MODERN DIFFERENTIAL EQUATIONS:

Theory, Applications, Technology

Martha L. Abell and James P. Braselton
Georgia Southern University

SAUNDERS COLLEGE PUBLISHING
Harcourt Brace College Publishers

Forth Worth   Philadelphia   San Diego   New York   Orlando   Austin
San Antonio   Toronto   Montreal   London   Sydney   Tokyo

Text Typeface: Times Roman
Compositor: York Graphic Services
Executive Editor: Jay Ricci
Senior Developmental Editor: Alexa Epstein
Managing Editor: Carol Field
Project Editor: Nancy Lubars
Copy Editor: Colleen Craney
Manager of Art and Design: Carol Bleistine
Senior Art Director: Joan Wendt
Text Designer: OX & Company
Cover Designer: Kathryn Needle
Text Artwork: Graficon, Inc.
Director of EDP: Tim Frelick
Production Manager: Joanne Cassetti
Product Manager: Nick Agnew

Cover Credit: © 1995 Lawrence Englesberg/Picture Perfect™

Printed in the United States of America

ISBN 0-03-098337-1
Library of Congress Catalog Card Number: 95-070697

5678901234   039   10 987654321

This book was printed on acid-free recycled content paper, containing **MORE THAN 10% POSTCONSUMER WASTE**

PREFACE

Computer algebra systems and sophisticated graphing calculators have changed the ways in which we learn and teach ordinary differential equations. Instead of focusing students' attention only on a sequence of solution methods, we want them to use their minds to understand what solutions mean and how differential equations can be used to answer pertinent questions.

Interestingly, this metamorphosis in the teaching of differential equations has occurred relatively "overnight" and has coincided with our professional careers at Georgia Southern University. Our interest in the use of technology in the mathematics classroom began in 1990 when we started to use computer laboratories and demonstrations in our calculus, differential equations, and applied mathematics courses. Over the past five years we have learned some ways of how to and how not to use technology in the mathematics curriculum. In the early stages, we simply wanted to show students how they could solve more difficult problems by using a computer algebra system so that they could be exposed to the technology. However, we soon realized that we were missing the great opportunity of allowing students to discover aspects of the subject matter on their own. We revised our materials (through our *In Touch with Technology* projects) to include experimental problems and thought-provoking questions in which students are asked to make conjectures and investigate supporting evidence. We also developed applications projects called *Differential Equations at Work*, not only to emphasize technology, but also to improve the problem-solving and communication skills of our students. To preserve the "wow" aspects of technology, we continue to use it to observe solutions in classroom demonstrations through such things as animating the motion of springs and pendulums. These demonstrations not only grab the attention of students, but also help them to make the connection between a formula and what it represents.

In presenting our findings to colleagues around the country, we quickly found out that others were interested in our work. As a result, we decided to develop a differential equations textbook to share this work with those who share our desire to improve mathematics education. This book is a culmination of years of "trials and tribulation" as the differential equations students at Georgia Southern can attest. Our hope is that its use will inspire students to open their eyes to the exciting discoveries that differential equations offer.

This book is designed to serve as a text for a beginning course in differential equations. Usually, introductory differential equations courses are taken by students who have successfully completed a first-year calculus course and this text is written at a level readable for them.

Technology

The benefits of incorporating technology into mathematics courses are well-known. Some of the advantages include enhancing the ability to solve a variety of problems, helping students work examples, supporting varied, realistic, and illuminating applications, exploiting and improving geometric intuition, encouraging mathematical experiments, teaching approximation, showing the mathematical significance of the computer revolution, and making higher-level mathematics accessible to students. In addition, technology is implemented throughout this text to promote the following goals in the learning of differential equations:

1. Solving problems: using different methods to solve problems and generalize solutions;
2. Reasoning: exploiting computer graphics to develop spatial reasoning through visualization;
3. Analyzing: finding the most reasonable solution to real problems or observing changes in the solution under changing conditions;
4. Communicating mathematics: developing written, verbal, and visual skills to communicate mathematical ideas; and
5. Synthesizing: making inferences and generalizations, evaluating outcomes, classifying objects, and controlling variables.

Students who develop these skills will succeed not only in differential equations, but also in subsequent courses and in the workforce.

Applications

Applications in this text are taken from a variety of fields, especially biology, physics, chemistry, engineering, and economics, and are documented by references. These applications can be found in many of the examples and exercises, in separate sections and chapters of the text, and in the *Differential Equations at Work* at the end of most chapters. Many of these applications are well-suited to exploration with technology because they incorporate real data. In particular, obtaining closed form solutions is not necessarily ''easy'' (or always possible). These applications, even if not formally discussed in class, show students that differential equations is an exciting and interesting subject with extensive application in many fields.

Style

To keep the text as flexible as possible, addressing the needs of both audiences with different mathematical backgrounds and instructors with varying preferences, *Modern Differential Equations: Theory, Applications, Technology* is written in an easy-to-read, yet mathematically precise, style. It contains all topics usually included in standard differential equations texts. Definitions, theorems, and proofs are concise but worded precisely for mathematical accuracy. Generally, theorems are proved if the proof is instructive or has ''teaching value''; these proofs are optional. In other cases, proofs of theorems are developed in the exercises or omitted. Theorems and definitions are boxed for easy reference; key terms are highlighted in boldface. Figures are used frequently to clarify material with a graphical interpretation.

CONTENT

The highlights of each chapter are described briefly below. Several suggestions for course syllabi follow the preface.

Chapter 1: After introducing preliminary definitions, we discuss direction fields not only for first-order differential equations, but also for systems of equations. In this presentation, we establish a basic understanding of solutions and their graphs. At the end of the chapter, we give an overview of some of the applications covered later in the text to point out the usefulness of the topic and some of the reasons we have for studying differential equations.

Chapter 2: In addition to discussing the standard techniques for solving several types of first-order differential equations (separable equations, homogeneous equations, exact equations, and linear equations), we introduce several numerical methods (Euler's Method, Improved Euler's Method, Runge-Kutta Method) and discuss the existence and uniqueness of solutions to first-order initial-value problems. Throughout the chapter, we encourage students to build an intuitive approach to the solution process by matching a graph to a solution without actually solving the equation.

Chapter 3: Not only do we cover most standard applications of first-order equations in Chapter 3 (orthogonal trajectories, population growth and decay, Newton's law of cooling, free-falling bodies), but also we present many that are not (due to their computational difficulty) through our *In Touch with Technology* problems and *Differential Equations at Work*.

Chapter 4: This chapter emphasizes the methods for solving homogeneous and non-homogeneous higher-order differential equations. It also stresses the Principle of Superposition and the differences between the properties of solutions to linear and nonlinear equations.

Chapter 5: Several applications of higher-order equations are presented. The distinctive presentation illustrates the motion of spring-mass systems and pendulums graphically to help students understand what solutions represent and to make the applications more meaningful.

Chapter 6: The topic of higher-order equations with nonconstant coefficients is covered. After discussing Cauchy-Euler equations, reviewing power series, and introducing series methods of solution to differential equations, the chapter culminates in a discussion of several special equations and the properties of their solutions—equations important in many areas of applied mathematics and physics.

Chapter 7: Laplace transforms are important in many areas of engineering and exhibit intriguing mathematical properties as well. Throughout the chapter, we point out the importance of initial conditions and forcing functions on initial-value problems.

Chapter 8: The study of systems of differential equations is perhaps the most exciting of all of the topics covered in the text. Although we direct most of our attention to solving systems of linear first-order equations with constant coefficients, technology allows us to investigate systems of nonlinear equations and observe phase planes.

Chapter 9: Several applications discussed earlier in the text are extended to more than one dimension and solved using systems of differential equations, in an effort to reinforce the understanding of these important problems. Numerous applications involving nonlinear systems are discussed as well.

PEDAGOGICAL FEATURES

Examples

Throughout the text, numerous examples are given, with thorough explanations and a substantial amount of detail. Solutions to more difficult examples are constructed with the help of graphics calculators or a computer algebra system and are indicated by an icon.

"Think about it!"

Many examples are followed by a question indicated by a 🔧. Generally, basic knowledge about the behavior of functions is sufficient to answer the question. Many of these questions encourage students to use technology. Others focus on the graph of a solution. Thus, "Think about it!" questions help students determine when to use technology and make this text more interactive.

In Touch with Technology

Many students entering their first differential course have had substantial experience with various sophisticated calculators and computer algebra systems. The text provides instructors with several ways to take advantage of technology's capabilities.

Nearly every section of the text contains *In Touch with Technology*. In these subsections, students use a graphics calculator and/or computer algebra system to study concepts of the section, to solve more advanced problems, or to discover new relationships. *In Touch with Technology* problems cover a wide range of difficulty: some use technology to solve problems similar to those in the section, others involve computations too tedious or time-consuming to do by hand. Still others are more difficult and, thus, are well suited for a lab project or other activity, if desired. (Specific suggestions regarding particular *In Touch with Technology* and *Differential Equations at Work* exercises are contained in the *Instructor's Resource Manual*.)

Technology can also be used to explore many of the applications and more difficult examples and exercises marked with 💾 throughout the text, and the problems in the subsections *Differential Equations at Work*. Solutions, partial solutions, or hints to those exercises indicated by an asterisk are contained in the *Student Resource Manual*.

Differential Equations at Work

Differential Equations at Work subsections describe detailed economics, biology, physics, chemistry, and engineering problems documented by references. These problems include real data when available and require students to provide answers based on different conditions. Students must analyze the problem and make decisions about the best way to solve it, including the appropriate use of technology. *Differential Equations at Work* can be assigned as projects requiring a written report, for group work, or for discussion in class.

Differential Equations at Work also illustrate how differential equations are used in the real world. Students are often reluctant to believe that the subject matter in calculus, linear algebra, and differential equations classes relates to subsequent courses and to their careers. Each *Differential Equations at Work* illustrates how the material discussed in the course is used in **real life.**

The problems are not connected to a specific section of the text; they require students to draw different mathematical skills and concepts together to solve a problem. Because each *Differential Equations at Work* is cumulative in nature, students must combine mathematical concepts, techniques, and experiences from previous chapters and math courses. Detailed hints regarding the use of technology in solving the problems encountered in *Differential Equations at Work* are included in the *Student Resource Manual;* suggestions for instructors are contained in the *Instructor's Resource Manual*.

Exercises

Numerous exercises, ranging in level from easy to difficult, are included in each section of the text. In particular, the exercise sets for topics that students find most difficult are rich and varied. The abundant "routine" exercises encourage students to master basic techniques. Most sections also contain interesting *mathematical* and *applied* problems to show that mathematics and its applications are both interesting and relevant. Instructors will find that they can assign a large number of problems, if desired, yet still have plenty for review in addition to those found in the review section at the end of each chapter. Answers to most odd-numbered exercises are included at the end of the text; detailed solutions to those exercises marked with an * are included in the *Student Resource Manual*.

Chapter Summary and Review Exercises

Each chapter ends with a chapter summary highlighting important concepts, key terms and formulas, and theorems. The Review Exercises following the chapter summary offer students extra practice on the topics in that chapter. These exercises are arranged by section so that students having difficulty can turn to the appropriate material for review.

Figures

One of technology's benefits is the ease of graphing and plotting. This text provides an abundance of figures and graphs, especially for solutions to examples. In addition, students are encouraged to develop spatial visualization and reasoning skills, to interpret graphs, and to discover and explore concepts from a graphical point of view. To ensure accuracy, the figures and graphs have been completely computer-generated. The *Instructor's Resource Manual* and *Student Resource Manual* show typical graphical output using *Mathematica* and *Maple* for exercise solutions.

Historical Material

Nearly every topic is motivated by either an application or an appropriate historical note. We have also included photos of many famous mathematicians and descriptions of the mathematics they discovered.

SUPPLEMENTS

The following aids prepared by the authors for instructors and students are available from the publisher:

The *Instructor's Resource Manual* contains detailed solutions to the exercises, plus *Mathematica* and *Maple* code and output for the *In Touch with Technology* and *Differential Equations at Work* problems. Suggestions regarding specific *In Touch with Technology* and *Differential Equations at Work* are also included.

The *Student Resource Manual* contains detailed solutions to selected exercises (those marked with an *) and problems and explanations of prerequisite techniques needed for a standard differential equations course, but not necessarily mastered by students enrolled in the course. The *Student Resource Manual* also contains solutions, partial solutions, or hints to *In Touch with Technology* and *Differential Equations at Work* problems and substantial guidance for users of *Mathematica* and *Maple* software.

ACKNOWLEDGMENTS

The development of this text has involved a thorough program of reviewing, for both pedagogical and topical content as well as for accuracy. Many colleagues were involved in this process. Their contributions are gratefully acknowledged:

Josefina Alvarez, New Mexico State University
Sidney Birnbaum, California State Polytechnic University, Pomona
Philip Crooke, Vanderbilt University
Steve R. Dunbar, University of Nebraska
Richard S. Falk, Rutgers University
Ronald B. Guenther, Oregon State University
Kenneth A. Heimes, Iowa State University
Dar-Veig Ho, Georgia Institute of Technology
Helmut Knaust, University of Texas, El Paso
David O. Lomen, University of Arizona
Marie Vanisko, Carroll College

We owe particular thanks for the careful comments and suggestions from those who class-tested this project during its development:

Lyle Cochran, Fresno Pacific College
Stuart Davidson, Georgia Southern University
Marie Vanisko, Carroll College

The accuracy of the examples and answers at the back of the book was carefully checked during manuscript, galleys, and page proof stages by Stuart Davidson (Georgia Southern University), Tom Richards (University of North Dakota), Marie Vanisko (Carroll College). Their careful and diligent work is greatly appreciated. Any remaining errors are our sole responsibility and we would greatly appreciate hearing about them.

This text would not have been possible without the substantial efforts of many at Saunders College Publishing. In particular, we would like to express our gratitude to

Jay Ricci, Executive Editor,
Alexa Epstein, Senior Developmental Editor,
Nancy Lubars, Project Editor,
Joan Wendt, Senior Art Director,
Katy Needle, Associate Art Director,
Joanne Cassetti, Production Manager, and
Nick Agnew, Product Manager.

We would also like to thank our families, particularly Imogene H. Abell, Lori Braselton, and Ada Braselton, for enduring with us the pressures of meeting a deadline and for graciously accepting our demanding work schedules. We could not have completed this task without their care and understanding.

Martha Abell and James Braselton
Statesboro, Georgia
June, 1995

TO THOSE WHO HAVE MEANT SO MUCH:

In honor of

Imogene Holcomb Abell	Evelyn Albrecht Farnsworth
Nelle Holcomb	Lorraine Maruth Braselton
Margaret Holcomb	Ada Doris Braselton
Mildred Holcomb	

In memory of

William J. Abell, Jr.	W. Emmett Braselton
Louise Chambless Abell	Doris Jacob Braselton

COURSE
SUGGESTIONS

*M*odern *Differential Equations: Theory, Applications, Technology* is a versatile and flexible text that can be used to teach differential equations in a variety of ways, depending upon student needs and instructor preferences. We list several suggestions for topics to be covered in a one-quarter course, a two-quarter course sequence, or a one-semester course. However, many other syllabi are possible.

The topics in Sections 1.1–1.3; 2.1–2.5; 4.1–4.4 (with less emphasis on differential operators, if Section 4.5 is not to be covered), 4.5 or 4.6, 4.7 form a core for nearly every syllabus. Beyond that, the instructor has many options. Some instructors may prefer to discuss many different topics on a less rigorous level; other instructors may want to concentrate on one of Chapters 6, 7, or 8. In addition, some topics, such as those contained in applications chapters, are well-suited for exploration in a computer laboratory setting.

One-Quarter Course Options

Option 1: First-Order Equations, Applications of First-Order Equations, Second-Order Equations, Applications of Second-Order Equations, Cauchy-Euler Equations, Power Series Solutions
SECTIONS 1.1–1.5; 2.1–2.5; 3.1–3.4; 4.1–4.4, 4.5 or 4.6, 4.7; 5.1–5.5; 6.1–6.5

Option 2: First-Order Equations, Applications of First-Order Equations, Second-Order Equations, Applications of Second-Order Equations, Cauchy-Euler Equations, Systems of Equations
SECTIONS 1.1–1.4; 2.1–2.5; 3.1–3.4; 4.1–4.4, 4.5 or 4.6, 4.7; 5.1–5.5; 6.1; 8.1–8.4; 8.7

Option 3: First-Order Equations, Applications of First-Order Equations, Second-Order Equations, Applications of Second-Order Equations, Power Series Solutions, Laplace Transforms
SECTIONS 1.1–1.5; 2.1–2.5; Two sections from 3.1–3.4; 4.1–4.4, 4.5 or 4.6, 4.7; Two sections from 5.1–5.5; 6.1–6.5, 7.1–7.7

Option 4: First-Order Equations (including numerical methods), Second-Order Equations, Laplace Transforms, Systems of Differential Equations (including numerical methods)
SECTIONS 1.1–1.5; 2.1–2.6; 4.1–4.4, 4.5 or 4.6, 4.7; 7.1–7.6; 8.1–8.7

Two-Quarter Sequence Option

First Quarter: First-Order Equations, Applications of First-Order Equations, Second-Order Equations, Applications of Second-Order Equations, Cauchy-Euler Equations, Power Series Solutions
SECTIONS 1.1–1.5; 2.1–2.6; 3.1–3.4; 4.1–4.4, 4.5 or 4.6, 4.7; 5.1–5.5; 6.1–6.3

Second Quarter: Laplace Transforms, Systems of Differential Equations (including numerical methods), Applications of Systems
SECTIONS 6.4–6.5; 7.1–7.7; 8.1–8.8; 9.1–9.4

One-Semester Course Options

Option 1: First-Order Equations, Applications of First-Order Equations, Second-Order Equations, Applications of Second-Order Equations, Cauchy-Euler Equations, Power Series Solutions, Systems of Differential Equations (brief discussion)
SECTIONS 1.1–1.5; 2.1–2.5; 3.1–3.4; 4.1–4.4, 4.5 or 4.6, 4.7; 5.1–5.5; 6.1–6.5; 8.1–8.4

Option 2: First-Order Equations, Applications of First-Order Equations, Second-Order Equations, Applications of Second-Order Equations, Cauchy-Euler Equations, Laplace Transforms, Systems of Differential Equations (brief discussion)
SECTIONS 1.1–1.5; 2.1–2.5; Two sections from 3.1–3.4; 4.1–4.4, 4.5 or 4.6, 4.7; Two sections from 5.1–5.5; 6.1; 7.1–7.7; 8.1–8.4

Option 3: First-Order Equations (including numerical methods), Second-Order Equations, Cauchy-Euler Equations, Laplace Transforms, Systems of Differential Equations; Systems of Differential Equations (brief discussion with numerical methods)
SECTIONS 1.1–1.5; 2.1–2.6; Two sections from 3.1–3.4; 4.1–4.4, 4.5 or 4.6, 4.7; Two sections from 5.1–5.5; 6.1; 7.1–7.6 (7.7-optional); 8.1–8.4, 8.8

Option 4: First-Order Equations, Applications of First-Order Equations, Second-Order Equations, Applications of Second-Order Equations, Cauchy-Euler Equations, Systems of Differential Equations, Applications of Systems of Differential Equations
SECTIONS 1.1–1.5; 2.1–2.5; Two sections from 3.1–3.4; 4.1–4.4, 4.5 or 4.6, 4.7; Two sections from 5.1–5.5; 6.1; 8.1–8.7; 9.1–9.4

CONTENTS

1

INTRODUCTION TO DIFFERENTIAL EQUATIONS

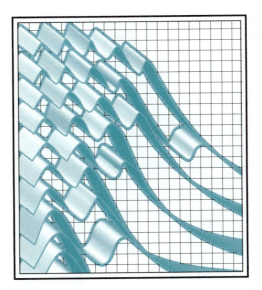

The purpose of ***Modern Differential Equations: Theory, Applications, Technology*** is twofold. First, we introduce and discuss the topics covered in an undergraduate course in ordinary differential equations. Second, we indicate how certain technologies such as computer algebra systems and graphics calculators are used to enhance the study of differential equations, not only by eliminating some of the computational difficulties that arise in the study of differential equations, but also by overcoming some of the visual limitations associated with the solutions of differential equations.

The advantages of using technology like graphics calculators and computer algebra systems in the study of differential equations are numerous, but perhaps the most useful is that of being able to produce the graphics associated with solu-

Gottfried Wilhelm Leibniz
(1646–1716) (North Wind
Picture Archives)

Isaac Newton (1642–1727)
(North Wind Picture Archives)

Leonhard Euler (1707–1783)
(North Wind Picture Archives)

tions of differential equations. This is particularly beneficial in the discussion of applications because many physical situations are modeled with differential equations. For example, in Chapter 5 we will see that the motion of a pendulum can be modeled by a differential equation. When we solve the problem of the motion of a pendulum, we use technology to watch the pendulum move. The same is true for the motion of a mass attached to the end of a spring, as well as many other problems. In having this ability to use technology, the study of differential equations becomes much more meaningful as well as interesting.

Although Chapter 1 is short in length, the vocabulary introduced will be used throughout the text. Even though, to a large extent, this chapter may be read quickly, subsequent chapters will take advantage of the terminology and techniques discussed here.

The formal introduction of differential equations begins with German scientist Gottfried Wilhelm Leibniz (1646–1716) and British scientist Isaac Newton (1642–1727), the inventors of calculus. Although we learn in integral calculus that the area under the graph of a smooth positive function is given by a definite integral, both Leibniz and Newton were more concerned with solving differential equations than finding areas. Many of the methods of solution we present in this text are due to the great Swiss mathematician Leonhard Euler (1707–1783). Subsequently, many problems, like determining the motion of a string, not only lead to ordinary and partial differential equations but to other areas of mathematics as well. Indeed, differential equations is full of rich and exciting applications; interesting applications are included throughout the text to motivate discussions, make the study of differential equations more interesting and pertinent to the real world, and indicate how some people use differential equations beyond this course. However, *mathematics* is also interesting in its own right; mathematical applications are also included throughout the text.

1.1 Definitions and Concepts

We begin our study of differential equations by explaining what a differential equation is. From our experience in calculus, we are familiar with some differential equations. For example, suppose that the acceleration of a falling object is

$$a(t) = -32,$$

measured in ft/s². Using the fact that the derivative of the velocity function $v(t)$ (measured in ft/s) is the acceleration function $a(t)$, we can solve the equation

$$v'(t) = a(t) \qquad \text{or} \qquad \frac{dv}{dt} = a(t)$$

to find the velocity of the object (within an arbitrary constant) with

$$v(t) = \int a(t)\, dt = \int (-32)\, dt = -32t + C.$$

Since the equation $dv/dt = a(t)$ is an example of a differential equation, we see that we already know how to solve many differential equations of the form $dy/dx = f(x)$ for the unknown function $y(x)$ through integration. In this course, we will extend our knowledge of differential equations.

Definition 1.1 Differential Equation

A **differential equation** is an equation that contains the derivatives or differentials of one or more dependent variables with respect to one or more independent variables. If the equation contains only ordinary derivatives (of one or more dependent variables) with respect to a single independent variable, the equation is called an **ordinary differential equation.**

Dependent and Independent Variables: We point out the differences between dependent and independent variables through two examples from calculus. Parametric equations are used many times in calculus. For example, parametric equations of the unit circle are

$$x = \cos t, \quad y = \sin t, \quad 0 \le t \le 2\pi.$$

In this case, the values of x and y *depend* on the choice of t. Therefore, t is the independent variable while x and y are the dependent variables. In multivariable calculus, the graph of

$$z = x^2 + y^2$$

is a paraboloid. Here, the value of z *depends* on the choices of x and y, so the independent variables are x and y, and z is the dependent variable.

Common Notation for Ordinary Derivatives: If y represents a function of x, then all of the following represent the derivative of y with respect to x, provided that the derivative exists:

$$y', \quad y'(x), \quad \frac{dy}{dx}, \quad Dy.$$

Another common notation used by scientists and engineers to denote differentiation with respect to t is

$$\dot{y}.$$

Similarly, all of the following represent the second derivative of y with respect to x, provided that the second derivative exists:

$$y'', \quad y''(x), \quad \frac{d^2y}{dx^2}, \quad D^2y.$$

Scientists and engineers frequently use

$$\ddot{y}$$

to represent the second derivative of y with respect to t.

For derivatives of order four and higher, we also use the notation

$$y^{(4)} = \frac{d^4y}{dx^4}, \quad y^{(5)} = \frac{d^5y}{dx^5}, \quad \cdots, \quad y^{(n)} = \frac{d^ny}{dx^n}.$$

Example 1

Determine which of the following are examples of ordinary differential equations:

(a) $\dfrac{dy}{dx} = \dfrac{x^2}{y^2 \cos y}$ (b) $\dfrac{dy}{dx} + \dfrac{du}{dx} = u + x^2y$ (c) $(y-1)\,dx + x\cos y\,dy = 0$

(d) $x^2y'' + xy' + (x^2 - n^2)y = 0$

Solution   All of the equations are ordinary differential equations. Notice that $(y-1)\,dx + x\cos y\,dy = 0$ includes the differentials dx and dy while the equations in (a), (b), and (d) involve derivatives (like dy/dx, du/dx, and $y'' = d^2y/dx^2$). Because of this, we say that the equation in (c) is in **differential form.** However, we can put it in **derivative form** (so that it involves derivatives) by solving for dy/dx or dx/dy to obtain either $\dfrac{dy}{dx} = -\dfrac{y-1}{x\cos y}$ or $\dfrac{dx}{dy} = -\dfrac{x\cos y}{y-1}$.

If the equation contains partial derivatives of one or more dependent variables, the equation is called a **partial differential equation.**

Definition 1.2 Partial Differential Equation

A **partial differential equation** is an equation that contains the partial derivatives or differentials of one or more dependent variables with respect to more than one independent variable.

Common Notation for Partial Derivatives: If u represents a function of x and y, both

$$\frac{\partial u}{\partial x} \quad \text{and} \quad u_x$$

represent the partial derivative of u with respect to x, provided the derivative exists. Similarly, both

$$\frac{\partial u}{\partial y} \quad \text{and} \quad u_y$$

represent the partial derivative of u with respect to y, provided the derivative exists. The expressions

$$\frac{\partial}{\partial y}\left(\frac{\partial u}{\partial x}\right) = \frac{\partial^2 u}{\partial y \partial x} \quad \text{and} \quad u_{xy}$$

represent the partial derivative of u with respect to x and then with respect to y, provided that both derivatives exist. The expressions

$$\frac{\partial}{\partial x}\left(\frac{\partial u}{\partial x}\right) = \frac{\partial^2 u}{\partial x^2} \quad \text{and} \quad u_{xx}$$

and

$$\frac{\partial}{\partial y}\left(\frac{\partial u}{\partial y}\right) = \frac{\partial^2 u}{\partial y^2} \quad \text{and} \quad u_{yy}$$

represent the second derivatives of u with respect to x and y, respectively, provided these derivatives exist.

Example 2

Determine which of the following are examples of partial differential equations:

(a) $u\dfrac{\partial u}{\partial t} = \dfrac{\partial u}{\partial x}$ (b) $uu_x + u = u_{yy}$ (c) $\dfrac{\partial^2 u}{\partial x^2} + \dfrac{\partial^2 u}{\partial y^2} = 0$ (d) $\dfrac{\partial^2 u}{\partial t^2} = \dfrac{\partial^2 u}{\partial x^2}$

(e) $\dfrac{\partial u}{\partial t} = \dfrac{\partial^2 u}{\partial x^2}$

Solution   All of these equations are partial differential equations. In fact, the equations in (c), (d), and (e) are well known and called **Laplace's equation,** the **wave equation,** and the **heat equation,** respectively.

Generally, given a differential equation, our goal in this course will be to construct a solution or a numerical approximation of the solution. Differential equations can be categorized into groups of equations that may be solved in similar ways. We have seen that some differential equations are *ordinary* differential equations and others are *partial* differential equations. We extend this classification system with the following definition.

Definition 1.3 Order

The **order** of a differential equation is the order of the highest-order derivative appearing in the equation.

Example 3

Determine the order of each of the following differential equations:
(a) $dy/dx = x^2/y^2 \cos y$ (b) $u_{xx} + u_{yy} = 0$ (c) $(dy/dx)^4 = y + x$ (d) $y^3 + dy/dx = 1$

Solution (a) The order of this equation is one because it includes only one first-order derivative, dy/dx. (b) This equation is classified as second-order because the highest-order derivatives, both u_{xx}, representing $\partial^2 u/\partial x^2$, and u_{yy}, representing $\partial^2 u/\partial y^2$, are of order two. Hence, Laplace's equation is a second-order partial differential equation. (c) This is a first-order equation because the highest-order derivative is the first derivative. Raising that derivative to the fourth power does not affect the order of the equation. The expressions

$$(dy/dx)^4 \quad \text{and} \quad d^4y/dx^4$$

do not represent the same quantities: $(dy/dx)^4$ represents the derivative of y with respect to x, dy/dx, raised to the fourth power; d^4y/dx^4 represents the fourth derivative of y with respect to x. (d) Again, we have a first-order equation because the highest-order derivative is the first derivative.

The next level of classification is based on the following definition.

Definition 1.4 Linear Differential Equation

An ordinary differential equation (of order n) is called **linear** if it is of the form

$$a_n(x)\frac{d^ny}{dx^n} + a_{n-1}(x)\frac{d^{n-1}y}{dx^{n-1}} + \cdots + a_2(x)\frac{d^2y}{dx^2} + a_1(x)\frac{dy}{dx} + a_0(x)y = f(x),$$

where the functions $a_j(x)$, $j = 0, 1, \ldots, n$, and $f(x)$ are given and $a_n(x)$ is not the zero function.

If the equation under consideration is not of this form, the equation is said to be **nonlinear.** Therefore, some of the properties that lead to classifying an equation as nonlinear are *powers of the dependent variable* (or one of its derivatives) and *functions of the dependent variable*. A similar classification is followed for partial differential equations. In this case, the coefficients in a linear partial differential equation are functions of the independent variables.

Example 4

Determine which of the following differential equations are linear: (a) $dy/dx = x^3$
(b) $\frac{d^2u}{dx^2} + u = e^x$ (c) $(y-1)dx + x\cos(y)dy = 0$ (d) $\frac{d^3y}{dx^3} + y\frac{dy}{dx} = x$
(e) $\frac{dy}{dx} + x^2y = x$ (f) $\frac{d^2x}{dt^2} + \sin x = 0$ (g) $u_{xx} + yu_y = 0$ (h) $u_{xx} + uu_y = 0$

Solution (a) This equation is linear because the nonlinear term x^3 is the function $f(x)$ of the independent variable in the general formula for a linear differential equation. (b) This equation is also linear. Using u as the name of the dependent variable does not affect the linearity. (c) If y is the *dependent* variable, solving for dy/dx gives us

$$\frac{dy}{dx} = \frac{1-y}{x \cos y}.$$

Because the right-hand side of this equation includes a nonlinear function of y, the equation is nonlinear (in y). However, if x is the *dependent* variable, solving for dx/dy yields

$$\frac{dx}{dy} = \frac{\cos y}{1-y} x.$$

This equation is linear in the dependent variable x. (d) The coefficient of the term dy/dx is y instead of an expression involving only the independent variable x. Hence, this equation is nonlinear in the dependent variable y. (e) This equation is linear. The term x^2 is the coefficient function. (f) This equation, known as the **pendulum equation** because it models the motion of a pendulum, is nonlinear because it involves a non-linear function of x, $\sin x$. Note that x is the dependent variable in this case; t is the independent variable. (g) This partial differential equation is linear because the coefficient of u_y is a function of one of the independent variables. (h) In this case, there is a product of u and one of its derivatives, so the equation is nonlinear.

In the same manner that we consider systems of equations in algebra, we can also consider systems of differential equations. For example, if x and y represent functions of t, we will learn in Chapter 8 to solve the **system of linear equations**

$$\begin{cases} x' = ax + by \\ y' = cx + dy \end{cases}$$

where a, b, c, and d represent constants and differentiation is with respect to t. We will see that systems of differential equations arise naturally in many physical situations that are modeled with more than one equation and involve more than one dependent variable. In addition, we will see that it is often useful to write a differential equation with order greater than one as a system of first-order equations.

Example 5

Write the nonlinear second-order equation $\dfrac{d^2x}{dt^2} + \sin x = 0$, which can be used to model the motion of a pendulum, as a system of first-order equations.

Solution   Let $y = \dfrac{dx}{dt} = x'$. Then, $y' = \dfrac{d^2x}{dt^2} = -\sin x$ so the second-order equation $\dfrac{d^2x}{dt^2} + \sin x = 0$ is equivalent to the system of first-order equations

$$\begin{cases} x' = y \\ y' = -\sin x \end{cases}$$

EXERCISES 1.1

For each of the following equations, determine (a) if the equation is an ordinary differential equation or partial differential equation; (b) the order of the ordinary differential equation; and (c) if the equation is linear or nonlinear.

1. $\dfrac{d^2y}{dx^2} + \dfrac{dy}{dx} - 2y = x^3$

2. $y\dfrac{dy}{dx} + y^4 = \sin x$

3. $\dfrac{\partial^2 y}{\partial t^2} = c^2 \dfrac{\partial^2 y}{\partial x^2}$

4. $y''' - 2y'' + 5y' + y = e^x$

5. $\left(\dfrac{dy}{dx}\right)^2 + y = 0$

6. $x^2 \dfrac{d^2y}{dx^2} + x\dfrac{dy}{dx} + 2y = 0$

7. $\dfrac{1}{c^2}\dfrac{\partial^2 z}{\partial t^2} = \dfrac{\partial^2 z}{\partial x^2} + \dfrac{\partial^2 z}{\partial y^2}$

8. $uu_x + u_t = 0$

9. $x\left(\dfrac{d^2y}{dx^2}\right)^4 + 2y = 2x$

10. $\dfrac{d^2x}{dt^2} + 2\sin x = \sin 2t$

11. $u_t + uu_x = \sigma u_{xx}$, σ constant

12. $(2x - 1)\, dx - dy = 0$

13. $(2x - y)\, dx - dy = 0$

14. $\dfrac{\partial u}{\partial x}\dfrac{\partial u}{\partial y} = u$

15. $(2x - y)\, dx - y\, dy = 0$

16. In 1840, the Belgian mathematician-biologist Pierre F. Verhulst (1804–1849) developed the **logistic equation** $dP/dt = rP - aP^2$ where r and a are positive constants to predict the population $P(t)$ in certain countries. Is this equation linear or nonlinear? Determine the order of the equation.

17. The current $I(t)$ in an **L-R circuit,** which contains a resistor, an inductor, and a voltage source, satisfies the differential equation $RI + L\dfrac{dI}{dt} = E(t)$, where R and L are constants representing the resistance and the inductance and $E(t)$ is the voltage source. Is this equation linear or nonlinear? Determine the order of the equation.

18. Suppose that the head of a drum is rectangular. The displacement u of the drumhead at the point (x, y) satisfies the **wave equation** $u_{tt} = c^2(u_{xx} + u_{yy})$ where c is a constant. Is this equation linear or nonlinear?

19. The **Korteweg-de Vries (KdV) equation** $u_t + u_{xxx} - 6uu_x = 0$ is studied in fluid dynamics. Is this equation linear or nonlinear?

20. The **diffusion equation** in cylindrical coordinates is $u_t = u_{rr} + \dfrac{u_r}{r}$. Is this equation linear or nonlinear?

21. Write each of the following second-order equations as a system of first-order equations.

(a) $\dfrac{d^2x}{dt^2} - \dfrac{dx}{dt} - 6x = 0$

(b) $4\dfrac{d^2x}{dt^2} + 4\dfrac{dx}{dt} + 37x = 0$

(c) $L\dfrac{d^2x}{dt^2} + g\sin x = 0$

(d) $\dfrac{d^2x}{dt^2} - \mu(1 - x^2)\dfrac{dx}{dt} + x = 0$

(e) $t\dfrac{d^2x}{dt^2} + (b - t)\dfrac{dx}{dt} - ax = 0$

1.2 Solutions of Differential Equations

When faced with a differential equation, our goal is frequently, but not always, to determine a solution to the equation. This means that we want to find an expression that satisfies the equation when substituted into the equation.

Definition 1.5 Solution

Do not let the notation
$F(x, y, y', y'', \ldots, y^{(n)}) = 0$
confuse you. It represents an nth order ordinary differential equation. For example, we can write the second-order differential equation
$x^2 y'' + xy' + y = \cos x$ *as*
$F(x, y, y', y'') = 0$, *where*
$F(x, y, y', y'') = x^2 y'' + xy' + y - \cos x$.

A **solution** of the nth-order ordinary differential equation

$$F(x, y, y', y'', \ldots, y^{(n)}) = 0$$

on the interval $a < x < b$ is a function $\phi(x)$ that is continuous on $a < x < b$ and has all the derivatives present in the differential equation such that

$$F(x, \phi, \phi', \phi'', \ldots, \phi^{(n)}) = 0$$

on $a < x < b$.

We will always assume that F is a real-valued function and we will only search for real-valued solutions $\phi(x)$.

In later chapters, we will discuss methods for solving differential equations. Here, in order to understand what is meant to be a solution, we give both the equation and a solution, and we verify the solution.

Example 1

Verify that the given function is a solution to the corresponding differential equation: (a) $\dfrac{dy}{dx} = 3y$, $y(x) = e^{3x}$ (b) $\dfrac{d^2 u}{dx^2} + 16u = 0$, $u(x) = \cos 4x$
(c) $x^2 y'' - 3xy' + 4y = 0$, $x > 0$, $y(x) = x^2 \ln x$

Solution   (a) Differentiating y we have $dy/dx = 3e^{3x}$ so that substitution into the equation yields

$$\frac{dy}{dx} = 3y$$
$$3e^{3x} = 3e^{3x}.$$

 Graph dy/dx and $3y$ to show that they are equivalent.

(b) Two derivatives are required in this case: $\dfrac{du}{dx} = -4 \sin 4x$ and

$\dfrac{d^2 u}{dx^2} = -16 \cos 4x$. Therefore,

$$\frac{d^2 u}{dx^2} + 16u = -16 \cos 4x + 16 \cos 4x = 0.$$

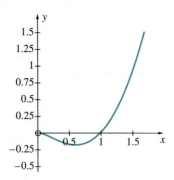

Figure 1.1 Graph of $y(x) = x^2 \ln x$

(c) Differentiating with the product rule we find that

$$y' = x^2\left(\frac{1}{x}\right) + 2x \ln x = x + 2x \ln x \text{ and } y'' = 1 + 2x\left(\frac{1}{x}\right) + 2 \ln x = 3 + 2 \ln x.$$

Therefore,

$$x^2 y'' - 3xy' + 4y = x^2(3 + 2 \ln x) - 3x(x + 2x \ln x) + 4x^2 \ln x = 0.$$

Notice that the solution $y(x) = x^2 \ln x$ is defined only for $x > 0$ because the domain of $\ln x$ is $(0, \infty)$, as shown in Figure 1.1, so $y(x) = x^2 \ln x$ is a solution only for $x > 0$.

Verify that both $y(t) = e^{-t/2}(\cos 2t - \sin 2t)$ and $y(t) = e^t$ are solutions to the differential equation $4y''' + 13y' - 17y = 0$. (Note that $y' = dy/dt$.)

We have some experience with solving differential equations from integral calculus where we were faced with solving equations like $dy/dx = f(x)$ to find an antiderivative of a function $f(x)$. This experience showed us that the equation $dy/dx = f(x)$ has (infinitely) many solutions because a constant of integration must be included in an antiderivative. (Why?) Familiar techniques for solving equations of this type include u-substitutions, trigonometric substitutions, partial fractions, and integration by parts. Again, in this course, we encounter problems of this nature, so we review these methods of integration now.

Just as you would not use a calculator to compute $2 + 3 = 5$, you would not use a computer algebra system to evaluate integrals like $\int \sin x \, dx = -\cos x + C$

or $\int xe^x \, dx = xe^x - e^x + C$. However, computer algebra systems can save you a lot of time if you use them to help you evaluate complicated integrals.

Example 2

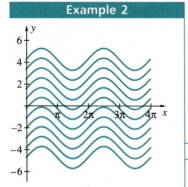

Figure 1.2 Graph of $y = \sin x + C$ for various values of C

Solve the following differential equations.

(a) $\dfrac{dy}{dx} = \cos x$ (b) $\dfrac{dy}{dx} = \dfrac{x}{\sqrt{x^2 + 1}}$ (c) $\dfrac{dy}{dx} = \dfrac{1}{16 + x^2}$ (d) $\dfrac{dy}{dx} = xe^x$

(e) $\dfrac{dy}{dx} = \dfrac{1}{4 - x^2}$

Solution   In each case, we integrate the indicated function and graph the solution for several values of the constant of integration. Each solution contains a constant of integration so there are (infinitely) many solutions to each equation.

(a) $y = \displaystyle\int \cos x \, dx = \sin x + C$ (See Figure 1.2.)

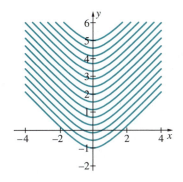

Figure 1.3 Graph of $y = \sqrt{x^2 + 1} + C$ for various values of C

(b) To evaluate $y = \displaystyle\int \frac{x}{\sqrt{x^2 + 1}}\, dx$ we let $u = x^2 + 1$ so that $du = 2x\, dx$. Then,

$$y = \frac{1}{2}\int \frac{2x}{\sqrt{x^2 + 1}}\, dx = \frac{1}{2}\int \frac{1}{\sqrt{u}}\, du = \frac{1}{2}\int u^{-1/2}\, du = u^{1/2} + C$$
$$= \sqrt{x^2 + 1} + C. \quad \text{(See Figure 1.3.)}$$

(c) In order to integrate $y = \displaystyle\int \frac{1}{16 + x^2}\, dx$, we use a trigonometric substitution.

Letting $x = 4\tan\theta$, $-\dfrac{\pi}{2} < \theta < \dfrac{\pi}{2}$ so that $dx = 4\sec^2\theta\, d\theta$ gives us

$$y = \int \frac{1}{16 + (4\tan\theta)^2} 4\sec^2\theta\, d\theta$$

$$= \int \frac{1}{16 + 16\tan^2\theta} 4\sec^2\theta\, d\theta \qquad \underbrace{=}_{1 + \tan^2\theta = \sec^2\theta} \qquad \int \frac{1}{16\sec^2\theta} 4\sec^2\theta\, d\theta$$

$$= \frac{1}{4}\int d\theta = \frac{1}{4}\theta + C \qquad \underbrace{=}_{\frac{x}{4} = \tan\theta \Rightarrow \theta = \tan^{-1}\frac{x}{4}} \qquad \frac{1}{4}\tan^{-1}\left(\frac{x}{4}\right) + C.$$

 Graph $y = \dfrac{1}{4}\tan^{-1}\left(\dfrac{x}{4}\right) + C$ *if* $C = -1, 0,$ *and* 1.

(d) To evaluate $y = \displaystyle\int xe^x\, dx$ by hand we use the integration by parts formula, $\displaystyle\int u\, dv = uv - \int v\, du$, with $u = x$, $dv = e^x\, dx$, $du = dx$, and $v = e^x$. This gives

$$y = \int xe^x\, dx = \underbrace{xe^x}_{uv} - \underbrace{\int e^x\, dx}_{vdu} = xe^x - e^x + C.$$

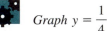 *Find* C *if* $y(1) = 2$.

(e) Use partial fractions to evaluate this integral. First, find the partial fraction decomposition of $\dfrac{1}{4 - x^2}$, which is determined by finding the constants A and B that satisfy the equation $\dfrac{1}{(2 - x)(2 + x)} = \dfrac{A}{2 - x} + \dfrac{B}{2 + x}$. These values are $A = B = \dfrac{1}{4}$. (Why?) Thus,

$$y = \frac{1}{4}\int \left(\frac{1}{2 - x} + \frac{1}{2 + x}\right) dx = \frac{1}{4}[-\ln|2 - x| + \ln|2 + x|] + C$$

$$= \frac{1}{4}\ln\left|\frac{2 + x}{2 - x}\right| + C. \quad \text{(See Figure 1.4.)}$$

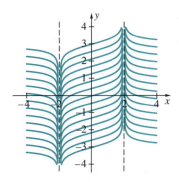

Figure 1.4 Graph of $y = \dfrac{1}{4}\ln\left|\dfrac{2 + x}{2 - x}\right| + C$

Explain why the solution is valid only if $x < -2$, $-2 < x < 2$, or $x > 2$.

In Example 2, each solution is given as a function $y = y(x)$ of the independent variable. In these cases, the solution is said to be **explicit.** In solving some differential equations, however, we can only find an equation involving x and y that the solution satisfies. In this case, we say that we have found an **implicit** solution. Also, in Example 2, the explicit solution of each differential equation involved an arbitrary constant C, so a solution exists for each choice of C. We call this collection of solutions a **family of solutions** of the differential equation.

*We will see that given an **arbitrary** differential equation, constructing an explicit or implicit solution is nearly always impossible. Consequently, although mathematicians were first concerned with finding analytic (explicit or implicit) solutions to differential equations, they have since (frequently) turned their attention to addressing properties of the solution and finding algorithms to approximate solutions.*

| **Example 3** | Verify that the equation $2x^2 + y^2 - 2xy + 5x = 0$ satisfies the differential equation |

$$\frac{dy}{dx} = \frac{2y - 4x - 5}{2y - 2x}.$$

| **Solution** | We use implicit differentiation to compute $y' = \dfrac{dy}{dx}$ if |

$2x^2 + y^2 - 2xy + 5x = 0$:

$$4x + 2y\frac{dy}{dx} - 2x\frac{dy}{dx} - 2y + 5 = 0$$

$$\frac{dy}{dx}(2y - 2x) = 2y - 4x - 5$$

$$\frac{dy}{dx} = \frac{2y - 4x - 5}{2y - 2x}.$$

The equation satisfies the differential equation $\dfrac{dy}{dx} = \dfrac{2y - 4x - 5}{2y - 2x}$.

Although we cannot solve $2x^2 + y^2 - 2xy + 5x = 0$ for y as a *function* of x (see Figure 1.5), we can determine the corresponding y value(s) for a given value of x. For example, if $x = -1$, then

$$2 + y^2 + 2y - 5 = y^2 + 2y - 3 = (y + 3)(y - 1) = 0.$$

Therefore, the points $(-1, -3)$ and $(-1, 1)$ lie on the graph of $2x^2 + y^2 - 2xy + 5x = 0$. (See Figure 1.6.)

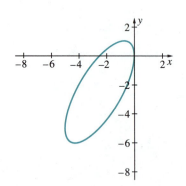

Figure 1.5 Graph of $2x^2 + y^2 - 2xy + 5x = 0$

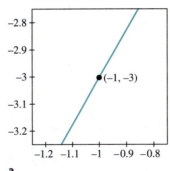 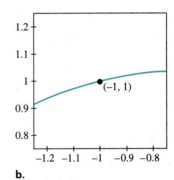

a.   **b.**

Figure 1.6 Notice that near the points $(-1, -3)$ and $(-1, 1)$ the implicit solution *looks* like a function. In fact, if we zoom in near the points $(-1, -3)$ and $(-1, 1)$, we see what *appears* to be the graph of a function.

   Find an equation of the line tangent to the graph of $2x^2 + y^2 - 2xy + 5x = 0$ at the points $(-1, -3)$ and $(-1, 1)$.

When a solution cannot be obtained from a family of solutions for any choice of the unknown constant(s), we say that the solution is **singular.**

Example 4	A family of solutions of $dy/dx = -2xy^2$ is $y(x) = 1/(x^2 + k)$ where k is an arbitrary constant. Determine if the constant function $y = 0$ is a solution by substitution into the differential equation. If $y = 0$ is a solution, can it be obtained from the given family of solutions?

*We call the solution $y = 0$ the **trivial** solution.*

Solution   If $y = 0$, then $dy/dx = 0$ and $-2xy^2 = 0$. Hence, $y = 0$ is a solution. However, in order to obtain this solution from $y(x) = 1/(x^2 + k)$, we must have $1/(x^2 + k) = 0$, which is impossible. Therefore, $y = 0$ is a singular solution.

   *Is $y = 0$ a solution to **every** differential equation?*

Most differential equations have more than one solution.

Example 5	Verify that the given solution which depends on an arbitrary constant satisfies the differential equation. (a) Solution: $y = C \sin x$ (b) Solution: $y = C_1 \sin x + C_2 \cos x$ Differential equation: $\dfrac{d^2y}{dx^2} + y = 0$ Differential equation: $\dfrac{d^2y}{dx^2} + y = 0$

Solution   (a) Differentiating, we obtain $dy/dx = C \cos x$ and $d^2y/dx^2 = -C \sin x$. Therefore,

$$\frac{d^2y}{dx^2} + y = -C \sin x + C \sin x = 0.$$

Some of the members of the family of solutions $y = C \sin x$ are graphed in Figure 1.7. Notice that these functions are the sine function with various amplitudes.

(b) In the same manner as in (a), $dy/dx = C_1 \cos x - C_2 \sin x$ and $d^2y/dx^2 = -C_1 \sin x - C_2 \cos x$. Substituting into the differential equation, we have

$$\frac{d^2y}{dx^2} + y = -C_1 \sin x - C_2 \cos x + C_1 \sin x + C_2 \cos x = 0.$$

We graph some of the members of this family of solutions in Figure 1.8.

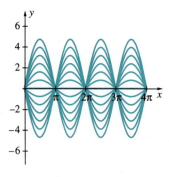

Figure 1.7 Graph of $y = C \sin x$ for various values of C

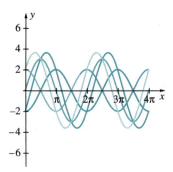

Figure 1.8

IN TOUCH WITH TECHNOLOGY

Throughout *Modern Differential Equations: Theory, Applications, Technology* we use graphs of solutions of differential equations. In some cases, we are able to predict what the graph of a solution should look like. If the graph of our proposed solution does not appear as predicted, we know that we either made a mistake in constructing our proposed solution or that our conjecture as to the general shape of the graph of the solution is wrong. In other cases, we will find that it is easier to examine the graph of a solution than it is to examine the solution (if we are able to construct one in the first place).

In Exercises 1–3, **(a)** verify that the indicated function is a solution of the given differential equation; and **(b)** graph the solution on the indicated interval(s).

1. $xy' + y = \cos x$, $y = \dfrac{\sin x}{x}$; $[-2\pi, 0) \cup (0, 2\pi]$

2. $16y'' + 24y' + 153y = 0$, $y = e^{-3x/4} \cos 3x$; $\left[0, \dfrac{3}{2}\pi\right]$

3. $x^3 \dfrac{d^3y}{dx^3} + x^2 \dfrac{d^2y}{dx^2} + x\dfrac{dy}{dx} - 40y = 0$,
$y = x^{-1} \sin(3\ln x)$; $(0, \pi]$

4. Verify that

(a) $\begin{cases} x(t) = e^{-t}\left[\dfrac{100\sqrt{3}}{3}\sin\sqrt{3}t + 20\cos\sqrt{3}t\right] \\[3mm] y(t) = e^{-t}\left[\dfrac{-40\sqrt{3}}{3}\sin\sqrt{3}t + 20\cos\sqrt{3}t\right] \end{cases}$

is a solution of the system of differential

equations $\begin{cases} x'(t) = 4y(t) \\ y'(t) = -x(t) - 2y(t) \end{cases}$. **(b)** Graph $x(t)$, $y(t)$,

and the parametric equations $\begin{cases} x = x(t) \\ y = y(t) \end{cases}$ for $0 \le t \le 2\pi$.

5. (a) Show that $(x^2 + y^2)^2 = 5xy$ is an implicit solution of

$$[4x(x^2 + y^2) - 5y]\,dx + [4y(x^2 + y^2) - 5x]\,dy = 0.$$

(b) Graph $(x^2 + y^2)^2 = 5xy$ on the rectangle $[-2, 2] \times [-2, 2]$.

(c) Approximate all points on the graph of $(x^2 + y^2)^2 = 5xy$ with x-coordinate 1.

(d) Approximate all points on the graph of $(x^2 + y^2)^2 = 5xy$ with y-coordinate -0.319.

EXERCISES 1.2

In Exercises 1–12, verify that each of the given functions is a solution to the corresponding differential equation. (A, B, and C represent constants.)

1. $\dfrac{dy}{dx} + 2y = 0$, $y(x) = e^{-2x}$, $y(x) = 5e^{-2x}$

2. $\dfrac{dy}{dx} + xy = 0$, $y(x) = e^{-x^2/2}$

3. $\dfrac{dy}{dx} + y = \sin x$, $y(x) = e^{-x} - \dfrac{1}{2}\cos x + \dfrac{1}{2}\sin x$

4. $\dfrac{d^2y}{dx^2} - \dfrac{dy}{dx} - 12y = 0$, $y(x) = e^{4x}$, $y(x) = e^{-3x}$

5. $\dfrac{d^2y}{dx^2} + 9\dfrac{dy}{dx} = 0$, $y(x) = A + Be^{-9x}$

6. $\dfrac{d^2x}{dt^2} + 3\dfrac{dx}{dt} - 10x = 0$, $x(t) = Ae^{2t} + Be^{-5t}$

7. $\dfrac{d^2x}{dt^2} + x = t\cos t - \cos t$, $x(t) = A\cos t + B\sin t + \dfrac{t^2}{4}\sin t - \dfrac{t}{2}\sin t + \dfrac{t}{4}\cos t$

8. $\dfrac{d^2y}{dx^2} - 12\dfrac{dy}{dx} + 40y = 0$, $y(x) = e^{6x}\cos 2x$, $y(x) = e^{6x}\sin 2x$

9. $\dfrac{d^3y}{dx^3} - 4\dfrac{dy}{dx} = 0$, $y(x) = A + Be^{2x} + Ce^{-2x}$

10. $\dfrac{d^3y}{dx^3} - 2\dfrac{d^2y}{dx^2} = 0$, $y(x) = A + Bx + Ce^{2x}$

11. $x^2\dfrac{d^2y}{dx^2} - 12x\dfrac{dy}{dx} + 42y = 0$, $y(x) = Ax^6 + Bx^7$

12. $x^2\dfrac{d^2y}{dx^2} + 3x\dfrac{dy}{dx} + 5y = 0$,
$y(x) = x^{-1}(A\cos(2\ln x) + B\sin(2\ln x))$

In Exercises 13–17, verify that the given equation satisfies the differential equation. Use the equation to determine y for the given value of x. Graph each equation.

13. $\dfrac{dy}{dx} = \dfrac{-x}{y}$, $x^2 + y^2 = 16$; $x = 0$

14. $3y(x^2 + y)\,dx + x(x^2 + 6y)\,dy = 0$, $x^3y + 3xy^2 = 8$; $x = 2$

15. $\dfrac{dy}{dx} = -\dfrac{2y}{x} - 3$, $x^3 + x^2y = 100$; $x = 1$

16. $y \cos x \, dx + (2y + \sin x) \, dy = 0$, $y^2 + y \sin x = 1$; $x = 0$

17. $\left(\dfrac{y}{x} + \cos y\right) dx + (\ln x - x \sin y) \, dy = 0$,

$y \ln x + x \cos y = 0$; $x = 1$

In Exercises 18–27, use integration to find a solution of the differential equation.

18. $\dfrac{dy}{dx} = (x^2 - 1)(x^3 - 3x)^3$

*19. $\dfrac{dy}{dx} = x \sin(x^2)$

20. $\dfrac{dy}{dx} = \dfrac{x}{\sqrt{x^2 - 16}}$

21. $\dfrac{dy}{dx} = \dfrac{1}{x \ln x}$

22. $\dfrac{dy}{dx} = x \ln x$

*23. $\dfrac{dy}{dx} = xe^{-x}$

24. $\dfrac{dy}{dx} = \dfrac{-2(x + 5)}{(x + 2)(x - 4)}$

*25. $\dfrac{dy}{dx} = \dfrac{x - x^2}{(x + 1)(x^2 + 1)}$

26. $\dfrac{dy}{dx} = \dfrac{\sqrt{x^2 - 16}}{x}$

*27. $\dfrac{dy}{dx} = (4 - x^2)^{3/2}$

28. The differential equation $\dfrac{dS}{dt} + \dfrac{3}{t + 100} S = 0$, where $S(t)$ is the number of pounds of salt in a particular tank at time t, is used to approximate the amount of salt in the tank containing a salt-water mixture in which pure water is allowed to flow into the tank while the mixture is allowed to flow out of the tank. If $S(t) = \dfrac{15{,}000{,}000}{(t + 100)^3}$, show that S satisfies $\dfrac{dS}{dt} + \dfrac{3}{t + 100} S = 0$. What is the initial amount of salt in the tank? As $t \to \infty$, what happens to the amount of salt in the tank?

29. The displacement (measured from $x = 0$) of a mass attached to the end of a spring at time t is given by $x(t) = 3 \cos 4t + \frac{9}{4} \sin 4t$. Show that x satisfies the ordinary differential equation $x'' + 16x = 0$. What is the initial position of the mass? What is the initial velocity of the mass?

30. Show that $u(x, y) = \ln \sqrt{x^2 + y^2}$ satisfies **Laplace's equation** $u_{xx} + u_{yy} = 0$.

31. The temperature in a thin rod of length 2π after t minutes at a position x between 0 and 2π is given by $u(x, t) = 3 - e^{-16kt} \cos 4x$. Show that u satisfies $u_t = ku_{xx}$. What is the initial temperature ($t = 0$) at $x = \pi$? What happens to the temperature at each point in the wire as $t \to \infty$?

32. The displacement u of a string of length 1 at time t and position x where x is measured from $x = 0$ is given by $u(x, t) = \sin \pi x \cos t$. Show that u satisfies $\pi^2 u_{tt} = u_{xx}$. What is the value of u at the endpoints $x = 0$ and $x = 1$ for all values of t?

*33. Find the value(s) of m so that $y = x^m$ is a solution of

$$x^2 y'' - 2xy' + 2y = 0.$$

34. Find the value(s) of m so that $y = e^{mx}$ is a solution of

$$y'' - 3y' - 18y = 0.$$

*35. Use the fact that $\dfrac{d}{dx}(e^{2x}y) = e^{2x}\dfrac{dy}{dx} + 2e^{2x}y$ and integration to solve $e^{2x}\dfrac{dy}{dx} + 2e^{2x}y = e^x$.

36. Use the fact that $\dfrac{d}{dx}(e^x y) = e^x\dfrac{dy}{dx} + e^x y$ and integration to solve $e^x\dfrac{dy}{dx} + e^x y = xe^x$.

37. The **time independent Schrödinger equation** is given by

$$-\frac{h^2}{2m}\frac{d^2\psi(x)}{dx^2} + U(x)\psi(x) = E\psi(x).$$

If $U(x) = 0$, find conditions on E so that $\psi(x) = A \sin\left(\dfrac{n\pi x}{L}\right)$ is a solution of the time independent Schrödinger equation.

38. Use implicit differentiation to show that $-\dfrac{1}{x} + \dfrac{2}{x^2} + \dfrac{1}{y} - \dfrac{1}{y^2} = C$ is an implicit solution of the differential equation $\dfrac{dy}{dx} = \dfrac{(x-4)y^3}{x^3(y-2)}$. Is $y = 0$ a solution of this differential equation? Is $y = 0$ a singular solution?

39. Show that $x + \dfrac{x^2}{y} = C$ is an implicit solution of the differential equation $\dfrac{dy}{dx} = \dfrac{y^2 + 2xy}{x^2}$. Is $y = 0$ a solution of this differential equation? Is $y = 0$ a singular solution?

40. Verify that $\begin{cases} x(t) = e^t(\sin t - 3\cos t) \\ y(t) = e^t(\cos t - 2\sin t) \end{cases}$ is a solution of the linear system of differential equations $\begin{cases} x'(t) = -2y(t) \\ y'(t) = x(t) + 2y(t) \end{cases}$.

41. Show that if $y_1(x)$ and $y_2(x)$ are solutions of the equation $ay'' + by' + cy = 0$, where a, b, and c are constants, then $C_1 y_1(x) + C_2 y_2(x)$ is also a solution for any numbers C_1 and C_2.

42. (a) Show that if $y_1(x)$ is a solution of the equation $ay'' + by' + cy = 0$ and $y_p(x)$ is a solution of $ay'' + by' + cy = f(x)$, $f(x) \neq 0$, then $y_1(x) + y_p(x)$ is a solution of $ay'' + by' + cy = f(x)$. **(b)** If $y_1(x)$ and $y_2(x)$ are solutions of the equation $ay'' + by' + cy = f(x)$, $f(x) \neq 0$, is $y_1(x) + y_2(x)$ also a solution? Explain.

1.3 Initial- and Boundary-Value Problems

In many applications, we are not only given a differential equation to solve but we are given one or more conditions that must be satisfied by the solution(s) as well. For example, if we want to find an antiderivative of the function $f(x) = 3x^2 - 4x$, we solve the differential equation

$$\frac{dy}{dx} = 3x^2 - 4x$$

which through integration yields

$$y(x) = x^3 - 2x^2 + C.$$

We call this a **general solution** because it involves an arbitrary constant and all solutions to the equation can be obtained from it. On the other hand, if we want to find the solution that passes through the point $(1, 4)$ we must find a solution that satisfies the **auxiliary condition** $y(1) = 4$. Substitution into the general solution $y(x) = x^3 - 2x^2 + C$ yields

$$y(1) = (1)^3 - 2(1)^2 + C = 4$$

so that $C = 5$. Therefore, the member of the family of solutions $y(x) = x^3 - 2x^2 + C$ that satisfies $y(1) = 4$ is

$$y(x) = x^3 - 2x^2 + 5.$$

In Figure 1.9, we graph several members of the family of solutions by substituting various values of C into the general solution in addition to the solution of the problem

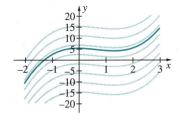

Figure 1.9 Graph of $y(x) = x^3 - 2x^2 + C$ for various values of C together with the graph of $y(x) = x^3 - 2x^2 + 5$

$$\begin{cases} \dfrac{dy}{dx} = 3x^2 - 4x \\ y(1) = 4 \end{cases}.$$

Notice that this first-order equation requires one auxiliary condition to eliminate the unknown coefficient in the general solution. (The number of conditions *typically* equals the order of the equation.) We call the auxiliary condition of a first-order equation the **initial condition** because it indicates the initial value of the dependent variable. Problems that involve a first-order equation and an initial condition are called (first-order) **initial-value problems.**

Example 1

Consider the first-order equation

$$\frac{dv}{dt} = 32 - v$$

that is solved to determine the velocity at time t, $v(t)$, of an object subjected to air resistance equivalent to the instantaneous velocity of the object. A general solution to this equation is $v(t) = 32 + ce^{-t}$, where c is a constant. If the initial velocity of the object is $v(0) = 0$, determine the solution that satisfies this initial condition.

Solution   Substituting into the general solution, we have $v(0) = 32 + c = 0$. Hence, $c = -32$, and the solution to the initial-value problem is $v(t) = 32 - 32e^{-t}$.

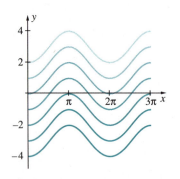

Figure 1.10 Graph of $v(t) = 32 - 32e^{-t}$

Determine $\lim_{t \to \infty} v(t)$. Does the graph of $v(t)$ in Figure 1.10 agree with this result? What does your result mean about the velocity of the object?

Example 2

Solve the initial-value problem $\dfrac{dy}{dx} = \sin x$, $y(0) = 2$.

Solution   First, we integrate to find that $y(x) = -\cos x + C$. Substitution of the initial condition into this equation gives us $y(0) = -\cos 0 + C = -1 + C = 2$, so $C = 3$ and the solution is $y(x) = -\cos x + 3$.

Figure 1.11 shows the graph of $y(x) = -\cos x + C$ for various values of C. Identify the graph of the solution that satisfies the initial condition $y(0) = 2$.

Figure 1.11

The following examples distinguish between initial- and boundary-value problems that involve higher-order equations.

Example 3

The second-order differential equation $x'' + x = 0$ can be used to model the motion of a mass attached to the end of a spring, where $x(t)$ represents the displacement of the mass from the equilibrium position $x = 0$ at time t. A general solution to this differential equation is $x(t) = A \cos t + B \sin t$. This is a second-order equation so we need two auxiliary conditions to determine the two unknown constants. Suppose that the initial position of the mass is $x(0) = 0$ and the initial velocity is $x'(0) = 1$. This is an initial-value problem because we have two auxiliary conditions given at the same value of t, namely $t = 0$. Use these initial conditions to determine the solution of this problem.

Solution We calculate $x'(t) = -A \sin t + B \cos t$ because we need the first derivative of the general solution. Substitution yields $x(0) = A = 0$ and $x'(0) = B = 1$. Hence, the solution is $x(t) = \sin t$. We graph this solution in Figure 1.12. Notice that the graph passes through the point $(0, 0)$ and the slope of the tangent line appears to be 1 at this point. If these properties (obtained from the initial conditions) did not hold, then we would know there was a mistake in our solution process.

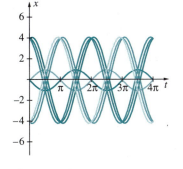

Figure 1.12

Figure 1.12 shows the graph of $x(t) = A \cos t + B \sin t$ for various values of A and B. Identify the solution that satisfies $x(0) = 0$ and $x'(0) = 1$.

Example 4

The shape of a bendable beam of length 1 unit that is subjected to a compressive force at one end can be described by the graph of the solution $y(x)$ of the differential equation $\dfrac{d^2y}{dx^2} + \dfrac{\pi^2}{4}y = 0$, $0 < x < 1$. If the height of the beam above the x-axis is known at the endpoints $x = 0$ and $x = 1$, we have a boundary-value problem because we know the value of y at the boundary (endpoints) of the beam. Use the boundary conditions $y(0) = 0$ and $y(1) = 2$ to find the shape of the beam if a general solution of the differential equation is $y(x) = A \cos \dfrac{\pi x}{2} + B \sin \dfrac{\pi x}{2}$.

Solution Applying the condition $y(0) = 0$ to the general solution yields

$$y(0) = A \cos 0 + B \sin 0 = A = 0.$$

Similarly, $y(1) = 2$ indicates that

$$y(1) = B \sin \frac{\pi}{2} = B = 2,$$

so the solution of the boundary-value problem is $y(x) = 2 \sin \dfrac{\pi x}{2}$, $0 < x < 1$.

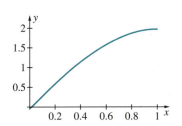

Figure 1.13 Graph of $y(x) = 2 \sin \dfrac{\pi x}{2}$

Use the graph in Figure 1.13 to check that the boundary conditions are satisfied. Does the graph pass through the points $(0, 0)$ and $(1, 2)$?

20

Chapter 1   Introduction to Differential Equations

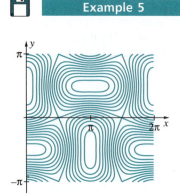

Figure 1.14   Graph of
$\cos^2 x + 2\cos x \sin y - \sin^2 y = C$ for various values of C

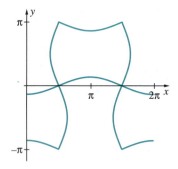

Figure 1.15   Graph of
$\cos^2 x + 2\cos x \sin y - \sin^2 y = 0$

Example 5

(a) Show that $\cos^2 x + 2\cos x \sin y - \sin^2 y = C$ is a general solution of

$$y' = \frac{\cos x \sin x + \sin x \sin y}{\cos x \cos y - \cos y \sin y}.$$

(b) Solve the initial-value problems

(i) $\begin{cases} y' = \dfrac{\cos x \sin x + \sin x \sin y}{\cos x \cos y - \cos y \sin y} \\ y(\pi/2) = 0 \end{cases}$ and   (ii) $\begin{cases} y' = \dfrac{\cos x \sin x + \sin x \sin y}{\cos x \cos y - \cos y \sin y} \\ y(3.14159) \approx 0.427079 \end{cases}$.

Does either initial-value problem have a unique solution?

Solution   (a) We see that $\cos^2 x + 2\cos x \sin y - \sin^2 y = C$ (see Figure 1.14) is a general solution of the equation by implicitly differentiating:

$$-2\cos x \sin x - 2\sin x \sin y + 2y' \cos x \cos y - 2y' \sin y \cos y = 0$$

$$-2(\cos x \sin x + \sin x \sin y) + 2(\cos x \cos y - \cos y \sin y)y' = 0$$

$$y' = \frac{\cos x \sin x + \sin x \sin y}{\cos x \cos y - \cos y \sin y}.$$

(b) We individually apply the initial conditions $y(\pi/2) = 0$ and $y(3.14159) \approx 0.427079$ and solve for C. For both (i) and (ii), we obtain the equation $\cos^2 x + 2\cos x \sin y - \sin^2 y = 0$ (see Figure 1.15).

The initial condition $y(\pi/2) = 0$ *does not* determine a unique solution to the problem because when we zoom in near the point $(\pi/2, 0)$ (see Figure 1.16(a)), the graph is not the graph of a function.

On the other hand, the initial condition $y(3.14159) \approx 0.427079$ *does* determine a unique solution because when we zoom in near the point $(3.14159, 0.427079)$ (see Figure 1.16(b)), the graph looks like the graph of a function.

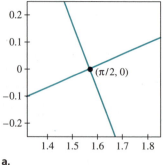

a.

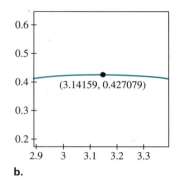

b.

Figure 1.16

Is there a solution to the differential equation that satisfies the boundary conditions $y(1.57079) \approx 0$ and $y(3.14159) \approx 0.427079$? The boundary conditions $y(1.57079) \approx 0$ and $y(3.14159) \approx 0$?

IN TOUCH WITH TECHNOLOGY

Often the calculus and algebra encountered in solving differential equations can be tedious, if not completely overwhelming or impossible. Today many sophisticated calculators and computer algebra systems are capable of performing the integration and algebraic simplification encountered when solving many differential equations. Having access to these tools can be a great advantage: a large number of problems can be solved relatively quickly, we make conjectures as to the general form of a solution to different forms of differential equations, and these tools allow us to check and verify our work.

1. Solve $\dfrac{dy}{dx} = \sin^4 x$, $y(0) = 0$ and graph the resulting solution on the interval $[0, 4\pi]$.

2. A general solution of $y^{(4)} + \dfrac{25}{2}y'' - 5y' + \dfrac{629}{16}y = 0$

   is given by

   $$y = e^{-x/2}(C_1 \cos 3x + C_2 \sin 3x) + e^{x/2}(C_3 \cos 2x + C_4 \sin 2x),$$

   where C_1, C_2, C_3, and C_4 are constants. Solve the initial-value problem

   $$\begin{cases} y^{(4)} + \dfrac{25}{2}y'' - 5y' + \dfrac{629}{16}y = 0 \\ y(0) = 0,\ y'(0) = 1,\ y''(0) = -1,\ y'''(0) = 1 \end{cases}$$

   and graph the resulting solution.

 3. A general solution of the system $\begin{cases} x' = 4y \\ y' = -4x \end{cases}$ is
   $\begin{cases} x = C_2 \sin 4t - C_1 \cos 4t \\ y = C_1 \sin 4t + C_2 \cos 4t \end{cases}$, where C_1 and C_2 are constants. Solve the initial-value problem

   $$\begin{cases} x' = 4y \\ y' = -4x \\ x(0) = 4,\ y(0) = 0 \end{cases}$$

and then graph $x(t)$, $y(t)$, and the parametric equations $\begin{cases} x = x(t) \\ y = y(t) \end{cases}$.

4. A general solution of the system $\begin{cases} x' = -5x + 4y \\ y' = 2x + 2y \end{cases}$ is
   $\begin{cases} x = C_1 e^{3t} - 4C_2 e^{-6t} \\ y = 2C_1 e^{3t} + C_2 e^{-6t} \end{cases}$, where C_1 and C_2 are constants. Solve the initial-value problem

   $$\begin{cases} x' = -5x + 4y \\ y' = 2x + 2y \\ x(0) = 4,\ y(0) = 0 \end{cases}$$

   and graph $x(t)$, $y(t)$, and the parametric equations $\begin{cases} x = x(t) \\ y = y(t) \end{cases}$.

5. A general solution of $\dfrac{d^2x}{dt^2} + \dfrac{1}{4}x = 0$ is

   $x(t) = C_1 \cos \dfrac{t}{2} + C_2 \sin \dfrac{t}{2}$. Is it possible to find C_1 and C_2 so that the solution satisfies the following boundary conditions? If so, confirm your results with a graph.

   (a) $x(0) = 2$, $x'(3\pi) = 1$
   (b) $x(0) = 2$, $x'(3\pi) = 9/10$
   (c) $x(0) = 1$, $x'(2\pi) = 2$
   (d) $x(0) = 1$, $x(3\pi) = 2$
   (e) $x(\pi) + 2x'(2\pi) = 0$
   (f) If a boundary-value problem has at least one solution and we change the boundary conditions slightly, how many solutions does the resulting boundary-value problem have? (Experiment with several different boundary conditions, if necessary.)

In Exercises 1–12, use the indicated initial or boundary conditions with the indicated general solution to determine the solution(s) to the given initial-value or boundary-value problem.

1. $\dfrac{dy}{dx} + 2y = 0$, $y(0) = 2$, $y(x) = Ae^{-2x}$

2. $\dfrac{dy}{dx} + y = \sin x$, $y(0) = -1$,

$\quad y(x) = Ae^{-x} - \dfrac{1}{2}\cos x + \dfrac{1}{2}\sin x$

3. $\dfrac{d^2y}{dx^2} - \dfrac{dy}{dx} - 12y = 0$, $y(0) = 0$, $y'(0) = 1$,

$\quad y(x) = Ae^{4x} + Be^{-3x}$

4. $\dfrac{d^2y}{dx^2} + 9\dfrac{dy}{dx} = 0$, $y(0) = 2$, $y'(0) = -1$,

$\quad y(x) = A + Be^{-9x}$

*5. $\dfrac{d^2y}{dx^2} + 9y = 0$, $y(0) = 0$, $y'(0) = 1$,

$\quad y(x) = A\cos 3x + B\sin 3x$

6. $\dfrac{d^2y}{dx^2} + 9y = 0$, $y(0) = 0$, $y'(0) = 0$,

$\quad y(x) = A\cos 3x + B\sin 3x$

7. $\dfrac{d^2y}{dx^2} - 9y = 0$, $y(0) = 0$, $y(\ln 2) = 63$,

$\quad y(x) = Ae^{3x} + Be^{-3x}$

8. $\dfrac{d^2y}{dx^2} + 9y = 0$, $y(0) = 0$, $y(\pi) = 1$,

$\quad y(x) = A\cos 3x + B\sin 3x$

*9. $\dfrac{d^3y}{dx^3} - 2\dfrac{d^2y}{dx^2} = 0$, $y(0) = 0$, $y'(0) = 1$, $y''(0) = 3$,

$\quad y(x) = A + Bx + Ce^{2x}$

10. $\dfrac{d^3y}{dx^3} - 4\dfrac{dy}{dx} = 0$, $y(0) = 1$, $y'(0) = -1$, $y''(0) = 0$,

$\quad y(x) = A + Be^{2x} + Ce^{-2x}$

11. $x^2\dfrac{d^2y}{dx^2} - 12x\dfrac{dy}{dx} + 42y = 0$, $y(1) = 1$, $y'(1) = -1$,

$\quad y(x) = Ax^6 + Bx^7$

12. $x^2\dfrac{d^2y}{dx^2} + 3x\dfrac{dy}{dx} + 5y = 0$, $\quad y(1) = 0$, $\quad y'(1) = 1$,

$\quad y(x) = x^{-1}(A\cos(2\ln x) + B\sin(2\ln x))$

In Exercises 13–16, solve the initial-value problem. Graph the solution on an appropriate interval.

13. $\dfrac{dy}{dx} = 4x^3 - x + 2$, $y(0) = 1$

14. $\dfrac{dy}{dx} = \sin 2x - \cos 2x$, $y(0) = 0$

*15. $\dfrac{dy}{dx} = \dfrac{1}{x^2}\cos\left(\dfrac{1}{x}\right)$, $y\left(\dfrac{2}{\pi}\right) = 1$

16. $\dfrac{dy}{dx} = \dfrac{\ln x}{x}$, $y(1) = 0$

17. A general solution of $y'' + 4y = 0$ is $y = A\cos 2x + B\sin 2x$. Show that the boundary-value problem

$$\begin{cases} y'' + 4y = 0 \\ y(0) = 0,\ y(\pi) = 0 \end{cases}$$

has infinitely many solutions.

18. A general solution of $y'' + y = 0$ is $y = A\cos x + B\sin x$. Determine the value(s) of k so that the boundary-value problem

$$\begin{cases} y'' + y = 0 \\ y(0) = 0,\ y(\pi) = k \end{cases}$$

has **(a)** infinitely many solutions and **(b)** no solutions.

19. If a general solution of $y'' + 36y = 0$ is $y = A\cos 6x + B\sin 6x$, determine if the boundary-value problem

$$\begin{cases} y'' + 36y = 0 \\ y(0) = 0,\ y\left(\dfrac{\pi}{2}\right) = -2 \end{cases}$$

has a solution.

20. The shape of a long, thin horizontal beam is described by $y(x) = -s(x)$ where s satisfies the differential equation $EI\dfrac{d^4s}{dx^4} = w$. In this equation E and I are constants that depend on the material of which the beam is made as well as the size and shape of the beam, and w is the constant weight per unit length of the beam. If

a general solution of this equation is

$$s(x) = A + Bx + Cx^2 + Dx^3 + \frac{w}{24EI}x^4,$$

find values of A, B, C, and D that satisfy the boundary conditions $s(0) = 0$, $s'(0) = 0$, $s(1) = 0$, $s''(1) = 0$.

21. The velocity of a falling object of mass m that is subjected to air resistance proportional to the instantaneous velocity v of the object is found by solving the initial-value problem

$$\begin{cases} m\dfrac{dv}{dt} = mg - cv & \text{(Note that } c > 0 \text{ is the} \\ & \text{proportionality constant.)} \\ v(0) = v_0 \end{cases}$$

(a) If a general solution to $m\dfrac{dv}{dt} = mg - cv$ is

$v(t) = \dfrac{mg}{c} + Ke^{-ct/m}$, find the solution of this initial-value problem.

(b) Determine $\lim_{t\to\infty} v(t)$.

22. The number of cells, $P(t)$, in a bacteria colony after t hours is determined by solving the initial-value problem

$$\begin{cases} \dfrac{dP}{dt} = kP \\ P(0) = P_0 \end{cases}$$

(a) If a general solution of $\dfrac{dP}{dt} = kP$ is $P(t) = Ce^{kt}$, use the initial condition to find C.

(b) Find the value of k so that the population doubles in 8 hours.

1.4 Direction Fields

Direction Fields • The Method of Isoclines • Phase Planes

Direction Fields

We generate the direction field for the equation $dy/dx = f(x, y)$ on the rectangle $a \le x \le b$ and $c \le y \le d$ by dividing $[a, b]$ into n equal subintervals and $[c, d]$ into m equal subintervals and then constructing the direction field for $(n + 1) \times (m + 1)$ points in this rectangle.

The geometrical interpretation of solutions to first-order differential equations of the form $dy/dx = f(x, y)$ is important to the basic understanding of problems of this type. Suppose that a solution to this equation is a function $y = \psi(x)$. If (x, y) is a point on the graph of $y = \psi(x)$, the slope of the line tangent to $y = \psi(x)$ at the point (x, y) is given by $f(x, y)$ because $dy/dx = f(x, y)$.

A set of (short) line segments representing the tangent lines can be constructed for a large number of points in the plane. This collection of line segments is known as the **direction field** of the differential equation and it provides a great deal of information concerning the behavior of the family of solutions. By determining the slope of the tangent line for a large number of points in the plane, the *shape* of the graphs of the solutions can be seen without having a formula for them. We will find direction fields especially useful when the differential equation is nonlinear or is being used to solve an applied problem.

Example 1

Determine the slope of the tangent line to the solution of the differential equation $dy/dx = x - y$ at the points $A(1, 0)$, $B(1, 1)$, $C(-1, 0)$, $D(1, -1)$, and $E(-1, -1)$. Graph a short line segment of the tangent line at each point with the corresponding slope.

TABLE 1.1

Point	Slope
$A(1, 0)$	$m = 1$
$B(1, 1)$	$m = 0$
$C(-1, 0)$	$m = -1$
$D(1, -1)$	$m = 2$
$E(-1, -1)$	$m = 0$

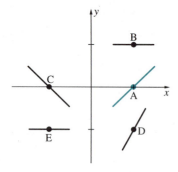

Figure 1.17

Solution   At the point $A(1, 0)$, the slope of the tangent line is $dy/dx = \underbrace{x}_{x=1} - \underbrace{y}_{y=0} = 1 - 0 = 1$. The value of dy/dx at the points B, C, and D is computed in the same way. See Table 1.1.

Line segments are drawn in Figure 1.17 at each point A, B, C, D, and E. To illustrate how these line segments are constructed, consider the line segment with slope 1 at the point $A(1, 0)$. At this point, we know that $dy/dx = 1 > 0$ so the y-coordinate increases one unit as the x-coordinate increases one unit. On the other hand, at $C(-1, 0)$, $dy/dx = -1 < 0$ so the y-coordinate decreases one unit as the x-coordinate increases one unit.

A general solution of the differential equation $dy/dx = x - y$ is $y = x - 1 + Ce^{-x}$. In Figure 1.18, we display the direction field along with several solutions of the differential equation. Notice how the solutions follow the line segments in the direction field.

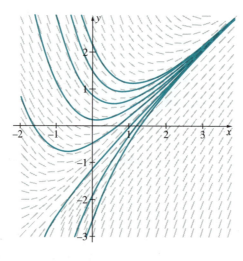

Figure 1.18   Graphs of the direction field and several solutions of $\dfrac{dy}{dx} = x - y$

For example, the slope of the tangent line to the solution that passes through $(0, 1)$ is $dy/dx = -1$. As the x-coordinate increases, the slope of the tangent along the solution curve approaches zero and is zero when the solution intersects the line $y = x$. (Why?) The slope of the tangent line along this solution then becomes positive and increases as we continue to move from left to right along the curve.

To better understand the behavior of solutions to the equation $dy/dx = x - y$, we note that $dy/dx > 0$ when $x > y$ and $dy/dx < 0$ when $x < y$. Thus, the slope of the tangent lines to the solution is positive at points (x, y) on solutions satisfying $x > y$, and is negative at points (x, y) satisfying $x < y$ as we see in the direction field.

The Method of Isoclines

Another way to generate the direction field associated with $dy/dx = f(x, y)$ is to graph several members of the family of curves $f(x, y) = c$, where c is a constant, which are called the **isoclines** of the differential equation. For each value of c, the tangent lines to solutions that intersect the isocline have equal slope c because $dy/dx = c$ along the isocline. After determining and sketching several isoclines, short line segments (representing tangent lines to solutions) of equal slope can be drawn along each isocline. The direction field is quickly generated by this technique, called the **method of isoclines.**

 Example 2   Graph the direction field associated with the differential equation $\dfrac{dy}{dx} = -\dfrac{x}{4y}$.

Solution   We graph the direction field in Figure 1.19. From $dy/dx = -x/4y$, we see that $dy/dx > 0$ when $y < 0$ and $x > 0$ (the fourth quadrant) or when $y > 0$ and $x < 0$ (the second quadrant). Similarly, $dy/dx < 0$ when $x > 0$ and $y > 0$ (the first quadrant) or when $x < 0$ and $y < 0$ (the third quadrant). From the direction field, we see that vectors in the second and fourth quadrants have positive slope while those in the first and third have negative slope. We verify these properties by graphing several members of the general solution, which is given by the implicit function $x^2 + 4y^2 = C$.

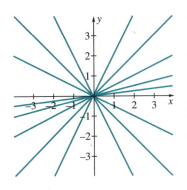 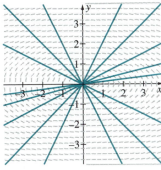 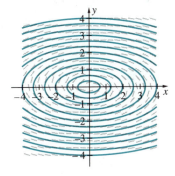

Figure 1.19

(a) Graph of $-\frac{x}{4y} = c$ for $c =$ $-2, -1, -\frac{1}{2}, -\frac{1}{4}, -\frac{1}{8}, 0, \frac{1}{8},$ $\frac{1}{4}, \frac{1}{2}, 1,$ and 2

(b) Direction field associated with $\dfrac{dy}{dx} = \dfrac{-x}{4y}$ together with several isoclines

(c) Graphs of $x^2 + 4y^2 = C$ for various values of C together with the direction field.

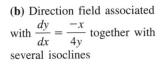

Show that $x^2 + 4y^2 = C$ is a general solution of $\dfrac{dy}{dx} = \dfrac{-x}{4y}$.

Technology allows us to graph solutions of equations and associated direction fields that would be nearly impossible by traditional methods, as shown in the following example.

Example 3

Graph the direction field associated with the differential equation

$$\frac{dy}{dx} = \frac{\cos y - y \cos x}{x \sin y + \sin x - 1}.$$

Solution   Verify that a general solution of $\dfrac{dy}{dx} = \dfrac{\cos y - y \cos x}{x \sin y + \sin x - 1}$ is the implicit function $y \sin x - x \cos y - y = C$. We graph the solution for various values of C and the direction field associated with $\dfrac{dy}{dx} = \dfrac{\cos y - y \cos x}{x \sin y + \sin x - 1}$ in Figure 1.20. Notice that the behavior of the solution depends very strongly on the initial condition: some solutions are closed paths while others are not.

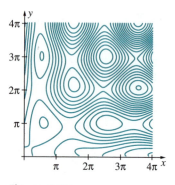
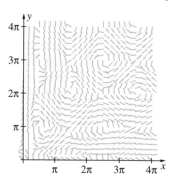
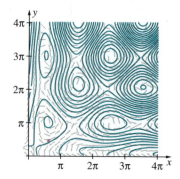

Figure 1.20

(a) Graph of
$y \sin x - x \cos y - y = C$

(b) Direction field of
$$\frac{dy}{dx} = \frac{\cos y - y \cos x}{x \sin y + \sin x - 1}$$

(c) Graphs of $y \sin x -$
$x \cos y - y = C$ and direction
field of $\dfrac{dy}{dx} = \dfrac{\cos y - y \cos x}{x \sin y + \sin x - 1}$

Phase Planes

Graphing utilities are particularly useful in graphing the direction field associated with a system of two first-order ordinary differential equations

$$\begin{cases} \dfrac{dx}{dt} = f(x, y) \\[2mm] \dfrac{dy}{dt} = g(x, y) \end{cases}.$$

Solutions of this system have the form $(x(t), y(t))$ so that a point $(x(t_0), y(t_0))$ in the xy-plane is generated for each value of t, $t = t_0$. This means that solutions form directed curves in the *xy*-plane where the collection of curves is called the **phase plane** for the system. In order to find the slope of the tangent line at points in the phase plane, we consider

$$\frac{dy}{dx} = \frac{g(x, y)}{f(x, y)}$$

because

$$\frac{dy}{dx} = \frac{dy/dt}{dx/dt}.$$

In addition to finding the slope of the tangent line, we must also determine the orientation of the curve. Therefore, the line segments in the direction field are directed so the direction field of a system of this form is made up of *vectors* that indicate the orientation of the solution curves. Although we will wait until Chapter 8 to explain how these vectors are determined, for now we will take advantage of the capabilities of a computer algebra system to generate direction fields associated with systems of differential equations.

Example 4

Graph the direction field associated with the second-order equation

$$9\frac{d^2x}{dt^2} - 6\frac{dx}{dt} + 10x = 0.$$

Solution Let $y = \dfrac{dx}{dt}$. Then, the second-order equation

$9\dfrac{d^2x}{dt^2} - 6\dfrac{dx}{dt} + 10x = 0$ is equivalent to the system of first-order equations

$\begin{cases} x' = y \\ y' = -\dfrac{10}{9}x + \dfrac{2}{3}y \end{cases}$. The direction field associated with this system is shown in

Figure 1.21. Notice that from the system of equations that $x' = y$. Therefore, in

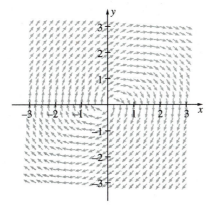

Figure 1.21 Direction field associated with
$$\begin{cases} x' = y \\ y' = -\dfrac{10}{9}x + \dfrac{2}{3}y \end{cases}$$
for $-3 \le x \le 3$ and $-3 \le y \le 3$

quadrants where $y > 0$, $x' > 0$ and x increases. This occurs in the first and second quadrants, which we can see in Figure 1.21. Similarly, in the third and fourth quadrants $x' = y < 0$, so x decreases. The vectors in these quadrants illustrate this behavior.

The phase plane associated with a system of differential equations can help us interpret the behavior of solutions to the system, especially when the equations are nonlinear and/or are being used to solve an applied problem. For example, in Chapter 8 we will see that the nonlinear system of equations

$$\begin{cases} S'(t) = -\lambda SI + \mu - \mu S \\ I'(t) = \lambda SI - \gamma I + \mu I \\ S(0) = S_0, \, I(0) = I_0 \end{cases},$$

can be used to model the spread of certain diseases, like measles, chickenpox, mumps, or scarlet fever, through a population. In the system, $S(t)$ and $I(t)$ represent the *percentage* of the population susceptible to and infected with the disease, respectively; λ, μ, and γ are constants (which we will determine at the end of Chapter 8) and the conditions $S(0) = S_0$ and $I(0) = I_0$ represent the percentage of the population initially susceptible to and infected with the disease. Because S and I represent *percentages,* we note that for all values of t, we must have that $S(t) + I(t) \le 1$. Once λ, μ, and γ have been determined, a graphing utility can be used to construct graphs of various solutions along with the direction field for the system. As an illustration, Figure 1.22 shows the graph of several solutions to the system along with the direction field associated with the system if $\lambda = 0.54203$, $\mu = 0.0142857$, and $\gamma = 0.0526316$.

Use Figure 1.22 to estimate $\lim_{t \to \infty} S(t)$ and $\lim_{t \to \infty} I(t)$ if $0 < I(0) < 1$. What are these limits if $I(0) = 0$ or $I(0) = 1$?

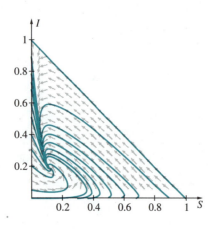

Figure 1.22   Typical direction field and solution curves associated with the nonlinear system
$$\begin{cases} S'(t) = -\lambda SI + \mu - \mu S \\ I'(t) = \lambda SI - \gamma I - \mu I \end{cases}$$

EXERCISES 1.4

In Exercises 1–10, sketch several isoclines for the given differential equation. Use the isoclines to sketch several members of the direction field.

1. $\dfrac{dy}{dx} = x - 2$ **2.** $\dfrac{dy}{dx} = 2x$

3. $\dfrac{dy}{dx} = xy$ **4.** $\dfrac{dy}{dx} = \dfrac{y}{x}$

5. $\dfrac{dy}{dx} = x - y^2$ **6.** $\dfrac{dy}{dx} = x^2 - y$

7. $\dfrac{dy}{dx} = x^2 + y^2$ **8.** $\dfrac{dy}{dx} = y^2 - x^2$

9. $\dfrac{dy}{dx} = \dfrac{1}{x + 2y}$ **10.** $\dfrac{dy}{dx} = \sqrt{x^2 + y^2}$

In Exercises 11–14, solve the differential equation and graph several solutions together with the direction field for the equation.

11. $\dfrac{dy}{dx} = \dfrac{1}{1 + x^2}$ **12.** $\dfrac{dy}{dx} = \dfrac{1}{1 - x}$

13. $\dfrac{dy}{dx} = \dfrac{1}{1 + x}$ **14.** $\dfrac{dy}{dx} = x \sin x$

In Exercises 15–20, verify that the indicated function or implicit function is a general solution to the differential equation. Graph several solutions and the direction field for the equation.

15. $\dfrac{dy}{dx} = x^2 \cos x + \dfrac{2y}{x}$, $y = Cx^2 + x^2 \sin x$

16. $\dfrac{dy}{dx} = \dfrac{6x - 3xy}{x^2 + 1}$, $y = C(x^2 + 1)^{-3/2} + 2$

***17.** $\dfrac{dy}{dx} = \dfrac{\sec^2 x - 2xy}{x^2 + 2y}$, $x^2y - \tan x + y^2 = C$

18. $\dfrac{dy}{dx} = -\dfrac{2x \cos y + 3x^2y}{x^3 - x^2 \sin y - y}$, $x^2 \cos y + x^3y - \dfrac{y^2}{2} = C$

***19.** $\dfrac{dy}{dx} = \dfrac{y - y^2 - x^2}{x}$, $\tan^{-1}\left(\dfrac{y}{x}\right) + x = C$

20. $\dfrac{dy}{dx} = \dfrac{-1}{xy^2}$, $y = (-3 \ln|x| + C)^{1/3}$

 21. Graph the direction field associated with **(a)** the *linear* equation $y' = -y \sin x$ and **(b)** the *nonlinear* equation $y' = -\dfrac{\sin x}{y}$.

22. **(a)** Graph the direction field associated with the *linear* equation $y' = x - y$ and the direction field associated with the *nonlinear* equation $y' = x - y^2$ for $0 \le x \le 4$ and $-2 \le y \le 2$. Use the direction fields to graph various solutions to each equation. **(b)** Comment on the following statement: "Suppose that $y' = f(x, y)$ is a linear equation and we are given an initial condition $y(x_0) = y_0$. If we slightly change the initial condition, then the corresponding solution also slightly changes. On the other hand, if $y' = f(x, y)$ is a nonlinear equation and we are given an initial condition $y(x_0) = y_0$ which we then slightly change, the corresponding solution may change considerably." *Hint:* Compare with Exercise 21.

23. Figure 1.23 shows the graph of various solutions of the first-order nonlinear equation $y' = 2 \sin(xy)$. Use Figure 1.23 to sketch the direction field associated with the equation. Compare your results with your classmates.

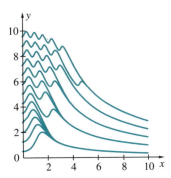

Figure 1.23

 24. Figure 1.24 shows the direction field associated with the first-order nonlinear equation $y' = \dfrac{1}{2} \sin(x - y)$. Use the direction field to graph the solutions that satisfy the initial conditions $y(0) = 2$, $y(0) = 4$, $y(0) = 6$, and $y(0) = 8$. Compare your results with your classmates. What do you notice as you move further away from your starting point?

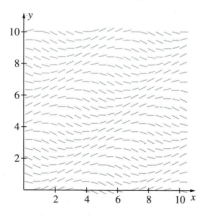

Figure 1.24

25. Consider the systems

$$\begin{cases} \dfrac{dx}{dt} = -y \\[2mm] \dfrac{dy}{dt} = x \end{cases} \quad \text{and} \quad \begin{cases} \dfrac{dx}{dt} = -y \\[2mm] \dfrac{dy}{dt} = -x \end{cases}$$

(a) For given initial conditions, do you *think* that solutions of the systems are similar or different? Why?
(b) Figure 1.25 shows the direction field associated with each system. Use the direction fields to graph the solutions that satisfy the initial conditions
(i) $x(0) = 0.5$, $y(0) = 0$; **(ii)** $x(0) = -0.25$, $y(0) = 0$; **(iii)** $x(0) = 0$, $y(0) = 0.75$; and **(iv)** $x(0) = 0$, $y(0) = -0.5$.
(c) How do your graphs affect your conjecture in (a)?

26. Consider the systems

$$\begin{cases} \dfrac{dx}{dt} = \dfrac{1}{2}x \\[2mm] \dfrac{dy}{dt} = y \end{cases} \quad \text{and} \quad \begin{cases} \dfrac{dx}{dt} = -\dfrac{1}{2}x \\[2mm] \dfrac{dy}{dt} = -y \end{cases}$$

(a) For given initial conditions, do you *think* that solutions of the systems are similar or different? Why?
(b) Figure 1.26 shows the direction field associated with each system. Use the direction fields to graph the solutions that satisfy the initial conditions **(i)** $x(0) = 0.5$, $y(0) = 0.25$; **(ii)** $x(0) = -0.25$, $y(0) = -0.5$; **(iii)** $x(0) = -0.5$, $y(0) = 0.75$; and **(iv)** $x(0) = 0.75$, $y(0) = -0.5$. **(c)** How do your graphs affect your conjecture in (a)? **(d)** For each system, use your graphs to calculate $\lim_{t\to\infty} x(t)$ and $\lim_{t\to\infty} y(t)$, as long as not both $x(t)$ and $y(t)$ are the zero function.

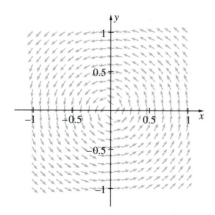

Figure 1.25
(a) Direction field associated with

$$\begin{cases} \dfrac{dx}{dt} = -y \\[2mm] \dfrac{dy}{dt} = x \end{cases}$$

(b) Direction field associated with

$$\begin{cases} \dfrac{dx}{dt} = -y \\[2mm] \dfrac{dy}{dt} = -x \end{cases}$$

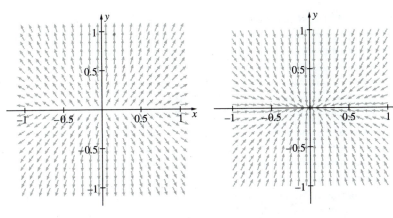

Figure 1.26
(a) Direction field associated with
$$\begin{cases} \dfrac{dx}{dt} = \dfrac{1}{2}x \\ \dfrac{dy}{dt} = y \end{cases}$$

(b)   Direction field associated with
$$\begin{cases} \dfrac{dx}{dt} = -\dfrac{1}{2}x \\ \dfrac{dy}{dt} = -y \end{cases}$$

1.5 The Origins of Differential Equations

Population • Ideal Capacitors and Inductors • L-R-C Electric Circuits • Mechanics • Spring-Mass Systems • Temperature in a Thin Rod or Wire

In the previous sections we have introduced some of the terminology used in this text. In particular, we have described several differential equations, determined how to classify them, and stated some solutions. Now, we turn our attention briefly to indicating how differential equations arise and, consequently, why it is necessary and interesting to study them. The applications discussed here and their origins will be discussed in more detail later in the text.

Population

Let $P(t)$ represent the number of individuals in a certain population. The population under consideration might include cells, humans, or radioactive material. If we assume that the rate of change of the size of the population is proportional to the number in the current population, then the solution of the differential equation

$$\frac{dP}{dt} = kP,$$

where k is a constant (representing the rate of growth or decay of P), yields the size of the population $P(t)$. This model was first studied by Thomas Malthus (1766–1834), a British economist who noticed that the size of many populations grow at a rate proportional to the size of the current population.

If $k > 0$, then $dP/dt > 0$, so $P(t)$ is an increasing function. (Why?) In this case, the differential equation corresponds to the growth of the size of a population, such as the growth of a colony of bacteria cells in a culture. A solution to a problem based on this model is shown in Figure 1.27(a). Notice how the size of the population increases without bound. If $k < 0$, then $dP/dt < 0$, so $P(t)$ is a decreasing function. In this case, the differential equation corresponds to the decay of the size of a population, such as the decay of a radioactive element like carbon-14 in fossils. The graph of a solution to a differential equation of this form is shown in Figure 1.27(b). We see that the size of the population decreases as t increases.

The model $dP/dt = kP$ does not accurately predict the number of individuals in many populations. To improve the approximation, Pierre F. Verhulst (1804–1849), a Belgian mathematician, introduced the model

$$\frac{dP}{dt} = k(P)P$$

in which the growth rate $k(P)$ depends on the population. In many cases, we let $k(P) = $ birth rate $-$ death rate (however, we can consider other factors). One choice for the function $k(P)$ is

$$k(P) = r - aP.$$

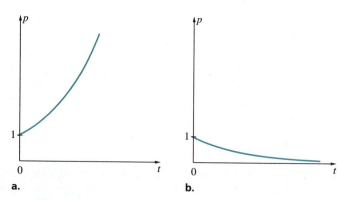

a.   **b.**

Figure 1.27

(a) Exponential Growth
   Typical solution of
   $\dfrac{dP}{dt} = kP$, $k > 0$

(b) Exponential Decay
   Typical solution of
   $\dfrac{dP}{dt} = kP$, $k < 0$

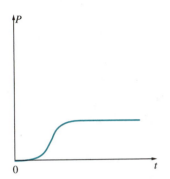

Figure 1.28 A typical solution of the logistic equation

This model is called the *logistic equation* and leads to a limiting size of the population, which is expected in most cases of population growth. In this equation, r and a are positive constants. This equation can be written as $dP/dt = rP - aP^2$ where the term $(-P^2)$ represents an inhibitive factor, so the population under these assumptions is not allowed to grow out of control as it was in the Malthus model. The graph of a solution to the logistic equation for certain values of r, a, and an initial population is shown in Figure 1.28. Notice that the population increases for all values of t. However, the size of the population approaches a limit in this case.

In many natural settings, two (or more) competing populations are present in one environment. This situation was first stated mathematically by Vito Volterra (1860–1940), an Italian mathematician who developed a model to study the interaction between sharks and food fish in the Adriatic Sea before and after World War I. Other examples of populations of species that can be used with this model, called the *predator-prey model,* include foxes and rabbits as well as ladybugs and aphids. We are considering two populations, so we have a differential equation to model each population. This yields the nonlinear system of differential equations

$$\begin{cases} \dfrac{dx}{dt} = x(a - by) \\ \dfrac{dy}{dt} = y(-c + dx) \end{cases},$$

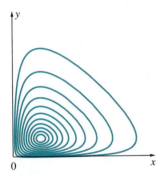

Figure 1.29 Graphs of typical solutions to the predator-prey system

where $x(t)$ and $y(t)$ represent the prey and predator populations, respectively, at time t and a, b, c, and d are positive constants. In this model, the xy terms indicate an interaction between the two populations (like a shark eating a fish or a fox eating a rabbit). This interaction causes $x(t)$ to decrease (at the rate b) while it causes $y(t)$ to increase (at the rate d). The constants a and c represent the rate of growth of $x(t)$ and the rate of death of $y(t)$, respectively. We graph solutions $(x(t), y(t))$ to this system of equations in the xy-plane for various initial populations in Figure 1.29.

Ideal Capacitors and Inductors

Capacitors and inductors are used in electrical circuits. The word *ideal* indicates that no resistance is involved. A **capacitor** is made up of two metallic conductors that are located close together but do not actually touch. If an electric charge is applied to one of the metal plates, then a charge of $-Q$ appears on the opposite plate and a voltage V arises across the capacitor. The quantities of Q (charge), C (capacitance), and V (voltage) are related by the equation $Q = CV$. Let V_A and V_B represent the voltage at terminals A and B, respectively, and let the difference $V_A - V_B$ be represented by V_{AB}. Then, the current $I_{A \to B}$ through the capacitor pictured in Figure 1.30 is determined by

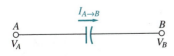

Figure 1.30

the differential equation $I_{A \to B} = C \dfrac{d}{dt} V_{AB}$.

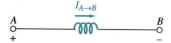

Figure 1.31

In a similar manner, we consider the ideal **inductor,** which is a two-terminal circuit element. In this case, L represents the inductance. For the situation described in Figure 1.31, $V_{AB} = L\dfrac{d}{dt}I$ where I represents the current that passes through the inductor.

More complicated circuits that involve resistors, capacitors, and inductors are described mathematically using these basic relationships. They are briefly described as follows as well as in Chapters 3, 5, and 9.

L-R-C Electric Circuits

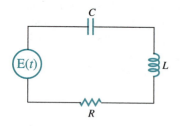

Figure 1.32

An *L-R-C* circuit, pictured in Figure 1.32, consists of an inductor (L = inductance), a resistor (R = resistance), and a capacitor (C = capacitance) in series with a supplied voltage $E(t)$, such as a battery. In an *L-R-C* circuit, we want to find the current $I(t)$ (in amperes) in the circuit and the charge $Q(t)$ (in coulombs) on the capacitor where $dQ/dt = I$. These quantities are found by modeling this situation according to Kirchhoff's law, which states that the sum of the voltage drops across the elements of the simple loop, the inductor, resistor, and capacitor, equals the applied voltage $E(t)$. Using the voltage drops in Table 1.2, we see that the differential equation

$$L\frac{dI}{dt} + RI + \frac{1}{C}Q = E(t)$$

or, after differentiating both sides of the equation,

$$LI'' + RI' + \frac{1}{C}Q' = E'(t)$$

models the circuit by summing the voltage drops over the elements. Writing this equation in terms of I yields

$$LI'' + RI' + \frac{1}{C}I = E'(t).$$

TABLE 1.2   Voltage Drops in an *L-R-C* Circuit

Circuit Element	Voltage Drop
Inductor	$L\dfrac{dI}{dt}$
Resistor	RI
Capacitor	$\dfrac{1}{C}Q$

Mechanics

Newton's second law of motion states that the rate of change of momentum of a body with respect to time equals the resultant force acting on the body. This principle, which is written mathematically as

$$ma = F \qquad \text{or} \qquad m\frac{dv}{dt} = F,$$

is used to model many problems in mechanics, such as determining the position and velocity of a free-falling body, finding the escape velocity of a rocket, and determining the motion of a mass attached to the end of a spring. Let m be the mass of an object such as a ball, g the gravitational constant, c a positive constant and $v(t)$ the velocity of the object at time t. Suppose that the object is dropped from the roof of a building. If we assume that the air resistance is equivalent to c times the instantaneous velocity of the object, then $v(t)$ is found by solving the differential equation

$$m\frac{dv}{dt} = mg - cv,$$

which is derived with $m\dfrac{dv}{dt} = F$ where F is the sum of the forces acting on the object.

Notice that on the right-hand side of this differential equation, the force due to air resistance (cv) acts *against* the force due to gravity (mg).

Spring-Mass Systems

Suppose that a mass is attached to the end of an elastic spring that is suspended from a fixed support such as a ceiling. When the mass is attached it stretches the spring by an amount s. Hooke's law states that the spring exerts an upward restoring force that is proportional to s. This is stated mathematically as $F = ks$, where k is called the spring constant. When the mass comes to rest, we say that it is at its equilibrium position. Let $x(t)$ represent the position of the mass measured from its equilibrium at time t. If the spring-mass system is subjected to an external force, $F(t)$, and a damping force proportional to x' (the velocity), the differential equation

$$mx'' + cx' + kx = F(t),$$

which is obtained by Newton's second law, is solved to find the displacement of the mass. In Figure 1.33, we graph the spring along with the graph of the negative of the solution, $-x(t)$. The displacement of the spring is given by the corresponding position $(t, -x(t))$ on the graph of the solution.

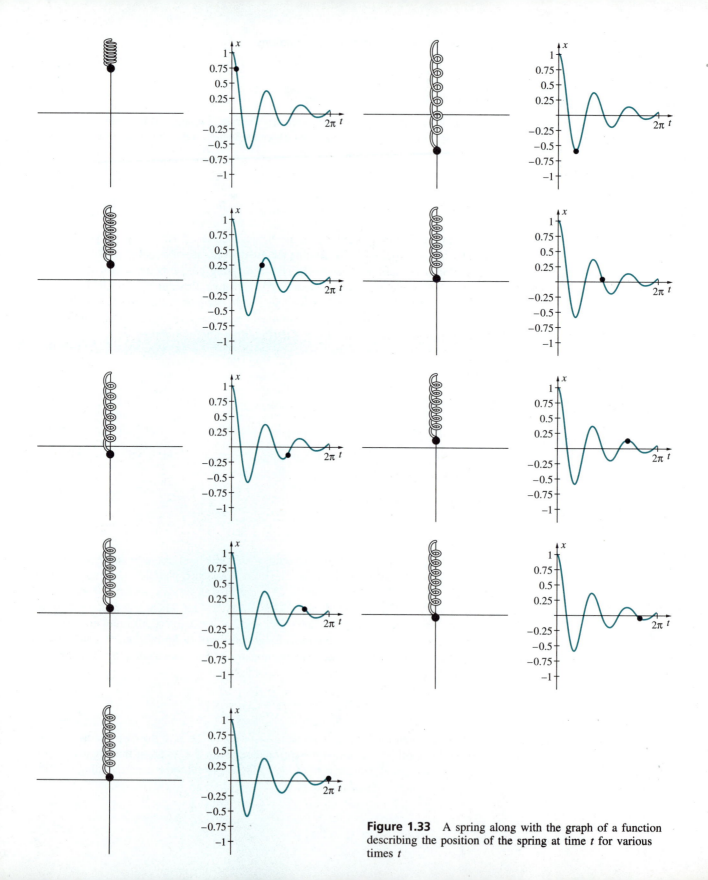

Figure 1.33 A spring along with the graph of a function describing the position of the spring at time t for various times t

Figure 1.34

Temperature in a Thin Rod or Wire

Let $u(x, t)$ represent the temperature at position x and time t in a homogeneous thin rod or wire of length L (see Figure 1.34). If we assume that the temperature flows only in the x-direction from warmer positions to cooler positions, then the partial differential equation

$$c^2 \frac{\partial^2 u}{\partial x^2} = \frac{\partial u}{\partial t}, \qquad 0 < x < L, \qquad t > 0$$

where c^2 is the thermal diffusivity of the wire, is used to find $u(x, t)$. The constant c^2 differs for various materials and describes the wire's ability to conduct heat. An interesting property associated with this problem is that after a sufficient period of time (based on the composition of the wire), the temperature at position x no longer depends on t. At this point, we say that we have reached the **steady-state temperature** $s(x)$. In Figure 1.35, we graph $u(x, t)$ for various values of t. Notice that these solutions approach a linear function as t increases so that eventually the temperature at position x can be approximated without knowing the time t.

In our attempt to describe situations, we observe that some models are easier to verify than others. Often, we will see that applications from many areas of physics and engineering are based on physical laws. In other cases, we will greatly simplify models by making various assumptions. These assumptions can considerably affect the accuracy of the model.

For example, although we might be able to use the logistic equation

$$\frac{dP}{dt} = rP - aP^2$$

to model the population growth of the United States through the year 1950, we probably would not want to use this equation to model the population of a species that is affected by harvesting (like hunting, fishing, or disease). If we wish to take a constant

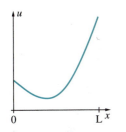
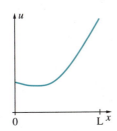
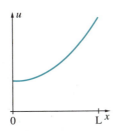
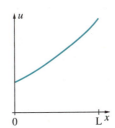
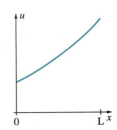

Figure 1.35

harvest rate, H, into consideration, then we might instead modify the logistic equation and use the equation

$$\frac{dP}{dt} = rP - aP^2 - H$$

to model the population.*

IN TOUCH WITH TECHNOLOGY

 1. (Competing Species) The system of equations

$$\begin{cases} \dfrac{dx}{dt} = x(a - b_1 x - b_2 y) \\[2mm] \dfrac{dy}{dt} = y(c - d_1 x - d_2 y) \end{cases}$$

where a, b_1, b_2, c, d_1, and d_2 represent positive constants, can be used to model the population of two species, represented by $x(t)$ and $y(t)$, competing for common food supply.

(a) Figure 1.36 shows the direction field for the system if $a = 1$, $b_1 = 2$, $b_2 = 1$, $c = 1$, $d_1 = 0.75$, and

$d_2 = 2$. **(i)** Use the direction field to graph various solutions if both $x(0)$ and $y(0)$ are not zero. **(ii)** Use the direction field and your graphs to approximate $\lim_{t\to\infty} x(t)$ and $\lim_{t\to\infty} y(t)$.

(b) Figure 1.37 shows the direction field for the system if $a = 1$, $b_1 = 1$, $b_2 = 1$, $c = 0.67$, $d_1 = 0.75$, and $d_2 = 1$. **(i)** Use the direction field to graph various solutions if both $x(0)$ and $y(0)$ are not zero. **(ii)** Use the direction field and your graphs to determine the fate of the species with population $y(t)$. What happens to the species with population $x(t)$?

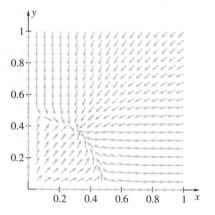

Figure 1.36

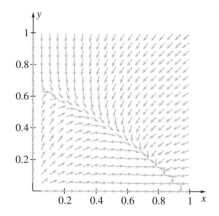

Figure 1.37

* David A. Sanchez, ''Populations and Harvesting,'' *Mathematical Modeling: Classroom Notes in Applied Mathematics,* Murray S. Klamkin, Editor, SIAM (1987) pp. 311–313.

1. The number of people $P(t)$ (in thousands) in a community after t years is determined by solving the initial-value problem $\begin{cases} \dfrac{dP}{dt} = kP \\ P(0) = 5 \end{cases}$. **(a)** Show that $P(t) = Ce^{kt}$ is a general solution of $dP/dt = kP$ and use the initial condition to find C. **(b)** If the population triples in 6 years, what is the value of k?

2. The number of bacteria $P(t)$ in a colony after t hours is determined by solving the initial-value problem $\begin{cases} \dfrac{dP}{dt} = kP \\ P(0) = P_0 \end{cases}$. **(a)** Use the result of Exercise 1 (a) and the initial condition to find C. **(b)** Find the value of k so that the population is $4P_0$ after 10 hours.

*3. The population $P(t)$ (in thousands) in a town grows at a rate proportional to the population at any time, so $P(t)$ satisfies $dP/dt = kP$ with general solution $P(t) = Ce^{kt}$. If the initial population of 10 increases to 15 in 3 years, we have the auxiliary conditions $P(0) = 10$ and $P(3) = 15$. Use these conditions to find $P(t)$.

4. The population of birds on an island is found to grow at a rate equal to $\frac{1}{10}$ the population at any time, so $P(t)$ satisfies the equation $dP/dt = \frac{1}{10}P$ with general solution $P(t) = Ce^{t/10}$. Find $P(t)$ if the initial population is 500.

*5. The amount $P(t)$ (in milligrams) of a radioactive isotope disintegrates at a rate proportional to the amount present, so P satisfies the differential equation $dP/dt = kP$ with general solution $P(t) = Ce^{kt}$. After 50 days, the initial amount of 1000 mg is reduced to 920 mg, so we have the auxiliary conditions $P(0) = 1000$ and $P(50) = 920$. Use these conditions to find $P(t)$.

6. The amount $P(t)$ of carbon-14 in a fossil satisfies the differential equation $dP/dt = kP$ with general solution $P(t) = Ce^{kt}$. If the half-life of carbon-14 is 5568 years, then $P(5568) = \frac{1}{2}P_0$ where $P(0) = P_0$. Use this information to find a formula for $P(t)$ that depends on P_0. Find the value of t when $P(t) = \frac{1}{4}P_0$.

*7. After spring break, two students with a flu virus return to an isolated college campus of 2500 students. If the students do not leave the campus during the duration of the virus, then the number of students $P(t)$ that acquire the virus is found by solving the logistic equation $dP/dt = kP(2500 - P)$ where $P(0) = 2$. A general solution to this problem after application of the initial condition is $P(t) = \dfrac{2500}{1 + 1249e^{-2500kt}}$. If 25 students have the virus after 2 days, then we have the condition $P(2) = 25$. Use this condition to find k.

8. The solution of the logistic equation $dP/dt = P(r - aP)$ with initial condition $P(0) = P_0$ is $P(t) = \dfrac{rP_0}{aP_0 + (r - aP_0)e^{-rt}}$. Find $\lim_{t \to \infty} P(t)$ assuming that $r > 0$.

9. A falling object of mass $m = 2$ slugs $\left(\dfrac{\text{lb} - \text{s}^2}{\text{ft}}\right)$ is subjected to air resistance that is equivalent to twice the instantaneous velocity. If the object is dropped with no initial velocity, its velocity is found by solving the initial-value problem $\begin{cases} 2\dfrac{dv}{dt} = 64 - 2v \\ v(0) = 0 \end{cases}$ where a general solution to $2\dfrac{dv}{dt} = 64 - 2v$ is $v(t) = 32 + Ke^{-t}$. Use the initial condition to find K. What is the velocity of the object at $t = 1$ second?

10. Suppose that the object in Exercise 9 is released with a downward velocity of 8 ft/s. Use the general solution $v(t) = 32 + Ke^{-t}$ with the initial velocity $v(0) = 8$ to find $v(t)$. What is the velocity of the object at $t = 2$ seconds?

11. Determine a differential equation that the current $I(t)$ of an L-R-C circuit in which $E(t) = 0$, $L = 0.2$ henry, $R = 300$ ohms, and $C = 10^{-5}$ farad satisfies.

12. If the L-R-C circuit in Exercise 11 has no initial current and an initial charge of 2×10^{-6}, state the two initial conditions on I that must be used to find $I(t)$.

13. An object of mass $m = 4\dfrac{\text{lb} - \text{s}^2}{\text{ft}}$ is attached to a spring with spring constant $k = 16$ lb/ft. If there is no damping ($c = 0$) and no external force, $F(t) = 0$, set up the differential equation that is satisfied by the displacement of the object on the spring.

14. An object of mass $m = 1\dfrac{\text{lb} - \text{s}^2}{\text{ft}}$ is attached to a spring with spring constant $k = 13$ lb/ft. If the motion of the object is subjected to a damping force equivalent to $4x'$ and there is no external force, find a differential equation satisfied by the displacement of the object on the spring.

15. The temperature in a wire of length 1 is found to be $u(x, t) = e^{-\pi^2 t} \cos \pi x$. Show that u satisfies the heat equation $\dfrac{\partial^2 u}{\partial x^2} = \dfrac{\partial u}{\partial t}$.

16. Find the steady-state temperature in a thin rod in which the temperature is given by $u(x, t) = e^{-4t} \cos 2x + x + 10$ by calculating $\lim_{t \to \infty} u(x, t)$.

CHAPTER 1 SUMMARY
Concepts and Formulas

Section 1.1

Differential Equation

An equation that contains the derivative or differentials of one or more dependent variables with respect to one or more independent variables.

Ordinary Differential Equation

If a differential equation contains only ordinary derivatives (of one or more dependent variables) with respect to a single independent variable, the equation is called an **ordinary differential equation.**

Partial Differential Equation

A differential equation that contains the partial derivatives or differentials of one or more independent variables with respect to more than one independent variable.

Linear Ordinary Differential Equation

A **linear ordinary differential equation** is of the form

$$a_n(x)\frac{d^n y}{dx^n} + a_{n-1}(x)\frac{d^{n-1}y}{dx^{n-1}} + \cdots$$
$$+ a_2(x)\frac{d^2 y}{dx^2} + a_1(x)\frac{dy}{dx} + a_0(x)y = f(x)$$

Order of an Equation

The order of highest derivative in the differential equation is called the **order** of the equation.

Section 1.2

Solution

A **solution** of a differential equation on a given interval is a function that is continuous on the interval and has all the necessary derivatives that are present in the differential equation such that when substituted into the equation yields an identity for all values on the interval.

Explicit Solution

A solution given as a function of the independent variable.

Implicit Solution

A solution given as a relation $f(x, y) = 0$.

Trivial Solution

$y(x) = 0$.

Section 1.3

General Solution of an Nth Order Linear Equation

A solution that depends on n arbitrary constants and includes all solutions of the equation.

Section 1.4

Direction Field

A collection of line segments that indicate the slope of the tangent line to the solution(s) of a differential equation.

CHAPTER 1 REVIEW EXERCISES

In Exercises 1–5, determine (a) if the equation is an ordinary differential equation or partial differential equation; (b) the order of the ordinary differential equation; and (c) if the equation is linear or nonlinear.

1. $y' = y$

2. $au_x + u_t = 0$, a constant

3. $\dfrac{d^2y}{dx^2} + 2\dfrac{dy}{dx} + y = 0$

4. $m\ddot{x} + kx = \sin t$, m and k constants

5. $\dfrac{\partial\phi}{\partial x}\dfrac{\partial^2\phi}{\partial x^2} = \dfrac{\partial^2\phi}{\partial y^2}$

In Exercises 6–13, verify that the given function is a solution to the corresponding differential equation. (A and B denote constants.)

6. $\dfrac{dy}{dx} + y\cos x = 0$, $y = e^{-\sin x}$

7. $\dfrac{dy}{dx} - y = \sin x$, $y = \dfrac{1}{2}(e^x - \cos x - \sin x)$

8. $y'' + 4y' - 5y = 0$, $y = e^{-5x}$, $y = e^x$

9. $y'' - 6y' + 45y = 0$, $y = e^{3x}[\cos 6x - \sin 6x]$

10. $x^2y'' - xy' - 16y = 0$, $y = Ax^5 + Bx^{-3}$

11. $x^2y'' + 3xy' + 2y = 0$,
    $y = x^{-1}[\cos(\ln x) - \sin(\ln x)]$

12. $\dfrac{d^2y}{dx^2} + 2\dfrac{dy}{dx} + 2y = x$, $y = \dfrac{1}{2}(x - 1)$

13. $y'' - 7y' + 12y = 2$, $y = Ae^{3x} + Be^{4x} + \dfrac{1}{6}$

In Exercises 14 and 15, verify that the given implicit function satisfies the differential equation.

14. $(2x - 3y)dx + (2y - 3x)dy = 0$, $x^2 - 3xy + y^2 = 1$

*15. $(y\cos(xy) + \sin x)dx + x\cos(xy)dy = 0$,
    $\sin(xy) - \cos x = 0$

In Exercises 16–19, find a solution of the differential equation.

16. $\dfrac{dy}{dx} = xe^{-x^2}$

*17. $\dfrac{dy}{dx} = x^2\sin x$

18. $\dfrac{dy}{dx} = \dfrac{2x^2 - x + 1}{(x - 1)(x^2 + 1)}$

*19. $\dfrac{dy}{dx} = \dfrac{x^2}{\sqrt{x^2 - 1}}$

In Exercises 20–22, use the indicated initial or boundary conditions with the given general solution to determine the solution(s) to the given problem.

20. $\dfrac{dy}{dx} + 2y = x^2$, $y(0) = 1$,
    $y = \dfrac{1}{4} - \dfrac{1}{2}x + \dfrac{1}{2}x^2 + Ae^{-2x}$

*21. $y'' + 4y = x$, $y(0) = 1$, $y\left(\dfrac{\pi}{4}\right) = \dfrac{\pi}{16}$,
    $y = \dfrac{x}{4} + A\cos 2x + B\sin 2x$

22. $x^2y'' + 5xy' + 4y = 0$, $y(1) = 1$, $y'(1) = 0$,
    $y = Ax^{-2} + Bx^{-2}\ln x$

In Exercises 23 and 24, solve the initial-value problem. Graph the solution on an appropriate interval.

23. $\dfrac{dy}{dx} = \cos^2 x\sin x$, $y(0) = 0$

24. $\dfrac{dy}{dx} = \dfrac{4x - 9}{3(x - 3)^{2/3}}$, $y(0) = 0$

25. The temperature on the surface of a steel ball at time t is given by $u(t) = 70e^{-kt} + 30$ (in °F) where k is a positive constant. Show that u satisfies the first-order equation $du/dt = -k(u - 30)$. What is the initial temperature ($t = 0$) on the surface of the ball? What happens to the temperature as $t \to \infty$?

26. The displacement (measured from $x = 0$) of a mass attached to the end of a spring at time t is given by $x(t) = \dfrac{1}{4}e^{-t}\left(\cos\sqrt{35}t + \dfrac{9}{\sqrt{35}}\sin\sqrt{35}t\right)$. Show that x satisfies the ordinary differential equation $x'' + 2x' + 36x = 0$. What is the initial displacement of the mass? What is the initial velocity of the mass?

27. For a particular wire of length 1 unit, the temperature u at time t hours at a position of x feet from the end $(x = 0)$ of the wire is estimated by $u(x, t) = e^{-\pi^2 kt} \sin \pi x - e^{-4\pi^2 kt} \sin 2\pi x$. Show that u satisfies the heat equation $u_t = ku_{xx}$. What is the initial temperature $(t = 0)$ at $x = 1$? What happens to the temperature at each point in the wire as $t \to \infty$?

28. The height u of a long string at time t and position x where x is measured from the middle of the string $(x = 0)$ is given by $u(x, t) = \sin x \cos 2t$. Show that u satisfies the wave equation $u_{tt} = 4u_{xx}$. What is the initial height $(t = 0)$ at $x = 0$?

29. Show that $u(x, y) = \tan^{-1}(y/x)$ satisfies Laplace's equation $u_{xx} + u_{yy} = 0$.

2

FIRST-ORDER ORDINARY DIFFERENTIAL EQUATIONS

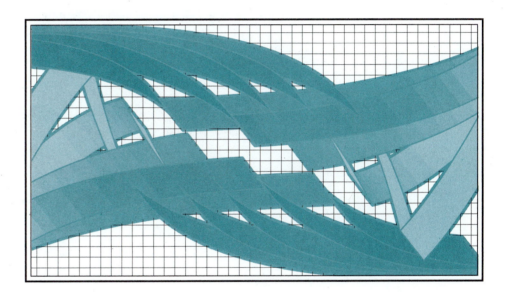

We will devote a considerable amount of time in this course to developing explicit, implicit, numerical, and graphical solutions of differential equations. In this chapter we discuss first-order ordinary differential equations and some methods used to construct explicit, implicit, numerical, and graphical solutions of them. Several of the equations and methods of solution discussed here will be used in later chapters of the text.

2.1 Separation of Variables

In Chapter 1 we used integration to solve problems of the form $dy/dx = f(x)$. This equation can be written as $dy = f(x)\, dx$ so that the expressions involving x and y are separated. This *separation of variables* technique can be carried out to solve similar problems.

Definition 2.1 Separable Differential Equation

A first-order differential equation that can be written in the form $g(y)y' = f(x)$ or $g(y)\, dy = f(x)\, dx$ is called a **separable differential equation.**

Separable differential equations are solved by collecting all the terms involving y on one side of the equation and all the terms involving x on the other side of the equation and then integrating each side of the equation. Rewriting $g(y)y' = f(x)$ in the form

$$g(y)\frac{dy}{dx} = f(x)$$

yields $g(y)\, dy = f(x)\, dx$, so that

$$\int g(y)\, dy = \int f(x)\, dx + C,$$

where C is a constant. This technique, called **separation of variables,** was first discovered by Leibniz in 1691.

 Explain why $\int g(y)\, dy$ and $\int f(x)\, dx$ differ by a constant.

Example 1 Show that the following equations are separable and solve them: (a) $\dfrac{dy}{dx} = 2y$

(b) $\dfrac{dy}{dx} = -\dfrac{x}{y}$ (c) $\dfrac{dy}{dx} = \dfrac{2y^{1/2} - 2y}{x}$

Solution (a) The equation is separable because we can write it as

$$\frac{dy}{y} = 2dx.$$

Integrating the left-hand side with respect to y and the right-hand side with respect to x gives us

$$\ln|y| = 2x + C_1,$$

where C_1 is a constant. Whenever possible, we *like* to obtain an explicit formula for the solution; in this case, we need to solve the equation $\ln|y| = 2x + C_1$ for y. Using the exponential function, we have

$$|y| = e^{2x+C_1} = e^{2x}e^{C_1},$$

or

$$y = \pm e^{2x}e^{C_1}.$$

Notice that e^{C_1} is a positive constant regardless of the value of C_1. This means that $y = \pm e^{2x}e^{C_1} = \pm e^{C_1}e^{2x}$ represents positive and negative constant multiples of the function e^{2x}. We can let $k = \pm e^{C_1}$ so that the solution is

$$y = ke^{2x},$$

where k takes on positive and negative values. If we allow k to take on the value $k = 0$, then $y = 0$ is obtained from $y = ke^{2x}$ and is a solution of the differential equation, which we lost when we divided by y in the separation of variables process. This shows us that functions that cause the divisor to be zero in the separation process may be solutions to the differential equation. We must check these functions on an individual basis by substitution into the differential equation as we did here.

 Show that e^{C_1} is positive for all numbers C_1.

(b) In this case, we see that the equation is separable by expressing it as

$$x\,dx = -y\,dy.$$

Integration yields

$$\frac{x^2}{2} = -\frac{y^2}{2} + C_1.$$

Multiplying by 2 and simplifying gives us $x^2 + y^2 = 2C_1$. If we let $k = 2C_1$, then the solutions satisfy the equation

$$x^2 + y^2 = k,$$

so that an implicit solution is circles centered at the origin of radius $\sqrt{k}$, as shown in Figure 2.1.

 Is $x^2 + y^2 = 0$ an implicit solution of the equation? Explain.

(c) The equation $\dfrac{dy}{dx} = \dfrac{2y^{1/2} - 2y}{x}$ is separable because it can be written in the form

$$\frac{dy}{2y^{1/2} - 2y} = \frac{dx}{x}.$$

To solve the equation, we integrate both sides and simplify. We rewrite this equation as

$$\int \frac{1}{2y^{1/2}}\frac{dy}{(1 - y^{1/2})} = \int \frac{dx}{x} + C_1.$$

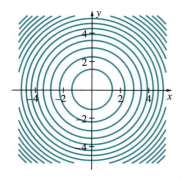

Figure 2.1 Graph of $x^2 + y^2 = k$ for various values of k

To evaluate the integral on the left-hand side, let $u = 1 - y^{1/2}$ so $du = \dfrac{-dy}{2y^{1/2}}$. We then obtain

$$\int \frac{-du}{u} = \int \frac{dx}{x} + C_1$$

so that $-\ln|u| = \ln|x| + C_1$. Recall that $-\ln|u| = \ln|u|^{-1} = \ln\dfrac{1}{|u|}$, so we have

$$\ln\frac{1}{|u|} = \ln|x| + C_1.$$

Simplification leads to

$$\frac{1}{|u|} = e^{\ln|x| + C_1} = C|x|$$

where $C = e^{C_1}$. Resubstituting we find that

$$\frac{1}{|1 - y^{1/2}|} = C|x| \qquad \text{or} \qquad x = \pm\frac{1}{C(1 - y^{1/2})}$$

is a general solution of the equation $\dfrac{dy}{dx} = \dfrac{2y^{1/2} - 2y}{x}$. We graph several members of this family of solutions in Figure 2.2. Notice that we must choose $0 \le y < 1$ or $y > 1$ so that these solutions are defined.

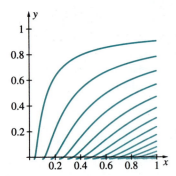

Figure 2.2

Identify the curve for which the value of C is approximately (a) 47.4132; and (b) 1.61884.

A separable differential equation, like other differential equations, can be stated along with an initial condition. An initial-value problem involving a separable equation is solved through the following steps:

1. Find a general solution of the differential equation using separation of variables.
2. Use the initial condition to determine the unknown constant in the general solution.

Example 2

Solve the initial-value problems (a) $\dfrac{dy}{dx} = e^{2x+y}$, $y(0) = 0$

(b) $y \cos x \, dx - (1 + y^2) \, dy = 0$, $y(0) = -1$

Solution   (a) If we write the equation as $dy/dx = e^{2x}e^{y}$, we see that the variables are separated as

$$e^{-y} \, dy = e^{2x} \, dx.$$

Integration gives us $-e^{-y} = \dfrac{e^{2x}}{2} + C_1$, and simplifying results in

$$e^{-y} = -\dfrac{e^{2x}}{2} - C_1.$$

Application of the natural logarithm yields

$$-y = \ln\left(-\dfrac{e^{2x}}{2} - C_1\right) \quad \text{or} \quad y = -\ln\left(-\dfrac{e^{2x}}{2} - C_1\right).$$

To find the value of C_1 so that the solution satisfies the condition $y(0) = 0$, we use the equation $e^{-y} = -\dfrac{e^{2x}}{2} - C_1$. Applying $y(0) = 0$ yields $1 = -\dfrac{1}{2} - C_1$ so $C_1 = -\dfrac{3}{2}$ and so the solution to the initial-value problem is

$$y = -\ln\left(-\dfrac{e^{2x}}{2} + \dfrac{3}{2}\right).$$

The domain of the natural logarithm is $(0, +\infty)$, so this solution is valid only for the values of x such that $3 - e^{2x} > 0$, which is equivalent to $e^{2x} < 3$ and has solution $x < \dfrac{\ln(3)}{2} \approx 0.5493$. Notice that the graph of the solution shown in Figure 2.3 passes through the point $(0, 0)$. If it had not, we would check for mistakes in our solution procedure.

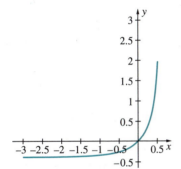

Figure 2.3

(b) Unlike the previous examples, this equation is in differential form because dx and dy appear as multiples in the equation instead of in the form dy/dx. The separation of variables is relatively straightforward, however, with

$$\cos x \, dx = \dfrac{1 + y^2}{y} \, dy.$$

Notice that the right-hand side can be written as $\dfrac{1}{y} + \dfrac{y^2}{y} = \dfrac{1}{y} + y$. Then integration gives us

$$\sin x + C_1 = \ln|y| + \dfrac{1}{2}y^2.$$

By substituting $y(0) = -1$ into this equation, we find that $C_1 = \dfrac{1}{2}$, so the implicit solution is given by $\sin x + \dfrac{1}{2} = \ln|y| + \dfrac{1}{2}y^2$. Notice that the graph of the solution, which is shown in Figure 2.4, passes through the point $(0, -1)$ as required by the initial condition.

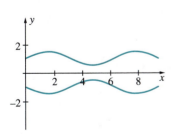

Figure 2.4

 Solve the initial-value problem $y \cos x \, dx - (1 + y^2) \, dy = 0$, $y(0) = 1$.

Example 3

Find a point on the graph of the solution to the initial-value problem

$$\begin{cases} \dfrac{dy}{dx} = \dfrac{1 - \cos x}{1 - \sin y} \\ y(0) = 0 \end{cases}$$

at which $dy/dx = 1$.

Solution   To solve the problem, we must find a point (x, y) that satisfies $\dfrac{dy}{dx} = \dfrac{1 - \cos x}{1 - \sin y} = 1$ and is on the graph of the solution to the initial-value problem $\begin{cases} \dfrac{dy}{dx} = \dfrac{1 - \cos x}{1 - \sin y} \\ y(0) = 0 \end{cases}$.

The equation $\dfrac{dy}{dx} = \dfrac{1 - \cos x}{1 - \sin y}$ is separable. Integrating and applying the initial condition yields the solution $y + \cos y = x - \sin x + 1$, which we graph together with $\dfrac{1 - \cos x}{1 - \sin y} = 1$ in Figure 2.5.

From the graph, we see that $dy/dx = 1$ near the point $(1.25, 0.35)$. A better approximation of the point of intersection is obtained by Newton's method and is found to be $(1.22682, 0.343977)$. This result is confirmed by graphing the solution to the initial-value problem together with the line $y = (x - 1.22682) + 0.343977$ as shown in Figure 2.6.

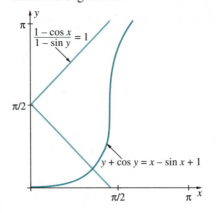

Figure 2.5   Graph of $y + \cos y = x - \sin x + 1$ and $\dfrac{1 - \cos x}{1 - \sin y} = 1$

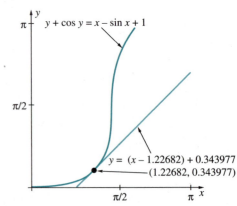

Figure 2.6

IN TOUCH WITH TECHNOLOGY

In many cases computer algebra systems can be used to perform the integration and algebraic simplification associated with a separable diffferential equation and can sometimes solve them. Similarly, initial-value problems can often be solved quickly with a computer algebra system.

In Exercises 1–4, find a general solution of the given equation and graph various solutions on the indicated interval.

 1. $\dfrac{dy}{dx} = y \cos x$; $[0, 4\pi]$

2. $\dfrac{dy}{dx} = \sqrt{1 - y^2} \sin x$; $[0, 4\pi]$

 3. $\cos x \ dx = \dfrac{1 + y^2}{y^2} \ dy$; $[0, 10]$

4. $e^y \cos y \ dy = \dfrac{x^2}{\sqrt{9 - x^2}} \ dx$; $[-2, 2]$

5. Solve the initial-value problem

$$\begin{cases} \dfrac{dy}{dx} = \dfrac{x^2}{\sqrt{9 - x^2}e^y \cos y} \\ y(0) = 0 \end{cases}$$

and graph the solution on an appropriate interval.

6. How does the graph of the solution to the initial-value problem

$$\begin{cases} y' = \dfrac{cy}{x^2} \\ y(1) = 1 \end{cases}$$

change as c changes from -2 to 2?

7. (a) Assume that $y(t) > 0$ and $\displaystyle\int_1^t f(t) \ dt$ exists for $t \geq 1$. Show that the solution to the initial-value problem

$$\begin{cases} \dfrac{dy}{dt} = yf(t) \\ y(1) = 2 \end{cases}$$

is

$$y(t) = 2e^{\int_1^t f(u) \, du}.$$

(b) Find three functions $f(t)$ so that **(i)** $y(t)$ is periodic; **(ii)** $\lim_{t\to\infty} y(t) = 0$; and **(iii)** $\lim_{t\to\infty} y(t) = \infty$. Confirm your results by graphing each solution.
(c) If $c > 0$ is given, is it possible to choose $f(t)$ so that $\lim_{t\to\infty} f(t) = c$? Explain.

EXERCISES 2.1

In Exercises 1–42, solve each equation. (A computer algebra system or table of integrals may be useful in evaluating some integrals.)

1. $\dfrac{dy}{dx} = \dfrac{6x^2}{7y^3}$

2. $\dfrac{1}{2}x^{-1/2} \ dx + y^2 \ dy = 0$

*3. $\dfrac{dy}{dx} = \dfrac{3y^7}{x^8}$

4. $\dfrac{dy}{dx} = \dfrac{1}{x^2(8 + 9y^2)}$

5. $(6 + 4x^3) \ dx + \left(5 + \dfrac{9}{y^8}\right) dy = 0$

6. $\left(\dfrac{6}{x^9} - \dfrac{6}{x^3} + x^7\right) dx + (9 + y^{-2} - 4y^8) \ dy = 0$

7. $4 \sinh 4y \ dy = 6 \cosh 3x \ dx$

8. $\dfrac{1}{3}x^{-2/3} \ dx = (y^5 - 6y^2 + 8) \ dy$

*9. $(x^2 + 2\sqrt{x}) \ dx = \dfrac{-1}{2}y^{-5/2} \ dy$

10. $\dfrac{3}{x^2} \ dx = \left(\dfrac{1}{\sqrt{y}} + \sqrt{y}\right) dy$

11. $3 \sin x \, dx - 4 \cos y \, dy = 0$

12. $\cos y \, dy = 8 \sin 8x \, dx$

*13. $(5x^5 - 4 \cos x) \, dx + (2 \cos 9y + 2 \sin 7y) \, dy = 0$

14. $9 \cosh 9y \, dy = 4 \sinh 4x \, dx$

15. $(\cosh 6x + 5 \sinh 4x) \, dx + 20 \sinh y \, dy = 0$

16. $\dfrac{dy}{dx} = e^{2y} e^{10x}$

*17. $(10 + 7e^{-3x}) \, dx - (e^y - 8y^3) \, dy = 0$

18. $\sin^2 x \, dx = \cos^2 y \, dy$

19. $(3 \sin x - \sin 3x) \, dx = (\cos 4y - 4 \cos y) \, dy$

20. $\dfrac{dy}{dx} = \dfrac{\sec^2 x}{\sec y \tan y}$

21. $\left(2 - \dfrac{5}{y^2}\right) dy + 4 \cos^2 x \, dx = 0$

22. $\dfrac{dy}{dx} = \dfrac{x^3}{y\sqrt{(1 - y^2)(x^4 + 9)}}$

*23. $\tan y \sec^2 y \, dy + \cos^3 2x \sin 2x \, dx = 0$

24. $\dfrac{dy}{dx} = \dfrac{1 + 2e^y}{e^y x \ln x}$

25. $x \sin(x^2) \, dx = \dfrac{\cos \sqrt{y}}{\sqrt{y}} \, dy$

26. $\dfrac{x - 2}{x^2 - 4x + 3} \, dx = \left(1 - \dfrac{1}{y}\right)^2 \dfrac{1}{y^2} \, dy$

*27. $\dfrac{\cos y}{(1 - \sin y)^2} \, dy = \sin^3 x \cos x \, dx$

28. $\dfrac{dy}{dx} = \dfrac{(5 - 2 \cos x)^3 \sin x \cos^4 y}{\sin y}$

29. $\dfrac{\sqrt{\ln x}}{x} \, dx = \dfrac{e^{3/y}}{y^2} \, dy$

30. $\dfrac{dy}{dx} = \dfrac{5^{-x}}{y^2}$

*31. $\dfrac{dy}{dx} = \dfrac{2x^5}{6 \ln(3y)}$

32. $\dfrac{4x - 15}{(2x + 3)(4x - 1)} \, dx - \dfrac{17y + 22}{(y - 8)(9y + 7)} \, dy = 0$

*33. $\dfrac{dy}{dx} = \dfrac{(5x + 23)(y + 1)(7y + 1)}{(x + 3)(3x + 1)(1 - 14y)}$

34. $x^3 \ln x \, dx + \dfrac{2y + 2}{y^2 - 2y + 1} \, dy = 0$

35. $\dfrac{dy}{dx} = \dfrac{16 \sin^2 x \cos^2 x}{\sin^{5/2} y \cos^3 y}$

36. $\dfrac{dy}{dx} = \dfrac{\sin^7 x}{8 \cos^4 y}$

*37. $\dfrac{dy}{dx} = \dfrac{y^3 + y}{(y^3 + 4y^2)\sqrt{9 - 4x^2}}$

38. $\dfrac{dy}{dx} = \dfrac{y^{-2} e^{4y}}{4x^2 + 4x + 2}$

*39. $\dfrac{dx}{(1 - x^2)^{3/2}} = e^{3y} \sin 4y \, dy$

40. $x \cos^5(x^2) \, dx = e^{2y} \cos^{-1}(e^y) \, dy$

*41. $\dfrac{dy}{dx} = \dfrac{\sin 4x}{y^2 e^{2x+y}}$

42. $\dfrac{dy}{dx} = \dfrac{\cos 7x \sin 3x}{9 e^{3y} \cos(6y)}$

In Exercises 43–54, solve the initial-value problem. Graph the solution on an appropriate interval.

43. $\dfrac{dy}{dx} = x^3, \; y(0) = 4$

44. $\dfrac{dy}{dx} = \cos x, \; y\left(\dfrac{\pi}{2}\right) = -1$

*45. $dx = \cos x \, dy, \; y(0) = 2$

46. $\sin^2 y \, dy = dx, \; y(0) = 0$

47. $\dfrac{dy}{dx} = \dfrac{\sqrt{x}}{y}, \; y(0) = 2$

48. $\dfrac{dy}{dx} = \sqrt{\dfrac{y}{x}}, \; y(1) = 2$

*49. $\dfrac{dy}{dx} = \dfrac{e^x}{y + 1}, \; y(0) = -2$

50. $\dfrac{dy}{dx} = e^{x-y}, \; y(0) = 0$

*51. $\dfrac{dy}{dx} = \dfrac{y}{\ln y}, \; y(0) = e$

52. $\dfrac{dy}{dx} = x \sin(x^2), \; y(\sqrt{\pi}) = 0$

53. $\dfrac{dy}{dx} = \dfrac{1}{1 + x^2}$, $y(0) = 1$

54. $\dfrac{dy}{dx} = \dfrac{\sin x}{\cos y + 1}$, $y(0) = 0$

*55. Find an equation of the curve that passes through the point $(0, 0)$ and has slope $-\dfrac{y - 2}{x - 2}$ at each point (x, y) on the curve.

56. A differential equation of the form $dy/dx = f(ax + by + k)$ is separable if $b = 0$. However, if $b \neq 0$, the substitution $u(x) = ax + by + k$ as the new dependent variable yields a separable equation. Use this transformation to solve **(a)** $dy/dx = (x + y - 4)^2$; and **(b)** $dy/dx = (3y + 1)^4$.

57. Let $\omega > 0$. **(a)** Show that the system

$$\begin{cases} \dfrac{dx}{dt} = x(1 - r) - \omega y \\ \dfrac{dy}{dt} = y(1 - r) + \omega x \end{cases}, \ r = \sqrt{x^2 + y^2}$$

can be rewritten as the system

$$\begin{cases} \dfrac{dr}{dt} = r(1 - r) \\ \dfrac{d\theta}{dt} = \omega \end{cases}$$

by changing to polar coordinates $\begin{cases} x = r \cos \theta \\ y = r \sin \theta \end{cases}$.

(b) Show that the solution to

$$\begin{cases} \dfrac{dr}{dt} = r(1 - r) \\ \dfrac{d\theta}{dt} = \omega \\ r(0) = r_0, \ \theta(0) = \theta_0 \end{cases}$$

is

$$\begin{cases} r(t) = \dfrac{r_0 e^t}{(1 - r_0) + r_0 e^t} \\ \theta(t) = \omega t + \theta_0 \end{cases}$$

and the solution to

$$\begin{cases} \dfrac{dx}{dt} = x(1 - r) - \omega y \\ \dfrac{dy}{dt} = y(1 - r) + \omega x \end{cases}, \ r = \sqrt{x^2 + y^2}$$

is

$$\begin{cases} x = r(t) \cos \theta(t) \\ y = r(t) \sin \theta(t) \end{cases}.$$

(c) How does the solution change for various initial conditions? (*Hint:* First determine how ω and θ_0 affect the solution. Then, determine how r_0 affects the solution. Try graphing the solution if $\omega = 2$, $\theta_0 = 0$, and $r_0 = \frac{1}{2}$, 1, and $\frac{3}{2}$. What happens if you increase ω? What happens if you increase θ_0?)

In Exercises 58–61, without actually solving the equations, match each differential equation in Group A with the graph of its direction field and two solutions in Group B.

Group A

58. $\dfrac{dy}{dx} = \dfrac{y^4 + 1}{x^4 + 1}$ **59.** $\dfrac{dy}{dx} = \dfrac{x^4 + 1}{y^4 + 1}$

60. $\dfrac{dy}{dx} = (x^4 + 1)(y^4 + 1)$ **61.** $\dfrac{dy}{dx} = \dfrac{1}{(x^4 + 1)(y^4 + 1)}$

Group B

(a)   **(b)**

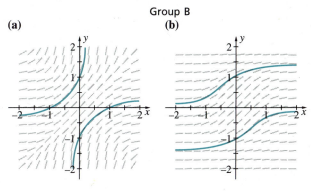

(c)   **(d)**

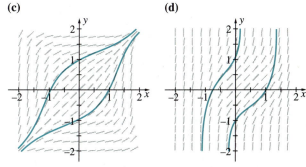

62. (Lasers) The variables that characterize a two-level laser are listed in the following table.

ϕ—the total number of photons in the optical resonator	l—length of the resonator	N_1—number of atoms per unit volume in Level 1	N_2—number of atoms per unit volume in Level 2
V—volume of optical resonator	n—the total inversion given by $(N_2 - N_1)V$	n_i—initial inversion	n_t—the total inversion at the threshold
t_c—the decay time constant for photons in the passive resonator.	c—the phase velocity of the light wave	ν—light wave frequency	h—magnetic field

Suppose that a light wave of frequency ν and intensity I_ν propagates through an atomic medium with N_2 atoms per unit volume in Level 2 and N_1 in Level 1. If $N_2 > N_1$, then the medium is amplifying as is the case with lasers. On the other hand, if $N_1 > N_2$, then the medium is absorbing. These two situations are pictured in Figure 2.7.

Optical resonators are used to build up large field intensities with moderate power inputs. A measure of this property is the quality factor Q where

$$Q = \omega \times \frac{\text{field energy stored by resonator}}{\text{power dissipated by resonator}}.$$

A technique called "Q-switching" is used to generate short and intense bursts of oscillations from lasers. This is done by lowering the quality factor Q during the pumping so that the inversion $N_2 - N_1$ builds up to a high value without oscillation. When the inversion reaches its peak, the factor Q is suddenly restored to its ordinary value. At this point, the laser medium is well above its threshold, which causes a large buildup of the oscillation and a simultaneous exhaustion of the inversion by stimulated transitions from Level 2 to Level 1. This process converts most of the energy that was stored by atoms pumped into Level 2 into photons, which are now in the optical resonator. These photons bounce back and forth between the reflectors with a fraction $1 - R$ escaping from the resonator with each bounce. This causes a decay of the pulse with a photon lifetime

$$t_c \approx \frac{nl}{c(1 - R)}.$$

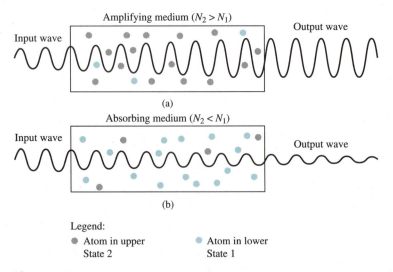

Amplifying medium ($N_2 > N_1$)

Input wave   Output wave

(a)

Absorbing medium ($N_2 < N_1$)

Input wave   Output wave

(b)

Legend:
- Atom in upper State 2
- Atom in lower State 1

Figure 2.7

The quantities of ϕ and n are related by the differential equation

$$\frac{d\phi}{dn} = \frac{n_t}{2n} - \frac{1}{2}.$$

(a) Solve this equation to determine the total number of photons in the optical resonator ϕ as a function of n (where ϕ depends on an arbitrary constant).

(b) Assuming that the initial values of ϕ and n are ϕ_0 and n_0 at $t = 0$, show that ϕ can be simplified to obtain

$$\phi = \frac{n_t}{2} \ln \frac{n}{n_0} + \frac{1}{2}(n_0 - n) + \phi_0$$

(c) If the instantaneous power output of the laser is given by

$$P = \frac{\phi h \nu}{t_c},$$

show that the maximum power output occurs when $n = n_t$. (*Hint:* Solve $\partial P/\partial n = 0$.)

*63. (Destruction of Microorganisms) Microorganisms can be removed from fluids by mechanical methods such as filtration, centrifugation, or flotation. However, they can also be destroyed by heat, chemical agents, or electromagnetic waves. The fermentation industry is interested in improving this process of sterilizing media. An important component in the design of a sterilizer is the kinetics of the death of microorganisms.

The destruction of microorganisms by heat indicates loss of viability as opposed to physical destruction. This destruction follows the rate of reaction

$$\frac{dN}{dt} = -kN$$

where k is the reaction constant (min^{-1}) and is a function of temperature; N is the number of viable organisms; and t is time.*

Microbiologists use the term **decimal reduction time**, D, to indicate the time of exposure to heat during which the original number of viable microbes is reduced by one-tenth. If $N(0) = N_0$, find $N(t)$. Use the fact that $N(D) = \frac{1}{10}N_0$ to find D in terms of k.

64. (Pollution) Under normal atmospheric conditions, the density of soot particles, $N(t)$, satisfies the differential equation

$$\frac{dN}{dt} = -k_c N^2 + k_d N,$$

where k_c, called the **coagulation constant,** is a constant that relates how well particles stick together; and k_d, called the **dissociation constant,** is a constant that relates to how well particles fall apart. Both of these constants depend on temperature, pressure, particle size, and other external forces.†

(a) Rewrite the equation $dN/dt = -k_c N^2 + k_d N$ in the form

$$\frac{1}{-k_c N^2 + k_d N} dN = dt$$

and use partial fractions to show that

$$N(t) = \frac{e^{k_d t}}{\frac{k_c}{k_d} e^{k_d t} + C}, \text{ where } C \text{ represents an arbitrary constant.}$$

(b) Find C so that $N(t)$ satisfies the condition $N(t_0) = N_0$.

The following table lists typical values of k_c and k_d.

k_c	k_d
163	5
125	26
95	57
49	85
300	26

---

* S. Aiba, A. E. Humphrey, N. F. Millis, *Biochemical Engineering,* Second Edition, Academic Press, 1973, pp. 240–242.

† Chr. Feldermann, H. Jander, and H. Gg. Wagner, ''Soot Particle Coagulation of Premixed Ethylene/Air Flames at 10 bar,'' *International Journal of Research in Physical Chemistry and Chemical Physics,* Volume 186, Part II, 1994, pp. 127–140.

(c) For each pair of values in the previous table, sketch the graph of $N(t)$ if $N(0) = N_0$ for $N_0 = 0.01, 0.05, 0.1, 0.5, 0.75, 1, 1.5,$ and 2. Regardless of the initial condition $N(0) = N_0$, what do you notice in each case? Do pollution levels seem to be more sensitive to k_c or k_d? Does your result make sense? Why?

(d) Show that if $k_d > 0$, $\lim_{t\to\infty} N(t) = k_d/k_c$. Why is the assumption that $k_d > 0$ reasonable?

(e) For each pair in the table, calculate $\lim_{t\to\infty} N(t) = k_d/k_c$. Which situation results in the highest pollution levels? How could the situation be changed?

2.2 Homogeneous Equations

Substitutions are used often in integral calculus to transform integrals into forms that can be computed easily. Similarly, with some differential equations, we may perform substitutions that transform a given differential equation into an equation that is easier to solve.

For example, the equation $(y - x)\,dy + (x + y)\,dx = 0$ is not a separable differential equation. However, if we let $y = ux$, then $dy = u\,dx + x\,du$. Substituting these expressions into $(y - x)\,dy + (x + y)\,dx = 0$ and simplifying results in

$$(ux - x)(u\,dx + x\,du) + (x + ux)\,dx = 0$$
$$x(u^2 + 1)\,dx + x^2(u - 1)\,du = 0.$$

Notice that the transformed equation $x(u^2 + 1)\,dx + x^2(u - 1)\,du = 0$ is separable, so we separate the variables:

$$x(u^2 + 1)\,dx = x^2(1 - u)\,du$$
$$\frac{dx}{x} = \frac{1 - u}{u^2 + 1}\,du$$

and integrate both sides of this equation to obtain

$$\ln|x| = \tan^{-1} u - \frac{1}{2}\ln(1 + u^2) + C.$$

Notice that absolute value is not needed in the term $\frac{1}{2}\ln(u^2 + 1)$ because $u^2 + 1 > 0$ for all u. Now, $u = y/x$ because $y = ux$, so resubstitution gives us

$$\ln|x| = \tan^{-1}(y/x) - \frac{1}{2}\ln(1 + (y/x)^2) + C$$

as a general solution of $(y - x)\,dy + (x + y)\,dx = 0$.

Notice the breaks in the graph along the y-axis shown in Figure 2.8. Why do they occur?

The equation $(y - x)\,dy + (x + y)\,dx = 0$ is called a *homogeneous equation*. We can always reduce a homogeneous equation to a separable equation by a suitable substitution.

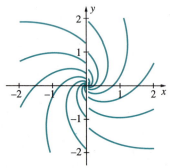

Figure 2.8 Graph of $\ln|x| = \tan^{-1}\!\left(\dfrac{y}{x}\right) - \dfrac{1}{2}\ln\!\left(1 + \left(\dfrac{y}{x}\right)^2\right) + C$ for several values of C

Definition 2.2 Homogeneous Differential Equation

A first-order differential equation that can be written in the form $M(x, y)\, dx + N(x, y)\, dy = 0$ where $M(tx, ty) = t^n M(x, y)$ and $N(tx, ty) = t^n N(x, y)$ is called a **homogeneous differential equation** (of degree n).

Example 1

Show that the equation $(x^2 + yx)\, dx - y^2\, dy = 0$ is homogeneous.

Solution Let $M(x, y) = x^2 + yx$ and $N(x, y) = -y^2$. The equation $(x^2 + yx)\, dx - y^2\, dy = 0$ is homogeneous of degree 2 because

$$M(tx, ty) = (tx)^2 + (ty)(tx) = t^2(x^2 + yx) = t^2 M(x, y)$$

and

$$N(tx, ty) = -t^2 y^2 = t^2 N(x, y).$$

Homogeneous equations are reduced to separable equations by either of the substitutions

$$y = ux \quad \text{or} \quad x = vy.$$

Use the substitution $y = ux$ if $N(x, y)$ is less complicated than $M(x, y)$, and use $x = vy$ if $M(x, y)$ is less complicated than $N(x, y)$. If a difficult integration problem is encountered after a substitution is made, try the other substitution to see if it yields an easier problem. As with the separation of variables technique, this technique was also discovered by Leibniz.

Example 2

Solve the equation $(x^2 - y^2)\, dx + xy\, dy = 0$.

Solution In this case, let $M(x, y) = x^2 - y^2$ and $N(x, y) = xy$. Then, $M(tx, ty) = t^2 M(x, y)$ and $N(tx, ty) = t^2 N(x, y)$, which means that $(x^2 - y^2)\, dx + xy\, dy = 0$ is a homogeneous equation of degree 2.

We let $y = ux$ because $N(x, y) = xy$ is less complicated than $M(x, y) = x^2 - y^2$. Then, $dy = u\, dx + x\, du$. With this substitution, we have

$$(x^2 - y^2)\, dx + xy\, dy = 0$$
$$(x^2 - (ux)^2)\, dx + x(ux)(u\, dx + x\, du) = 0$$
$$(x^2 - u^2 x^2)\, dx + x^2 u(u\, dx + x\, du) = 0$$
$$x^2\, dx - u^2 x^2\, dx + u^2 x^2\, dx + ux^3\, du = 0$$
$$x^2\, dx + x^3 u\, du = 0$$

which results in the separated equation

$$\frac{1}{x} \, dx = -u \, du.$$

Integration yields

$$\ln|x| = -\frac{1}{2} u^2 + C$$

which is simplified with $u = y/x$ to obtain

$$\ln|x| = -\frac{1}{2}(y/x)^2 + C \qquad \text{or} \qquad x = Ke^{-y^2/(2x^2)}$$

where $K = \pm e^C$.

The same result is obtained with the subtitution $x = vy$. If $x = vy$, $dx = v \, dy + y \, dv$. Substituting into the equation and simplifying yields

$$
\begin{aligned}
0 &= (x^2 - y^2) \, dx + xy \, dy \\
&= (v^2y^2 - y^2)(v \, dy + y \, dv) + vyy \, dy \\
&= v^3y^2 \, dy - y^2v \, dy + v^2y^3 \, dv - y^3 \, dv + vy^2 \, dy \\
&= (v^3y^2 - y^2v + vy^2) \, dy + (v^2y^3 - y^3) \, dv \\
&= y^2v^3 \, dy + y^3(v^2 - 1) \, dv.
\end{aligned}
$$

Dividing this equation by y^3v^3 yields the separable differential equation $\dfrac{dy}{y} + \dfrac{(v^2 - 1) \, dv}{v^3} = 0$. We solve this equation by rewriting it in the form $\dfrac{dy}{y} = \dfrac{(1 - v^2) \, dv}{v^3} = \left(\dfrac{1}{v^3} - \dfrac{1}{v} \right) dv$ and integrating. This yields

$$\ln|y| = -\frac{1}{2v^2} - \ln|v| + C_1$$

which can be simplified as $\ln|vy| = -\dfrac{1}{2v^2} + C_1$, so

$$vy = Ce^{-1/(2v^2)}, \text{ where } C = \pm e^{C_1}.$$

Because $x = vy$, $v = x/y$. Resubstituting into the equation $vy = Ce^{-1/(2v^2)}$ yields

$$x = Ce^{-y^2/(2x^2)}$$

as a general solution of the equation $(x^2 - y^2) \, dx + xy \, dy = 0$.

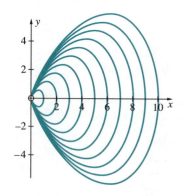

Figure 2.9

Figure 2.9 shows the graph of $x = Ce^{-y^2/(2x^2)}$ for various positive values of C. Sketch the graph of $x = Ce^{-y^2/(2x^2)}$ for $C = -1, -2, \ldots, -5$.

We solve an initial-value problem in the following example.

Example 3

Solve $(x^2 e^{y/x} - y^2)\, dx + xy\, dy = 0$ subject to the initial condition $y(1) = 0$.

Solution   We identify $M(x, y) = x^2 e^{y/x} - y^2$ and $N(x, y) = xy$. Then,

$$M(tx, ty) = (tx)^2 e^{(ty)/(tx)} - (ty)^2 = t^2(x^2 e^{y/x} - y^2) = t^2 M(x, y)$$

and

$$N(tx, ty) = (tx)(ty) = t^2 xy = t^2 N(x, y).$$

The equation $(x^2 e^{y/x} - y^2)\, dx + xy\, dy = 0$ is homogeneous of degree 2.

Assume that $y = ux$. Then $dy = u\, dx + x\, du$ and substituting into the equation results in

$$(x^2 e^{(ux)/x} - (ux)^2)\, dx + x(ux)(u\, dx + x\, du) = 0.$$

Simplifying, we obtain the separable equation

$$x^2(e^u - u^2)\, dx + x^2 u^2 dx + x^3 u\, du = 0$$
$$x^2 e^u\, dx + x^3 u\, du = 0$$
$$u e^{-u}\, du = \frac{-1}{x}\, dx.$$

Applying integration by parts, a table of integrals, or a computer algebra system we obtain

$$\int u e^{-u}\, du = -u e^{-u} - \int -e^{-u}\, du = -u e^{-u} - e^{-u} + C_1$$

and $\displaystyle \int \frac{-1}{x}\, dx = -\ln|x| + C_2$. Therefore, a general solution of $u e^{-u}\, du = \dfrac{-1}{x}\, dx$ is

$$u e^{-u} + e^{-u} = \ln|x| + C.$$

Because $y = ux$, $u = y/x$. Substituting into the solution yields

$$\left(1 + \frac{y}{x}\right) e^{-y/x} = \ln|x| + C$$

as a general solution of $(x^2 e^{y/x} - y^2)\, dx + xy\, dy = 0$.

To find the value of C that satisfies the initial condition $y(1) = 0$, we replace x with 1 and y with 0 so that $1 = 0 + C$, then solve for C to find that $C = 1$. Thus, the solution to the initial-value problem is $(1 + y/x)e^{-y/x} = \ln|x| + 1$. We see that the initial condition is satisfied by the solution because the graph of the solution obtained here, which is shown in Figure 2.10, intersects the x-axis at $(1, 0)$.

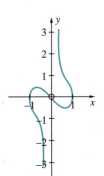

Figure 2.10   Graph of $(1 + y/x)e^{-y/x} = \ln|x| + 1$

Is it possible to find a solution of the equation that satisfies the initial condition $y(0) = 1$?

In addition to verifying that $M(tx, ty) = t^n M(x, y)$ and $N(tx, ty) = t^n N(x, y)$, there are other ways to determine if an equation is homogeneous or not. For example, solving the differential equation $4xy^2\, dx + (x^3 + y^3)\, dy = 0$ for dy/dx, we obtain

$$\frac{dy}{dx} = \frac{-4xy^2}{x^3 + y^3} = \frac{\dfrac{1}{x^3}(-4xy^2)}{\dfrac{1}{x^3}(x^3 + y^3)} = \frac{-4\left(\dfrac{y}{x}\right)^2}{1 + \left(\dfrac{y}{x}\right)^3} = F\left(\frac{y}{x}\right),$$

where $F(t) = -4t^2/(1 + t^3)$. Similarly, we can find that

$$\frac{dy}{dx} = \frac{-4xy^2}{x^3 + y^3} = \frac{\dfrac{1}{y^3}(-4xy^2)}{\dfrac{1}{y^3}(x^3 + y^3)} = \frac{-4\dfrac{x}{y}}{\left(\dfrac{x}{y}\right)^3 + 1} = G\left(\frac{x}{y}\right),$$

where $G(t) = -4t/(t^3 + 1)$. This indicates (although we have not shown it in general) that an equation is homogeneous if we can write it in either of the forms $dy/dx = F(y/x)$ or $dy/dx = G(x/y)$.

Example 4

Find values of a_0, a_1, a_2, and a_3, none of which are zero, so that the nontrivial solutions to $\dfrac{dy}{dx} = \dfrac{a_0 x + a_1 y}{a_2 x + a_3 y}$ are periodic.

Solution   This equation is homogeneous because $\dfrac{dy}{dx} = \dfrac{a_0 x + a_1 y}{a_2 x + a_3 y} =$

$$\frac{a_0 + a_1 \dfrac{y}{x}}{a_2 + a_3 \dfrac{y}{x}} = F\left(\frac{y}{x}\right), \text{ where } F(t) = \frac{a_0 + a_1 t}{a_2 + a_3 t}. \text{ Letting } y = ux \text{ leads to}$$

$$-x(a_3 u^2 + (a_2 - a_1)u - a_0)\, dx - x^2(a_3 u + a_2)\, du = 0$$

$$\frac{a_3 u + a_2}{a_3 u^2 + (a_2 - a_1)u - a_0}\, du = -\frac{1}{x}\, dx.$$

Integrating and substituting $u = y/x$ yields a general solution of the equation:

$$\ln \sqrt{a_0 + (a_1 - a_2)u - a_3 u^2} -$$

$$\frac{a_1 + a_2}{\sqrt{(a_2 - a_1)^2 + 4a_0 a_3}} \tanh^{-1}\left(\frac{a_2 - a_1 + 2a_3 u}{\sqrt{(a_2 - a_1)^2 + 4a_0 a_3}}\right) = -\ln x + C$$

$$\ln \sqrt{a_0 + (a_1 - a_2)\frac{y}{x} - a_3 \frac{y^2}{x^2}} -$$

$$\frac{a_1 + a_2}{\sqrt{(a_2 - a_1)^2 + 4a_0 a_3}} \tanh^{-1}\left(\frac{(a_2 - a_1)x + 2a_3 y}{x\sqrt{(a_2 - a_1)^2 + 4a_0 a_3}}\right) = -\ln x + C.$$

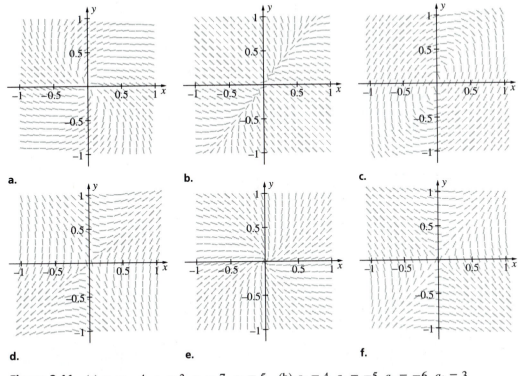

Figure 2.11 (a) $a_0 = -4$, $a_1 = 3$, $a_2 = 7$, $a_3 = 5$ (b) $a_0 = 4$, $a_1 = -5$, $a_2 = -6$, $a_3 = 3$ (c) $a_0 = 3$, $a_1 = -1$, $a_2 = 1$, $a_3 = -2$ (d) $a_0 = 3$, $a_1 = -1$, $a_2 = 1$, $a_3 = 2$ (e) $a_0 = -2$, $a_1 = -7$, $a_2 = -7$, $a_3 = 2$ (f) $a_0 = -5$, $a_1 = 1$, $a_2 = 2$, $a_3 = -4$

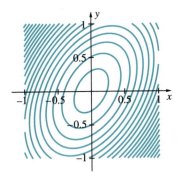

Figure 2.12 Graph of $\frac{3}{2}x^2 - xy + y^2 = C$ for various values of C

This solution is complicated and nearly impossible to analyze analytically so we proceed graphically. We begin by generating *random* values of a_0, a_1, a_2, and a_3, none of which are zero, and then graphing the direction field associated with the equation $\dfrac{dy}{dx} = \dfrac{a_0 x + a_1 y}{a_2 x + a_3 y}$. Several of the results we obtain are shown in Figure 2.11. (Use each direction field to graph several solutions to each equation.)

From these graphs, we see that the choices $a_0 = 3$, $a_1 = -1$, $a_2 = 1$, and $a_3 = -2$ *may* lead to periodic solutions. A general solution of the equation $\dfrac{dy}{dx} = \dfrac{3x - y}{x - 2y}$ is $\dfrac{3}{2}x^2 - xy + y^2 = C$ (why?), which is a family of ellipses as shown in Figure 2.12. We see that the choices $a_0 = 3$, $a_1 = -1$, $a_2 = 1$, and $a_3 = -2$ yield periodic solutions to the equation.

Show that the nontrivial solutions to $\dfrac{dy}{dx} = \dfrac{a_0 x + a_1 y}{a_2 x + a_3 y}$ *are periodic if*

$a_2 = -a_1$ *and* $a_1^2 + a_0 a_3 < 0$.

IN TOUCH WITH TECHNOLOGY

In the same manner that computer algebra systems and graphing utilities can be used to help find solutions of separable equations, they can also be used to help solve and graph solutions of homogeneous equations.

1. Solve the equation
$(x^{1/3}y^{2/3} + x) \, dx + (x^{2/3}y^{1/3} + y) \, dy = 0$. Graph several solutions.

2. Solve the initial-value problem $\begin{cases} y' = \dfrac{y^2 - x^2}{xy} \\ y(4) = 0 \end{cases}$ and

graph the solution for $0 < x \le 4$.

3. (a) The **sine integral function,** $\mathrm{Si}(x)$, is defined by

$$\mathrm{Si}(x) = \int_0^x \frac{\sin t}{t} \, dt.$$

(i) Evaluate $\mathrm{Si}'(x)$ and $\lim_{x \to 0} \mathrm{Si}(x)$. **(ii)** Graph $\mathrm{Si}(x)$ on the interval $[0, 6\pi]$. **(iii)** Approximate the maximum value of $\mathrm{Si}(x)$. **(iv)** Can you predict $\lim_{x \to \infty} \mathrm{Si}(x)$? **(b)** Graph the direction field associated with the equation $y \sin(x/y) \, dx - (x + x \sin(x/y)) \, dy = 0$. **(c)** Solve the initial-value problem

$$\begin{cases} y \sin(x/y) \, dx - (x + x \sin(x/y)) \, dy = 0 \\ y(1) = 2 \end{cases}$$

and graph the solution.

EXERCISES 2.2

In Exercises 1–10, determine if the differential equation is homogeneous. If so, determine its degree.

1. $(x + 3y) \, dx - 4x \, dy = 0$

2. $(y^2 - x^2) \, dx + xy \, dy = 0$

3. $\sqrt{x^2 + xy} \, dy - xy \, dx = 0$

4. $dx + (x + y) \, dy = 0$

5. $\cos\left(\dfrac{x}{x + y}\right) dx + e^{2y/x} \, dy = 0$

6. $y \ln\left(\dfrac{x}{y}\right) dx + \dfrac{x^2}{x + y} \, dy = 0$

7. $2 \ln x \, dx - \ln(4y^2) \, dy = 0$

8. $\left(\dfrac{2}{x} + \dfrac{1}{y}\right) dx + \dfrac{x}{y^2} \, dy = 0$

9. $\dfrac{\sin 2x}{\cos 2y} \, dx + \left(\dfrac{\ln y}{\ln x}\right) dy = 0$

10. $\sqrt{x^2 + 1} \, dx + y \, dy = 0$

In Exercises 11–40, solve each equation.

11. $2x \, dx + (y - 3x) \, dy = 0$

12. $y \, dx + (x - 2y) \, dy = 0$

*13. $(2y - 3x) \, dx + x \, dy = 0$

14. $(x - y) \, dx + (4y - x) \, dy = 0$

15. $(xy - y^2) \, dx + x(x - 3y) \, dy = 0$

16. $(x^2 + xy - y^2) \, dx + xy \, dy = 0$

*17. $(x^2 + xy + y^2) \, dx - xy \, dy = 0$

18. $(30xy + 15y^2) \, dx + (4x^2 - 6xy) \, dy = 0$

19. $(x^3 + y^3) \, dx - xy^2 \, dy = 0$

20. $(5xy + 20y^2) \, dx - x(3x + 2y) \, dy = 0$

*21. $\dfrac{dy}{dx} = \dfrac{x + 4y}{4x + y}$

22. $\sqrt{xy} \, dx + (x + \sqrt{xy} + y) \, dy = 0$

23. $(x - y) \, dx + x \, dy = 0$

24. $(4x + 3y) \, dx + 2y \, dy = 0$

*25. $y \, dx + (xy + y) \, dy = 0$

26. $x^2 \, dx + (xy + x^2) \, dy = 0$

27. $(2x^2 - 7xy + 5y^2) \, dx + xy \, dy = 0$

28. $12xy \, dx + (20x^2 - 26xy + 8y^2) \, dy = 0$

*29. $(y + 2\sqrt{x^2 + y^2})\,dx - x\,dy = 0$

30. $4xy\,dx - (x^2 + y^2)\,dy = 0$

31. $y^2\,dx = (xy - 4x^2)\,dy$

32. $x\,dy = (\sqrt{xy} - y)\,dx$

*33. $y\,dx - (3\sqrt{xy} + x)\,dy = 0$

34. $(x^2 - xy)\,dx = (x^2 - y^2)\,dy$

35. $(x^2 - y^2)\,dy + (y^2 + xy)\,dx = 0$

36. $xy\,dy - (x^2 e^{-y/x} + y^2)\,dx = 0$

*37. $\dfrac{dx}{dy} = \dfrac{2y}{x}e^{-x/y} + \dfrac{x}{y}$

38. $x\,dy - y\ln\!\left(\dfrac{y}{x}\right)^2 dx = 0$

39. $x(\ln x - \ln y)\,dy = y\,dx$

40. $x^2 y e^{x/y}\,dx - (x^3 e^{x/y} + y^3)\,dy = 0$

In Exercises 41–47, solve the initial-value problem.

*41. $\dfrac{dy}{dx} = \dfrac{4y^2 - x^2}{2xy}$, $y(1) = 1$

42. $(x + y)\,dx - x\,dy = 0$, $y(1) = 1$

43. $x\,dy - (y + \sqrt{x^2 + y^2})\,dx = 0$, $y(1) = 0$

44. $(x^2 + y^2\sqrt{x^2 + y^2})\,dx - xy\sqrt{x^2 + y^2}\,dy = 0$,
$y(1) = 1$

*45. $(y^3 - x^3)\,dx - xy^2\,dy = 0$, $y(1) = 3$

46. $xy^3\,dx - (x^4 + y^4)\,dy = 0$, $y(1) = 1$

47. $y^4\,dx + (x^4 - xy^3)\,dy = 0$, $y(1) = 2$

48. First-order equations of the form

$$(a_1 x + b_1 y + c_1)\,dx + (a_2 x + b_2 y + c_2)\,dy = 0$$

can be transformed into a homogeneous equation with a transformation. If $a_2/a_1 \neq b_2/b_1$, the transformation $\begin{cases} x = X + h \\ y = Y + k \end{cases}$, where (h, k) satisfies the linear system $\begin{cases} a_1 h + b_1 k + c_1 = 0 \\ a_2 h + b_2 k + c_2 = 0 \end{cases}$, reduces this equation to a homogeneous equation in the variables X and Y. If $a_2/a_1 = b_2/b_1 = k$, the transformation $z = a_1 x + b_1 y$ reduces this equation to a homogeneous equation in the variables x and z. Use this transformation to solve the following equations.

(a) $(x - 2y + 1)\,dx + (4x - 3y - 6)\,dy = 0$

(b) $(5x + 2y + 1)\,dx + (2x + y + 1)\,dy = 0$

(c) $(3x - y + 1)\,dx - (6x - 2y - 3)\,dy = 0$

(d) $(2x + 3y + 1)\,dx + (4x + 6y + 1)\,dy = 0$

*49. Find an equation of the curve that passes through the point $(\sqrt{e}, \sqrt{e})$ and has slope $y/x + x/y$ at each point (x, y) on the curve.

In Exercises 50–53, without actually solving the homogeneous equations, match each equation in Group A with the graph of its direction field in Group B.

Group A

50. $\dfrac{dy}{dx} = \dfrac{xe^{y/x}}{y}$

51. $\dfrac{dy}{dx} = \dfrac{ye^{y/x}}{x}$

52. $\dfrac{dy}{dx} = \dfrac{ye^{x/y}}{x}$

53. $\dfrac{dy}{dx} = \dfrac{xe^{x/y}}{y}$

Group B

(a)

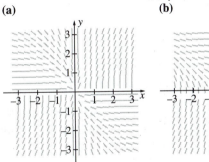

(b)

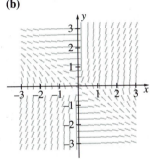

(c)

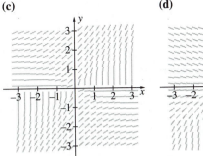

(d)

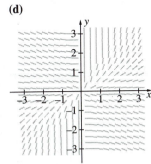

54. Show that if the differential equation $dy/dx = f(x, y)$ is homogeneous, the equation can be written as $dy/dx = F(y/x)$. (*Hint:* Let $t = 1/x$.)

55. If $M(x, y)\,dx + N(x, y)\,dy = 0$ is a homogeneous equation, show that the change of variables $x = r\cos\theta$ and $y = r\sin\theta$ transform the homogeneous equation into a separable equation.

56. Equations of the form

$$f(xy' - y) = g(y')$$

are called **Clairaut equations** after the French mathematician Alexis Clairaut (1713–1765) who studied these equations in 1734. Solutions to this equation are determined by differentiating each side of the equation with respect to x.

(a) Use the chain rule to show that the derivative of $f(xy' - y)$ is

$$f'(xy' - y)(xy'' + y' - y') = f'(xy' - y)(xy'')$$

where $'$ denotes differentiation with respect to the argument of the function, x.

(b) Show that the equation

$$f'(xy' - y)(xy'') = g'(y')y''$$

which is equivalent to

$$[f'(xy' - y)x - g'(y')]y'' = 0$$

is obtained by differentiating both sides of the Clairaut equation with respect to x.

(c) This result indicates that $y'' = 0$ or $f'(xy' - y)x - g'(y') = 0$. If $y'' = 0$, $y' = c$ where c is a constant. Substitute $y' = c$ into the differential equation $f(xy' - y) = g(y')$, to find

that a general solution is $f(xc - y) = g(c)$. If $f'(xy' - y)x - g'(y') = 0$, this equation can be used along with $f(xy' - y) = g(y')$ to determine another solution by eliminating y'. This is called the *singular solution* of the Clairaut equation.

57. Use the following steps to solve the Clairaut equation $xy' - (y')^3 = y$.

(a) Place the equation in the appropriate form to find that $f(x) = x$ and $g(x) = x^3$.

(b) Use the form of a general solution to find that $xc - y = c^3$ and solve this equation for y.

(c) Find the singular solution by differentiating $xy' - (y')^3 = y$ with respect to x to obtain $xy'' + y' - 3(y')^2 y'' = y'$, which can be simplified to $[x - 3(y')^2]y'' = 0$. Since $y'' = 0$ was used to find the general solution, solve $x - 3(y')^2 = 0$ for y' to obtain $y' = (x/3)^{1/2}$. Substitute this expression for y' into $xy' - y = (y')^3$ to obtain a relationship between x and y.

In Exercises 58–61, solve the Clairaut equation.

58. $xy' - y - 2(xy' - y)^2 = y' + 1$

59. $xy' - y - 1 = (y')^2 - y'$

60. $1 + y - xy' = \ln(y')$

61. $1 - 2(xy' - y) = (y')^{-2}$

2.3   Exact Equations

Many first-order differential equations cannot be reduced to separable differential equations by a suitable substitution, unlike homogeneous equations. For example, the first-order differential equation

$$(\sin y + y \cos x)\, dx + (\sin x + x \cos y)\, dy = 0$$

is neither separable nor can be reduced to a separable equation by an appropriate substitution. Nevertheless, a general solution of

$$(\sin y + y \cos x)\, dx + (\sin x + x \cos y)\, dy = 0$$

can be calculated, as we will see in this section.

Definition 2.3 Exact Differential Equation

A first-order differential equation that can be written in the form

$$M(x,y)\, dx + N(x, y)\, dy = 0$$

where

$$M(x, y)\, dx + N(x, y)\, dy = \frac{\partial f}{\partial x}(x, y)\, dx + \frac{\partial f}{\partial y}(x, y)\, dy$$

for some function $f(x, y)$ is called an **exact differential equation.**

In calculus we learn that the **total differential** of the function $f(x, y)$ is

$$df = \frac{\partial f}{\partial x}(x, y)\, dx + \frac{\partial f}{\partial y}(x, y)\, dy.$$

Therefore, the equation $M(x, y)\, dx + N(x, y)\, dy = 0$ is exact if there exists a function $f(x, y)$ such that $M(x, y)\, dx + N(x, y)\, dy$ is the total differential of $f(x, y)$.

In multivariable calculus, we learned that if f, $\frac{\partial f}{\partial x}$, $\frac{\partial f}{\partial y}$, $\frac{\partial^2 f}{\partial x \partial y}$, and $\frac{\partial^2 f}{\partial y \partial x}$ are

continuous on an open region R, then $\frac{\partial^2 f}{\partial x \partial y} = \frac{\partial^2 f}{\partial y \partial x}$ on R. Hence, if

$M(x, y)\, dx + N(x, y)\, dy = 0$ is exact and $M(x, y)\, dx + N(x, y)\, dy$ is the total differential of $f(x, y)$,

$$\frac{\partial N}{\partial x} = \frac{\partial}{\partial x}\left(\frac{\partial f}{\partial y}\right) = \frac{\partial}{\partial y}\left(\frac{\partial f}{\partial x}\right) = \frac{\partial M}{\partial y}.$$

In fact, we can prove the following. (See Exercise 55.)

Theorem 2.1 Test for Exactness

The first-order differential equation $M(x, y)\, dx + N(x, y)\, dy = 0$ is exact if and only if $\dfrac{\partial M}{\partial y} = \dfrac{\partial N}{\partial x}.$

Example 1

Show that the equation $2xy^3\, dx + (1 + 3x^2y^2)\, dy = 0$ is exact and that the equation $x^2y\, dx + 5xy^2\, dy = 0$ is not exact.

Solution The equation $2xy^3\, dx + (1 + 3x^2y^2)\, dy = 0$ is an exact equation because

$$\frac{\partial}{\partial y}(2xy^3) = 6xy^2 = \frac{\partial}{\partial x}(1 + 3x^2y^2).$$

On the other hand, the equation $x^2y\, dx + 5xy^2\, dy = 0$ is not exact because

$$\frac{\partial}{\partial y}(x^2y) = x^2 \neq 5y^2 = \frac{\partial}{\partial x}(5xy^2).$$

If the equation $M(x, y)\, dx + N(x, y)\, dy = 0$ is exact, we can find a function $f(x, y)$ such that $M(x, y) = \dfrac{\partial f}{\partial x}(x, y)$ and $N(x, y) = \dfrac{\partial f}{\partial y}(x, y)$. Then the differential equation becomes

$$M(x, y)\, dx + N(x, y)\, dy = 0$$
$$df = 0.$$

A general solution of the equation is

$$f(x, y) = C,$$

where C is a constant.

Example 2

Find a general solution of $(\sin y + y \cos x)\, dx + (\sin x + x \cos y)\, dy = 0$.

Solution   The equation is exact because

$$\frac{\partial}{\partial y}(\sin y + y \cos x) = \cos y + \cos x = \frac{\partial}{\partial x}(\sin x + x \cos y).$$

Let $f(x, y)$ be a function with $\dfrac{\partial f}{\partial x} = \sin y + y \cos x$ and $\dfrac{\partial f}{\partial y} = \sin x + x \cos y$. Integrating $\dfrac{\partial f}{\partial x} = \sin y + y \cos x$ with respect to x results in

$$\int (\sin y + y \cos x)\, dx = x \sin y + y \sin x + g(y),$$

where $g(y)$ denotes an arbitrary function of y. We must include this arbitrary function $g(y)$, because the derivative of a function of y with respect to x is zero. That is, the general form of the function $f(x, y)$ whose partial derivative with respect to x is $\partial f/\partial x = \sin y + y \cos x$ is given by

$$f(x, y) = x \sin y + y \sin x + g(y).$$

Because we are looking for a function $f(x, y)$ that satisfies

$$\frac{\partial f}{\partial y} = \sin x + x \cos y,$$

and differentiating $f(x, y) = x \sin y + y \sin x + g(y)$ with respect to y results in

$$\frac{\partial f}{\partial y} = \sin x + x \cos y + g'(y),$$

it must be true that

$$\sin x + x \cos y + g'(y) = \sin x + x \cos y$$

which indicates that $g'(y) = 0$. This means that $g(y) = k$ for some constant k. Thus,

$$f(x, y) = x \sin y + y \sin x + k.$$

Therefore, the implicit function $x \sin y + y \sin x + k = C_1$ or $x \sin y + y \sin x = C$, where $C = C_1 - k$ represents an arbitrary constant, is a general solution of $(\sin y + y \cos x) \, dx + (\sin x + x \cos y) \, dy = 0$. (Note that there is no need to include an arbitrary constant in $f(x, y)$, because it is included in $f(x, y) = C$.) Several members of the family of solutions are graphed in Figure 2.13 by graphing several level curves of the function $f(x, y) = x \sin y + y \sin x$.

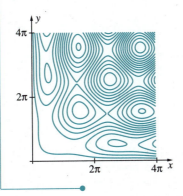

Figure 2.13 We graph various solutions to the equation by graphing several level curves of the function $f(x, y) = x \sin y + y \sin x$.

Example 3

Solve $2x \sin y \, dx + (x^2 \cos y - 1) \, dy = 0$ subject to $y(0) = \frac{1}{2}$.

Solution The equation is exact because

$$\frac{\partial}{\partial y}(2x \sin y) = 2x \cos y = \frac{\partial}{\partial x}(x^2 \cos y - 1).$$

Let $f(x, y)$ be a function with $\dfrac{\partial f}{\partial x} = 2x \sin y$ and $\dfrac{\partial f}{\partial y} = x^2 \cos y - 1$. Integrating $\dfrac{\partial f}{\partial x}$ with respect to x yields

$$f(x, y) = \int 2x \sin y \, dx = x^2 \sin y + g(y).$$

Notice that the arbitrary function $g(y)$ serves as a "constant" of integration with respect to x. From the differential equation, we have

$$\frac{\partial f}{\partial y} = x^2 \cos y - 1$$

and differentiating $f(x, y) = x^2 \sin y + g(y)$ with respect to y gives us

$$\frac{\partial f}{\partial y} = x^2 \cos y + g'(y).$$

Thus,

$$x^2 \cos y - 1 = x^2 \cos y + g'(y)$$
$$g'(y) = -1$$

and $g(y) = -y + C_1$ so that substitution into $f(x, y) = x^2 \sin y + g(y)$ yields

$$f(x, y) = x^2 \sin y - y + C_1.$$

A general solution of the exact equation is then $x^2 \sin y - y + C_1 = C$. Simplifying, we have

$$x^2 \sin y - y = k,$$

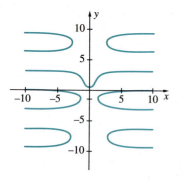

Figure 2.14

where $k = C - C_1$ is a constant. Our solution must satisfy $y(0) = \frac{1}{2}$, so we must find the solution that passes through the point $(0, \frac{1}{2})$. Substituting $x = 0$ and $y = \frac{1}{2}$ into the general solution, we obtain $0^2 \sin(\frac{1}{2}) - \frac{1}{2} = k$ so that $k = -\frac{1}{2}$ and the solution is $x^2 \sin y - y = -\frac{1}{2}$. The solution is graphed in Figure 2.14. We see that the graph passes through the point $(0, \frac{1}{2})$, as required by the initial condition.

Solving the Exact Differential Equation
$M(x, y)dx + N(x, y)dy = 0$

① Assume that $M(x, y) = \dfrac{\partial f}{\partial x}(x, y)$ and $N(x, y) = \dfrac{\partial f}{\partial y}(x, y)$.

② Integrate $M(x, y)$ with respect to x. (Add an arbitrary function of y, $g(y)$.)

③ Differentiate the result in step 2 with respect to y and set the result equal to $N(x, y)$. Solve for $g'(y)$.

④ Integrate $g'(y)$ with respect to y to obtain an expression for $g(y)$. (There is no need to include an arbitrary constant.)

⑤ Substitute $g(y)$ into the result obtained in step 2 for $f(x, y)$.

⑥ A general solution is $f(x, y) = C$, where C is a constant.

⑦ Apply the initial condition if given.

A similar algorithm can be stated so that in step 2, $N(x, y)$ is integrated with respect to y as we show in Example 4.

Example 4

Solve $\left(e^{y/x} - \dfrac{y}{x}e^{y/x} + \dfrac{1}{1+x^2}\right) dx + e^{y/x}\, dy = 0$.

Solution This equation is exact because $\dfrac{\partial}{\partial y}\left(e^{y/x} - \dfrac{y}{x}e^{y/x} + \dfrac{1}{1+x^2}\right) =$

$-\dfrac{y}{x^2}e^{y/x} = \dfrac{\partial}{\partial x}(e^{y/x})$. Let $f(x, y)$ be a function such that $\dfrac{\partial f}{\partial x} = e^{y/x} - \dfrac{y}{x}e^{y/x} +$

$\dfrac{1}{1+x^2}$ and $\dfrac{\partial f}{\partial y} = e^{y/x}$. Integrating $\dfrac{\partial f}{\partial y}$ with respect to y because it is a less compli-

cated expression than $\dfrac{\partial f}{\partial x}$ gives us

$$f(x, y) = \int e^{y/x}\, dy = \dfrac{1}{1/x}e^{y/x} + g(x) = xe^{y/x} + g(x)$$

where $g(x)$ is an arbitrary function of x. Differentiating $f(x, y)$ with respect to x leads to

$$\dfrac{\partial f}{\partial x} = e^{y/x} + x\left(-\dfrac{y}{x^2}e^{y/x}\right) + g'(x) = e^{y/x} - \dfrac{y}{x}e^{y/x} + g'(x),$$

so $g'(x) = \dfrac{1}{1+x^2}$. This implies that $g(x) = \tan^{-1} x$, so $f(x, y) = xe^{y/x} + \tan^{-1} x$.

Therefore, a general solution of the exact equation is $xe^{y/x} + \tan^{-1} x = C$.

Figure 2.15 shows the graph of $x \cos y + 4x - \sin y = C$ for various values of C. Identify the curve(s) corresponding to $C = 5$.

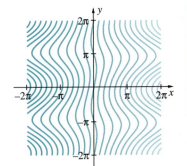

Figure 2.15 Graph of $x \cos y + 4x - \sin y = C$ for various values of C

Up to now, all initial-value problems that we have considered have had a *unique* solution. However, that need not be true. (This topic is considered again in Section 2.5.)

Example 5

Find a value of y_0 so that there is *not* a unique solution to the initial-value problem

$$\begin{cases} (\cos x \cos y - \sin x)\, dx + (\cos y - \sin x \sin y)\, dy = 0 \\ \qquad\qquad\qquad\qquad y(0) = y_0 \end{cases}$$

Solution The equation $(\cos x \cos y - \sin x)\, dx + (\cos y - \sin x \sin y)\, dy = 0$ is exact because

$$\dfrac{\partial}{\partial y}(\cos x \cos y - \sin x) = -\cos x \sin y = \dfrac{\partial}{\partial x}(\cos y - \sin x \sin y).$$

A general solution is found to be $\cos x + \sin x \cos y + \sin y = C$, which we graph for various values of C in Figure 2.16.

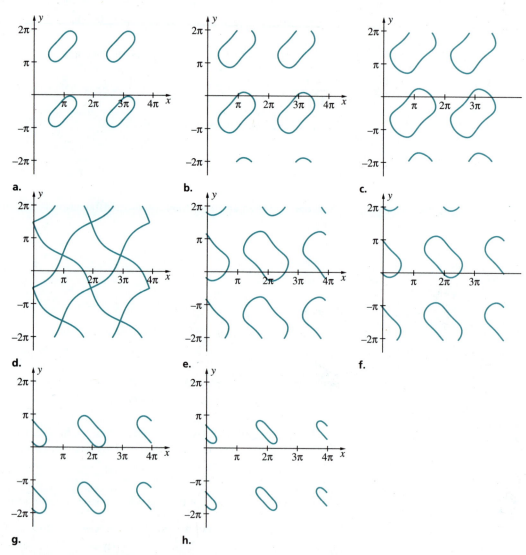

Figure 2.16   (a) $C = -\frac{3}{2}$ (b) $C = -1$ (c) $C = -\frac{1}{2}$ (d) $C = 0$ (e) $C = \frac{1}{2}$ (f) $C = 1$
(g) $C = \frac{3}{2}$ (h) $C = 9/5$

From the graphs, we see that if $C = 0$, it is possible to find y_0 so that there is *not* a unique solution to the initial-value problem. Indeed an exact value of y_0 is $3\pi/2$. (Why?) See Figure 2.17.)

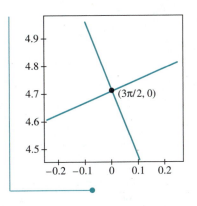

Figure 2.17 The solution is *not unique* near this point because when we zoom in we see the graph of more than one function passing through the point. (See Example 5 in Section 1.3.)

IN TOUCH WITH TECHNOLOGY

As with other types of equations, technology is useful in solving exact equations or in performing the steps necessary to solve an exact equation.

1. Find a general solution of the equation
$(2x - y^2 \sin(xy)) \, dx + (\cos(xy) - xy \sin(xy)) \, dy = 0$.
Graph various solutions on the rectangle $[0, 3\pi] \times [0, 3\pi]$.

2. Find a general solution of the equation
$(-1 + e^{xy}y + y \cos(xy)) \, dx +$
$\qquad\qquad (1 + e^{xy}x + x \cos(xy)) \, dy = 0$.

Graph several solutions.

3. (a) Find a general solution for each of the following differential equations:
(i) $(2x + 2y) \, dx + (2x + 2y) \, dy = 0$;
(ii) $(1.8x + 2y) \, dx + (2x + 2y) \, dy = 0$; and
(iii) $(2x + 1.9y) \, dx + (1.9x + 2y) \, dy = 0$.

(b) Graph the direction field for each equation along with several solutions.

(c) Comment on the statement: "If we slightly change a differential equation, the solutions also slightly change."

4. How does the graph of the solution to the initial-value problem

$$\begin{cases} [\sin(cy) - yc \sin(cx)] \, dx + \\ \qquad\qquad [xc \cos(cy) + \cos(cx)] \, dy = 0 \\ y(0) = 1 \end{cases}$$

change as c takes on values from -2 to 2?

EXERCISES 2.3

In Exercises 1–10, determine if the equation is exact.

1. $\left(y^2 - \dfrac{y}{2\sqrt{x}}\right) dx + (2xy - \sqrt{x} + 1) \, dy = 0$

2. $\dfrac{x}{\sqrt{x^2 + y^2}} \, dx + \dfrac{y}{\sqrt{x^2 + y^2}} \, dy = 0$

*3. $y \cos(xy) \, dx + x \cos(xy) \, dy = 0$

4. $(y \sec^2 x + 2x)\, dx + (\tan x)\, dy = 0$

5. $3xy^2\, dy + y^3\, dx = 0$

6. $(x - y \sin x)\, dx + (y^6 + \cos x)\, dy = 0$

*7. $(y \sin 2x)\, dx - (\sqrt{y} + \cos 2x)\, dy = 0$

8. $(e^{2x} + y)\, dx - (e^y - x)\, dy = 0$

9. $\ln(xy)\, dx + \dfrac{x}{y}\, dy = 0$

10. $e^{xy}\, dx + \dfrac{x}{y}e^{xy}\, dy = 0$

In Exercises 11–30, solve each equation.

11. $3x^2\, dx - dy = 0$

12. $-dx + 3y^2\, dy = 0$

*13. $y^2\, dx + 2xy\, dy = 0$

14. $\dfrac{3x^2}{y}\, dx - \dfrac{x^3}{y^2}\, dy = 0$

15. $(2x + y^3)\, dx + (3xy^2 + 4)\, dy = 0$

16. $-\dfrac{1}{y}\, dx + \left(\dfrac{x}{y^2} + 3y^2\right) dy = 0$

*17. $2xy\, dx + (x^2 + y^2)\, dy = 0$

18. $2xy^3\, dx + (1 + 3x^2y^2)\, dy = 0$

19. $\sin^2 y\, dx + x \sin 2y\, dy = 0$

20. $(3x^2 + 3y^2)\, dx + 6xy\, dy = 0$

*21. $\dfrac{y + y^2}{(y - x)^2}\, dx - \dfrac{x + x^2}{(x - y)^2}\, dy = 0$

22. $(3x^2y + 3y^2 - 1)\, dx + (x^3 + 6xy)\, dy = 0$

23. $-2xy^2 \sin(x^2)\, dx + 2y \cos(x^2)\, dy = 0$

24. $(2x - y^2 \sin(xy))\, dx + (\cos(xy) - xy \sin(xy))\, dy = 0$

*25. $(1 + y^2 \cos(xy))\, dx + (xy \cos(xy) + \sin(xy))\, dy = 0$

26. $ye^{xy}(\cos(xy) + \sin(xy))\, dx + xe^{xy}(\cos(xy) + \sin(xy))\, dy = 0$

27. $((3 + x)\cos(x + y) + \sin(x + y))\, dx + (3 + x)\cos(x + y)\, dy = 0$

28. $\dfrac{2x^2y \cos(x^2) - y \sin(x^2)}{x^2}\, dx + \dfrac{2xy + \sin(x^2)}{x}\, dy = 0$

*29. $\dfrac{e^{y/x}(x - y)}{x}\, dx + e^{y/x}\, dy = 0$

30. $\dfrac{e^{x/y}\left(x^2 \cos\left(\frac{y}{x}\right) + y^2 \sin\left(\frac{y}{x}\right)\right)}{x^2 y}\, dx -$
$\dfrac{e^{x/y}\left(x^2 \cos\left(\frac{y}{x}\right) + y^2 \sin\left(\frac{y}{x}\right)\right)}{xy^2}\, dy = 0$

In Exercises 31–40, solve the initial-value problem. Graph the solution on an appropriate region.

31. $2xy^2\, dx + 2x^2y\, dy = 0,\; y(1) = 1$

32. $\left(1 + \dfrac{y}{x^2}\right) dx - \dfrac{1}{x}\, dy = 0,\; y(2) = 1$

*33. $(2xy + 3x^2)\, dx + (x^2 - 1)\, dy = 0,\; y(0) = 1$

34. $(1 + 5x - y)\, dx - (x + 2y)\, dy = 0,\; y(0) = 0$

35. $(e^y - 2xy)\, dx + (xe^y - x^2)\, dy = 0,\; y(0) = 0$

36. $(2xe^{x^2}y + 2xe^{-y})\, dx + (e^{x^2} - x^2e^{-y} + 1)\, dy = 0,\; y(0) = 0$

*37. $(y^2 - 2 \sin 2x)\, dx + (1 + 2xy)\, dy = 0,\; y(0) = 1$

38. $(\cos^2 x - \sin^2 x + y)\, dx + (\sec y \tan y + x)\, dy = 0,\; y(0) = 0$

39. $\left(\dfrac{1}{1 + x^2} - y^2\right) dx - 2xy\, dy = 0,\; y(0) = 0$

40. $\left(\dfrac{2x}{1 + x^2} + y\right) dx + (e^y + x)\, dy = 0,\; y(0) = 0$

In Exercises 41–44, without actually solving the equations, match each equation in Group A with the graph of some of its solutions in Group B.

Group A

41. $\dfrac{dy}{dx} = \dfrac{e^{y-x}(\cos x - \sin x)}{\cos y + \sin y}$

42. $\dfrac{dy}{dx} = \dfrac{-e^{x-y}(\cos x + \sin x)}{\cos y - \sin y}$

43. $\dfrac{dy}{dx} = \dfrac{-e^y \cos x - e^x \cos y}{e^y \sin x - e^x \sin y}$

44. $\dfrac{dy}{dx} = \dfrac{e^x \sin x + e^y \sin y}{e^y \cos y - e^x \cos x}$

Group B

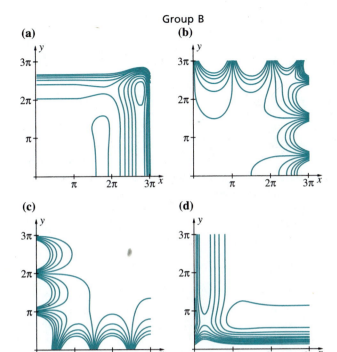

(a)

(b)

(c)

(d)

*45. Show that a separable equation of the form
$g(y)\,dy - h(x)\,dx = 0$ is exact.

46. (Integrating Factors) If the differential equation
$M(x, y)\,dx + N(x, y)\,dy = 0$ is not exact, multiplying
it by an *appropriate* function $\mu(x, y)$ yields an exact
equation. To find $\mu(x, y)$, we use the fact that if the
equation $\mu(x, y)M(x, y)\,dx + \mu(x, y)N(x, y)\,dy = 0$
is exact, $[\mu M]_y = [\mu N]_x$.

(a) Use the product rule to show that μ must satisfy
the differential equation
$M\mu_y - N\mu_x + (M_y - N_x)\mu = 0.$

(b) Use this equation to show that if $\mu = \mu(x)$,
μ satisfies the differential equation
$$\frac{d\mu}{dx} = \frac{M_y - N_x}{N}\mu, \text{ where } \frac{M_y - N_x}{N} \text{ is a function}$$
of x only.

(c) Show that if $\mu = \mu(x)$,
$$\mu(x) = \exp\left(\int \frac{1}{N(x, y)}\left[\frac{\partial M(x, y)}{\partial y} - \frac{\partial N(x, y)}{\partial x}\right]dx\right).$$

(d) If $\mu = \mu(y)$, find a differential equation that μ
must satisfy and a restriction on $\dfrac{N_x - M_y}{M}$. Show
that
$$\mu(y) = \exp\left(\int \frac{1}{M(x, y)}\left[\frac{\partial N(x, y)}{\partial x} - \frac{\partial M(x, y)}{\partial y}\right]dy\right).$$

In Exercises 47–54, use an integrating factor to solve each
differential equation. (See Exercise 46.)

47. $x^2y\,dx + x^3\,dy = 0$

48. $y(2e^x + 4x)\,dx + 3(e^x + x^2)\,dy = 0$

*49. $y\,dx + (2x - ye^y)\,dy = 0$

50. $(2xy + y^2)\,dx - x^2\,dy = 0$

51. $(y + 2x^2)\,dx + (x^2y - x)\,dy = 0$

52. $(5xy + 4y^2 + 1)\,dx + (x^2 + 2xy)\,dy = 0$

*53. $(2xy^2 + y)\,dx + (2x^3 - x)\,dy = 0$

54. $(2x + \tan y)\,dx + (x - x^2\tan y)\,dy = 0$

55. Suppose that $M(x, y)\,dx + N(x, y)\,dy = 0$ is an equation for which $\partial M/\partial y = \partial N/\partial x$.

(a) Let $g(y) = \int \left(N(x, y) - \dfrac{\partial}{\partial y}\int M(x, y)\,dx\right)dy.$

Show that g is a function of y. $\bigg($Hint: Show that

$N(x, y) - \dfrac{\partial}{\partial y}\int M(x, y)\,dx$ is a function of y by

showing that $\dfrac{\partial}{\partial x}\left(N(x, y) - \dfrac{\partial}{\partial y}\int M(x, y)\,dx\right) =$

$0.\bigg)$

(b) Let $f(x, y) = g(y) + \int M(x, y)\,dx.$ Show that
$$M(x, y)\,dx + N(x, y)\,dy = \frac{\partial f}{\partial x}\,dx + \frac{\partial f}{\partial y}\,dy.$$

2.4 Linear Equations

In the previous sections, we have seen that some first-order equations may be classified
as separable equations, others as homogeneous, and others as exact equations. (Of
course, most first-order differential equations are neither separable, homogeneous, nor

First-order linear equations are particularly important because as long as the necessary integrals can be evaluated, an explicit solution can be produced as shown here. If the integration cannot be carried out, often the solution can be approximated numerically by taking advantage of numerical integration techniques.

exact.) Calculating explicit or implicit closed-form solutions of most first-order equations may be a formidable task, at best. However, first-order linear equations $a_1(x)\dfrac{dy}{dx} + a_0(x)y = f(x)$, which we can rewrite in the form

$$\frac{dy}{dx} + p(x)y = q(x),$$

can always be solved, so we discuss their method of solution in this section.

In Exercise 46 in Section 2.3, we saw that some equations of the form $M(x, y)\, dx + N(x, y)\, dy = 0$ can be transformed into exact equations by multiplying by an appropriate function $\mu(x,y)$, which is called an **integrating factor.** Let us see if we can transform the linear equation $dy/dx + p(x)y = q(x)$ into an exact equation by multiplying it by an appropriate function $\mu(x, y)$. First, we rewrite the first-order linear equation $dy/dx + p(x)y = q(x)$ in differential form:

$$dy + (p(x)y - q(x))\, dx = 0.$$

If it is possible to transform the equation into an exact equation, we would like our choice of $\mu(x, y)$ to be as simple as possible. For example, we would like $\mu(x, y)$ to be either a function of only x or a function of only y. If μ is a function of x, multiplying the equation by $\mu(x)$ results in

$$\mu(x)\, dy + \mu(x)(p(x)y - q(x))\, dx = 0$$

and in order for this equation to be exact, we must have

$$\frac{\partial \mu}{\partial x} = \frac{\partial}{\partial y}[\mu(x)(p(x)y - q(x))].$$

Because μ, p, and q are functions of x, $\dfrac{\partial}{\partial y}[\mu(x)p(x)y] = \mu(x)p(x)\dfrac{\partial}{\partial y}[y] = \mu(x)p(x)$ and $\dfrac{\partial}{\partial y}[q(x)] = 0$, so we can rewrite this equation as

$$\mu'(x) = \mu(x)p(x) \qquad \text{or} \qquad \frac{d\mu(x)}{dx} = \mu(x)p(x),$$

which is a separable equation. Solving this equation for μ yields $\mu(x) = e^{\int p(x)\, dx}$. Thus, multiplying the equation $dy/dx + p(x)y = q(x)$ by $\mu(x) = e^{\int p(x)\, dx}$ yields the exact equation

$$e^{\int p(x)\, dx}\frac{dy}{dx} + e^{\int p(x)\, dx}p(x)y = e^{\int p(x)\, dx}q(x).$$

Is it possible to transform the equation $\dfrac{dy}{dx} + p(x)y = q(x)$ into an exact equation by multiplying it by a function of y?

Although this equation is exact and we can use the methods discussed in Section 2.3 to solve it, we find an easier method by noticing that by the product rule and the Fundamental Theorem of Calculus,

$$\frac{d}{dx}(e^{\int p(x)\,dx}y) = e^{\int p(x)\,dx}\frac{dy}{dx} + e^{\int p(x)\,dx}p(x)y.$$

Using this relationship, we rewrite the equation as

$$\frac{d}{dx}(e^{\int p(x)\,dx}y) = e^{\int p(x)\,dx}q(x).$$

Integrating we obtain

$$e^{\int p(x)\,dx}y = \int e^{\int p(x)\,dx}q(x)\,dx$$

and dividing by $e^{\int p(x)\,dx}$ yields a general solution of $dy/dx + p(x)y = q(x)$:

$$y = \frac{\int e^{\int p(x)\,dx}q(x)\,dx}{e^{\int p(x)\,dx}} = e^{-\int p(x)\,dx}\int e^{\int p(x)\,dx}q(x)\,dx,$$

where a *constant of integration* must be included when $\int e^{\int p(x)\,dx}q(x)\,dx$ is evaluated.

The term $\mu(x) = e^{\int p(x)\,dx}$ is called an **integrating factor** for the linear equation $dy/dx + p(x)y = q(x)$ and is useful because $\mu(x)\left(\frac{dy}{dx} + p(x)y\right) = \frac{d}{dx}[\mu(x)y]$ as we saw in the derivation above.

A general solution of the first-order ordinary differential equation $\frac{dy}{dx} + p(x)y = q(x)$ is found by solving

$$\frac{d}{dx}[\mu(x)y] = \mu(x)q(x)$$

for y where $\mu(x) = e^{\int p(x)\,dx}$.

Example 1   Find a general solution of $x\frac{dy}{dx} + y = x\sin x$.

Solution   First, we place the equation in the form $\frac{dy}{dx} + p(x)y = q(x)$. Dividing the equation by x yields

$$\frac{dy}{dx} + \frac{1}{x}y = \sin x,$$

so $p(x) = \dfrac{1}{x}$ and $q(x) = \sin x$. An integrating factor is

$$\mu(x) = e^{\int 1/x\,dx} = e^{\ln|x|} = x, \text{ for } x > 0,$$

and

$$\frac{d}{dx}(xy) = x\frac{dy}{dx} + y = x \sin x$$

so

$$xy = \int x \sin x\,dx.$$

Using the integration by parts formula, $\int u\,dv = uv - \int v\,du$, with $u = x$ and $dv = \sin x\,dx$, we obtain $du = dx$ and $v = -\cos x$ so

$$xy = \int x \sin x\,dx = -x\cos x + \int \cos x\,dx = -x\cos x + \sin x + C.$$

Therefore, a general solution of the equation $x\dfrac{dy}{dx} + y = x\sin x$ for $x > 0$ is

$$y = \frac{-x\cos x + \sin x + C}{x}.$$ In Figure 2.18, we graph several solutions for $x > 0$ (we assumed that $x > 0$ in finding $\mu(x) = x$). Note that if we want to solve the equation for $x < 0$, we let $e^{\int 1/x\,dx} = e^{\ln|x|} = -x$ for $x < 0$.

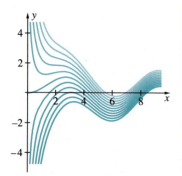

Figure 2.18 Graph of $y = \dfrac{-x\cos x + \sin x + C}{x}$ for various values of C.

Example 2

Find a general solution of $\dfrac{dy}{dx} - \dfrac{4x}{x^2+1}y = (1+x^2)^3 e^x$.

Solution   We begin by computing the integrating factor $\mu(x) = e^{\int -4x/(x^2+1)\,dx}$:

$$\mu(x) = e^{\int -4x/(x^2+1)\,dx} = e^{-2\ln(x^2+1)} = (x^2+1)^{-2} = 1/(x^2+1)^2.$$

Notice that we do not have to include absolute value signs with the natural logarithm, because $x^2 + 1 > 0$ for all x. To solve $\dfrac{d}{dx}[\mu(x)y] = \mu(x)(1+x^2)^3 e^x$, we compute

$$\int \mu(x)(1+x^2)^3 e^x\,dx = \int \frac{1}{(1+x^2)^2}(1+x^2)^3 e^x\,dx = \int (1+x^2)e^x\,dx$$

using integration by parts:

$$\int (1 + x^2)e^x \, dx = (1 + x^2)e^x - 2\int xe^x \, dx$$
$$= (1 + x^2)e^x - 2xe^x + 2e^x + C$$
$$= (x^2 - 2x + 3)e^x + C.$$

Thus, a general solution of the equation is given by

$$y(x) = \frac{(x^2 - 2x + 3)e^x + C}{(x^2 + 1)^{-2}} = (x^2 + 1)^2[(x^2 - 2x + 3)e^x + C].$$

As with other types of equations, we solve initial-value problems by first finding a general solution of the equation and then applying the initial condition to determine the value of the constant.

Example 3

Solve the initial-value problem $\dfrac{dy}{dx} + 5x^4y = x^4$, $y(0) = -7$.

Solution   We begin by solving the linear equation $\dfrac{dy}{dx} + 5x^4y = x^4$ using the integrating factor $e^{\int 5x^4 \, dx} = e^{x^5}$. Then, the equation can be written as

$$\frac{d}{dx}(e^{x^5}y) = x^4 e^{x^5}$$

so that integration of both sides of the equation yields

$$e^{x^5}y = \frac{1}{5}e^{x^5} + C.$$

A general solution is

$$y = \frac{1}{5} + Ce^{-x^5}.$$

We find the unknown constant C by substituting the initial condition $y(0) = -7$. This gives $-7 = \frac{1}{5} + C$, so $C = -\frac{36}{5}$. Therefore, the solution to the initial-value problem is

$$y = \frac{1}{5} - \frac{36}{5}e^{-x^5}.$$

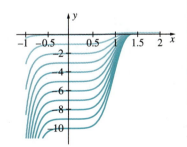

Figure 2.19

Figure 2.19 shows the graph of $y = \dfrac{1}{5} + Ce^{-x^5}$ for various values of C. Identify the graph of the solution that satisfies the condition $y(0) = -7$.

In some cases, we must rewrite an equation in order to place it in the form of a linear first-order ordinary differential equation.

Example 4

Solve $\dfrac{dt}{dr} = \dfrac{1}{\sin t - r \tan t}$, $0 < t < \pi/2$.

Solution   Notice that this equation is neither separable, exact, nor homogeneous. In fact, if t is the *dependent* variable, the equation is nonlinear (in t). (Why?) However, solving the equation for $\dfrac{dr}{dt}$ yields

$$\frac{dr}{dt} = \sin t - r \tan t \qquad \text{or} \qquad \frac{dr}{dt} + (\tan t)r = \sin t,$$

which is a *linear* equation in the dependent variable r. With $p(t) = \tan t$, the integrating factor is

$$\mu(t) = e^{\int \tan t\, dt} = e^{-\ln|\cos t|} = \frac{1}{\cos t}, \quad 0 < t < \pi/2.$$

Then,

$$\frac{d}{dt}[\mu(t)r] = \mu(t) \sin t$$

$$\frac{d}{dt}\left[\frac{1}{\cos t}r\right] = \frac{\sin t}{\cos t}$$

$$\frac{1}{\cos t}r = -\ln|\cos t| + C$$

$$r = -(\cos t)\ln(\cos t) + C \cos t.$$

 Solve the initial-value problem $\begin{cases} \dfrac{dt}{dr} = \dfrac{1}{\sin t - r \tan t} \\ t(1) = \pi/4 \end{cases}$. *Graph the solution on an appropriate interval.*

Example 5

If a drug is introduced into the bloodstream in dosages of $D(t)$ and is removed at a rate proportional to the concentration, the concentration $C(t)$ at time t is given by

$$\begin{cases} \dfrac{dC}{dt} = D(t) - kC \\ C(0) = 0 \end{cases},$$

where $k > 0$ is the constant of proportionality.*

---

* J. D. Murray, *Mathematical Biology*, Springer-Verlag, 1990, pp. 645–649.

(a) Solve this initial-value problem.

(b) Suppose that over a 24-hour period, a drug is introduced into the bloodstream at a rate of $24/t_0$ for exactly t_0 hours and then stopped so that $D_{t_0}(t) = \begin{cases} 24/t_0, & \text{if } 0 \le t \le t_0 \\ 0, & \text{if } t > t_0 \end{cases}$. What is the total dosage and average dosage over a 24-hour period?

(c) Calculate and then graph $C(t)$ on the interval $[0, 30]$ if $k = 0.05, 0.10, 0.15,$ and 0.20 for $t_0 = 4, 8, 12, 16,$ and 20. How does increasing t_0 affect the concentration of the drug in the bloodstream? increasing k?

Solution   (a) The solution to this linear initial-value problem is

$$C(t) = e^{-kt} \int_0^t e^{ks} D(s) \, ds.$$

(b) In each case, the total dosage over a 24-hour period is $\displaystyle\int_0^{24} D_{t_0}(t) \, dt =$

$\displaystyle\int_0^{t_0} \frac{24}{t_0} \, dt = 24$; the average dosage is

$$\frac{1}{24 - 0} \int_0^{24} D_{t_0}(t) \, dt = \frac{1}{24} \int_0^{t_0} \frac{24}{t_0} \, dt = 1.$$

(c) To compute $C(t) = e^{-kt} \displaystyle\int_0^t e^{ks} D_{t_0}(s) \, ds$, we must keep in mind that $D_{t_0}(t)$ is a piecewise defined function:

$$C(t) = e^{-kt} \int_0^t e^{ks} D_{t_0}(s) \, ds = \begin{cases} e^{-kt} \displaystyle\int_0^t e^{ks} \frac{24}{t_0} \, ds, & \text{if } 0 \le t \le t_0 \\ e^{-kt} \displaystyle\int_0^{t_0} e^{ks} \frac{24}{t_0} \, ds, & \text{if } t > t_0 \end{cases}$$

$$= \begin{cases} \dfrac{24}{kt_0}(1 - e^{-kt}), & \text{if } 0 \le t \le t_0 \\ \dfrac{24}{kt_0}(e^{-k(t-t_0)} - e^{-kt}), & \text{if } t > t_0 \end{cases}.$$

We graph $C(t)$ on the interval $[0, 30]$ if $k = 0.05, 0.10, 0.15,$ and 0.20 for $t_0 = 4, 8, 12, 16,$ and 20 in Figure 2.20. From the graphs, we see that as t_0 is increased, the maximum concentration level decreases and occurs at later times, while increasing k increases the rate at which the drug is removed from the bloodstream.

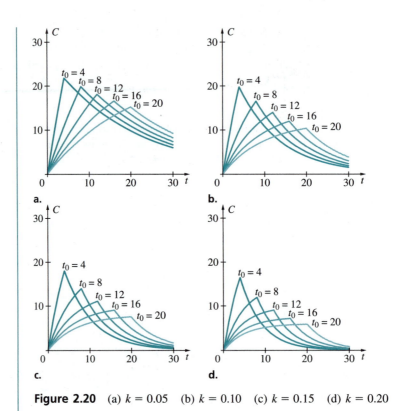

Figure 2.20   (a) $k = 0.05$ (b) $k = 0.10$ (c) $k = 0.15$ (d) $k = 0.20$

IN TOUCH WITH TECHNOLOGY

One advantage of using technology is that often when a large number of problems are solved, their solutions are compared, and conjectures about general patterns may be discovered and tested. Many computer algebra systems are capable of solving a variety of linear equations, particularly those that are frequently encountered in an elementary differential equations course.

1. Find a general solution of the equation
   $$x\frac{dy}{dx} + y = x\cos x.$$ Graph various solutions on the rectangle $[0, 2\pi] \times [-10, 10]$.

2. Compare the solutions of $\dfrac{dy}{dx} + y = f(x)$ subject to $y(0) = 0$, where $f(x) = x$, $\sin x$, $\cos x$, and e^x.

3. Compare the solutions of $\dfrac{dy}{dx} + ky = x$, $y(0) = 1$ for $k = -2, -1, 0, 1, 2$.

4. Compare the solutions of $\dfrac{dy}{dx} + y = x$, $y(0) = k$ for $k = -2, -1, 0, 1, 2$.

5. (a) Graph the direction field for the equation $y' = y - x$. (b) Solve and graph the solution to the initial-value problems (i) $\begin{cases} y' = y - x \\ y(0) = 1 \end{cases}$; (ii) $\begin{cases} y' = y - x \\ y(0) = 1.1 \end{cases}$; and (iii) $\begin{cases} y' = y - x \\ y(0) = 0.9 \end{cases}$. (c) Comment on the statement: "If we slightly change the initial conditions in a linear initial-value problem, the solution also slightly changes."

6. **(Antibiotic Production)** When you are injured or sick, your doctor may prescribe antibiotics to prevent or cure infections. In the journal article "Changes in the Protein Profile of *Streptomyces Griseus* during a Cycloheximide Fermentation" we see that production of the antibiotic cycloheximide by *Streptomyces* is typical of antibiotic production. During the production of cycloheximide, the mass of *Streptomyces* grows relatively quickly and produces little cycloheximide. After approximately 24 hours, the mass of *Streptomyces* remains relatively constant and cycloheximide accumulates. However, once the level of cycloheximide reaches a certain level, extracellular cycloheximide is degraded **(feedback inhibited)**. One approach to alleviating this problem and to maximize cycloheximide production is to continuously remove extracellular cycloheximide. The rate of growth of *Streptomyces* can be described by the equation

$$\frac{dX}{dt} = \mu_{max}\left(1 - \frac{X}{X_{max}}\right)X,$$

where X represents the mass concentration in g/L, μ_{max} is the maximum specific growth rate, and X_{max} represents the maximum mass concentration.*

(a) Find the solution to the initial-value problem

$$\begin{cases} \dfrac{dX}{dt} = \mu_{max}\left(1 - \dfrac{X}{X_{max}}\right)X \\ X(0) = 1 \end{cases}$$

by first converting the equation $\frac{dX}{dt} = \mu_{max}\left(1 - \frac{X}{X_{max}}\right)X$ to a linear equation with the substitution $y = X^{-1}$. (See Exercise 48.)

Experimental results have shown that $\mu_{max} = 0.3 \text{ hr}^{-1}$ and $X_{max} = 10$ g/L.

(b) Substitute these values into the result obtained in (a). (i) Graph $X(t)$ on the interval $[0, 24]$. (ii) Find the mass concentration at the end of 4, 8, 12, 16, 20, and 24 hours.

The rate of accumulation of cycloheximide is the difference between the rate of synthesis and the rate of degradation:

$$\frac{dP}{dt} = R_s - R_d.$$

It is known that $R_d = K_d P$, where $k_d \approx 5 \times 10^{-3} \text{ h}^{-1}$, so $dP/dt = R_s - R_d$ is equivalent to $dP/dt = R_s - K_d P$. Furthermore,

$$R_s = Q_{po}EX\left(1 + \frac{P}{K_I}\right)^{-1},$$

where Q_{po} represents the specific enzyme activity with value $Q_{po} \approx 0.6$ g CH/g protein $\cdot$ h and K_I represents the inhibition constant. E represents the intracellular concentration of an enzyme which we will assume is constant. For large values of K_I and t, $X(t) \approx 10$ and $\left(1 + \frac{P}{K_I}\right)^{-1} \approx 1$. Thus, $R_s \approx 10Q_{po}E$ so

$$\frac{dP}{dt} = 10Q_{po}E - K_d P.$$

(c) Solve the initial-value problem

$$\begin{cases} \dfrac{dP}{dt} = 10Q_{po}E - K_d P \\ p(24) = 0 \end{cases}$$

(d) Graph $\frac{1}{E}P(t)$ on the interval $[24, 1000]$.

(e) What happens to the net accumulation of the antibiotic as time increases?

(f) If instead the antibiotic is removed from the solution so that no degradation occurs, what happens to the net accumulation of the antibiotic as time increases?

---

* Kevin H. Dykstra and Henry Y. Wang, "Changes in the Protein Profile of *Streptomyces Griseus* during a Cycloheximide Fermentation," *Biochemical Engineering V,* Annals of the New York Academy of Sciences, Volume 56, New York Academy of Sciences (1987), pp. 511–522.

In Exercises 1–24, solve each equation.

1. $\dfrac{dy}{dx} + \dfrac{1}{x}y = x$

2. $\dfrac{dy}{dx} + \dfrac{1}{x}y = \sin x$

*3. $\dfrac{dy}{dx} + \dfrac{1}{x}y = e^x$

4. $\dfrac{dy}{dx} + \dfrac{1}{x}y = xe^{-x}$

5. $\dfrac{dy}{dx} - \dfrac{2x}{1+x^2}y = 2x$

6. $\dfrac{dy}{dx} - \dfrac{2x}{1+x^2}y = x^2$

*7. $dy = \left(2x + \dfrac{xy}{x^2 - 1}\right)dx$

8. $\dfrac{dy}{dx} + y\cot x = \cos x$

9. $\dfrac{dy}{dx} - y\tan x = e^{-2x}$

10. $dy = \left(2x + \dfrac{2x-6}{x^2 - 6x + 10}y\right)dx$

*11. $\dfrac{dy}{dx} - \dfrac{3x}{x^2 - 4}y = x^2$

12. $\dfrac{dy}{dx} - \dfrac{10x^2 - 1}{x(10x^2 + 7x + 1)}y = \dfrac{1}{10x^2 + 7x + 1}$

13. $\dfrac{dy}{dx} - \dfrac{4x}{4x^2 - 9}y = x^3$

14. $\dfrac{dy}{dx} - \dfrac{16x}{16x^2 + 25}y = x$

*15. $\dfrac{dy}{dx} - \dfrac{9x}{9x^2 + 49}y = x$

16. $\dfrac{dy}{dx} + (2\cot x)y = \cos x$

17. $\dfrac{dy}{dx} + xy = x^3$

18. $\dfrac{dy}{dx} - xy = x$

*19. $\dfrac{dy}{dx} = \dfrac{1}{y^2 + x}$

20. $\dfrac{dx}{dy} - x = y$

21. $y\,dx - (x + 3y^2)\,dy = 0$

22. $\dfrac{dx}{dy} = \dfrac{3xy^2}{1 - y^3}$

*23. $\dfrac{dp}{dt} = t^3 + \dfrac{p}{t}$

24. $\dfrac{dv}{ds} + v = e^{-s}$

In Exercises 25–34, solve the initial-value problem. Graph the solution on an appropriate interval.

25. $\dfrac{dy}{dx} - y = 4e^x,\ y(0) = 4$

26. $\dfrac{dy}{dx} + y = e^{-x},\ y(0) = -1$

*27. $\dfrac{dy}{dx} + 3x^2 y = e^{-x^3},\ y(0) = 2$

28. $\dfrac{dy}{dx} + 2xy = 2x,\ y(0) = -1$

29. $\dfrac{dy}{dx} + \dfrac{y}{x} = \dfrac{\cos x}{x},\ y\left(\dfrac{\pi}{2}\right) = \dfrac{4}{\pi},\ x > 0$

30. $\dfrac{dy}{dx} + \dfrac{y}{x} = 2e^x,\ y(1) = -1,\ x > 0$

31. $\dfrac{dy}{dx} + \dfrac{e^x}{e^x + 1}y = \dfrac{x}{e^x + 1},\ y(0) = 1$

32. $\dfrac{dy}{dx} + \dfrac{2x}{x^2 + 4}y = \dfrac{2x}{x^2 + 4},\ y(0) = -4$

33. $\dfrac{dx}{dt} = x + t + 1,\ x(0) = 2$

34. $\dfrac{d\theta}{dt} = e^{2t} + 2\theta,\ \theta(0) = 0$

35. The equation $\dfrac{dy}{dx} + p(x)y = 0$ is called a **homogeneous first-order linear equation** because $q(x) = 0$. **(a)** Show that $y = 0$ (the trivial solution) is a solution. **(b)** Show that if $y = y_1(x)$ is a solution and k is a constant, $y = ky_1(x)$ is also a solution. **(c)** Show that if $y = y_1(x)$ and $y = y_2(x)$ are solutions, $y = y_1(x) + y_2(x)$ is also a solution.

36. If $y = y_1(x)$ satisfies the homogeneous equation $\dfrac{dy}{dx} + p(x)y = 0$ and $y = y_2(x)$ satisfies the nonhomogeneous equation $\dfrac{dy}{dx} + p(x)y = r(x)$, show that $y = y_1(x) + y_2(x)$ satisfies the nonhomogeneous equation $\dfrac{dy}{dx} + p(x)y = r(x)$.

37. (a) Show that if $y = y_1(x)$ is a solution of $\dfrac{dy}{dx} + p(x)y = r(x)$ and $y = y_2(x)$ is a solution of $\dfrac{dy}{dx} + p(x)y = q(x)$, then $y = y_1(x) + y_2(x)$ is a solution of $\dfrac{dy}{dx} + p(x)y = r(x) + q(x)$.

(b) Use the result obtained in **(a)** to solve

$$\frac{dy}{dx} + 2y = e^{-x} + \cos x.$$

Without actually solving the problems in Exercises 38–40, match each initial-value problem in Group A with the graph of its solution in Group B.

Group A

38. $y' - xy = 1$, $y(0) = 1$

39. $y' - \sin(2\pi x)y = 1$, $y(0) = 1$

40. $y' - \dfrac{1}{x^2 + 1}y = 1$, $y(0) = 1$

Group B

(a)

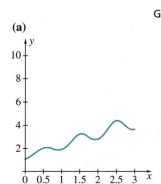

(b)

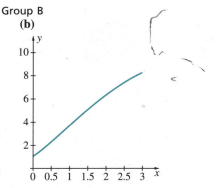

(c)

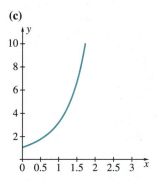

*41. **(Algae Growth)** When wading in a river or stream, you may notice that microorganisms like algae are frequently found on rocks. Similarly, if you have a swimming pool, you may notice that without maintaining appropriate levels of chlorine and algaecides, small patches of algae take over the pool surface, sometimes overnight. Underwater surfaces are attrac-

tive environments for microorganisms because water movement removes wastes and provides a continuous supply of nutrients. On the other hand, the organisms must spread over the surface without being washed away. If conditions become unfavorable, they must be able to free themselves from the surface and recolonize on a new surface.

The rate at which cells accumulate on a surface is proportional to the rate of growth of the cells and the rate at which the cells attach to the surface. An equation describing this situation is given by

$$\frac{dN}{dt} = \mu(N + A),$$

where N represents the cell density, μ the growth rate, A the attachment rate, and t time.*

(a) If the attachment rate, A, is constant, solve the

initial-value problem $\begin{cases} \dfrac{dN}{dt} = \mu(N + A) \\ N(0) = 0 \end{cases}$ for N

and then solve the result for μ.

(b) In a colony of cells it was observed that $A = 3$. The number of cells, N, at the end of t hours is shown in the following table. Estimate the growth rate at the end of each hour.

t	N	μ
1	3	
2	9	
3	21	
4	45	

(c) Using the growth rate obtained in **(b)**, estimate the number of cells at the end of 24 hours and 36 hours.

42. (Dialysis) The primary purpose of the kidney is to remove waste products like urea, creatinine, and excess fluid, from blood. When the kidneys are not working properly, wastes accumulate in the blood; when toxic levels are reached, death is certain. The leading causes of chronic kidney failure in the United

* Douglas E. Caldwell, ''Microbial Colonization of Solid-Liquid Interfaces,'' *Biochemical Engineering V,* Annals of the New York Academy of Sciences, Volume 56, New York Academy of Sciences (1987) pp. 274–280.

States are hypertension (high blood pressure) and diabetes mellitus. In fact, one-quarter of all patients requiring **kidney dialysis** have diabetes. Fortunately, kidney dialysis removes waste products from the blood of patients with improperly working kidneys. During the hemodialysis process, the patient's blood is pumped through a **dialyser,** usually at a rate of 1 to 3 deciliters per minute. The patient's blood is separated from the "cleaning fluid" by a semipermeable membrane, which permits wastes (but not blood cells) to diffuse to the cleaning fluid; the cleaning fluid contains some substances beneficial to the body which diffuse to the blood. The cleaning fluid, called the **dialysate,** is flowing in the **opposite** direction as the blood, usually at a rate of 2 to 6 deciliters per minute. Waste products from the blood diffuse to the dialysate through the membrane at a rate proportional to the difference in concentration of the waste products in the blood and dialysate. If we let $u(x)$ represent the concentration of wastes in blood and $v(x)$ represent the concentration of wastes in the dialysate, where x is the distance along the dialyser; Q_D represent the flow rate of the dialysate through the machine; and Q_B represent the flow rate of the blood through the machine, then

$$\begin{cases} Q_B u' = -k(u-v) \\ -Q_D v' = k(u-v) \end{cases},$$

where k is the proportionality constant.*

We let L denote the length of the dialyser and the initial concentration of wastes in the blood is $u(0) = u_0$, while the initial concentration of wastes in the dialysate is $v(L) = 0$. Then, we must solve the initial-value problem

$$\begin{cases} Q_B u' = -k(u-v) \\ -Q_D v' = k(u-v) \\ u(0) = u_0, v(L) = 0 \end{cases}.$$

(a) Show that the solution to
$$\begin{cases} Q_B u' = -k(u-v) \\ -Q_D v' = k(u-v) \\ u(0) = u_0, v(L) = 0 \end{cases} \quad \text{is}$$

$$u(x) = u_0 \frac{Q_B e^{\alpha x} - Q_D e^{\alpha L}}{e^{\alpha x}(Q_B - Q_D e^{\alpha L})},$$

$$v(x) = u_0 \frac{e^{\alpha L} - e^{\alpha x}}{e^{\alpha x} \dfrac{Q_D}{Q_B}(e^{\alpha L} - 1)}$$

where $\alpha = \dfrac{k}{Q_B} - \dfrac{k}{Q_D}$. First add the equations
$$\begin{cases} Q_B u' = -k(u-v) \\ -Q_D v' = k(u-v) \end{cases}$$ to obtain the linear equation
$$\frac{d}{dx}(u-v) = -\frac{k}{Q_B}(u-v) + \frac{k}{Q_D}(u-v),$$

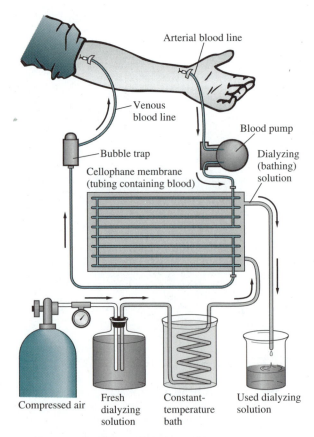

Arterial blood line

Venous blood line

Blood pump

Bubble trap

Cellophane membrane (tubing containing blood)

Dialyzing (bathing) solution

Compressed air

Fresh dialyzing solution

Constant-temperature bath

Used dialyzing solution

A diagram of a kidney dialysis machine

---

* D. N. Burghess and M. S. Borrie, *Modeling with Differential Equations,* Ellis Horwood Limited, pp. 41–45. Joyce M. Black and Esther Matassarin-Jacobs, *Luckman and Sorensen's Medical-Surgical Nursing: A Psychophysiologic Approach,* Fourth Edition, W. B. Saunders Company (1993) pp. 1509–1519, 1775–1808.

then let $z = u - v$. Next solve for z and subsequently solve for u and v.

In healthy adults, typical urea nitrogen levels are 11 to 23 milligrams per deciliter (1 deciliter = 100 milliliters) while serum creatinine levels range from 0.6 to 1.2 milligrams per deciliter, and the total volume of blood is 4 to 5 liters (1 liter = 1000 milliliters).

(b) Suppose that hemodialysis is performed on a patient with urea nitrogen level of 34 mg/dl and serum creatinine level of 1.8 using a dialyser with $k = 2.25$ and $L = 1$. If the flow rate of blood, Q_B, is 2 dl/minute while the flow rate of the dialysate, Q_D, is 4 dl/minute, will the level of wastes in the patient's blood reach normal levels after dialysis is performed? For what waste levels would dialysis have to be performed twice?

(c) The **amount of waste removed** is given by $\int_0^L k[u(x) - v(x)]\, dx$. Show that

$$\int_0^L k[u(x) - v(x)]\, dx = Q_B[u_0 - u(L)].$$

(d) The **clearance of a dialyser,** CL, is given by $CL = \dfrac{Q_B}{u_0}[u_0 - u(L)]$. Use the solution obtained in **(a)** to show that

$$CL = Q_B \frac{1 - e^{-\alpha L}}{1 - \dfrac{Q_B}{Q_D} e^{-\alpha L}}.$$

Typically, hemodialysis is performed 3 to 4 hours at a time 3 or 4 times per week. In some cases, a kidney transplant can free patients from the restrictions of dialysis. Of course, transplants have other risks not necessarily faced by those on dialysis; the number of available kidneys also affects the number of transplants performed. For example, in 1991 over 130,000 patients were on dialysis while only 7000 kidney transplants had been performed.

43. Find restrictions on p and q so that the first-order linear equation $\dfrac{dy}{dx} + p(x)y = q(x)$ is exact.

Some differential equations involve functions with jump discontinuities at $x = a$. In these cases, we must solve the differential equation for each piece of the function and match up the pieces so that the solution is continuous at $x = a$. Use this procedure to solve the initial-value problems in Exercises 44–47.

44. $\dfrac{dy}{dx} + y = q(x)$ where
$$q(x) = \begin{cases} 4, & 0 \le x \le 2 \\ 0, & x > 2 \end{cases}, \quad y(0) = 0.$$

*45. $\dfrac{dy}{dx} + y = q(x)$ where
$$q(x) = \begin{cases} x, & 0 \le x \le 1 \\ 0, & x > 1 \end{cases}, \quad y(0) = 1.$$

46. $\dfrac{dy}{dx} + p(x)y = 0$ where
$$p(x) = \begin{cases} 1, & 0 \le x \le 2 \\ -1, & x > 2 \end{cases}, \quad y(0) = 2.$$

*47. $\dfrac{dy}{dx} + p(x)y = 0$ where
$$p(x) = \begin{cases} 2, & 0 \le x \le 1 \\ 4, & x > 1 \end{cases}, \quad y(0) = 1.$$

48. A **Bernoulli equation** is a nonlinear equation of the form

$$y' + p(x)y = q(x)y^n.$$

and is named for the Swiss mathematician Jacques Bernoulli (1654–1705). Both Bernoulli and Leibniz solved the Bernoulli equation; Jean Bernoulli (1667–1748) solved this equation by reducing it to a linear equation with the substitution

$$w = y^{1-n}.$$

Equations of this form can be expressed as first-order linear equations if an appropriate substitution is made. Notice that if $n = 0$, the equation is first-order linear; if $n = 1$, the equation can be written as $y' + [p(x) - q(x)]y = 0$, which is also a first-order linear equation.

(a) Show that for $n \ne 0, 1$, differentiating $w = y^{1-n}$ with respect to y results in $\dfrac{dw}{dx} = (1 - n)y^{-n}\dfrac{dy}{dx}$.

(b) Show that substituting $\dfrac{dw}{dx} = (1 - n)y^{-n}\dfrac{dy}{dx}$ into the equation $y' + p(x)y = q(x)y^n$ yields

$$\frac{y^n}{1 - n}\frac{dw}{dx} + p(x)y^n w = q(x)y^n.$$

(c) Show that multiplying by $\dfrac{1-n}{y^n}$ yields the first-order linear equation

$$\frac{dw}{dx} + (1-n)p(x)w = (1-n)q(x)$$

which can be solved for $w(x)$. Once $w(x)$ is found, $y(x)$ is found by using the relationship $w = y^{1-n}$ or $y = w^{1/(1-n)}$.

49. Solve the Bernoulli equation $\dfrac{dy}{dx} + \dfrac{y}{x} = \dfrac{1}{xy^2}$, where $p(x) = \dfrac{1}{x}$, $q(x) = \dfrac{1}{x}$, and $n = -2$ using the following steps.

(a) Substitute $w = y^{1-(-2)} = y^3$ into this equation to obtain the equation $\dfrac{dw}{dx} + \dfrac{3w}{x} = \dfrac{3}{x}$.

(b) Solve the first-order linear equation $\dfrac{dw}{dx} + \dfrac{3w}{x} = \dfrac{3}{x}$ to find that $w = 1 + Cx^{-3}$.

(c) Use the substitution $w = y^3$ to show that the solution of $\dfrac{dy}{dx} + \dfrac{y}{x} = \dfrac{1}{xy^2}$ is $y = (1 + Cx^{-3})^{1/3}$.

In Exercises 50–55, solve the Bernoulli equation.

50. $y' + y = xy^2$ *51. $y' - \dfrac{1}{2}y = \dfrac{x}{y}$

52. $y' - \dfrac{1}{x}y = y^3 \sin x$ 53. $y' - \dfrac{1}{2x}y = y^3 \cos x$

54. $y' + 3y = \sqrt{y}\sin x$ *55. $y' - 2y = \dfrac{\cos x}{\sqrt{y}}$

56. Equations of the form $y = xf(y') + g(y')$ are called **Lagrange equations.** These equations are solved by making the substitution

$$p = y'(x).$$

(a) Differentiate $y = xf(y') + g(y')$ with respect to x to obtain

$$y' = xf'(y')y'' + f(y') + g'(y')y''.$$

(b) Substitute p into the equation to obtain

$$p = xf'(p)\frac{dp}{dx} + f(p) + g'(p)\frac{dp}{dx}$$
$$= f(p) + \frac{dp}{dx}[xf'(p) + g'(p)].$$

(c) Solve this equation for $\dfrac{dx}{dp}$ to obtain the linear equation

$$\frac{dx}{dp} = \frac{xf'(p) + g'(p)}{p - f(p)}$$

which is equivalent to

$$\frac{dx}{dp} + \frac{f'(p)}{f(p) - p}x = \frac{g'(p)}{p - f(p)}.$$

This linear first-order equation can be solved for x in terms of p. Then, $x(p)$ can be used with $y = xf(p) + g(p)$ to obtain an equation for y.

57. Solve the equation $y = -xy' + \dfrac{1}{5}(y')^5$ using the following steps. In this case, $f(y') = -y'$ and $g(y') = \dfrac{1}{5}(y')^5$.

(a) Differentiate the equation with respect to x to obtain $y' = -y' - xy'' + (y')^4y''$.

(b) Substitute $y' = p$ to obtain $p = -p - x\dfrac{dp}{dx} + (p)^4\dfrac{dp}{dx}$. Simplify this equation to obtain

$$\frac{dp}{dx} = \frac{2p}{p^4 - x}.$$

(c) Rewrite this equation as $\dfrac{dx}{dp} = \dfrac{p^4 - x}{2p}$ and solve this first-order linear equation for x to obtain $x = \dfrac{1}{9}p^4 + Cp^{-1/2}$.

(d) Substitute $y' = p$ and $x = \dfrac{1}{9}p^4 + Cp^{-1/2}$ into the differential equation $y = -xy' + \dfrac{1}{5}(y')^5$ to obtain a formula for y.

(e) Graph the solution curves for various values of C.

In Exercises 58–60, solve the Lagrange equation.

58. $y = x\left(\dfrac{dy}{dx}\right)^2 + 3\left(\dfrac{dy}{dx}\right)^2 - 2\left(\dfrac{dy}{dx}\right)^3$

*59. $y = x\left(\dfrac{dy}{dx} + 1\right) + \left(2\dfrac{dy}{dx} + 1\right)$

60. $y = x\left(2 - \dfrac{dy}{dx}\right) + \left(2\left(\dfrac{dy}{dx}\right)^2 + 1\right)$

61. Euler was the first mathematician to take advantage of integrating factors to solve linear differential equa-

tions.* Euler used the following steps to solve the equation

$$\frac{dz}{dv} - 2z + \frac{z}{v} = \frac{1}{v}.$$

(a) Multiply the equation by the integrating factor $e^{-2v}v$.

(b) Show that $\dfrac{d}{dv}(e^{-2v}vz) = e^{-2v}v\dfrac{dz}{dv} - 2e^{-2v}vz + e^{-2v}z.$

(c) Express the equation as $\dfrac{d}{dv}(e^{-2v}vz) = e^{-2v}$ and solve this equation for z.

2.5 Theory of First-Order-Equations: A Brief Discussion

We have previously seen that initial-value and boundary-value problems may have none, one, or many solutions. In order to understand the types of initial-value problems which yield a *unique* solution, the following theorem is stated. The proof is omitted but may be found in more advanced differential equations and analysis textbooks.†

Theorem 2.2 Existence and Uniqueness

Consider the initial-value problem

$$y' = f(x, y), \; y(x_0) = y_0.$$

If f and $\dfrac{\partial f}{\partial y}$ are continuous functions on the rectangular region R: $a < x < b$, $c < y < d$ containing the point (x_0, y_0), then there exists an interval $|x - x_0| < h$ centered at x_0 on which there exists one and only one solution to the differential equation that satisfies the initial condition.

If the condition that $\dfrac{\partial f}{\partial y}$ is continuous on the rectangular region R: $a < x < b$, $c < y < d$ containing the point (x_0, y_0) is not included in Theorem 2.2, then we can say that at least one solution exists. (We call this more relaxed theorem an *existence theorem* because the uniqueness of the solution is not guaranteed.)

---

* Victor J. Katz, *A History of Mathematics: An Introduction,* HarperCollins, 1993, p. 503.

† C. Corduneanu, *Principles of Differential and Integral Equations,* Chelsea Publishing Co., New York, 1977, pp. 19–24 or T. Apostol, *Mathematical Analysis,* Second Edition, Addison-Wesley, Reading, MA, 1974, p. 181.

Example 1

Solve the initial-value problem

$$\begin{cases} \dfrac{dy}{dx} = \dfrac{x}{y} \\ y(0) = 0 \end{cases}.$$

Does this result contradict the Existence and Uniqueness Theorem?

Solution   This equation is solved by separation of variables. After rewriting the equation as $y\,dy = x\,dx$, integration yields the family of solutions $y^2 - x^2 = C$. Application of the initial condition gives us $0^2 - 0^2 = C$, so $C = 0$ and the solutions that pass through $(0, 0)$ satisfy $y^2 - x^2 = 0$. Thus, we have two solutions to the initial-value problem, $y = x$ and $y = -x$.

Although more than one solution satisfies this initial-value problem, the Existence and Uniqueness Theorem is not contradicted because the function x/y is not continuous at the point $(0, 0)$: the requirements of the hypotheses of the theorem are not met.

Example 2

Verify that the initial-value problem $\dfrac{dy}{dx} = y$, $y(0) = 1$ has a unique solution.

Solution   In this case, $f(x, y) = y$, $x_0 = 0$, and $y_0 = 1$. Hence, both f and $\dfrac{\partial f}{\partial y}$ are continuous on all rectangular regions containing the point $(x_0, y_0) = (0, 1)$. By the Existence and Uniqueness Theorem, there exists a unique solution to the differential equation that satisfies the initial condition $y(0) = 1$. We verify this by solving the initial-value problem. This equation is separable and equivalent to $\dfrac{dy}{y} = dx$. A general solution is given by $y = Ce^x$ and the solution that satisfies the initial condition $y(0) = 1$ is $y = e^x$.

The Existence and Uniqueness Theorem gives sufficient, but not necessary, conditions for the existence of a unique solution of an initial-value problem. If an initial-value problem does not satisfy the hypotheses of the theorem, we cannot conclude that a unique solution does not exist. In fact, the problem may have a unique solution, no solution, or many solutions.

Example 3

Find a solution to the initial-value problems   (a) $\begin{cases} \dfrac{dy}{dx} = \dfrac{x}{y^{2/3}} \\ y(0) = 0 \end{cases}$ and

(b) $\begin{cases} 2(xy^3 - x^5y)\,dx + (x^6 - x^2y^2)\,dy = 0 \\ y(0) = 0 \end{cases}$, if possible.

| Solution |

The Existence and Uniqueness Theorem does not guarantee the existence of a solution to either problem because both $\dfrac{x}{y^{2/3}}$ and $-\dfrac{2(xy^3 - x^5 y)}{x^6 - x^2 y^2}$ are discontinuous at the point $(0, 0)$ specified by the initial condition.

(a) The equation $\dfrac{dy}{dx} = \dfrac{x}{y^{2/3}}$ is separable and has solution

$$\frac{3}{5} y^{5/3} = \frac{1}{2} x^2 + C_1$$

$$y = \left(\frac{5}{6} x^2 + C\right)^{3/5}$$

Application of the initial condition leads to the *unique* solution $y = (5/6)^{3/5} x^{6/5}$, which is graphed in Figure 2.21.

(b) The equation $2(xy^3 - x^5 y)\, dx + (x^6 - x^2 y^2)\, dy = 0$ is more difficult to solve. Dividing the equation by $(x^4 + y^2)^2$ leads to an exact equation that has general solution $x^2 y = C(x^4 + y^2)$. Applying the initial condition $y(0) = 0$ results in the identity $0 = 0$, which means that $x^2 y = C(x^4 + y^2)$ is a solution to the initial-value problem for *any* value of C. There are infinitely many solutions to the initial-value problem so the solution is *nonunique*. Several solutions are graphed in Figure 2.22.

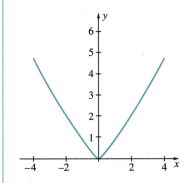

Figure 2.21

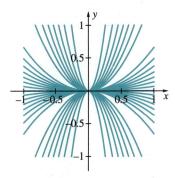

Figure 2.22

Are the hypotheses of the Existence and Uniqueness Theorem satisfied for the initial-value problem $\dfrac{dy}{dx} = \sqrt{y^2 - 1}$, $y(0) = 1$? *If* $\displaystyle\int \dfrac{dy}{\sqrt{y^2 - 1}} = \ln|y + \sqrt{y^2 - 1}| + C$, *does a unique solution to this problem exist? Does more than one solution exist?*

IN TOUCH WITH TECHNOLOGY

1. Find all points $(x_0, 0)$ in the rectangle $R = \{(x, y) \mid 0 \le x \le 2\pi, -\pi \le y \le \pi\}$ so that the initial-value problem

$$\begin{cases} \dfrac{dy}{dx} = \dfrac{\cos x \sin x + \sin x \sin y}{\cos x \cos y - \cos y \sin y} \\ y(x_0) = 0 \end{cases}$$

has *more than one* solution. Confirm your results graphically.

2. Find all points (x_0, y_0) in the rectangle $R = \{(x, y) \mid 0 \le x \le 4\pi, 0 \le y \le 4\pi\}$ so that the initial-value problem

$$\begin{cases} \dfrac{dy}{dx} = \dfrac{\cos y - y \cos x}{x \sin y + \sin x - 1} \\ y(x_0) = y_0 \end{cases}$$

has *no* solutions. *Hint:* Graph the direction field associated with the equation.

EXERCISES 2.5

1. Does the Existence and Uniqueness Theorem guarantee a unique solution to the following initial-value problems on some interval? Explain.

   (a) $\dfrac{dy}{dx} + x^2 = y^2$, $y(0) = 0$

   (b) $\dfrac{dy}{dx} + x^2 = y^{-2}$, $y(0) = 0$

   (c) $\dfrac{dy}{dx} = y + \dfrac{1}{1-x}$, $y(1) = 0$

2. According to the Existence and Uniqueness Theorem, the initial-value problem $\dfrac{dy}{dx} = |y|$, $y(1) = 0$ has a solution. (a) Must the solution be unique? (b) Solve the problem by hand. Is the solution unique? *Hint:* Separate the initial-value problem into two parts. First solve the problem for $y \ge 0$ and then solve it for $y < 0$.

*3. The Existence and Uniqueness Theorem implies that at least one solution to the initial-value problem $\dfrac{dy}{dx} = y^{1/5}$, $y(0) = 0$ exists. (a) Must the solution be unique according to the theorem? (b) Solve the problem by hand. Is the solution unique? Note that $y = 0$ satisfies the differential equation as well as the initial condition.

4. Show that $y = 0$ and $y = \dfrac{x^4}{16}$ both satisfy the initial-value problem $\dfrac{1}{x}\dfrac{dy}{dx} = \sqrt{y}$, $y(0) = 0$. Does this contradict the Existence and Uniqueness Theorem?

5. Show that $y = 0$ and $y = x|x|$ both satisfy the initial-value problem $\dfrac{dy}{dx} = 2\sqrt{|y|}$, $y(0) = 0$. Does this contradict the Existence and Uniqueness Theorem?

6. Does the initial-value problem $\dfrac{dy}{dx} = 4x^2 - xy^2$, $y(2) = 1$ have a unique solution on an interval containing $x = 2$?

*7. Does the initial-value problem $\dfrac{dy}{dx} = y\sqrt{x}$, $y(1) = 1$ have a unique solution on an interval containing $x = 1$? Verify your result by solving the problem.

8. Does the initial-value problem $\dfrac{dy}{dx} = 6y^{2/3}$, $y(1) = 0$ have a unique solution on an interval containing $x = 1$? Solve the problem using separation of variables. Is $y = 0$ obtained through this method? Is $y = 0$ a solution?

9. Does the initial-value problem $y' = \sin y - \cos x$, $y(\pi) = 0$ have a unique solution on an interval containing $x = \pi$?

10. Does the initial-value problem $xy' = y$, $y(0) = 1$ have a unique solution on an interval containing $x = 0$? Verify your response by solving the initial-value problem.

*11. Show that $y = \sec x$ satisfies the initial-value problem $y' = y \tan x$, $y(0) = 1$. What is the largest open interval containing $x = 0$ over which $y = \sec x$ is a solution? Explain.

12. Show that $y = \tan x$ satisfies the initial-value problem $y' = 1 + y^2$, $y(0) = 0$. What is the largest open interval containing $x = 0$ over which $y = \tan x$ is a solution? Explain.

13. Using the Existence and Uniqueness Theorem, determine if $\dfrac{dy}{dx} = \sqrt{y^2 - 1}$ has a unique solution passing through the point **(a)** $(0, 2)$; **(b)** $(4, -1)$; **(c)** $(0, \tfrac{1}{2})$; **(d)** $(2, 1)$ is guaranteed.

14. Using the Existence and Uniqueness Theorem, determine if $\dfrac{dy}{dx} = \sqrt{25 - y^2}$ has a unique solution passing through the point **(a)** $(-4, 3)$; **(b)** $(0, 5)$; **(c)** $(3, -6)$; **(d)** $(4, -5)$.

 2.6 **Numerical Approximations of Solutions of First-Order Equations**

Euler's Method • **Improved Euler's Method** • **Errors** • **The Runge-Kutta Method**

Euler's Method

In many cases, we cannot obtain a formula for the solution to an initial-value problem of the form

$$\begin{cases} y' = f(x, y) \\ y(x_0) = y_0 \end{cases}.$$

Often, we can approximate the solution using a numerical method like **Euler's method** that is based on tangent line approximations, which we discuss now.

Let h represent a small change or **stepsize** in the independent variable x. We approximate the value of y at the sequence of x-values, $x_1, x_2, x_3, \ldots$, where

$$x_1 = x_0 + h$$
$$x_2 = x_1 + h = x_0 + 2h$$
$$x_3 = x_2 + h = x_0 + 3h$$
$$\vdots$$
$$x_n = x_{n-1} + h = x_0 + nh$$

The slope of the tangent line to the graph of y at these values of x is found with the differential equation $y' = f(x, y)$. For example, at $x = x_0$, the slope of the tangent line is $f(x_0, y(x_0)) = f(x_0, y_0)$. An equation of the tangent line to the graph of y at the point $(x_0, y(x_0))$ is

$$y - y_0 = f(x_0, y_0)(x - x_0) \qquad \text{or} \qquad y = f(x_0, y_0)(x - x_0) + y_0.$$

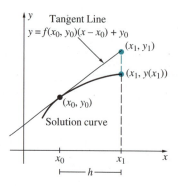

Figure 2.23

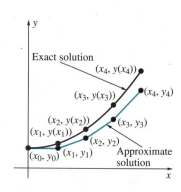

Figure 2.24

Using this line to find the value of y at $x = x_1$ (which we call y_1) then yields

$$y_1 = f(x_0, y_0)(x_1 - x_0) + y_0 = hf(x_0, y_0) + y_0.$$

Therefore, we obtain the approximate value of y at $x = x_1$. In Figure 2.23, we illustrate the difference between the actual value of y at $x = x_1$, $y(x_1)$, and the approximate value of y at $x = x_1$, y_1.

We have the point (x_1, y_1) and use this point to estimate the value of y when $x = x_2$. Using a similar procedure, we approximate the tangent line at $x = x_1$ with

$$y - y_1 = f(x_1, y_1)(x - x_1) \qquad \text{or} \qquad y = f(x_1, y_1)(x - x_1) + y_1.$$

Then, at $x = x_2$,

$$y_2 = f(x_1, y_1)(x_2 - x_1) + y_1 = hf(x_1, y_1) + y_1.$$

Continuing with this procedure, we see that at $x = x_n$, where $x_n = x_0 + nh$,

$$y_n = hf(x_{n-1}, y_{n-1}) + y_{n-1}.$$

Using this formula, we obtain a sequence of points of the form (x_n, y_n) $(n = 1, 2, \ldots)$ where y_n is the approximate value of $y(x_n)$. We show several points of this type in Figure 2.24 along with the actual values of y.

Euler's Method

The solution of the initial-value problem

$$y' = f(x, y), \, y(x_0) = y_0$$

is approximated at the sequence of points (x_n, y_n) $(n = 1, 2, \ldots)$, where y_n is the approximate value of $y(x_n)$ by computing

$$y_n = hf(x_{n-1}, y_{n-1}) + y_{n-1} \, (n = 1, 2, \ldots),$$

where $x_n = x_0 + nh$ and h is the selected stepsize.

Example 1

Use Euler's method with $h = 0.1$ and $h = 0.05$ to approximate the solution of $y' = xy$, $y(0) = 1$ on $0 \le x \le 1$. Determine the exact solution and compare the results.

Solution First, we note that $f(x, y) = xy$, $x_0 = 0$, and $y_0 = 1$. With $h = 0.1$, we have the formula

$$y_n = hf(x_{n-1}, y_{n-1}) + y_{n-1} = 0.1x_{n-1}y_{n-1} + y_{n-1}.$$

Then, for $x_1 = x_0 + h = 0.1$, we have

$$y_1 = 0.1x_0y_0 + y_0 = 0.1(0)(1) + 1 = 1.$$

Similarly, for $x_2 = x_0 + 2h = 0.2$,

$$y_2 = 0.1x_1y_1 + y_1 = 0.1(0.1)(1) + 1 = 1.01.$$

In Table 2.1, we show the results of this sequence of approximations. From this, we see that $y(1)$ is approximately 1.54711.

TABLE 2.1 Euler's method with $h = 0.1$

x_n	y_n	x_n	y_n
0.0	1.0	0.6	1.15873
0.1	1.0	0.7	1.22825
0.2	1.01	0.8	1.31423
0.3	1.0302	0.9	1.41937
0.4	1.06111	1.0	1.54711
0.5	1.10355		

For $h = 0.05$, we use

$$y_n = hf(x_{n-1}, y_{n-1}) + y_{n-1} = 0.05x_{n-1}y_{n-1} + y_{n-1}$$

to obtain the values given in Table 2.2. With this stepsize, the approximate value of $y(1)$ is 1.59594.

The exact solution to the initial-value problem, which is found with separation of variables, is $y = e^{x^2/2}$, so the exact value of $y(1)$ is $e^{1/2} \approx 1.64872$. The smaller value of h, therefore, yields a better approximation. We graph the approximations obtained with $h = 0.1$ and $h = 0.05$ as well as the graph of $y = e^{x^2/2}$ in Figure 2.25. Notice from these graphs that the approximation is more accurate when h is decreased.

TABLE 2.2   Euler's method with $h = 0.05$

x_n	y_n	x_n	y_n	x_n	y_n
0.0	1.0	0.35	1.05361	0.70	1.2523
0.05	1.0	0.40	1.07204	0.75	1.29613
0.10	1.0025	0.45	1.09348	0.80	1.34474
0.15	1.00751	0.50	1.11809	0.85	1.39853
0.20	1.01507	0.55	1.14604	0.90	1.45796
0.25	1.02522	0.60	1.17756	0.95	1.52357
0.30	1.03803	0.65	1.21288	1.00	1.59594

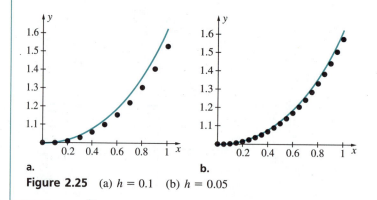

a.

b.

Figure 2.25   (a) $h = 0.1$ (b) $h = 0.05$

Improved Euler's Method

Euler's method is improved by using an average slope over each interval. Using the tangent line approximation of the curve through (x_0, y_0), $y = f(x_0, y_0)(x - x_0) + y_0$, we find the approximate value of y at $x = x_1$, which we now call y_1^*. Then,

$$y_1^* = hf(x_0, y_0) + y_0$$

and with the differential equation $y' = f(x, y)$, we find that the approximate slope of the tangent line at $x = x_1$ is $f(x_1, y_1^*)$. The average of the two slopes, $f(x_0, y_0)$ and $f(x_1, y_1^*)$, is $\dfrac{f(x_0, y_0) + f(x_1, y_1^*)}{2}$, and an equation of the line through (x_0, y_0) with slope $\dfrac{f(x_0, y_0) + f(x_1, y_1^*)}{2}$ is

$$y = \frac{f(x_0, y_0) + f(x_1, y_1^*)}{2}(x - x_0) + y_0.$$

We illustrate the determination of this average slope and the position of these lines in Figure 2.26.

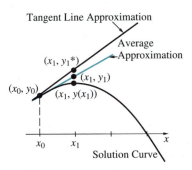

Tangent Line Approximation

Average Approximation

(x_1, y_1^*)

(x_1, y_1)

(x_0, y_0)

$(x_1, y(x_1))$

x_0 x_1 x

Solution Curve

Figure 2.26

At $x = x_1$, the approximate value of y is given by

$$y_1 = \frac{f(x_0, y_0) + f(x_1, y_1^*)}{2}(x_1 - x_0) + y_0 = \frac{f(x_0, y_0) + f(x_1, y_1^*)}{2}h + y_0.$$

Continuing in this manner, the approximation in each step in the improved Euler's method depends on the following two calculations:

$$y_n^* = hf(x_{n-1}, y_{n-1}) + y_{n-1}$$

$$y_n = \frac{f(x_{n-1}, y_{n-1}) + f(x_n, y_n^*)}{2}h + y_{n-1},$$

where $x_n = x_0 + nh$.

Improved Euler's Method

The solution of the initial-value problem

$$y' = f(x, y), \; y(x_0) = y_0$$

is approximated at the sequence of points (x_n, y_n) $(n = 1, 2, \ldots)$, where y_n is the approximate value of $y(x_n)$ by computing at each step the two calculations:

$$y_n^* = hf(x_{n-1}, y_{n-1}) + y_{n-1}$$

$$y_n = \frac{f(x_{n-1}, y_{n-1}) + f(x_n, y_n^*)}{2}h + y_{n-1} \qquad (n = 1, 2, \ldots),$$

where $x_n = x_0 + nh$ and h is the selected stepsize.

Example 2

Use the improved Euler's method to approximate the solution of $y' = xy$, $y(0) = 1$ on $0 \le x \le 1$ for $h = 0.1$. Compare the results to the exact solution.

Solution In this case, $f(x, y) = xy$, $x_0 = 0$, and $y_0 = 1$. Therefore, we use the equations

$$y_n^* = hx_{n-1}y_{n-1} + y_{n-1}$$

$$y_n = \frac{x_{n-1}y_{n-1} + x_n y_n^*}{2}h + y_{n-1}$$

for $n = 1, 2, \ldots, 10$. For example, if $n = 1$, we have

$$y_1^* = hx_0 y_0 + y_0 = (0.1)(0)(1) + 1 = 1$$

$$y_1 = \frac{x_0 y_0 + x_1 y_1^*}{2} h + y_0 = \frac{(0)(1) + (0.1)(1)}{2}(0.1) + 1 = 1.005.$$

Then,

$$y_2^* = hx_1 y_1 + y_1 = (0.1)(0.1)(1.005) + 1.005 = 1.01505$$

$$y_2 = \frac{x_1 y_1 + x_2 y_2^*}{2} h + y_1 = \frac{(0.1)(1.005) + (0.2)(1.01505)}{2}(0.1) + 1.005$$

$$= 1.0201755.$$

In Table 2.3, we list the approximations obtained with this improved method and compare them to those obtained with Euler's method in Example 1 (see Figure 2.27). Which method yields a better approximation?

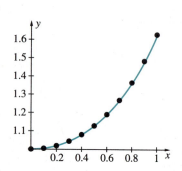

Figure 2.27

TABLE 2.3   Improved Euler's method with $h = 0.1$

x_n	y_n (IEM)	y_n (EM)	Actual Value
0.0	1.0	1.0	1.0
0.1	1.005	1.0	1.00501
0.2	1.0201755	1.01	1.0202
0.3	1.0459859	1.0302	1.04603
0.4	1.083223	1.06111	1.08329
0.5	1.1330513	1.10355	1.13315
0.6	1.1970687	1.15873	1.19722
0.7	1.277392	1.22825	1.27762
0.8	1.3767731	1.31423	1.37713
0.9	1.4987552	1.41937	1.49930
1.0	1.6478813	1.54711	1.64872

Errors

When approximating the solution of an initial-value problem, there are several sources of error. One of these sources is **round-off error** because computers and calculators use only a finite number of digits in all calculations. Of course, the error is compounded as rounded values are used in subsequent calculations. Therefore, one way to minimize round-off error is to reduce the number of calculations.

Another source of error is **truncation error,** which is due to the fact that we use an approximate formula. We begin our discussion of error by considering the error associated with Euler's method, which uses only the first two terms of the Taylor series expansion.

Recall from calculus the Taylor formula with remainder:

$$y(x) = y(x_0) + \frac{dy}{dx}(x_0)(x - x_0) + \frac{1}{2!}\frac{d^2y}{dx^2}(x_0)(x - x_0)^2 + \cdots$$

$$+ \frac{1}{n!}\frac{d^ny}{dx^n}(x_0)(x - x_0)^n + \frac{1}{(n+1)!}\frac{d^{n+1}y}{dx^{n+1}}(c)(x - x_0)^{n+1}$$

where c is a number between x and x_0. The remainder term is

$$R_n(x) = \frac{1}{(n+1)!}\frac{d^{n+1}y}{dx^{n+1}}(c)(x - x_0)^{n+1},$$

so the accuracy of the approximation obtained by using the first n terms of the Taylor series depends on the size of $R_n(x)$. For Euler's method, the remainder is

$$R_1(x) = \frac{1}{2!}\frac{d^2y}{dx^2}(c)(x - x_0)^2,$$

so at $x = x_1 = x_0 + h$, the remainder is

$$R_1(x_1) = R(x_0 + h) = \frac{1}{2!}\frac{d^2y}{dx^2}(c)h^2.$$

Therefore, a **bound** on the error is given by

$$|R_1(x_1)| \le \frac{h^2}{2} \max_{x_0 \le x \le x_1} \left|\frac{d^2y}{dx^2}(x)\right|.$$

Example 3

Find a bound for the local truncation error when Euler's method is used to approximate the solution of the initial-value problem $y' = xy$, $y(0) = 1$ at $x_1 = 0.1$.

Solution We found in Example 1 that the exact solution of this problem is $y = e^{x^2/2}$. Therefore, a bound on the error is

$$|R_1(x_1)| \le \frac{(0.1)^2}{2} \max_{x_0 \le x \le x_1} \left|\frac{d^2y}{dx^2}(x)\right|$$

where $\dfrac{dy}{dx} = xe^{x^2/2}$ and $\dfrac{d^2y}{dx^2} = e^{x^2/2} + x^2e^{x^2/2}$.

Then,

$$|R_1(x_1)| \le \frac{(0.1)^2}{2} \max_{x_0 \le x \le x_1} \left|\frac{d^2y}{dx^2}(x)\right| \le \frac{(0.1)^2}{2} \max_{0 \le x \le 0.1} \left|e^{x^2/2} + x^2 e^{x^2/2}\right|$$

$$\le (0.005)[\max_{0 \le x \le 0.1} |e^{x^2/2}| + \max_{0 \le x \le 0.1} |x^2 e^{x^2/2}|]$$

$$\le (0.005)[e^{(0.1)^2/2} + (0.1)^2 e^{(0.1)^2/2}] \approx 0.005075.$$

Notice that the value of $y = e^{x^2/2}$ at $x = 0.1$ is 1.005012521 and the approximate value obtained with Euler's method is $y_1 = 1$. Therefore, the error is $1.005012521 - 1.0 = 0.005012521$, which is less than the bound $|R_1(x_1)| \le 0.005075$.

The Runge-Kutta Method

In an attempt to improve on the approximation obtained with Euler's method as well as to avoid the analytic differentiation of the function $f(x, y)$ to obtain y'', y''', ..., we introduce the Runge-Kutta method. We begin with the Runge-Kutta method of order two.

Suppose that we know the value of y at x_n. We now use the point (x_n, y_n) to approximate the value of y at a nearby value $x = x_n + h$ by assuming that

$$y_{n+1} = y_n + Ak_1 + Bk_2$$

where

$$k_1 = hf(x_n, y_n) \quad \text{and} \quad k_2 = hf(x_n + ah, y_n + bk_1).$$

We also use the Taylor series expansion of y to obtain another representation of $y_{n+1} = y(x_n + h)$:

$$y(x_n + h) = y(x_n) + hy'(x_n) + h^2\frac{y''(x_n)}{2!} + \cdots = y_n + hy'(x_n) + h^2\frac{y''(x_n)}{2!} + \cdots.$$

Now, because

$$y_{n+1} = y_n + Ak_1 + Bk_2 = y_n + Ahf(x_n, y_n) + Bhf(x_n + ah, y_n + bhf(x_n, y_n)),$$

we wish to determine values of A, B, a, and b such that these two representations of y_{n+1} agree. Notice that is we let $A = 1$ and $B = 0$, then the relationships match up to order h. However, we can choose these parameters more wisely so that agreement occurs up through terms of order h^2. This is accomplished by considering the Taylor expansion of a function F of two variables about (x_0, y_0), which is given by

$$F(x, y) = F(x_0, y_0) + \frac{\partial F}{\partial x}(x_0, y_0)(x - x_0) + \frac{\partial F}{\partial y}(x_0, y_0)(y - y_0) + \cdots.$$

In our case, we have

$$f(x_n + ah, y_n + bfh(x_n, y_n)) =$$
$$f(x_n, y_n) + ah\frac{\partial f}{\partial x}(x_n, y_n) + bhf(x_n, y_n)\frac{\partial f}{\partial x}(x_n, y_n) + O(h^2).$$

The power series is then substituted into the following expression and simplified to yield

$$y_{n+1} = y_n + Ahf(x_n, y_n) + Bhf(x_n + ah, y_n + bhf(x_n, y_n))$$
$$= y_n + (A + B)hf(x_n, y_n) + aBh^2\frac{\partial f}{\partial x}(x_n, y_n) + bBh^2f(x_n, y_n)\frac{\partial f}{\partial x}(x_n, y_n) + O(h^3).$$

Comparing the above expression to the following power series obtained directly from the Taylor series of y given by

$$y(x_n + h) = y(x_n) + hf(x_n, y_n) + \frac{h^2}{2}\frac{\partial f}{\partial x}(x_n, y_n) + \frac{h^2}{2}f(x_n, y_n)\frac{\partial f}{\partial y}(x_n, y_n) + O(h^3)$$

or

$$y_{n+1} = y_n + hf(x_n, y_n) + \frac{h^2}{2}\frac{\partial f}{\partial x}(x_n, y_n) + \frac{h^2}{2}f(x_n, y_n)\frac{\partial f}{\partial y}(x_n, y_n) + O(h^3),$$

we see that A, B, a, and b must satisfy the following system of nonlinear equations:

$$A + B = 1, \quad aA = \frac{1}{2}, \quad \text{and} \quad bB = \frac{1}{2}.$$

Choosing $a = b = 1$, the Runge-Kutta method of order two uses the equations

$$y_{n+1} = y(x_n + h) = y_n + \frac{1}{2}hf(x_n, y_n) + \frac{1}{2}hf(x_n + h, y_n + hf(x_n, y_n))$$

$$= y_n + \frac{1}{2}(k_1 + k_2)$$

where $k_1 = hf(x_n, y_n)$ and $k_2 = hf(x_n + h, y_n + k_1)$. Notice that this method is equivalent to the improved Euler method.

Runge-Kutta Method of Order Two

The solution of the initial-value problem

$$y' = f(x, y), \ y(x_0) = y_0$$

is approximated at the sequence of points (x_n, y_n) $(n = 1, 2, \ldots)$ where y_n is the approximate value of $y(x_n)$ by computing at each step

$$y_{n+1} = y(x_n + h) = y_n + \frac{1}{2}hf(x_n, y_n) + \frac{1}{2}hf(x_n + h, \ y_n + hf(x_n, y_n))$$

$$(n = 0, 1, \ldots)$$

$$= y_n + \frac{1}{2}(k_1 + k_2)$$

where $k_1 = hf(x_n, y_n)$, $k_2 = hf(x_n + h, \ y_n + k_1)$, $x_n = x_0 + nh$, and h is the selected stepsize.

Example 4

Use the Runge-Kutta method of order two with $h = 0.1$ to approximate the solution of the initial-value problem $y' = xy$, $y(0) = 1$ on $0 \le x \le 1$.

Solution   In this case, $f(x, y) = xy$, $x_0 = 0$, and $y_0 = 1$. Therefore, on each step we use the three equations

$$k_1 = hf(x_n, y_n) = 0.1x_ny_n$$
$$k_2 = hf(x_n + h, \ y_n + k_1) = 0.1(x_n + 0.1)(y_n + k_1)$$

$$y_{n+1} = y_n + \frac{1}{2}(k_1 + k_2).$$

For example, if $n = 0$, then

$$k_1 = 0.1x_0y_0 = 0.1(0)(1) = 0$$
$$k_2 = 0.1(x_0 + 0.1)(y_0 + k_1) = 0.1(0.1)(1) = 0.01$$

$$y_1 = y_0 + \frac{1}{2}(k_1 + k_2) = 1 + \frac{1}{2}(0.01) = 1.005.$$

Therefore, the Runge-Kutta method of order two approximates that the value of y at $x = 0.1$ is 1.005. Similarly, if $n = 1$, then

$$k_1 = 0.1x_1y_1 = 0.1(0.1)(1.005) = 0.01005$$
$$k_2 = 0.1(x_1 + 0.1)(y_1 + k_1) = 0.1(0.2)(1.01505) = 0.020301$$

$$y_2 = y_1 + \frac{1}{2}(k_1 + k_2) = 1.005 + \frac{1}{2}(0.01005 + 0.020301) = 1.0201755.$$

In Table 2.4, we display the results obtained for the other values on $0 \le x \le 1$ using the Runge-Kutta method of order two.

TABLE 2.4 Runge-Kutta Method of Order Two with $h = 0.1$

x_n	y_n (RK)	Actual Value
0.0	1.0	1.0
0.1	1.005	1.00501
0.2	1.0201755	1.0202
0.3	1.0459859	1.04603
0.4	1.083223	1.08329
0.5	1.1330513	1.13315
0.6	1.1970687	1.19722
0.7	1.277392	1.27762
0.8	1.3767731	1.37713
0.9	1.4987552	1.4993
1.0	1.6478813	1.64874

The terms of the power series expansions used in the derivation of the Runge-Kutta method of order two can be made to match up to order four. These computations are rather complicated, so they are not discussed here. However, after much work, the approximation at each step is found to be made with

$$y_{n+1} = y_n + \frac{h}{6}[k_1 + 2k_2 + 2k_3 + k_4], \ n = 0, 1, 2, \ldots$$

where $k_1 = f(x_n, y_n)$, $k_2 = f\left(x_n + \frac{h}{2}, y_n + \frac{hk_1}{2}\right)$, $k_3 = f\left(x_n + \frac{h}{2}, y_n + \frac{hk_2}{2}\right)$, and $k_4 = f(x_{n+1}, y_n + hk_3)$.

Runge-Kutta Method of Order Four

The solution of the initial-value problem

$$y' = f(x, y), \ y(x_0) = y_0$$

is approximated at the sequence of points (x_n, y_n) ($n = 1, 2, \ldots$) where y_n is the approximate value of $y(x_n)$ by computing at each step

$$y_{n+1} = y_n + \frac{h}{6}[k_1 + 2k_2 + 2k_3 + k_4] \ (n = 0, 1, \ldots)$$

where $k_1 = f(x_n, y_n)$, $k_2 = f\left(x_n + \dfrac{h}{2}, \; y_n + \dfrac{hk_1}{2}\right)$, $k_3 = f\left(x_n + \dfrac{h}{2}, \; y_n + \dfrac{hk_2}{2}\right)$, $k_4 = f(x_{n+1}, \; y_n + hk_3)$, $x_n = x_0 + nh$, and h is the selected stepsize.

Example 5

Use the fourth-order Runge-Kutta method with $h = 0.1$ to approximate the solution of $y' = xy$, $y(0) = 1$ on $0 \le x \le 1$.

Detailed discussions of error analysis as well as other numerical methods can be found in most numerical analysis texts.*

Solution   With $f(x, y) = xy$, $x_0 = 0$, and $y_0 = 1$, the formulas are

$$k_1 = f(x_n, y_n) = x_n y_n,$$

$$k_2 = f\left(x_n + \frac{h}{2}, \; y_n + \frac{hk_1}{2}\right) = \left(x_n + \frac{0.1}{2}\right)\left(y_n + \frac{0.1 k_1}{2}\right),$$

$$k_3 = f\left(x_n + \frac{h}{2}, \; y_n + \frac{hk_2}{2}\right) = \left(x_n + \frac{0.1}{2}\right)\left(y_n + \frac{0.1 k_2}{2}\right),$$

$$k_4 = f(x_{n+1}, \; y_n + hk_3) = x_{n+1}(y_n + 0.1 k_3)$$

$$y_{n+1} = y_n + \frac{h}{6}[k_1 + 2k_2 + 2k_3 + k_4] = y_n + \frac{0.1}{6}[k_1 + 2k_2 + 2k_3 + k_4].$$

For $n = 0$, we have $k_1 = x_0 y_0 = (0)(1) = 0$, $k_2 = \left(x_0 + \dfrac{0.1}{2}\right)\left(y_0 + \dfrac{0.1 k_1}{2}\right) =$

$(0.05)(1) = 0.05$, $k_3 = \left(x_0 + \dfrac{0.1}{2}\right)\left(y_0 + \dfrac{0.1 k_2}{2}\right) = (0.05)(1 + 0.0025) =$

0.050125,   and $k_4 = x_1(y_0 + 0.1 k_3) = (0.1)(1 + 0.0050125) = 0.10050125$. Therefore,

$$y_1 = y_0 + \frac{0.1}{6}[k_1 + 2k_2 + 2k_3 + k_4]$$

$$= 1 + \frac{0.1}{6}[0 + 0.05 + 0.050125 + 0.10050125] = 1.005012521.$$

In Table 2.5, we list the results for the Runge-Kutta method of order four to five decimal places. Notice that this method yields the most accurate approximation of the methods used to this point.

---

* See, for example, Richard L. Burden and J. Douglas Faires, *Numerical Analysis,* Third Edition, PWS Publishers, 1985.

TABLE 2.5 Fourth-Order Runge-Kutta Method with $h = 0.1$

x_n	y_n (RK–order 4)	y_n (IEM)	Actual Value
0.0	1.0	1.0	1.0
0.1	1.00501	1.005	1.00501
0.2	1.0202	1.0201755	1.0202
0.3	1.04603	1.0459859	1.04603
0.4	1.08329	1.083223	1.08329
0.5	1.13315	1.1330513	1.13315
0.6	1.19722	1.1970687	1.19722
0.7	1.27762	1.277392	1.27762
0.8	1.37713	1.3767731	1.37713
0.9	1.4993	1.4987552	1.4993
1.0	1.64872	1.6478813	1.64874

IN TOUCH WITH TECHNOLOGY

As indicated in our examples, computers and computer algebra systems are of great use in implementing numerical techniques. In addition to being able to implement the algorithms illustrated above, many computer algebra systems contain built-in commands that you can use to implement various numerical methods.

1. Graph the solution to the initial-value problem
$$\begin{cases} \dfrac{dy}{dx} = \sin(2x - y) \\ y(0) = 0.5 \end{cases}$$
on the interval $[0, 15]$.

2. Graph the solution of $y' = \sin(xy)$ subject to the initial condition $y(0) = i$ on the interval $[0, 7]$ for $i = 0.5$, 1.0, 1.5, 2.0, and 2.5. In each case, approximate the value of the solution if $x = 0.5$.

3. (a) Graph the direction field associated with
$$\frac{dy}{dx} = x^2 + y^2 \text{ for } -1 \le x \le 1 \text{ and } -1 \le y \le 1.$$

   (b) Graph the solution to the initial-value problem
$$\begin{cases} \dfrac{dy}{dx} = x^2 + y^2 \\ y(0) = 0 \end{cases}$$
on the interval $[-1, 1]$.

   (c) Approximate $y(1)$.

In Exercises 1–8, use Euler's method with $h = 0.1$ and $h = 0.05$ to approximate the solution at the given value of x.

1. $y' = 4y + 3x + 2$, $y(0) = 1$, $x = 1$
2. $y' = 4x - y + 1$, $y(0) = 0$, $x = 1$
3. $y' - x = y^2 - 1$, $y(0) = 1$, $x = 1$
4. $y' + x = 5y^{1/2}$, $y(0) = 1$, $x = 1$
5. $y' = \sqrt{xy} + 5y$, $y(1) = 1$, $x = 2$
6. $y' = xy^{1/3} - y$, $y(1) = 1$, $x = 2$
7. $y' = \sin y$, $y(0) = 1$, $x = 1$
8. $y' = \sin(y - x)$, $y(0) = 0$, $x = 1$

In Exercises 9–16, use the improved Euler's method with $h = 0.1$ and $h = 0.05$ to approximate the solution of the corresponding exercise above at the given value of x. Compare these results with those obtained in Exercises 1–8.

9. Exercise 1
10. Exercise 2
11. Exercise 3
12. Exercise 4
13. Exercise 5
14. Exercise 6
15. Exercise 7
16. Exercise 8

In Exercises 17–24, use the Runge-Kutta method with $h = 0.1$ and $h = 0.05$ to approximate the solution of the corresponding exercise above. Compare these results with those obtained in Exercises 1–16.

17. Exercise 1
18. Exercise 2
19. Exercise 3
20. Exercise 4
21. Exercise 5
22. Exercise 6
23. Exercise 7
24. Exercise 8

CHAPTER 2 SUMMARY
Concepts and Formulas

Section 2.1

Separable Differential Equation

A differential equation that can be written in the form $g(y)y' = f(x)$ or $g(y)\,dy = f(x)\,dx$ is called a **separable differential equation.**

Section 2.2

Homogeneous Differential Equation

A differential equation that can be written in the form $M(x, y)\,dx + N(x, y)\,dy = 0$, where

$$M(tx, ty) = t^n M(x, y) \quad \text{and} \quad N(tx, ty) = t^n N(x, y)$$

is called a **homogeneous differential equation (of degree n).**

Section 2.3

Exact Differential Equation

A differential equation that can be written in the form

$$M(x, y)\,dx + N(x, y)\,dy = 0$$

where

$$M(x, y)\,dx + N(x, y)\,dy = \frac{\partial f}{\partial x}(x, y)\,dx + \frac{\partial f}{\partial y}(x, y)\,dy$$

for some function $f(x, y)$ is called an **exact differential equation.**

Section 2.4

First-Order Linear Differential Equation
A differential equation that can be written in the form
$\frac{dy}{dx} + p(x)y = q(x)$ is called a **first-order linear differential equation.**

Integrating Factor
An **integrating factor** for the first-order linear equation
$\frac{dy}{dx} + p(x)y = q(x)$ is $e^{\int p(x)\,dx}$.

Section 2.5

Existence and Uniqueness Theorem
If f and $\frac{\partial f}{\partial y}$ are continuous functions on the rectangular region R: $a < x < b$, $c < y < d$ containing the point (x_0, y_0), then there exists an interval $|x - x_0| < h$ centered at x_0 on which there exists one and only one solution to the initial-value problem
$$y' = f(x, y),\ y(x_0) = y_0.$$

Section 2.6

Euler Method
The approximate value of y at $x = x_n$ is
$$y_n = hf(x_{n-1}, y_{n-1}) + y_{n-1}$$
where $x_n = x_0 + nh$.

Improved Euler's Method
The approximate value of y at $x = x_n$ is computed with
$$y_n^* = hf(x_{n-1}, y_{n-1}) + y_{n-1}$$
$$y_n = \frac{f(x_{n-1}, y_{n-1}) + f(x_n, y_n^*)}{2} h + y_{n-1}.$$

The Runge-Kutta Method of Order Two
The approximate value of y at $x = x_n$ is $y_{n+1} = y_n + \frac{1}{2}(k_1 + k_2)$ where $k_1 = hf(x_n, y_n)$ and $k_2 = hf(x_n + h, y_n + k_1)$.

The Runge-Kutta Method of Order Four
The approximate value of y at $x = x_n$ is
$$y_{n+1} = y_n + \frac{h}{6}[k_1 + 2k_2 + 2k_3 + k_4],\ n = 0, 1, 2, \dots$$
where $k_1 = f(x_n, y_n)$, $k_2 = f\left(x_n + \frac{h}{2}, y_n + \frac{hk_1}{2}\right)$,
$k_3 = f\left(x_n + \frac{h}{2}, y_n + \frac{hk_2}{2}\right)$, and $k_4 = f(x_{n+1}, y_n + hk_3)$.

CHAPTER 2 REVIEW EXERCISES

In Exercises 1–23, solve each equation.

*1. $\frac{dy}{dx} = \frac{2x^5}{5y^2}$

2. $\cos 4x\,dx - 8 \sin y\,dy = 0$

3. $\frac{dy}{dx} = \frac{\sinh x}{2 \cosh y}$

4. $\frac{dy}{dx} = \frac{e^{8y}}{x}$

*5. $\frac{dy}{dx} = \frac{e^{5x}}{y^4}$

6. $(-x^{-5} + x^{-3})\,dx - (2y^4 - 6y^9)\,dy = 0$

7. $\frac{dy}{dx} = \frac{y}{e^{2x} \ln y}$

8. $\frac{dy}{dx} = \frac{(4 - 7x)(2y - 3)}{(x - 1)(2x - 5)}$

*9. $\frac{dy}{dx} = \frac{\cos 7x \sin(3x)}{e^y \cos 6y}$

10. $3x\,dx + (x - 4y)\,dy = 0$

11. $(y - x)\,dx + (x + y)\,dy = 0$

12. $(y^2 - x^2)\,dx + (x + y)\,dy = 0$

*13. $(2y - x)\,dx + (4y^2 - 2xy)\,dy = 0$

14. $\dfrac{dy}{dx} = \dfrac{y^2 + x^2}{xy}$

15. $\dfrac{dy}{dx} = \dfrac{5xy}{y^2 + x^2}$

16. $(x^2 - y)\,dx + (y - x)\,dy = 0$

*17. $(x^2 y + \sin x)\,dx + \left(\dfrac{1}{3}x^3 - \cos y\right) dy = 0$

18. $(\tan y - x)\,dx + (x \sec^2 y + 1)\,dy = 0$

19. $(x \ln y)\,dx + \left(\dfrac{x^2}{2y} + 1\right) dy = 0$

20. $y' + y = 5$ ***21.** $y' + xy = x$

22. $\dfrac{dx}{dy} + \dfrac{x}{y} = y^2$ **23.** $y' + \dfrac{1}{x}y = \cos x$

Solve the following Bernoulli equations. (See Exercise 48, Section 2.4.)

24. $\dfrac{dy}{dx} - y = xy^3$ ***25.** $\dfrac{dy}{dx} + y = e^x y^{-2}$

Solve the following Clairaut equations. (See Exercise 57, Section 2.2.)

26. $y = xy' + 3y'^4$ ***27.** $y - xy' = 2\ln(y')$

Solve the following Lagrange equations. (See Exercise 56, Section 2.4.)

28. $y - 2xy' = -2(y')^3$ ***29.** $y - 2xy' = -4(y')^2$

In Exercises 30–35, solve each initial-value problem.

30. $(2x - y - 2)\,dx + (2y - x)\,dy = 0$, $y(0) = 1$

*31. $\cos(x - y)\,dx + [1 - \cos(x - y)]\,dy = 0$, $y(\pi) = \pi$

32. $(ye^{xy} - 2x)\,dx + xe^{xy}\,dy = 0$, $y(0) = 0$

*33. $(\sin y - y \cos x)\,dx + (x \cos y - \sin x)\,dy = 0$, $y(\pi) = 0$

34. $y^2\,dx + (2xy - 2\cos y \sin y)\,dy = 0$, $y(0) = \pi$

*35. $\left(\dfrac{y}{x} + \ln y\right) dx + \left(\dfrac{x}{y} + \ln x\right) dy = 0$, $y(1) = 1$

For each of the following initial-value problems, use **(a)** Euler's method, **(b)** the improved Euler's method, and **(c)** the Runge-Kutta method with $h = 0.05$ to approximate the solution to the initial-value problem on the indicated interval.

36. $y' = y^2 - x$, $y(0) = 0$, $[0, 1]$

37. $y' = \sqrt{x - y}$, $y(1) = 1$, $[1, 2]$

38. $y' = x + y^{1/3}$, $y(1) = 1$, $[1, 2]$

39. $y' = \sin(xy)$, $y(0) = 1$, $[0, 1]$

40. Does the Existence and Uniqueness Theorem guarantee a unique solution to the following initial-value problems?

(a) $\dfrac{dy}{dx} = xy^3$, $y(0) = 0$

*(b) $\dfrac{dy}{dx} = xy^{-3}$, $y(0) = 0$

(c) $\dfrac{dy}{dx} = -\dfrac{y}{x - 2}$, $y(2) = 0$

41. (Higher-Order Methods With Taylor Series Expansion) Consider the initial-value problem

$$y' = f(x, y), \; y(x_0) = y_0.$$

Recall that the Taylor series expansion of y about $x = x_0$ is given by

$$y(x) = y(x_0) + y'(x_0)(x - x_0) + \dfrac{y''(x_0)}{2!}(x - x_0)^2 +$$
$$\dfrac{y'''(x_0)}{3!}(x - x_0)^3 + \cdots.$$

We know the value of y at the initial value of $x = x_0$ so we use this value to approximate y at $x_1 = x_0 + h$, which is near x_0, in the following manner. We first evaluate the Taylor series at $x_1 = x_0 + h$ to yield

$$y(x_0 + h) = y(x_0) + y'(x_0)h + \dfrac{y''(x_0)}{2!}h^2 +$$
$$\dfrac{y'''(x_0)}{3!}h^3 \cdots.$$

Substituting $y' = f(x, y)$, $y'' = \dfrac{df}{dx}(x, y)$, and $y''' = \dfrac{d^2 f}{dx^2}(x, y)$ into this expansion, using $y(x_0) = y_0$ and calling this new value y_1, we have

$$y_1 = y(x_0 + h) = y(x_0) + f(x_0, y_0)h +$$
$$\frac{1}{2!}\frac{df}{dx}(x_0, y_0)h^2 + \frac{1}{3!}\frac{d^2f}{dx^2}(x_0, y_0)h^3 + \cdots$$

$$= y_0 + f(x_0, y_0)h + \frac{1}{2!}\frac{df}{dx}(x_0, y_0)h^2 +$$
$$\frac{1}{3!}\frac{d^2f}{dx^2}(x_0, y_0)h^3 + \cdots.$$

Hence, the initial point (x_0, y_0) is used to determine y_1. A first-order approximation is obtained from this series by disregarding the terms of order h^2 and higher. In other words, we determine y_1 from

$$y_1 = y_0 + f(x_0, y_0)h.$$

We next use the point (x_1, y_1) to approximate the value of y at $x_2 = x_1 + h$. Calling this value y_2, we have

$$y(x) = y(x_1) + y'(x_1)(x - x_1) +$$
$$\frac{y''(x_1)}{2!}(x - x_1)^2 + \cdots$$

so that an approximate value of $y(x_2)$ is

$$y_2 = y(x_1) + f(x_1, y_1)h = y_1 + f(x_1, y_1)h.$$

Continuing this procedure, we have that the approximate value of y at $x = x_n = x_{n-1} + h$ is given by

$$y_n = y(x_{n-1}) + f(x_{n-1}, y_{n-1})h$$
$$= y_{n-1} + f(x_{n-1}, y_{n-1})h.$$

If we use the first three terms of the Taylor series expansion instead, then we obtain the approximation

$$y_n = y(x_{n-1}) + f(x_{n-1}, y_{n-1})h + \frac{df}{dx}(x_{n-1}, y_{n-1})\frac{h^2}{2}$$
$$= y_{n-1} + f(x_{n-1}, y_{n-1})h + \frac{df}{dx}(x_{n-1}, y_{n-1})\frac{h^2}{2}.$$

We call the approximation which uses the first three terms of the expansion the **three-term Taylor method.**

(a) Use the three-term Taylor method with $h = 0.1$ to approximate the solution of the initial-value problem $y' = xy$, $y(0) = 1$ on $0 \leq x \leq 1$.

A four-term approximation is derived in a similar manner to be

$$y_n = y(x_{n-1}) + f(x_{n-1}, y_{n-1})h +$$
$$\frac{df}{dx}(x_{n-1}, y_{n-1})\frac{h^2}{2} + \frac{d^2f}{dx^2}(x_{n-1}, y_{n-1})\frac{h^3}{6}$$

$$= y_{n-1} + f(x_{n-1}, y_{n-1})h + \frac{df}{dx}(x_{n-1}, y_{n-1})\frac{h^2}{2} +$$
$$\frac{d^2f}{dx^2}(x_{n-1}, y_{n-1})\frac{h^3}{6}.$$

(b) Use the four-term method with $h = 0.1$ to approximate the solution of $y' = 1 + y + x^2$, $y(0) = 0$ on $0 \leq x \leq 1$.

42. **(Running Shoes)** The rate of change of energy absorption in a running shoe is given by

$$\frac{dE}{dt} = F\frac{du}{dt},$$

where

$$F(t) = \frac{F_0}{2}[1 - \cos(\omega t)]$$

represents the force magnitude and pulse duration exerted on the shoe and

$$\frac{du}{dt} = \frac{u_0\omega}{2}\sin(\omega t - \delta)$$

represents the rate of change of the vertical displacement of the midsole.[*] The constant F_0 represents the maximum magnitude of the input force in Newtons (N), ω represents the frequency of the input profile in radians per second (rad/s), u_0 represents the maximum rate of change of the vertical displacement of the midsole in meters (m), δ represents the phase angle between $F(t)$ and $u(t)$ in radians (rad), and t represents time in seconds (s). Thus, the rate of change of energy absorbed by the shoe is given by

$$\frac{dE}{dt} = \frac{F_0 u_0\omega}{4}[1 - \cos(\omega t)]\sin(\omega t - \delta).$$

(a) Find the energy absorbed by the shoe from $t = t_1$ to $t = t_2$.

---

[*] John F. Swigart, Arthur G. Erdman, and Patrick J. Cain, "An Energy-Based Method for Testing Cushioning Durability of Running Shoes," *Journal of Applied Biomechanics*, Volume 9 (1993) pp. 47–65.

(b) The **maximum energy absorbed by the shoe,** ME, is found by calculating the energy absorbed if $t_1 = \dfrac{\delta}{\omega}$ and $t_2 = \dfrac{\pi + \delta}{\omega}$. Show that

$$ME = \frac{F_0 u_0}{2}\left(1 + \frac{\pi}{4}\sin\delta\right).$$

*43. (Fermentation) In the fermentation industry, one of the goals of the molecular biologist is to control the environment and regulate the fermentation. To achieve meaningful environmental control, fermentation research must be carried out on fully monitored environmental systems, the environmental observations must be correlated with existing knowledge of cellular control mechanisms, and environmental control conditions must be reproduced through continuous computer monitoring, analysis, and feedback control of the fermentation environment.

One component of environmental control is the measurement of dissolved oxygen with a steam-sterilizable dissolved oxygen sensor made up of a polymer membrane-covered electrode. Suppose that the electrode is immersed in a liquid medium and that the oxygen is reduced according to the chemical reaction

$$O_2 + 2H_2O + 4e \rightarrow 4OH^-$$

The rate of change, $\dfrac{d\overline{C}}{dt}$, in dissolved oxygen concentration at a particular point in a fermentor vessel is given by

$$\frac{d\overline{C}}{dt} = k_L a(C^* - \overline{C}) - Q_{O_2} X$$

where C^* is the concentration of dissolved oxygen that is in equilibrium with partial pressure $\overline{p}$ in bulk gas phase, $\overline{C}$ is the concentration of dissolved oxygen in bulk liquid, $k_L a$ is the volumetric oxygen-transfer coefficient, Q_{O_2} is the specific rate of oxygen uptake (microbial respiration), and X is the cell mass concentration. If there are no cells present, the equation becomes

$$\frac{d\overline{C}}{dt} = k_L a(C^* - \overline{C}).\text{*}$$

(a) If $\overline{C}(0) = 0$, determine $\overline{C}(t)$ as a function of C^* and $k_L a$.

(b) Suppose that $\overline{C}_p$ is the concentration of dissolved oxygen that corresponds to the sensor reading and k_p is the sensor constant that depends on the conductance of the membrane and the liquid film outside the sensor. If $\overline{C}_p$ satisfies $d\overline{C}_p/dt = k_p(\overline{C} - \overline{C}_p)$ and $\overline{C}_p(0) = 0$, determine $\overline{C}_p(t)$ as a function of k_p, $k_L a$, and C^*. By knowing k_p in advance and by observing $\overline{C}_p$ experimentally, the value of $k_L a$ can be estimated with the solution of this initial-value problem. Thus, the amount of dissolved oxygen can be determined.

Differential Equations at Work:
Modeling the Spread of a Disease

- Suppose that a disease is spreading among a population of size N. In some diseases, like chicken pox, once an individual has had the disease, the individual becomes immune to the disease. In other diseases, like most venereal diseases, once an individual has had the disease and recovers from the disease, the individual does not become immune to the disease; subsequent encounters can lead to recurrence of the infection.

---

* S. Aiba, A. E. Humphrey, N. F. Millis, *Biochemical Engineering,* Second Edition, Academic Press, 1973, pp. 317–336.

Let $S(t)$ denote the percent of the population susceptible to a disease at time t, $I(t)$ the percent of the population infected with the disease, and $R(t)$ the percent of the population unable to contract the disease. For example, $R(t)$ could represent the percent of persons who have had a particular disease, recovered, and have subsequently become immune to the disease.

In order to model the spread of various diseases, we begin by making several assumptions and introducing some notation.

1. Susceptible and infected individuals die at a rate proportional to the number of susceptible and infected individuals with proportionality constant μ called the **daily death removal rate;** the number $1/\mu$ is the **average lifetime** or **life expectancy.**
2. The constant λ represents the **daily contact rate.** On average, an infected person will spread the disease to λ people per day.
3. Individuals recover from the disease at a rate proportional to the number infected with the disease with proportionality constant γ. The constant γ is called the **daily recovery removal rate;** the **average period of infectivity** is $1/\gamma$.
4. The **contact number** $\sigma = \lambda/(\gamma + \mu)$ represents the average number of contacts an infected person has with both susceptible and infected persons.

If a person becomes susceptible to a disease after recovering from it (like gonorrhea, meningitis, and streptococcal sore throat), then the percent of persons susceptible to becoming infected with the disease, $S(t)$, and the percent of people in the population infected with the disease, $I(t)$, can be modeled by the system

$$\begin{cases} S'(t) = -\lambda IS + \gamma I + \mu - \mu S \\ I'(t) = \lambda IS - \gamma I - \mu I \\ S(0) = S_0, \ I(0) = I_0, \ S(t) + I(t) = 1 \end{cases}.$$

This model is called an **SIS** (susceptible-infected-susceptible) model since once an individual has recovered from the disease, the individual again becomes susceptible to the disease.*

Since $S(t) = 1 - I(t)$, we can write $I'(t) = \lambda IS - \gamma I + \mu I$ as

$$I'(t) = \lambda I(1 - I) - \gamma I - \mu I$$

and thus we need to solve the initial-value problem

$$\begin{cases} I'(t) = [\lambda - (\gamma + \mu)]I - \lambda I^2 \\ I(0) = I_0 \end{cases}.$$

1. Convert the Bernoulli equation $I'(t) = \lambda I(1 - I) - \gamma I - \mu I$ to a linear equation and solve the result with the substitution $y = I^{-1}$.
2. **(a)** Show that the solution to the initial-value problem

$$\begin{cases} I'(t) = [\lambda - (\gamma + \mu)]I - \lambda I^2 \\ I(0) = I_0 \end{cases}$$

---

* Herbert W. Hethcote, "Three Basic Epidemiological Models," *Applied Mathematical Ecology*, edited by Simon A. Levin, Thomas G. Hallan, and Louis J. Gross, Springer-Verlag (1989), pp. 119–143.

is

$$I(t) = \begin{cases} \dfrac{e^{(\lambda+\mu)(\sigma-1)t}}{\sigma[e^{(\lambda+\mu)(\sigma-1)t} - 1]/(\sigma - 1) + 1/I_0}, & \text{if } \sigma \neq 1 \\[2em] \dfrac{1}{\lambda t + 1/I_0}, & \text{if } \sigma = 1 \end{cases}$$

and graph various solutions if **(b)** $\lambda = 3.6$, $\gamma = 2$, and $\mu = 1$; **(c)** $\lambda = 3.6$, $\gamma = 2$, and $\mu = 2$. In each case, find the contact number. How does the contact number affect $I(t)$ for large values of t?

3. Evaluate $\lim_{t \to \infty} I(t)$.

The incidence of some diseases, like measles, rubella, and gonorrhea, oscillate seasonally. To model these diseases, we may wish to replace the constant contact rate, λ, by a periodic function $\lambda(t)$.

4. Graph various solutions if **(a)** $\lambda(t) = 5 - 2\sin(6t)$, $\gamma = 1$, and $\mu = 4$; and **(b)** $\lambda(t) = 5 - 2\sin(6t)$, $\gamma = 1$, and $\mu = 2$. In each case, calculate the average contact number. How does the average contact number affect $I(t)$ for large values of t?

5. Explain why diseases like gonorrhea, meningitis, and streptococcal sore throat continue to persist in the population. Do you think there is any way to completely eliminate these diseases from the population? Why or why not?

3

APPLICATIONS OF FIRST-ORDER DIFFERENTIAL EQUATIONS

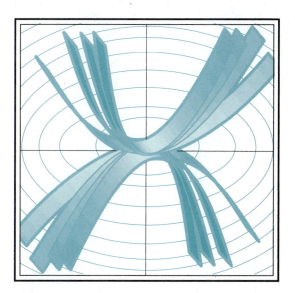

When a space shuttle is launched from the Kennedy Space Center, the minimum initial velocity needed for the shuttle to escape the Earth's atmosphere is determined by solving a first-order ordinary differential equation. The same can be said for finding the flow of electromagnetic forces, the temperature of a cup of coffee, the population of a species, and numerous other applications. In this chapter, we show how these problems can be expressed as first-order equations. We will focus our attention on setting up problems and explaining the meaning of the subsequent solutions because the techniques for solving the first-order equations were discussed in Chapter 2.

3.1   Orthogonal Trajectories

We begin our discussion with a topic that is encountered in the study of electromagnetic fields and heat flow. In algebra, we learn that two lines L_1 and L_2 with slopes m_1 and m_2, respectively, are **orthogonal** (or perpendicular) if their slopes satisfy the relationship $m_1 = -1/m_2$. This relationship can be extended to curves.

Definition 3.1 Orthogonal Curves

Two curves C_1 and C_2 are **orthogonal** (or perpendicular) at a point of intersection if their respective tangent lines to the curves at that point are orthogonal (or perpendicular). (See Figure 3.1.)

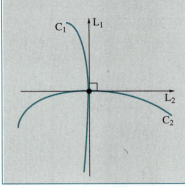

Figure 3.1   Orthogonal Curves C_1 and C_2

Example 1

Verify that the graphs of $y = x$ and $y = \sqrt{1 - x^2}$ are orthogonal at the point $\left(\frac{\sqrt{2}}{2}, \frac{\sqrt{2}}{2}\right)$.

Solution   First note that the point $\left(\frac{\sqrt{2}}{2}, \frac{\sqrt{2}}{2}\right)$ lies on the graph of both $y = x$ and $y = \sqrt{1 - x^2}$. The derivatives of these functions are $y' = 1$ and $y' = -x/\sqrt{1 - x^2}$, respectively. The slope of the line tangent to $y = x$ at the point with x-coordinate $x = \frac{\sqrt{2}}{2}$ is 1. Substitution of $x = \frac{\sqrt{2}}{2}$ into $y' = -x/\sqrt{1 - x^2}$ yields

$$\frac{-\frac{\sqrt{2}}{2}}{1 - \left(\frac{\sqrt{2}}{2}\right)^2} = -1$$

as the slope of the tangent line at the point with x-coordinate $x = \frac{\sqrt{2}}{2}$. Thus, these curves are orthogonal at the point $\left(\frac{\sqrt{2}}{2}, \frac{\sqrt{2}}{2}\right)$ because the slopes of the lines tangent to their graphs at that point are negative reciprocals. We graph these two curves in Figure 3.2 along with the tangent line to $y = \sqrt{1 - x^2}$ at $\left(\frac{\sqrt{2}}{2}, \frac{\sqrt{2}}{2}\right)$ to illustrate that the two are orthogonal.

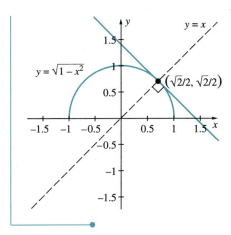

Figure 3.2

The next step in our discussion of orthogonal curves is to determine the set of orthogonal curves to a given family of curves. If we are given a family of curves F, then we would like to find a set of curves that are orthogonal to each curve in F. We refer to this set of orthogonal curves as the **family of orthogonal trajectories**.

Suppose that a family of curves is defined as $F(x, y) = c$ and that the slope of the tangent line at any point on these curves is

$$\frac{dy}{dx} = f(x, y)$$

$\left(\text{which can be obtained by differentiating } F(x, y) = c \text{ with respect to } x \text{ and solving for } \frac{dy}{dx}\right)$. Then the slope of the tangent line on the orthogonal trajectory is

$\dfrac{dy}{dx} = \dfrac{-1}{f(x - y)}$ so the family of orthogonal trajectories is found by solving the first-order equation

$$\frac{dy}{dx} = -\frac{1}{f(x, y)}.$$

*Always make sure to take the **negative reciprocal** before solving the differential equation for the orthogonal trajectories. (Otherwise, the original family of curves will be recovered!)*

<table>
<tr><td>Example 2</td><td>Determine the family of orthogonal trajectories to the family of curves $y = cx^2$.</td></tr>
</table>

Solution   First we must find the slope of the tangent line at any point on the parabola $y = cx^2$. Differentiating with respect to x gives us

$$\frac{dy}{dx} = 2cx.$$

Note that we need to express the slope of the curves solely in terms of x and $y = y(x)$. From $y = cx^2$, we have $c = \dfrac{y}{x^2}$. Substitution into $\dfrac{dy}{dx} = 2cx$ then yields

$$\frac{dy}{dx} = 2cx = 2\left(\frac{y}{x^2}\right)x = \frac{2y}{x}$$

on the parabolas. We must solve

$$\frac{dy}{dx} = -\frac{x}{2y}$$

to determine the orthogonal trajectories. This equation is separable, so we write it as

$$2y\, dy = -x\, dx,$$

integrate both sides, and see that an implicit solution can be expressed as

$$y^2 + \frac{x^2}{2} = k,$$

which we recognize as a family of ellipses. In Figure 3.3, we graph the family of parabolas $y = cx^2$, the family of ellipses $y^2 + \dfrac{x^2}{2} = k$, and the two families of curves together. Notice that the graphs appear orthogonal, confirming the results we obtained.

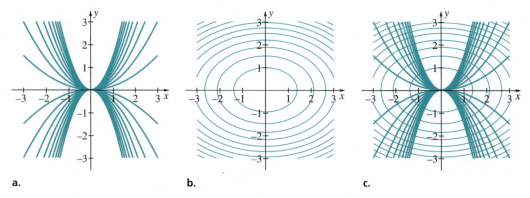

a.   b.   c.

Figure 3.3   (a) Graph of $y = cx^2$ (b) Graph of $y^2 + \frac{1}{2}x^2 = k$ (c) Graphs of $y = cx^2$ and $y^2 + \frac{1}{2}x^2 = k$

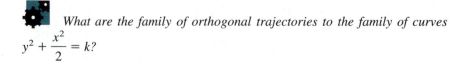

What are the family of orthogonal trajectories to the family of curves $y^2 + \dfrac{x^2}{2} = k$?

Example 3

Determine the family of orthogonal trajectories of the family of ellipses

$$x^2 - xy + y^2 = c^2.$$

Solution We first determine the differential equation satisfied by the family of ellipses. Implicit differentiation yields

$$2x - x\frac{dy}{dx} - y + 2y\frac{dy}{dx} = 0$$

$$\frac{dy}{dx}(2y - x) = y - 2x$$

$$\frac{dy}{dx} = \frac{y - 2x}{2y - x}.$$

Notice that we do not have to eliminate c as in Example 2 because c does not appear in this equation after differentiation. The family of orthogonal trajectories satisfies

$$\frac{dy}{dx} = \frac{-1}{\dfrac{y - 2x}{2y - x}} = \frac{x - 2y}{y - 2x}.$$

Rewriting this equation as $(y - 2x)\,dy + (2y - x)\,dx = 0$, we see that this is a homogeneous first-order equation. Letting $y = vx$, we obtain $dy = v\,dx + x\,dv$ and substitute into the equation $(y - 2x)\,dy + (2y - x)\,dx = 0$ to obtain a separable equation:

$$(vx - 2x)(v\,dx + x\,dv) + (2vx - x)\,dx = 0$$

$$\frac{dx}{x} = \frac{(2 - v)\,dv}{v^2 - 1}.$$

Integrating both sides of this equation results in $\ln|x| = \frac{1}{2}\ln|1 - v| - \frac{3}{2}\ln|v + 1| + C$, and substituting $v = \frac{y}{x}$ yields $\ln|x| = \frac{1}{2}\ln\left|1 - \frac{y}{x}\right| - \frac{3}{2}\ln\left|\frac{y}{x} + 1\right| + C$. Multiplying this equation by 2 and applying properties of logarithms we obtain $(x + y)^3/(x - y) = C$.

In Figure 3.4, we graph members of the family of curves $x^2 - xy + y^2 = c^2$, members of $(x + y)^3/(x - y) = C$, and the two families together to observe that they are orthogonal.

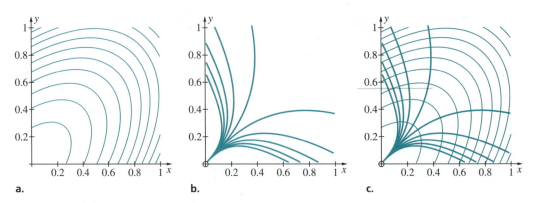

Figure 3.4   (a) Graph of $x^2 - xy + y^2 = c^2$ (b) Graph of $(x + y)^3/(x - y) = C$
(c) Graphs of $x^2 - xy + y^2 = c^2$ and $(x + y)^3/(x - y) = C$

Graph several members of the family of curves $x^2 - xy + y^2 = c^2$ and $(x + y)^3/(x - y) = C$ in Quadrant 3. Hint: What is the relationship between the level curves of $f(x, y) = (x + y)^3/(x - y)$ and $g(x, y) = (x - y)/(x + y)^3$?

Example 4

Let $T(x, y)$ represent the temperature at the point (x, y). The curves given by $T(x, y) = c$ (where c is constant) are called **isotherms.** The orthogonal trajectories are curves along which heat will flow. Determine the isotherms if the curves of heat flow are given by $y^2 + 2xy - x^2 = c$.

Solution   We begin by finding the slope of the tangent line at each point on the heat flow curves $y^2 + 2xy - x^2 = c$ using implicit differentiation to obtain $\dfrac{dy}{dx} = \dfrac{x - y}{x + y}$. Hence, the orthogonal trajectories satisfy the differential equation $\dfrac{dy}{dx} = -\dfrac{x + y}{x - y}$.

Writing this equation in differential form as

$$(x + y)\, dx + (x - y)\, dy = 0,$$

we see that it is exact because $\dfrac{\partial}{\partial y}(x + y) = 1$ and $\dfrac{\partial}{\partial x}(x - y) = 1$. We solve this equation by integrating $x + y$ with respect to x and then differentiating the result with respect to y to find that the family of orthogonal trajectories (isotherms) is given by $\dfrac{x^2}{2} + xy - \dfrac{y^2}{2} = k$ or $x^2 + 2xy - y^2 = C$, where $C = 2k$. These two families of hyperbolas are shown in Figure 3.5.

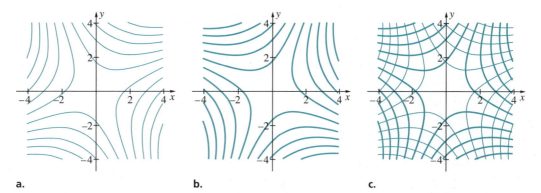

Figure 3.5 (a) Curves of Heat Flow   Graph of $y^2 + 2xy - x^2 = c$ (b) Isotherms   Graph of $x^2 + 2xy - y^2 = C$ (c) Graphs of $y^2 + 2xy - x^2 = c$ and $x^2 + 2xy - y^2 = C$

IN TOUCH WITH TECHNOLOGY

 1. Find the family of orthogonal trajectories for the family of curves $y = x - 1 + ce^{-x}$. Graph several members of both families of curves on the same set of axes.

2. (a) Determine the orthogonal trajectories of the family of curves $y^2 = 2cx + 2c^2$. **(b)** Graph several members of both families of curves on the same set of axes. **(c)** What is your reaction to the graphs?

3. (a) Sketch several members of the family of curves $f(x, y) = c$ for which $y' = \sin(xy)$. **(b)** Graph several members of the family of orthogonal trajectories associated with this family of curves.

4. (a) Sketch several members of the family of curves $f(x, y) = c$ for which $y' = x^2 + y^2$. **(b)** Graph several members of the family of orthogonal trajectories associated with this family of curves.

EXERCISES 3.1

In Exercises 1–20, determine the orthogonal trajectories of the given family of curves. (Graph the orthogonal trajectories and curves simultaneously for each of the exercises in 1–20.)

1. $y + 2x = c$

2. $y = cx - 1$

*3. $y = e^{cx}$

4. $y = ce^x$

5. $y^2 = x^2 + c$

6. $y^2 = -x^2 + c$

*7. $y^2 = x^2 + cx$

8. $y = cx^3$

*9. $y = x^2 + cx$ (*Hint:* See Exercise 48 in Section 2.4.)

10. $x^2 + y^2 = cx^3$

11. $y = (x - c)^2$

12. $y = c \cos x$

*13. $2y^3 - 3x^2 = C$

14. $y^2 = -x^2 + 2cx$

15. $y = e^x + c$

16. $y = \ln(x + c)$

*17. $y = \dfrac{x}{1 + cx}$

18. $y = \dfrac{1 - cx}{1 + cx}$ **19.** $e^y - e^{-y} = cx$

20. $y - x = ce^x$

*21. A family of curves is **self-orthogonal** if the family of orthogonal trajectories is the same as the original family of curves. Is $y^2 - 2cx = c^2$ a self-orthogonal family of curves (parabolas)?

22. Find a value of c so that the two families of curves $y = k_1 x^2 + c$ and $x^2 + 2y^2 = k_2 + y$, where k_1 and k_2 are constants, are orthogonal.

23. Suppose that an electrical current is flowing in a wire along the z-axis. Then, the equipotential lines in the xy-plane are concentric circles centered at the origin. If the electric lines of force are the orthogonal trajectories of these circles, find the electric lines of force.

24. The path along which a fluid particle flows is called a **streamline** and the orthogonal trajectories are called **equipotential lines.** If the streamlines are $y = k/x$, find the equipotential lines.

*25. (Oblique Trajectories) Let ℓ_1 and ℓ_2 denote two lines, not perpendicular to each other, with slopes m_1 and m_2, respectively; and let θ denote the angle between them as shown in Figure 3.6. Then,

$$\tan \theta = \frac{m_2 - m_1}{1 + m_2 m_1}.$$

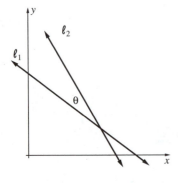

Figure 3.6

(a) Show that $m_2 = \dfrac{m_1 + \tan \theta}{1 - m_1 \tan \theta}$.

(b) Suppose we are given a family of curves that satisfies the differential equation $dy/dx = f(x, y)$. Use **(a)** to show that if we want to find a family of curves that intersects this family at a constant angle θ, we must solve the differential equation

$$\frac{dy}{dx} = \frac{f(x, y) \pm \tan \theta}{1 \mp f(x, y) \tan \theta}.$$

Hint: The picture in Figure 3.7 might help.

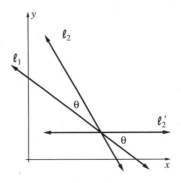

Figure 3.7

(c) Find a family of curves that intersects the family of curves $x^2 - y^2 = c$ at an angle of $\frac{\pi}{4}$. Graph several members of both families to confirm your result.

(d) Find a family of curves that intersects the family of curves $x^2 + y^2 = c^2$ at an angle of $\frac{\pi}{6}$. Graph several members of both families to confirm your result.

Population Growth and Decay

The Malthus Model • The Logistic Equation

Many interesting problems involving populations can be solved through the use of first-order differential equations. These include the determination of the number of cells in a bacteria culture, the number of citizens in a country, and the amount of radioactive substance remaining in a fossil. We begin our discussion by solving a population problem.

The Malthus Model

Suppose that the rate at which a population of size $y(t)$ at time t changes is proportional to the amount present. Mathematically, this statement is represented as the first-order initial-value problem

$$\begin{cases} \dfrac{dy}{dt} = ky, \\ y(0) = y_0 \end{cases}$$

where y_0 is the initial population. This is known as the **Malthus model** and is due to the work of the English clergyman and economist Thomas R. Malthus. If $k > 0$, then the population increases (growth), while the population decreases (decay) if $k < 0$.

We solve the Malthus model for all values of k and y_0, which enables us to refer to the solution in other problems without solving the differential equation again. Rewriting $\dfrac{dy}{dt} = ky$ in the form $\dfrac{dy}{y} = k\,dt$, integrating, and simplifying results in:

$$\int \frac{dy}{y} = \int k\,dt$$
$$\ln|y| = kt + C_1$$
$$y = Ce^{kt} \qquad (C = e^{C_1}).$$

Notice that $|y| = y$ because the population y is greater than or equal to zero. To find C, we apply the initial condition, obtaining $y_0 = y(0) = Ce^{k \cdot 0} = C$. Thus, the solution to the initial-value problem $\dfrac{dy}{dt} = ky$, $y(0) = y_0$ is

$$y = y_0 e^{kt}.$$

Example 1

Forms of a given element with different numbers of neutrons are called **nuclides.** Some nuclides are not stable. For example, potassium–40 (^{40}K) naturally decays to reach argon–40 (^{40}Ar). This decay that occurs in some nuclides was first observed, but not understood, by Henri Becquerel (1852–1908) in 1896. Marie Curie, however, began studying this decay in 1898, named it **radioactivity,** and discovered the radioactive substances polonium and radium. Marie Curie (1867–1934), along with her husband Pierre Curie (1859–1906) and Henri Becquerel, received the Nobel Prize in Physics in 1903 for their work on radioactivity. Marie Curie subsequently received the Nobel Prize in Chemistry in 1910 for discovering polonium and radium. Given a sample of ^{40}K of sufficient size, after 1.2×10^9 years approximately half of the sample will have decayed to ^{40}Ar. The **half-life** of a nuclide is the time

a.   b.

(a) Marie (1867–1934) and Pierre (1859–1906) Curie in their Paris laboratory (1896).
(b) Henri Becquerel (1852–1908) discovered radioactivity in 1896. Marie Curie, Pierre Curie, and Henri Becquerel shared the Nobel Prize in Physics in 1903 for their work on radioactivity. (*a*, AIP Emilio Segré Visual Archives; *b*, North Wind Picture Archives)

for half the nuclei in a given sample to decay. (See Table 3.1.) We see that the rate of decay of a nuclide is proportional to the amount present because the half-life of a given nuclide is constant and independent of the sample size.

    If the half-life of polonium ^{209}Po is 100 years, determine the percentage of the original amount of ^{209}Po that remains after 50 years.

TABLE 3.1   Half-Life of Various Nuclides

Element	Nuclide	Half-Life	Element	Nuclide	Half-Life
Aluminum	^{26}Al	7.4×10^5 Years	Polonium	^{209}Po	100 Years
Beryllium	^{10}Be	1.51×10^6 Years	Polonium	^{210}Po	138 Days
Carbon	^{14}C	5730 Years	Radon	^{222}Rn	3.82 Days
Chlorine	^{36}Cl	3.01×10^5 Years	Radium	^{226}Ra	1700 Years
Iodine	^{131}I	8.05 Days	Thorium	^{230}Th	75,000 Years
Potassium	^{40}K	1.2×10^9 Years	Uranium	^{238}U	4.51×10^9 Years

Solution   Let y_0 represent the original amount of ^{209}Po that is present. The amount present after t years is $y(t) = y_0 e^{kt}$. Using $y(100) = \frac{1}{2}y_0$ and $y(100) = y_0 e^{100k}$, we solve $y_0 e^{100k} = \frac{1}{2}y_0$ for e^k:

$$e^{100k} = \tfrac{1}{2}$$
$$(e^k)^{100} = \tfrac{1}{2}$$
$$e^k = (\tfrac{1}{2})^{1/100}.$$

Hence,

$$y(t) = y_0 e^{kt} = y_0(e^k)^t = y_0(\tfrac{1}{2})^{t/100}.$$

In order to determine the percentage of y_0 that remains, we evaluate

$$y(50) = y_0(\tfrac{1}{2})^{50/100} \approx 0.7071 y_0.$$

Therefore, 70.71% of the original amount of ^{209}Po remains after 50 years.

In Example 1, we determine the percentage of the original amount of ^{209}Po that remains even though we do not know the value of y_0, the initial amount of ^{209}Po. Instead of letting $y(t)$ represent the *amount* of the substance present after time t, we can let it represent the *fraction* (or *percent*) of y_0 that remains after time t. In doing so, we use the initial condition $y(0) = 1 = 1.00$ to indicate that 100% of y_0 is present at $t = 0$.

Example 2

The wood of an Egyptian sarcophagus (burial case) is found to contain 63% of the carbon-14 found in a present day sample. What is the age of the sarcophagus?

Sarcophagus A Roman sarcophagus in the Antalya Museum, Perge, Turkey. (Borys Malkin/Anthro-Photo)

Solution   From Table 3.1, we see that the half-life of carbon-14 is 5730 years. Let $y(t)$ be the percent of carbon-14 in the sample after t years. Then $y(0) = 1$. Now, $y(t) = y_0 e^{kt}$, so $y(5730) = e^{5730k} = 0.5$. Solving for k yields:

$$\ln(e^{5730k}) = \ln(.5)$$
$$5730k = \ln(.5)$$
$$k = \frac{\ln(.5)}{5730} = \frac{-\ln 2}{5730}.$$

Thus, $y(t) = e^{kt} = e^{\frac{-\ln 2}{5730}t} = 2^{-t/5730}$. (An alternate approach to obtain the solution is to solve $e^{5730k} = 0.5$ for e^k instead of for k as we did in Example 1. This yields $e^k = (.5)^{1/5730} = (\tfrac{1}{2})^{1/5730} = 2^{-1/5730}$. Substitution of this expression into $y(t) = y_0 e^{kt} = y_0(e^k)^t$ gives the same solution as was found previously.)

In this problem, we must find the value of t for which $y(t) = 0.63$. Solving this equation results in:

$$2^{-t/5730} = 0.63 = \tfrac{63}{100}$$
$$\ln(2^{-t/5730}) = \ln \tfrac{63}{100}$$
$$\frac{-t}{5730} \ln 2 = \ln \tfrac{63}{100}$$
$$t = \frac{-5730 \ln \tfrac{63}{100}}{\ln 2} = \frac{5730(\ln 100 - \ln 63)}{\ln 2} \approx 3819.48.$$

We conclude that the sarcophagus is approximately 3819 years old.

We can use the Malthus model to predict the size of a population at a given time if the rate of growth of the population is proportional to the present population.

Example 3

Suppose that the number of cells in a bacteria culture doubles after three days. Determine the number of days required for the initial population to triple.

Solution   In this case, $y(0) = y_0$, so the population is given by $y(t) = y_0 e^{kt}$ and $y(3) = 2y_0$ because the population doubles after three days. Substituting this value into $y(t) = y_0 e^{kt}$, we have

$$y(3) = y_0 e^{3k} = 2y_0.$$

Solving for e^k, we find that $e^{3k} = 2$ or $(e^k)^3 = 2$. Therefore, $e^k = 2^{1/3}$. Substitution into $y(t) = y_0 e^{kt}$ then yields

$$y(t) = y_0(e^k)^t = y_0 2^{t/3}.$$

We find when the population triples by solving $y(t) = y_0 2^{t/3} = 3y_0$ for t. This yields $2^{t/3} = 3$ or $\dfrac{t}{3} \ln 2 = \ln 3$. Therefore, the population triples in $t = \dfrac{3 \ln 3}{\ln 2} \approx 4.755$ days.

Determine the number of days required for the culture of bacteria considered in Example 3 to reach nine times its initial size.

To observe some of the limitations of the Malthus model, we consider a population problem in which the rate of growth of the population does *not* exclusively depend on the present population.

Example 4

The population of the United States was recorded as 5.3 million in 1800. Use the Malthus model to approximate the population for years after 1800 if $k = 0.03$. Compare these results to the actual population. Is this a good approximation for years after 1800?

Solution   In this example, $k = 0.03$, $y_0 = 5.3$, and our model for the population of the United States at time t (where t is the number of years from 1800) is $y(t) = 5.3e^{0.03t}$. In order to compare this model with the actual population of the United States, census figures for the population of the United States for various years are listed in Table 3.2 along with the corresponding value of $y(t)$. A graph of $y(t)$ with the corresponding points is shown in Figure 3.8.

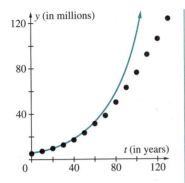

Figure 3.8

TABLE 3.2

Year (t)	Actual Population (in millions)	Value of $y(t) = 5.3e^{0.03t}$	Year (t)	Actual Population (in millions)	Value of $y(t) = 5.3e^{0.03t}$
1800 (0)	5.30	5.30	1870 (70)	38.56	43.28
1810 (10)	7.24	7.15	1880 (80)	50.19	58.42
1820 (20)	9.64	9.66	1890 (90)	62.98	78.86
1830 (30)	12.68	13.04	1900 (100)	76.21	106.45
1840 (40)	17.06	17.60	1910 (110)	92.23	143.70
1850 (50)	23.19	23.75	1920 (120)	106.02	193.97
1860 (60)	31.44	32.06	1930 (130)	123.20	261.83

Although the model appears to approximate the data for several years after 1800, the accuracy of the approximation diminishes over time because the population of the United States does not exclusively increase at a rate proportional to the population. Another model that better approximates the population would take other factors into account.

The Logistic Equation

Because the approximation obtained with the Malthus model is less than desirable in the previous example, we see that another model is needed. The **logistic equation** (or **Verhulst equation**) is the equation

$$y'(t) = (r - ay(t))y(t),$$

where r and a are constants, subject to the condition $y(0) = y_0$. This equation was first introduced by the Belgian mathematician Pierre Verhulst to study population growth. The logistic equation differs from the Malthus model in that the term $(r - ay(t))$ is not constant. This equation can be written as $\dfrac{dy}{dt} = (r - ay)y = ry - ay^2$ where the term $(-y^2)$ represents an inhibitive factor. Under these assumptions the population is neither allowed to grow out of control nor grow or decay constantly as it was with the Malthus model.

The logistic equation is separable. For convenience, we write the equation as

$$y' = (r - ay)y \qquad \text{or} \qquad \frac{dy}{dt} = (r - ay)y$$

Separating variables and using partial fractions to integrate with respect to y, we have

$$\frac{dy}{(r-ay)y} = dt$$

$$\left(\frac{a/r}{r-ay} + \frac{1/r}{y}\right) dy = dt$$

$$\left(\frac{a}{r-ay} + \frac{1}{y}\right) dy = r\,dt$$

$$-\ln|r-ay| + \ln|y| = rt + c$$

We solve this expression for y using the properties of logarithms:

$$\ln\left|\frac{y}{r-ay}\right| = rt + c$$

$$\frac{y}{r-ay} = \pm e^{rt+c} = Ke^{rt} \qquad (K = \pm e^c)$$

$$y = r\left(\frac{1}{K}e^{-rt} + a\right)^{-1}.$$

Applying the initial condition $y(0) = y_0$ and solving for K, we find that

$$\frac{y_0}{r-ay_0} = K.$$

After substituting this value into the general solution and simplifying, the solution can be written as

$$y = \frac{ry_0}{ay_0 + (r-ay_0)e^{-rt}}.$$

Notice that $\lim_{t\to\infty} y(t) = r/a$ because $\lim_{t\to\infty} e^{-rt} = 0$ if $r > 0$. This makes the solution to the logistic equation different from that of the Malthus model in that the solution to the logistic equation approaches a finite nonzero limit as $t \to \infty$ while that of the Malthus model approaches either infinity or zero as $t \to \infty$.

Use a computer algebra system to solve the logistic equation. If the result you obtain is not in the same form as that given above, show (by hand) that the two are the same.

Example 5

Use the logistic equation to approximate the population of the United States using $r = 0.03$, $a = 0.0001$, and $y_0 = 5.3$. Compare this result with the actual census values given in Table 3.2. Use the model obtained to predict the population of the United States in the year 2000.

| **Solution** | We substitute the indicated values of r, a, and y_0 into

$y = \dfrac{ry_0}{ay_0 + (r - ay_0)e^{-rt}}$ to obtain the approximation of the population of the United States at time t, where t represents the number of years since 1800,

$$y(t) = \frac{0.03 \cdot 5.3}{0.0001 \cdot 5.3 + (0.03 - 0.0001 \cdot 5.3)e^{-.03t}} = \frac{0.159}{0.00053 + 0.02947e^{-0.3t}}.$$

In Table 3.3, we compare the approximation of the population of the United States given by the approximation $y(t)$ with the actual population obtained from census figures. Note that this model appears to more closely approximate the population over a longer period of time than the Malthus model did (Example 4), as we can see in Figure 3.9. To predict the population of the United States in the year 2000 with this model, we evaluate $y(200)$ to obtain $y(200) = \dfrac{0.159}{0.00053 + 0.02947e^{-0.3 \cdot 200}} \approx$

263.66. Thus, we predict that the population will be approximately 263.66 million in the year 2000. Note that projections of the population of the United States in the year 2000 made by the Bureau of the Census range from 259.57 million to 278.23 million, so the approximation obtained with the logistic equation seems reasonable.

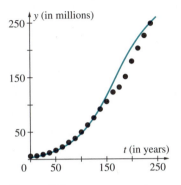

Figure 3.9

TABLE 3.3

Year (t)	Actual Population (in millions)	Value of $y(t)$	Year (t)	Actual Population (in millions)	Value of $y(t)$
1800 (0)	5.30	5.30	1900 (100)	76.21	79.61
1810 (10)	7.24	7.11	1910 (110)	92.23	98.33
1820 (20)	9.64	9.52	1920 (120)	106.02	119.08
1830 (30)	12.68	12.71	1930 (130)	123.20	141.14
1840 (40)	17.06	16.90	1940 (140)	132.16	163.59
1850 (50)	23.19	22.38	1950 (150)	151.33	185.45
1860 (60)	31.44	29.44	1960 (160)	179.32	205.82
1870 (70)	38.56	38.42	1970 (170)	203.30	224.05
1880 (80)	50.19	49.63	1980 (180)	226.54	239.78
1890 (90)	62.98	63.33	1990 (190)	248.71	252.94

IN TOUCH WITH TECHNOLOGY

1. Consider the Malthus population model with $k = 0.01$, 0.05, 0.1, 0.5, and 1.0 using $y_0 = 1$. Solve the model, plot the solution with these values, and compare the results. How does the value of k affect the solution?

2. Consider the logistic equation with $r = 0.01, 0.05, 0.1$, 0.5, and 1.0 using $y_0 = 1$ and $a = 1$. Solve the model, plot the solution with these values, and compare the results. How does the value of r affect the solution?

3. We may use graphing utilities to investigate the behavior of solutions of the logistic equation under different conditions. (a) Use a graphing utility to investigate the behavior of the solution if values of r, a, and y_0 are varied. Under what conditions does the limit of the solution approach (b) a nonzero number as $t \to \infty$; and (c) zero as $t \to \infty$? (*Hint:* Consider $0 < r \leq 1, r > 1$; $0 < a < 1, a > 1$.)

4. (**Harvesting**) If we wish to model a population of size $P(t)$ at time t and consider a constant harvest rate h (like hunting, fishing, or disease), then we might modify the logistic equation and use the equation

$$\frac{dP}{dt} = rP - aP^2 - h$$

to model the population under consideration.* Assume that $h \geq \dfrac{r^2}{4a}$.

(a) Show that a general solution of
$$\frac{dP}{dt} = rP - aP^2 - h \text{ is}$$
$$P(t) = \frac{1}{2a}\left[r + \sqrt{4ah - r^2} \tan\left(\frac{1}{2}(C - t)\sqrt{4ah - r^2}\right)\right].$$

(b) Suppose that for a certain species it is found that $r = 0.03$, $a = 0.0001$, $h = 2.26$, and $C = -1501.85$. At what time will the species become extinct?

(c) If $r = 0.03$, $a = 0.0001$, and $P(0) = 5.3$, graph $P(t)$ if $h = 0$, $h = 0.5$, $h = 1.0$, $h = 1.5$, $h = 2.0$, $h = 2.25$ and $h = 2.5$.

(d) What is the maximum allowable harvest rate to assure that the species survives?

(e) Generalize your result from (d). For arbitrary a and r, what is the maximum allowable harvest rate that assures survival of the species?

5. Shortly after the Chernobyl accident in the Soviet Union in 1986, several nations reported that the level of ^{131}I in milk was five times that considered safe for human consumption. Make a table of the level of ^{131}I in milk as a multiple of that considered safe for human consumption for the first three weeks following the accident. After how long did the milk become safe for human consumption?

EXERCISES 3.2

Solve the following problems. Unless otherwise stated, use the Malthus model.

1. Suppose that a culture of bacteria has initial population of $n = 100$. If the population doubles every three days, determine the number of bacteria present after

30 days. How much time is required for the population to reach 4250 in number?

2. Suppose that the population in a yeast culture triples every seven days. What is the population after 35 days? How much time is required for the population to be 10 times the initial population?

---

* David A Sanchez, "Populations and Harvesting," *Mathematical Modeling: Classroom Notes in Applied Mathematics,* Murray S. Klamkin, Editor, SIAM (1987) pp. 311–313.

*3. Suppose that two-thirds of the cells in a culture remain after one day. Use this information to determine the number of days until only one-third of the initial population remains.

4. Consider a radioactive substance with half-life 10 days. If there are initially 5000 grams of the substance, how much remains after 365 days?

5. Suppose that the half-life of an element is 1000 hours. If there are initially 100 grams, how much remains after one hour? How much remains after 500 hours?

6. Suppose that the population of a small town is initially 5000. Due to the construction of an interstate highway, the population doubles over the next year. If the rate of growth is proportional to the current population, when will the population reach 25,000? What is the population after five years?

*7. Suppose that mold grows at a rate proportional to the amount present. If there are initially 500 grams of mold and six hours later there are 600 grams, determine the amount of mold present after one day. When is the amount of mold 1000 grams?

8. Suppose that the rabbit population on a small island grows at a rate proportional to the number of rabbits present. If this population doubles after 100 days, when does the population triple?

9. In a chemical reaction, chemical A is converted to chemical B at a rate proportional to the amount of chemical A present. If half of chemical A remains after five hours, when does $\frac{1}{6}$ of the initial amount of chemical A remain? How much of the initial amount remains after 15 hours?

10. If 90% of the initial amount of a radioactive element remains after one day, what is the half-life of the element?

*11. If $y(t)$ represents the percent of a radioactive element that is present at time t and the values of $y(t_1)$ and $y(t_2)$ are known, show that the half-life H is given by
$$H = \frac{(t_2 - t_1)\ln 2}{\ln[y(t_1)] - \ln[y(t_2)]}.$$

12. The half-life of carbon-14 is 5730 years. If the original amount of carbon-14 in a particular living organism is 20 grams and that found in a fossil of that organism is 0.01 grams, determine the approximate age of the fossil.

13. After 10 days 800 grams of a radioactive element remain, and after 15 days 560 grams remain. What is the half-life of this element?

14. After one week, 10% of the initial amount of a radioactive element decays. How much decays after two weeks? When does half of the original amount decay?

*15. Determine the percentage of the original amount of ^{226}Ra that remains after 100 years.

16. If an artifact contains 40% of the amount of ^{230}Th as a present day sample, what is the age of the artifact?

17. On an archeological dig, scientists find an ancient tool near a fossilized human bone. If the tool and fossil contain 65% and 60% of the amount of carbon-14 as that in present day samples, respectively, determine if the tool could have been used by the human.

18. A certain group of people with initial population 10,000 grows at a rate proportional to the number present. The population doubles in five years. In how many years will the population triple?

*19. Solve the logistic equation by viewing it as a Bernoulli equation. (See Exercise 2.4.)

20. What is the limiting population ($\lim_{t \to \infty} y(t)$) of the U. S. population using the result obtained in Example 5?

21. Solve the logistic equation if $r = \dfrac{1}{100}$ and $a = \dfrac{1}{10^8}$ given that $y(0) = 100{,}000$. Find $y(25)$. What is the limiting population?

22. Five college students with the flu return to an isolated campus of 2500 students. If the rate at which this virus spreads is proportional to the number of infected students y as well as the number not infected $2500 - y$, solve the initial value problem $dy/dt = ky(2500 - y)$, $y(0) = 5$ to find the number of infected students after t days if 25 students have the virus after one day. How many students have the flu after five days?

*23. One student in a college organization of 200 members proceeds to spread a rumor. If the rate at which this rumor spreads is proportional to the number of students y that know about the rumor as well as the number that do not know, then solve the initial-value problem to find the number of students informed of the rumor after t days if 50 students are informed after one day. How many students know the rumor after two days? Will all of the students eventually be informed of the rumor? (See Exercise 22.)

24. Suppose that glucose enters the bloodstream at the constant rate of r grams per minute while it is removed at a rate proportional to the amount y present at any time. Solve the initial-value problem

$dy/dt = r - ky$, $y(0) = y_0$ to find $y(t)$. What is the eventual concentration of glucose in the bloodstream according to this model?

25. What is the concentration of glucose in the bloodstream after 10 minutes if $r = 5$ grams per minute and $k = 5$ and the initial concentration is $y(0) = 500$? After 20 minutes? Does the concentration appear to reach its limiting value quickly or slowly? (See Exercise 24.)

26. Suppose that we deposit a sum of money in a money market fund that pays interest at an annual rate k, and let $S(t)$ represent the value of the investment at time t. If the compounding takes place continuously, then the rate at which the value of the investment changes is the interest rate times the value of the investment, $dS/dt = kS$. Use this equation to find $S(t)$ if $S(0) = S_0$.

*27. Banks use different methods to compound interest. If the interest rate is k, and if interest is compounded m times per year, then $S(t) = S_0\left(1 + \dfrac{k}{m}\right)^{mt}$. When interest is compounded continuously, then $m \to \infty$. Compare $\lim_{m\to\infty} S_0\left(1 + \dfrac{k}{m}\right)^{mt}$ to the formula obtained in Exercise 26.

28. (Dating works of art) We can determine if a work of art is more than 100 years old by determining if the lead-bearing materials contained in the work were manufactured within the last 100 years. The half-life of lead-210 (^{210}Pb) is 22 years while the half-life of radium-226 is 1700 years. Let SF denote the ratio of ^{210}Pb to ^{226}Ra per unit mass of lead. The approximate value of SF for works of art created in the last 80 years is 100. Then, the quantity of lead $\dfrac{1 - \text{Ra}}{\text{Po}}$ at time t is given by

$$\frac{1 - \text{Ra}}{\text{Po}} = \frac{(SF - 1)e^{-\lambda t}}{(SF - 1)e^{-\lambda t} + 1},$$

where λ is the disintegration constant for ^{210}Pb.* On the other hand, for very old paintings $\dfrac{1 - \text{Ra}}{\text{Po}} \cong 0$.

(a) Determine the disintegration constant λ for ^{210}Pb where the amount of ^{210}Pb at time t is $y = y_0 e^{\lambda t}$.

(b) Graph $\dfrac{1 - \text{Ra}}{\text{Po}}$ for $0 \le t \le 250$ using $SF = 100$.

The following table shows the ratio of $\dfrac{1 - \text{Ra}}{\text{Po}}$ for various famous paintings.

Painting	Po^{210} concentration (dpm/g of Pb)	Ra^{226} concentration (dpm/g of Pb)	$\dfrac{1 - \text{Ra}}{\text{Po}}$
Washing of Feet	12.6	0.26	0.98
Woman Reading Music	10.3	0.30	0.97
Woman Playing Mandolin	8.2	0.17	0.98
Woman Drinking	8.3	0.1	0.99
Disciples of Emmaus	8.5	0.8	0.91
Boy Smoking	4.8	0.31	0.94
Lace Maker	1.5	1.4	0.07
Laughing Girl	5.2	6.0	−0.15

* Bernard Keisch, "Dating Works of Art through Their Natural Radioactivity: Improvements and Applications," *Science,* Volume 160, April 26, 1968, pp. 413–415.

The last two paintings, *Lace Maker* and *Laughing Girl* were painted by the Dutch painter Jan Vermeer who lived from 1632 to 1675.
(c) Determine if it is likely that the first six paintings

were also painted by Vermeer (which would make them very valuable!). If not, approximate when they were painted.

3.3 Newton's Law of Cooling

First-order linear differential equations can be used to solve a variety of problems that involve temperature. For example, a medical examiner can find the time of death in a homicide case, a chemist can determine the time required for a plastic mixture to cool to a hardening temperature, and an engineer can design the cooling and heating system of a manufacturing facility. Although distinct, each of these problems depends on a basic principle, Newton's law of cooling, which is used to develop the differential equation associated with each problem.

Newton's law of cooling states that the rate at which the temperature $T(t)$ changes in a cooling body is proportional to the difference between the temperature of the body and the constant temperature T_s of the surrounding medium. This situation is represented as the first-order initial-value problem

$$\frac{dT}{dt} = k(T - T_s), \quad T(0) = T_0$$

where T_0 is the initial temperature of the body and k is the constant of proportionality. The equation

$$\frac{dT}{dt} = k(T - T_s)$$

is separable and separating variables gives us

$$\frac{dT}{T - T_s} = k \, dt,$$

so $\ln|T - T_s| = kt + C$. Using the properties of natural logarithms and simplifying yields

$$T = C_1 e^{kt} + T_s$$

where $C_1 = e^c$. Applying the initial condition implies that $T_0 = C_1 + T_s$, so $C_1 = T_0 - T_s$. Therefore, the solution of the equation is

$$T = (T_0 - T_s)e^{kt} + T_s.$$

Notice that if $k < 0$, $\lim_{t \to \infty} e^{kt} = 0$. Therefore, $\lim_{t \to \infty} T(t) = T_s$, so the temperature of the body approaches that of its surroundings.

 If $k < 0$ and $T > T_s$, use the differential equation $\frac{dT}{dt} = k(T - T_s)$ to determine if $\frac{dT}{dt} > 0$ or $\frac{dT}{dt} < 0$. What does this mean? What if $T < T_s$?

A pie is removed from a 350°F oven and placed to cool in a room with temperature 75°F. In 15 minutes the pie has a temperature of 150°F. Determine the time required to cool the pie to a temperature of 80°F.

Solution   In this example, $T_0 = 350$ and $T_s = 75$. Substituting these values into $T = (T_0 - T_s)e^{kt} + T_s$, we obtain $T(t) = (350 - 75)e^{kt} + 75 = 275e^{kt} + 75$. To solve the problem we must find k or e^k. We know that $T(15) = 150$ so $T(15) = 275e^{15k} + 75 = 150$. Solving this equation for e^k gives us:

$$275e^{15k} = 75$$
$$e^{15k} = \tfrac{3}{11}$$
$$(e^k)^{15} = \tfrac{3}{11}$$
$$e^k = (\tfrac{3}{11})^{1/15}.$$

Thus, $T(t) = 275(e^k)^t + 75 = 275(\tfrac{3}{11})^{t/15} + 75$.

To find the value of t for which $T(t) = 80$, we solve the equation $275(\tfrac{3}{11})^{t/15} + 75 = 80$ for t:

$$275(\tfrac{3}{11})^{t/15} = 5$$
$$(\tfrac{3}{11})^{t/15} = \tfrac{1}{55}$$
$$\ln[(\tfrac{3}{11})^{t/15}] = \ln(\tfrac{1}{55}) = -\ln 55$$

$$\frac{t}{15}\ln(\tfrac{3}{11}) = -\ln 55$$

$$t = \frac{-15 \ln 55}{\ln(\tfrac{3}{11})} = \frac{-15 \ln 55}{\ln 3 - \ln 11} \approx 46.264.$$

Thus, the pie will reach a temperature of 80°F after approximately 46 minutes. An interesting problem associated with this example is to determine if the pie ever reaches room temperature. From the formula $T(t) = 275(\tfrac{3}{11})^{t/15} + 75$, we note that $275(\tfrac{3}{11})^{t/15} > 0$, so $T(t) = 275(\tfrac{3}{11})^{t/15} + 75 > 75$. Therefore, the pie never actually reaches room temperature according to our model. However, we see its temperature approaches 75° as t increases because $\lim_{t \to \infty}[275(\tfrac{3}{11})^{t/15} + 75] = 75$.

In the investigation of a homicide, the time of death is important. Newton's law of cooling can be used to approximate this time. For example, the normal body temperature of most healthy people is 98.6°F. Suppose that when a body is discovered at noon, its temperature is 82°F. Two hours later it is 72°F. If the temperature of the surroundings is 65°F, what was the approximate time of death?

Solution   This problem is solved like the previous example. Let $T(t)$ denote the temperature of the body at time t where $T(0)$ represents the temperature of the body when it is discovered and $T(2)$ represents the temperature of the body two hours after it is discovered. In this case we have that $T_0 = 82$ and $T_s = 65$, and substituting these values into $T = (T_0 - T_s)e^{kt} + T_s$ yields $T(t) = (82 - 65)e^{kt} + 65 =$

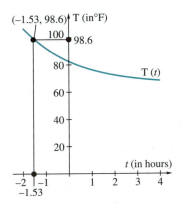

Figure 3.10 Graph of $T(t) = 17(\frac{7}{17})^{t/2} + 65$

$17e^{kt} + 65$. Using $T(2) = 72$, we solve the equation $T(2) = 17e^{2k} + 65 = 72$ for e^k to find $e^k = (\frac{7}{17})^{1/2}$ so $T(t) = 17(e^k)^t + 65 = 17(\frac{7}{17})^{t/2} + 65$. To find the value of t for which $T(t) = 98.6$, we solve the equation $17(\frac{7}{17})^{t/2} + 65 = 98.6$ for t and obtain $t = \dfrac{2\ln(1.97647)}{\ln 7 - \ln 17} = -1.53569$.

This result means that the time of death occurred approximately 1.53 hours before being discovered, as we observe in Figure 3.10. Therefore, the time of death was approximately 10:30 A.M. because the body was discovered at noon.

In each of the previous cases, the temperature of the surroundings was assumed to be constant. However, this does not have to be the case. For example, determining the temperature inside a building over the span of a 24-hour day is more complicated because the outside temperature varies. If we assume that a building has no heating or air conditioning system, the differential equation that needs to be solved to find the temperature $u(t)$ at time t inside the building is

$$\frac{du}{dt} = k(C(t) - u(t))$$

where $C(t)$ is a function that describes the outside temperature and $k > 0$ is a constant that depends on the insulation of the building. According to this equation, if $C(t) > u(t)$, then $\dfrac{du}{dt} > 0$, which implies that u increases. On the other hand, if $C(t) < u(t)$, then $\dfrac{du}{dt} < 0$, which means that u decreases.

Example 3

Suppose that during the month of April in Atlanta, Georgia, the outside temperature in degrees Fahrenheit is given by $C(t) = 70 - 10\cos\dfrac{\pi t}{12}, 0 \le t \le 24$. (This implies that the average value of $C(t)$ is 70°F.) Determine the temperature in a building that has an initial temperature of 60°F if $k = \frac{1}{4}$.

Solution   The initial-value problem that we must solve is

$$\begin{cases} \dfrac{du}{dt} = k[70 - 10\cos\dfrac{\pi t}{12} - u] \\ u(0) = 60 \end{cases}.$$

The differential equation can be solved if we write it as

$$\frac{du}{dt} + ku = k\left[70 - 10\cos\frac{\pi t}{12}\right]$$ and use an integrating factor. This gives us

$$u(t) = \frac{10}{9 + \pi^2}\left(63 + 7\pi^2 - 9\cos\frac{\pi t}{12} - 3\pi\sin\frac{\pi t}{12}\right) + C_1 e^{-t/4}.$$

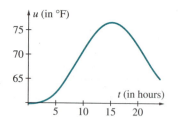

Figure 3.11

We then apply the initial condition $u(0) = 60$ to determine the arbitrary constant C_1 and obtain the solution

$$u(t) = \frac{10}{9 + \pi^2}\left(63 + 7\pi^2 - 9\cos\frac{\pi t}{12} - 3\pi \sin\frac{\pi t}{12}\right) - \frac{10\pi^2}{9 + \pi^2}e^{-t/4}$$

where $k = \frac{1}{4}$, which is graphed in Figure 3.11. From the graph, we see that the temperature reaches its maximum (approximately 72°) near $t = 15.5$ hours, which corresponds to 3:30 P.M.

At what time during the day is the temperature in the building increasing (decreasing) at the fastest rate?

In many situations, a heating or cooling system is installed to control the temperature in a building. Another factor in determining the temperature, which we ignored in prior calculations, is the generation of heat from the occupants of the building, including people and machinery. We investigate the inclusion of these factors next.

IN TOUCH WITH TECHNOLOGY

The temperature $u(t)$ inside a building can be based on three factors: (1) the heat produced by people or machinery inside the building; (2) the heating (or cooling) produced by the furnace (or air conditioning system); (3) the temperature outside the building based on Newton's law of cooling. If the rate at which these factors affect (increase or decrease) the temperature is given by $A(t)$, $B(t)$, and $C(t)$, respectively, the differential equation that models this situation is

$$\frac{du}{dt} = k(C(t) - u(t)) + A(t) + B(t),$$

where the constant $k > 0$ depends on the insulation of the building.

1. Find the temperature (and the maximum temperature) in the building with $k = \frac{1}{4}$ if the initial temperature is 70°, and **(a)** $A(t) = 0.25$, $C(t) = 75$, and $B(t) = 0$; **(b)** $A(t) = 0.25$, $C(t) = 70 - 10\cos\frac{\pi t}{12}$, and $B(t) = 0$; **(c)** $A(t) = 1$, $C(t) = 70 - 10\cos\frac{\pi t}{12}$ and $B(t) = 0$.

If a heating or cooling system is considered, we must model the system with an appropriate function. Of course, the system could run constantly, but we know that most are controlled by a thermostat. Suppose that the desired temperature is u_d. Then $B(t) = k_d(u_d - u(t))$, where k_d is a constant approximately equal to two for most systems.

2. Determine the temperature (and the maximum temperature) in a building with $k = \frac{1}{4}$ and initial temperature 70° if **(a)** $A(t) = 0.25$, $B(t) = 1.75(68 - u(t))$, and $C(t) = 70 - 10\cos\frac{\pi t}{12}$; **(b)** $A(t) = 1$, $B(t) = 1.75(68 - u(t))$, and $C(t) = 70 - 10\cos\frac{\pi t}{12}$; **(c)** $A(t) = 0.25$, $B(t) = 1.75(68 - u(t))$, and $C(t) = 80 - 10\cos\frac{\pi t}{12}$.

3. If $A(t) = 0.25$, $B(t) = 1.75(u_d - u(t))$, and $C(t) = 70 - 10\cos\frac{\pi t}{12}$; determine the value of u_d needed so that the average temperature in the building over a 24-hour period is 70°.

Solve the following problems related to Newton's law of cooling.

1. A hot cup of tea is initially 100°C when poured. How long does it take for the tea to reach a temperature of 50°C if it is 80°C after 15 minutes and the room temperature is 30°C?

2. Suppose that the tea in Exercise 1 is allowed to cool at room temperature for 20 minutes. It is then placed in a cooler with temperature 15°C. What is the temperature of the tea after 60 minutes if it is 60°C after 10 minutes in the cooler?

*3. A can of orange soda is removed from a refrigerator having temperature 40°F. If the can is 50°F after five minutes, how long does it take for the can to reach a temperature of 60°F if the surrounding temperature is 75°F?

4. Suppose that a container of tea is placed in a refrigerator at 35°F to cool. If the tea is initially 75°F and it has a temperature of 70°F after one hour, then when does the tea reach 55°F?

5. Determine the time of death if a corpse is 79°F when discovered at 3:00 P.M. and 68°F three hours later. Assume that the temperature of the surroundings is 60°F. (Normal body temperature is 98.6°F.)

6. At the request of his children, a father makes homemade popsicles. At 2:00 P.M., one of the children asks if the popsicles are frozen (0°C), at which time the father tests the temperature of a popsicle and finds it to be 5°C. If the father placed the popsicles with a temperature of 15°C in the freezer at 12:00 P.M. and the temperature of the freezer is −2°C, when will the popsicles be frozen?

*7. A thermometer that reads 90°F is placed in a room with temperature 70°F. After three minutes, the thermometer reads 80°F. What does the thermometer read after five minutes?

8. A thermometer is placed outdoors at 80°F. After two minutes, the thermometer reads 68°F, and after five minutes, it reads 72°F. What was the initial temperature reading of the thermometer?

9. A casserole is placed in a microwave oven to defrost. It is then placed in a conventional oven at 300°F and bakes for 30 minutes at which time its temperature is 150°F. If after baking an additional 30 minutes its temperature is 200°F, what was the temperature of the casserole when it was removed from the microwave?

10. A bottle of wine at room temperature (70°F) is placed in ice to chill at 32°F. After 20 minutes, the temperature of the wine is 58°F. When will its temperature be 50°F?

*11. When a cup of coffee is poured its temperature is 200°F. Two minutes later, its temperature is 170°F. If the temperature of the room is 68°F, when is the temperature of the coffee 140°F?

12. After dinner, a couple orders two cups of coffee. Upon being served, the gentleman immediately pours one container of cream into his cup of coffee. His companion waits four minutes before adding the same amount of cream to her cup. Which person has the hotter cup of coffee when they both take a sip of coffee after she adds the cream to her cup? (Assume that the cream's temperature is less than that of the coffee.) Explain.

13. Suppose that during the month of February in Washington, D.C., the outside temperature in °F is given by

$$C(t) = 40 - 5\cos\frac{\pi t}{12}, \quad 0 \le t \le 24.$$ Determine the temperature in a building that has an initial temperature of 50° if $k = \frac{1}{4}$. (Assume that the building has no heating or air conditioning system.)

14. Suppose that during the month of August in Savannah, GA, the outside temperature in °F is given by

$$C(t) = 85 - 10\cos\frac{\pi t}{12}, \quad 0 \le t \le 24.$$ Determine the temperature in a building that has an initial temperature of 60°F if $k = \frac{1}{4}$. (Assume that the building has no heating or air conditioning system.)

*15. Suppose that during the month of October in Los Angeles, CA, the outside temperature in °F is given by $C(t) = 70 - 5\cos\frac{\pi t}{12}, 0 \le t \le 24.$ Find the temperature in a building that has an initial temperature of 65°F if $k = \frac{1}{4}$. (Assume that the building has no heat or air conditioning system.)

16. Suppose that during the month of January in Cincinnati, OH, the outside temperature in °F is given by

$C(t) = 20 - 5 \cos \dfrac{\pi t}{12}, 0 \le t \le 24$. Find the temperature in a building that has an initial temperature of
40°F if $k = \frac{1}{4}$. (Assume that the building has no heat
or air conditioning system.)

17. **(Mixture Problem)** Suppose that a tank contains V_0
gallons of a brine solution, a mixture of dissolved salt
and water. A brine solution of concentration S_1
pounds per gallon is allowed to flow into the tank at a
rate R_1 gallons per minute while at the same time a
well-stirred mixture flows out of the tank at the rate of
R_2 gallons per minute. If $y(t)$ represents the amount of
salt in the tank at time t and $y(0) = y_0$, show that $y(t)$
is found by solving the initial-value problem

$$\frac{dy}{dt} = \left(S_1 \frac{\text{lb}}{\text{gal}}\right)\left(R_1 \frac{\text{gal}}{\text{min}}\right) - \left(\frac{y(t)}{V(t)} \frac{\text{lb}}{\text{gal}}\right)\left(R_2 \frac{\text{gal}}{\text{min}}\right),$$
$$y(0) = y_0$$

where $V(t)$, the volume of solution in the tank at any
time t, is found by solving the initial-value problem

$$\frac{dV}{dt} = R_1 - R_2, \; V(0) = V_0.$$

What can be said of $V(t)$ if $R_1 = R_2$? What if $R_1 > R_2$? What if $R_1 < R_2$?

18. **(See Exercise 17)** A tank contains 100 gallons of a
brine solution in which 20 lb of salt is initially dissolved. **(a)** Water (containing no salt) is then allowed
to flow into the tank at a rate of 4 gal/min and the
well-stirred mixture flows out of the tank at an equal
rate of 4 gal/min. Determine the amount of salt $y(t)$ at
any time t. What is the eventual concentration of the
brine solution in the tank? **(b)** If instead of water a
brine solution with concentration 2 lb/gal flows into
the tank at a rate of 4 gal/min, what is the eventual
concentration of the brine solution in the tank?

*19. A tank contains 200 gallons of a brine solution in
which 10 lb of salt is initially dissolved. A brine solution with concentration 2 lb/gal is then allowed to
flow into the tank at a rate of 4 gal/min and the well-
stirred mixture flows out of the tank at a rate of 3 gal/
min. Determine the amount of salt $y(t)$ at any time t. If
the tank can hold a maximum of 400 gallons, what is
the concentration of the brine solution in the tank
when the volume reaches this maximum?

20. A tank contains 300 gallons of a brine solution in
which 30 lb of salt is initially dissolved. A brine solution with concentration 4 lb/gal is then allowed to
flow into the tank at a rate of 3 gal/min and the well-
stirred mixture flows out of the tank at a rate of 4 gal/
min. Determine the amount of salt $y(t)$ at any time t.
What is the concentration of the brine solution after
10 minutes? What is the eventual concentration of the
brine solution in the tank? For what values of t is the
solution defined? Why?

3.4 Free-Falling Bodies

The motion of some objects can be determined through the solution of a first-order
equation. We begin by explaining some of the theory that is needed to set up the
differential equation that models the situation.

Newton's Second Law of Motion

The rate at which the momentum of a body changes with respect to time is equal to
the resultant force acting on the body.

The body's momentum is defined as the product of its mass and velocity, so this
statement is modeled as

$$\frac{d}{dt}(mv) = F,$$

where m and v represent the body's mass and velocity, respectively, and F is the sum of the forces acting on the body. The mass m of the body is constant, so differentiation leads to the differential equation

$$m\frac{dv}{dt} = F.$$

If the body is subjected to the force due to gravity, then its velocity is determined by solving the differential equation

$$m\frac{dv}{dt} = mg \qquad \text{or} \qquad \frac{dv}{dt} = g$$

where $g \approx 32$ ft/s^2 (English system) or $g \approx 9.8$ m/s^2 (international system). See the summary of units in Table 3.4.

TABLE 3.4 Units that are useful in solving problems associated with Newton's second law of motion.

	English	International
Mass	slug $\left(\dfrac{\text{lb} - \text{s}^2}{\text{ft}}\right)$	kilogram (kg)
Force	pound (lb)	Newton $\left(\dfrac{\text{m} - \text{kg}}{\text{s}^2}\right)$
Distance	foot (ft)	meter (m)
Time	second (s)	second (s)

This differential equation is applicable only when the resistive force due to the medium (such as air resistance) is ignored. If this offsetting resistance is considered, we must discuss all of the forces acting on the object. Mathematically, we write the equation as

$$m\frac{dv}{dt} = \sum (\text{forces acting on the object})$$

where the direction of motion is taken to be the positive direction.

We use a force diagram in Figure 3.12 to set up the differential equation that models the situation.

Air resistance acts against the object as it falls and g acts in the same direction of the motion. We state the differential equation in the form:

$$m\frac{dv}{dt} = mg + (-F_R) \qquad \text{or} \qquad m\frac{dv}{dt} = mg - F_R,$$

where F_R represents this resistive force. Note that down is assumed to be the positive direction. The resistive force is typically proportional to the body's velocity (v) or a

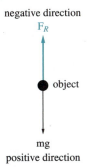

negative direction

F_R

object

mg

positive direction

Figure 3.12 Force Diagram

power of its velocity. Hence, the differential equation is linear or nonlinear based on the resistance of the medium taken into account.

Example 1

(a) Determine the velocity and the distance traveled by an object with mass $m = 1$ slug that is thrown downward with an initial velocity of 2 ft/s from a height of 1000 feet. Assume that the object is subjected to air resistance that is equivalent to the instantaneous velocity of the object. (b) Determine the time at which the object strikes the ground and its velocity when it strikes the ground.

Solution   (a) First, we set up the initial-value problem to determine the velocity of the object. The air resistance is equivalent to the instantaneous velocity, so $F_R = v$. The formula $m\dfrac{dv}{dt} = mg - F_R$ then gives us

$$\frac{dv}{dt} = 32 - v,$$

and imposing the initial velocity $v(0) = 2$ yields the initial-value problem

$$\begin{cases} \dfrac{dv}{dt} = 32 - v \\ v(0) = 2 \end{cases},$$

which can be solved through several methods. We choose to solve it as a linear first-order equation and use an integrating factor. (It also can be solved by separating variables.) With the integrating factor e^t, we have $\dfrac{d}{dt}(e^t v) = 32e^t$. Integrating both sides gives us $e^t v = 32e^t + C$ so

$$v = 32 + Ce^{-t}$$

and applying the initial velocity gives us $v(0) = 32 + Ce^0 = 32 + C = 2$ so $C = -30$. Therefore, the velocity of the object is

$$v = 32 - 30e^{-t}.$$

Notice that the velocity of the object cannot exceed 32 ft/sec, called the **limiting velocity,** which is found by evaluating $\lim_{t \to \infty} v(t)$.

To determine the distance traveled at time t, $s(t)$, we solve the first-order equation

$$\frac{ds}{dt} = v = 32 - 30e^{-t}$$

with initial condition $s(0) = 0$. This differential equation is solved by integrating both sides of the equation to obtain $s = 32t + 30e^{-t} + C_2$. Application of $s(0) = 0$

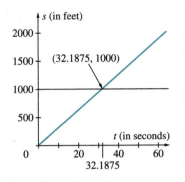

Figure 3.13 Graph of
$s = 32t + 30e^{-t} - 30$

then gives us $s(0) = 32(0) + 30e^0 + C_2 = 30 + C_2 = 0$, so $C_2 = -30$, and the distance traveled by the object is given by

$$s = 32t + 30e^{-t} - 30.$$

(b) The object strikes the ground when $s(t) = 1000$. Therefore, we must solve $s = 32t + 30e^{-t} - 30 = 1000$ for t. The roots of this equation can be approximated with numerical methods like Newton's method. From the graph of this function, shown in Figure 3.13, we see that $s(t) = 1000$ near $t = 35$. Numerical methods show that $s = 32t + 30e^{-1} - 30 = 1000$ when $t \approx 32.1875$.

The velocity at the point of impact is found to be 32.0 ft/sec by evaluating the derivative at the time at which the object strikes the ground, given by $s'(32.1875)$.

Example 2

Suppose that the object in Example 1 of mass 1 slug is thrown downward with an initial velocity of 2 ft/s and that the object is attached to a parachute, increasing this resistance so that it is given by v^2. Find the velocity at any time t and determine the limiting velocity of the object.

Solution   This situation is modeled by the initial-value problem

$$\begin{cases} \dfrac{dv}{dt} = 32 - v^2 \\[2mm] v(0) = 2 \end{cases},$$

We solve the differential equation by separating the variables and using partial fractions:

$$\frac{dv}{32 - v^2} = dt$$

$$\frac{dv}{(4\sqrt{2} + v)(4\sqrt{2} - v)} = dt$$

$$\frac{1}{8\sqrt{2}}\left[\frac{1}{v + 4\sqrt{2}} - \frac{1}{v - 4\sqrt{2}}\right] dv = dt$$

$$\ln|v + 4\sqrt{2}| - \ln|v - 4\sqrt{2}| = 8\sqrt{2}t + C$$

$$\ln\left|\frac{v + 4\sqrt{2}}{v - 4\sqrt{2}}\right| = 8\sqrt{2}t + C$$

$$\left|\frac{v + 4\sqrt{2}}{v - 4\sqrt{2}}\right| = \tilde{C}e^{8\sqrt{2}t} \qquad (\tilde{C} = e^C)$$

$$\frac{v + 4\sqrt{2}}{v - 4\sqrt{2}} = Ke^{8\sqrt{2}t} \qquad (K = \pm\tilde{C}).$$

Solving for v, we find that $v + 4\sqrt{2} = Ke^{8\sqrt{2}t}(v - 4\sqrt{2})$ or $(1 - Ke^{8\sqrt{2}t})v = -4\sqrt{2}(Ke^{8\sqrt{2}t} + 1)$, so

$$v = \frac{-4\sqrt{2}(Ke^{8\sqrt{2}t} + 1)}{1 - Ke^{8\sqrt{2}t}}.$$

Application of the initial condition yields $K = \dfrac{1 + 2\sqrt{2}}{1 - 2\sqrt{2}}$. The limiting velocity of the object is found with L'Hopital's rule to be $\lim_{t\to\infty} v(t) = \dfrac{-4\sqrt{2}K}{-K} = 4\sqrt{2}$ ft/sec.

In Example 1, the limiting velocity is 32 ft/sec, so the parachute causes the velocity of the object to be reduced. (See Figure 3.14.) This shows that the object does not have to endure as great an impact as it would without the help of the parachute.

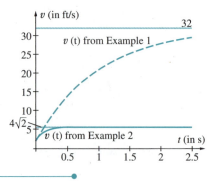

Figure 3.14   The velocity functions from Example 1 and Example 2. Notice how the different forces due to air resistance affect the velocity of the object.

Example 3

Determine a solution (for the velocity and the height) of the differential equation that models the motion of an object of mass m when directed upward with an initial velocity of v_0 from an initial position s_0, assuming that the air resistance equals cv where c is constant.

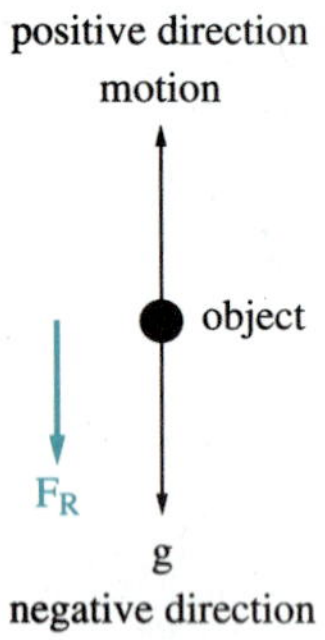

positive direction
motion

object

F_R

g

negative direction

Figure 3.15   By drawing a force diagram, we see that g and F_R are in the negative direction.

Solution   The motion of the object is upward so g and F_R act against the upward motion of the object as shown in Figure 3.15.

Therefore, the differential equation that must be solved in this case is the linear equation $\dfrac{dv}{dt} = -g - \dfrac{c}{m}v$. We solve the initial-value problem

$$\begin{cases} \dfrac{dv}{dt} = -g - \dfrac{c}{m}v \\ v(0) = v_0 \end{cases},$$

by first rewriting the equation $\dfrac{dv}{dt} = -g - \dfrac{c}{m}v$ as $\dfrac{dv}{dt} + \dfrac{c}{m}v = -g$ and then calculating the integrating factor $e^{\int \frac{c\,dt}{m}} = e^{ct/m}$. Multiplying each side of the equa-

tion by $e^{ct/m}$ gives us $e^{ct/m}\dfrac{dv}{dt} + \dfrac{c}{m}ve^{ct/m} = -ge^{ct/m}$ so that $\dfrac{d}{dt}(e^{ct/m}v) = -ge^{ct/m}$.

Integrating we obtain $e^{ct/m}v = -\dfrac{gm}{c}e^{ct/m} + C$ and, consequently, the general

solution of $\dfrac{dv}{dt} = -g - \dfrac{c}{m}v$ is $v(t) = -\dfrac{gm}{c} + Ce^{-ct/m}$.

Applying the initial condition $v(0) = v_0$ and solving for C yields

$C = \dfrac{cv_0 + gm}{c}$ so that the solution to the initial-value problem $\begin{cases} \dfrac{dv}{dt} = -g - \dfrac{c}{m}v \\ v(0) = v_0 \end{cases}$ is

$$v(t) = -\frac{gm}{c} + \frac{cv_0 + gm}{c}e^{-ct/m}.$$

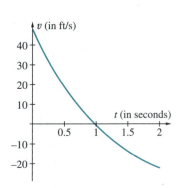

Figure 3.16

For example, the velocity function for the case with $m = \frac{1}{128}$ slugs, $c = \frac{1}{160}$, $g = 32$ ft/s^2 and $v_0 = 48$ ft/s is $v(t) = 88e^{-4t/5} - 40$. This function is graphed in Figure 3.16. Notice where $v(t) = 0$. This value of t represents the time at which the object reaches its maximum height and begins to fall towards the ground.

Similarly, this function can be employed to investigate numerous situations without solving the differential equation each time.

The height function $s(t)$, which represents the distance above the ground at time t, is determined by integrating the velocity function:

$$s(t) = \int v(t)\,dt = \int \left(-\frac{gm}{c} + \frac{cv_0 + gm}{c}e^{-ct/m}\right)dt$$
$$= -\frac{gm}{c}t - \frac{cmv_0 + gm^2}{c^2}e^{-ct/m} + C.$$

If the initial height is given by $s(0) = s_0$, solving for C results in $C = \dfrac{gm^2 + c^2s_0 + cmv_0}{c^2}$, so that

$$s(t) = -\frac{gm}{c}t - \frac{cmv_0 + gm^2}{c^2}e^{-ct/m} + \frac{gm^2 + c^2s_0 + cmv_0}{c^2}.$$

The height and velocity functions are shown in Figure 3.17 using the parameters $m = \dfrac{1}{128}$ slugs, $c = \dfrac{1}{160}$, $g = 32$ ft/s^2, and $v_0 = 48$ ft/s as well as $s_0 = 0$.

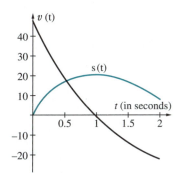

Figure 3.17 Graph of $v(t) =$ $88e^{-4t/5} - 40$ (in blue) and $s(t) = 110 - 40t - 110e^{-4t/5}$ (in black)

The time at which the object reaches its maximum height occurs when the derivative of the position is equal to zero. From Figure 3.17 we see that $s'(t) = v(t) = 0$ when $t \approx 1$. Solving $s'(t) = 0$ for t yields the better approximation $t \approx 0.985572$.

Weight and Mass: Notice that in the English system, *pounds* describe *force*. Therefore, when the weight W of an object is given, we must calculate its *mass* with the relationship $W = mg$ or $m = \dfrac{W}{g}$. On the other hand, in the International system, the mass of the object (in kilograms) is typically given.

    We now combine several of the topics discussed in this section to solve the following problem.

Example 4

A 32 lb object is dropped from a height of 50 feet above the surface of a small pond. While the object is in the air, the force due to air resistance is v. However, when the object is in the pond it is subjected to a buoyancy force equivalent to $6v$. Determine how much time is required for the object to reach a depth of 25 feet in the pond.

> **Solution**   The mass of this object is found using the relationship $W = mg$, where W is the weight of the object. With this, we find that $32 \text{ lb} = m(32 \text{ ft/s}^2)$, so $m = 1\dfrac{\text{lb} - \text{s}^2}{\text{ft}}$ (slug).

    This problem must be broken into two parts: an initial-value problem for the object above the pond, and an initial-value problem for the object below the surface of the pond. Using techniques discussed in previous examples, the initial-value problem above the pond's surface is found to be

$$\begin{cases} \dfrac{dv}{dt} = 32 - v \\[2mm] v(0) = 0 \end{cases}.$$

However, to determine the initial-value problem that yields the velocity of the object beneath the pond's surface, the velocity of the object when it reaches the surface must be known. Hence, the velocity of the object above the surface must be determined first.

    The equation $\dfrac{dv}{dt} = 32 - v$ is separable and rewriting it yields $\dfrac{dv}{32 - v} = dt$. Integrating and applying the initial condition results in $v(t) = 32 - 32e^{-t}$. In order to find the velocity when the object hits the pond's surface we must know the time at which the object has fallen 50 ft. Thus, we find the distance traveled by the object by solving $\dfrac{ds}{dt} = v(t)$, $s(0) = 0$, obtaining $s(t) = 32e^{-t} + 32t - 32$. From the graph of $s(t)$ shown in Figure 3.18, we see that the value of t at which the object has traveled 50 feet appears to be approximately 2.5 seconds.

    A more accurate value of the time at which the object hits the surface is $t \approx 2.47864$. The velocity at this time is then determined by substitution into the velocity function resulting in $v(2.47864) \approx 29.3166$. Note that this value is the initial velocity of the object when it hits the surface of the pond.

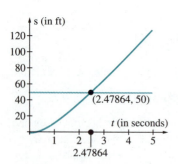

Figure 3.18   Graph of $s(t) = 32e^{-t} + 32t - 32$

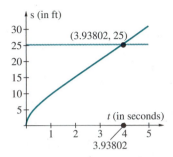

Figure 3.19 Graph of $s(t) = 3.99722 - 3.99722e^{-6t} + \frac{16}{3}t$

Thus, the initial-value problem that determines the velocity of the object beneath the surface of the pond is given by $\dfrac{dv}{dt} = 32 - 6v$, $v(0) = 29.3166$. The solution of this initial-value problem is $v(t) = \dfrac{16}{3} + 23.9833e^{-t}$, and solving

$$\frac{ds}{dt} = \frac{16}{3} + 23.9833e^{-t}, \quad s(0) = 0 \text{ we obtain}$$

$$s(t) = 3.99722 - 3.99722e^{-6t} + \frac{16}{3}t$$

which gives the depth of the object at time t. From the graph of this function, shown in Figure 3.19, we see that the object is 25 feet beneath the surface of the pond after approximately 4 seconds.

A more accurate approximation of the time at which the object is 25 feet beneath the pond's surface is $t \approx 3.93802$.

Finally, the time required for the object to reach the pond's surface is added to the time needed for it to travel 25 feet beneath the surface to see that approximately 6.41667 seconds are required for the object to travel from a height of 50 feet above the pond to a depth of 25 feet below the surface.

IN TOUCH WITH TECHNOLOGY

Solve the following falling-bodies problems.

1. Suppose that a falling body is subjected to air resistance assumed to be $F_R = cv$. Use the values of $c = 0.5$, 1, and 2; and plot the velocity function with $m = 1$, $g = 32$, and $v_0 = 0$. How does the value of c affect the velocity?

2. Compare the effects that air resistance has on the velocity of a falling object of mass $m = 0.5$ that is released with an initial velocity of $v_0 = 16$. Consider $F_R = 16v^2$ and $F_R = 16v$.

3. Compare the effects that air resistance has on the velocity of a falling object of mass $m = 0.5$ that is released with an initial velocity of $v_0 = 0$. Consider $F_R = 16v^3$ and $F_R = 16\sqrt{v}$.

4. Consider the velocity and height functions found in Example 3 with $m = \frac{1}{128}$, $c = \frac{1}{160}$, and $g = 32$ ft/s².

   (a) Suppose that on the first toss, the object is thrown with $v_0 = 48$ ft/s from an initial height of $s_0 = 0$ ft and on the second toss with $v_0 = 36$ ft/s and $s_0 = 6$. On which toss does the object reach the greater maximum height?

   (b) If $s_0 = 0$ ft, compare the effect that the initial velocities $v_0 = 48$ ft/s, $v_0 = 64$ ft/s, and $v_0 = 80$ ft/s have on the height function.

   (c) If $v_0 = 48$ ft/s, compare the effect that the initial heights $s_0 = 0$ ft, $s_0 = 10$ ft, and $s_0 = 20$ ft have on the height function.

5. A woman weighing 125 lb falls from an airplane at an altitude of 4000 ft and opens her parachute after 5 sec. If the force due to air resistance is $F_R = v$ before she opens her parachute and $F_R = 10v$ afterwards, how long does it take for the woman to reach the ground?

6. Consider the problem discussed in Example 4. Instead of a buoyancy force equivalent to $6v$, suppose that when the object is in the pond, it is subjected to a buoyancy force equivalent to $6v^2$. Determine how much time is required for the object to reach a depth of 25 feet in the pond.

1. A rock that weighs 32 lb is dropped from rest from the edge of a cliff. **(a)** Find the velocity of the rock at time t if the air resistance is equivalent to the instantaneous velocity v. **(b)** What is the velocity of the rock at $t = 2$ seconds?

2. An object that weighs 4 lb is dropped from the top of a tall building. **(a)** Find the velocity of the object at time t if the air resistance is equivalent to the instantaneous velocity v. **(b)** What is the velocity of the object at $t = 2$ seconds? How does this compare to the result in Exercise 1?

*3. An object weighing 1 lb is thrown downward with an initial velocity of 8 ft/s. **(a)** Find the velocity of the object at time t if the air resistance is equivalent to twice the instantaneous velocity. **(b)** What is the velocity of the object at $t = 1$ sec?

4. An object weighing 16 lb is dropped from a tall building. **(a)** Find the velocity of the object at time t if the air resistance is equivalent to twice the instantaneous velocity. **(b)** What is the velocity of the object at $t = 1$ sec? How does this compare to the result in Exercise 3?

5. A ball of weight 4 oz is tossed into the air with an initial velocity of 64 ft/s. **(a)** Find the velocity of the object at time t if the air resistance is equivalent to $\frac{1}{16}$ the instantaneous velocity. **(b)** When does the ball reach its maximum height?

6. A tennis ball weighing 8 oz is hit vertically into the air with an initial velocity of 128 ft/s. **(a)** Find the velocity of the object at time t if the air resistance is equivalent to half of the instantaneous velocity. **(b)** When does the ball reach its maximum height?

*7. A rock of weight 0.5 lb is dropped (with zero initial velocity) from a height of 300 feet. If the air resistance is equivalent to $\frac{1}{64}$ times the instantaneous velocity, find the velocity and distance traveled by the object at time t. Does the rock hit the ground before 4 seconds elapse?

8. An object of weight 0.5 lb is thrown downward with an initial velocity of 16 ft/s from a height of 300 feet. If the air resistance is equivalent to $\frac{1}{64}$ times the instantaneous velocity, find the velocity of and distance traveled by the object at time t. Compare these results to those in Exercise 7.

9. An object of mass 10 kg is dropped from a great height. **(a)** If the object is subjected to air resistance equivalent to 10 times the instantaneous velocity, find the velocity. **(b)** What is the limiting velocity of the object?

10. Suppose that an object of mass 1 kg is thrown with a downward initial velocity of 5 m/s and is subjected to an air resistance equivalent to the instantaneous velocity. **(a)** Find the velocity of the object and the distance fallen at time t. **(b)** How far does the object drop after 5 seconds?

*11. A projectile of mass 100 kg is launched vertically from ground level with an initial velocity of 100 m/s. **(a)** If the projectile is subjected to air resistance equivalent to $\frac{1}{10}$ times the instantaneous velocity, determine the velocity of and the height of the projectile at any time t. **(b)** What is the maximum height attained by the projectile?

12. In a carnival game, a participant uses a mallet to project an object up a 20-meter pole. If the mass of the object is 1 kg and the object is subjected to resistance equivalent to $\frac{1}{10}$ times the instantaneous velocity, determine if an initial velocity of 20 m/s causes the object to reach the top of the pole.

13. Assuming that air resistance is ignored, find the velocity and height functions if an object with mass m is thrown vertically up into the air with an initial velocity v_0 from an initial height s_0.

14. Use the results of Exercise 13 to find the velocity and height functions if $m = \frac{1}{128}$, $g = 32$ ft/s², $v_0 = 48$ ft/s, and $s_0 = 0$ ft. What is the shape of the height function? What is the maximum height attained by the object? When does the object reach its maximum height? When does the object hit the ground? How do these results compare to those of Example 3?

*15. Consider the situation described in Exercise 13. What is the velocity of the object when it hits the ground, assuming that the object is thrown from ground level?

16. Consider the situation described in Exercise 13. If the object reaches its maximum height after T seconds, when does the object hit the ground assuming that the object is thrown from ground level?

17. Suppose that an object of mass 10 kg is thrown vertically into the air with an initial velocity v_0 m/s. If the limiting velocity is -19.6 m/s, what can be said about c in the force due to air resistance $F_R = cv$ acting on the object?

18. If the limiting velocity of an object of mass m which is thrown vertically into the air with an initial velocity v_0 m/s is -9.8 m/s, what can be said about the relationship between m and c in the force due to air resistance $F_R = cv$ acting on the object?

*19. A parachutist weighing 192 lb falls from a plane (that is, $v_0 = 0$). When the parachutist's speed is 60 ft/s, the parachute is opened and the parachutist is then subjected to air resistance equivalent to $F_R = 3v^2$. Find the velocity $v(t)$ of the parachutist. What is the limiting velocity of the parachutist?

20. A parachutist weighing 60 kg falls from a plane (that is, $v_0 = 0$) and is subjected to air resistance equivalent to $F_R = 10v$. After one minute, the parachute is opened so that the parachutist is subjected to air resistance equivalent to $F_R = 100v$. **(a)** What is the parachutist's velocity when the parachute is opened? **(b)** What is the parachutist's velocity $v(t)$ after the parachute is opened? **(c)** What is the parachutist's limiting velocity? How does this compare to the limiting velocity if the parachute does not open?

21. **(Escape Velocity)** Suppose that a rocket is launched from the earth's surface. At a great (radial) distance r from the center of the earth, the rocket's acceleration is not the constant g. Instead, according to Newton's law of gravitation, $a = \dfrac{k}{r^2}$, where k is the constant of proportionality ($k > 0$ if the rocket is falling towards the earth; $k < 0$ if the rocket is moving away from the earth). **(a)** If $a = -g$ at the earth's surface (when $r = R$), find k and show that the rocket's velocity is found by solving the initial-value problem $\dfrac{dv}{dt} = -\dfrac{gR^2}{r^2}$, $v(0) = v_0$. **(b)** Show that $\dfrac{dv}{dt} = v\dfrac{dv}{dr}$ so that the solution to the initial-value problem $v\dfrac{dv}{dr} = -\dfrac{gR^2}{r^2}$, $v(R) = v_0$ is $v^2 = \dfrac{2gR^2}{r} + v_0^2 - 2gR$. **(c)** Compute $\lim_{r\to\infty} v^2$. If $v > 0$ (so that the rocket does not fall to the ground), show that the minimum value of v_0 for which this is true (even for very large values of r) is $v_0 = \sqrt{2gR}$. This value is called the **escape velocity**

and signifies the minimum velocity required so that the rocket does not return to the earth.

22. If $R \approx 3960$ miles and $g \approx 32$ ft/s$^2 \approx 0.006$ mi/s^2, use the results of Exercise 21 to compute the escape velocity of the earth.

*23. **(See Exercise 21)** Determine the minimum initial velocity needed to launch the lunar module (used on early space missions) from the surface of the earth's moon given that the moon's radius is $R \approx 1080$ miles and the acceleration of gravity of the moon is 16.5% of that of the earth.

24. **(See Exercise 21)** Compare the earth's escape velocity to those of Venus and Mars if for Venus $R \approx 3800$ miles and acceleration of gravity is 85% of that of the earth; and for Mars $R \approx 2100$ and acceleration of gravity is 38% of that of the earth. Which planet has the largest escape velocity? Which has the smallest?

25. In an electric circuit with one loop that contains a resistor R, a capacitor C, and a voltage source $E(t)$, the charge Q on the capacitor is found by solving the initial-value problem $R\dfrac{dQ}{dt} + \dfrac{1}{C}Q = E(t)$, $Q(0) = Q_0$. Solve this problem to find $Q(t)$ if $E(t) = E_0$ where E_0 is constant. Find $I(t) = Q'(t)$ where $I(t)$ is the current at any time t.

26. If in the R–C circuit described in Exercise 25 $C = 10^{-6}$ farads, $R = 4000$ ohms, and $E(t) = 200$ volts, find $Q(t)$ and $I(t)$ if $Q(0) = 0$. What eventually happens to the charge and the current as $t \to \infty$?

*27. An object that weighs 48 lb is released from rest at the top of a plane metal slide that is inclined 30° to the horizontal. Air resistance (pounds) is numerically equal to one-half the velocity (ft/sec), and the coefficient of friction is $\mu = \frac{1}{4}$. Using Newton's second law of motion by summing the forces along the surface of the slide, we find the following forces:

(a) the component of the weight parallel to the slide: $F_1 = 48 \sin 30° = 24$;

(b) the component of the weight perpendicular to the slide: $N = 48 \cos 30° = 24\sqrt{3}$;

(c) the frictional force (against the motion of the object): $F_2 = -\mu N = -\frac{1}{4}(24\sqrt{3}) = -6\sqrt{3}$; and

(d) the force due to air resistance (against the motion of the object): $F_3 = -\frac{1}{2}v$.

Because the mass of the object is $m = \frac{48}{32} = \frac{3}{2}$, we find that the velocity of the object satisfies the initial-value problem $m\dfrac{dv}{dt} = F_1 + F_2 + F_3$ or

$$\frac{3}{2}\frac{dv}{dt} = 24 - 6\sqrt{3} - \frac{1}{2}v, \; v(0) = 0.$$

Solve this problem for $v(t)$. Determine the distance traveled by the object at time t, $x(t)$, if $x(0) = 0$.

28. A boat weighing 150 lb with a single rider weighing 170 lb is being towed in a particular direction at a rate of 20 mph. At $t = 0$, the tow rope is cut and the rider begins to row in the same direction, exerting a constant force of 12 lb in the direction that the boat is moving. The resistance is equivalent to twice the instantaneous velocity (ft/sec). The forces acting on the boat are $F_1 = 12$ in the direction of motion and the force due to resistance in the opposite direction, $F_2 = -2v$. Because the total weight (boat and rider) is 320 lb, $m = \dfrac{320}{32} = 10$. Therefore, the velocity satisfies the differential equation $m\dfrac{dv}{dt} = F_1 + F_2$ or

$10\dfrac{dv}{dt} = 12 - 2v$ with initial velocity 20 mph, which

is equivalent to $v(0) = 20\dfrac{\text{mi}}{\text{hr}} \times \dfrac{5280\ \text{ft}}{1\ \text{mi}} \times$

$\dfrac{1\ \text{hr}}{3600\ \text{sec}} = \dfrac{88}{3}$ ft/sec. Find $v(t)$ and the distance traveled by the boat, $x(t)$, if $x(0) = 0$.

CHAPTER 3 SUMMARY
Concepts and Formulas

Section 3.1

Orthogonal Curves

Two lines L_1 and L_2, with slopes m_1 and m_2, respectively, are **orthogonal** (or perpendicular) if their slopes satisfy the relationship $m_1 = -1/m_2$. Two curves C_1 and C_2 are **orthogonal** (or perpendicular) at a point if their respective tangent lines to the curves at that point are perpendicular.

Section 3.2

Malthus model

The initial-value problem $\dfrac{dy}{dt} = ky$, $y(0) = y_0$ has solution

$$y = y_0 e^{kt}.$$

Logistic equation (or Verhulst equation)

The initial-value problem $y'(t) = (r - ay(t))y(t)$, $y(0) = y_0$ has solution

$$y = \frac{ry_0}{ay_0 + (r - ay_0)e^{-rt}}.$$

Section 3.3

Newton's law of cooling

The initial-value problem $\dfrac{dT}{dt} = k(T - T_s)$, $T(0) = T_0$ has solution

$$T = (T_0 - T_s)e^{kt} + T_s.$$

Section 3.4

Newton's second law of motion

The rate at which the momentum of a body changes with respect to time is equal to the resultant force acting on the body: $\dfrac{d}{dt}(mv) = F$.

Falling Body Problem

The velocity of the falling body is found by solving the differential equation determined with

$$m\frac{dv}{dt} = \sum (\text{forces acting on the object}).$$

CHAPTER 3 REVIEW EXERCISES

In Exercises 1–4, find the family of orthogonal trajectories to the given family of curves.

1. $4x - y = c$

2. $y - x^2 = c$

*3. $y = x - 1 + ce^{-x}$

4. $y(1 - cx) = 1 + cx$

5. The initial population in a bacteria culture is y_0. Suppose that after four days the population is $3y_0$. When is the population $5y_0$?

6. Suppose that a culture contains 200 cells. After one day the culture contains 600 cells. How many cells does the culture contain after two days?

*7. What percentage of the original amount of the element ^{226}Ra remains after 50 years?

8. If an artifact contains 10% of the amount of carbon-14 as that of a present day sample, how old is the artifact?

9. Suppose that in an isolated population of 1000 people, 250 initially have a virus. If after one day 500 have the virus, how many days are required for three-fourths of the population to acquire the virus?

10. The **Gompertz equation** given by

$\dfrac{dy}{dt} = y(r - a \ln y)$ is used by actuaries to predict certain populations. If $y(0) = y_0$, then find $y(t)$. Find $\lim_{t \to \infty} y(t)$ if $a > 0$.

*11. A bottle that contains water with a temperature of 40°F is placed on a tennis court with temperature 90°F. After 20 minutes, the water is 65°F. What is the water's temperature after 30 minutes?

12. A can of diet cola at room temperature of 70°F is placed in a cooler with temperature 40°F. After 30 minutes the can is 60°F. When is the can 45°F?

13. A frozen turkey breast is placed in a microwave oven to defrost. It is then placed in a conventional oven at 325°F and bakes for one hour, at which time its temperature is 100°F. If after baking an additional 45 minutes its temperature is 150°F, what was the temperature of the turkey when it was removed from the microwave?

14. Suppose that during the month of July in Statesboro, GA, the outside temperature in °F is given by $C(t) =$

$85 - 10 \cos \dfrac{\pi t}{12}$, $0 \le t \le 24$. Find the temperature in

a parked car that has an initial temperature of 70°F if $k = \frac{1}{4}$. (Assume that the car has no heat or air conditioning system.)

*15. A rock weighing 4 lb is dropped from rest from a large height and is subjected to air resistance equivalent to $F_R = v$. Find the velocity $v(t)$ of the rock at any time t. What is the velocity of the rock after three seconds? How far has the rock fallen after three seconds?

16. A container of waste weighing 6 lb is accidentally released from an airplane at an altitude of 1000 ft with an initial velocity of 6 ft/s. If the container is subjected to air resistance equivalent to $F_R = \dfrac{2v}{3}$, find the velocity $v(t)$ of the container at any time t. What is the velocity of the container after five seconds? How far has the container fallen after five seconds? Approximately when does the container hit the ground?

17. An object of mass 5 kg is thrown vertically in the air from ground level with an initial velocity of 40 m/s. If the object is subjected to air resistance equivalent to $F_R = 5v$, find the velocity of the object at any time t. When does the object reach its maximum height? What is its maximum height?

18. A ball weighing 0.75 lb is thrown vertically in the air with an initial velocity of 20 ft/s. If the ball is subjected to air resistance equivalent to $F_R = \dfrac{v}{64}$, find the velocity of the object at any time t. When does the object reach its maximum height? What is its maximum height if it is thrown from an initial height of 5 ft?

*19. A parachutist weighing 128 lb falls from a plane ($v_0 = 0$). When the parachutist's speed is 30 ft/s, the parachute is opened and the parachutist is then subjected to air resistance equivalent to $F_R = 2v^2$. Find the velocity $v(t)$ of the parachutist. What is the limiting velocity of the parachutist?

20. A relief package weighing 256 lb is dropped from a plane ($v_0 = 0$) over a war-ravaged area and is subjected to air resistance equivalent to $F_R = 16v$. After 2 seconds the parachute opens and the package is then subjected to air resistance equivalent to $F_R = 4v^2$. Find the velocity $v(t)$ of the package. What is the limiting velocity of the package? Compare this to the limiting velocity if the parachute does not open.

21. Atomic waste is placed in sealed canisters and dumped in the ocean. It has been determined that the seal will not break and leak the waste when the canister hits the bottom of the ocean as long as the velocity of the canister is less than 12 m/s when it hits the bottom. Using Newton's second law, the velocity satisfies the equation $m\dfrac{dv}{dt} = W - B - kv$, where $v(0) = 0$, W is the weight of the canister, B is the buoyancy force, and the drag is given by $-kv$. Solve this first-order linear equation for $v(t)$ and then integrate to find the position $y(t)$. If $W = 2254$ Newtons, $B = 2090$ Newtons, and $k = 0.637$ kg/sec, determine the time at which the velocity is 12 m/s. Determine the depth H of the ocean so that the seal will not break when the canister hits the bottom.

22. According to the **Law of Mass Action,** if the temperature is constant, then the velocity of a chemical reaction is proportional to the product of the concentrations of the substances that are reacting. The reaction $A + B \rightarrow M$ combines a moles per liter of substance A and b moles per liter of substance B. If $y(t)$ is the number of moles per liter that have reacted after time t, the reaction rate is $\dfrac{dy}{dt} = k(a - y)(b - y)$. Find $y(t)$ if $y(0) = 0$. Find $\lim_{t\to\infty} y(t)$ if $a > b$ and if $b > a$.

23. Solve the following equations for $r(\theta)$. Graph the polar equation that results.

   (a) $r\dfrac{dr}{d\theta} + 4 \sin 2\theta = 0$, $r(0) = 2$

   *(b) $\dfrac{dr}{d\theta} - 2 \sec \theta \tan \theta = 0$, $r(0) = 4$

   (c) $\dfrac{dr}{d\theta} - 6 \sin \dfrac{\theta}{2} \cos \dfrac{\theta}{2} = 0$, $r(0) = 0$

24. A cylindrical tank 1.50 meters high stands on its circular base of radius $r = 0.50$ meters and is initially filled with water. At the bottom of the tank, there is a hole of radius $r = 0.50$ cm that is opened at some in-

stant so that draining starts due to gravity. According to **Torricelli's Law,** $v = 0.600\sqrt{2gh}$ where $g = 980$ cm/s^2 and h is the height of the water. By determining the rate at which the volume changes, we find that $\dfrac{dh}{dt} = -\dfrac{0.600A\sqrt{2g}}{B}\sqrt{h}$, where A is the cross-sectional area of the outlet and B is the cross-sectional area of the tank. In this case, $A = 0.500^2\pi$ cm^2 and $B = 50.0^2\pi$ cm^2, so $\dfrac{dh}{dt} = -0.00266\sqrt{h}$. Find $h(t)$ if $h(0) = 150$.

*25. (Fishing) Consider a population of fish with size at time t given by $x(t)$. Suppose the fish are harvested at a rate of $h(t)$. If the fish are sold at price p and δ is the interest rate, the present value P of the harvest is given by the improper integral.

$$P = \int_0^\infty e^{-\delta t}\left[p - \dfrac{c}{qx(t)}\right]h(t)\ dt,*$$

where c and q are constants related to the cost of the effort of catching fish (q is called the **catchability**).
   (a) Evaluate P if $\delta = 0.05$, $x(t) = 1$, and $h(t) = \frac{1}{2}$.

   If we assume that the harvesting rate $h(t)$ is proportional to the population of the fish, then

$$h(t) = qEx(t),$$

   where E represents the **effort** in catching the fish, and, under certain assumptions, the size of the population of the fish $x(t)$ satisfies the differential equation

$$\dfrac{dx}{dt} = (r - ax(t))x(t) - h(t)$$

$$\dfrac{dx}{dt} = (r - ax(t))x(t) - qEx(t).$$

   This equation can be rewritten in the form

$$\dfrac{dx}{dt} = [(r - qE) - ax(t)]x(t).$$

   (b) Solve the equation $\dfrac{dx}{dt} = [(r - qE) - ax(t)]x(t)$

---

* J. N. Kapur, ''Some Problems in Biomathematics,'' *International Journal of Mathematical Education in Science and Technology,* Volume 9, Number 3 (August 1978) pp. 287–306; and Colin W. Clark, ''Bioeconomic Modeling and Resource Management,'' *Applied Mathematical Ecology,* edited by Simon A. Levin, Thomas G. Hallam, and Louis J. Goss, Springer-Verlag, New York (1980) pp. 11–57. For more information, see Robert M. May, John R. Beddington, Colin W. Clark, Sidney J. Holt, and Richard M. Laws ''Management of Multispecies Fisheries,'' *Science,* Volume 205, Number 4403 (July 20, 1979) pp. 267–277.

and find the solution that satisfies the initial condition $x(0) = x_0$. What is $h(t)$?

    Suppose that $x_0 = 1$, $r = 1$, and $a = \frac{1}{2}$.

(c) Graph $x(t)$ if there is no harvesting for $0 \le t \le 10$. (*Hint:* If there is no harvesting, $qE = 0$.)

(d) Graph $x(t)$ and $h(t)$ for $0 \le t \le 20$ using $qE = 0$, $0.1, 0.2, \ldots, 2$. What is the maximum sustainable harvest rate? In other words, what is the highest rate at which the fish can be harvested without becoming extinct? At what rate should the fish be harvested to produce the largest overall harvest? How does this result compare to (a)?

    In 1965, the values of r, a, p, c, and q for the Antarctic whaling industry were determined to be $r = 0.05$, $a = 1.25 \times 10^{-7}$, $p = 7000$, $c = 5000$, and $q = 1.3 \times 10^{-5}$. Assume that $t = 0$ corresponds to the year 1965 and that $x(0) = 78{,}000$. Assume that a typical firm expects a return of 10% on their investment, so that $\delta = 0.10$.

(e) Approximate

$$P = \int_0^\infty e^{-\delta t}\left[p - \frac{c}{qx(t)}\right]h(t)\, dt$$

if **(i)** $E = 5000$ and **(ii)** $E = 7000$. What happens to the whale population in each case? What advice would you give to the whaling industry?

(f) Approximate

$$P = \int_0^\infty e^{-\delta t}\left[p - \frac{c}{qx(t)}\right]h(t)\, dt$$

using the values in the following table.

E	Approximation of $P = \int_0^\infty e^{-\delta t}\left[p - \frac{c}{qx(t)}\right]h(t)\, dt$
1000	
1500	
2000	
2500	
3000	
3500	
4000	
4500	

What value of E produces the maximum profit? What happens to the whale population in this case?

(g) Some reports have indicated that the optimal stock level of whales should be around 227,500. How does this number compare to the maximum number of whales that the environment can sustain? *Hint:* Evaluate $\lim_{t\to\infty} x(t)$ if there is no harvesting. How can the whaling industry make a profit and maintain this number of whales in the ocean?

Differential Equations at Work:

Mathematics of Finance

• Suppose that P dollars are invested in an account at an annual rate of $r\%$ compounded continuously. To find the balance of the account $x(t)$ at time t, we must solve the initial-value problem

$$\begin{cases} \dfrac{dx}{dt} = rx \\ x(0) = P \end{cases}$$

for x.

1. Show that $x(t) = Pe^{rt}$.

2. If \$1000 is deposited into an account with an annual interest rate of 8% compounded continuously, what is the balance of the account at the end of 5, 10, 15, and 20 years?

If we allow additions or subtractions of sums of money from the account, the problem becomes more complicated. Suppose that an account like a savings account, home mortgage loan, student loan, or car loan, has an initial balance of P dollars and r denotes the interest rate per compounding period. Let $p(t)$ denote the money flow per unit time and $\delta = \ln(1 + r)$. Then the balance of the account at time t, $x(t)$, satisfies the initial-value problem

$$\begin{cases} \dfrac{dx}{dt} - \delta x = p(t) \\ x(0) = P \end{cases} \quad \overset{*}{}$$

3. Show that the balance of the account at time t_0 is given by

$$x(t_0) = Pe^{\delta t_0} + e^{\delta t_0} \int_0^{t_0} p(t)e^{-\delta t}\, dt.$$

4. Suppose that the initial balance of a student loan is \$12,000 and that monthly payments are made in the amount of \$130. If the annual interest rate, compounded monthly, is 9%, then $P = 12{,}000$, $p(t) = -130$, $r = 0.09/12 = 0.0075$, and $\delta = \ln(1 + r) = \ln(1.0075)$. **(a)** Show that the balance of the loan at time t, in months, is given by

$$x(t) = 17398.3 - 5398.25 \cdot 1.0075^t.$$

(b) Graph $x(t)$ on the interval $[0, 180]$, corresponding to the loan balance for the first fifteen years. **(c)** How long will it take to pay off the loan?

5. Suppose that the initial balance of a home mortgage loan is \$80,000 and that monthly payments are made in the amount of \$599. If the annual interest rate, compounded monthly, is 8%, how long will it take to pay off the mortgage loan?

6. Suppose that the initial balance of a home mortgage loan is \$80,000 and that monthly payments are made in the amount of \$599. If the annual interest rate, compounded monthly, is 8%, how long will it take to pay off the mortgage loan if the monthly payment is increased at an annual rate of 3%, which corresponds to a monthly increase of $\frac{1}{4}$%?

7. If an investor invests \$250 per month in an account paying an annual interest rate of 10%, compounded monthly, how much will the investor have accumulated at the end of 10, 20, and 30 years?

* Thoddi C. T. Kotiah, "Difference and differential equations in the mathematics of finance," *International Journal of Mathematics in Education, Science, and Technology,* Volume 22, Number 5 (1991) pp. 783–789.
Edward W. Herold, "Inflation Mathematics for the Professional," *Mathematical Modeling: Classroom Notes in Applied Mathematics,* Edited by Murray S. Klamkin, SIAM (1987) pp. 206–209.

8. Suppose an investor begins by investing $250 per month in an account paying an annual interest rate of 10%, compounded monthly, and in addition increases the amount invested by 6% each year for an increase of $\frac{1}{2}$% each month. How much will the investor have accumulated at the end of 10, 20, and 30 years?

9. Suppose that a 25-year-old investor begins investing $250 per month in an account paying an annual interest rate of 10%, compounded monthly. If at the age of 35 the investor stops making monthly payments, what will be the account balance when the investor reaches 45, 55, and 65 years of age? Suppose that a 35-year-old friend begins investing $250 per month in an account paying an annual interest rate of 10%, compounded monthly, at the same time the first investor stops. Who has a larger investment at the age of 65?

10. If you are given a choice between saving $150 a month beginning when you first start working and continuing until you retire, or saving $300 per month beginning 10 years after you first start working and continuing until you retire, which should you do to help assure a financially secure retirement?

From Exercises 8 and 9, we see that consistent savings beginning at an early age can help assure that a large sum of money will accumulate by the age of retirement. How much money does a person need to have accumulated to help assure a financially secure retirement?

For example, corporate pension plans and social security generally provide a relatively small portion of living expenses, especially for those with above-average incomes. In addition, inflation erodes the buying power of the dollar; large sums of money today will not necessarily be large sums of money tomorrow.

As an illustration, we see that if inflation were to average a modest 3% per year for the next 20 years and a person has living expenses of $20,000 annually, then after 20 years the person would have living expenses of

$$\$20{,}000 \cdot 1.03^{20} \approx \$36{,}122.$$

Let t denote years and suppose that a person's after-tax income as a function of years is given by $I(t)$ and $E(t)$ represents living expenses. Here $t = 0$ might represent the year a person enters the work force. Generally, during working years $I(t) > E(t)$; during retirement years when $I(t)$ represents income from sources like corporate pension plans and social security, $I(t) < E(t)$.

Suppose that an account has an initial balance of S_0 and the after-tax return on the account is r%. We assume that the amount deposited into the account each year is $I(t) - E(t)$. What is the balance of the account at year t, $S(t)$? S must satisfy the initial-value problem

$$\begin{cases} \dfrac{dS}{dt} = rS(t) + I(t) - E(t) \\ S(0) = S_0 \end{cases}.$$

11. Show that the balance of the account at time $t = t_0$ is

$$S(t_0) = e^{rt_0}\left\{ S_0 + \int_0^{t_0} [I(t) - E(t)]e^{-rt}\, dt \right\}.$$

Assume that inflation averages an annual rate of i. Then, in terms of $E(0)$,

$$E(t) = E(0)e^{it}.$$

Similarly, during working years we will assume that annual raises are received at an annual rate of j. Then, in terms of $I(0)$,

$$I(t) = I(0)e^{jt}.$$

On the other hand, during retirement years, we will assume that $I(t)$ is given by a fixed sum, F, like a corporate pension or annuity, and a portion indexed to the inflation rate, V, like social security. Thus, during retirement years

$$I(t) = F + Ve^{i(t-T)},$$

where T denotes the number of working years. Therefore,

$$I(t) = \begin{cases} I(0)e^{jt}, & 0 \le t \le T \\ F + Ve^{i(t-T)}, & t > T \end{cases}.$$

12. Suppose that a person has an initial income of $I(0) = 20{,}000$ and receives annual average raises of 5%, so that $j = 0.05$ and that initial living expenses are $E(0) = 18{,}000$. Further, we will assume that inflation averages 3%, so that $i = 0.03$ while the after-tax return on the investment is 6% so that $r = 0.06$. Upon retirement after T years of work, we will assume that the person receives a fixed pension equal to 20% of his living expenses at that time so that

$$F = 0.2 \cdot 18{,}000 \cdot e^{0.03T}$$

while social security provides 30% of his living expenses at that time so that

$$V = 0.3 \cdot 18{,}000 \cdot e^{0.03T}.$$

(a) Find the smallest value of T so that the balance in the account is zero after 30 years of retirement. Sketch a graph of S for this value of T.

(b) Find the smallest value of T so that the balance in the account is never zero. Sketch a graph of S for this value of T.

13. What is the relationship between the results you obtained in Exercises 9 and 10 and that obtained in Exercise 12?

14. (a) How would you advise a person 22 years of age first entering the workforce to prepare for a financially secure retirement? (b) How would you advise a person 50 years of age with no savings who hopes to retire at 65 years of age? (c) When should you start saving for retirement?

4

HIGHER-ORDER DIFFERENTIAL EQUATIONS

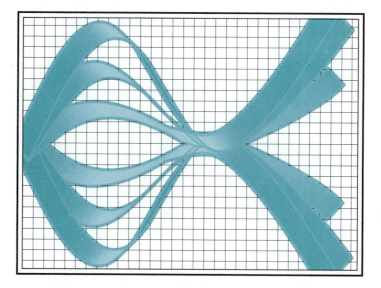

In Chapters 2 and 3 we saw that first-order differential equations can be used to model a variety of physical situations. However, there are many physical situations that need to be modeled by higher-order differential equations.

For example, in 1735 Daniel Bernoulli's (1700–1782) study of the vibrations of an elastic beam led to the fourth-order differential equation

$$k^4 \frac{d^4 y}{dx^4} = y,$$

which describes the displacement of the *simple modes*. This equation can be rewritten in the form

$$y - k^4 \frac{d^4 y}{dx^4} = 0.$$

Both Bernoulli and Euler realized that a solution to the equation is $y = e^{x/k}$, but that other solutions to the equation must also exist.*

4.1 Introduction

The *n*th-Order Ordinary Linear Differential Equation • Linear Dependence and Linear Independence • The Wronskian

The *n*th-Order Ordinary Linear Differential Equation

In order to develop the methods needed to solve higher-order differential equations, we must state several important definitions and theorems. We begin by introducing the types of higher-order equations that we will be solving in this chapter by restating the following definition that was given in Chapter 1.

Definition 4.1 *n*th-order Ordinary Linear Differential Equation

An ordinary differential equation of the form

$$a_n(x)y^{(n)}(x) + a_{n-1}(x)y^{(n-1)}(x) + \cdots + a_1(x)y'(x) + a_0(x)y(x) = g(x),$$

where $a_n(x) \not\equiv 0$, is called an **nth-order ordinary linear differential equation**. If $g(x)$ is identically the zero function, the equation is said to be **homogeneous**; if $g(x)$ is not the zero function, the equation is said to be **nonhomogeneous**; and if the functions $a_i(x)$, $i = 0, 1, 2, \ldots, n$ are constants, the equation is said to have **constant coefficients**. An nth-order equation accompanied by the conditions

$$y(x_0) = y_0, \ y'(x_0) = y_0', \ldots, y^{(n-1)}(x_0) = y_0^{(n-1)}$$

where $y_0, y_0', \ldots, y_0^{(n-1)}$ are constants is called an **nth-order initial-value problem**.

Example 1

Classify the linear differential equations:

(a) $x^2y'' - 2xy' + 4y = 0$ (b) $y''' - 8y'' + 10y' - 3y = \cos x$
(c) $y^{(4)} - y = 0$.

Solution   (a) second-order homogeneous constant coefficient   (b) third-order nonhomogeneous constant coefficient   (c) fourth-order homogeneous constant coefficient

---

* Victor J. Katz, *A History of Mathematics: An Introduction*, HarperCollins, 1993, p. 504.

The following theorem gives sufficient conditions for the existence of a unique solution to the nth-order initial-value problem.

Theorem 4.1 Existence of a Unique Solution

If $a_n(x)$, $a_{n-1}(x)$, . . ., $a_1(x)$, $a_0(x)$ and $g(x)$ are continuous throughout an interval I and $a_n(x) \neq 0$ for all x in the interval I, then for every x_0 in I there is a unique solution to the initial-value problem

$$\begin{cases} a_n(x)y^{(n)}(x) + a_{n-1}(x)y^{(n-1)}(x) + \cdots + a_1(x)y'(x) + a_0(x)y(x) = g(x) \\ y(x_0) = y_0 \\ \vdots \\ y^{(n-1)}(x_0) = y_0^{(n-1)} \end{cases}$$

on I where y_0, y_0', . . ., $y_0^{(n-1)}$ represent arbitrary constants.

The notation $y^{(n-1)}(x_0) = y_0^{(n-1)}$ in this theorem means the $(n-1)$st derivative of y evaluated at $x = x_0$ where n is the order of the differential equation. For example, if the differential equation is fifth order, we have five initial conditions

$y(x_0) = y_0$, $y'(x_0) = y_0'$, $y''(x_0) = y_0''$, $y'''(x_0) = y_0'''$, and $y^{(4)}(x_0) = y_0^{(4)}$

where y_0, y_0', y_0'', y_0''', and $y_0^{(4)}$ are constants.

The proof of this theorem is well beyond the scope of this text but can be found in advanced differential equations texts.*

Linear Dependence and Linear Independence

Now that we have conditions that indicate the existence of solutions, we become familiar with the properties of the functions that form the solution. We will see that *general* solutions to nth-order ordinary linear differential equations require n solutions with the following property.

Definition 4.2 Linearly Dependent and Linearly Independent

Let $S = \{f_1(x), f_2(x), f_3(x), . . ., f_{n-1}(x), f_n(x)\}$ be a set of n functions. S is **linearly dependent** on an interval I if there are constants $c_1, c_2, . . ., c_n$, not all zero, so that

$$c_1 f_1(x) + c_2 f_2(x) + \cdots + c_{n-1}f_{n-1}(x) + c_n f_n(x) = 0$$

for every value of x in the interval I.
S is **linearly independent** if S is not linearly dependent.

Thus, $S = \{f_1(x), f_2(x), f_3(x), . . ., f_{n-1}(x), f_n(x)\}$ is linearly independent if

$$c_1 f_1(x) + c_2 f_2(x) + \cdots + c_{n-1}f_{n-1}(x) + c_n f_n(x) = 0$$

implies that $c_1 = c_2 = \cdots = c_n = 0$.

---

* For example, see Chapter 2 of C. Corduneanu, *Principles of Differential and Integral Equations*, Chelsea Publishing Company, 1971.

Example 2

Classify the sets of functions as linearly independent or linearly dependent.
(a) $S = \{2x, 4x\}$ (b) $S = \{x, 3x - 6, 1\}$
(c) $S = \{\cos 2x, \sin 2x, \sin x \cos x\}$ (d) $S = \{1, x, x^2\}$

Solution   (a) We try to find constants c_1 and c_2, not both zero, that satisfy

$$c_1(2x) + c_2(4x) = (2c_1 + 4c_2)x = 0.$$

Many choices for these constants satisfy the equation. For example, if we choose $c_1 = 2$ and $c_2 = -1$, substitution yields $c_1(2x) + c_2(4x) = (2)(2x) + (-1)(4x) = 0$. At least one of these constants is not zero, so the set $S = \{2x, 4x\}$ is linearly dependent.
(b) Here, we must find constants c_1, c_2, and c_3 such that

$$c_1 x + c_2(3x - 6) + c_3(1) = (c_1 + 3c_2)x + (c_3 - 6c_2) = 0.$$

Equating each of the coefficients to zero leads to the system of equations $\{c_1 + 3c_2 = 0, c_3 - 6c_2 = 0\}$. This system has infinitely many solutions of the form $\{c_1 = -3c_2, c_3 = 6c_2, c_2 \text{ arbitrary}\}$. With the choice $c_1 = -3$, $c_2 = 1$, and $c_3 = 6$, the equation is satisfied. Therefore, the set $S = \{x, 3x - 6, 1\}$ is linearly dependent.
(c) For this set of functions, we apply the identity $\sin 2x = 2 \sin x \cos x$. Doing this, we consider the equation

$$c_1 \cos 2x + c_2 \sin 2x + c_3 \sin x \cos x =$$
$$c_1 \cos 2x + 2c_2 \sin x \cos x + c_3 \sin x \cos x = 0.$$

The choices of $c_1 = 0$, $c_2 = 1$, and $c_3 = -2$ lead to a solution. At least one of the constants is not zero, so the set $S = \{\cos 2x, \sin 2x, \sin x \cos x\}$ is linearly dependent.
(d) For the set $S = \{1, x, x^2\}$, we consider the equation $c_1(1) + c_2 x + c_3 x^2 = 0$. The only constants that satisfy the equation are $c_1 = c_2 = c_3 = 0$, so the set is linearly independent.

The Wronskian

According to the definition of linear dependence, a set of two functions is linearly dependent if and only if the two functions are constant multiples of each other. Of course, as the number of functions increases, the problem of determining if a set of functions is linearly independent or linearly dependent becomes more difficult. Fortunately, better techniques exist for making this determination.

Definition 4.3 Wronskian

Let $S = \{f_1(x), f_2(x), f_3(x), \ldots, f_{n-1}(x), f_n(x)\}$ be a set of n functions for which each is differentiable at least $n - 1$ times. The **Wronskian** of S, denoted by

$$W(S) = W(f_1(x), f_2(x), f_3(x), \ldots, f_{n-1}(x), f_n(x)),$$

is the determinant

$$W(S) = \begin{vmatrix} f_1(x) & f_2(x) & \cdots & f_n(x) \\ f_1'(x) & f_2'(x) & \cdots & f_n'(x) \\ \vdots & \vdots & \vdots & \vdots \\ f_1^{(n-1)}(x) & f_2^{(n-1)}(x) & \cdots & f_n^{(n-1)}(x) \end{vmatrix}.$$

Even though most computer algebra systems can quickly find and simplify the Wronskian of a set of functions (which is particularly useful if you have not had linear algebra), the following example illustrates how to compute the Wronskian for a set of two or three functions by hand.

Example 3

Compute the Wronskian for each of the following sets of functions.
(a) $S = \{\sin x, \cos x\}$ (b) $S = \{e^x, e^{2x}\}$
(c) $S = \{1, x, x^2\}$ (d) $S = \{2x, 4x\}$

Solution The 2×2 determinant $\begin{vmatrix} a_{11} & a_{12} \\ a_{21} & a_{22} \end{vmatrix}$ is computed by calculating $a_{11}a_{22} - a_{12}a_{21}$.

(a) $W(S) = \begin{vmatrix} \sin x & \cos x \\ \dfrac{d}{dx}(\sin x) & \dfrac{d}{dx}(\cos x) \end{vmatrix} = \begin{vmatrix} \sin x & \cos x \\ \cos x & -\sin x \end{vmatrix} = -\sin^2 x - \cos^2 x = -1.$

(b) $W(S) = \begin{vmatrix} e^x & e^{2x} \\ \dfrac{d}{dx}(e^x) & \dfrac{d}{dx}(e^{2x}) \end{vmatrix} = \begin{vmatrix} e^x & e^{2x} \\ e^x & 2e^{2x} \end{vmatrix} = 2e^{3x} - e^{3x} = e^{3x}.$

(c) The 3×3 determinant $\begin{vmatrix} a_{11} & a_{12} & a_{13} \\ a_{21} & a_{22} & a_{23} \\ a_{31} & a_{32} & a_{33} \end{vmatrix}$ can be computed in several equivalent ways. For example,

$$\begin{vmatrix} a_{11} & a_{12} & a_{13} \\ a_{21} & a_{22} & a_{23} \\ a_{31} & a_{32} & a_{33} \end{vmatrix} = a_{11}\begin{vmatrix} a_{22} & a_{23} \\ a_{32} & a_{33} \end{vmatrix} - a_{12}\begin{vmatrix} a_{21} & a_{23} \\ a_{31} & a_{33} \end{vmatrix} + a_{13}\begin{vmatrix} a_{21} & a_{22} \\ a_{31} & a_{32} \end{vmatrix}.$$

Also, we have $\begin{vmatrix} a_{11} & a_{12} & a_{13} \\ a_{21} & a_{22} & a_{23} \\ a_{31} & a_{32} & a_{33} \end{vmatrix} = a_{11} \begin{vmatrix} a_{22} & a_{23} \\ a_{32} & a_{33} \end{vmatrix} - a_{21} \begin{vmatrix} a_{12} & a_{13} \\ a_{32} & a_{33} \end{vmatrix} + a_{31} \begin{vmatrix} a_{12} & a_{13} \\ a_{22} & a_{23} \end{vmatrix}$

which is used here:

$$W(S) = \begin{vmatrix} 1 & x & x^2 \\ \dfrac{d}{dx}(1) & \dfrac{d}{dx}(x) & \dfrac{d}{dx}(x^2) \\ \dfrac{d^2}{dx^2}(1) & \dfrac{d^2}{dx^2}(x) & \dfrac{d^2}{dx^2}(x^2) \end{vmatrix}$$

$$= \begin{vmatrix} 1 & x & x^2 \\ 0 & 1 & 2x \\ 0 & 0 & 2 \end{vmatrix}$$

$$= 1 \begin{vmatrix} 1 & 2x \\ 0 & 2 \end{vmatrix} - 0 \begin{vmatrix} x & x^2 \\ 0 & 2 \end{vmatrix} + 0 \begin{vmatrix} x & x^2 \\ 1 & 2x \end{vmatrix} = 2.$$

(d) $W(S) = \begin{vmatrix} 2x & 4x \\ \dfrac{d}{dx}(2x) & \dfrac{d}{dx}(4x) \end{vmatrix} = \begin{vmatrix} 2x & 4x \\ 2 & 4 \end{vmatrix} = 8x - 8x = 0.$

In (a)–(c) of the previous example, the Wronskian is not 0 while in (d) the Wronskian is 0. Moreover, in Example 3 we know that the sets of functions in (a)–(c) are linearly independent while the set of functions in (d) is linearly dependent. In fact, we can use the Wronskian to determine if a set of functions is linearly dependent or linearly independent.

Theorem 4.2

Let $S = \{f_1(x), f_2(x), f_3(x), \ldots, f_{n-1}(x), f_n(x)\}$ be a set of n functions each differentiable at least $n - 1$ times on an interval I. If $W(S) \neq 0$ for at least one value of x in the interval I, S is linearly independent.

PROOF OF THEOREM 4.2

We use a proof by contradiction to prove this theorem for two functions. Suppose that $W(y_1(a), y_2(a)) \neq 0$ for $a \in I$ and $y_1(x)$ and $y_2(x)$ are linearly dependent functions that are each differentiable on I. By the definition of linear dependence, there are constants c_1 and c_2 (not both zero) such that $c_1 y_1(x) + c_2 y_2(x) = 0$. Both of these functions are differentiable, so we can take the derivative of the linear combination to obtain $c_1 y_1'(x) + c_2 y_2'(x) = 0$. Hence, we have the system of equations $\begin{cases} c_1 y_1(x) + c_2 y_2(x) = 0 \\ c_1 y_1'(x) + c_2 y_2'(x) = 0 \end{cases}$.

Then, because $y_1(x)$ and $y_2(x)$ are linearly dependent, the system has a solution in which at least one of these constants is not zero. In linear algebra courses, we learn that this homogeneous system of equations has a nontrivial solution if and only if $\begin{vmatrix} y_1(x) & y_2(x) \\ y_1'(x) & y_2'(x) \end{vmatrix} = 0$ for all x on I. However, this contradicts the assumption $W(y_1(a), y_2(a)) \neq 0$ for $a \in I$. Therefore, $y_1(x)$ and $y_2(x)$ are linearly independent.

The proof of the general case with n functions follows the same steps as those given above for two functions. By assuming that the n functions are linearly dependent, there are constants (not all zero) such that $c_1 f_1(x) + c_2 f_2(x) + \cdots + c_n f_n(x) = 0$. This equation can be differentiated $(n - 1)$ times to obtain the homogeneous system of n equations

$$\begin{cases} c_1 f_1(x) + c_2 f_2(x) + \cdots + c_n f_n(x) = 0 \\ c_1 f_1'(x) + c_2 f_2'(x) + \cdots + c_n f_n'(x) = 0 \\ c_1 f_1''(x) + c_2 f_2''(x) + \cdots + c_n f_n''(x) = 0. \\ \quad\quad\quad\quad\quad \vdots \\ c_1 f_1^{(n-1)}(x) + c_2 f_2^{(n-1)}(x) + \cdots + c_n f_n^{(n-1)}(x) = 0 \end{cases}$$

Because $\{c_1 = 0, c_2 = 0, \ldots, c_n = 0\}$, if $W(f_1(x), f_2(x) \ldots, f_n(x)) \neq 0$, our assumption that the n functions are linearly dependent is incorrect. Therefore, the n functions are linearly independent. • • •

Example 4

Use the Wronskian to classify each of the following sets of functions as linearly independent or linearly dependent.
(a) $S = \{e^{-x}, e^{2x}\}$ (b) $S = \{2 \tan^2 x, \sec^2 x - 1\}$
(c) $S = \{e^x, xe^x, x^2 e^x\}$

Solution   (a) In this case, $W(S) = \begin{vmatrix} e^{-x} & e^{2x} \\ -e^{-x} & 2e^{2x} \end{vmatrix} = 2e^x + e^x = 3e^x \neq 0.$
Hence, the set $S = \{e^{-x}, e^{2x}\}$ is linearly independent.
(b) In computing the Wronskian for S we take advantage of the identity $\tan^2 x + 1 = \sec^2 x$:

$$W(S) = \begin{vmatrix} 2 \tan^2 x & \sec^2 x - 1 \\ 4 \tan x \sec^2 x & 2 \sec^2 x \tan x \end{vmatrix} = 4 \sec^2 x \tan^3 x - 4 \sec^4 x \tan x + $$
$$4 \tan x \sec^2 x$$

$$= 4 \sec^2 x \tan x (\tan^2 x - \sec^2 x + 1) = 4 \sec^2 x \tan x (\tan^2 x - (\tan^2 x + 1) + 1)$$
$$= 4 \sec^2 x \tan x (\tan^2 x - \tan^2 x - 1 + 1) = 0.$$

Therefore, the set $S = \{2 \tan^2 x, \sec^2 x - 1\}$ is linearly dependent. Note that we could have answered this question more quickly by using $\tan^2 x = \sec^2 x - 1$ with $2c_1 \tan^2 x + c_2(\sec^2 x - 1) = 2c_1(\sec^2 x - 1) + c_2(\sec^2 x - 1) = 0.$ Therefore, any choice of $c_2 = -2c_1$ satisfies the equation.

(c) Here, we compute the determinant of a 3×3 matrix. In this case,

$$W(S) = \begin{vmatrix} e^x & xe^x & x^2e^x \\ \dfrac{d}{dx}(e^x) & \dfrac{d}{dx}(xe^x) & \dfrac{d}{dx}(x^2e^x) \\ \dfrac{d^2}{dx^2}(e^x) & \dfrac{d^2}{dx^2}(xe^x) & \dfrac{d^2}{dx^2}(x^2e^x) \end{vmatrix} = \begin{vmatrix} e^x & xe^x & x^2e^x \\ e^x & (x+1)e^x & (x^2+2x)e^x \\ e^x & (x+2)e^x & (x^2+4x+2)e^x \end{vmatrix}$$

$$= e^x \begin{vmatrix} (x+1)e^x & (x^2+2x)e^x \\ (x+2)e^x & (x^2+4x+2)e^x \end{vmatrix} - xe^x \begin{vmatrix} e^x & (x^2+2x)e^x \\ e^x & (x^2+4x+2)e^x \end{vmatrix} +$$

$$x^2e^x \begin{vmatrix} e^x & (x+1)e^x \\ e^x & (x+2)e^x \end{vmatrix}$$

$$= e^x[(x+1)(x^2+4x+2)e^{2x} - (x+2)(x^2+2x)e^{2x}] -$$
$$xe^x[(x^2+4x+2)e^{2x} - (x^2+2x)e^{2x}] + x^2e^x[(x+2)e^{2x} - (x+1)e^{2x}]$$

$$= 2e^{3x}.$$

We conclude that S is linearly independent because the Wronskian of S is not identically zero.

Example 5   Is it possible to choose values of $c_1, \ldots, c_5$, not all of which are zero, so that

$$f(x) = c_1 + c_2x^{-3/2} + c_3x^{-1/2} + c_4x^{1/2} + c_5x^{3/2}$$

is the zero function?

Solution   Let $S = \{1, x^{-3/2}, x^{-1/2}, x^{1/2}, x^{3/2}\}$. Then,

$$W(S) = \begin{vmatrix} 1 & x^{-3/2} & x^{-1/2} & x^{1/2} & x^{3/2} \\ \dfrac{d}{dx}(1) & \dfrac{d}{dx}(x^{-3/2}) & \dfrac{d}{dx}(x^{-1/2}) & \dfrac{d}{dx}(x^{1/2}) & \dfrac{d}{dx}(x^{3/2}) \\ \dfrac{d^2}{dx^2}(1) & \dfrac{d^2}{dx^2}(x^{-3/2}) & \dfrac{d^2}{dx^2}(x^{-1/2}) & \dfrac{d^2}{dx^2}(x^{1/2}) & \dfrac{d^2}{dx^2}(x^{3/2}) \\ \dfrac{d^3}{dx^3}(1) & \dfrac{d^3}{dx^3}(x^{-3/2}) & \dfrac{d^3}{dx^3}(x^{-1/2}) & \dfrac{d^3}{dx^3}(x^{1/2}) & \dfrac{d^3}{dx^3}(x^{3/2}) \\ \dfrac{d^4}{dx^4}(1) & \dfrac{d^4}{dx^4}(x^{-3/2}) & \dfrac{d^4}{dx^4}(x^{-1/2}) & \dfrac{d^4}{dx^4}(x^{1/2}) & \dfrac{d^4}{dx^4}(x^{3/2}) \end{vmatrix} = \frac{27}{4}x^{-10} \neq 0$$

so S is a linearly independent set of functions; it is not possible to choose values of $c_1, \ldots, c_5$, not all of which are zero, so that $f(x) = c_1 + c_2x^{-3/2} + c_3x^{-1/2} + c_4x^{1/2} + c_5x^{3/2}$ is the zero function.

Is it possible to choose c_1, c_2, c_3, and c_4 so that

$$y(x) = c_3c_4 + \frac{c_2^2 c_3}{24c_1}x^{-3/2} + \left(\frac{c_2 c_3}{2} - \frac{c_2^2 c_4}{8c_1}\right)x^{-1/2} + \frac{3c_1 c_3 + c_2 c_4}{2}x^{1/2} - \frac{c_1 c_4}{2}x^{3/2}$$

is the zero function?

IN TOUCH WITH TECHNOLOGY

Using traditional pencil-and-paper to compute the Wronskian of a set of functions quickly becomes tedious and time-consuming. Fortunately, computer algebra systems are capable of computing derivatives and evaluating the determinant encountered when calculating the Wronskian of a set of functions (like that encountered in Example 5). Moreover, the graphing capabilities can be used to help determine whether the resulting expression is or is not zero. Compute the Wronskian of each of the following sets S to classify each set as linearly independent or linearly dependent. If necessary, graph the Wronskian to see if it is the zero function.

1. $S = \{1 - 2\sin^2 x,\ \cos 2x\}$

2. $S = \{1, x, \frac{1}{2}(3x^2 - 1), \frac{1}{2}(5x^3 - 3x), \frac{1}{8}(35x^4 - 30x^2 + 3)\}$

3. $S = \{\sin x,\ \sin 2x,\ \sin 3x,\ \sin 4x\}$

4. $S = \{e^x, xe^x, x^2 e^x, x^3 e^x, x^4 e^x\}$

5. Is it possible to choose values of $c_1, \ldots, c_7$, not all zero, so that

$$f(x) = c_1 + c_2 x^{-5/2} + c_3 x^{-3/2} + c_4 x^{-1/2} + c_5 x^{1/2} + c_6 x^{3/2} + c_7 x^{5/2}$$

is the zero function?

6. Use the symbolic manipulation capabilities of a computer algebra system to help you prove the following theorem. Suppose that $f(x)$ is a differentiable function on an interval I and that $f(x) \neq 0$ for all x in I. **(a)** Prove that $f(x)$ and $xf(x)$ are linearly independent on I. **(b)** Prove that $f(x)$, $xf(x)$, and $x^2 f(x)$ are linearly independent on I. **(c)** Generalize your result.

EXERCISES 4.1

In Exercises 1–14, calculate the Wronskian of the indicated set of functions. Classify each set of functions as linearly independent or linearly dependent.

1. $S = \{x, 4x - 1\}$

2. $S = \{x, e^x\}$

*3. $S = \{e^{-6x}, e^{-4x}\}$

4. $S = \{\cos 2x, \sin 2x\}$

5. $S = \{e^{-3x}\cos 3x, e^{-3x}\sin 3x\}$

6. $S = \{e^{5x}\cos 4x, e^{5x}\sin 4x\}$

*7. $S = \{3x^2, x, 2x - 2x^2\}$

8. $S = \{\cos 2x, \sin x, 1\}$

9. $S = \{e^{-x}, e^{3x}, xe^{3x}\}$

10. $S = \{e^x, e^{-2x}, e^{-4x}\}$

*11. $S = \{e^x, e^{-5x}\sin x, e^{-5x}\cos x\}$

12. $S = \{e^{3x}, e^{-2x}, e^{5x}, xe^{5x}\}$

13. $S = \{e^{-3x}, e^{-x}, e^{-4x}\cos 3x, e^{-4x}\sin 3x\}$

14. $S = \{e^{-x}\cos 2x, e^{-x}\sin 2x, e^{2x}\cos 5x, e^{2x}\sin 5x\}$

In Exercises 15–20, show that $y = c_1 y_1(x) + c_2 y_2(x)$ satisfies the given differential equation and that $\{y_1(x), y_2(x)\}$ is a linearly independent set.

15. $y = c_1 e^x + c_2 e^{-x}$, $y'' - y = 0$

16. $y = c_1 e^x + c_2 e^{-3x}$, $y'' + 2y' - 3y = 0$

*17. $y = c_1 e^{-x} + c_2 x e^{-x}$, $y'' + 2y' + y = 0$

18. $y = c_1 e^{2x} + c_2 x e^{2x}$, $y'' - 4y' + 4y = 0$

19. $y = c_1 \cos 2x + c_2 \sin 2x$, $y'' + 4y = 0$

20. $y = e^{3x}(c_1 \cos 4x + c_2 \sin 4x)$, $y'' - 6y' + 25y = 0$

In Exercises 21–30, use the given solution on the indicated interval to find the solution to the initial-value problem.

21. $\begin{cases} y'' - y' - 2y = 0 \\ y(0) = -1, \ y'(0) = -5 \end{cases}$
$y = c_1 e^{-x} + c_2 e^{2x}$, $-\infty < x < \infty$

22. $\begin{cases} y'' - 7y' + 12y = 0 \\ y(0) = 0, \ y'(0) = -1 \end{cases}$
$y = c_1 e^{3x} + c_2 e^{4x}$, $-\infty < x < \infty$

*23. $\begin{cases} y'' + 4y = 0 \\ y(0) = -1, \ y'(0) = 2 \end{cases}$
$y = c_1 \cos 2x + c_2 \sin 2x$, $-\infty < x < \infty$

24. $\begin{cases} y'' + 4y' + 13y = 0 \\ y(0) = 2, \ y'(0) = -1 \end{cases}$
$y = e^{-2x}(c_1 \cos 3x + c_2 \sin 3x)$, $-\infty < x < \infty$

25. $\begin{cases} y'' - 8y' + 16y = 0 \\ y(1) = 0, \ y'(1) = -e^4 \end{cases}$
$y = c_1 e^{4x} + c_2 x e^{4x}$, $-\infty < x < \infty$

26. $\begin{cases} y''' + y' = 0 \\ y(\pi) = 3, \ y'(\pi) = 2, \ y''(\pi) = 1 \end{cases}$
$y = c_1 + c_2 \cos x + c_3 \sin x$, $-\infty < x < \infty$

*27. $\begin{cases} x^2 y'' + 7xy' - 7y = 0 \\ y(1) = 2, \ y'(1) = -22 \end{cases}$ $y = c_1 x^{-7} + c_2 x$, $x > 0$

28. $\begin{cases} x^2 y'' + 6xy' + 4y = 0 \\ y(1) = 4, \ y'(1) = -4 \end{cases}$, $y = c_1 x^{-1} + c_2 x^{-4}$, $x > 0$

29. $\begin{cases} y'' + y = 2 \cos x \\ y(0) = 1, \ y'(0) = 1 \end{cases}$
$y = c_1 \cos x + c_2 \sin x + x \sin x$, $-\infty < x < \infty$

30. $\begin{cases} y'' - y = 4e^x \\ y(0) = 1, \ y'(0) = -1 \end{cases}$
$y = c_1 e^x + c_2 e^{-x} + 2x e^x$, $-\infty < x < \infty$

*31. Use the Wronskian to show that if y_1 and y_2 are solutions of the first-order differential equation

$$y' + p(x)y = 0,$$

then y_1 and y_2 are linearly dependent.

32. Consider the hyperbolic trigonometric functions
$$\cosh x = \frac{e^x + e^{-x}}{2} \quad \text{and} \quad \sinh x = \frac{e^x - e^{-x}}{2}. \quad \text{Show}$$
that

(a) $\dfrac{d}{dx}(\cosh x) = \sinh x$.

(b) $\dfrac{d}{dx}(\sinh x) = \cosh x$.

(c) $\cosh^2 x - \sinh^2 x = 1$.

(d) $\cosh x$ and $\sinh x$ are linearly independent functions.

4.2 Fundamental Set of Solutions

The Principle of Superposition • Fundamental Set of Solutions • Existence of a Fundamental Set of Solutions • Reduction of Order

The Principle of Superposition

We will see that to find a general solution of the nth-order linear equation

$$a_n(x)y^{(n)}(x) + \cdots + a_1(x)y'(x) + a_0(x)y(x) = g(x)$$

we must find a general solution of the corresponding homogeneous equation

$$a_n(x)y^{(n)}(x) + \cdots + a_1(x)y'(x) + a_0(x)y(x) = 0$$

because of the following *very* important property of linear homogeneous equations.

Theorem 4.3 Principle of Superposition

If $S = \{f_1(x), f_2(x), \ldots, f_{k-1}(x), f_k(x)\}$ is a set of k solutions of the equation

$$a_n(x)y^{(n)}(x) + \cdots + a_1(x)y'(x) + a_0(x)y(x) = 0$$

and $\{c_1, c_2, \ldots, c_{k-1}, c_k\}$ is a set of k constants, then

$$f(x) = c_1 f_1(x) + c_2 f_2(x) + \cdots + c_{k-1}f_{k-1}(x) + c_k f_k(x)$$

is also a solution of $a_n(x)y^{(n)}(x) + \cdots + a_1(x)y'(x) + a_0(x)y(x) = 0$.

The expression $f(x) = c_1 f_1(x) + c_2 f_2(x) + \cdots + c_{k-1}f_{k-1}(x) + c_k f_k(x)$ is called a **linear combination** of the functions in the set $S = \{f_1(x), f_2(x), \ldots, f_{k-1}(x), f_k(x)\}$.

Example 1

Show that $y(x) = c_1 e^{-x} + c_2 \cos x$ is a solution of $y''' + y'' + y' + y = 0$.

Solution The functions e^{-x} and $\cos x$ are both solutions to the equation because

$$\frac{d^3}{dx^3}(e^{-x}) + \frac{d^2}{dx^2}(e^{-x}) + \frac{d}{dx}(e^{-x}) + e^{-x} = -e^{-x} + e^{-x} - e^{-x} + e^{-x} = 0$$

and

$$\frac{d^3}{dx^3}(\cos x) + \frac{d^2}{dx^2}(\cos x) + \frac{d}{dx}(\cos x) + \cos x = \sin x - \cos x - \sin x + \cos x = 0$$

so by the Principle of Superposition, $y(x) = c_1 e^{-x} + c_2 \cos x$ is also a solution of $y''' + y'' + y' + y = 0$.

It is *very* important to remember that the Principle of Superposition is *only* valid for linear homogeneous differential equations.

Example 2

The functions $y_1(x) = c_1 x^2 + c_2 x - \dfrac{c_2^2}{12c_1}$, $y_2(x) = c_3 x^{1/2}$, and $y_3(x) = c_4 x^{3/2}$, where c_1, c_2, c_3, and c_4 are constants, are solutions to the nonlinear differential equation

$$\frac{4}{3}(y' - xy'')^2 + 2yy'' - (y')^2 = 0.$$

According to the Principle of Superposition, any linear combination of a set of solutions to a *linear homogeneous equation* is also a solution. Is the linear combination $y = y_1 + y_2 + y_3$ a solution to this *nonlinear equation*?

Solution   If $y(x) = y_1(x) + y_2(x) + y_3(x)$,

$$\frac{4}{3}(y' - xy'')^2 + 2yy'' - (y')^2 = c_3 c_4 + \frac{c_2^2 c_3}{24 c_1} x^{-3/2} + \left(\frac{c_2 c_3}{2} - \frac{c_2^2 c_4}{8 c_1} \right) x^{-1/2}$$

$$+ \frac{3 c_1 c_3 + c_2 c_4}{2} x^{1/2} - \frac{c_1 c_4}{2} x^{3/2},$$

which is not the zero function because the set of functions $\{1, x^{-3/2}, x^{-1/2}, x^{1/2}, x^{3/2}\}$ is linearly independent. (Why?) Thus, we see that no nontrivial linear combination of y_1, y_2, and y_3 is a solution to the nonlinear equation.

Fundamental Set of Solutions

To find a general solution of an nth-order linear homogeneous equation, we will see that we must obtain a collection of n linearly independent solutions to the equation. We therefore state the following definition.

Definition 4.4 Fundamental Set of Solutions

A set $S = \{f_1(x), f_2(x), f_3(x), \ldots, f_{n-1}(x), f_n(x)\}$ of n linearly independent solutions of the nth-order linear homogeneous equation

$$a_n(x) y^{(n)}(x) + a_{n-1}(x) y^{(n-1)}(x) + \cdots + a_1(x) y'(x) + a_0(x) y(x) = 0$$

is called a **fundamental set of solutions** of the equation.

An important consequence of the Principle of Superposition is that any linear combination of the functions in a fundamental set of solutions of an nth-order homogeneous linear differential equation is also a solution of the differential equation.

Corollary

If $S = \{f_1(x), f_2(x), \ldots, f_{n-1}(x), f_n(x)\}$ is a fundamental set of solutions of the equation

$$a_n(x) y^{(n)}(x) + \cdots + a_1(x) y'(x) + a_0(x) y(x) = 0$$

and $\{c_1, c_2, \ldots, c_{n-1}, c_n\}$ is a set of n constants, then

$$f(x) = c_1 f_1(x) + c_2 f_2(x) + \cdots + c_{n-1} f_{n-1}(x) + c_n f_n(x)$$

is also a solution of $a_n(x) y^{(n)}(x) + \cdots + a_1(x) y'(x) + a_0(x) y(x) = 0$.

Example 3

Show that $S = \{e^{-5x}, e^{-x}\}$ is a fundamental set of solutions of the equation

$$y'' + 6y' + 5y = 0.$$

Solution S is linearly independent because

$$W(S) = \begin{vmatrix} e^{-5x} & e^{-x} \\ -5e^{-5x} & -e^{-x} \end{vmatrix} = -e^{-6x} + 5e^{-6x} = 4e^{-6x} \neq 0.$$

In addition, we must verify that each function is a solution of the differential equation. Because

$$\frac{d^2}{dx^2}(e^{-5x}) + 6\frac{d}{dx}(e^{-5x}) + 5e^{-5x} = 25e^{-5x} - 30e^{-5x} + 5e^{-5x} = 0$$

and

$$\frac{d^2}{dx^2}(e^{-x}) + 6\frac{d}{dx}(e^{-x}) + 5e^{-x} = e^{-x} - 6e^{-x} + 5e^{-x} = 0,$$

we conclude that S is a fundamental set of solutions of the equation
$y'' + 6y' + 5y = 0$.

Example 4

Determine if $S = \{e^x, e^{2x}\}$ is a fundamental set of solutions of the differential equation $y'' - y = 0$.

Solution The Wronskian of S is $W(S) = \begin{vmatrix} e^x & e^{2x} \\ e^x & 2e^{2x} \end{vmatrix} = 2e^{3x} - e^{3x} = e^{3x} \neq 0,$

so S is linearly independent. However, when we investigate if the functions satisfy the differential equation, we run into a problem. For $y = e^x$ we have

$$\frac{d^2}{dx^2}(e^x) - (e^x) = e^x - e^x = 0.$$

On the other hand,

$$\frac{d^2}{dx^2}(e^{2x}) - (e^{2x}) = 4e^{2x} - e^{2x} = 3e^{2x} \neq 0,$$

so $y = e^{2x}$ is *not* a solution to the differential equation. Therefore, S is not a fundamental set of solutions.

Show that both $S = \{e^{-x}, e^x\}$ and $S = \{\sinh x, \cosh x\}$ are fundamental sets of solutions of $y'' - y = 0$.

Example 5

Determine if $S = \{e^x, e^{2x}\}$ is a fundamental set of solutions of the differential equation $y''' - 5y'' + 8y' - 4y = 0$.

Solution   In Example 4 we saw that these functions are linearly independent. However, this set cannot form a fundamental set of solutions of the third-order equation $y''' - 5y'' + 8y' - 4y = 0$ because there are only two functions in S. In order to have a fundamental set of this equation, we must have three linearly independent solutions in the set. You should verify that $S = \{e^x, e^{2x}, xe^{2x}\}$ is a fundamental set of solutions of $y''' - 5y'' + 8y' - 4y = 0$.

Existence of a Fundamental Set of Solutions

The following two theorems tell us that under reasonable conditions, the nth-order linear homogeneous equation

$$a_n(x)y^{(n)}(x) + a_{n-1}(x)y^{(n-1)}(x) + \cdots + a_1(x)y'(x) + a_0(x)y(x) = 0$$

has a fundamental set of n solutions.

Theorem 4.4

If $a_i(x)$ is continuous on an open interval I for $i = 0, 1, \ldots, n$, and $a_n(x) \neq 0$ for all x in the interval I, then the nth-order linear homogeneous equation

$$a_n(x)y^{(n)}(x) + a_{n-1}(x)y^{(n-1)}(x) + \cdots + a_1(x)y'(x) + a_0(x)y(x) = 0$$

has a fundamental set of n solutions.

Theorem 4.5

Any set of $n + 1$ solutions of the nth-order linear homogeneous equation

$$a_n(x)y^{(n)}(x) + a_{n-1}(x)y^{(n-1)}(x) + \cdots + a_1(x)y'(x) + a_0(x)y(x) = 0$$

is linearly dependent.

See the exercises for a discussion of the proof of these two theorems.

Is every set of n functions a fundamental set of solutions of an nth-order homogeneous equation? Can a fundamental set of this equation contain more than n functions?

Thus, if $S = \{f_1(x), f_2(x), \ldots, f_{n-1}(x), f_n(x)\}$ is a fundamental set of solutions of the linear homogeneous equation

$$a_n(x)y^{(n)}(x) + \cdots + a_1(x)y'(x) + a_0(x)y(x) = 0$$

and $f(x)$ is *any* solution to the equation, then the set of functions $S = \{f_1(x), f_2(x), \ldots, f_{n-1}(x), f_n(x), f(x)\}$ is linearly dependent. There are $n + 1$ constants $\{c_1, c_2, \ldots, c_{n-1}, c_n, c_{n+1}\}$, not all of which are zero, so that

$$c_1 f_1(x) + c_2 f_2(x) + \cdots + c_{n-1}f_{n-1}(x) + c_n f_n(x) + c_{n+1}f(x) = 0.$$

If $c_{n+1} = 0$, we obtain $c_1 f_1(x) + c_2 f_2(x) + \cdots + c_{n-1}f_{n-1}(x) + c_n f_n(x) = 0$, which implies that S is linearly dependent, so $c_{n+1} \neq 0$. Thus,

$$f(x) = -\frac{1}{c_{n+1}}(c_1 f_1(x) + c_2 f_2(x) + \cdots + c_{n-1}f_{n-1}(x) + c_n f_n(x))$$ and we see that $f(x)$

can be written as a linear combination of the functions in S.

We can summarize the results of these theorems by saying that in order to find *all* solutions to an nth-order linear homogeneous ordinary differential equation, we must determine a set of n linearly independent solutions. (Notice that the number of linearly independent solutions equals the order of the differential equation.)

Definition 4.5 General Solution

If $S = \{f_1(x), f_2(x), \ldots, f_{n-1}(x), f_n(x)\}$ is a fundamental set of solutions of the nth-order linear homogeneous equation

$$a_n(x)y^{(n)}(x) + \cdots + a_1(x)y'(x) + a_0(x)y(x) = 0,$$

a **general solution** of the equation is

$$y(x) = c_1 f_1(x) + c_2 f_2(x) + \cdots + c_{n-1}f_{n-1}(x) + c_n f_n(x),$$

where $\{c_1, c_2, \ldots, c_{n-1}, c_n\}$ is a set of n arbitrary constants.

In other words, if we have a fundamental set of solutions S, a general solution of the differential equation is formed by taking the linear combination of the functions in S. Notice that because the constants are arbitrary, we obtain the trivial solution, $y = 0$, if we let $c_1 = c_2 = \cdots = c_n = 0$. The trivial solution satisfies every homogeneous linear differential equation. In addition, any constant multiple of a solution of a homogeneous linear differential equation is also a solution of the equation. (This is seen by letting all but one of the constants $\{c_1, c_2, \ldots, c_{n-1}, c_n\}$ be zero.)

Example 6 Show that $y = c_1 e^{-5x} + c_2 e^{-x}$ is a general solution of $y'' + 6y' + 5y = 0$.

Solution In Example 3, we showed that $S = \{e^{-5x}, e^{-x}\}$ is a fundamental set of solutions of $y'' + 6y' + 5y = 0$. If follows that $y = c_1 e^{-5x} + c_2 e^{-x}$ is a general solution of $y'' + 6y' + 5y = 0$.

Show that the solution to the initial-value problem $\begin{cases} y'' + 3y' + 2y = 0 \\ y(0) = a, \ y'(0) = b \end{cases}$ *is*

$$y(x) = (2a + b)e^{-x} - (a + b)e^{-2x}.$$

Regardless of the choices of a and b, calculate $\lim_{x \to \infty} y(x).$

Example 7   Show that $y = c_1 e^{3x} + c_2 x e^{3x}$ is a general solution of $y'' - 6y' + 9y = 0.$

Solution   You should verify that $S = \{e^{3x}, xe^{3x}\}$ is a fundamental set of solutions of $y'' - 6y' + 9y = 0$ and, hence, $y = c_1 e^{3x} + c_2 x e^{3x}$ is a general solution of $y'' - 6y' + 9y = 0.$

 Show that the solution to the initial-value problem $\begin{cases} y'' - 6y' + 9y = 0 \\ y(0) = a, \ y'(0) = b \end{cases}$ *is*

$$y(x) = ae^{3x} + (b - 3a)xe^{3x}.$$

Find conditions on a and b so that $y'(x) \neq 0$ for all $x > 0$. For these choices of a and b, describe the shape of the graph of the solution. Confirm your results by graphing the solution for several choices of a and b.

Reduction of Order

In the next section, we learn how to find solutions of homogeneous equations with constant coefficients. In doing so, we will find it necessary to determine a second linearly independent solution from a known solution. We illustrate this procedure, called **reduction of order,** by considering a second-order equation.

Suppose we have the equation

$$y'' + p(x)y' + q(x)y = 0,$$

and that $y = f(x)$ is a solution. We know from our previous discussion that in order to find a general solution of this second-order equation, we must have two linearly independent solutions. We must determine a second linearly independent solution. We accomplish this by attempting to find a solution of the form

$$y = v(x)f(x)$$

and solving for $v(x)$. Differentiating with the product rule, we obtain

$$y' = f'v + v'f \quad \text{and} \quad y'' = f''v + 2v'f' + fv''.$$

For convenience, we have omitted the argument of these functions. We now substitute y, y', and y'' into the equation $y'' + p(x)y' + q(x)y = 0$. This gives us

$$y'' + p(x)y' + q(x)y = f''v + 2v'f' + fv'' + p(x)(f'v + fv') + q(x)vf$$
$$= \underbrace{[f'' + p(x)f' + q(x)f]}_{= 0}v + fv'' + 2v'f' + p(x)fv'$$
$$= fv'' + (2f' + p(x)f)v'.$$

Therefore, we have the equation

$$fv'' + (2f' + p(x)f)v' = 0$$

which can be written as a first-order equation by letting $w = v'$. Making this substitution gives us the linear first-order equation

$$fw' + (2f' + p(x)f)w = 0 \qquad \text{or} \qquad f\frac{dw}{dx} + (2f' + p(x)f)w = 0$$

which is separable, so we obtain the separated equation

$$\frac{dw}{w} = \left(-\frac{2f'}{f} - p\right) dx.$$

We solve this equation by integrating both sides of the equation to yield

$$\ln|w| = \ln\left(\frac{1}{f^2}\right) - \int p(x)\,dx.$$

This means that $w = \dfrac{1}{f^2}e^{-\int p(x)\,dx}$, so we have the formula $\dfrac{dv}{dx} = \dfrac{1}{f^2}e^{-\int p(x)\,dx}$ or

$$v(x) = \int \frac{e^{-\int p(x)\,dx}}{[f(x)]^2}\,dx.$$

If $y = f(x)$ is a known solution of the differential equation $y'' + p(x)y' + q(x)y = 0$, we can obtain a second linearly independent solution of the form $y(x) = f(x)v(x)$ where $v(x) = \int \dfrac{e^{-\int p(x)\,dx}}{[f(x)]^2}\,dx.$

We leave the proof that $y_1(x) = f(x)$ and $y_2(x) = f(x)v(x) = f(x)\displaystyle\int \frac{e^{-\int p(x)\,dx}}{[f(x)]^2}\,dx$ are linearly independent as an exercise.

Example 8

Determine a second linearly independent solution to the differential equation $y'' + 6y' + 9y = 0$ given that $y = e^{-3x}$ is a solution.

Solution First we identify the functions $p(x) = 6$ and $f(x) = e^{-3x}$. Then we determine the function $v(x)$ such that $y(x) = f(x)v(x)$ with the formula

$$v(x) = \int \frac{e^{-\int p(x)\,dx}}{[f(x)]^2}\,dx = \int \frac{e^{-\int 6\,dx}}{[e^{-3x}]^2}\,dx = \int \frac{e^{-6x}}{e^{-6x}}\,dx = \int dx = x.$$

A second linearly independent solution is $y = f(x)v(x) = xe^{-3x}$.

Example 9

Determine a second linearly independent solution to the differential equation

$$4x^2\frac{d^2y}{dx^2} + 8x\frac{dy}{dx} + y = 0, \ x > 0, \ \text{if} \ y = x^{-1/2} \ \text{is a solution.}$$

Solution   In this case, we must divide by $4x^2$ in order to obtain an equation of the form $y'' + p(x)y' + q(x)y = 0$. This gives us the equation $\frac{d^2y}{dx^2} + \frac{2}{x}\frac{dy}{dx} + \frac{1}{4x^2}y = 0$. Therefore, $p(x) = 2/x$ and $f(x) = x^{-1/2}$. Using the formula for v, we obtain

$$v(x) = \int \frac{e^{-\int p(x)\,dx}}{[f(x)]^2}\,dx = \int \frac{e^{-\int 2/x\,dx}}{[x^{-1/2}]^2}\,dx = \int \frac{e^{-2\ln(x)}}{x^{-1}}\,dx = \int x^{-1}\,dx = \ln x, \ x > 0.$$

A second solution is $y = f(x)v(x) = x^{-1/2}\ln x$.

IN TOUCH WITH TECHNOLOGY

Many computer algebra systems can compute a general solution of nth-order homogeneous equations with constant coefficients and some nth-order homogeneous equations with nonconstant coefficients as long as the order of the equation is less than or equal to 4. For each of the following exercises, use a computer algebra system to compute a general solution of each equation. Use the results to determine and graph a fundamental set of solutions for each equation.

1. $y^{(4)} + 2y''' + 5y'' + 8y' + 4y = 0$

2. $16y^{(4)} + 72y'' - 48y' + 85y = 0$

3. $x^4y^{(4)} + 6x^3y''' + 2x^2y'' - 4xy' + 4y = 0$

4. $x^4y^{(4)} + 4x^3y''' + 7x^2y'' + 3xy' + 5y = 0$

5. (a) Find conditions on the constants c_1, c_2, c_3, and c_4 so that $y(x) = c_1 + c_2 \tan(c_3 + c_4 \ln x)$ is a solution of the nonlinear second-order equation $xy'' - 2yy' = 0$. **(b)** Is the Principle of Superposition valid for this equation? Explain.

As with first-order equations, most computer algebra systems contain built-in commands to generate numerical solutions of higher-order initial-value problems. Use the numerical capabilities of a computer algebra system to help you with the following problems.

6. Graph the solution to the initial-value problem
$$\begin{cases} y'' = \sin(x^2y') \\ y(0) = 0, \ y'(0) = 1 \end{cases}$$ on the interval $[0, 10]$.

 7. The **Van-der-Pol equation,** which arises in the study of nonlinear damping, is the equation

$$x'' + \mu(x^2 - 1)x' + x = 0.$$

(a) Graph the solution on the interval $[0, 15]$ for various values of μ (try $\frac{1}{8}$, $\frac{1}{4}$, $\frac{1}{2}$, 1, $\frac{3}{2}$, 2, 3, 5, 7, and 9) if $x(0) = 1$ and $x'(0) = 0$. **(b)** How does the solution look in these cases? Compare the graphs of these solutions to the graph of the solution of the initial-value problem $\begin{cases} x'' + x = 0 \\ x(0) = 1, \ x'(0) = 0 \end{cases}$, which most computer algebra systems can solve exactly.

In Exercises 1–10, show that S is a fundamental set of solutions for the given equation. (See Exercises 4.1 to verify linear independence.)

1. $S = \{e^{-6x}, e^{-4x}\}$, $y'' + 10y' + 24y = 0$

2. $S = \{\cos 2x, \sin 2x\}$; $y'' + 4y = 0$

*3. $S = \{e^{-3x} \cos 3x, e^{-3x} \sin 3x\}$; $y'' + 6y' + 18y = 0$

4. $S = \{e^{5x} \cos 4x, e^{5x} \sin 4x\}$; $y'' - 10y' + 41y = 0$

5. $S = \{e^{-x}, e^{3x}, xe^{3x}\}$; $y''' - 5y'' + 3y' + 9y = 0$

6. $S = \{e^{-x}, e^{-2x}, e^{-4x}\}$; $y''' + 7y'' + 14y' + 8y = 0$

*7. $S = \{e^x, e^{-5x} \sin x, e^{-5x} \cos x\}$;
$y''' + 9y'' + 16y' - 26y = 0$

8. $S = \{e^{3x}, e^{-2x}, e^{5x}, xe^{5x}\}$;
$y^{(4)} - 11y''' + 29y'' + 35y' - 150y = 0$

9. $S = \{e^{-3x}, e^{-x}, e^{-4x} \cos 3x, e^{-4x} \sin 3x\}$;
$y^{(4)} + 12y''' + 60y'' + 124y' + 75y = 0$

10. $S = \{e^{-x} \cos 2x, e^{-x} \sin 2x, e^{2x} \cos 5x, e^{2x} \sin 5x\}$;
$y^{(4)} - 2y''' + 26y'' + 38y' + 145y = 0$

11. Let $ay'' + by' + cy = 0$ be a homogeneous second-order equation with constant coefficients and let m_1 and m_2 be the solutions of the equation $am^2 + bm + c = 0$.

(a) If $m_1 \neq m_2$ and both m_1 and m_2 are real, show that $\{e^{m_1 x}, e^{m_2 x}\}$ is a fundamental set of solutions of $ay'' + by' + cy = 0$.

(b) If $m_1 = m_2$, show that $\{e^{m_1 x}, xe^{m_1 x}\}$ is a fundamental set of solutions of $ay'' + by' + cy = 0$.

(c) If $m_1 = \alpha + i\beta$, $\beta \neq 0$, and $m_2 = \overline{m_1} = \alpha - i\beta$, show that $\{e^{\alpha x} \cos \beta x, e^{\alpha x} \sin \beta x\}$ is a fundamental set of solutions of $ay'' + by' + cy = 0$.

In Exercises 12–21, use reduction of order with the solution $y_1(x)$ to find a second linearly independent solution of the given differential equation.

12. $y_1(x) = e^{3x}$, $y'' - 5y' + 6y = 0$

13. $y_1(x) = e^{-2x}$, $y'' + 6y' + 8y = 0$

14. $y_1(x) = e^{2x}$, $y'' - 4y' + 4y = 0$

*15. $y_1(x) = e^{-5x}$, $y'' + 10y' + 25y = 0$

16. $y_1(x) = \cos 2x$, $y'' + 4y = 0$

17. $y_1(x) = \sin 7x$, $y'' + 49y = 0$

18. $y_1(x) = x^{-4}$, $x^2 y'' + 4xy' - 4y = 0$

*19. $y_1(x) = x^{-2}$, $x^2 y'' + 6xy' + 6y = 0$

20. $y_1(x) = x$, $x^2 y'' + 3xy' + y = 0$

21. $y_1(x) = x^{-1}$, $x^2 y'' - xy' + y = 0$

22. Find a and b so that $S = \{e^{-x} \cos 2x, e^{-x} \sin 2x\}$ is a fundamental set of solutions for $y'' + ay' + by = 0$.

*23. Find a, b, and c so that $S = \{e^{-2x}, e^{-x}, e^x\}$ is a fundamental set of solutions for $y''' + ay'' + by' + cy = 0$.

24. Show that if $y_1(x)$ and $y_2(x)$ are linearly independent solutions of $p_0(x)y'' + p_1(x)y' + p_2(x)y = 0$, then so are the functions $y(x) = y_1(x) + y_2(x)$ and $y(x) = y_1(x) - y_2(x)$.

25. Prove Theorem 4.3 (the Principle of Superposition) and the subsequent corollary.

26. If the Principle of Superposition ever valid for linear nonhomogeneous equations? Explain.

*27. Suppose that $f(x)$ is a solution to the equation $y'' + p(x)y' + q(x)y = 0$. Show that $f(x)$ and the solution $f(x) \int \dfrac{e^{-\int p(x)\,dx}}{[f(x)]^2}\,dx$ obtained by reduction of order are linearly independent. (*Hint:* Use the Wronskian.)

28. **(Existence of a Fundamental Set of Solutions)** (a) Use Theorem 4.1 (Existence of a Unique Solution) to prove Theorems 4.4 and 4.5 for second-order equations using the following steps. By Theorem 4.1, the initial-value problems

$$\begin{cases} a_2(x)y'' + a_1(x)y' + a_0(x)y = 0 \\ y(x_0) = 1, \ y'(x_0) = 0 \end{cases}$$

and

$$\begin{cases} a_2(x)y'' + a_1(x)y' + a_0(x)y = 0 \\ y(x_0) = 0, \ y'(x_0) = 1 \end{cases}$$

have unique solutions $y_1(x)$ and $y_2(x)$, respectively. **(i)** Show that $\{y_1(x), y_2(x)\}$ is a fundamental set of solutions for $a_2(x)y'' + a_1(x)y' + a_0(x)y = 0$. **(ii)** If $y_3(x)$ is any other solution, suppose that $y_3(x_0) = c_1$ and $y_3'(x_0) = c_2$. Use the fact that the solution to the initial-value problem

$$\begin{cases} a_2(x)y'' + a_1(x)y' + a_0(x)y = 0 \\ y(x_0) = c_1, \ y'(x_0) = c_2 \end{cases}$$

is both $y_3(x)$ and $c_1y_1(x) + c_2y_2(x)$ (why?) so $y_3(x) = c_1y_1(x) + c_2y_2(x)$ to show that $\{y_1(x), y_2(x), y_3(x)\}$ is linearly dependent.

(b) Indicate how to generalize the result in **(a)** for higher-order equations.

29. Show that a general solution of $y'' - k^2y = 0$ is $y = c_1 \cosh kx + c_2 \sinh kx$, where $\cosh kx = \dfrac{e^{kx} + e^{-kx}}{2}$ and $\sinh kx = \dfrac{e^{kx} - e^{-kx}}{2}$.

4.3 Solutions of Linear Homogeneous Equations with Constant Coefficients

The Characteristic Equation • Second-Order Linear Homogeneous Equations with Constant Coefficients • Higher-Order Linear Homogeneous Equations with Constant Coefficients

We now turn our attention to solving linear homogeneous equations with constant coefficients.

Euler was the first to develop a general method for solving the linear homogeneous differential equation with constant coefficients

$$a_ny^{(n)} + \cdots + a_1y' + a_0y = 0$$

by factoring the corresponding **characteristic equation**

$$a_np^n + \cdots + a_1p + a_0 = 0.$$

For example, to solve the equation $y - k^4\dfrac{d^4y}{dx^4} = 0$ *proposed by Bernoulli in 1735, Euler first wrote the corresponding characteristic equation*

$$1 - k^4p^4 = 0,$$

factored it to obtain

$$(1 - kp)(1 + kp)(1 + k^2p^2) = 0$$

with solutions $-1/k$, $1/k$, *and* $\pm i/k$ *and then said that a general solution of the equation is*

$$y = Ae^{-x/k} + Be^{x/k} + C \sin \frac{x}{k} + D \cos \frac{x}{k}.$$

Thus, Euler produced a general solution of $y - k^4\dfrac{d^4y}{dx^4} = 0$ *in 1739, four years after the equation was proposed by Bernoulli. Unfortunately, Euler did not explain how he arrived at this solution. In fact, in [Katz]* we learn that Bernoulli further factored*

$$(1 - kp)(1 + kp)(1 + k^2p^2) = 0$$

---

* Victor J. Katz, *A History of Mathematics: An Introduction*, HarperCollins, 1993, p. 506.

as

$$(1 - kp)(1 + kp)(1 + ikp)(1 - ikp) = 0$$

with solutions $-1/k$, $1/k$, and $\pm i/k$ and deduced that another general solution of the equation is

$$y = Ae^{-x/k} + Be^{x/k} + Ce^{xi/k} + De^{-xi/k}.$$

*The fact that these two general solutions do not **look** the same bothered Bernoulli until Euler convinced him that $2 \cos x$ and $e^{ix} + e^{-ix}$ are the same because they satisfy the same differential equation, and subsequently that*

$$e^{ix} = \cos x + i \sin x$$

and

$$e^{-ix} = \cos x - i \sin x.$$

We will find these formulas useful later in the section.

The Characteristic Equation

Now that we understand the meaning of a general solution of a linear homogeneous differential equation, we want to explain how to find general solutions of linear homogeneous differential equations with constant coefficients. Suppose that we wish to find a general solution of $7y'' + 3y' - 4y = 0$. Looking back on our previous examples in Section 4.2, we see that some second-order linear homogeneous differential equations with constant coefficients have a general solution of the form $y = c_1 e^{m_1 x} + c_2 e^{m_2 x}$, where c_1 and c_2 represent arbitrary constants and m_1 and m_2 are constants. If $y = e^{mx}$, where m is a number to be determined, is a solution of $7y'' + 3y' - 4y = 0$, differentiating twice yields $y' = me^{mx}$ and $y'' = m^2 e^{mx}$ and substituting into the equation $7y'' + 3y' - 4y = 0$ results in

$$7 \cdot m^2 e^{mx} + 3 \cdot me^{mx} - 4e^{mx} = 0$$
$$e^{mx}(7m^2 + 3m - 4) = 0.$$

Thus, we see that the solutions of $7y'' + 3y' - 4y = 0$ depend on the solutions of the equation $7m^2 + 3m - 4 = 0$. In fact, solving this equation for m results in

$$7m^2 + 3m - 4 = (m + 1)(7m - 4) = 0$$

so that $m = -1$ or $m = \frac{4}{7}$. Substituting these values of m into $y = e^{mx}$, we obtain that two solutions of $7y'' + 3y' - 4y = 0$ are $y = e^{-x}$ and $y = e^{4x/7}$. You should verify that these two functions are linearly independent, which implies that $\{e^{-x}, e^{4x/7}\}$ is a fundamental set of solutions of $7y'' + 3y' - 4y = 0$ and, thus, a general solution is given by $y = c_1 e^{-x} + c_2 e^{4x/7}$, where c_1 and c_2 denote arbitrary constants.

In fact, we will see that solutions of any nth-order linear homogeneous differential equation with constant coefficients can be constructed in the same way. We state the following definition.

Definition 4.6 Characteristic Equation

The equation

$$a_n m^n + a_{n-1} m^{n-1} + \cdots + a_1 m + a_0 = 0$$

is called the **characteristic equation** of the nth-order linear homogeneous differential equation with constant coefficients

$$a_n y^{(n)}(x) + a_{n-1} y^{(n-1)}(x) + \cdots + a_1 y'(x) + a_0 y(x) = 0.$$

Second-Order Linear Homogeneous Equations with Constant Coefficients

Let us begin our investigation by considering the second-order linear homogeneous equation with constant coefficients

$$ay'' + by' + cy = 0.$$

Suppose that

$$y = e^{mx}$$

is a solution to this equation. We would like to determine the value(s) of m, if any, which lead to a solution. We accomplish this by differentiating $y = e^{mx}$ twice, which yields $y' = m e^{mx}$ and $y'' = m^2 e^{mx}$. Substituting into the differential equation $ay'' + by' + cy = 0$,

$$ay'' + by' + cy = am^2 e^{mx} + bm e^{mx} + c e^{mx} = e^{mx}(am^2 + bm + c) = 0.$$

The value(s) of m that leads to a solution of the differential equation is found by solving the characteristic equation

$$am^2 + bm + c = 0$$

because $e^{mx} \neq 0$ for all values of x. (Why?) Of course, we know that the roots of this quadratic equation depend on the values of a, b, and c. We state these possibilities with the corresponding solutions in the following theorem.

Theorem 4.6 Solving Second-Order Equations with Constant Coefficients

Let $ay'' + by' + cy = 0$ be a homogeneous second-order equation with constant real coefficients and let m_1 and m_2 be the solutions of the equation $am^2 + bm + c = 0$.

(a) If $m_1 \neq m_2$ and both m_1 and m_2 are real, a general solution of $ay'' + by' + cy = 0$ is

$$y = c_1 e^{m_1 x} + c_2 e^{m_2 x}.$$

(b) If $m_1 = m_2$, a general solution of $ay'' + by' + cy = 0$ is

$$y = c_1 e^{m_1 x} + c_2 x e^{m_1 x}.$$

$\overline{m_1}$ is the complex conjugate of m_1:

$$\overline{m_1} = \overline{\alpha + i\beta} = \alpha - i\beta.$$

(c) If $m_1 = \alpha + i\beta$, $\beta \neq 0$, and $m_2 = \overline{m_1} = \alpha - i\beta$, a general solution of $ay'' + by' + cy = 0$ is

$$y = c_1 e^{\alpha x} \cos \beta x + c_2 e^{\alpha x} \sin \beta x = e^{\alpha x}(c_1 \cos \beta x + c_2 \sin \beta x).$$

The proof of Theorem 4.6 is discussed after the following examples, which illustrate each of these three situations.

Example 1

Solve $y'' + 3y' - 4y = 0$.

Solution The characteristic equation of $y'' + 3y' - 4y = 0$ is $m^2 + 3m - 4 = (m + 4)(m - 1) = 0$ with roots $m = -4$ and $m = 1$. Therefore, a general solution of $y'' + 3y' - 4y = 0$ is

$$y(x) = c_1 e^{-4x} + c_2 e^x.$$

(a) Determine the solution to $y'' + 3y' - 4y = 0$ that satisfies $y(0) = a$ and $y'(0) = b$. (b) Find conditions on a and b so that the limit as $x \to \infty$ of the resulting solution is 0. Confirm your result with a graph.

Example 2

Solve $y'' + 2y' + y = 0$.

Solution The characteristic equation of $y'' + 2y' + y = 0$ is $m^2 + 2m + 1 = (m + 1)^2 = 0$. The solution of the characteristic equation is $m_1 = m_2 = -1$, so a general solution of $y'' + 2y' + y = 0$ is $y(x) = c_1 e^{-x} + c_2 x e^{-x}$.

Find conditions on b, if any, so that the solution which satisfies the initial-value problem $\begin{cases} y'' + 2y' + y = 0 \\ y(0) = 0, y'(0) = b \end{cases}$ attains a relative maximum at $x = 1$. (See Figure 4.1.)

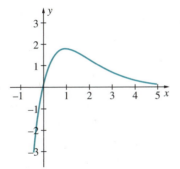

Figure 4.1 Graph of a solution of $y'' + 2y' + y = 0$ that attains a relative maximum if $x = 1$ and satisfies $y(0) = 0$

Solve $y'' + 4y' + 20y = 0$ subject to $y(0) = 3$ and $y'(0) = -1$.

Solution   The   characteristic   equation   of $y'' + 4y' + 20y = 0$ is $m^2 + 4m + 20 = 0$ with roots $m = -2 \pm 4i$. Therefore, a general solution of $y'' + 4y' + 20y = 0$ is

$$y(x) = e^{-2x}(c_1 \cos 4x + c_2 \sin 4x),$$

because the solutions of the characteristic equation are complex conjugates. To find the solution for which $y(0) = 3$ and $y'(0) = -1$, we first calculate

$$y' = 2e^{-2x}(-c_1 \cos 4x + 2c_2 \cos 4x - 2c_1 \sin 4x - c_2 \sin 4x)$$

with the product rule and then evaluate both

$$y(0) = c_1 \qquad \text{and} \qquad y'(0) = 2(2c_2 - c_1)$$

obtaining the system of equations $\begin{cases} c_1 = 3 \\ 2(2c_2 - c_1) = -1 \end{cases}$. Substituting $c_1 = 3$ into the

second equation results in $c_2 = \frac{5}{4}$. Thus, the solution of $y'' + 4y' + 20y = 0$ for which $y(0) = 3$ and $y'(0) = -1$ is $y(x) = e^{-2x}(3 \cos 4x + \frac{5}{4} \sin 4x)$.

Figure 4.2   Graph of $y(x) = e^{-2x}\left(3 \cos 4x + \dfrac{5}{4} \sin 4x\right)$

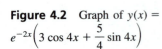

In Example 3, evaluate $\lim_{x \to \infty} y(x)$. *Does this limit depend on the initial conditions? Does this result agree with the graph in Figure 4.2?*

Now we explain the results that were given for the three cases associated with the equation $ay'' + by' + cy = 0$ in Theorem 4.6 by considering the roots of the character-istic equation $am^2 + bm + c = 0$, which are $m_{1,2} = \dfrac{-b \pm \sqrt{b^2 - 4ac}}{2a}$.

PROOF OF THEOREM 4.6:

(a) The roots m_1 and m_2 satisfy the characteristic equation so the functions $e^{m_1 x}$ and $e^{m_2 x}$ are solutions of $ay'' + by' + cy = 0$. In addition, we verify that the solutions are linearly independent by computing the Wronskian:

$$W = \begin{vmatrix} e^{m_1 x} & e^{m_2 x} \\ m_1 e^{m_1 x} & m_2 e^{m_2 x} \end{vmatrix} = m_2 e^{(m_1 + m_2)x} - m_1 e^{(m_1 + m_2)x} = (m_2 - m_1)e^{(m_1 + m_2)x} \neq 0$$

because $m_2 \neq m_1$ and $e^{(m_1 + m_2)x} \neq 0$. Therefore, the functions are linearly independent, so $\{e^{m_1 x}, e^{m_2 x}\}$ is a fundamental set of solutions and $y = c_1 e^{m_1 x} + c_2 e^{m_2 x}$ is a general solution.

(b) If $m_1 = m_2$, we must have $m_{1,2} = \dfrac{-b \pm \sqrt{b^2 - 4ac}}{2a} = -\dfrac{b}{2a}$. Because m_1 satis-fies the characteristic equation, $y_1(x) = e^{m_1 x} = e^{-(b/2a)x}$ is a solution to the differential equation. We use this solution to obtain a second linearly independent solution through reduction of order with the formula

$$y_2(x) = y_1(x) \int \frac{e^{-\int (b/a)\,dx}}{[y_1(x)]^2}\,dx = e^{(b/2a)x} \int \frac{e^{-(b/a)x}}{[e^{-(b/2a)x}]^2}\,dx$$

$$= e^{-(b/2a)x} \int \frac{e^{-(b/a)x}}{e^{-(b/a)x}}\,dx = e^{(b/2a)x} \int dx$$

$$= xe^{-(b/2a)x} = xe^{m_1 x}.$$

Thus, $\{e^{m_1 x}, xe^{m_1 x}\}$ is a fundamental set of solutions, so a general solution is
$y = c_1 e^{m_1 x} + c_2 x e^{m_1 x}$.

(c) Finally we consider the case with complex conjugate roots $m_{1,2} = \alpha \pm \beta i$. In order
to do this, however, we use Euler's formula $e^{i\theta} = \cos\theta + i\sin\theta$, mentioned in the
section introduction, which can be obtained through the use of the Maclaurin series

$$e^x = \sum_{k=0}^{\infty} \frac{x^k}{k!} = 1 + x + \frac{x^2}{2!} + \frac{x^3}{3!} + \frac{x^4}{4!} + \frac{x^5}{5!} + \cdots,$$

$$\cos x = \sum_{k=0}^{\infty} \frac{(-1)^k x^{2k}}{(2k)!} = 1 - \frac{x^2}{2!} + \frac{x^4}{4!} - \cdots,$$

and

$$\sin x = \sum_{k=0}^{\infty} \frac{(-1)^k x^{2k+1}}{(2k+1)!} = x - \frac{x^3}{3!} + \frac{x^5}{5!} - \cdots.$$

We derive Euler's formula using these Maclaurin series and substitution:

$$e^{i\theta} = \sum_{k=0}^{\infty} \frac{(i\theta)^k}{k!} = 1 + i\theta + \frac{i^2\theta^2}{2!} + \frac{i^3\theta^3}{3!} + \frac{i^4\theta^4}{4!} + \frac{i^5\theta^5}{5!} + \cdots$$

$$= 1 + i\theta - \frac{\theta^2}{2!} - \frac{i\theta^3}{3!} + \frac{\theta^4}{4!} + \frac{i\theta^5}{5!} + \cdots$$

$$= \left(1 - \frac{\theta^2}{2!} + \frac{\theta^4}{4!} + \cdots\right) + i\left(\theta - \frac{\theta^3}{3!} + \frac{\theta^5}{5!} + \cdots\right) = \cos\theta + i\sin\theta.$$

Using the properties of sine and cosine, Euler's formula also implies that
$e^{-i\theta} = \cos\theta - i\sin\theta$. The roots of the characteristic equation are $m_{1,2} = \alpha \pm \beta i$,
so $y_1 = e^{(\alpha + \beta i)x} = e^{\alpha x}(\cos\beta x + i\sin\beta x)$ and $y_2 = e^{(\alpha - \beta i)x} = e^{\alpha x}(\cos\beta x - i\sin\beta x)$
are both solutions to the differential equation. By the Principle of Superposition, any
linear combination of y_1 and y_2 is also a solution. For example, $z_1 = \frac{1}{2}(y_1 + y_2) =$
$e^{\alpha x}\cos\beta x$ and $z_2 = -\frac{i}{2}(y_1 - y_2) = e^{\alpha x}\sin\beta x$ are both solutions. We found in Exercise 11, Section 4.2 that $\{e^{\alpha x}\cos\beta x, e^{\alpha x}\sin\beta x\}$ is a fundamental set of solutions, so a
general solution is $y = e^{\alpha x}(c_1 \cos\beta x + c_2 \sin\beta x)$. ● ● ●

Higher-Order Linear Homogeneous Equations with Constant Coefficients

As with second-order equations, a general solution of the nth-order linear homogeneous differential equation with constant coefficients is determined by the solutions of its characteristic equation. In order to explain the process of finding a general solution of a higher-order equation, we state the following definition.

Definition 4.7 Multiplicity

Suppose that the characteristic equation $a_n m^n + a_{n-1}m^{n-1} + \cdots + a_1 m + a_0 = 0$ can be written in factored form as

$$a_n(m - m_1)^{k_1}(m - m_2)^{k_2}\cdots(m - m_r)^{k_r} = 0.$$

Then the roots of the equation are $m = m_1$, $m = m_2$, ..., and $m = m_r$, where the roots have **multiplicity** $k_1, k_2, \ldots, k_r$, respectively, and $k_1 + k_2 + \cdots + k_r = n$.

Example 4

Determine the roots and corresponding multiplicity of the characteristic equation of each of the following higher-order differential equations.
(a) $y''' + 4y'' + 4y' = 0$
(b) $y^{(4)} - 16y = 0$
(c) $y^{(4)} + 4y''' + 6y'' + 4y' + y = 0$

Solution   (a) The corresponding characteristic equation is $m^3 + 4m^2 + 4m = 0$. Factoring we have $m^3 + 4m^2 + 4m = m(m^2 + 4m + 4) = m(m + 2)^2 = 0$. Therefore, $m = 0$ is a root of multiplicity 1, and $m = -2$ is a root of multiplicity 2.
(b) In this case, the characteristic equation is $m^4 - 16 = 0$ and is factored as $(m^2 - 4)(m^2 + 4) = (m - 2)(m + 2)(m - 2i)(m + 2i) = 0$. The four roots $m = 2$, $m = -2$, $m = 2i$, and $m = -2i$ each have multiplicity 1.
(c) Factoring the characteristic equation $m^4 + 4m^3 + 6m^2 + 4m + 1 = 0$ yields

$$m^4 + 4m^3 + 6m^2 + 4m + 1 = (m + 1)^4 = 0.$$

Therefore, $m = -1$ is a root of multiplicity 4.

We state the following rules for finding a general solution of an nth-order linear homogeneous equation with constant coefficients because of the many situations that are encountered.

Rules for Determining a General Solution of a Higher-Order Equation

① Let m be a real root of the characteristic equation

$$a_n m^n + a_{n-1} m^{n-1} + \cdots + a_1 m + a_0 = 0$$

of an nth-order homogeneous linear differential equation with real constant coefficients. Then, e^{mx} is the solution associated with the root m.

If m is a root of multiplicity k where $k \geq 2$ of the characteristic equation, then the k solutions associated with m are

$$e^{mx}, \; xe^{mx}, \; x^2 e^{mx}, \ldots, \; x^{k-1} e^{mx}.$$

② Suppose that m and $\overline{m}$ represent the complex conjugate pair $\alpha \pm \beta i$. Then the two solutions associated with these two roots are

$$e^{\alpha x} \cos \beta x \qquad \text{and} \qquad e^{\alpha x} \sin \beta x.$$

If the values $\alpha \pm \beta i$ are each a root of multiplicity k of the characteristic equation, then the other solutions associated with this pair are

$$xe^{\alpha x} \cos \beta x, \; xe^{\alpha x} \sin \beta x, \; x^2 e^{\alpha x} \cos \beta x, \; x^2 e^{\alpha x} \sin \beta x, \ldots,$$
$$x^{k-1} e^{\alpha x} \cos \beta x, \; x^{k-1} e^{\alpha x} \sin \beta x.$$

A general solution to the nth order differential equation is the linear combination of the solutions obtained for all values of m.

Note that if $m_1, m_2, \ldots, m_r$ are the roots of the equation of multiplicity k_1, $k_2, \ldots, k_r$, respectively, then $k_1 + k_2 + \cdots + k_r = n$, where n is the order of the differential equation.

We now show how to use these rules to find a general solution. The key to the process is identifying each root of the characteristic equation and the associated solution(s).

Example 5

Solve the following equations: (a) $y''' - y' = 0$; (b) $y''' - 3y' + 2y = 0$; and (c) $y^{(4)} - y = 0$.

Solution   (a) In this case, the characteristic equation is $m^3 - m = 0$. Factoring, we have

$$m^3 - m = m(m^2 - 1) = m(m - 1)(m + 1) = 0.$$

The three distinct roots are

$$m_1 = 0, \qquad m_2 = 1, \qquad \text{and} \qquad m_3 = -1$$

and the corresponding solutions are

$$y = e^0 = 1, \qquad y = e^x, \qquad \text{and} \qquad y = e^{-x},$$

respectively. (See Table 4.1.)

TABLE 4.1

Root	Multiplicity	Corresponding Solution
$m = 0$	$k = 1$	$y = 1$
$m = 1$	$k = 1$	$y = e^x$
$m = -1$	$k = 1$	$y = e^{-x}$

A general solution of this differential equation is the linear combination of these functions,

$$y = c_1 + c_2 e^x + c_3 e^{-x}.$$

(b) The characteristic equation for this differential equation is

$$m^3 - 3m + 2 = 0.$$

The left-hand side of this equation is not as easy to factor as that of the previous problem. If we do not have access to a computer algebra system we try to guess* a root and then use division, if necessary, to find the other roots. By inspection, we see that $m = 1$ is a solution of the equation. Using this, we find that the equation can be written as

$$m^3 - 3m + 2 = (m - 1)^2(m + 2) = 0.$$

The roots are $m = 1$ of multiplicity 2 and $m = -2$ of multiplicity 1. The solutions corresponding to $m = 1$ are $y = e^x$ and $y = xe^x$, while the solution corresponding to $m = -2$ is $y = e^{-2x}$. (See Table 4.2.) A general solution is, therefore, $y = c_1 e^x + c_2 xe^x + c_3 e^{-2x}$. (Notice that the sum of the multiplicities of the roots is 3, which matches the order of the differential equation.)

TABLE 4.2

Root	Multiplicity	Corresponding Solution(s)
$m = 1$	$k = 2$	$y = e^x, y = xe^x$
$m = -2$	$k = 1$	$y = e^{-2x}$

(c) Here, the characteristic equation is

$$m^4 - 1 = (m^2 - 1)(m^2 + 1) = (m - 1)(m + 1)(m^2 + 1) = 0$$

with roots $m = 1$, $m = -1$, $m = i$, and $m = -i$, all of multiplicity 1. The solutions corresponding to the roots $m = 1$ and $m = -1$ are $y = e^x$, and $y = e^{-x}$, while the

---

* By the rational root theorem, the possible rational solutions to the equation are ± 2 and ± 1.

solutions corresponding to the complex conjugate pair $\pm i$ are $y = \cos x$ and $y = \sin x$. (See Table 4.3.) Therefore, a general solution is

$$y = c_1 e^x + c_2 e^{-x} + c_3 \cos x + c_4 \sin x.$$

TABLE 4.3

Roots	Multiplicity	Corresponding Solution(s)
$m = 1$	$k = 1$	$y = e^x$
$m = -1$	$k = 1$	$y = e^{-x}$
$m = \pm i$	$k = 1, k = 1$	$y = \cos x,\ y = \sin x$

Figure 4.3 shows the graph of $y = c_1 e^x + c_2 e^{-x} + c_3 \cos x + c_4 \sin x$ for various values of c_1, c_2, c_3, and c_4. Identify those graphs for which (a) $c_1 = c_3 = c_4 = 0$; (b) $c_1 = c_2 = 0$; and (c) $c_3 = c_4 = 0$.

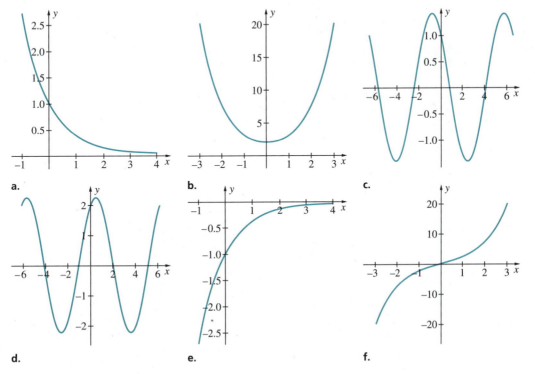

a.

b.

c.

d.

e.

f.

Figure 4.3 (a)–(f)

Example 6

Determine a general solution of the eighth-order homogeneous equation with constant coefficients if the roots and corresponding multiplicities of the characteristic equation are: $m_1 = 2 - 3i$, $k_1 = 2$; $m_2 = 2 + 3i$, $k_2 = 2$; $m_3 = -5$, $k_3 = 1$; $m_4 = 2$, $k_4 = 3$.

Solution   The solutions that correspond to the complex conjugate pair $2 \pm 3i$ are $y = e^{2x} \cos 3x$ and $y = e^{2x} \sin 3x$. These roots are repeated so the other solutions corresponding to the pair are $y = xe^{2x} \cos 3x$ and $y = xe^{2x} \sin 3x$. For the single root $m_3 = -5$, the corresponding solution is $y = e^{-5x}$. Finally, the root of multiplicity three $m_4 = 2$ yields the three solutions $y = e^{2x}$, $y = xe^{2x}$, and $y = x^2e^{2x}$. Therefore, a general solution is

$$y = e^{2x}(c_1 \cos 3x + c_2 \sin 3x) + xe^{2x}(c_3 \cos 3x + c_4 \sin 3x) + c_5 e^{-5x} + c_6 e^{2x} +$$
$$c_7 xe^{2x} + c_8 x^2 e^{2x}.$$

In the previous examples, we have been interested in finding a general solution of a higher-order equation. Here, we consider finding a solution to an nth-order initial-value problem.

Example 7

Solve $9y''' - 3y'' + 31y' - 37y = 0$ subject to the initial conditions $y(0) = 0$, $y'(0) = -1$, and $y''(0) = 1$.

Solution   In this case, the characteristic equation is

$$9m^3 - 3m^2 + 31m - 37 = 0$$

with three roots $m = 1$, $m = -\frac{1}{3} + 2i$, and $m = -\frac{1}{3} - 2i$. Each root has multiplicity one, so a general solution of the equation is $y(x) = c_1 e^x + e^{-x/3}(c_2 \cos 2x + c_3 \sin 2x)$. To find the values of c_1, c_2, and c_3 that satisfy the initial conditions, we calculate $y(0) = c_1 + c_2 = 0$, $y'(0) = c_1 - \frac{1}{3}c_2 + 2c_3 = -1$, and $y''(0) = c_1 - \frac{35}{9} c_2 - \frac{4}{3}c_3 = 1$. Solving this system of equations for c_1, c_2, and c_3 yields $c_1 = \frac{3}{52}$, $c_2 = -\frac{3}{52}$, and $c_3 = -\frac{7}{13}$. Therefore, the desired solution is

$$y(x) = \frac{3}{52}e^x + e^{-x/3}\left(-\frac{3}{52} \cos 2x - \frac{7}{13} \sin 2x\right),$$

which is graphed in Figure 4.4. What is $\lim_{x \to \infty} y(x)$?

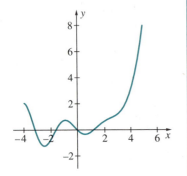

Figure 4.4   Graph of
$y(x) = \dfrac{3}{52}e^x +$
$e^{-x/3}\left(-\dfrac{3}{52} \cos 2x - \dfrac{7}{13} \sin 2x\right)$

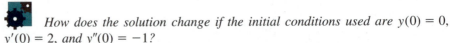
How does the solution change if the initial conditions used are $y(0) = 0$, $y'(0) = 2$, and $y''(0) = -1$?

Example 8

(a) If a is a positive constant, find conditions on the constant b so that $y(x)$ satisfies

$$\begin{cases} 5.69889y''' + 5.21742y'' - 3.39914y' - 5.6932y = 0 \\ \quad\quad y(0) = 0, y'(0) = a, y''(0) = b \end{cases}$$

and $\lim_{x\to\infty} y(x) = 0$. (b) For this function, find and classify the first critical point on the interval $[0, +\infty)$.

Solution A general solution of the equation $5.69889y''' + 5.21742y'' - 3.39914y' - 5.69232y = 0$ is

$$y(x) = e^{-0.91693x}(c_1 \cos 0.496889x + c_2 \sin 0.496889x) + c_3 e^{0.918345x}$$

while the solution to the initial-value problem is

$$y(x) = e^{-0.91693x}[(-0.507273a - 0.276615b) \cos 0.496889x - (-0.138893a + 1.02168b) \sin 0.496889x] + (0.507273a + 0.276615b)e^{0.918345x}.$$

We see that $\lim_{x\to\infty} y(x) = 0$ if $0.507273a + 0.276615b = 0$ which leads to $b = -1.83386a$.

To find and classify the first critical point of $y(x)$, we compute

$$y'(x) = ae^{-0.91693x}(\cos 0.496889x - 1.84534 \sin 0.496889x)$$

and graph $e^{-0.91693x}(\cos 0.496889x - 1.84534 \sin 0.496889x)$ in Figure 4.5 to locate the first zero of y'.

From the graph, we see that $y'(x) = 0$ near 1 and, numerically, we obtain the critical number $x = 0.999432$. At this critical number $y(0.999432) = 0.383497a$ so by the first derivative test, $(0.999432, 0.383497a)$ is a local maximum. To see that $(0.999432, 0.383497a)$ is the *absolute* maximum, we graph $y(x)$ for various choices of a in Figure 4.6.

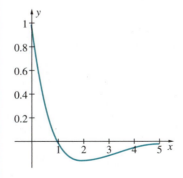

Figure 4.5 Graph of $\dfrac{1}{a}y'(x) =$ $e^{-0.91693x}(\cos 0.496889x - 1.84534 \sin 0.496889x)$

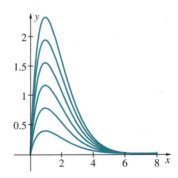

Figure 4.6 Graph of $y(x)$ for various choices of a

IN TOUCH WITH TECHNOLOGY

Computer algebra systems have built-in functions that can be used to find exact solutions of nth-order linear homogeneous differential equations with constant coefficients as long as n is smaller than 5. In cases when the roots of the characteristic equation are symbolically complicated, when approximations are desired, or when n is greater than 4, the roots of the characteristic equation can be approximated by taking advantage of the numerical approximation capabilities of the system being used.

1. Find a general solution of

(a) $y'' + 2y' + 17y = 0$

(b) $y''' + 2y'' + 5y' - 26y = 0$

(c) $1.90413y'' + 0.267791y' + 3.58554y = 0$

(d) $0.9y''' + 18.78y' - 0.2987y = 0$

(e) $8.9y^{(4)} - 2.5y'' + 32.0y' + 0.773y = 0$.

Graph the solution for various initial conditions.

2. Solve each of the following initial-value problems. Verify that your result satisfies the initial conditions by graphing it on an appropriate interval.

(a) $\begin{cases} y''' + 3y'' + 2y' + 6y = 0 \\ y(0) = 0,\ y'(0) = 1,\ y''(0) = -1 \end{cases}$

(b) $\begin{cases} y^{(4)} - 8y''' + 30y'' - 56y' + 49y = 0 \\ y(0) = 1,\ y'(0) = 2,\ y''(0) = -1,\ y'''(0) = -1 \end{cases}$

(c) $\begin{cases} 0.31y''' + 11.2y'' - 9.8y' + 5.3y = 0 \\ y(0) = -1,\ y'(0) = -1,\ y''(0) = 0 \end{cases}$.

3. Use a computer algebra system to find a general solution of

$$y - k^4 \frac{d^4 y}{dx^4} = 0.$$

Show that the result you obtain is equivalent to

$$y = Ae^{-x/k} + Be^{x/k} + C \sin \frac{x}{k} + D \cos \frac{x}{k}.$$

4. Find conditions on a and b, if possible, so that the solution to the initial-value problem

$$\begin{cases} y'' + 4y' + 3y = 0 \\ y(0) = a,\ y'(0) = b \end{cases}$$

has (a) neither local maxima nor local minima; (b) exactly one local maximum; and (c) exactly one local minimum on the interval $[0, \infty)$.

5. Show that if $y_0 \neq 0$ the problem

$$\begin{cases} 4.02063y''' - 0.224975y'' + 4.486y' - 2.48493y = 0 \\ y(0) = 0,\ y'(0) = y_0,\ y(4) = 0 \end{cases}$$

has a unique solution.

6. (a) Show that the roots of the characteristic equation

of $\dfrac{d^2 x}{dt^2} + a_1 \dfrac{dx}{dt} + a_0 x = 0$ are

$$m_{1,2} = \frac{-a_1 \pm \sqrt{a_1^2 - 4a_0}}{2}.$$

(b) Graph $a_0 = \frac{1}{4}a_1^2$ and $a_1 = 0$, using the horizontal axis to represent a_0 and the vertical axis to represent a_1. *Randomly* generate three points on the graph of $a_0 = \frac{1}{4}a_1^2$, three points in the region $a_0 < \frac{1}{4}a_1^2$, and three points in the region $a_0 > \frac{1}{4}a_1^2$.

(c) Write the second-order homogeneous equation $\dfrac{d^2 x}{dt^2} + a_1 \dfrac{dx}{dt} + a_0 x = 0$ as a system by letting $y = x' = \dfrac{dx}{dt}$. (See Section 1.4.)

(d) For each pair of points obtained in (b), graph the phase plane associated with the system for $-1 \leq x \leq 1$ and $-1 \leq y \leq 1$. Compare your results with your classmates.

(e) How does the phase plane associated with the system change as the roots of the characteristic equation change?

7. How do the graphs of the solutions to the equations

$y'' - 2\lambda y' + (\lambda^2 + 4)y = 0$, $y'' - 2y' + (\lambda^2 + 1)y = 0$, and $y'' + 2y' + (\lambda^2 + 1)y = 0$

subject to the conditions $y(0) = 1$ and $y'(0) = 0$ change as λ goes from -1 to 1?

In Exercises 1–42 find a general solution of each equation. (A computer algebra system may be useful in solving some of the higher-order equations.)

1. $y'' + 8y' + 12y = 0$

2. $y'' - 4y' - 12y = 0$

*3. $y'' + 3y' - 4y = 0$

4. $y'' + y' - 72y = 0$

5. $y'' + 16y = 0$

6. $4y'' + 9y = 0$

*7. $y'' + 7y = 0$

8. $y'' + 8y = 0$

9. $y'' - 6y' + 25y = 0$

10. $y'' - 8y' + 20y = 0$

*11. $y'' + 6y' + 18y = 0$

12. $4y'' + 21y' + 5y = 0$

13. $7y'' + 4y' - 3y = 0$

14. $y'' + 4y' + 4y = 0$

*15. $y'' - 6y' + 9y = 0$

16. $y''' - 4y' = 0$

17. $y''' - 10y'' + 25y' = 0$

18. $y''' - 6y'' = 0$

*19. $8y''' + y'' = 0$

20. $y''' + 7y'' + 17y' + 15y = 0$

*21. $y''' + 7y'' + 24y' - 32y = 0$

22. $y''' - 8y'' + 5y' + 14y = 0$

23. $y''' - 6y'' - 7y' + 60y = 0$

24. $y''' + y'' - 16y' + 20y = 0$

*25. $y''' + 12y'' + 36y' + 32y = 0$

26. $y''' + 11y'' + 24y' - 36y = 0$

27. $y''' + 2y'' - 15y' - 36y = 0$

28. $y^{(4)} - 9y'' = 0$

*29. $8y^{(4)} + y' = 0$

30. $y^{(4)} - 16y = 0$

31. $9y^{(4)} + 4y''' = 0$

32. $y^{(4)} - 6y''' - y'' + 54y' - 72y = 0$

*33. $y^{(4)} - 5y''' - 10y'' + 80y' - 96y = 0$

34. $y^{(4)} + y''' - 6y'' - 4y' + 8y = 0$

*35. $y^{(4)} + 7y''' + 6y'' - 32y' - 32y = 0$

36. $y^{(4)} - 15y''' + 85y'' - 215y' + 204y = 0$

*37. $y^{(4)} + y''' + 10y'' - 52y' + 40y = 0$

38. $y^{(4)} - 6y''' + 18y'' - 54y' + 81y = 0$

39. $y^{(4)} + 2y''' - 2y'' + 8y = 0$

40. $y^{(4)} + 8y''' + 32y'' + 64y' + 64y = 0$

*41. $y^{(4)} + 32y'' + 256y = 0$

42. $y^{(5)} + 25y''' = 0$

In Exercises 43–56, solve the initial-value problem. Graph the solution on an appropriate interval.

43. $y'' - y' = 0$, $y(0) = 3$, $y'(0) = 2$

44. $3y'' - y' = 0$, $y(0) = 0$, $y'(0) = 7$

*45. $y'' + y' - 12y = 0$, $y(0) = 0$, $y'(0) = 7$

46. $y'' - 7y' + 12y = 0$, $y(0) = 3$, $y'(0) = -2$

47. $2y'' - 7y' - 4y = 0$, $y(0) = 0$, $y'(0) = 1$

48. $y'' - 7y' + 10y = 0$, $y(0) = 1$, $y'(0) = 5$

49. $y'' + 36y = 0$, $y(0) = 2$, $y'(0) = -6$

50. $y'' + 100y = 0$, $y(0) = 1$, $y'(0) = 10$

*51. $y'' - 2y' + y = 0$, $y(0) = 4$, $y'(0) = 0$

52. $y'' + 4y' + 4y = 0$, $y(0) = 1$, $y'(0) = 3$

*53. $y'' + 2y' + 5y = 0$, $y(0) = 1$, $y'(0) = 3$

54. $y'' + 4y' + 20y = 0$, $y(0) = 2$, $y'(0) = 0$

55. $y''' - 2y'' = 0$, $y(0) = 1$, $y'(0) = 2$, $y''(0) = 0$

56. $y''' + 49y' = 0$, $y(0) = -1$, $y'(0) = 0$, $y''(0) = -2$

57. A **Ricatti equation,** named for the Italian mathematician Jacopo Francesco Ricatti (1676–1754), is a nonlinear first-order equation of the form

$$y' + a(x)y^2 + b(x)y + c(x) = 0.$$

We make the substitution $y(x) = \dfrac{w'(x)}{w(x)} \dfrac{1}{a(x)}$ to solve Ricatti equations.

(a) Show that

$$y'(x) = \frac{w''(x)}{a(x)w(x)} - \frac{(w'(x))^2}{a(x)(w(x))^2} - \frac{a'(x)w'(x)}{(a(x))^2 w(x)}.$$

(b) Substitute y and y' into the Ricatti equation to obtain the second-order equation

$$\frac{w''(x)}{a(x)w(x)} - \frac{(w'(x))^2}{a(x)(w(x))^2} - \frac{a'(x)w'(x)}{(a(x))^2 w(x)} +$$
$$\frac{(w'(x))^2}{a(x)(w(x))^2} + \frac{b(x)w'(x)}{a(x)w(x)} + c(x) = 0.$$

Multiply this equation by $a(x)w(x)$ to obtain

$$w''(x) - \frac{a'(x)w'(x)}{a(x)} + b(x)w'(x) +$$

$a(x)c(x)w(x) = 0$ and simplify the result to

obtain the second-order equation

$$w'' - \left(\frac{a'(x)}{a(x)} - b(x)\right)w' + a(x)c(x)w = 0.$$

58. Convert the Ricatti equation $y' + (x^4 + x^2 + 1)y^2 + \dfrac{2(1 - x + x^2 - 2x^3 + x^4)}{1 + x^2 + x^4}y + \dfrac{1}{x^4 + x^2 + 1} = 0$ to a second-order equation by using the following steps.

(a) Note that $a(x) = x^4 + x^2 + 1$,

$$b(x) = \frac{2(1 - x + x^2 - 2x^3 + x^4)}{1 + x^2 + x^4}, \text{ and}$$

$$c(x) = \frac{1}{x^4 + x^2 + 1}. \text{ Let } y(x) = \frac{w'(x)}{w(x)}\frac{1}{a(x)} \text{ so}$$

that the second-order equation is

$$w'' - \left(\frac{4x^3 + 2x}{x^4 + x^2 + 1} - \frac{2(1 - x + x^2 - 2x^3 + x^4)}{1 + x^2 + x^4}\right)w' + (x^4 + x^2 + 1)\frac{1}{x^4 + x^2 + 1}w = 0,$$

which simplifies to $w'' - 2w' + w = 0$.

(b) Show that $w(x) = C_1 e^x + C_2 x e^x$ is a general solution of $w'' - 2w' + w = 0$.

(c) Use $y(x) = \dfrac{w'(x)}{w(x)}\dfrac{1}{a(x)}$ to show that

$$y(x) = \frac{C_1 e^x + C_2(e^x + xe^x)}{(x^4 + x^2 + 1)(C_1 e^x + C_2 x e^x)} \text{ satisfies}$$

the Ricatti equation.

In Exercises 59–64 solve the given Ricatti equation. (See Exercises 57 and 58.)

59. $y' + \dfrac{1}{x^2 + 1}y^2 - \dfrac{2(-6 + x - 6x^2)}{x^2 + 1}y + 45(x^2 + 1) = 0$

60. $y' + \sin x y^2 + (2 + \cot x)y + 61 \csc x = 0$

61. $y' + x^2 \cos x y^2 - \dfrac{x \tan x + 2x - 2}{x}y + \dfrac{2 \sec x}{x^2} = 0$

62. $y' + \dfrac{\sin x}{x}y^2 - \dfrac{1 - 2x - x \cot x}{x}y + 26x \csc x = 0$

*63. $y' + \dfrac{x}{x^2 + 4}y^2 - \dfrac{x^2 - 4}{x(x^2 + 4)}y + \dfrac{4(x^2 + 4)}{x} = 0$

64. $y' + x \tan 4x + \dfrac{1 + 10x + 4x \csc 4x \sec 4x}{x}y + \dfrac{41 \cot 4x}{x} = 0$

*65. Find a differential equation for which the characteristic equation has roots with corresponding multiplicities $m_1 = 2 - 3i$, $k_1 = 2$; $m_2 = 2 + 3i$, $k_2 = 2$; $m_3 = -5$, $k_3 = 1$; $m_4 = 2$, $k_4 = 3$.

66. Determine if it is possible to find values of a_3, a_2, a_1, and a_0 so that a general solution of

$$y^{(4)} + a_3 y''' + a_2 y'' + a_1 y' + a_0 y = 0$$

has the indicated form. In each case, either find such values or explain why no such numbers exist.

(a) $y(x) = c_1 e^{-2x} + c_2 e^{-x} + c_3 e^x + c_4 e^{2x}$

(b) $y(x) = c_1 e^{-2x} + c_2 x e^{-2x} + c_3 x^2 e^{-2x} + c_4 x^3 e^{-2x}$

(c) $y(x) = c_1 x e^x + c_2 x^2 e^x + c_3 x^3 e^x + c_4 x^4 e^x$

(d) $y(x) = c_1 e^{-x} + c_2 e^x + c_3 \cos x + c_4 \sin x$

(e) $y(x) = c_1 \cos x + c_2 \sin x + c_3 x \cos x + c_4 x \sin x$

(f) $y(x) = c_1 x \cos x + c_2 x \sin x + c_3 x^2 \cos x + c_4 x^2 \sin x$

67. Consider the homogeneous equation $y'' + p(x)y' + q(x)y = 0$. (a) If $y_1(x)$ satisfies the equation and c is a constant, show that $y(x) = cy_1(x)$ satisfies the equation. (b) If $y_1(x)$ and $y_2(x)$ are both solutions of the equation, show that $y(x) = y_1(x) + y_2(x)$ is a solution.

68. Use the substitution $x = e^t$ to solve (a) $3x^2 y'' - 2xy' + 2y = 0$, $x > 0$ and (b) $x^2 y'' - xy' + y = 0$, $x > 0$. $\left(\text{Hint: Show that if } x = e^t, \dfrac{dy}{dx} = \dfrac{1}{x}\dfrac{dy}{dt} \text{ and } \dfrac{d^2 y}{dx^2} = \dfrac{1}{x^2}\left(\dfrac{d^2 y}{dt^2} - \dfrac{dy}{dt}\right).\right)$

69. (a) Use the Maclaurin series $e^x = \displaystyle\sum_{k=0}^{\infty} \dfrac{x^k}{k!}$ and the Maclaurin series for $\sin x$ and $\cos x$ to prove $e^{-i\theta} = \cos \theta - i \sin \theta$. (b) Use (a) and trigonometric identities to prove that $e^{i\theta} = \cos \theta + i \sin \theta$.

70. Show that a general solution of the differential equation $ay'' + 2by' + cy = 0$ where $b^2 - ac > 0$ can be written as

$$y = e^{-bx/a}\left[c_1 \cosh \frac{x\sqrt{b^2 - ac}}{a} + c_2 \sinh \frac{x\sqrt{b^2 - ac}}{a}\right].$$

71. Express the solution to each differential equation in terms of the hyperbolic trigonometric functions. (See Exercise 70.)

(a) $y'' + 6y' + 2y = 0$

(b) $y'' - 5y' + 6y = 0$

(c) $y'' - 6y' - 16y = 0$

(d) $y'' - 16y = 0$

72. Show that the boundary-value problem $\begin{cases} y'' + 2y' + 5y = 0 \\ y(0) = 0, \, y(\pi/2) = 0 \end{cases}$ has infinitely many solutions,

the boundary-value problem $\begin{cases} y'' + 2y' + 5y = 0 \\ y(0) = 0, \, y(\pi/4) = 0 \end{cases}$

has no nontrivial solutions, and that the boundary-value problem $\begin{cases} y'' + 2y' + 5y = 0 \\ y(0) = 1, \, y(\pi/4) = 0 \end{cases}$ has one (nontrivial) solution.

*73. Use factoring to solve each of the following nonlinear equations. For which equations, if any, is the Principle of Superposition valid? Explain.

(a) $(y'')^2 - 5y''y + 4y^2 = 0$

(b) $(y'')^2 - 2y''y + y^2 = 0$

74. Complete the following table.

Differential Equation	Characteristic Equation	Roots of Characteristic Equation	General Solution
$y^{(n)} = 0$			
	$(m - k)^n$		
		$m_{1,2} = \alpha \pm i\beta$, $\beta \neq 0$	
			$y = e^{\alpha x}[(c_{1,1} + c_{1,2}x + \cdots c_{1,n-1}x^{n-1}) \cos \beta x + (c_{2,1} + c_{2,2}x + \cdots c_{2,n-1}x^{n-1}) \sin \beta x]$

4.4 Introduction to Solving Nonhomogeneous Equations with Constant Coefficients

All instructors will want to discuss the introduction to solving nonhomogeneous equations with constant coefficients. However, instructors who do not plan to cover the material in Section 4.5 (The Annihilator Method) may want to omit the discussion of operator notation and annihilators.

General Solution of a Nonhomogeneous Equation • **Operator Notation** • **Annihilators of Familiar Functions**

General Solution of a Nonhomogeneous Equation

In the previous section, we learned how to solve nth-order linear homogeneous equations with real constant coefficients. These techniques are also useful in solving some nonhomogeneous equations of the form

$$a_n y^{(n)}(x) + a_{n-1} y^{(n-1)}(x) + \cdots + a_1 y'(x) + a_0 y(x) = g(x).$$

Before describing how to obtain solutions of some nonhomogeneous equations, we first describe what is meant by a general solution of a nonhomogeneous equation.

Definition 4.8 Particular Solution

A **particular solution,** $y_p(x)$, of the nonhomogeneous differential equation $a_n y^{(n)}(x) + a_{n-1} y^{(n-1)}(x) + \cdots + a_1 y'(x) + a_0 y(x) = g(x)$ is a specific function that contains no arbitrary constants and satisfies the differential equation.

Example 1

Verify that $y_p(x) = \dfrac{3}{5}\sin x$ is a particular solution of $y'' - 4y = -3\sin x$.

Solution   First we compute $y_p'(x) = \frac{3}{5}\cos x$ and $y_p''(x) = -\frac{3}{5}\sin x$. Substituting into $y'' - 4y$ results in:

$$y_p'' - 4y_p = -\frac{3}{5}\sin x - 4 \cdot \frac{3}{5}\sin x = -3\sin x.$$

We conclude that $y_p(x) = \frac{3}{5}\sin x$ is a particular solution of $y'' - 4y = -3\sin x$ because y_p satisfies the equation $y'' - 4y = -3\sin x$ and contains no arbitrary constants.

 Show that $y_p(x) = \dfrac{1}{5}e^{-2x}(15 - 10e^{4x} + 3e^{2x}\sin x)$ *is also a particular solution of* $y'' - 4y = -3\sin x$.

In this example, we see that $y_p(x) = \frac{3}{5}\sin x$ is a particular solution of $y'' - 4y = -3\sin x$. The **corresponding homogeneous equation** of $y'' - 4y = -3\sin x$ is $y'' - 4y = 0$ with general solution $y_h(x) = c_1 e^{-2x} + c_2 e^{2x}$. Let $y(x) = y_h(x) + y_p(x)$. Then,

$$
\begin{aligned}
y'' - 4y &= (y_h(x) + y_p(x))'' - 4(y_h(x) + y_p(x)) \\
&= y_h'' + y_p'' - 4y_h - 4y_p \\
&= \underbrace{y_h'' - 4y_h}_{0} + \underbrace{y_p'' - 4y_p}_{-3\sin x} = -3\sin x.
\end{aligned}
$$

We see that $y(x) = y_h(x) + y_p(x)$ is a solution of the nonhomogeneous equation $y'' - 4y = -3\sin x$. In fact, it is not hard to show that if $y(x)$ is *any* solution to the equation, then there are constants c_1 and c_2 so that $y(x) = y_h(x) + y_p(x)$. (See Exercise 31.)

More generally, if $y_p(x)$ is a particular solution of the nonhomogeneous equation

$$a_n y^{(n)}(x) + a_{n-1}y^{(n-1)}(x) + \cdots + a_1 y'(x) + a_0 y(x) = g(x)$$

and

$$y_h(x) = c_1 f_1(x) + c_2 f_2(x) + \cdots + c_n f_n(x)$$

is a general solution of the corresponding homogeneous equation

$$a_n y^{(n)}(x) + a_{n-1}y^{(n-1)}(x) + \cdots + a_1 y'(x) + a_0 y(x) = 0,$$

every solution, $y(x)$, of the nonhomogeneous equation can be written in the form

$$y(x) = y_h(x) + y_p(x),$$

for some choice of $c_1, c_2, \ldots, c_n$. (See Exercise 32.)

To prove this theorem for second-order equations, consider

$$a_2(x)y''(x) + a_1(x)y'(x) + a_0(x)y(x) = g(x)$$

and assume that $\{f_1(x), f_2(x)\}$ is a fundamental set of solutions for the corresponding homogeneous equation

$$a_2(x)y''(x) + a_1(x)y'(x) + a_0(x)y(x) = 0.$$

If both $y(x)$ and $y_p(x)$ are solutions to

$$a_2(x)y''(x) + a_1(x)y'(x) + a_0(x)y(x) = g(x),$$

let $y_h(x) = y(x) - y_p(x)$ so that substitution into the nonhomogeneous equation yields

$$a_2(x)y_h''(x) + a_1(x)y_h'(x) + a_0(x)y_h(x) =$$
$$a_2(x)[y''(x) - y_p''(x)] + a_1(x)[y'(x) - y_p'(x)] + a_0(x)[y(x) - y_p'(x)]$$
$$= [a_2(x)y''(x) + a_1(x)y'(x) + a_0(x)y(x)] - [a_2(x)y_p''(x) + a_1(x)y_p'(x) + a_0(x)y_p'(x)]$$
$$= g(x) - g(x) = 0.$$

Thus, $y_h(x) = y(x) - y_p(x)$ is a solution to the homogeneous equation $a_2(x)y''(x) + a_1(x)y'(x) + a_0(x)y(x) = 0$. There are constants c_1 and c_2 so that $y_h(x) = c_1 f_1(x) + c_2 f_2(x)$, which indicates that

$$y(x) - y_p(x) = \underbrace{c_1 f_1(x) + c_2 f_2(x)}_{y_h(x)} \qquad \text{or} \qquad y(x) = \underbrace{c_1 f_1(x) + c_2 f_2(x)}_{y_h(x)} + y_p(x).$$

This means that every solution to the nonhomogeneous equation can be written as the sum of a general solution to the corresponding homogeneous equation and a particular equation of the nonhomogeneous equation. This leads us to the following definition.

Definition 4.9 General Solution of a Nonhomogeneous Equation

A **general solution to the linear nonhomogeneous nth-order differential equation**

$$a_n(x)y^{(n)}(x) + a_{n-1}(x)y^{(n-1)}(x) + \cdots + a_1(x)y'(x) + a_0(x)y(x) = g(x)$$

is

$$y(x) = y_h(x) + y_p(x)$$

where $y_h(x)$ is a general solution of the corresponding homogeneous equation

$$a_n(x)y^{(n)}(x) + a_{n-1}(x)y^{(n-1)}(x) + \cdots + a_1(x)y'(x) + a_0(x)y(x) = 0,$$

and $y_p(x)$ is a particular solution to the nonhomogeneous equation.

Example 2

Let $y_p(x) = \dfrac{1}{2}x^3 - \dfrac{3}{2}x^2 + \dfrac{3}{2}x$ and $y_h(x) = e^{-x}(c_1 \cos x + c_2 \sin x)$. Show that $y(x) = y_h(x) + y_p(x)$ is a general solution of $y'' + 2y' + 2y = x^3$.

Solution   We first show that $y_p(x) = \frac{1}{2}x^3 - \frac{3}{2}x^2 + \frac{3}{2}x$ is a particular solution of $y'' + 2y' + 2y = x^3$. Calculating $y_p' = \frac{3}{2}x^2 - 3x + \frac{3}{2}$, $y_p'' = 3x - 3$, and $y_p'' + 2y_p' + 2y_p$ gives us:

$$y_p'' + 2y_p' + 2y_p = (3x - 3) + 2\left(\frac{3}{2}x^2 - 3x + \frac{3}{2}\right) + 2\left(\frac{1}{2}x^3 - \frac{3}{2}x^2 + \frac{3}{2}x\right)$$

$$= 3x - 3 + 3x^2 - 6x + 3 + x^3 - 3x^2 + 3x = x^3.$$

The associated homogeneous equation of $y'' + 2y' + 2y = x^3$ is $y'' + 2y' + 2y = 0$ with characteristic equation $m^2 + 2m + 2 = 0$ and roots $m = -1 \pm i$. Thus, a general solution of $y'' + 2y' + 2y = 0$ is $y_h(x) = e^{-x}(c_1 \cos x + c_2 \sin x)$, so that

$$y(x) = y_h(x) + y_p(x)$$

$$= e^{-x}(c_1 \cos x + c_2 \sin x) + \frac{1}{2}x^3 - \frac{3}{2}x^2 + \frac{3}{2}x$$

is a general solution of $y'' + 2y' + 2y = x^3$.

 If $y_1(x)$ and $y_2(x)$ are nontrivial solutions of $y'' + 2y' + 2y = x^3$, is $y_1(x) + y_2(x)$ also a solution?

Operator Notation

The nth-order derivative of a function y is given in **operator notation** by

$$D^n y = \frac{d^n y}{dx^n}.$$

Thus, the left-hand side of the nth-order linear homogeneous differential equation with real constant coefficients

$$a_n y^{(n)} + a_{n-1} y^{(n-1)} + \cdots + a_1 y' + a_0 y = 0$$

can be expressed in operator notation as

$$a_n y^{(n)} + a_{n-1} y^{(n-1)} + \cdots + a_1 y' + a_0 y = a_n D^n y + a_{n-1} D^{n-1} y + \cdots + a_1 Dy + a_0 y$$
$$= (a_n D^n + a_{n-1} D^{n-1} + \cdots + a_1 D + a_0)y.$$

The linear differential equation with real constant coefficients

$$a_n y^{(n)} + a_{n-1} y^{(n-1)} + \cdots + a_1 y' + a_0 y = g(x)$$

can be written in operator form as

$$(a_n D^n + a_{n-1} D^{n-1} + \cdots + a_1 D + a_0)y = g(x).$$

Definition 4.10 Nth-Order Linear Differential Operator with Constant Coefficients

The expression

$$P = p(D) = a_n D^n + a_{n-1} D^{n-1} + \cdots + a_1 D + a_0,$$

where $a_0, a_1, \ldots,$ and a_n are numbers and $a_n \neq 0$, is called an **nth-order linear differential operator with constant coefficients**.

Example 3

Write the following differential equations in operator form.
(a) $y' - 6y = 0$ (b) $y'' + 2y' - 8y = 0$
(c) $y'' + 4y = \cos x$ (d) $y''' + 3y'' + 3y' + y = e^{2x}$

Solution   (a) $y' - 6y = Dy - 6y = (D - 6)y = 0$
(b) $y'' + 2y' - 8y = D^2 y + 2Dy - 8y = (D^2 + 2D - 8)y = 0$
(c) $y'' + 4y = D^2 y + 4y = (D^2 + 4)y = \cos x$
(d) $y''' + 3y'' + 3y' + y = D^3 y + 3D^2 y + 3Dy + y = (D^3 + 3D^2 + 3D + 1)y = e^{2x}$.

Notice that when we place the differential equation in operator form, we end up with an equation of the form $p(D)y = g(x)$ where $p(D)$ is a function of the differential operator D. Functions of this type will be of great use to us in solving nonhomogeneous equations in Section 4.5.

Before we move on, we discuss some of the important properties of the linear differential operator with constant coefficients.

Property 1

Let $P = p(D)$ be a linear differential operator with constant coefficients. Then,

$$P[f(x) + g(x)] = P[f(x)] + P[g(x)].$$ • • •

Example 4

Calculate $P[f(x) + g(x)]$ and $P[f(x)] + P[g(x)]$ if $P = p(D) = D^2 + 2D - 1$, $f(x) = e^{2x}$, and $g(x) = \sin x$.

Solution   Calculating $P[f(x) + g(x)]$ yields

$$
\begin{aligned}
P[f(x) + g(x)] &= (D^2 + 2D - 1)[e^{2x} + \sin x] \\
&= D^2[e^{2x} + \sin x] + 2D[e^{2x} + \sin x] - [e^{2x} + \sin x] \\
&= D[2e^{2x} + \cos x] + 2[2e^{2x} + \cos x] - [e^{2x} + \sin x] \\
&= 4e^{2x} - \sin x + 4e^{2x} + 2\cos x - e^{2x} - \sin x \\
&= 7e^{2x} - 2\sin x + 2\cos x.
\end{aligned}
$$

We obtain the same results calculating $P[f(x)] + P[g(x)]$:

$$
\begin{aligned}
P[f(x)] + P[g(x)] &= (D^2 + 2D - 1)[e^{2x}] + (D^2 + 2D - 1)[\sin x] \\
&= D^2[e^{2x}] + 2D[e^{2x}] - [e^{2x}] + D^2[\sin x] + 2D[\sin x] - [\sin x] \\
&= D[2e^{2x}] + 2[2e^{2x}] - [e^{2x}] + D[\cos x] + 2\cos x - \sin x \\
&= 4e^{2x} + 4e^{2x} - e^{2x} - \sin x + 2\cos x - \sin x \\
&= 7e^{2x} - 2\sin x + 2\cos x.
\end{aligned}
$$

Property 2

The product of two linear differential operators with constant coefficients P_1 and P_2 is defined by $P_1P_2[f(x)] = P_1(P_2[f(x)])$. ● ● ●

A consequence of Property 2 is that $(P_1P_2)[f(x)] = P_1(P_2[f(x)])$ and consequently $P_1P_2[f(x)] = P_2P_1[f(x)]$, as illustrated in the following example.

Example 5

Compute $P_1P_2[f(x)]$ and $P_2P_1[f(x)]$ if $P_1 = p_1(D) = 3D - 1$, $P_2 = p_2(D) = D + 1$, and $f(x) = e^{2x} + \sin x$.

Solution   First we compute $P_1P_2[f(x)]$:

$$
\begin{aligned}
P_1(P_2[f(x)]) &= (3D - 1)[(D + 1)(e^{2x} + \sin x)] \\
&= (3D - 1)[2e^{2x} + \cos x + e^{2x} + \sin x] \\
&= 3D[3e^{2x} + \cos x + \sin x] - [3e^{2x} + \cos x + \sin x] \\
&= 3[6e^{2x} - \sin x + \cos x] - [3e^{2x} + \cos x + \sin x] \\
&= 15e^{2x} - 4\sin x + 2\cos x.
\end{aligned}
$$

We obtain the same results when we compute $P_2P_1[f(x)]$:

$$
\begin{aligned}
P_2(P_1[f(x)]) &= (D + 1)[(3D - 1)(e^{2x} + \sin x)] \\
&= (D + 1)[3D(e^{2x} + \sin x) - (e^{2x} + \sin x)] \\
&= (D + 1)[3(2e^{2x}) + 3\cos x - e^{2x} - \sin x] \\
&= D[5e^{2x} + 3\cos x - \sin x] + [5e^{2x} + 3\cos x - \sin x] \\
&= 15e^{2x} - 4\sin x + 2\cos x.
\end{aligned}
$$

Property 3

Linear differential operators with constant coefficients can be treated as polynomials in D. ● ● ●

Example 6

Treat the differential operators $P_1 = p_1(D) = 3D - 1$ and $P_2 = p_2(D) = D + 1$ as polynomials to calculate $P_1 P_2[f(x)]$ if $f(x) = e^{2x} + \sin x$.

Solution If we treat these operators as polynomials, we multiply the operators to yield $P_1 P_2 = (3D - 1)(D + 1) = 3D^2 - D + 3D - 1 = 3D^2 + 2D - 1$. Therefore, we have the following:

$$
\begin{aligned}
P_1 P_2[f(x)] &= (3D - 1)(D + 1)[e^{2x} + \sin x] = (3D^2 + 2D - 1)[e^{2x} + \sin x] \\
&= 3D^2[e^{2x} + \sin x] + 2D[e^{2x} + \sin x] - [e^{2x} + \sin x] \\
&= 3D[2e^{2x} + \cos x] + 2[2e^{2x} + \cos x] - [e^{2x} + \sin x] \\
&= 3[4e^{2x} - \sin x] + 4e^{2x} + 2 \cos x - e^{2x} - \sin x \\
&= 15e^{2x} - 4 \sin x + 2 \cos x.
\end{aligned}
$$

Notice that the same result is obtained for $P_1 P_2[f(x)]$ as was obtained when the product was interpreted as the compositions $P_1(P_2[f(x)])$ and $P_2(P_1[f(x)])$.

Now that we are familiar with some of the properties of linear differential operators with constant coefficients, we can use them to solve nonhomogeneous differential equations.

Definition 4.11 Annihilator

The linear differential operator $p(D)$ is said to **annihilate** a function $f(x)$ if $p(D)[f(x)] = 0$ for all x. In this case, $p(D)$ is called an **annihilator** of $f(x)$.

Example 7

Show that the following operators annihilate the corresponding function.
(a) $p(D) = D^3$; $f(x) = x^2$
(b) $p(D) = D - 5$; $f(x) = e^{5x}$
(c) $p(D) = D^2 + 16$; $f(x) = \cos 4x$

Solution (a) $D^3(x^2) = D^2(2x) = D(2) = 0$
(b) $(D - 5)(e^{5x}) = D(e^{5x}) - 5(e^{5x}) = 5e^{5x} - 5e^{5x} = 0$
(c) $(D^2 + 16)(\cos 4x) = D^2(\cos 4x) + 16 \cos 4x$
$$= D(-4 \sin 4x) + 16 \cos 4x = -16 \cos 4x + 16 \cos 4x = 0.$$

Annihilators of Familiar Functions

We now take advantage of the techniques of the previous section to find annihilators of familiar functions.

The differential equation $y^{(n)} = 0$ is expressed as $D^n y = 0$ in differential operator notation and has general solution $y = c_1 + c_2 x + c_3 x^2 + \cdots + c_{n-1} x^{n-1}$. Therefore, D^n annihilates the functions $1, x, x^2, \ldots, x^{n-1}$ as well as any linear combination of these functions, because each of the functions $1, x, x^2, \ldots, x^{n-1}$ satisfies the differential equation $D^n y = 0$.

Example 8

Find a differential operator that annihilates the indicated function.
(a) $f(x) = x^3$
(b) $f(x) = x^3 + x - 1$

Solution   (a) Because the power of x is 3, the operator D^4 annihilates $f(x) = x^3$. We can verify this with

$$D^4(x^3) = D^3(3x^2) = D^2(6x) = D(6) = 0.$$

(b) Again in this case, the highest power of x in the polynomial is 3. Therefore, D^4 annihilates $f(x) = x^3 + x - 1$. As in (a), we can verify this with

$$D^4(x^3 + x + 1) = D^3(3x^2 + 1) = D^2(6x) = D(6) = 0.$$

 Show that D^5, D^6, . . . annihilate the functions $f(x) = x^3$ and $f(x) = x^3 + x - 1$.

A general solution of the differential equation $(D - k)^n y = 0$ is $y = c_1 e^{kx} + c_2 x e^{kx} + c_3 x^2 e^{kx} + \cdots + c_{n-1} x^{n-1} e^{kx}$ so the differential operator $(D - k)^n$ annihilates the functions $e^{kx}, x e^{kx}, x^2 e^{kx}, \ldots, x^{n-1} e^{kx}$.

Example 9

Find a differential operator that annihilates the given function.
(a) $f(x) = e^{-3x}$ (b) $f(x) = x e^{4x}$
(c) $f(x) = x^4 e^x$ (d) $f(x) = x^2 e^{-x} + e^{-x}$

Solution   (a) e^{-3x} satisfies $y' + 3y = (D + 3)y = 0$, so the operator $(D + 3)$ annihilates $f(x) = e^{-3x}$.
(b) The function $f(x) = x e^{4x}$ is the second linearly independent solution of the second-order differential equation with repeated root $m = 4$ of the characteristic equation. Therefore it is a solution of $(D - 4)^2 y = 0$ and is annihilated by $(D - 4)^2$.
(c) In this case, the function $f(x) = x^4 e^x$ is the fifth linearly independent solution of the fifth-order differential equation with repeated root $m = 1$ of the characteristic equation $(m - 1)^5$ which corresponds to the differential equation $(D - 1)^5 y = 0$. The operator $(D - 1)^5$ annihilates $f(x) = x^4 e^x$.
(d) First, consider the function $x^2 e^{-x}$ which we know is the third linearly independent solution of the differential equation with characteristic equation $(m + 1)^3 = 0$. This corresponds to the differential equation $(D + 1)^3 y = 0$. Note also that the function e^{-x} satisfies this differential equation. Therefore, $(D + 1)^3$ annihilates $f(x) = x^2 e^{-x} + e^{-x}$.

If the characteristic equation of a differential equation has complex conjugate roots $m_{1,2} = \alpha \pm i\beta$, $\beta \neq 0$, we can factor the characteristic equation as $[m - (\alpha + i\beta)][m - (\alpha - i\beta)] = 0$. Expanding the left-hand side of this equation, we obtain

$$[m - (\alpha + i\beta)][m - (\alpha - i\beta)] = m^2 - (\alpha + i\beta)m - (\alpha - i\beta)m + (\alpha + i\beta)(\alpha - i\beta)$$
$$= m^2 - 2\alpha m + (\alpha^2 + \beta^2)$$

which corresponds to the differential equation $y'' - 2\alpha y' + (\alpha^2 + \beta^2)y = [D^2 - 2\alpha D + (\alpha^2 + \beta^2)]y = 0$. Hence, the differential operator $[D^2 - 2\alpha D + (\alpha^2 + \beta^2)]$ annihilates the functions $e^{\alpha x}\cos\beta x$ and $e^{\alpha x}\sin\beta x$.

If the complex roots are repeated with multiplicity n, they correspond to the characteristic equation

$$[m - (\alpha + i\beta)]^n[m - (\alpha - i\beta)]^n = ([m - (\alpha + i\beta)][m - (\alpha - i\beta)])^n$$
$$= (m^2 - 2\alpha m + (\alpha^2 + \beta^2))^n = 0 \quad .$$

This is the characteristic equation of the differential equation of order $2n$ which can be written in operator form as $(D^2 - 2\alpha D + (\alpha^2 + \beta^2))^n y = 0$. Therefore, the differential operator $(D^2 - 2\alpha D + (\alpha^2 + \beta^2))^n$ annihilates the functions $e^{\alpha x}\cos\beta x$, $e^{\alpha x}\sin\beta x$, $xe^{\alpha x}\cos\beta x$, $xe^{\alpha x}\sin\beta x$, ..., $x^{n-1}e^{\alpha x}\cos\beta x$, $x^{n-1}e^{\alpha x}\sin\beta x$.

Example 10

Determine an operator that annihilates the indicated function.
(a) $f(x) = \cos x$ (b) $f(x) = e^{2x}\cos x$
(c) $f(x) = xe^{2x}\cos x$ (d) $f(x) = x^2 e^{2x}\cos x$
(e) $f(x) = x^2 e^{2x}\cos x + e^{2x}\cos x$

Solution   (a) In this case, we have a function of the form $e^{\alpha x}\cos\beta x$ with $\alpha = 0$ and $\beta = 1$. The differential operator $(D^2 + 1)$ annihilates $f(x) = \cos x$, which makes sense because $\cos x$ is a solution of $(D^2 + 1)y = y'' + y = 0$.
(b) For the function $f(x) = e^{2x}\cos x$, $\alpha = 2$ and $\beta = 1$. Therefore, the differential operator $(D^2 - 4D + 5)$ annihilates $f(x) = e^{2x}\cos x$.
(c) The function $f(x) = xe^{2x}\cos x$ is a solution of a differential equation with repeated roots of the corresponding characteristic equation. Also, notice from part (b) that $(D^2 - 4D + 5)$ annihilates $f(x) = e^{2x}\cos x$. Hence, $(D^2 - 4D + 5)^2$ annihilates $f(x) = xe^{2x}\cos x$.
(d) In this case, the function $f(x) = x^2 e^{2x}\cos x$ corresponds to a solution of a differential equation with the complex conjugate pair $2 \pm i$ as roots of multiplicity 3. Therefore, $(D^2 - 4D + 5)^3$ annihilates $f(x) = x^2 e^{2x}\cos x$.
(e) We notice that $f(x) = x^2 e^{2x}\cos x + e^{2x}\cos x$ is made up of two functions that satisfy the differential equation with characteristic roots $2 \pm i$. Therefore, $(D^2 - 4D + 5)^3$ annihilates $f(x) = x^2 e^{2x}\cos x + e^{2x}\cos x$.

To construct an annihilator of a *sum* of functions, like those given in Table 4.4, we compute the *product* of the annihilators.

TABLE 4.4   Annihilators of Familiar Functions

Functions	Annihilator
$1, x, x^2, \ldots, x^{n-1}$	D^n
$e^{kx}, xe^{kx}, \ldots, x^{n-1}e^{kx}$	$(D-k)^n$
$e^{\alpha x}\cos\beta x, xe^{\alpha x}\cos\beta x, \ldots, x^{n-1}e^{\alpha x}\cos\beta x,$ $e^{\alpha x}\sin\beta x, xe^{\alpha x}\sin\beta x, \ldots, x^{n-1}e^{\alpha x}\sin\beta x$	$[D^2 - 2\alpha D + (\alpha^2 + \beta^2)]^n$

Property 4

Let $P_1 = p_1(D)$ and $P_2 = p_2(D)$ be linear differential operators with constant coefficients. If P_1 annihilates $f(x)$ and P_2 annihilates $g(x)$, then P_1P_2 (or P_2P_1) annihilates $af(x) + bg(x)$, where a and b are constants. • • •

PROOF OF PROPERTY 4

$$P_1P_2[af(x) + bg(x)] = P_1P_2[af(x)] + P_1P_2[bg(x)]$$
$$= aP_1P_2[f(x)] + bP_1P_2[g(x)]$$
$$= aP_2P_1[f(x)] + bP_1P_2[g(x)]$$
$$= aP_2(P_1[f(x)]) + bP_1(P_2[g(x)])$$
$$= aP_2(0) + bP_1(0) = 0. \; • • •$$

We can use this property to determine the operator that annihilates any function of the forms discussed earlier. (Note that we can extend this property to include the linear combination of more than two functions.)

Example 11

Show that the differential operator $D(D^2 + 4)$ annihilates $f(x) = \sin 2x - 2\cos 2x + 1$.

Solution   Note that $(D^2 + 4)$ annihilates $\sin 2x$ and $\cos 2x$ while D annihilates 1. Verifying this, we have

$$D(D^2 + 4)[\sin 2x - 2\cos 2x + 1]$$
$$= D(D^2 + 4)[\sin 2x] - 2D(D^2 + 4)[\cos 2x] + D(D^2 + 4)[1]$$
$$= D(0) - 2D(0) + (D^2 + 4)D[1]$$
$$= 0 - 2\cdot 0 + (D^2 + 4)(0) = 0.$$

| Example 12 | Determine a differential operator that annihilates the indicated function. |

(a) $f(x) = x + \cos x$ (b) $g(x) = 3x^2 - e^{-x} \sin 4x$
(c) $h(x) = x \cos x + 4$ (d) $k(x) = x^2 e^x - x^4 + x$

Solution   (a) Let $f(x) = f_1(x) + f_2(x)$ where $f_1(x) = x$ and $f_2(x) = \cos x$. The operator D^2 annihilates $f_1(x) = x$ and $(D^2 + 1)$ annihilates $f_2(x) = \cos x$. Therefore, $D^2(D^2 + 1)$ or $(D^2 + 1)D^2$ annihilates $f(x) = x + \cos x$.

(b) Let $g(x) = g_1(x) + g_2(x)$ where $g_1(x) = 3x^2$ and $g_2(x) = -e^{-x} \sin 4x$. Then, D^3 annihilates $g_1(x) = 3x^2$ and $(D^2 + 2D + 17)$ annihilates $g_2(x) = -e^{-x} \sin 4x$. Hence, $D^3(D^2 + 2D + 17)$ or $(D^2 + 2D + 17)D^3$ annihilates $g(x) = 3x^2 - e^{-x} \sin 4x$.

(c) Let $h(x) = h_1(x) + h_2(x)$ where $h_1(x) = x \cos x$ and $h_2(x) = 4$. Because $(D^2 + 1)^2$ annihilates $h_1(x) = x \cos x$ and D annihilates $h_2(x) = 4$, $(D^2 + 1)^2 D$ or $D(D^2 + 1)^2$ annihilates $h(x) = x \cos x + 4$.

(d) Let $k(x) = k_1(x) + k_2(x) + k_3(x)$ where $k_1(x) = x^2 e^x$, $k_2(x) = -x^4$, and $k_3(x) = x$. The operator $(D - 1)^3$ annihilates $k_1(x) = x^2 e^x$ and D^5 annihilates $k_2(x) = -x^4$. Because D^5 also annihilates $k_3(x) = x$, the operator $(D - 1)^3 D^5$ or $D^5(D - 1)^3$ annihilates $k(x) = x^2 e^x - x^4 + x$.

Verify that the operators in Example 12 annihilate the given functions. (A computer algebra system will be helpful.)

IN TOUCH WITH TECHNOLOGY

1. (a) Find a general solution of $4y'' + 4y' + 37y = 0$.
(b) Find values of A and B so that $y_p(x) = xe^{-x/2}(A \cos 3x + B \sin 3x)$ is a particular solution of $4y'' + 4y' + 37y = e^{-x/2} \cos 3x$. **(c)** Use the results of **(a)** and **(b)** to solve the initial-value problem
$$\begin{cases} 4y'' + 4y' + 37y = e^{-x/2} \cos 3x \\ y(0) = 0, \; y'(0) = 1 \end{cases}.$$
Verify that your result satisfies the initial conditions by graphing it on an appropriate interval.

2. (a) Find conditions on ω so that $y_p(x) = A \cos \omega x + B \sin \omega x$, where A and B are constants to be determined, is a particular solution of $y'' + y = \cos \omega x$.
(b) For the value(s) of ω obtained in (a), show that every solution of the nonhomogeneous equation $y'' + y = \cos \omega x$ is bounded (and periodic).

3. (a) Find conditions on ω so that $y_p(x) = x(A \cos \omega x + B \sin \omega x)$, where A and B are constants to be determined, is a particular solution of $y'' + 4y = \sin \omega x$.
(b) For the value(s) of ω obtained in (a), show that every solution of the nonhomogeneous equation $y'' + 4y = \sin \omega x$ is unbounded.

The following exercise illustrates how we can use operator notation to solve linear homogeneous systems of equations. (This topic discussed in more detail in Chapter 8.)

4. (a) Graph the direction field associated with the
system $\begin{cases} 3x' = -x + y \\ y' = -3x - \dfrac{1}{3}y \end{cases}$
for $-3 \le x \le 3$ and $-3 \le y \le 3$.

(b) Find a general solution of this system as follows. First, rewrite the system in operator notation as

$$\begin{cases} (3D+1)x - y = 0 \\ 3x + \left(D + \dfrac{1}{3}\right)y = 0 \end{cases}$$

where $D = \dfrac{d}{dt}$.

Apply $D + \frac{1}{3}$ to the first equation to obtain the system

$$\begin{cases} \left(D + \dfrac{1}{3}\right)(3D+1)x - \left(D + \dfrac{1}{3}\right)y = 0 \\ \\ 3x + \left(D + \dfrac{1}{3}\right)y = 0 \end{cases}$$

Eliminate y from the system by adding these two equations and solve for x. Use the relationship

$$3x' = -x + y \text{ or } y = 3x' + x$$

to solve for y.

(c) Find the solution that satisfies the initial conditions **(i)** $x(0) = 0$, $y(0) = 2$, **(ii)** $x(0) = 2$, $y(0) = 0$, and

(iii) $x(0) = 2$, $y(0) = 2$. In each case, graph $x(t)$, $y(t)$, and the parametric equations $\begin{cases} x = x(t) \\ y = y(t) \end{cases}$ for $0 \le t \le 5\pi$. Display the graphs of the parametric equations together with the direction field. Are the vectors in the direction field tangent to the graphs of the solutions?

(d) Show that the sum of the solutions obtained in **(i)** and **(ii)** is the same as that obtained in **(iii)**. Is the Principle of Superposition valid for this linear homogeneous system of equations? Explain.

(e) Is the Principle of Superposition valid for linear homogeneous systems of equations? Explain.

Find an annihilator of each of the following functions. Use the differentiation capabilities of your technology to verify your result.

5. $x^8 e^{2x}$

6. $x^3 e^x \sin 4x$

7. $x^6 e^{-2x} + 3e^{4x} \sin x$

8. $2x^2 e^{3x} \cos 5x + e^{-8x}x^3 \cos 2x + x$

EXERCISES 4.4

In Exercises 1–8, **(a)** show that the given function is a particular solution of the given nonhomogeneous differential equation. **(b)** Find a general solution by solving the corresponding homogeneous equation.

1. $3y'' + 2y' - 8y = -5$, $y_p(x) = \dfrac{5}{8}$

2. $y'' + y' = 2$, $y_p(x) = 2x$

*3. $10y'' - 9y' - 9y = -3x$, $y_p(x) = \dfrac{x}{3} - \dfrac{1}{3}$

4. $y'' - 7y' + 10y = x$, $y_p(x) = \dfrac{x}{10} + \dfrac{7}{100}$

5. $y'' + 10y' + 41y = -e^{3x}$, $y_p(x) = -\dfrac{1}{80}e^{3x}$

6. $y'' - 3y' = xe^{3x}$, $y_p(x) = -\dfrac{1}{9}xe^{3x} + \dfrac{1}{6}x^2 e^{3x}$

*7. $y'' - 2y' + y = 3\cos x$, $y_p(x) = -\dfrac{3}{2}\sin x$

8. $y'' + 6y' + 13y = 2x^{-2x}\sin x$,
$$y_p(x) = -\dfrac{1}{5}e^{-2x}\cos x + \dfrac{2}{5}e^{-2x}\sin x$$

In Exercises 9–28 find a differential operator that annihilates each function.

9. 3

10. $4x$

11. $2x^2$

12. e^{-8x}

13. e^{4x}

14. $\sin 8x$

15. $\cos 2x$

16. $\cos 5x + 1$

*17. $10 - \sin 3x$

18. $2xe^{-x}$

19. $x^2 e^{8x}$

20. $x^2 e^{2x} + x + 1$

*21. $x^3 e^{3x}$

22. $e^{5x} \cos 2x$

23. $e^{-x} \cos 3x$

24. $xe^{3x} \sin x$

*25. $11x^2 e^{8x} \sin 9x$

26. $-4xe^{2x} - 5x^5 \cos 4x$

*27. $3x^4 e^{-4x} - 2xe^{-2x} \sin 2x$

28. $-3 + 4x^2 - 2x^2 e^{-3x} - xe^{3x} \cos 2x + e^{4x} \sin 4x$

29. Let $P = p(D)$ be a differential operator of the form

$$p(D) = a_n D^n + a_{n-1} D^{n-1} + \cdots + a_1 D + a_0$$

Show that $P[f(x) + g(x)] = P[f(x)] + P[g(x)]$.

30. If P_1 and P_2 are two linear differential operators with constant coefficients, show that $P_1 P_2[f(x)] = P_2 P_1[f(x)]$.

31. Let $y_p(x) = \frac{3}{5} \sin x$. Show that if $y(x)$ is any solution of $y'' - 4y = -3 \sin x$ then $y(x) - y_p(x)$ is a solution of $y'' - 4y = 0$. Explain why there are constants c_1 and c_2 so that $y(x) - y_p(x) = c_1 e^{-2x} + c_2 e^{2x}$ and, thus, $y(x) = c_1 e^{-2x} + c_2 e^{2x} + y_p(x)$.

32. Show that if $y_p(x)$ is a particular solution of the nonhomogeneous equation

$$a_n y^{(n)}(x) + a_{n-1} y^{(n-1)}(x) + \cdots$$
$$+ a_1 y'(x) + a_0 y(x) = g(x)$$

and

$$y_h(x) = c_1 f_1(x) + c_2 f_2(x) + \cdots + c_n f_n(x)$$

is a general solution of the corresponding homogeneous equation

$$a_n y^{(n)}(x) + a_{n-1} y^{(n-1)}(x) + \cdots$$
$$+ a_1 y'(x) + a_0 y(x) = 0$$

(assume that $\{f_1(x), f_2(x), \ldots, f_n(x)\}$ is a fundamental set of solutions for the corresponding homogeneous equation), then every solution of the nonhomogeneous equation, $y(x)$, can be written in the form

$$y(x) = y_h(x) = y_p(x),$$

for some choice of $c_1, c_2, \ldots, c_n$.

(**Variable-Coefficient Operators**) As with differential equations, differential operators can have coefficients that are not constant. For example, we can write the differential equation $x^2 y'' + xy' + y = 0$ as $(x^2 D^2 + xD + 1)y = 0$.

We can also consider products of operators with variable coefficients such as

$$(xD + 2)(xD - 3) = xD(xD - 3) + 2(xD - 3)$$
$$= xD(xD) - xD3 + 2xD - 6$$
$$= x(xD^2 + D) + 2xD - 6 = x^2 D^2 + 3xD - 6.$$

In Exercises 33–40, write the differential equation in the form $a_n(x)y^{(n)} + \cdots + a_2(x)y'' + a_1(x)y' + a_0(x)y = 0$.

33. $(3x^2 D^2 - xD + 8)y = 0$

34. $(2x^2 D^2 + 5xD - 10)y = 0$

*35. $(xD + 6)(D - 1)y = 0$

36. $(xD - 10)(xD + 4)y = 0$

37. $(4xD + 3)(D - 1)y = 0$

38. $(x^2 D + 1)(xD - 1)y = 0$

39. $xD(x^2 D^2 + D + 10x)y = 0$

40. $(xD - 1)(D^2 + 2xD + 1)y = 0$

In Exercises 41–45, show that the given function satisfies the differential equation.

41. $(x^2 D^2 - 2xD - 4)y = 0$; $y = c_1 x^4 + c_2 x^{-1}$

42. $(9x^2 D^2 + 15xD + 1)y = 0$; $y = x^{-1/3}(c_1 + c_2 \ln x)$

43. $(x^2 D^2 - xD + 5)y = 0$;
$y = x(c_1 \cos(2 \ln x) + c_2 \sin(2 \ln x))$

44. $(x^2 D^2 - 5xD + 9)y = 2x^3$;
$y = x^3(c_1 + c_2 \ln x + (\ln x)^2)$

45. $(x^2 D^2 - 4xD + 6)y = 4x - 6$;
$y = c_1 x^3 + c_2 x^2 + 2x - 1$

46. If the operators P_1 and P_2 have variable coefficients, does the relationship $P_1 P_2[f(x)] = P_2 P_1[f(x)]$ always hold?

47. Are the differential operators $P_1 = xD$ and $P_2 = Dx$ equivalent? Why or why not?

4.5   Nonhomogeneous Equations with Constant Coefficients: Using the Annihilator Method

Another approach, the method of undetermined coefficients (using the Principle of Superposition), is discussed in the next section. Both methods lead to the same result. Your instructor may prefer the method of undetermined coefficients and choose to omit this section or cover this section and choose to omit the next. Of course, some instructors will discuss both approaches.

In this section we show how annihilators are used to solve some nonhomogeneous linear differential equations with constant coefficients.

The nonhomogeneous linear nth-order differential equation with constant coefficients can be expressed as $p(D)y = g(x)$. When $g(x)$ is a function of one of the forms listed in Table 4.5, another differential operator, $q(D)$, which annihilates $g(x)$, can be determined.

TABLE 4.5

$g(x)$
$1, x, x^2, \ldots, x^{n-1}$
$e^{kx}, xe^{kx}, \ldots, x^{n-1}e^{kx}$
$e^{\alpha x}\cos\beta x, xe^{\alpha x}\cos\beta x, \ldots, x^{n-1}e^{\alpha x}\cos\beta x,$
$e^{\alpha x}\sin\beta x, xe^{\alpha x}\sin\beta x, \ldots, x^{n-1}e^{\alpha x}\sin\beta x$

If the differential operator $q(D)$ annihilates $g(x)$, applying $q(D)$ to the nonhomogeneous equation yields $q(D)p(D)y = q(D)g(x) = 0$. A particular solution is found by solving the homogeneous equation $q(D)q(D)y = 0$.

> **Procedure to Solve $p(D)y = g(x)$ if $g(x)$ is a linear combination of the functions**
> $1, x, x^2, \ldots, e^{kx}, xe^{kx}, x^2e^{kx}, \ldots, e^{\alpha x}\cos\beta x, xe^{\alpha x}\cos\beta x,$
> $x^2e^{\alpha x}\cos\beta x, \ldots, e^{\alpha x}\sin\beta x, xe^{\alpha x}\sin\beta x, x^2e^{\alpha x}\sin\beta x, \ldots.$

① Determine an operator $q(D)$ that annihilates $g(x)$: $q(D)(g(x)) = 0$.
② Apply the operator to both sides of the differential equation: $q(D)(p(D)y) = q(D)(g(x)) = 0$.
③ Solve the homogeneous equation $q(D)(p(D)y) = 0$.
④ Find the solution $y_h(x)$ of the homogeneous equation corresponding to the original equation, $p(D)y = 0$.
⑤ Eliminate the terms of the homogeneous solution $y_h(x)$ from the general solution obtained in step 3. The function that remains is the correct form of a particular solution.
⑥ Solve for the unknown coefficients in the particular solution to obtain $y_p(x)$ by substitution of $y_p(x)$ into $p(D)y = y(x)$.
⑦ A general solution of the nonhomogeneous equation is $y(x) = y_h(x) + y_p(x)$.

| Example 1 |

Solve the nonhomogeneous equation $y'' + y = x$.

Solution First, we write the differential equation in operator form $(D^2 + 1)y = x$ and note that the operator D^2 annihilates the function $g(x) = x$. Applying this annihilator to both sides of the equation, we obtain the homogeneous equation

$$D^2(D^2 + 1)y = 0,$$

which has characteristic equation $m^2(m^2 + 1) = 0$. A general solution of this equation is $y(x) = b_1 + b_2x + b_3 \cos x + b_4 \sin x$, where b_1, b_2, b_3, and b_4 are arbitrary constants. Solving the corresponding homogeneous equation $y'' + y = 0$, we find that $y_h(x) = c_1 \cos x + c_2 \sin x$. Eliminating these terms from $y(x) = b_1 + b_2x + b_3 \cos x + b_4 \sin x$ indicates that a particular solution has the form $y_p(x) = A + Bx$, where A and B are constants to be determined by substituting $y_p(x)$ into the original equation $y'' + y = x$. Now, $y_p'(x) = B$ and $y_p''(x) = 0$ so $y'' + y = 0 + A + Bx = A + Bx = x$. Equating the coefficients of like terms, we find that $A = 0$ and $B = 1$. Therefore, $y_p(x) = A + Bx = 0 + (1)x = x$ and a general solution of the nonhomogeneous equation is $y(x) = y_h(x) + y_p(x) = c_1 \cos x + c_2 \sin x + x$.

| Example 2 |

Solve $y''' + y' = \cos x + x$.

Solution We begin by determining an operator that annihilates the function $g(x) = \cos x + x$. Because $(D^2 + 1)$ annihilates $\cos x$ and D^2 annihilates x, the operator $D^2(D^2 + 1)$ annihilates $g(x) = \cos x + x$. Applying this operator to both sides of $y''' + y' = \cos x + x$ in operator form yields $D^2(D^2 + 1)(D^3 + D)y = D^2(D^2 + 1)(\cos x + x) = 0$. This homogeneous equation has characteristic equation $m^2(m^2 + 1)(m^3 + m) = m^3(m^2 + 1)^2 = 0$. Therefore, it has general solution

$$y(x) = b_1 + b_2x + b_3x^2 + b_4 \cos x + b_5 \sin x + b_6x \cos x + b_7x \sin x.$$

Now, we solve the corresponding homogeneous equation $y''' + y' = 0$ to see that

$$y_h(x) = c_1 + c_2 \cos x + c_3 \sin x.$$

Comparing the two solutions

$$y_h(x) = c_1 + c_2 \cos x + c_3 \sin x \qquad \text{and}$$
$$y(x) = b_1 + b_2x + b_3x^2 + b_4 \cos x + b_5 \sin x + b_6x \cos x + b_7x \sin x,$$

we recognize that $y_p(x) = Ax + Bx^2 + Cx \cos x + Ex \sin x$ after eliminating the terms of $y_h(x)$ from $y(x)$. Differentiating $y_p(x)$, we have

$$y_p'(x) = A + 2Bx + C \cos x - Cx \sin x + E \sin x + Ex \cos x,$$
$$y_p''(x) = 2B - 2C \sin x - Cx \cos x + 2E \cos x - Ex \sin x, \qquad \text{and}$$
$$y_p'''(x) = -3C \cos x + Cx \sin x - 3E \sin x - Ex \cos x.$$

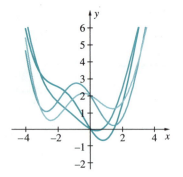

Figure 4.7

Substitution into $y''' + y' = \cos x + x$ then gives us $-2C \cos x - 2E \sin x + 2Bx + A = \cos x + x$. Hence, $A = 0$, $B = \frac{1}{2}$, $C = -\frac{1}{2}$, and $E = 0$, so a particular solution of the nonhomogeneous equation is $y_p(x) = \frac{1}{2}x^2 - \frac{1}{2}x \cos x$. This means that a general solution is

$$y(x) = y_h(x) + y_p(x) = c_1 + c_2 \cos x + c_3 \sin x + \frac{1}{2}x^2 - \frac{1}{2}x \cos x.$$

Figure 4.7 shows the graph $y(x) = y_h(x) + y_p(x)$ in Example 2 for various values of c_1, c_2 and c_3. Identify the graph of $y_p(x) = \frac{1}{2}x^2 - \frac{1}{2}x \cos x$.

As with other differential equations, we solve initial-value problems that involve nonhomogeneous equations by first finding a general solution to the nonhomogeneous equation and then using the initial conditions to find the values of the unknown coefficients so that the initial conditions are satisfied.

Example 3

Solve the initial-value problem $\begin{cases} y'' + y = x \\ y(0) = 1, \ y'(0) = 0 \end{cases}$

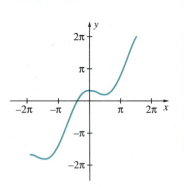

Figure 4.8 Graph of $y(x) = \cos x - \sin x + x$

Solution A general solution of the nonhomogeneous equation $y'' + y = x$ was found to be $y(x) = c_1 \cos x + c_2 \sin x + x$ in Example 1. Differentiating we obtain $y'(x) = -c_1 \sin x + c_2 \cos x + 1$. Substituting the initial conditions yields $y(0) = c_1 \cos(0) + c_2 \sin(0) + 0 = c_1 = 1$ and $y'(0) = -c_1 \sin(0) + c_2 \cos(0) + 1 = c_2 + 1 = 0$. Hence, $c_1 = 1$ and $c_2 = -1$, so the solution of the initial-value problem is $y(x) = \cos x - \sin x + x$, which is graphed in Figure 4.8. Do the initial conditions appear to be satisfied?

 Solve the initial-value problem $\begin{cases} y'' + y = x \\ y(0) = a, \ y'(0) = b \end{cases}$. Calculate $\lim_{x \to \infty} y(x)$. How does the choice of different initial conditions affect the limit?

Example 4

Solve $y'' + 2y' - 3y = 4e^x - \sin x$ subject to the conditions $y(0) = 0$ and $y'(0) = 1$.

Solution First, we notice that $(D - 1)$ annihilates e^x and $(D^2 + 1)$ annihilates $\sin x$. Hence, $(D - 1)(D^2 + 1)$ annihilates $g(x) = 4e^x - \sin x$. Applying this operator to both sides of the equation $y'' + 2y' - 3y = 4e^x - \sin x$ gives us

$$(D - 1)(D^2 + 1)(D^2 + 2D - 3)y = (D - 1)^2(D^2 + 1)(D + 3)y = 0$$

in operator form. The solution of this equation with characteristic equation $(m - 1)^2(m^2 + 1)(m + 3) = 0$ is

$$y(x) = b_1 e^{-3x} + b_2 e^x + b_3 x e^x + b_4 \cos x + b_5 \sin x.$$

The solution to the corresponding homogeneous equation $y'' + 2y' - 3y = 0$ with characteristic equation $m^2 + 2m - 3 = (m + 3)(m - 1) = 0$ is

$$y_h(x) = c_1 e^{-3x} + c_2 e^x.$$

Eliminating the terms of $y_h(x)$ from $y(x) = b_1 e^{-3x} + b_2 e^x + b_3 x e^x + b_4 \cos x + b_5 \sin x$, we see that a particular solution is $y_p(x) = A x e^x + B \cos x + C \sin x$. Differentiating $y_p(x)$ yields

$$y_p'(x) = A e^x + A x e^x - B \sin x + C \cos x \qquad \text{and}$$
$$y_p''(x) = 2A e^x + A x e^x - B \cos x - C \sin x.$$

Substitution into the equation $y'' + 2y' - 3y = 4e^x - \sin x$ then gives us

$$4A e^x + (2C - 4B) \cos x + (-4C - 2B) \sin x = 4e^x - \sin x.$$

Therefore, $4A = 4$, $2C - 4B = 0$, and $-4C - 2B = -1$ which implies that $A = 1$, $B = \frac{1}{10}$, and $C = \frac{1}{5}$. Hence, a particular solution is $y_p(x) = x e^x + \frac{1}{10} \cos x + \frac{1}{5} \sin x$ and a general solution of $y'' + 2y' - 3y = 4e^x - \sin x$ is

$$y(x) = y_h(x) + y_p(x) = c_1 e^{-3x} + c_2 e^x + x e^x + \frac{1}{10} \cos x + \frac{1}{5} \sin x.$$

Because $y(0) = c_1 + c_2 + \frac{1}{10} = 0$ and $y'(0) = c_2 - 3c_1 + \frac{6}{5} = 1$, $c_2 = \frac{-1}{8}$ and $c_1 = \frac{1}{40}$ so that the solution of the initial-value problem is

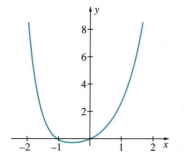

Figure 4.9 Graph of $y(x) = -\frac{1}{8}e^x + \frac{1}{40}e^{-3x} + x e^x + \frac{1}{10} \cos x + \frac{1}{5} \sin x$

$$y(x) = -\frac{1}{8}e^x + \frac{1}{40}e^{-3x} + x e^x + \frac{1}{10} \cos x + \frac{1}{5} \sin x. \text{ (See Figure 4.9.)}$$

Of course, boundary-value problems may have none, one, or more than one solution.

Example 5

Show that the boundary-value problem

$$\begin{cases} 4y'' + 4y' + 37y = \cos 3x \\ \qquad\qquad y(0) = y(\pi) \end{cases}$$

has infinitely many solutions.

Solution   The operator $D^2 + 9$ annihilates $\cos 3x$ and a general solution of $(D^2 + 9)(4D^2 + 4D + 37)y = 0$ is

$$y = e^{-x/2}(b_1 \cos 3x + b_2 \sin 3x) + b_3 \cos 3x + b_4 \sin 3x,$$

and a general solution of the corresponding homogeneous equation

$$4y'' + 4y' + 37y = 0 \text{ is } y_h(x) = e^{-x/2}(c_1 \cos 3x + c_2 \sin 3x).$$

A particular solution to the nonhomogeneous equation has the form $y_p(x) = A \cos 3x + B \sin 3x$. Substituting $y_p(x)$ into the nonhomogeneous equation, solving for A and B, and simplifying the result we find that a general solution of the non-homogeneous equation is

$$ y = \cos(3x)\left[c_1 e^{-x/2} + \frac{1}{145}\right] + \sin(3x)\left[c_2 e^{-x/2} + \frac{12}{145}\right]. $$

To satisfy the boundary conditions, we first calculate $y(0) = c_1 + \dfrac{1}{145}$ and

$y(\pi) = -\dfrac{1}{145} - c_1 e^{-\pi/2}$ and then solve $c_1 + \dfrac{1}{145} = -\dfrac{1}{145} - c_1 e^{-\pi/2}$ for c_1

which leads to $c_1 = \dfrac{-2e^{\pi/2}}{145(1 + e^{\pi/2})} \approx -0.01141927$.

We conclude that if $c_1 = \dfrac{-2e^{\pi/2}}{145(1 + e^{\pi/2})}$ and c_2 is *any* real number, y is a solution to the boundary-value problem. These results are confirmed by the graph of several solutions to the boundary-value problem in Figure 4.10.

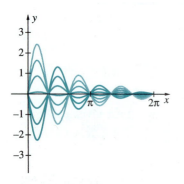

Figure 4.10 Graph of several solutions of
$$\begin{cases} 4y'' + 4y' + 37y = \cos 3x \\ y(0) = y(\pi) \end{cases}$$

 Solve the boundary-value problem $\begin{cases} 4y'' + 4y' + 37y = \cos 3x \\ y(0) = 0, \; y(\pi/2) = 0 \end{cases}$.

IN TOUCH WITH TECHNOLOGY

As when calculating solutions of homogeneous equations, technology is useful in helping us calculate and visualize solutions of nonhomogeneous equations, particularly when the calculations encountered are symbolically complicated.

1. Find a general solution of each equation.

   (a) $y''' - y'' - 7y' + 15y = x^2 e^{-3x} + e^{2x} \cos 3x$

   (b) $y^{(4)} - 9y''' + 24y'' - 36y' + 80y = \cos 2x + 4 \sin 2x$

2. Solve the following initial-value problems. Graph the solution.

   (a) $\begin{cases} 2y''' - 6y'' + 18y' - 56y = e^{-2x} + 4e^{3x} \\ y''(0) = -1, \; y'(0) = 0, \; y(0) = 1 \end{cases}$

   (b) $\begin{cases} y^{(4)} - \frac{1}{4}y''' - 21y'' - \frac{50}{4}y' + 5y = x \\ y'''(0) = 0, \; y''(0) = 0, \; y'(0) = 0, \; y(0) = 1 \end{cases}$

3. Consider the initial-value problem

$$ \begin{cases} y''' + y'' + 4y' + 4y = e^{-x} \cos 2x \\ y(0) = a, \; y'(0) = b, \; y''(0) = c \end{cases}. $$

   (a) Find conditions on a, b, and c, if any, so that $y(x) \to 0$ as $x \to \infty$. Confirm your results by graphing a (the) solution that satisfies the conditions.

   (b) Is it possible to choose a, b, and c so that the solution behaves like $\cos 2x$ for positive large values of x? $\sin 2x$?

4. Show that the solution to the initial-value problem

$$\begin{cases} y'' + 4y = f(x) \\ y(0) = 0, \ y'(0) = 2 \end{cases}$$

is

$$y(x) = \sin(2x) + \frac{1}{2}\sin(2x)\int_0^x f(t)\cos 2t\, dt - \frac{1}{2}\cos(2x)\int_0^x f(t)\sin 2t\, dt$$

(as long as the integrals can be evaluated).

(a) Verify this result by using the Fundamental Theorem of Calculus to show that

$$y(x) = \sin(2x) + \frac{1}{2}\sin(2x)\int_0^x f(t)\cos 2t\, dt - \frac{1}{2}\cos(2x)\int_0^x f(t)\sin 2t\, dt$$

is the solution to

$$\begin{cases} y'' + 4y = f(x) \\ y(0) = 0, \ y'(0) = 2 \end{cases}.$$

(b) Show that if $f(x)$ is constant, the resulting solution is periodic and find its period. Confirm your result with a graph.

(c) If $f(x)$ is periodic, is the resulting solution periodic? As in (b), confirm your result with a graph.

5. The substitution $u = y'$ into the equation $f(y^{(n)}, y^{(n-1)}, \ldots, y') = 0$ yields a differential equation of lower order. Use the substitution $u = y'$ to solve the equations **(a)** $\dfrac{d^2y}{dx^2} + 5\left(\dfrac{dy}{dx}\right)^2 - 4\dfrac{dy}{dx} = 0$ and

(b) $\dfrac{d^2y}{dx^2} - (3y - 5)\dfrac{dy}{dx} = 0$. Graph the solutions to each for various values of the constants.

The following exercise illustrates how we can use operator notation together with the methods discussed in this section to solve linear nonhomogeneous systems of equations. (This topic is discussed in more detail in Chapter 8.)

6. (a) Find a general solution of the system
$$\begin{cases} x' = 2x - 5y + \cos 4t \\ y' = -4x - 2y + \sin 4t \end{cases}$$ using the following steps.
First, rewrite the system in operator notation as

$$\begin{cases} (D - 2)x + 5y = \cos 4t \\ 4x + (D + 2)y = \sin 4t \end{cases}, \quad \text{where } D = \frac{d}{dt}.$$

Apply $D + 2$ to the first equation and -5 to the second equation to obtain the system

$$\begin{cases} (D - 2)(D + 2)x + 5(D + 2)y = (D + 2)[\cos 4t] \\ -20x - 5(D + 2)y = -5\sin 4t \end{cases}.$$

Add these two equations to eliminate y. Solve the resulting second-order equation for x.

$$(D - 2)(D + 2)x - 20x = -9\sin 4t + 2\cos 4t$$

Use x and the relationship

$$y = \frac{1}{5}(x' - 2x - \cos 4t)$$

to solve for y. **(b)** Find the solution that satisfies the initial conditions **(i)** $x(0) = 0$, $y(0) = 2$, **(ii)** $x(0) = 2$, $y(0) = 0$, and **(iii)** $x(0) = 2$, $y(0) = 2$. In each case, graph $x(t)$, $y(t)$, and the parametric equations
$$\begin{cases} x = x(t) \\ y = y(t) \end{cases}$$ for $0 \le t \le 5\pi$. **(c)** Show that the sum of the solutions obtained in **(i)** and **(ii)** is not the same as that obtained in **(iii)**. **(d)** Is the Principle of Superposition valid for this linear nonhomogeneous system of equations? Explain. **(e)** Is the Principle of Superposition valid for linear nonhomogeneous systems of equations? Explain.

EXERCISES 4.5

In Exercises 1–42, find a general solution of each equation. (A computer algebra system may be useful in solving the higher-order equations.)

*1. $y'' - 5y' + 6y = e^x$

2. $y'' + 7y' + 12y = 2x^2$

3. $y'' + 14y' + 49y = e^{-7x}$

4. $y'' - 8y' + 16y = e^{4x}$

*5. $y'' - 6y' + 10y = 10$

6. $y'' + 4y' + 13y = e^{-x}$

7. $y'' + 4y' = xe^{-x}$

8. $y'' - 4y' = \cos x$

9. $9y'' + y = x$

10. $4y'' + 25y = 2e^{-x}$

*11. $y'' + 2y' + 17y = e^{-x}\cos 4x$

12. $y'' + 3y' - 10y = x + e^{-3x}$

13. $y'' - 2y' - 15y = 64e^{5x}$

14. $y'' + 3y' + 2y = e^{-x} + 3e^{-2x}$

*15. $y'' - 2y' + 26y = 29e^{-x} + 78x - 6$

16. $y'' - 4y' + 8y = xe^{2x} - 2e^{-2x}$

17. $y'' - 4y' + 5y = 25x$

18. $y'' + 10y' + 41y = -3e^{2x} - e^{-5x}\sin 4x$

*19. $y'' + 6y' + 34y = e^{-3x}\cos 5x + e^{5x}\sin 3x$

20. $y'' - 6y' + 13y = x^2 - xe^{3x}\sin 2x$

21. $y'' + 8y' + 20y = 2e^{-4x}\cos 2x - 8e^{-4x}\sin 2x$

22. $y''' + y'' = 4x + 8$

*23. $y''' + 16y' = 10x$

24. $y''' + 5y'' - 4y' - 20y = 4e^{-2x}$

25. $y''' + y'' - 16y' - 16y = 17e^{3x} - 28xe^{3x}$

26. $y''' - 7y'' + 14y' - 8y = e^{2x} + 3e^x$

*27. $y''' - 4y'' + 21y' - 34y = 2e^{-2x}$

28. $y''' + 5y'' + 11y' + 15y = x - e^{-3x}$

29. $y''' - y'' + 8y' + 60y = 60x^2 + 16x - 2$

30. $y''' - 3y'' + 4y' - 2y = e^x \sin x - 3e^x$

*31. $y^{(4)} - 13y'' + 36y = \cos x$

32. $y^{(4)} - 10y'' + 9y = xe^{3x} + 2e^{-x}$

33. $y^{(4)} - 5y''' + 22y'' - 80y' + 96y = 960$

34. $y^{(4)} - 10y''' + 34y'' - 48y' + 32y = e^{2x}$

*35. $y^{(4)} + 2y''' - 7y'' + 12y' + 72y = 144x + 24$

36. $y^{(4)} + 2y'' + 8y' + 5y = xe^{-x} + e^x \cos x$

37. $y^{(4)} - 32y'' + 256y = 256x^2 - 64$

38. $y^{(4)} - 8y'' + 16y = xe^{-2x} - e^{2x}$

*39. $y^{(4)} - 2y''' - y'' + 2y' + 10y = -x$

40. $y^{(4)} - 8y''' + 26y'' - 40y' + 25y = e^{2x}\cos x$

41. $y^{(4)} - 4y''' + 8y'' - 8y' + 4y = 4$

42. $y^{(4)} - 4y''' = 4x^3 - 1$

In Exercises 43–58, solve the initial-value problem. Graph the solution on an appropriate interval.

*43. $5y'' - 18y' + 9y = -5$, $y(0) = -\frac{5}{9}$, $y'(0) = \frac{12}{5}$

44. $5y'' - 2y' - 3y = -2x$, $y(0) = \frac{5}{9}$, $y'(0) = \frac{2}{3}$

45. $6y'' + 7y' + 2y = -1$, $y(0) = \frac{3}{2}$, $y'(0) = \frac{1}{6}$

46. $3y'' - 5y' - 2y = -1$, $y(0) = \frac{1}{4}$, $y'(0) = \frac{9}{2}$

*47. $y'' - y = x^3$, $y(0) = 0$, $y'(0) = 8$

48. $y'' + 4y = 4x^2$, $y(0) = 3$, $y'(0) = 2$

49. $y'' + y = -2\sin 2x$, $y(0) = 0$, $y'(0) = -1$

50. $y'' + 3y' = 3\cos x$, $y(0) = \frac{17}{10}$, $y'(0) = \frac{39}{10}$

*51. $y'' + 2y' - 3y = 3e^{-x}$, $y(0) = \frac{13}{4}$, $y'(0) = \frac{3}{4}$

52. $y'' + 4y' + 4y = 4x^2$, $y(0) = 0$, $y'(0) = 0$

53. $y'' + 2y' + y = x$, $y(0) = 1$, $y'(0) = 0$

54. $y'' - 4y' = 8x$, $y(0) = 0$, $y'(0) = 0$

*55. $y'' + 4y' = 8e^{2x}$, $y(0) = 1$, $y'(0) = 0$

56. $y'' - 3y' + 2y = -xe^{4x}$, $y(0) = \frac{5}{36}$, $y'(0) = \frac{25}{18}$

57. $y'' + 6y' + 10y = -4$, $y(0) = \frac{3}{5}$, $y'(0) = 1$

58. $y'' - 8y' + 17y = -4$, $y(0) = \frac{14}{17}$, $y'(0) = 0$

59. Show that the differential equation $ay'' + by' + cy = k$, where a, b, c, and k are constants such that $c, k \neq 0$, has the particular solution $y_p(x) = k/c$.

60. Find a particular solution to $a_n y^{(n)} + a_{n-1}y^{(n-1)} + \cdots + a_1 y' + a_0 y = k$, where $a_n, a_{n-1}, \ldots, a_1, a_0$ are constants such that $a_0, k \neq 0$.

4.6 Nonhomogeneous Equations with Constant Coefficients: The Method of Undetermined Coefficients

Note that in subsequent sections, examples involving nonhomogeneous equations are typically solved with the method of this section.

We now present an approach that can be used instead of the annihilator method to determine the form of a particular solution of a nonhomogeneous differential equation. We begin the discussion with several examples.

Consider the nonhomogeneous second-order linear differential equation

$$y'' + 4y' + 3y = e^x.$$

A general solution to this nonhomogeneous equation is $y(x) = y_h(x) + y_p(x)$ where $y_h(x)$ is a solution of the corresponding homogeneous equation $y'' + 4y' + 3y = 0$. This equation has characteristic equation $m^2 + 4m + 3 = (m + 1)(m + 3) = 0$ so $y_h(x) = c_1 e^{-x} + c_2 e^{-3x}$.

Now we must select the proper form of a particular solution $y_p(x)$. Because $g(x) = e^x$, we are somewhat safe to assume that a particular solution is a multiple of e^x. Notice that the function we choose for $y_p(x)$ must satisfy $y'' + 4y' + 3y = e^x$, so we need to select a function that includes e^x in the function and its derivatives. We let $y_p(x) = Ae^x$ and attempt to find A. Substituting this function into $y'' + 4y' + 3y = e^x$ yields

$$y_p'' + 4y_p' + 3y_p = Ae^x + 4Ae^x + 3Ae^x = 8Ae^x = e^x.$$

Equating the coefficients of e^x gives us $8A = 1$ or $A = \dfrac{1}{8}$. Therefore, $y_p(x) = \dfrac{1}{8}e^x$, and a general solution is

$$y(x) = y_h(x) + y_p(x) = c_1 e^{-x} + c_2 e^{-3x} + \frac{1}{8}e^x.$$

Find a particular solution of $y'' + 4y = \sin x$ of the form $y_p(x) = A \cos x + B \sin x$.

Now consider the equation

$$y'' + 4y' + 3y = e^{-3x}.$$

We just saw that the solution of the corresponding homogeneous equation

$$y'' + 4y' + 3y = 0$$

is $y_h(x) = c_1 e^{-x} + c_2 e^{-3x}$. Therefore, if we assume that $y_p(x) = Ae^{-3x}$ as we did in the previous example, we have the derivatives $y_p'(x) = -3Ae^{-3x}$ and $y_p''(x) = 9Ae^{-3x}$. Substitution into $y'' + 4y' + 3y = e^{-3x}$ then yields

$$y'' + 4y' + 3y = 9Ae^{-3x} - 12Ae^{-3x} + 3Ae^{-3x} = 0 = e^{-3x}$$

which does *not* lead to determining the value of A. Of course, this should not surprise us because e^{-3x} is a solution of the corresponding homogeneous equation. Therefore, we *cannot* include this function in the particular solution. However, we can obtain a function that resembles e^{-3x} and includes e^{-3x} in $y_p(x)$, $y_p'(x)$, and $y_p''(x)$, by multiplying e^{-3x} by x. We let $y_p(x) = Axe^{-3x}$. The derivatives of this function are

$$y_p'(x) = Ae^{-3x} - 3Axe^{-3x} \quad \text{and} \quad y_p''(x) = -6Ae^{-3x} + 9Axe^{-3x}.$$

Substitution into $y'' + 4y' + 3y = e^{-3x}$ gives us

$$y_p'' + 4y_p' + 3y_p = -6Ae^{-3x} + 9Axe^{-3x} + 4(Ae^{-3x} - 3Axe^{-3x}) + 3Axe^{-3x}$$
$$= -2Ae^{-3x} = e^{-3x}.$$

Equating the coefficients of e^{-3x} then yields $-2A = 1$ or $A = -\frac{1}{2}$. Hence, $y_p(x) = -\frac{1}{2}xe^{-3x}$ and a general solution is

$$y(x) = y_h(x) + y_p(x) = c_1 e^{-x} + c_2 e^{-3x} - \frac{1}{2}xe^{-3x}.$$

*Show that it is **not** possible to find a particular solution of $y'' + 4y = \sin 2x$ of the form $y_p(x) = A\cos 2x + B\sin 2x$ but that it is possible to find a particular solution of the form $y_p(x) = x(A\cos 2x + B\sin 2x)$.*

These two examples illustrate the guesswork involved in choosing the form of a particular solution. In order to eliminate some of the guessing required to apply this method, we introduce an algorithm that can be used to determine the form of $y_p(x)$.

Consider the linear nonhomogeneous nth-order differential equation with constant coefficients

$$a_n y^{(n)}(x) + a_{n-1}y^{(n-1)}(x) + \cdots + a_1 y'(x) + a_0 y(x) = g(x)$$

where $a_n \neq 0$ and $g(x)$ is a linear combination of the functions $1, x, x^2, \ldots, e^{kx}, xe^{kx}, x^2 e^{kx}, \ldots,\ e^{\alpha x}\cos\beta x,\ xe^{\alpha x}\cos\beta x,\ x^2 e^{\alpha x}\cos\beta x, \ldots,\ e^{\alpha x}\sin\beta x,\ xe^{\alpha x}\sin\beta x, x^2 e^{\alpha x}\sin\beta x, \ldots$. A general solution of this differential equation is

$$y(x) = y_h(x) + y_p(x)$$

where $y_h(x)$ is a solution of the corresponding homogeneous equation

$$a_n y^{(n)}(x) + a_{n-1}y^{(n-1)}(x) + \cdots + a_1 y'(x) + a_0 y(x) = 0,$$

and $y_p(x)$ is a particular solution involving no arbitrary constants of the nonhomogeneous equation

$$a_n y^{(n)}(x) + a_{n-1}y^{(n-1)}(x) + \cdots + a_1 y'(x) + a_0 y(x) = g(x).$$

We learned how to solve homogeneous equations in Section 4.2, so we must learn how to find the form of a particular solution in order to solve nonhomogeneous equations with the method of undetermined coefficients.

Outline of the Method of Undetermined Coefficients to Solve
$a_n y^{(n)}(x) + a_{n-1}y^{(n-1)}(x) + \cdots + a_1 y'(x) + a_0 y(x) = g(x),$
where $g(x)$ is a linear combination of the functions
$1, x, x^2, \ldots, e^{kx}, xe^{kx}, x^2 e^{kx}, \ldots, e^{\alpha x}\cos\beta x, xe^{\alpha x}\cos\beta x, x^2 e^{\alpha x}\cos\beta x, \ldots,$ $e^{\alpha x}\sin\beta x, xe^{\alpha x}\sin\beta x, x^2 e^{\alpha x}\sin\beta x, \ldots$.

① Solve the corresponding homogeneous equation for $y_h(x)$.

② Determine the form of a particular solution $y_p(x)$. (See **Determining the Form of $y_p(x)$.**)

③ Determine the unknown coefficients in $y_p(x)$ by substituting $y_p(x)$ into the nonhomogeneous equation and equating the coefficients of like terms.

④ Form a general solution with $y(x) = y_h(x) + y_p(x)$.

Determining the Form of $y_p(x)$ (step 2):

Suppose that $g(x) = b_1 g_1(x) + b_2 g_2(x) + \cdots + b_j g_j(x)$ where $b_1, b_2, \ldots, b_j$ are constants and each $g_i(x)$, $i = 1, 2, \ldots, j$ is a function of the form x^m, $x^m e^{kx}$, $x^m e^{\alpha x} \cos \beta x$, or $x^m e^{\alpha x} \sin \beta x$.

(A) If $g_i(x) = x^m$, the associated set of functions is

$$S = \{x^m, x^{m-1}, \ldots, x^2, x, 1\}.$$

(B) If $g_i(x) = x^m e^{kx}$, the associated set of functions is

$$S = \{x^m e^{kx}, x^{m-1}e^{kx}, \ldots, x^2 e^{kx}, xe^{kx}, e^{kx}\}.$$

(C) If $g_i(x) = x^m e^{\alpha x} \cos \beta x$, or $g_i(x) = x^m e^{\alpha x} \sin \beta x$, the associated set of functions is

$$S = \{x^m e^{\alpha x} \cos \beta x, x^{m-1}e^{\alpha x} \cos \beta x, \ldots, x^2 e^{\alpha x} \cos \beta x, xe^{\alpha x} \cos \beta x, e^{\alpha x} \cos \beta x$$
$$x^m e^{\alpha x} \sin \beta x, x^{m-1}e^{\alpha x} \sin \beta x, \ldots, x^2 e^{\alpha x} \sin \beta x, xe^{\alpha x} \sin \beta x, e^{\alpha x} \sin \beta x\}.$$

For each function in $g(x)$, determine the associated set of functions S. If any of the functions in S appears in the homogeneous solution $y_h(x)$, multiply each function in S by x^r to obtain a new set S', where r is the smallest positive integer so that each function in S' is not a function in $y_h(x)$.

The correct form of a particular solution is obtained by taking the linear combination of all functions in the associated sets where repeated functions should appear only once in the form of the particular solution.

Example 1

For each of the following functions, determine the associated set of functions: (a) $f(x) = x^4$, (b) $f(x) = x^3 e^{-2x}$, and (c) $f(x) = x^2 e^{-x} \cos 4x$.

Solution (a) Using (A), we have $S = \{x^4, x^3, x^2, x, 1\}$.
(b) In this case we use (B):

$$S = \{x^3 e^{-2x}, x^2 e^{-2x}, xe^{-2x}, e^{-2x}\}.$$

(c) According to (C),

$$S = \{x^2 e^{-x} \cos 4x, xe^{-x} \cos 4x, e^{-x} \cos 4x, x^2 e^{-x} \sin 4x, xe^{-x} \sin 4x, e^{-x} \sin 4x\}.$$

Example 2

Solve the nonhomogeneous equations (a) $y'' + 5y' + 6y = 2e^x$, (b) $y'' + 5y' + 6y = 2x^2 + 3x$, (c) $y'' + 5y' + 6y = 3e^{-2x}$, and (d) $y'' + 5y' + 6y = 4 \cos x$.

Solution (a) The corresponding homogeneous equation $y'' + 5y' + 6y = 0$ has general solution $y_h(x) = c_1 e^{-2x} + c_2 e^{-3x}$. (Why?) Next, we determine the form of $y_p(x)$. Because $g(x) = 2e^x$, we choose $S = \{e^x\}$.

Notice that e^x is not a solution to the homogeneous equation, so we take $y_p(x)$ to be the linear combination of the functions in S. Therefore,

$$y_p(x) = Ae^x.$$

Substituting this solution into $y'' + 5y' + 6y = 2e^x$, we have

$$Ae^x + 5Ae^x + 6Ae^x = 12Ae^x = 2e^x.$$

Equating the coefficients of e^x then gives us $A = \dfrac{1}{6}$. A particular solution is $y_p(x) = \dfrac{1}{6}e^x$, so a general solution of $y'' + 5y' + 6y = 2e^x$ is

$$y(x) = y_h(x) + y_p(x) = c_1 e^{-2x} + c_2 e^{-3x} + \frac{1}{6}e^x.$$

(b) Here, we see that $g(x) = b_1 g_1(x) + b_2 g_2(x) = 2x^2 + 3x$. The set of functions associated with $g_1(x) = 2x^2$ is $S_1 = \{x^2, x, 1\}$ and that associated with $g_2(x) = 3x$ is $S_2 = \{x, 1\}$. Notice that none of these functions are solutions of the corresponding homogeneous equation. Notice also that S_1 and S_2 have two elements in common. If we take the linear combination of the functions x^2, x, and 1, then

$$y_p(x) = Ax^2 + Bx + C.$$

Substitution of the derivatives $y_p'(x) = 2Ax + B$ and $y_p''(x) = 2A$ into $y'' + 5y' + 6y = 2x^2 + 3x$ yields

$$y_p'' + 5y_p' + 6y_p = 6Ax^2 + (10A + 6B)x + (2A + 5B + 6C) = 2x^2 + 3x.$$

Therefore, $6A = 2$, $10A + 6B = 3$, and $2A + 5B + 6C = 0$. This system of equations has the solution $A = \dfrac{1}{3}$, $B = -\dfrac{1}{18}$, and $C = -\dfrac{7}{108}$, so

$$y_p(x) = \frac{1}{3}x^2 - \frac{1}{18}x - \frac{7}{108}.$$

A general solution is

$$y(x) = y_h(x) + y_p(x) = c_1 e^{-2x} + c_2 e^{-3x} + \frac{1}{3}x^2 - \frac{1}{18}x - \frac{7}{108}.$$

(c) In this case, we see that $g(x) = 3e^{-2x}$. Then the associated set is $S = \{e^{-2x}\}$. However, because e^{-2x} is a solution to the corresponding homogeneous equation, we must multiply this function by x^r so that it is no longer a solution. We multiply the element of S by $x^1 = x$ to obtain $S' = \{xe^{-2x}\}$ because xe^{-2x} is not a solution of $y'' + 5y' + 6y = 0$. Hence, $y_p(x) = Axe^{-2x}$. Differentiating $y_p(x)$ twice and substituting into the equation yields:

$$y_p'' + 5y_p' + 6y_p = -4Ae^{-2x} + 4Axe^{-2x} + 5(Ae^{-2x} - 2Axe^{-2x}) + 6Axe^{-2x}$$
$$= Ae^{-2x} = 3e^{-2x}.$$

Thus, $A = 3$ so $y_p(x) = 3xe^{-2x}$, and a general solution of $y'' + 5y' + 6y = 3e^{-2x}$ is

$$y(x) = y_h(x) + y_p(x) = c_1 e^{-2x} + c_2 e^{-3x} + 3xe^{-2x}.$$

(d) In this case, we see that $g(x) = 4 \cos x$, so the associated set is $S = \{\cos x, \sin x\}$. Because neither of these functions is a solution of the homogeneous equation, we take the particular solution to be the linear combination

$$y_p(x) = A \cos x + B \sin x.$$

We determine A and B by differentiating $y_p(x)$ twice and substituting into the nonhomogeneous equation

$$
\begin{aligned}
y_p'' + 5y_p' + 6y_p &= -A \cos x - B \sin x - 5A \sin x + 5B \cos x + 6A \cos x + 6B \sin x \\
&= (5A + 5B) \cos x + (-5A + 5B) \sin x \\
&= 4 \cos x.
\end{aligned}
$$

Hence, $5A + 5B = 4$ and $-5A + 5B = 0$, so $A = B = \frac{2}{5}$. Therefore,

$$y_p(x) = \frac{2}{5} \cos x + \frac{2}{5} \sin x,$$

and a general solution is

$$y(x) = y_h(x) + y_p(x) = c_1 e^{-2x} + c_2 e^{-3x} + \frac{2}{5} \cos x + \frac{2}{5} \sin x.$$

Example 3

Solve the following nonhomogeneous equations: (a) $y'' + 5y' + 6y = 4xe^{-2x} + e^{-2x}$; and (b) $y'' + 4y' + 13y = 4e^{-2x} \sin 3x$.

Solution   (a) As we saw in Example 2, $y_h(x) = c_1 e^{-2x} + c_2 e^{-3x}$ is a general solution of the corresponding homogeneous equation $y'' + 5y' + 6y = 0$. Next, we notice that $g(x) = b_1 g_1(x) + b_2 g_2(x) = 4xe^{-2x} + e^{-2x}$. The set associated with $g_1(x) + xe^{-2x}$ is $S_1 = \{xe^{-2x}, e^{-2x}\}$ and that associated with $g_2(x) = e^{-2x}$ is $S_2 = \{e^{-2x}\}$. Notice that both sets contain a function that appears in the corresponding homogeneous solution $y_h(x) = c_1 e^{-2x} + c_2 e^{-3x}$. We multiply both sets by x to obtain $S_1' = \{x^2 e^{-2x}, xe^{-2x}\}$ and $S_2' = \{xe^{-2x}\}$ so that neither set contains a solution to the corresponding homogeneous set. However, because S_1' and S_2' contain the common element xe^{-2x}, we let

$$y_p(x) = Ax^2 e^{-2x} + Bxe^{-2x}.$$

Notice that if we include the repeated function twice, we arrive at the same result:

$$y_p(x) = c_1 x^2 e^{-2x} + c_2 xe^{-2x} + c_3 xe^{-2x} = c_1 x^2 e^{-2x} + (c_2 + c_3) xe^{-2x}.$$

Differentiating $y_p(x)$ twice and substituting into $y'' + 5y' + 6y = 4xe^{-2x} + e^{-2x}$ yields

$$y_p'' + 5y_p' + 6y_p = (2A + B)e^{-2x} + 2Axe^{-2x} = 4xe^{-2x} + e^{-2x}.$$

Therefore, $2A + B = 1$ and $2A = 4$ so $A = 2$ and $B = -3$. This results in the particular solution

$$y_p(x) = 2x^2e^{-2x} - 3xe^{-2x},$$

and general solution

$$y(x) = y_h(x) + y_p(x) = c_1e^{-2x} + c_2e^{-3x} + 2x^2e^{-2x} - 3xe^{-2x}.$$

(b) As in (a), first find a general solution of the corresponding homogeneous equation $y'' + 4y' + 13y = 0$. Verify that a general solution of the corresponding homogeneous equation is

$$y_h(x) = c_1e^{-2x}\cos 3x + c_2e^{-2x}\sin 3x.$$

Because $g(x) = 4e^{-2x}\sin 3x$, the associated set of functions is $S = \{e^{-2x}\cos 3x, e^{-2x}\sin 3x\}$ but both of the functions in S are solutions to the homogeneous equation, so we multiply by x so that they are no longer solutions. Hence, $S' = \{xe^{-2x}\cos 3x, xe^{-2x}\sin 3x\}$ and we let

$$y_p(x) = Axe^{-2x}\cos 3x + Bxe^{-2x}\sin 3x.$$

This function has first and second derivatives

$$y_p'(x) = Ae^{-2x}\cos 3x + (3B - 2A)xe^{-2x}\cos 3x + Be^{-2x}\sin 3x + (-3A - 2B)e^{-2x}\sin 3x$$

and

$$y_p''(x) = (-4A + 6B)e^{-2x}\cos 3x + (-5A - 12B)xe^{-2x}\cos 3x + (6A - 4B)e^{-2x}\sin 3x + (12A - 5B)e^{-2x}\sin 3x$$

which when substituted into the equation $y'' + 4y' + 13y = 4e^{-2x}\sin 3x$ yield

$$y_p'' + 4y_p' + 13y_p = 6Be^{-2x}\cos 3x - 6Ae^{-2x}\sin 3x = 4e^{-2x}\sin 3x.$$

Equating the coefficients of like terms, we have $-6A = 4$ and $6B = 0$. Hence, $A = -\dfrac{2}{3}$ and $B = 0$, so

$$y_p(x) = -\frac{2}{3}xe^{-2x}\cos 3x.$$

Therefore, a general solution is

$$y(x) = y_h(x) + y_p(x) = c_1e^{-2x}\cos 3x + c_2e^{-2x}\sin 3x - \frac{2}{3}xe^{-2x}\cos 3x.$$

*What is the limit as $x \to \infty$ of **every** solution? Why doesn't the choice of c_1 and c_2 matter? (Hint: First, identify and use the graph of $y_p(x)$ in Figure 4.11 (or, generate your own) and then apply L'Hopital's rule, if necessary.)*

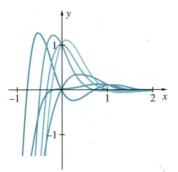

Figure 4.11 Graph of $y(x) = y_h(x) + y_p(x)$ for various values of c_1 and c_2

In order to solve an initial-value problem, first determine a general solution and then use the initial conditions to solve for the unknown constants in the general solution.

Example 4

Solve the initial-value problem $y''' + y'' = 12x^2$, $y(0) = 0$, $y'(0) = 1$, $y''(0) = 25$.

Solution First, we solve the corresponding homogeneous equation $y''' + y'' = 0$, which has characteristic equation $m^3 + m^2 = m^2(m + 1) = 0$ and thus, general solution

$$y_h(x) = c_1 + c_2 x + c_3 e^{-x}.$$

Next, we set up the set of functions associated with $g(x) = 12x^2$ which is $S = \{x^2, x, 1\}$. However, both of the functions x and 1 are solutions of $y''' + y'' = 0$, so we multiply the functions in S by x^2 so that the functions that result do not appear in $y_h(x)$. Hence, $S' = \{x^4, x^3, x^2\}$, so

$$y_p(x) = Ax^4 + Bx^3 + Cx^2.$$

The derivatives of $y_p(x)$ are $y_p'(x) = 4Ax^3 + 3Bx^2 + 2Cx$, $y_p''(x) = 12Ax^2 + 6Bx + 2C$, and $y_p'''(x) = 24Ax + 6B$. Substitution into $y''' + y'' = 12x^2$ then gives us

$$y_p''' + y_p'' = 12Ax^2 + (24A + 6B)x + (2C + 6B) = 12x^2,$$

so $12A = 12$, $24A + 6B = 0$, and $2C + 6B = 0$. This system of equations has the solution $A = 1$, $B = -4$, and $C = 12$, so

$$y_p(x) = x^4 - 4x^3 + 12x^2$$

and a general solution is

$$y(x) = y_h(x) + y_p(x) = c_1 + c_2 x + c_3 e^{-x} + x^4 - 4x^3 + 12x^2.$$

Figure 4.12 shows the graph of $y(x) = c_1 + c_2x + c_3e^{-x} + x^4 - 4x^3 + 12x^2$ for various values of c_1, c_2, and c_3. Identify the graph of the solution that satisfies the initial conditions $y(0) = 0$, $y'(0) = 1$ and $y''(0) = 25$.

To determine the unknown coefficients c_1, c_2, and c_3, we apply the initial conditions. First, we compute $y'(x) = c_2 - c_3e^{-x} + 4x^3 - 12x^2 + 24x$ and $y''(x) = c_3e^{-x} + 12x^2 - 24x + 24$. Applying the initial conditions we have, $y(0) = c_1 + c_3 = 0$, $y'(0) = c_2 - c_3 = 1$, and $y''(0) = c_3 + 24 = 25$, so $c_1 = -1$, $c_2 = 2$, and $c_3 = 1$. Therefore, the solution of the initial-value problem is

$$y(x) = y_h(x) + y_p(x) = -1 + 2x + e^{-x} + x^4 - 4x^3 + 12x^2. \text{ (See Figure 4.13.)}$$

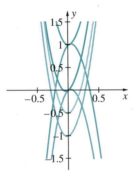

Figure 4.12

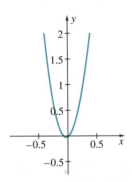

Figure 4.13 Graph of $y(x) = -1 + 2x + e^{-x} + x^4 - 4x^3 + 12x^2$

Example 5

If $\omega \geq 0$, solve the initial-value problem $\begin{cases} y'' + y = \cos \omega x \\ y(0) = 0, \ y'(0) = 1 \end{cases}$.

Solution   A general solution of the corresponding homogeneous equation $y'' + y = 0$ is $y_h(x) = c_1 \cos x + c_2 \sin x$. We see that if $\omega \neq 1$ we can find a particular solution of the nonhomogeneous equation of the form $y_p(x) = A \cos \omega x + B \sin \omega x$, while if $\omega = 1$ we can find a particular solution of the form $y_p(x) = x(A \cos x + B \sin x)$. Solving for a particular solution, forming a general solution, and applying the initial conditions yields the solution

$$y(x) = \begin{cases} \dfrac{1}{\omega^2 - 1}(\cos x - \cos \omega x) + \sin x, & \text{if } \omega \neq 1 \\[2mm] \dfrac{1}{2}(x + 2) \sin x, & \text{if } \omega = 1 \end{cases}$$

Notice that the behavior of the solution changes dramatically when $\omega = 1$: if $0 \leq \omega < 1$ or $\omega > 1$, the solution is periodic and bounded (why?) but if $\omega = 1$, the solution is unbounded. (See Figure 4.14.)

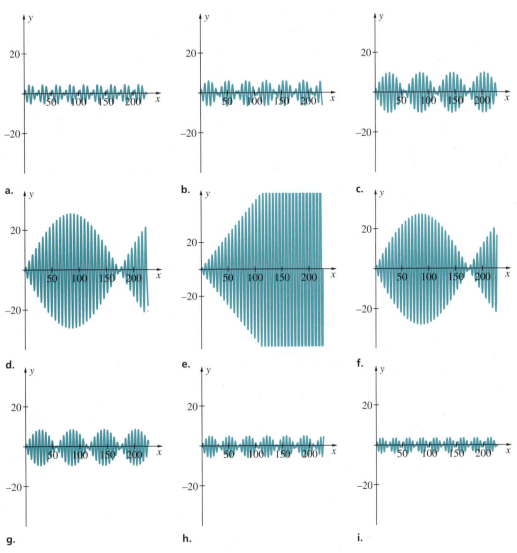

Figure 4.14   Graph of the solution of $\begin{cases} y'' + y = \cos \omega x \\ y(0) = 0, y'(0) = 1 \end{cases}$ for various values of ω

(a) $\omega = \frac{3}{4}$ (b) $\omega = \frac{23}{28}$ (c) $\omega = \frac{25}{28}$ (d) $\omega = \frac{27}{28}$ (e) $\omega = 1$ (f) $\omega = \frac{29}{28}$ (g) $\omega = \frac{31}{28}$
(h) $\omega = \frac{33}{28}$ (i) $\omega = \frac{5}{4}$

IN TOUCH WITH TECHNOLOGY

Technology can be used to determine the unknown coefficients in the particular solution. This is particularly useful when the differential equation and its solution are complicated, as is the case in the following problems.

1. Solve the following equations. In each case, graph the particular solution you obtain.

   (a) $y''' + 4y'' + 9y' + 36y = e^{-x} + 2\cos 3x$

   (b) $y''' + 6y'' + 21y' + 26y = e^{2x}\cos 3x + 10e^{-2x}$

2. Solve the following initial-value problems. In each case, graph the resulting solution.

   (a) $y''' + 11y'' + 32y' + 28y = xe^{-2x} + 3e^{-2x}$,
   $y''(0) = 0,\ y'(0) = -1,\ y(0) = 1$

   (b) $y''' + 2y'' + 16y' + 32y = 4e^{-2x} + x\cos 4x$,
   $y''(0) = 4,\ y'(0) = -2,\ y(0) = 0$

3. Determine if it is possible to find values of a_3, a_2, a_1, and a_0 so that a general solution of

   $$y^{(4)} + a_3 y''' + a_2 y'' + a_1 y' + a_0 y = \cos t$$

   has the indicated form. In each case, either find such values (and confirm your results) or explain why no such numbers exist.

   (a) $y(t) = c_1 e^{-2t} + c_2 e^{-t} + c_3 e^{t} + c_4 e^{2t} + \frac{1}{10}\cos t$

   (b) $y(t) = c_1 e^{-t} + c_2 e^{t} + c_3 \cos t + c_4 \sin t$
   $- \frac{3}{8}\cos t - \frac{1}{4}t\sin t$

(c) $y(t) = c_1 \cos t + c_2 \sin t + c_3 t \cos t$
$+ c_4 t \sin t + \frac{3}{8}\cos t + \frac{3}{8}\sin t - \frac{1}{8}t^2 \cos t$

(d) $y(t) = c_1 \cos t + c_2 \sin t + c_3 t \cos t$
$+ c_4 t \sin t + t^3 \sin t$

4. Show that the solution to the initial-value problem

$$\begin{cases} y'' + y' - 2y = f(x) \\ y(0) = 0,\ y'(0) = a \end{cases}$$

is

$$y(x) = \frac{1}{3}ae^{x} - \frac{1}{3}ae^{-2x} + \frac{1}{3}e^{x}\int_0^x f(t)e^{-t}\,dt -$$
$$\frac{1}{3}e^{-2x}\int_0^x f(t)e^{2t}\,dt$$

(as long as the integrals can be evaluated).

(a) Use the Fundamental Theorem of Calculus to verify this result.

(b) If possible, find a function $f(t)$ and a value of a so that (i) the solution is periodic; (ii) the solution approaches 0 as x approaches infinity; and (iii) the solution has no limit as x approaches infinity. Confirm your results graphically.

5. How does the solution to the initial-value problem

$$\begin{cases} x'' + 9x = \sin(ct) \\ x(0) = 0,\ x'(0) = 0 \end{cases}$$

change as c assumes value from 0 to 6?

EXERCISES 4.6

In Exercises 1–42, find a general solution of each equation.

1. $y'' + y' - 2y = -1$

2. $y'' - 7y' + 12y = -3$

*3. $5y'' + y' - 4y = -3$

4. $5y'' + 14y' - 3y = -3$

5. $y'' - 2y' - 8y = 32x$

6. $y'' + 2y' - 8y = -x$

*7. $16y'' - 8y' - 15y = 75x$

8. $2y'' + y' - y = -5x$

9. $y'' + 2y' + 26y = -338x$

10. $y'' + 5y' + 4y = x^2$

*11. $y'' + 3y' - 4y = -32x^2$

12. $y'' - 4y' + 4y = 4x^2$

13. $8y'' + 6y' + y = 5x^2$

14. $4y'' - 4y' + y = -5x^2$

*15. $y'' - 6y' + 8y = -256x^3$

16. $3y'' - 4y' - 4y = x^3$

17. $y'' - 2y' = 52 \sin 3x$

18. $y'' - y' - 2y = 2 \sin x$

*19. $y'' - 6y' + 13y = 25 \sin 2x$

20. $y'' + 4y' + 8y = -4 \sin 4x$

21. $y'' - 9y = 54x \sin 3x$

22. $y'' - 4y' + 3y = 2 \cos x$

*23. $y'' - 5y' + 6y = -78 \cos 3x$

24. $y'' - 3y' = -2x \cos x$

25. $y'' + 4y' + 4y = -32x^2 \cos 2x$

26. $y'' + 4y' - 5y = -3e^x$

*27. $y'' - y' - 20y = -2e^x$

28. $y'' - 8y' + 32y = -4e^{-3x}$

29. $y'' - 4y' - 5y = -648x^2e^{5x}$

30. $y'' - 6y' + 5y = -3xe^{4x}$

*31. $y'' - 7y' + 12y = -2x^3e^{4x}$

32. $y'' + 16y = 16x^3 + 4$

33. $y''' + 6y'' + 11y' + 6y = 2e^{-3x} - xe^{-x}$

34. $y''' - y'' + 9y' - 9y = x^3$

*35. $y''' + 10y'' + 34y' + 40y = xe^{-4x} + 2e^{-3x} \cos x$

36. $y''' + y'' - 17y' - 65y = e^{5x} - xe^{-3x} \sin 2x$

37. $y''' + 6y'' - 14y' - 104y = -111e^x$

38. $y^{(4)} + 10y''' + 29y'' + 20y' = e^x$

*39. $y^{(4)} - 6y''' + 13y'' - 24y' + 36y = 108x$

40. $y^{(4)} - 8y''' + 25y'' - 36y' + 20y = e^{2x} \cos x$

41. $y^{(4)} - 10y''' + 38y'' - 64y' + 40y = 153e^{-x}$

42. $y^{(4)} - 18y'' + 81y = e^{3x}$

In Exercises 43–61, solve the initial-value problem.

43. $y'' + 3y' = 18$, $y(0) = 0$, $y'(0) = 3$

44. $y'' - y = 4$, $y(0) = 0$, $y'(0) = 0$

*45. $y'' - 4y = 32x$, $y(0) = 0$, $y'(0) = 6$

46. $y'' + 2y' - 3y = -2$, $y(0) = \frac{2}{3}$, $y'(0) = 8$

47. $y'' + y' - 6y = 3x$, $y(0) = \frac{23}{12}$, $y'(0) = -\frac{3}{2}$

48. $y'' + 8y' + 16y = 4$, $y(0) = \frac{5}{4}$, $y'(0) = 0$

*49. $y'' + 4y = \frac{17}{4} \cos x$, $y(0) = \frac{1}{4}$, $y'(0) = 5$

50. $y'' + 7y' + 10y = xe^{-x}$, $y(0) = -\frac{15}{48}$, $y'(0) = \frac{9}{16}$

51. $y'' + 6y' + 25y = -1$, $y(0) = -\frac{1}{25}$, $y'(0) = 7$

52. $y'' + 2y' + 5y = 2 \sin 5x$, $y(0) = -\frac{1}{25}$, $y'(0) = \frac{8}{5}$

*53. $y'' - 3y' = -e^{3x} - 2x$, $y(0) = 0$, $y'(0) = \frac{8}{9}$

54. $y'' + 2y' + 5y = e^{-x} \cos 2x$, $y(0) = -\frac{1}{25}$,
    $y'(0) = \frac{8}{5}$

55. $y'' - y' = -3x - 4x^2e^{2x}$, $y(0) = -\frac{7}{2}$, $y'(0) = 0$

56. $y'' + y' - 2y = -2x^2$, $y(0) = \frac{3}{2}$, $y'(0) = 4$

*57. $y'' - 2y' = 2x^2$, $y(0) = 3$, $y'(0) = \frac{3}{2}$

58. $y''' - 4y' = 16e^x$, $y(0) = 0$, $y'(0) = 0$, $y''(0) = 4$

59. $y''' + 5y'' = 125x$, $y(0) = 0$, $y'(0) = 0$, $y''(0) = 0$

60. $y''' + 4y' = 16 \sin 2x$, $y(0) = 0$, $y'(0) = 0$, $y''(0) = 0$

*61. $y''' + 25y' = 325e^{-x}$, $y(0) = 0$, $y'(0) = 0$, $y''(0) = 0$

In Exercises 62–64, without actually solving the equations, match each initial-value problem in Group A with the graph of its solution in Group B.

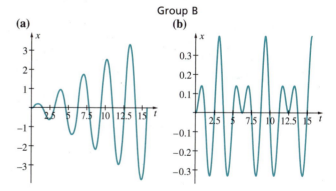

Group A

62. $\begin{cases} x'' + 4x = \cos t \\ x(0) = 0, \ x'(0) = 0 \end{cases}$
63. $\begin{cases} x'' + 4x = \cos 2t \\ x(0) = 0, \ x'(0) = 0 \end{cases}$

64. $\begin{cases} x'' + 4x = \cos 3t \\ x(0) = 0, \ x'(0) = 0 \end{cases}$

Group B

(a)

(b)

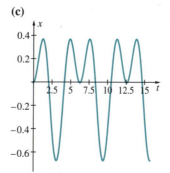

(c)

65. The method of undetermined coefficients can be used to solve first-order nonhomogeneous equations. Use this method to solve the following problems.

(a) $y' - 4y = x^2$

*(b) $y' + y = \cos 2x$

(c) $y' - y = e^{4x}$

66. (a) Suppose that $y_1(x)$ and $y_2(x)$ are solutions of $a\dfrac{d^2y}{dx^2} + b\dfrac{dy}{dx} + cy = f(x)$, where a, b, and c are positive constants. Show that

$$\lim_{x \to \infty} [y_2(x) - y_1(x)] = 0.$$

(b) Is the result of **(a)** true if $b = 0$?

(c) Suppose that $f(x) = k$ where k is a constant. Show that $\lim_{x \to \infty} y(x) = \dfrac{k}{c}$ for every solution $y(x)$ of $a\dfrac{d^2y}{dx^2} + b\dfrac{dy}{dx} + cy = k$.

(d) Determine the solution of $a\dfrac{d^2y}{dx^2} + b\dfrac{dy}{dx} = k$ and find $\lim_{x \to \infty} y(x)$.

(e) Determine the solution of $a\dfrac{d^2y}{dx^2} = k$ and find $\lim_{x \to \infty} y(x)$.

4.7 Nonhomogeneous Equations with Constant Coefficients: Variation of Parameters

Second-Order Equations • Higher-Order Nonhomogeneous Equations

Second-Order Equations

For the nonhomogeneous equation $y'' + \frac{1}{4}y = \sec\frac{x}{2} + \csc\frac{x}{2}, 0 < x < \pi$, the nonhomogeneous function is $g(x) = \sec\frac{x}{2} + \csc\frac{x}{2}$, which is not a function of terms of the form x^m, $x^m e^{kx}$, $x^m e^{\alpha x} \cos \beta x$, or $x^m e^{\alpha x} \sin \beta x$. Because the method of undetermined coefficients and the annihilator method are limited to equations involving these functions, we need to introduce another method, called **variation of parameters,** discovered by Lagrange, that can be used to find a particular solution of other equations such as this one.

A general solution to the corresponding homogeneous equation $y'' + \frac{1}{4}y = 0$ is $y_h(x) = c_1 \cos\frac{x}{2} + c_2 \sin\frac{x}{2}$.

In the method of variation of parameters, we try to find a particular solution of the form

$$y_p(x) = u_1(x) \cos\frac{x}{2} + u_2(x) \sin\frac{x}{2},$$

where we replace the constants c_1 and c_2 in $y_h(x)$ with the unknown functions $u_1(x)$ and $u_2(x)$. We arrive at the name of the method from this replacement because we *vary the parameters* c_1 and c_2 by allowing them to be *functions* of x instead of *constants*. (Notice that there should be many choices for $u_1(x)$ and $u_2(x)$ because we have two unknown functions and only one equation, the nonhomogeneous differential equation, to use in finding them.) We find possible choices for $u_1(x)$ and $u_2(x)$ by substitution of

$y_p(x)$ into the nonhomogeneous equation $y'' + \frac{1}{4}y = \sec\frac{x}{2} + \csc\frac{x}{2}$. Differentiating $y_p(x)$, we find that

$$y_p'(x) = -\frac{1}{2}u_1(x)\sin\frac{x}{2} + u_1'(x)\cos\frac{x}{2} + \frac{1}{2}u_2(x)\cos\frac{x}{2} + u_2'(x)\sin\frac{x}{2}.$$

To simplify the process of finding $u_1(x)$ and $u_2(x)$, we assume that

$$u_1'(x)\cos\frac{x}{2} + u_2'(x)\sin\frac{x}{2} = 0$$

which is our *first restriction* on $u_1(x)$ and $u_2(x)$. (In other words, of all functions $u_1(x)$ and $u_2(x)$ that lead to a particular solution to the nonhomogeneous differential equation, we look for functions that satisfy this condition.) Eliminating this expression from $y_p'(x)$ gives us

$$y_p'(x) = -\frac{1}{2}u_1(x)\sin\frac{x}{2} + \frac{1}{2}u_2(x)\cos\frac{x}{2},$$

so that the second derivative is

$$y_p''(x) = -\frac{1}{4}u_1(x)\cos\frac{x}{2} - \frac{1}{2}u_1'(x)\sin\frac{x}{2} - \frac{1}{4}u_2(x)\sin\frac{x}{2} + \frac{1}{2}u_2'(x)\cos\frac{x}{2}.$$

Next, we substitute $y_p(x)$ into the nonhomogeneous equation $y'' + \frac{1}{4}y = \sec\frac{x}{2} + \csc\frac{x}{2}$ to obtain

$$y_p'' + \frac{1}{4}y_p = -\frac{1}{4}u_1(x)\cos\frac{x}{2} - \frac{1}{2}u_1'(x)\sin\frac{x}{2} - \frac{1}{4}u_2(x)\sin\frac{x}{2}$$
$$+ \frac{1}{2}u_2'(x)\cos\frac{x}{2} + \frac{1}{4}\left[u_1(x)\cos\frac{x}{2} + u_2(x)\sin\frac{x}{2}\right]$$
$$= -\frac{1}{2}u_1'(x)\sin\frac{x}{2} + \frac{1}{2}u_2'(x)\cos\frac{x}{2}.$$

Then, because $y_p(x)$ satisfies the nonhomogeneous equation, we have a *second restriction* on $u_1(x)$ and $u_2(x)$,

$$-\frac{1}{2}u_1'(x)\sin\frac{x}{2} + \frac{1}{2}u_2'(x)\cos\frac{x}{2} = \sec\frac{x}{2} + \csc\frac{x}{2}.$$

This gives us the system of two equations

$$\begin{cases} u_1'(x)\cos\frac{x}{2} + u_2'(x)\sin\frac{x}{2} = 0 \\ -\frac{1}{2}u_1'(x)\sin\frac{x}{2} + \frac{1}{2}u_2'(x)\cos\frac{x}{2} = \sec\frac{x}{2} + \csc\frac{x}{2} \end{cases}$$

which we solve for the two functions $u_1'(x)$ and $u_2'(x)$. Multiplying the first equation by $\frac{1}{2}\sin\frac{x}{2}$, the second by $\cos\frac{x}{2}$, and adding the resulting equations yields

$$u_2'(x) = \frac{\cos\dfrac{x}{2}\left[\sec\dfrac{x}{2} + \csc\dfrac{x}{2}\right]}{\dfrac{1}{2}}$$

so that one choice for $u_2(x)$ is obtained through integration to be

$$u_2(x) = \int \frac{\cos\dfrac{x}{2}\left(\sec\dfrac{x}{2} + \csc\dfrac{x}{2}\right)}{\dfrac{1}{2}} = 2\int\left(1 + \frac{\cos\dfrac{x}{2}}{\sin\dfrac{x}{2}}\right)dx = 2x + 4\ln\left|\sin\dfrac{x}{2}\right|.$$

(Notice that any antiderivative of $u_2'(x)$ is a possible choice for $u_2(x)$. Here, we assume that the constant of integration is zero.) Similarly, if we mutliply the first equation in the system by $-\frac{1}{2}\cos\frac{x}{2}$, the second by $\sin\frac{x}{2}$, and we add the equations that result, we have

$$u_1'(x) = \frac{-\sin\dfrac{x}{2}\left(\sec\dfrac{x}{2} + \csc\dfrac{x}{2}\right)}{\dfrac{1}{2}}$$

so one possibility for $u_1(x)$ is

$$u_1(x) = \int \frac{-\sin\dfrac{x}{2}\left(\sec\dfrac{x}{2} + \csc\dfrac{x}{2}\right)}{\dfrac{1}{2}}\,dx = -2\int\left(\frac{\sin\dfrac{x}{2}}{\cos\dfrac{x}{2}} + 1\right)dx =$$

$$-2x + 4\ln\left|\cos\dfrac{x}{2}\right|.$$

Again, we have selected the antiderivative with constant of integration equal to zero.
Then by *variation of parameters*,

$$y_p(x) = u_1(x)y_1(x) + u_2(x)y_2(x)$$

$$= \cos\frac{x}{2}\left[-2x + 4\ln\left|\cos\frac{x}{2}\right|\right] + \sin\frac{x}{2}\left[2x + 4\ln\left|\sin\frac{x}{2}\right|\right]$$

is a particular solution of $y'' + \frac{1}{4}y = \sec\frac{x}{2} + \csc\frac{x}{2}$ and

$$y = y_h(x) + y_p(x)$$

$$= c_1\cos\frac{x}{2} + c_2\sin\frac{x}{2} + \cos\frac{x}{2}\left[-2x + 4\ln\left|\cos\frac{x}{2}\right|\right] + \sin\frac{x}{2}\left[2x + 4\ln\left|\sin\frac{x}{2}\right|\right]$$

is a general solution. (Note that $\cos \frac{x}{2} > 0$ and $\sin \frac{x}{2} > 0$ on $0 < x < \pi$, so the absolute value signs can be eliminated from the arguments of the natural logarithm functions.)

In general, to solve the second-order linear nonhomogeneous differential equation

$$a_2(x)y'' + a_1(x)y' + a_0(x)y = g(x),$$

where $y_h(x) = c_1 y_1(x) + c_2 y_2(x)$ is a general solution of the corresponding homogeneous equation $a_2(x)y'' + a_1(x)y' + a_0(x)y = 0$, we first divide the equation by $a_2(x)$ to rewrite it in the form

$$y'' + p(x)y' + q(x)y = f(x),$$

assume that a particular solution has a form similar to the general solution by varying the parameters c_1 and c_2, and let

$$y_p(x) = u_1(x)y_1(x) + u_2(x)y_2(x).$$

We need two equations in order to determine the two unknown functions $u_1(x)$ and $u_2(x)$ and we obtain them by substituting $y_p(x) = u_1(x)y_1(x) + u_2(x)y_2(x)$ into the nonhomogeneous differential equation $y'' + p(x)y' + q(x)y = f(x)$. Differentiating $y_p(x)$, we obtain

$$y_p'(x) = u_1(x)y_1'(x) + u_1'(x)y_1(x) + u_2(x)y_2'(x) + u_2'(x)y_2(x)$$

which can be simplified to

$$y_p'(x) = u_1(x)y_1'(x) + u_2(x)y_2'(x)$$

with the assumption that $u_1'(x)y_1(x) + u_2'(x)y_2(x) = 0$. (Notice that this assumption is made to eliminate the derivatives of $u_1(x)$ and $u_2(x)$ from this expression for $y_p'(x)$, which simplifies the process for finding $u_1(x)$ and $u_2(x)$.) Therefore, the *first equation* satisfied by $u_1(x)$ and $u_2(x)$ is

$$u_1'(x)y_1(x) + u_2'(x)y_2(x) = 0.$$

With our simplified expression for $y_p'(x)$, the second derivative is

$$y_p''(x) = u_1(x)y_1''(x) + u_1'(x)y_1'(x) + u_2(x)y_2''(x) + u_2'(x)y_2'(x).$$

Substitution into $y'' + p(x)y' + q(x)y = f(x)$ then yields

$$\begin{aligned} y_p'' + p(x)y_p' + q(x)y_p &= u_1(x)[y_1''(x) + p(x)y_1'(x) + q(x)y_1(x)] \\ &\quad + u_2(x)[y_2''(x) + p(x)y_2'(x) + q(x)y_2(x)] \\ &\quad + u_1'(x)y_1'(x) + u_2'(x)y_2'(x) \\ &= u_1'(x)y_1'(x) + u_2'(x)y_2'(x) = f(x) \end{aligned}$$

because $y_1(x)$ and $y_2(x)$ are solutions of the corresponding homogeneous equation. Therefore, our *second equation* for determining $u_1(x)$ and $u_2(x)$ is $u_1'(x)y_1'(x) + u_2'(x)y_2'(x) = f(x)$, so we have the system

$$\begin{cases} u_1'(x)y_1(x) + u_2'(x)y_2(x) = 0 \\ u_1'(x)y_1'(x) + u_2'(x)y_2'(x) = f(x) \end{cases}$$

which is written in matrix form as $\begin{pmatrix} y_1(x) & y_2(x) \\ y_1'(x) & y_2'(x) \end{pmatrix}\begin{pmatrix} u_1'(x) \\ u_2'(x) \end{pmatrix} = \begin{pmatrix} 0 \\ f(x) \end{pmatrix}$. In linear algebra, we learn that this system has a unique solution if and only if $\begin{vmatrix} y_1(x) & y_2(x) \\ y_1'(x) & y_2'(x) \end{vmatrix} \neq 0$.

Notice that this determinant is the Wronskian of the set $S = \{y_1(x), y_2(x)\}$, $W(S)$. We stated in Section 4.1 that $W(S) \neq 0$ if the functions $y_1(x)$ and $y_2(x)$ in the set S are linearly independent. Because $S = \{y_1(x), y_2(x)\}$ represents a fundamental set of solutions of the corresponding homogeneous equation, $W(S) \neq 0$. Hence, this system has a unique solution that can be found by Cramer's Rule to be

$$u_1'(x) = \frac{\begin{vmatrix} 0 & y_2(x) \\ f(x) & y_2'(x) \end{vmatrix}}{W(S)} = \frac{-y_2(x)f(x)}{W(S)}$$

and

$$u_2'(x) = \frac{\begin{vmatrix} y_1(x) & 0 \\ y_1'(x) & f(x) \end{vmatrix}}{W(S)} = \frac{y_1(x)f(x)}{W(S)}.$$

(Note: These formulas can be found with *elimination* if you are not familiar with Cramer's Rule.) The functions $u_1(x)$ and $u_2(x)$ are then found through integration.

Summary of Variation of Parameters for Second-Order Equations

Given the second-order equation $y'' + p(x)y' + q(x)y = f(x)$:

① Find a general solution $y_h(x) = c_1 y_1(x) + c_2 y_2(x)$ and fundamental set of solutions $S = \{y_1(x), y_2(x)\}$ of the corresponding homogeneous equation $y'' + p(x)y' + q(x)y = 0$.

② Let $u_1'(x) = \dfrac{-y_2(x)f(x)}{W(S)}$ and $u_2'(x) = \dfrac{y_1(x)f(x)}{W(S)}$.

③ Integrate to obtain $u_1(x)$ and $u_2(x)$.

④ A particular solution of $y'' + p(x)y' + q(x)y = f(x)$ is given by $y_p(x) = u_1(x)y_1(x) + u_2(x)y_2(x)$.

⑤ A general solution of $y'' + p(x)y' + q(x)y = f(x)$ is given by $y = y_h(x) + y_p(x)$.

Example 1

Solve $y'' - 2y' + y = e^x \ln x$, $x > 0$.

Solution   The corresponding homogeneous equation has the characteristic equation $m^2 - 2m + 1 = (m - 1)^2 = 0$ so $y_h(x) = c_1 \underbrace{e^x}_{y_1(x)} + c_2 \underbrace{xe^x}_{y_2(x)}$ and $S = \{e^x, xe^x\}$. Therefore, we compute

$$W(S) = \begin{vmatrix} e^x & xe^x \\ e^x & e^x + xe^x \end{vmatrix} = e^{2x} + xe^{2x} - xe^{2x} = e^{2x}.$$

Through the use of integration by parts, a table of integrals, or a computer algebra system

$$u_1(x) = \int \frac{-xe^x(e^x \ln x)}{e^{2x}}\,dx = -\int x \ln x\,dx = \frac{1}{4}x^2 - \frac{1}{2}x^2 \ln x$$

and

$$u_2(x) = \int \frac{e^x(e^x \ln x)}{e^{2x}}\,dx = \int \ln x\,dx = x \ln x - x.$$

Then a particular solution, which is assumed to have the form $y_p(x) = u_1(x)y_1(x) + u_2(x)y_2(x)$, is

$$y_p(x) = \left(\frac{1}{4}x^2 - \frac{1}{2}x^2 \ln x\right)e^x + (x \ln x - x)xe^x = \frac{1}{2}x^2 e^x \ln x - \frac{3}{4}x^2 e^x.$$

Therefore, a general solution is

$$y(x) = y_h(x) + y_p(x) = c_1 e^x + c_2 xe^x + \frac{1}{2}x^2 e^x \ln x - \frac{3}{4}x^2 e^x,\ x > 0.$$

Of course, the problems that were solved in the preceding sections by the method of undetermined coefficients or the annihilator method can be solved by variation of parameters as well.

Example 2

Solve the initial-value problem $y'' + 4y = \sin 2x,\ y(0) = 0,\ y'(0) = 1$ using variation of parameters.

Solution   The characteristic equation of the corresponding homogeneous equation has the complex conjugate roots $m = \pm 2i$, so $y_h(x) = c_1 \underbrace{\cos 2x}_{y_1(x)} + C_2 \underbrace{\sin 2x}_{y_2(x)}$

and $S = \{\cos 2x,\ \sin 2x\}$. Thus, $W(S) = \begin{vmatrix} \cos 2x & \sin 2x \\ -2\sin 2x & 2\cos 2x \end{vmatrix} = 2$. Through the use of a computer algebra system (or trigonometric identities), we obtain

$$u_1(x) = \int \frac{-\sin^2 2x}{2}\,dx = -\frac{x}{4} + \frac{\sin 4x}{16}$$

and

$$u_2(x) = \int \frac{\cos 2x \sin 2x}{2}\,dx = -\frac{1}{8}\cos^2 2x.$$

Simplifying (verify!) leads to the particular solution

$$y_p(x) = u_1(x)y_1(x) + u_2(x)y_2(x) = -\frac{1}{4}x \cos 2x$$

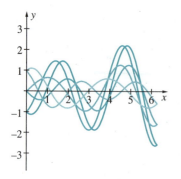

Figure 4.15

and, thus, a general solution is

$$y(x) = y_h(x) + y_p(x) = c_1 \cos 2x + c_2 \sin 2x - \frac{1}{4}x \cos 2x.$$

 Figure 4.15 shows the graph $y(x) = c_1 \cos 2x + c_2 \sin 2x - \frac{1}{4}x \cos 2x$ for various values of c_1 and c_2. Identify the graph of the solution that satisfies the initial conditions (a) $y(0) = 0$ and $y'(0) = 1$; and (b) $y(0) = 0$ and $y'(0) = -1$.

In order to solve the initial-value problem, we compute

$$y'(x) = -2c_1 \sin 2x + 2c_2 \cos 2x - \frac{1}{4}\cos 2x + \frac{1}{2}x \sin 2x.$$

We then evaluate $y(0) = c_1 = 0$ and $y'(0) = 2c_2 - \frac{1}{4} = 1$. Hence, $c_1 = 0$ and $c_2 = \frac{5}{8}$, so the solution is

$$y(x) = \frac{5}{8}\sin 2x - \frac{1}{4}x \cos 2x.$$

Note that if we had used the method of undetermined coefficients or the annihilator method, we should not expect a particular solution of the same form as that obtained through variation of parameters unless we simplify the original functions $u_1(x)$ and $u_2(x)$ obtained through integration. (Why?)

Example 3

For what value(s) of $\lambda > 0$, if any, does the boundary-value problem
$\begin{cases} y'' + \lambda^2 y = \sin 2x \\ y(0) = 0, y(\pi) = 0 \end{cases}$ have (a) one solution, (b) no solutions, and
(c) infinitely many solutions?

Solution   A fundamental set of solutions of the corresponding homogeneous equation is $S = \{\cos \lambda x, \sin \lambda x\}$ with $W(s) = \lambda$. By variation of parameters,

$$u_1(x) = \int \frac{-\sin \lambda x \sin 2x}{\lambda}\, dx$$

$$= \begin{cases} \dfrac{(2 + \lambda) \sin((2 - \lambda)x) - (2 - \lambda) \sin((2 + \lambda)x)}{2\lambda(\lambda^2 - 4)}, & \text{if } \lambda \neq 2 \\[2ex] \dfrac{\sin 4x - 4x}{16}, & \text{if } \lambda = 2 \end{cases}$$

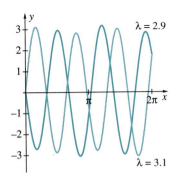

a.

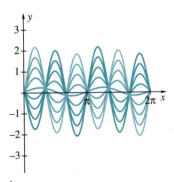

b.

Figure 4.16 (a) Graph of *the* solution to the boundary-value problem if $\lambda = 2.9$ and if $\lambda = 3.1$ (b) Graph of *several* solutions to the boundary-value problem if $\lambda = 3$

and

$$u_2(x) = \int \frac{\cos \lambda x \sin 2x}{\lambda} \, dx$$

$$= \begin{cases} \dfrac{(2 + \lambda) \cos((2 - \lambda)x) + (2 - \lambda) \cos((2 + \lambda)x)}{2\lambda(\lambda^2 - 4)}, & \text{if } \lambda \neq 2 \\[3mm] -\dfrac{\cos^2 2x}{8}, & \text{if } \lambda = 2 \end{cases}$$

so a general solution of the nonhomogeneous equation is

$$y(x) = c_1 \cos \lambda x + c_2 \sin \lambda x + u_1(x)y_1(x) + u_2(x)y_2(x)$$

$$= \begin{cases} c_1 \cos \lambda x + c_2 \sin \lambda x + \dfrac{1}{\lambda^2 - 4} \sin 2x, & \text{if } \lambda \neq 2 \\[3mm] c_1 \cos 2x + c_2 \sin 2x - \dfrac{1}{4}x \cos 2x, & \text{if } \lambda = 2 \end{cases}$$

Applying the boundary conditions yields the system of equations

$$\begin{cases} c_1 = 0 \\ c_1 \cos \lambda \pi + c_2 \sin \lambda \pi = 0 \end{cases}, \quad \text{if } \lambda \neq 2$$

and

$$\begin{cases} c_1 = 0 \\ c_1 - \dfrac{\pi}{4} = 0 \end{cases}, \quad \text{if } \lambda = 2.$$

The system $\begin{cases} c_1 = 0 \\ c_1 - \dfrac{\pi}{4} = 0 \end{cases}$ does not have a solution (why?) so the boundary-value problem does not have a solution if $\lambda = 2$. On the other hand, the solution to the system $\begin{cases} c_1 = 0 \\ c_1 \cos \lambda \pi + c_2 \sin \lambda \pi = 0 \end{cases}$ is $c_1 = 0$ and $c_2 = -\csc \lambda \pi$, provided that λ is not an integer, because $\sin \lambda \pi = 0$ if λ is an integer. But, if $\lambda \neq 2$ is an integer, then *any* value of c_2 is a solution to this system, so there are infinitely many solutions to the boundary-value problem if $\lambda \neq 2$ is an integer.

Thus, the boundary-value problem has no solution if $\lambda = 2$, a unique solution if λ is not an integer, and infinitely many solutions if $\lambda \neq 2$ is an integer. (See Figure 4.16.)

An advantage that the method of variation of parameters has over the method of undetermined coefficients (and the annihilator method) is that nonhomogeneous equations with *nonconstant coefficients* can be considered.

Example 4

A general solution of $x^2 y'' + xy' = 0$, $x > 0$ is $y_h(x) = c_1 + c_2 \ln x$. Solve the nonhomogeneous equation $x^2 y'' + xy' = 2x^2$, $x > 0$.

Solution   In this case, $S = \{y_1(x), y_2(x)\} = \{1, \ln x\}$, so $W(S) = \begin{vmatrix} 1 & \ln x \\ 0 & 1/x \end{vmatrix} =$ $1/x$. Recall that we divided the second-order equation $a_2(x)y'' + a_1(x)y' + a_0(x)y = g(x)$ by the coefficient function of y'' to place it in the form $y'' + p(x)y' + q(x)y = f(x)$ before deriving the formulas for $u_1(x)$ and $u_2(x)$. Doing this, we have $y'' + \dfrac{1}{x}y' = 2$, so $f(x) = 2$. Therefore, using integration by parts, we obtain

$$u_1(x) = \int \frac{-\ln x \,(2)}{\dfrac{1}{x}}\,dx = -2 \int x \ln x \, dx = -x^2 \ln x + \frac{1}{2}x^2,$$

and

$$u_2(x) = \int \frac{(1)(2)}{\dfrac{1}{x}}\,dx = 2 \int x \, dx = x^2,$$

so $\quad y_p(x) = u_1(x)y_1(x) + u_2(x)y_2(x) = (1)\left(-x^2 \ln x + \frac{1}{2}x^2\right) + (\ln x)(x^2) = \frac{1}{2}x^2,$

and a general solution to the nonhomogeneous equation is

$$y(x) = y_h(x) + y_p(x) = c_1 + c_2 \ln x + \frac{1}{2}x^2, \ x > 0.$$

Higher-Order Nonhomogeneous Equations

Higher-order nonhomogeneous linear equations can be solved through variation of parameters as well. In general, if we are given the nonhomogeneous equation

$$y^{(n)}(x) + a_{n-1}(x)y^{(n-1)}(x) + \cdots + a_1(x)y'(x) + a_0(x)y(x) = g(x)$$

and a fundamental set of solutions $\{y_1(x), y_2(x), \ldots, y_n(x)\}$ of the associated homogeneous equation

$$y^{(n)}(x) + a_{n-1}(x)y^{(n-1)}(x) + \cdots + a_1(x)y'(x) + a_0(x)y(x) = 0,$$

we generalize the method for second-order equations to find $u_1(x), u_2(x), \ldots, u_n(x)$ such that

$$y_p(x) = u_1(x)y_1(x) + u_2(x)y_2(x) + \cdots + u_n(x)y_n(x)$$

is a particular solution of the nonhomogeneous equation.

If

$$u_1'(x)y_1(x) + u_2'(x)y_2(x) + \cdots + u_n'(x)y_n(x) = 0,$$

then

$$y_p^{(m)}(x) = u_1(x)y_1^{(m)}(x) + u_2(x)y_2^{(m)}(x) + \cdots + u_n(x)y_n^{(m)}(x)$$

for $m = 0, 1, 2, \ldots, n-1$ and if

$$u_1'(x)y_1^{(m-1)}(x) + u_2'(x)y_2^{(m-1)}(x) + \cdots + u_n'(x)y_n^{(m-1)}(x) = 0$$

for $m = 1, 2, \ldots, n-1$, then

$$\begin{aligned} y_p^{(n)}(x) = &\; (u_1(x)y_1^{(n)}(x) + u_2(x)y_2^{(n)}(x) + \cdots + u_n(x)y_n^{(n)}(x)) \\ &+ (u_1'(x)y_1^{(n-1)}(x) + u_2'(x)y_2^{(n-1)}(x) + \cdots + u_n'(x)y_n^{(n-1)}(x)). \end{aligned}$$

Following a method similar to that of a second-order differential equation, we obtain the system of n equations

$$\begin{cases} u_1'(x)y_1(x) + u_2'(x)y_2(x) + \cdots + u_n'(x)y_n(x) = 0 \\ u_1'(x)y_1'(x) + u_2'(x)y_2'(x) + \cdots + u_n'(x)y_n'(x) = 0 \\ \qquad\qquad\qquad \vdots \\ y_1^{(n-1)}(x)u_1'(x) + y_2^{(n-1)}(x)u_2'(x) + \cdots + y_n^{(n-1)}(x)u_n'(x) = g(x) \end{cases}$$

which can be solved for $u_1'(x)$, $u_2'(x)$, $\ldots$, $u_n'(x)$ using Cramer's rule.

Let $W_m(y_1(x), y_2(x), \ldots, y_n(x))$ denote the determinant of the matrix obtained by replacing the mth column of

$$\begin{pmatrix} y_1(x) & y_2(x) & \cdots & y_n(x) \\ y_1'(x) & y_2'(x) & \cdots & y_n'(x) \\ \vdots & \vdots & \vdots & \vdots \\ y_1^{(n-1)}(x) & y_2^{(n-1)}(x) & \cdots & y_n^{(n-1)}(x) \end{pmatrix}$$

by the column $\begin{pmatrix} 0 \\ 0 \\ \vdots \\ 0 \\ g(x) \end{pmatrix}$. Then, by Cramer's rule,

$$u_i'(x) = \frac{g(x)W_i(y_1(x), y_2(x), \ldots, y_n(x))}{W(y_1(x), y_2(x), \ldots, y_n(x))},$$

for $i = 1, 2, \ldots, n$, and

$$u_i(x) = \int \frac{g(x)W_i(y_1(x), y_2(x), \ldots, y_n(x))}{W(y_1(x), y_2(x), \ldots, y_n(x))}\,dx$$

so

$$y_p(x) = u_1(x)y_1(x) + u_2(x)y_2(x) + \cdots + u_n(x)y_n(x)$$

is a particular solution of the nonhomogeneous equation. A general solution of the nonhomogeneous equation is given by $y(x) = y_h(x) + y_p(x)$, where $y_h(x)$ is a general solution of the corresponding homogeneous equation.

Example 5

Solve $y''' + 3y'' + 2y' = \cos x$.

Solution   You should verify that a general solution of the corresponding homogeneous equation is $y_h(x) = c_1 + c_2 e^{-x} + c_3 e^{-2x}$ and a fundamental set of solutions is $S = \{1, e^{-x}, e^{-2x}\}$. Therefore, we must solve the system

$$\begin{pmatrix} 1 & e^{-x} & e^{-2x} \\ 0 & -e^{-x} & -2e^{-2x} \\ 0 & e^{-x} & 4e^{-2x} \end{pmatrix} \begin{pmatrix} u_1'(x) \\ u_2'(x) \\ u_3'(x) \end{pmatrix} = \begin{pmatrix} 0 \\ 0 \\ \cos x \end{pmatrix}$$

where

$$W(S) = \begin{vmatrix} 1 & e^{-x} & e^{-2x} \\ 0 & -e^{-x} & -2e^{-2x} \\ 0 & e^{-x} & 4e^{-2x} \end{vmatrix} = -2e^{-3x}.$$

Using this with Cramer's Rule, we have

$$u_1'(x) = \frac{\begin{vmatrix} 0 & e^{-x} & e^{-2x} \\ 0 & -e^{-x} & -2e^{-2x} \\ \cos x & e^{-x} & 4e^{-2x} \end{vmatrix}}{-2e^{-3x}} = \frac{-e^{-3x}\cos x}{-2e^{-3x}} = \frac{1}{2}\cos x,$$

$$u_2'(x) = \frac{\begin{vmatrix} 1 & 0 & e^{-2x} \\ 0 & 0 & -2e^{-2x} \\ 0 & \cos x & 4e^{-2x} \end{vmatrix}}{-2e^{-3x}} = \frac{2e^{-2x}\cos x}{-2e^{-3x}} = -e^{x}\cos x,$$

and

$$u_3'(x) = \frac{\begin{vmatrix} 1 & e^{-x} & 0 \\ 0 & -e^{-x} & 0 \\ 0 & e^{-x} & \cos x \end{vmatrix}}{-2e^{-3x}} = \frac{-e^{-x}\cos x}{-2e^{-3x}} = \frac{1}{2}e^{2x}\cos x.$$

Integration then gives us $u_1(x) = \dfrac{1}{2}\sin x$, $u_2(x) = -\dfrac{1}{2}e^{x}(\cos x + \sin x)$, and

$u_3(x) = \dfrac{1}{10}e^{2x}(2\cos x + \sin x)$.

Because

$$y_p(x) = u_1(x)y_1(x) + u_2(x)y_2(x) + u_3(x)y_3(x),$$

we find through substitution into this equation and simplification that $y_p(x) = -\frac{3}{10}\cos x + \frac{1}{10}\sin x$ is a particular solution of the nonhomogeneous equation. A general solution is

$$y(x) = y_h(x) + y_p(x) = c_1 + c_2 e^{-x} + c_3 e^{-2x} - \frac{3}{10}\cos x + \frac{1}{10}\sin x.$$

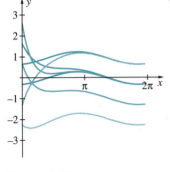

Figure 4.17

 Figure 4.17 shows the graph of $y(x) = y_h(x) + y_p(x)$ for various values of c_1, c_2, and c_3. Identify the graph of $y_p(x)$.

IN TOUCH WITH TECHNOLOGY

As in previous sections when we have used computer algebra systems and graphing utilities to help solve differential equations and graph the resulting solutions, computer algebra systems are particularly useful in carrying out the integration encountered when implementing the method of variation of parameters.

1. Find a general solution of each equation. In each case, graph the solution for several values of the arbitrary constants.

   (a) $y'' + 4y' + 13y = x\cos^2 3x$

   (b) $y'' - 4y = x^{-3}e^{-4x}$

   (c) $y'' - 10y' + 125y = e^{-6x}$

   (d) $y'' - 8y' + 32y = e^{4x}\cos 4x$

   (e) $y'' - 8y' + 20y = \cos 2x$

2. Consider the initial-value problem

   $$\begin{cases} 4y'' + 4y' + y = e^{-x/2} \\ y(0) = a,\ y'(0) = b \end{cases}.$$

   (a) Show that regardless of the choices of a and b, the limit as $x \to \infty$ of every solution is 0.

   (b) Find conditions on a and b, if possible, so that $y(x)$ has (i) no local minima or maxima; (ii) exactly one local minimum or local maximum; and (iii) two distinct local minima and maxima.

   (c) If possible, find a and b so that $y(x)$ has local extrema if $x = 1$ and $x = 3$. Graph $y(x)$ on the interval $[0, 5]$.

   (d) If possible, find a and b so that $y(x)$ has exactly one local extrema at $x = 3$. Graph $y(x)$ on the interval $[0, 6]$.

   (e) Is it possible to determine a and b so that $y(x)$ has local extrema if $x = x_0$ and $x = x_1$ ($x_0 \neq x_1$ both arbitrary)? Support your conclusion with several randomly generated values of x_0 and x_1.

3. Solve the equation

   $$e^{-2x}\left[y\frac{d^2y}{dx^2} - \left(\frac{dy}{dx}\right)^2 \right] - 2x(1 + x)y^2 = 0$$

   by making the substitution $y = e^{u(x)}$.
   tion for various values of the constan

EXERCISES 4.7

In Exercises 1–44, solve each differential equation using variation of parameters.

1. $y'' - 7y' + 10y = e^{3x}$

2. $y'' + 5y' + 6y = e^{-x}$

*3. $y'' + 16y = 2\cos 4x$

4. $y'' + y = 6x$

5. $y'' + 4y' + 4y = e^{-2x}$

6. $y'' - y' + \dfrac{1}{4}y = 16xe^{x/2}$

*7. $y'' - 4y' + 13y = 8e^{2x}\cos 3x$

8. $y'' + 4y' + 3y = 54x^3e^{-4x}$

9. $y'' + 4y' + 20y = 2xe^{-2x}$

10. $y'' + 4y = \sec 2x$

*11. $y'' + 16y = \csc 4x$

12. $y'' + 16y = \cot 4x$

13. $y'' + 2y' + 50y = e^{-x}\csc 7x$

14. $y'' + 6y' + 25y = e^{-3x}(\sec 4x + \csc 4x)$

*15. $y'' - 2y' + 26y = e^x(\sec 5x + \csc 5x)$

16. $y'' + 12y' + 37y = \dfrac{e^{-6x}\sin x}{x}, \ x > 0$

17. $y'' - 6y' + 34y = e^{3x}\tan 5x$

18. $y'' - 10y' + 34y = \dfrac{e^{-5x}\sin 3x}{x^2}, \ x > 0$

*19. $y'' - 12y' + 37y = e^{6x}\sec x$

20. $y'' - 8y' + 17 = e^{4x}\sec x$

21. $y'' - 9y = \dfrac{1}{1 + e^{3x}}$

22. $y'' - 25y = \dfrac{1}{1 - e^{5x}}$

*23. $y'' - y = 2\sinh x$

24. $y'' - 4y = 8\cosh x$

25. $y'' - 2y' + y = \dfrac{e^x}{x}, \ x > 0$

26. $y'' - 4y' + 4y = \dfrac{e^{2x}}{x^2}, \ x > 0$

*27. $y'' + 8y' + 16y = \dfrac{e^{-4x}}{x^4}, \ x > 0$

28. $y'' + 6y' + 9y = \dfrac{1}{xe^{3x}}, \ x > 0$

29. $y'' + 6y' + 9y = e^{-3x}\ln x, \ x > 0$

30. $y'' + 3y' + 2y = \cos(e^x)$

*31. $y'' + 4y' + 4y = e^{-2x}\sqrt{1 - x^2}, \ -1 \le x \le 1$

32. $y'' - 2y' + y = e^x\sqrt{1 - x^2}, \ -1 \le x \le 1$

33. $y'' - 10y' + 25y = e^{5x}\ln 2x$

34. $y'' - 4y' + 4y = e^{2x}\tan^{-1} x$

*35. $y'' + 8y' + 16y = \dfrac{e^{-4x}}{1 + x^2}$

36. $y'' + \dfrac{1}{4}y = \sec \dfrac{x}{2} + \csc \dfrac{x}{2}$

37. $y''' + y' = -2\sin x - \dfrac{\sin x}{\cos^2 x}$

38. $y''' + 4y' = \sec 2x$

*39. $y''' - 2y'' = -\dfrac{1 + 2x}{x^2}, \ x > 0$

40. $y''' - 3y'' + 3y' - y = \dfrac{e^x}{x}, \ x > 0$

41. $y''' - 4y'' - 11y' + 30y = e^{4x}$

42. $y''' + 3y'' - 10y' - 24y = e^{-3x}$

*43. $y''' - 13y' + 12y = \cos x$

44. $y''' - 6y'' + 32y = e^{-4x}$

In Exercises 45–54, solve the initial-value problem.

45. $y'' + y = \tan x, \ y(0) = 2, \ y'(0) = 1, \ -\dfrac{\pi}{2} < x < \dfrac{\pi}{2}$

46. $y'' + 9y = \dfrac{1}{2}\csc x, \ y\left(\dfrac{\pi}{4}\right) = \sqrt{2}, \ y'\left(\dfrac{\pi}{4}\right) = 0$

*47. $y'' + 5y' + 6y = -3 - \sin 4x, \ y(0) = -\dfrac{23}{50},$
$y'(0) = \dfrac{27}{25}$

48. $y'' + \dfrac{2}{3}y' = -3\sin 4x, \ y(0) = 0, \ y'(0) = \dfrac{27}{37}$

49. $y'' - 16y = \dfrac{16x}{e^{4x}}$, $y(0) = 0$, $y'(0) = 0$

50. $y'' - 6y' + 9y = 5x^3 e^{4x}$, $y(0) = 0$, $y'(0) = 0$

*51. $y'' - 2y' + y = -e^{-2x} \cos 3x$,

$y(0) = -\dfrac{18}{13}$, $y'(0) = -\dfrac{10}{13}$

52. $y'' - 6y' + 18y = -e^{3x} \sin 3x$, $y(0) = -\dfrac{1}{6}$,

$y'(0) = 0$

53. $y''' - y'' = 3x^2$, $y(0) = 0$, $y'(0) = 0$, $y''(0) = 0$

54. $y''' - y'' = 3x^2 - 3x^3$, $y(0) = 0$, $y'(0) = 0$, $y''(0) = 0$

55. Solve $y'' + 4y' + 3y = 65 \cos 2x$ by **(a)** the method of undetermined coefficients; **(b)** the method of variation of parameters. Which method is more easily applied?

56. Show that the solution of the initial-value problem

$$\begin{cases} y'' + a_1(x)y' + a_0(x)y = f(x) \\ y(x_0) = y_0, y'(x_0) = y_0' \end{cases}$$

can be written as $y(x) = u(x) + v(x)$ where u is the solution of

$$\begin{cases} u'' + a_1(x)u' + a_0(x)u = 0 \\ u(x_0) = y_0, u'(x_0) = y_0' \end{cases}$$

and v is the solution of

$$\begin{cases} v'' + a_1(x)v' + a_0(x)v = f(x) \\ v(x_0) = 0, v'(x_0) = 0 \end{cases}.$$

57. Let $y_1(x)$ and $y_2(x)$ be two linearly independent solutions of the second-order homogeneous linear differential equation

$$a_2(x)y'' + a_1(x)y' + a_0(x)y = 0.$$

What is a general solution of $a_2(x)y'' + a_1(x)y' + a_0(x)y = g(x)$? (*Hint:* In order to apply the formulas of variation of parameters, the lead coefficient must be 1; divide by $a_2(x)$ and apply the variation of parameters formula.)

58. Suppose that we attempt to solve the initial-value problem

$$\begin{cases} y'' + a_1(x)y' + a_0(x)y = f(x) \\ y(x_0) = 0, y'(x_0) = 0 \end{cases}$$

using variation of parameters. Show that the particular solution can be written as

$$y_p(x) = \int_{x_0}^{x} \frac{y_1(t)y_2(x) - y_1(x)y_2(t)}{y_1(t)y_2'(t) - y_1'(t)y_2(t)} f(t)\, dt$$

$$= \int_{x_0}^{x} \frac{\begin{vmatrix} y_1(t) & y_2(t) \\ y_1(x) & y_2(x) \end{vmatrix}}{W(y_1(t), y_2(t))} f(t)\, dt$$

where $\quad G(t, x) = \dfrac{\begin{vmatrix} y_1(t) & y_2(t) \\ y_1(x) & y_2(x) \end{vmatrix}}{W(y_1(t), y_2(t))}\quad$ is called

the **Green's function**; $y_1(x)$ and $y_2(x)$ are solutions to the corresponding homogeneous equation, $y'' + a_1(x)y' + a_0(x)y = 0$.

59. Use the results of Exercises 56 and 58 to show that the solution of

$$\begin{cases} y'' + a_1(x)y' + a_0(x)y = f(x) \\ y(x_0) = y_0, y'(x_0) = y_0' \end{cases}$$

can be written as

$$y(x) = u(x) + \int_{x_0}^{x} \frac{\begin{vmatrix} y_1(t) & y_2(t) \\ y_1(x) & y_2(x) \end{vmatrix}}{W(y_1(t), y_2(t))} f(t)\, dt.$$

In Exercises 60–63, use a Green's function to solve the initial-value problem.

60. $\begin{cases} y'' + y = \sin x \\ y(0) = -1, y'(0) = 1 \end{cases}$

61. $\begin{cases} y'' - 2y' + y = 2e^x \\ y(0) = 1, y'(0) = 0 \end{cases}$

62. $\begin{cases} y'' + 7y' + 10y = e^{-2x} \\ y(0) = 1, y'(0) = 0 \end{cases}$

*63. $\begin{cases} y'' - 4y' + 3y = 9x \\ y(0) = 0, y'(0) = 2 \end{cases}$

64. Given that $y = c_1 x^{-1} + c_2 x^{-1} \ln x$ is a general solution to $x^2 y'' + 3xy' + y = 0$, $x > 0$, use variation of parameters to solve $x^2 y'' + 3xy' + y = \ln x$, $x > 0$.

*65. Given that $y = c_1 \sin(2 \ln x) + c_2 \cos(2 \ln x)$ is a general solution to $x^2 y'' + xy' + 4y = 0$, $x > 0$, use variation of parameters to solve $x^2 y'' + xy' + 4y = x$, $x > 0$.

66. Given that $y = c_1 x^6 + c_2 x^{-1}$ is a general solution to $x^2 y'' - 4xy' - 6y = 0$, $x > 0$, use variation of parameters to solve $x^2 y'' - 4xy' - 6y = 2 \ln x$, $x > 0$.

Section 4.1

Nth-order ordinary linear differential equation
$$a_n(x)y^{(n)}(x) + \cdots + a_1(x)y'(x) + a_0(x)y(x) = g(x)$$

Homogeneous
$$a_n(x)y^{(n)}(x) + \cdots + a_1(x)y'(x) + a_0(x)y(x) = 0$$

Constant coefficients
$$a_n y^{(n)}(x) + \cdots + a_1 y'(x) + a_0 y(x) = g(x)$$

Linearly dependent and independent set of functions
If $S = \{f_1(x), f_2(x), f_3(x), \ldots, f_{n-1}(x), f_n(x)\}$, S is **linearly dependent** if there are constants $c_1, c_2, \ldots, c_n$, not all zero, so that
$$c_1 f_1(x) + c_2 f_2(x) + \cdots + c_n f_n(x) = 0.$$

S is **linearly independent** if S is not linearly dependent.

Wronskian
The **Wronskian** of
$$S = \{f_1(x), f_2(x), f_3(x), \ldots, f_{n-1}(x), f_n(x)\},$$
is the determinant
$$W(S) = \begin{vmatrix} f_1(x) & f_2(x) & \cdots & f_n(x) \\ f_1'(x) & f_2'(x) & \cdots & f_n'(x) \\ \vdots & \vdots & \vdots & \vdots \\ f_1^{(n-1)}(x) & f_2^{(n-1)}(x) & \cdots & f_n^{(n-1)}(x) \end{vmatrix}.$$

If $W(S) \neq 0$ for at least one value of x in the interval I, S is linearly independent.

Section 4.2

Principle of Superposition
Any linear combination of a set of solutions of the nth-order linear homogeneous equation is also a solution.

Fundamental set of solutions
A set $S = \{f_1(x), f_2(x), f_3(x), \ldots, f_{n-1}(x), f_n(x)\}$ of n linearly **independent** solutions of the nth-order linear homogeneous

equation. Every nth-order linear homogeneous equation has a fundamental set of solutions from which a general solution can be obtained.

General solution of a homogeneous equation
If $S = \{f_1(x), f_2(x), \ldots, f_{n-1}(x), f_n(x)\}$ is a fundamental set of solutions of the nth-order linear homogeneous equation
$$a_n(x)y^{(n)}(x) + \cdots + a_1(x)y'(x) + a_0(x)y(x) = 0,$$
a **general solution** of the equation is
$$f(x) = c_1 f_1(x) + c_2 f_2(x) + \cdots + c_{n-1} f_{n-1}(x) + c_n f_n(x),$$
where $\{c_1, c_2, \ldots, c_{n-1}, c_n\}$ is a set of n arbitrary constants.

Section 4.3

Characteristic equation
The equation
$$a_n m^n + a_{n-1} m^{n-1} + \cdots + a_1 m + a_0 = 0$$
is the **characteristic equation** of the linear homogeneous differential equation with constant coefficients
$$a_n y^{(n)}(x) + a_{n-1} y^{(n-1)}(x) + \cdots + a_1 y'(x) + a_0 y(x) = 0.$$

Solving second-order equations with constant coefficients
If $ay'' + by' + cy = 0$ and m_1 and m_2 are the solutions of the equation $am^2 + bm + c = 0$, then a general solution is
(a) $y = c_1 e^{m_1 x} + c_2 e^{m_2 x}$, if $m_1 \neq m_2$ and both m_1 and m_2 are real;
(b) $y = c_1 e^{m_1 x} + c_2 x e^{m_1 x}$, if $m_1 = m_2$; and
(c) $y = e^{\alpha x}(c_1 \cos \beta x + c_2 \sin \beta x)$, if $m_1 = \alpha + i\beta$, $\beta \neq 0$, and $m_2 = \overline{m_1} = \alpha - i\beta$.

Section 4.4

Particular solution
A **particular solution,** $y_p(x)$, of the differential equation $a_n y^{(n)}(x) + \cdots + a_1 y'(x) + a_0 y(x) = g(x)$ is a specific function, containing no arbitrary constants, that satisfies the equation.

General solution of a nonhomogeneous equation

A general solution to the nonhomogeneous equation is

$$y(x) = y_h(x) + y_p(x),$$

where $y_h(x)$ is a general solution of the corresponding homogeneous equation

$$a_n y^{(n)}(x) + \cdots + a_1 y'(x) + a_0 y(x) = 0,$$

and $y_p(x)$ is a particular solution of the nonhomogeneous equation.

Annihilator

The differential operator $p(D)$ is said to **annihilate** a function $f(x)$ if $p(D)[f(x)] = 0$ for all x. In this case, $p(D)$ is called an **annihilator** of $f(x)$.

Section 4.5

Solving nonhomogeneous equations with annihilators

1. Determine an operator $q(D)$ that annihilates $g(x)$.
2. Apply the operator to both sides of the differential equation $q(D)(p(D)y) = q(D)(g(x)) = 0$.
3. Solve the homogeneous equation $q(D)(p(D)y) = 0$.
4. Find a general solution $y_h(x)$ of the corresponding homogeneous equation $p(D)y = 0$.
5. Eliminate the terms of the homogeneous solution $y_h(x)$ from the general solution obtained in step 3. The function that remains is the correct form of a particular solution of the nonhomogeneous equation.
6. Solve for the unknown coefficients in the particular solution.
7. A general solution of the nonhomogeneous equation is $y(x) = y_h(x) + y_p(x)$.

Section 4.6

Method of undetermined coefficients

Method used to find a particular solution:
1. If $g_i(x) = x^m$, the associated set of functions is

$$S = \{x^m, x^{m-1}, \ldots, x^2, x, 1\}.$$

2. If $g_i(x) = x^m e^{kx}$, the associated set of functions is

$$S = \{x^m e^{kx}, x^{m-1} e^{kx}, \ldots, x^2 e^{kx}, x e^{kx}, e^{kx}\}.$$

3. If $g_i(x) = x^m e^{\alpha x} \cos \beta x$, or $g_i(x) = x^m e^{\alpha x} \sin \beta x$, the associated set of functions is

$$S = \{x^m e^{\alpha x} \cos \beta x, \ldots, x e^{\alpha x} \cos \beta x, e^{\alpha x} \cos \beta x, \\ x^m e^{\alpha x} \sin \beta x, \ldots, x e^{\alpha x} \sin \beta x, e^{\alpha x} \sin \beta x\}.$$

For each function in S, determine the associated set of functions. If any of the functions in S appear in the homogeneous solution $y_h(x)$, multiply each function in S by x^r to obtain a new set S' (r is the smallest positive integer so that each function in S' is not a function in $y_h(x)$). A particular solution is obtained by taking the linear combination of all functions in the associated sets where repeated functions should appear only once in the particular solution.

Section 4.7

Variation of parameters

A particular solution of $y'' + p(x)y' + q(x)y = f(x)$ is

$$y_p(x) = u_1(x)y_1(x) + u_2(x)y_2(x)$$

where $u_1(x) = \int \dfrac{-y_2(x)f(x)}{W(S)}\, dx$ and $u_2(x) = \int \dfrac{y_1(x)f(x)}{W(S)}\, dx$;
$S = \{y_1(x), y_2(x)\}$ is a fundamental set of solutions of the corresponding homogeneous equation.

CHAPTER 4 REVIEW EXERCISES

In Exercises 1–12, determine if the given set is linearly independent or linearly dependent.

1. $S = \{e^{5x}, 1\}$
*3. $S = \{\cos^2 x, \sin^2 x\}$
2. $S = \{x, \cos x\}$
4. $S = \{\cos^2 x, 2\sin^2 x - 2\}$

5. $S = \{x, x \ln x\}$
*7. $S = \{x, x - 1, 3x\}$
9. $S = \{x, \cos x, \sin x\}$
*11. $S = \{e^x, e^{-2x}, e^{-x}\}$

6. $S = \{x^{3/2}, x^2\}$
8. $S = \{x^2, x + 1, 5\}$
10. $S = \{\cos^2 x, 1, \sin^2 x\}$
12. $S = \{e^x, \cos 2x, 1\}$

In Exercises 13–18, verify that y is a general solution of the given differential equation.

13. $y'' - 7y' + 10y = 0$; $y = c_1 e^{5x} + c_2 e^{2x}$

14. $y'' - 4y' - 5y = 0$; $y = c_1 e^{5x} + c_2 e^{-x}$

*15. $y'' - y' - 2y = 0$; $y = c_1 e^{2x} + c_2 e^{-x}$

16. $y'' - 7y' + 6y = 0$; $y = c_1 e^x + c_2 e^{6x}$

17. $y'' - 2y' + 2y = 0$; $y = e^x(c_1 \sin x + c_2 \cos x)$

18. $2y'' - 6y' + 17y = 0$;

$$y = e^{3x/2}\left(c_1 \sin \frac{5x}{2} + c_2 \cos \frac{5x}{2}\right)$$

In Exercises 19 and 20, show that the function $y_1(x)$ satisfies the differential equation and find a second linearly independent solution.

*19. $y_1(x) = x + 1$, $y'' - \dfrac{2}{x+1}y' + \dfrac{2}{(x+1)^2}y = 0$

20. $y_1(x) = \dfrac{\sin x}{x}$, $y'' + \dfrac{2}{x}y' + y = 0$

In Exercises 21–61, find a general solution for each equation.

21. $y'' + 7y' + 10y = 0$ 22. $y'' - 4y' - 5y = 0$

*23. $6y'' + 5y' - 4y = 0$ 24. $6y'' + 11y' - 10y = 0$

25. $y'' + 2y' + y = 0$ 26. $y'' - 4y' + 4y = 0$

*27. $y'' + 3y' + 2y = 0$ 28. $y'' + 4y = 0$

29. $y'' - 10y' + 34y = 0$ 30. $y'' - 6y' + 34y = 0$

*31. $2y'' - 5y' + 2y = 0$ 32. $5y'' - 9y' + 4y = 0$

33. $15y'' - 11y' + 2y = 0$ 34. $4y'' - y' - 3y = 0$

*35. $20y'' + y' - y = 0$ 36. $20y'' + 7y' - 6y = 0$

37. $12y'' + 8y' + y = 0$ 38. $4y''' + 16y'' + 15y' = 0$

*39. $2y''' + 3y'' + y' = 0$ 40. $4y'' - 16y' + 17y = 0$

41. $9y'' + 36y' + 40y = 0$ 42. $4y'' + 8y' + 5y = 0$

*43. $9y'' + 12y' + 13y = 0$ 44. $9y'' + 12y' + 40y = 0$

45. $y'' - 2y' - 8y = -x$ 46. $3y'' + 5y' = -3x$

*47. $y'' + 5y' = 5x^2$ 48. $y'' - 4y' + 4y = 5x^2$

49. $y'' - 4y' = -3 \sin x$ 50. $y'' - 2y' = -2 \sin 3x$

*51. $y'' + 2y' + 5y = 3 \sin 2x$

52. $y'' - 9y = \cos 3x$

53. $y'' - 2y' = 2 \cos 4x$

54. $y'' - 3y' + 2y = -4e^{-2x}$

*55. $y'' - 6y' + 13y = 3e^{-2x}$

56. $y'' + 9y' + 20y = -2xe^x$

57. $y'' + 7y' + 12y = 3x^2 e^{-4x}$

58. $y''' + 3y'' - 9y' + 5y = e^x$

*59. $y''' - 12y' - 16y = e^{4x} - e^{-2x}$

60. $y^{(4)} + 6y''' + 18y'' + 30y' + 25y = e^{-x} \cos 2x + e^{-2x} \sin x$

61. $y^{(4)} + 4y''' + 14y'' + 20y' + 25y = x^2$

In Exercises 62–69, solve the initial-value problem.

62. $y'' + 5y' + 6y = 0$, $y(0) = 2$, $y'(0) = 0$

63. $y'' + 10y' + 16y = 0$, $y(0) = 0$, $y'(0) = 4$

64. $y'' + 16y = 0$, $y(0) = 0$, $y'(0) = -8$

*65. $y'' + 25y = 0$, $y(0) = 1$, $y'(0) = 0$

66. $y'' - 4y = x$, $y(0) = 2$, $y'(0) = 0$

67. $y'' + 3y' - 4y = e^x$, $y(0) = 0$, $y'(0) = 4$

68. $y'' + 9y = \sin 3x$, $y(0) = 6$, $y'(0) = 0$

69. $y'' + y = \cos x$, $y(0) = 0$, $y'(0) = 0$

In Exercises 70–75, use variation of parameters to solve the indicated differential equations and initial-value problems.

70. $y'' + 4y = \tan 2x$

71. $y'' + y = \csc x$

72. $y'' - 8y' + 16y = x^{-3}e^{4x}$

*73. $y'' - 8y' + 16y = x^{-3}e^{4x}$, $y(1) = 0$, $y'(1) = 0$

74. $y'' - 2y' + y = e^x \ln x$

75. $y'' - 2y' + y = e^x \ln x$, $y(1) = 0$, $y'(1) = 0$

76. Use the substitution $x = e^t$ to solve $2x^3 y''' - 4x^2 y'' - 20xy' = 0$, $x > 0$. (*Hint:* See Exercise 68 in Section 4.3.)

77. Show that the substitution $u = y'/y$ converts the equation $p_0(x)y'' + p_1(x)y' + p_2(x)y = 0$ to $p_0(x)(u' + u^2) + p_1(x)u + p_2(x) = 0$.

78. Use the substitution in the previous problem to solve $y'' - 2xy' + x^2 y = 0$. $\left(\text{*Hint:* You should obtain the differential equation } \dfrac{du}{dx} = -(u - x)^2. \text{ Make the substitution } v = u - x \text{ to solve this equation.}\right)$

*79. (**Abel's Formula**) Suppose that y_1 and y_2 are two solutions of $y'' + p(x)y' + q(x)y = 0$ on an interval I where $p(x)$ and $q(x)$ are continuous functions. Then, the Wronskian of y_1 and y_2 is

$$W(y_1, y_2)(x) = Ce^{-\int p(x)\, dx}.$$

Prove Abel's Formula by computing $\dfrac{d}{dx} W(y_1, y_2)(x)$ and using the relationship $y'' = -p(x)y' - q(x)y$ for y_1'' to obtain a first-order ordinary differential equation for W.

80. Can the Wronskian be zero at only one value of x on I? (*Hint:* Use Abel's formula.)

81. Use Abel's formula to find the Wronskian (within a constant multiple) associated with the following differential equations. Also, obtain a fundamental set of solutions and compute the Wronskian directly. Compare the results.
  (a) $y'' + 3y' - 4y = 0$ *(b) $y'' + 4y' + 13y = 0$
  (c) $y'' + 4y' + 4y = 0$ (d) $y'' + 9y = 0$

82. Use Abel's formula to find the Wronskian (within a constant multiple) associated with the following differential equations.
  (a) $xy'' + y' + xy = 0$ (b) $x^2 y'' - 5xy' + 5y = 0$
  (c) $x^2 y'' - xy' + 5y = 0$

83. Solve each of the following boundary-value problems.
  (a) $\begin{cases} y'' - y = 0 \\ y'(0) + 3y(0) = 0,\ y'(1) + y(1) = 1 \end{cases}$

  (b) $\begin{cases} y'' + \lambda y = 0 \\ y(0) = 0,\ y(p) = 0 \end{cases},\ p > 0$
  (*Hint:* Consider three cases: $\lambda = 0$, $\lambda < 0$, and $\lambda > 0$.)

  (c) $\begin{cases} y'' + \lambda y = 0 \\ y'(0) = 0,\ y'(p) = 0 \end{cases},\ p > 0$
  (*Hint:* Consider three cases: $\lambda = 0$, $\lambda < 0$, and $\lambda > 0$.)

  (d) $\begin{cases} y'' + 2y' - (\lambda - 1)y = 0 \\ y(0) = 0,\ y(2) = 0 \end{cases}$
  (*Hint:* Consider three cases: $\lambda = 0$, $\lambda < 0$, and $\lambda > 0$.)

84. (Testing for Diabetes) Diabetes mellitus affects approximately 12 million Americans; approximately one-half of these people are unaware that they have diabetes. Diabetes is a serious disease: it is the leading cause of blindness in adults, the leading cause of renal failure, and is responsible for approximately one-half of all nontraumatic amputations in the United States. In addition, people with diabetes have an increased rate of coronary artery disease and strokes. People at risk for developing diabetes include those who are obese; those suffering from excessive thirst, hunger, urination, and weight loss; women who have given birth to a baby of weight greater than nine pounds; those with a family history of diabetes; and those who are over 40 years of age.

People with diabetes cannot metabolize glucose because their pancreas produces an inadequate or ineffective supply of insulin. Subsequently, glucose levels rise. The body attempts to remove the excess glucose through the kidneys; the glucose acts as a diuretic, resulting in increased water consumption. Since some cells require energy, which is not being provided by glucose, fat and protein is broken down and ketone levels rise. Although there is no cure for diabetes at this time, many cases can be effectively managed by a balanced diet and insulin therapy in addition to maintaining an optimal weight.

Diabetes can be diagnosed by several tests. In the **fasting blood sugar** test, a patient fasts for at least four hours and the glucose level is measured. In a fasting state, the glucose level in normal adults ranges from 70 to 110 milligrams per milliliter. An adult in a fasting state with consistent readings of over 150 milligrams probably has diabetes. However, individuals vary greatly, so people with mild cases of diabetes might have fasting state glucose levels within the normal range. A highly accurate test that is frequently used to diagnose mild diabetes is the **glucose tolerance test** (GTT), which was developed by Dr. Rosevear and Dr. Molnar of the Mayo Clinic and Dr. Ackerman and Dr. Gatewood of the University of Minnesota. During the GTT, a blood and urine sample are taken from a patient in a fasting state to measure the glucose (G_0), hormone (H_0), and glycosuria levels, respectively. We assume that these values are equilibrium values. The patient is then given 100 grams of glucose. Blood and urine samples are then taken at 1, 2, 3, and 4 hour intervals. In a person without diabetes, glucose levels return to normal after two hours; in diabetics the blood sugar levels either take longer or never return to normal levels. Let G denote the cumulative level of glucose in the blood, $g = G - G_0$, H the cumulative level of hormones that affect insulin production (like glucagon, epinephrine, cortisone and thyroxin), and $h = H - H_0$. Notice that g and h represent the fluctuation of the cumulative levels of glucose and hormones from their equilibrium values. The relationship between the rate of change of glucose in the blood and the rate of change of the cumulative levels of the hormones in the blood which affect insulin production is

$$\begin{cases} g' = f_1(g, h) + J(t), \\ h' = f_2(g, h) \end{cases},$$

where $J(t)$ represents the **external** rate at which the blood glucose concentration is being increased.* If we assume that f_1 and f_2 are linear functions, then this system of equations becomes

$$\begin{cases} g' = -ag - bh + J(t), \\ h' = -ch + dg \end{cases},$$

where a, b, c, and d represent positive numbers.

(a) Show that if $g' = -ag - bh + J(t)$ then

$$h = \frac{1}{b}(-g' - ag + J) \text{ and}$$

$$h' = \frac{1}{b}(-g'' - ag' + J').$$

(b) Substitute $h = \frac{1}{b}(-g' - ag + J)$ and $h' = \frac{1}{b}(-g'' - ag' + J')$ into $h' = -ch + dg$ to obtain the second-order equation

$$\frac{1}{b}(-g'' - ag' + J') = -\frac{c}{b}(-g' - ag + J) + dg$$

$$g'' + (a + c)g' + (ac + bd)g = J' + cJ.$$

Since the glucose solution is consumed at $t = 0$, for $t > 0$ we have that

$$g'' + (a + c)g' + (ac + bd)g = 0.$$

(c) Show that the solutions of the characteristic equation of $g'' + (a + c)g' + (ac + bd)g = 0$ are

$$\frac{1}{2}(-a - c - \sqrt{(a - c)^2 - 4bd}) \text{ and}$$

$$\frac{1}{2}(-a - c + \sqrt{(a - c)^2 - 4bd}).$$

(d) Explain why it is reasonable to assume that glucose levels are periodic and subsequently that $(a - c)^2 - 4bd < 0$.

(e) If $(a - c)^2 - 4bd < 0$ and $t > 0$, show that a general solution of

$$g'' + (a + c)g' + (ac + bd)g = 0$$

is

$$g(t) = e^{-(a+c)t/2}\left[A \cos\left(\frac{t}{2}\sqrt{4bd - (a - c)^2}\right) + B \sin\left(\frac{t}{2}\sqrt{4bd - (a - c)^2}\right)\right]$$

and that

$$G(t) =$$
$$G_0 + e^{-(a+c)t/2}\left[A \cos\left(\frac{t}{2}\sqrt{4bd - (a - c)^2}\right) + B \sin\left(\frac{t}{2}\sqrt{4bd - (a - c)^2}\right)\right].$$

Let $\alpha = \frac{1}{2}(a + c)$ and $\omega = \frac{1}{2}\sqrt{4bd - (a - c)^2}$. Then we can rewrite the general solution obtained here as

$$G(t) = G_0 + e^{-\alpha t}[A \cos \omega t + B \sin \omega t].$$

Research has shown that lab results of $2\pi/\omega > 4$ indicate a mild case of diabetes.

(f) Suppose that you have given the GTT to four patients you suspect of having a mild case of diabetes. The results for each patient are shown in the following table. Which patients, if any, have a mild case of diabetes?

	Patient 1	Patient 2	Patient 3	Patient 4
G_0	80.00	90.00	100.00	110.00
$t = 1$	85.32	91.77	103.35	114.64
$t = 2$	82.54	85.69	98.26	105.89
$t = 3$	78.25	92.39	96.59	108.14
$t = 4$	76.61	91.13	99.47	113.76

* D. N. Burghess and M. S. Borrie, *Modeling with Differential Equations*, Ellis Horwood Limited, pp. 113–116. Joyce M. Black and Esther Matassarin-Jacobs, *Luckman and Sorensen's Medical-Surgical Nursing: A Psychophysiologic Approach*, Fourth Edition, W. B. Saunders Company, 1993, pp. 1775–1808.

Differential Equations at Work:
Modeling the Motion of a Skier

- During a sporting event, an athlete loses strength because the athlete has to perform work against many physical forces. Some of the forces acting on an athlete include gravity (the athlete must usually work against gravity when moving body parts; friction is created between the ground and the athlete) and aerodynamic drag (a pressure gradient exists between the front and the back of the athlete; the athlete must overcome the force of air friction). In downhill skiing, friction and drag affect the skier most because the skier is not working against gravity (why?). The distance traveled by a skier moving down a slope is given by

$$m\frac{d^2s}{dt^2} = F_g - F_\mu - D,$$

where m is the mass of the skier, s is the distance traveled by the skier at time t, F_g is the gravitational force, F_μ is the friction force between skis and snow, and D is the aerodynamic drag.

If the slope has constant angle α, we can rewrite the equation as

$$\frac{d^2s}{dt^2} = g[\sin\alpha - \mu\cos\alpha] - \frac{C_D A\rho}{2m}\left(\frac{ds}{dt}\right)^2,$$

where μ is the coefficient of friction, C_D the drag coefficient, A the projected area of the skier, ρ the air density, and $g \cong 9.81$ m $\cdot$ s^{-2} the gravitational constant.* Because g, α, μ, m, C_D, A, and ρ are constants, $g[\sin\alpha - \mu\cos\alpha]$ and $\frac{C_D A\rho}{2m}$ are constants. Thus, if we let $k^2 = g[\sin\alpha - \mu\cos\alpha]$ (assuming that $g[\sin\alpha - \mu\cos\alpha]$ is nonnegative) and $h^2 = \frac{C_D A\rho}{2m}$, we can rewrite this equation in the simpler form

$$\frac{d^2s}{dt^2} = k^2 - h^2\left(\frac{ds}{dt}\right)^2.$$

Remember that the relationship between displacement, s, and velocity, v, is

$$v = \frac{ds}{dt}.$$

Thus, we can find displacement by integrating velocity, if the velocity is known.

---

* Sauli Savolainen and Reijo Visuri, ''A Review of Athletic Energy Expenditure, Using Skiing as a Practical Example,'' *Journal of Applied Biomechanics,* Volume 10, Number 3 (August 1994) pp. 253–269.

1. Use the substitution $v = ds/dt$ to rewrite $d^2s/dt^2 = k^2 - h^2(ds/dt)^2$ as a first-order equation and find the solution that satisfies $v(0) = v_0$. (*Hint:* Use the method of partial fractions.)

2. Find $s(t)$ if $s(0) = 0$.

Thus, in this case, we see that we are able to express both s and v as functions of t. Typical values of the constants m, ρ, μ, and C_DA are shown in the following table.

Constant	Typical Value
m	75 kg
ρ	1.29 kg · m^{-3}
μ	0.06
C_DA	0.16 m^2

3. Use the values given in the previous table to complete the entries in the following table. For each value of α, graph $v(t)$ and $s(t)$ if $v(0) = 0$, 5, and 10 on the interval $[0, 40]$. In each case, calculate the maximum velocity achieved by the skier. What is the limit of the velocity achieved by the skier? How does changing the initial velocity affect the maximum velocity?

α	$k^2 = g[\sin \alpha - \mu \cos \alpha]$	$h^2 = \dfrac{C_DA\rho}{2m}$
$\alpha = 30° = \pi/6$ rad		
$\alpha = 40° = 2\pi/9$ rad		
$\alpha = 45° = \pi/4$ rad		
$\alpha = 50° = 5\pi/18$ rad		

4. For each value of α in the previous table, graph $v(t)$ and $s(t)$ if $v(0) = 80$ and 100 on the interval $[0, 40]$, if possible. Interpret your results.

Sometimes, it is more useful to express v (velocity) as a function of a different variable, like s (displacement). To do so, we write $dt/ds = 1/v$ because $ds/dt = v$. Multiplying $dv/dt = k^2 - h^2v^2$ by $1/v$ and simplifying leads to

$$\frac{1}{v}\frac{dv}{dt} = \frac{1}{v}(k^2 - h^2v^2)$$

$$\frac{dv}{dt}\frac{dt}{ds} = k^2v^{-1} - h^2v$$

$$\frac{dv}{ds} + h^2v = k^2v^{-1}.$$

Now let $w = v^2$. Then, $\dfrac{dw}{ds} = 2v\dfrac{dv}{ds}$ so $\dfrac{1}{2v}\dfrac{dw}{ds} = \dfrac{dv}{dx}$ and the equation becomes

$$\frac{1}{2v}\frac{dw}{dx} + h^2v = k^2v^{-1}.$$

Multiplying this equation by $2v$ and applying the substitution $w = v^2$ leads us to the first-order linear differential equation

$$\frac{dw}{ds} + 2h^2w = 2k^2.$$

5. Show that a solution of this differential equation is given by

$$w(s) = (k/h)^2 + Ce^{-2h^2s},$$

where C is an arbitrary constant. Resubstitute $w = v^2$ to obtain

$$[v(s)]^2 = (k/h)^2 + Ce^{-2h^2s}.$$

6. Show that the solution that satisfies the initial condition $v(0) = v_0$ is

$$[v(s)]^2 = \frac{-k^2 + k^2e^{2h^2s} + h^2v_0^2}{h^2e^{2hs}} = \left(\frac{k}{h}\right)^2(1 - e^{-2h^2s}) + v_0^2e^{-2h^2s}.$$

7. For each value of α in the previous table, graph $v(s)$ if $v(0) = 0$, 5, and 10 on the interval $(0, 1500]$. In each case, calculate the maximum velocity achieved by the skier. How does changing the initial velocity affect the maximum velocity? Compare your results to (3). Are your results consistent?

8. How do these results change if μ is increased or decreased? if C_DA is increased or decreased? How much does a 10% increase or decrease in friction decrease or increase the skiers velocity?

9. The values of the constants considered here for three skiers are listed in the following table.

Constant	Skier 1	Skier 2	Skier 3
m	75 kg	75 kg	75 kg
ρ	1.29 kg $\cdot$ m^{-3}	1.29 kg $\cdot$ m^{-3}	1.29 kg $\cdot$ m^{-3}
μ	0.05	0.06	0.07
C_DA	0.16 m^2	0.14 m^2	0.12 m^2

Suppose that these three skiers are racing down a slope with constant angle $\alpha = 30° = \pi/6$ rad. Who wins the race if the length of the slope if 400 meters? Who loses? Who wins the race if the length of the slope is 800 or 1200 meters? Who loses? Does the length of the slope matter as to who wins or loses? What if the angle of the slope is increased or decreased? Explain.

10. Which of the variables considered here affects the velocity of the skier the most? Which variable do you think is the easiest to change? How would you advise a group of skiers who want to increase their maximum attainable velocity by 10%?

5

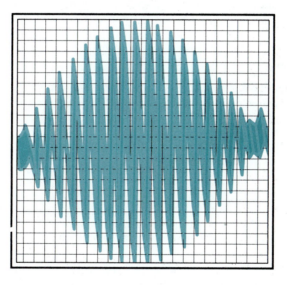

In Chapter 4, we discussed several techniques for solving higher-order differential equations. In this chapter, we illustrate how those methods can be used to solve some initial-value problems that model physical situations.

5.1 Simple Harmonic Motion

Suppose that an object of mass m is attached to an elastic spring that is suspended from a rigid support such as a ceiling or a horizontal rod. The object causes the spring to stretch to a distance s from its *natural length.* The position at which it comes to rest is called the *equilibrium position.* According to Hooke's law, the spring exerts a restoring force in the upward (opposite) direction that is proportional to the distance s that the spring is stretched. Mathematically, this is stated as

$$F = ks$$

where $k > 0$ is the constant of proportionality or **spring constant.** In Figure 5.1 we see that the spring has natural length b. When the object is attached to the spring, it is stretched s units past its natural length to the equilibrium position $x = 0$. When the system is in motion, the displacement from $x = 0$ at time t is given by $x(t)$.

By Newton's Second Law of Motion,

$$F = ma = m\frac{d^2x}{dt^2},$$

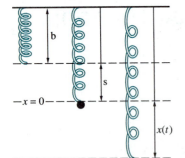

Figure 5.1   A Spring–Mass System

where m represents mass of the object and a represents acceleration. If we assume that there are no forces other than the force due to gravity acting on the mass, we determine the differential equation that models this situation by summing the forces acting on the **spring-mass system** with

$$m\frac{d^2x}{dt^2} = \sum (\text{forces acting on the system})$$
$$= -k(s + x) + mg$$
$$= -ks - kx + mg.$$

At equilibrium $ks = mg$, so after simplification we obtain the differential equation

$$m\frac{d^2x}{dt^2} = -kx \qquad \text{or} \qquad m\frac{d^2x}{dt^2} + kx = 0.$$

The two initial conditions that are used with this problem are the *initial position* $x(0) = \alpha$ and the *initial velocity* $\dfrac{dx}{dt}(0) = \beta$. The function $x(t)$, which describes the *displacement* of the object with respect to the equilibrium position at time t, is found by solving the initial-value problem

$$\begin{cases} m\dfrac{d^2x}{dt^2} + kx = 0 \\[2mm] x(0) = \alpha, \ \dfrac{dx}{dt}(0) = \beta \end{cases}$$

Based on the assumption made in deriving the differential equation, *positive values of $x(t)$ indicate that the mass is below the equilibrium position* and *negative values of $x(t)$*

indicate that the mass is above the equilibrium position. The units that are encountered in these problems are summarized in Table 5.1.

TABLE 5.1   Units Encountered When Solving Spring-Mass Systems

System	Force	Mass	Length	k (Spring Constant)	Time
English	pounds (lb)	slugs (lb-s²/ft)	feet (ft)	lb/ft	seconds (s)
Metric	Newtons (N)	kilograms (kg)	meters (m)	N/m	seconds (s)

Example 1

Determine the spring constant of the spring with natural length 10 in. that is stretched to a distance of 13 in. by an object weighing 5 lb.

Solution   Because the mass weighs 5 lb, $F = 5$ lb, and because displacement from the equilibrium position is $13 - 10 = 3$ in., $s = 3$ in. $\times \dfrac{1 \text{ ft}}{12 \text{ in.}} = \dfrac{1}{4}$ ft. Therefore,

$$F = ks$$

$$5 = \frac{1}{4}k$$

$$k = 20.$$

Notice that the spring constant is given in the units lb/ft because $F = 5$ lb and $s = \dfrac{1}{4}$ ft.

Example 2

An object of weight 16 lb stretches a spring 3 in. Determine the initial-value problem that models this situation if (a) the object is released from a point 4 in. below the equilibrium position with an upward initial velocity of 2 ft/s; (b) the object is released from rest 6 in. above the equilibrium position; (c) the object is released from the equilibrium position with a downward initial velocity of 8 ft/s.

Solution   We first determine the differential equation that models the spring-mass system (we use the same equation for all three parts of the problem). The information is given in English units so we use $g = 32$ ft/s². We must convert all measurements given in inches to feet. The object stretches the spring 3 in. so $s = 3$ in. $\times \dfrac{1 \text{ ft}}{12 \text{ in.}} = \dfrac{1}{4}$ ft. Also, because the mass weighs 16 lb, $F = 16$. According to Hooke's law, $16 = k \cdot \dfrac{1}{4}$, so $k = 64$ lb/ft. We then find the mass m of the object with $F = mg$ to find that $16 = m \cdot 32$ or $m = \dfrac{1}{2}$ slug. The differential equation used to find the displacement of the object at time t is

$$\frac{1}{2}\frac{d^2x}{dt^2} + 64x = 0.$$

The initial conditions for parts (a), (b), and (c) are then found.

(a) Because 4 in. $\times \dfrac{1 \text{ ft}}{12 \text{ in.}} = \dfrac{1}{3}$ ft and down is the positive direction, $x(0) = \dfrac{1}{3}$.

Notice, however, that the initial velocity is 2 ft/s in the upward (negative) direction. Hence, $\dfrac{dx}{dt}(0) = -2$.

(b) A position 6 in. $\left(\text{or } \dfrac{1}{2} \text{ ft}\right)$ above the equilibrium position (in the negative direction) corresponds to initial position $x(0) = -\dfrac{1}{2}$. Being released from rest indicates that $\dfrac{dx}{dt}(0) = 0$.

(c) Because the mass is released from the equilibrium position, $x(0) = 0$. Also, the initial velocity is 8 ft/s in the downward (positive) direction, so $\dfrac{dx}{dt}(0) = 8$.

Example 3

An object weighing 60 lb stretches a spring 6 in. Determine the function $x(t)$ that describes the displacement of the object if it is released from rest 12 in. below the equilibrium position.

Solution   First, the spring constant k is determined from the supplied information. By Hooke's law, $F = ks$, so we have $60 = k \cdot 0.5$. Therefore, $k = 120$ lb/ft. Next, the mass m of the object is determined using $F = mg$. In this case, $60 = m \cdot 32$, so $m = \dfrac{15}{8}$ slugs. Because $\dfrac{k}{m} = 64$ and 12 inches is equivalent to 1 foot, the initial-value problem that models the situation is

$$\begin{cases} \dfrac{d^2x}{dt^2} + 64x = 0 \\[2mm] x(0) = 1, \ \dfrac{dx}{dt}(0) = 0 \end{cases}$$

(Recall that we used m in the characteristic equations in Chapter 4. Here, we use r instead of m to avoid confusion with the mass of the object.) The characteristic equation that corresponds to the differential equation is $r^2 + 64 = 0$. It has solutions $r = \pm 8i$, so a general solution of the equation is $x(t) = c_1 \cos 8t + c_2 \sin 8t$.

To find the values of c_1 and c_2 that satisfy the initial conditions, we calculate $\dfrac{dx}{dt}(t) = -8c_1 \sin 8t + 8c_2 \cos 8t$. Then, $x(0) = c_1 = 1$ and $\dfrac{dx}{dt}(0) = 8c_2 = 0$. So $c_1 = 1$, $c_2 = 0$ and $x(t) = \cos 8t$.

Notice that $x(t) = \cos 8t$ indicates that the spring-mass system never comes to rest once it is set into motion. The solution is periodic so the mass moves vertically, retracing its motion, as shown in Figure 5.2. Motion of this type is called **simple harmonic motion.**

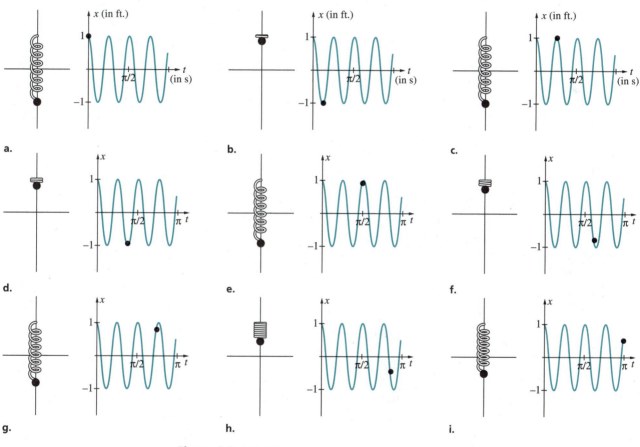

Figure 5.2   (a)–(i)

What is the maximum displacement of the object in Example 3 from the equilibrium position?

| Example 4 |

An object weighing 2 lb stretches a spring 1.5 in. (a) Determine the function $x(t)$ that describes the displacement of the object if it is released with a downward initial velocity of 32 ft/s from 12 in. above the equilibrium position. (b) At what value of t does the object first pass through the equilibrium position?

Solution   (a) We begin by determining the spring constant. Because the force $F = 2$ stretches the spring $\frac{3}{2}$ in. $\times \dfrac{1\ \text{ft}}{12\ \text{in.}} = \frac{1}{8}$ ft, k is found by solving $2 = k \cdot \frac{1}{8}$. Hence, $k = 16$ lb/ft. With $F = mg$, we have $m = \frac{2}{32} = \frac{1}{16}$ slug. The differential equation that models this situation is $\dfrac{1}{16}\dfrac{d^2x}{dt^2} + 16x = 0$ or $\dfrac{d^2x}{dt^2} + 256x = 0$. Because 12 inches is equivalent to 1 foot, the initial position above the equilibrium (in the negative direction) is $x(0) = -1$. The downward initial velocity (in the positive direction) is $\dfrac{dx}{dt}(0) = 32$. Therefore, we must solve the initial-value problem

$$\begin{cases} \dfrac{d^2x}{dt^2} + 256x = 0 \\ x(0) = -1,\ \dfrac{dx}{dt}(0) = 32 \end{cases}$$

Because the characteristic equation is $r^2 + 256 = 0$, the roots are $r = \pm 16i$. A general solution is

$$x(t) = c_1 \cos 16t + c_2 \sin 16t$$

with derivative $\dfrac{dx}{dt}(t) = -16c_1 \sin 16t + 16c_2 \cos 16t$. Application of the initial conditions then yields $x(0) = c_1 \cos 0 + c_2 \sin 0 = c_1 = -1$ and

$$\dfrac{dx}{dt}(0) = -16c_1 \sin 0 + 16c_2 \cos 0 = 16c_2 = 32,\ \text{so } c_2 = 2.$$

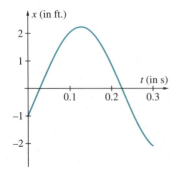

Figure 5.3

The position function is given by $x(t) = -\cos 16t + 2 \sin 16t$. (See Figure 5.3.) (b) In order to determine when the object first passes through its equilibrium position, we solve the equation $x(t) = -\cos 16t + 2 \sin 16t = 0$ or $\tan 16t = \frac{1}{2}$. Therefore, $t = \frac{1}{16}\tan^{-1}(.5) \approx 0.03$ second, which appears to be a reasonable approximation based on the graph of $x(t)$.

Approximate the second time the mass considered in Example 4 passes through the equilibrium position.

A general formula for the solution of the initial-value problem

$$\begin{cases} m\dfrac{d^2x}{dt^2} + kx = 0 \\ x(0) = \alpha, \quad \dfrac{dx}{dt}(0) = \beta \end{cases}$$

is

$$x(t) = \alpha \cos \omega t + \frac{\beta}{\omega} \sin \omega t \quad \text{where} \quad \omega = \sqrt{\frac{k}{m}}.$$

Through the use of the trigonometric identity $\cos (a + b) = \cos a \cos b - \sin a \sin b$, we can write $x(t)$ in terms of a cosine function with a phase shift. First, let

$$x(t) = A \cos (\omega t - \phi).$$

Then $x(t) = A \cos \omega t \cos \phi + A \sin \omega t \sin \phi$. Comparing the functions

$$x(t) = \alpha \cos \omega t + \frac{\beta}{\omega} \sin \omega t \quad \text{and} \quad x(t) = A \cos \omega t \cos \phi + A \sin \omega t \sin \phi$$

indicates that

$$A \cos \phi = \alpha \quad \text{and} \quad A \sin \phi = \frac{\beta}{\omega}.$$

Thus,

$$\cos \phi = \frac{\alpha}{A} \quad \text{and} \quad \sin \phi = \frac{\beta}{A\omega}.$$

Because $\cos^2 \phi + \sin^2 \phi = 1$, $\left(\dfrac{\alpha}{A}\right)^2 + \left(\dfrac{\beta}{A\omega}\right)^2 = 1$. Therefore, the amplitude of the solution is

$$A = \sqrt{\alpha^2 + \frac{\beta^2}{\omega^2}}$$

and

$$x(t) = \sqrt{\alpha^2 + \frac{\beta^2}{\omega^2}} \cos (\omega t - \phi),$$

where $\phi = \cos^{-1}\left(\dfrac{\alpha}{\sqrt{\alpha^2 + \dfrac{\beta^2}{\omega^2}}}\right)$ and $\omega = \sqrt{\dfrac{k}{m}}$.

Note that the period of $x(t)$ is

$$T = 2\frac{\pi}{\omega} = 2\pi\sqrt{\frac{m}{k}}.$$

In many cases, questions about the displacement function are more easily answered if the solution is written in this form.

How does an increase in the magnitude (absolute value) of the initial posi-tion and initial velocity affect the amplitude of the resulting motion of the spring-mass system? From experience, does this agree with the actual physical situation?

Example 5

A 4 kg mass stretches a spring 0.392 m. (a) Determine the displacement function if the mass is released from 1 m below the equilibrium position with a downward initial velocity of 10 m/s. (b) What is the maximum displacement of the mass? (c) What is the approximate period of the displacement function?

Solution   (a) Because the mass of the object (in metric units) is $m = 4$ kg, we use this with $F = mg$ to determine the force. We first compute

$$F = (4)(9.8) = 39.2 \text{ Newtons.}$$

We then find the spring constant with $39.2 = k \cdot 0.392$, so $k = 100$ N/m. The differential equation that models this spring-mass system is $4\dfrac{d^2x}{dt^2} + 100x = 0$ or $\dfrac{d^2x}{dt^2} + 25x = 0$. The initial position is $x(0) = 1$, while the initial velocity is $\dfrac{dx}{dt}(0) = 10$. We must solve the initial-value problem

$$\begin{cases} \dfrac{d^2x}{dt^2} + 25x = 0 \\ x(0) = 1, \; \dfrac{dx}{dt}(0) = 10 \end{cases}$$

either directly or with the general formula obtained above. Using the general formula with $\alpha = 1$, $\beta = 10$, and $\omega = \sqrt{100/4} = 5$, we have

$$x(t) = \sqrt{\alpha^2 + \frac{\beta^2}{\omega^2}} \cos(\omega t - \phi) = \sqrt{1^2 + \frac{10^2}{5^2}} \cos(5t - \phi)$$

$$= \sqrt{5} \cos(5t - \phi),$$

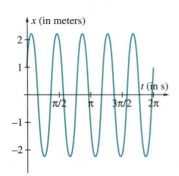

Figure 5.4 Simple Harmonic Motion

where $\phi = \cos^{-1}\dfrac{1}{\sqrt{5}} \approx 1.11$ rad, which we graph in Figure 5.4.

(b) From our knowledge of trigonometric functions, we know that the maximum value of $x(t) = \sqrt{5} \cos(5t - \phi)$ is $x = \sqrt{5}$. Therefore, the maximum displacement of the mass from its equilibrium position is $\sqrt{5} \approx 2.236$ meters.

(c) The period of this trigonometric function is $T = \dfrac{2\pi}{\omega} = \dfrac{2\pi}{5}$. The mass returns to its initial position every $\dfrac{2\pi}{5} \approx 1.257$ seconds.

IN TOUCH WITH TECHNOLOGY

1. An object with mass $m = 1$ slug is attached to a spring with spring constant $k = 4$ lb/ft. **(a)** Determine the displacement function of the object if $x(0) = \alpha$ and $x'(0) = 0$. Graph the solution for $\alpha = 1, 4, -2$. How does varying the value of α affect the solution? Does it change the values of t at which the mass passes through the equilibrium position? **(b)** Determine the displacement function of the object if $x(0) = 0$ and $x'(0) = \beta$. Graph the solution for $\beta = 1, 4, -2$. How does varying the value of β affect the solution? Does it change the values of t at which the mass passes through the equilibrium position?

2. An object of mass $m = 4$ slugs is attached to a spring with spring constant $k = 20$ lb/ft. If the object is released from 7 in. above its equilibrium with a downward initial velocity of 2.5 ft/s, find **(a)** the maximum displacement from the equilibrium position; **(b)** the time at which the object first passes through its equilibrium position; **(c)** the period of the motion.

3. If the spring in Problem 2 has the spring constant $k = 16$ lb/ft, what is the maximum displacement from the equilibrium position? How does this compare to the result in the previous exercise? Determine the maximum displacement if $k = 24$ lb/ft. Do these results agree with those that would be obtained with the general formula $A = \sqrt{\alpha^2 + \dfrac{\beta^2}{\omega^2}}$?

4. An object of mass $m = 3$ slugs is attached to a spring with spring constant $k = 15$ lb/ft. If the object is released from 9 in. below its equilibrium with a downward initial velocity of 1 ft/s, find **(a)** the maximum displacement from the equilibrium position; **(b)** the time at which the object first passes through its equilibrium position; **(c)** the period of the motion.

5. If the mass in Problem 4 is $m = 4$ slugs, find the maximum displacement from the equilibrium position. Compare this result to that obtained with $m = 3$ slugs.

EXERCISES 5.1

In Exercises 1–4, determine the mass m and the spring constant k for the given spring-mass system.

$$\begin{cases} m\dfrac{d^2x}{dt^2} + kx = 0 \\ x(0) = \alpha, \dfrac{dx}{dt}(0) = \beta \end{cases}$$

Interpret the initial conditions. (Assume the English system.)

1. $4\dfrac{d^2x}{dt^2} + 9x = 0$, $x(0) = -1$, $x'(0) = 0$

2. $2\dfrac{d^2x}{dt^2} + 128x = 0$, $x(0) = -0.5$, $x'(0) = 1$

*3. $\dfrac{1}{4}\dfrac{d^2x}{dt^2} + 16x = 0$, $x(0) = 0.75$, $x'(0) = -2$

4. $\dfrac{1}{25}\dfrac{d^2x}{dt^2} + 4x = 0$, $x(0) = -\dfrac{1}{4}$, $x'(0) = 1$

In Exercises 5–10, express the solution of the initial-value problem in the form $x(t) = \sqrt{\alpha^2 + \dfrac{\beta^2}{\omega^2}}\cos(\omega t - \phi)$. What is the period and amplitude of the solution?

5. $\dfrac{d^2x}{dt^2} + x = 0$, $x(0) = 3$, $x'(0) = -4$

6. $\dfrac{d^2x}{dt^2} + 4x = 0$, $x(0) = 1$, $x'(0) = 1$

*7. $\dfrac{1}{16}\dfrac{d^2x}{dt^2} + x = 0$, $x(0) = -2$, $x'(0) = 1$

8. $\dfrac{d^2x}{dt^2} + 256x = 0$, $x(0) = 2$, $x'(0) = 4$

9. $\dfrac{d^2x}{dt^2} + 9x = 0$, $x(0) = \dfrac{1}{3}$, $x'(0) = -1$

10. $10\dfrac{d^2x}{dt^2} + \dfrac{1}{10}x = 0$, $x(0) = -5$, $x'(0) = 1$

11. A 16 lb object stretches a spring 6 in. If the object is lowered 1 ft below the equilibrium position and released, determine the displacement of the object. What is the maximum displacement of the object? When does it occur?

12. A 4 lb weight stretches a spring 1 ft. A 16-lb weight is then attached to the spring, and it comes to rest in its equilibrium position. If it is then put into motion with a downward initial velocity of 2 ft/s, determine the displacement of the mass. What is the maximum displacement of the object? When does it occur?

*13. A 6 lb object stretches a spring 6 in. If the object is lifted 3 in. above the equilibrium position and released, determine the time required for the mass to return to its equilibrium position. What is the displacement of the object at $t = 5$ seconds? If the object is released from its equilibrium position with a downward initial velocity of 1 ft/s, determine the time required for the object to return to its equilibrium position.

14. A 16 lb weight stretches a spring 8 in. If the weight is lowered 4 in below the equilibrium position and released, find the time required for the weight to return to the equilibrium position. What is the displacement of the weight at $t = 4$ seconds? If the weight is released from its equilibrium position with an upward initial velocity of 2 ft/s, determine the time required for the weight to return to the equilibrium position.

15. Solve the initial-value problem $\dfrac{d^2x}{dt^2} + kx = 0$, $x(0) = -1$, $x'(0) = 0$ for values of $k = 1$, 4, and 9. Comment on the effect that k has on the resulting motion.

16. Solve the initial-value problem $\dfrac{d^2x}{dt^2} + \dfrac{4}{m}x = 0$, $x(0) = -1$, $x'(0) = 0$ for values of $m = 1$, 4, and 9. Comment on the effect that m has on the resulting motion.

*17. Suppose that a 1 lb object stretches a spring $\frac{1}{8}$ ft. The object is pulled downward and released from a position b ft beneath its equilibrium with an upward initial velocity of 1 ft/s. Determine the value of b so that the maximum displacement is 2 ft.

18. Suppose that a 70 gram mass stretches a spring 5 cm and that the mass is pulled downward and released from a position b units beneath its equilibrium with an upward initial velocity of 10 cm/s. Determine the value of b so that the mass first returns to its equilibrium at $t = 1$ s. ($g \approx 980$ cm/s^2)

19. The period of the motion of an undamped spring-mass system is $\frac{\pi}{2}$ seconds. Find the mass m if the spring constant is $k = 32$ lb/ft.

20. Find the period of the motion of an undamped spring-mass system if the mass of the object is 4 kg and the spring constant is $k = 0.25$ N/m.

21. If the motion of an object satisfies the initial-value problem
$$\begin{cases} m\dfrac{d^2x}{dt^2} + kx = 0 \\ x(0) = \alpha, \dfrac{dx}{dt}(0) = \beta \end{cases},$$
find the maximum velocity of the object.

22. Show that the solution to the initial-value problem
$$\begin{cases} m\dfrac{d^2x}{dt^2} + kx = 0 \\ x(0) = \alpha, \dfrac{dx}{dt}(0) = \beta \end{cases}$$
can be written as $x(t) = u(t) + v(t)$, where u and v satisfy the same differential equation as x, u satisfies the initial conditions $u(0) = \alpha$, $\dfrac{du}{dt}(0) = 0$, and v satisfies the initial conditions $v(0) = 0$, $\dfrac{dv}{dt}(0) = \beta$.

23. (**Archimedes' Principle**) Suppose that an object of mass m is submerged (either partially or totally) in a liquid of density ρ. Archimedes' Principle states that a body in liquid experiences a buoyant upward force equal to the weight of the liquid displaced by the body. The object is in equilibrium when the buoyant force of the displaced liquid equals the force of gravity on the object. (See Figure 5.5.) Consider the cylinder of radius r and height H of which h units of the height is submerged at equilibrium.

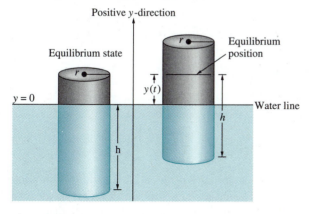

Figure 5.5

(a) Show that the weight of liquid displaced at equilibrium is $\pi r^2 h \rho$. Therefore, at equilibrium $\pi r^2 h \rho = mg$.

(b) Let $y(t)$ represent the vertical displacement of the cylinder from equilibrium. Show that when the cylinder is raised out of the liquid, the downward force is $\pi r^2 [h - y(t)] \rho$.

(c) Use Newton's second law of motion to show that
$$m \frac{d^2 y}{dt^2} = \pi r^2 [h - y(t)] \rho - mg. \quad \text{Simplify this}$$
equation to obtain a second-order equation that models this situation.

24. Determine if the cylinder can float in a deep pool of water $(\rho \approx 62.4 \text{ lb/ft}^3)$ using the given radius r, height H, and weight W: **(a)** $r = 3$ in., $H = 12$ in., $W = 5$ lb; **(b)** $r = 4$ in., $H = 8$ in., $W = 20$ lb; **(c)** $r = 6$ in., $H = 9$ in., $W = 50$ lb.

*25. Determine the motion of the cylinder of weight 512 lb, radius $r = 1$ ft, and height $H = 4$ ft if it is released with 3 ft of its height above the water $(\rho = 62.5 \text{ lb/ft}^3)$ with a downward initial velocity of 3 ft/s. What is the maximum displacement of the cylinder from its equilibrium?

26. Consider the cylinder of radius $r = 3$ in., height $H = 12$ in., and weight 10 lb. Show that the portion of the cylinder submerged in water of density (ρ) 62.5 lb/ft^3 is $h \approx 0.815$ ft ≈ 9.78 in. Find the motion of the cylinder if it is released with 1.22 in. of its height above the water with no initial velocity. (*Hint:* At $t = 0$, $h \approx$ 9.78 in., so 0.22 in. is above the water. Therefore, the initial position is $y(0) = 1$.)

5.2 Damped Motion

Because the differential equation derived in Section 5.1 disregarded all retarding forces acting on the motion of the mass, a more realistic model is needed. Studies in mechanics reveal that the resistive force due to **damping** is a function of the velocity of the motion. For $c > 0$, functions such as

$$F_R = c \frac{dx}{dt}, \quad F_R = c \left(\frac{dx}{dt} \right)^3, \quad \text{and } F_R = c \operatorname{sgn}\left(\frac{dx}{dt} \right), \quad \text{where } \operatorname{sgn}\left(\frac{dx}{dt} \right) = \begin{cases} 1, & \dfrac{dx}{dt} > 0 \\[2mm] 0, & \dfrac{dx}{dt} = 0 \\[2mm] -1, & \dfrac{dx}{dt} < 0 \end{cases}$$

can be used to represent the damping force. We follow procedures similar to those used in Section 5.1 to model simple harmonic motion, and to determine a differential equation that models the spring-mass system which includes damping. Assuming that $F_R = c\dfrac{dx}{dt}$, we have after summing the forces acting on the spring-mass system,

$$m\frac{d^2x}{dt^2} = -c\frac{dx}{dt} - kx \quad \text{or} \quad m\frac{d^2x}{dt^2} + c\frac{dx}{dt} + kx = 0.$$

The displacement function is found by solving the initial-value problem

$$\begin{cases} m\dfrac{d^2x}{dt^2} + c\dfrac{dx}{dt} + kx = 0 \\ x(0) = \alpha, \ \dfrac{dx}{dt}(0) = \beta \end{cases}.$$

From our experience with second-order ordinary differential equations with constant coefficients in Chapter 4, the solutions to initial-value problems of this type depend on the values of m, k, and c. Suppose we assume (as we did in Section 5.1) that solutions of the differential equation have the form $x(t) = e^{rt}$. Then the characteristic equation is $mr^2 + cr + k = 0$ with solutions

$$r = \frac{-c \pm \sqrt{c^2 - 4mk}}{2m}.$$

The solution depends on the value of the quantity $c^2 - 4mk$. In fact, problems of this type are classified by the value of $c^2 - 4mk$ as follows.

Case 1: $c^2 - 4mk > 0$

This situation is said to be **overdamped,** because the damping coefficient c is large in comparison with the spring constant k.

Case 2: $c^2 - 4mk = 0$

This situation is described as **critically damped,** because the resulting motion is oscillatory with a slight decrease in the damping coefficient c.

Case 3: $c^2 - 4mk < 0$

This situation is called **underdamped,** because the damping coefficient c is small in comparison with the spring constant k.

Example 1

An 8 lb object is attached to a spring of length 4 ft. At equilibrium, the spring has length 6 ft. Determine the displacement function $x(t)$ if $F_R = 2\dfrac{dx}{dt}$ and (a) the object is released from the equilibrium position with a downward initial velocity of 1 ft/sec; (b) the object released 6 in. above the equilibrium position with an initial velocity of 5 ft/sec in the downward direction.

Solution   Notice that $s = 6 - 4 = 2$ ft and that $F = 8$ lb. We find the spring constant with $8 = k \cdot 2$, so $k = 4$ lb/ft. Also, the mass of the object is $m = \frac{8}{32} = \frac{1}{4}$ slug. The differential equation that models this spring-mass system is

$$\frac{1}{4}\frac{d^2x}{dt^2} + 2\frac{dx}{dt} + 4x = 0 \text{ or } \frac{d^2x}{dt^2} + 8\frac{dx}{dt} + 16x = 0.$$

The corresponding characteristic equation is

$$r^2 + 8r + 16 = (r + 4)^2 = 0,$$

so $r = -4$ is a root of multiplicity two. A general solution is

$$x(t) = c_1 e^{-4t} + c_2 t e^{-4t}.$$

Differentiating yields

$$\frac{dx}{dt}(t) = (-4c_1 + c_2)e^{-4t} - 4c_2 t e^{-4t}.$$

(a) The initial conditions in this case are $x(0) = 0$ and $\frac{dx}{dt}(0) = 1$, so $c_1 = 0$ and $-4c_1 + c_2 = 1$. Thus, $c_2 = 1$ and the solution is $x(t) = te^{-4t}$, which is shown in Figure 5.6. Notice that $x(t)$ is always *positive*, so the object is always *below* the equilibrium position and approaches zero (the equilibrium position) as t approaches infinity. Because of the resistive force due to damping, the object is not allowed to pass through its equilibrium position.

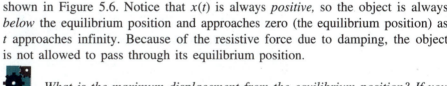

   What is the maximum displacement from the equilibrium position? If you were looking at this spring, would you perceive its motion?

(b) In this case, $x(0) = -\dfrac{1}{2}$ and $\dfrac{dx}{dt}(0) = 5$. When we apply these initial conditions, we find that $x(0) = c_1 = -\dfrac{1}{2}$ and $\dfrac{dx}{dt}(0) = -4c_1 + c_2 = -4\left(-\dfrac{1}{2}\right) + c_2 = 5$.

Hence, $c_2 = 3$, and the solution is $x(t) = -\dfrac{1}{2}e^{-4t} + 3te^{-4t}$. This function, which is graphed in Figure 5.7, indicates the importance of the initial conditions on the resulting motion. In this case, the displacement is negative (above the equilibrium position) initially, but the positive initial velocity causes the function to become positive (below the equilibrium position) before approaching zero.

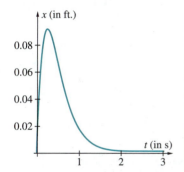

Figure 5.6   Critically Damped Motion

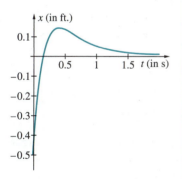

Figure 5.7   Critically Damped Motion

   If the object in Example 1 is released from any point below its equilibrium position with an upward initial velocity of 1 ft/sec, can it possibly pass through its equilibrium position? If the object is released from below its equilibrium position with any upward initial velocity, can it possibly pass through the equilibrium position?

Example 2

A 32 lb object stretches a spring 8 ft. If the resistive force due to damping is $F_R = 5\dfrac{dx}{dt}$, determine the displacement function if the object is released from 1 ft below the equilibrium position with (a) an upward velocity of 1 ft/sec; (b) an upward velocity of 6 ft/sec.

Solution   (a) Because $F = 32$ lb, the spring constant is found with $32 = k(8)$, so $k = 4$ lb/ft. Also, $m = \frac{32}{32} = 1$ slug. The differential equation that models this situation is $\dfrac{d^2x}{dt^2} + 5\dfrac{dx}{dt} + 4x = 0$. The initial position is $x(0) = 1$ and the initial velocity in (a) is $\dfrac{dx}{dt}(0) = -1$. The characteristic equation of the differential equation is

$$r^2 + 5r + 4 = (r + 1)(r + 4) = 0$$

with roots $r_1 = -1$ and $r_2 = -4$. A general solution is

$$x(t) = c_1 e^{-t} + c_2 e^{-4t} \quad \text{and} \quad \frac{dx}{dt}(t) = -c_1 e^{-t} - 4c_2 e^{-4t}.$$

(Because $c^2 - 4mk = 5^2 - 4(1)(4) = 9 > 0$, the system is overdamped.) Application of the initial conditions yields the system of equations $\begin{cases} c_1 + c_2 = 1 \\ -c_1 - 4c_2 = -1 \end{cases}$ with solution $\{c_1 = 1, c_2 = 0\}$, so the solution to the initial-value problem is $x(t) = e^{-t}$. The graph of $x(t)$ is shown in Figure 5.8(a). Notice that this solution is always positive and, due to the damping, approaches zero as t approaches infinity. Therefore, the object is always below its equilibrium position. (b) Using the initial velocity $\dfrac{dx}{dt}(0) = -6$, we solve the system $\begin{cases} c_1 + c_2 = 1 \\ -c_1 - 4c_2 = -6 \end{cases}$, which has solution $\{c_1 = -\frac{2}{3}, c_2 = \frac{5}{3}\}$. The solution of the initial-value problem is $x(t) = \frac{5}{3}e^{-4t} - \frac{2}{3}e^{-t}$. The graph of this function is shown in Figure 5.8(b). As in Example 1, these results indicate the importance of the initial conditions on the resulting motion. In this case, the displacement is positive (below its equilibrium) initially, but the larger negative initial velocity causes the function to become negative (above its equilibrium) before approaching zero. Therefore, we see that the initial velocity in part (b) causes the mass to pass through its equilibrium position.

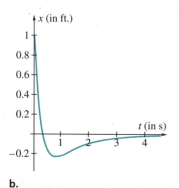

a.

b.

Figure 5.8   (a) and (b)

Can you find a minimum value for the upward initial velocity v_0 in Example 2 so that the object passes through its equilibrium position for all values greater than v_0?

Example 3	A 16 lb object stretches a spring 2 ft. Determine the displacement $x(t)$ if the resistive force due to damping is $F_R = \dfrac{1}{2}\dfrac{dx}{dt}$ and the object is released from the equilibrium position with a downward velocity of 1 ft/s.

Solution   Because $F = 16$ lb, the spring constant is determined with $16 = k \cdot 2$. Hence, $k = 8$ lb/ft. Also, $m = \frac{16}{32} = \frac{1}{2}$ slug. Therefore, the differential equation is $\dfrac{1}{2}\dfrac{d^2x}{dt^2} + \dfrac{1}{2}\dfrac{dx}{dt} + 8x = 0$ or $\dfrac{d^2x}{dt^2} + \dfrac{dx}{dt} + 16x = 0$. The initial position is $x(0) = 0$ and the initial velocity is $\dfrac{dx}{dt}(0) = 1$. We must solve the initial-value problem

$$\begin{cases} \dfrac{d^2x}{dt^2} + \dfrac{dx}{dt} + 16x = 0 \\[2mm] x(0) = 0,\ \dfrac{dx}{dt}(0) = 1 \end{cases}$$

A general solution of $\dfrac{d^2x}{dt^2} + \dfrac{dx}{dt} + 16x = 0$ is

$$x(t) = e^{-t/2}\left(c_1 \cos \frac{3\sqrt{7}\,t}{2} + c_2 \sin \frac{3\sqrt{7}\,t}{2}\right).$$

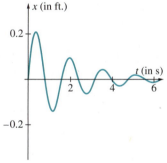

Notice that $c^2 - 4mk = \left(\dfrac{1}{2}\right)^2 - 4\left(\dfrac{1}{2}\right)(8) = -\dfrac{63}{4} < 0$, so the spring-mass system is underdamped. Because

$$\frac{dx}{dt}(t) = -\frac{1}{2}e^{-t/2}\left(c_1 \cos \frac{3\sqrt{7}\,t}{2} + c_2 \sin \frac{3\sqrt{7}\,t}{2}\right)$$
$$+ e^{-t/2}\frac{3\sqrt{7}}{2}\left(-c_1 \sin \frac{3\sqrt{7}\,t}{2} + c_2 \cos \frac{3\sqrt{7}\,t}{2}\right),$$

a.

application of the initial conditions yields $\left\{c_1 = 0,\ c_2 = \dfrac{2}{3\sqrt{7}}\right\}$. Therefore, the solution is $x(t) = \dfrac{2}{3\sqrt{7}}e^{-t/2}\sin \dfrac{3\sqrt{7}\,t}{2}$.

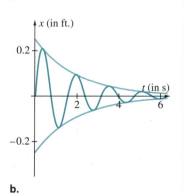

Solutions of this type have several interesting properties. First, the trigonometric component of the solution causes the motion to *oscillate*. Also, the exponential portion forces the solution to approach zero as t approaches infinity. These qualities are illustrated in the graph of $x(t)$ in Figure 5.9(a). Physically, the position of the mass in this case oscillates about the equilibrium position and eventually comes to rest in the equilibrium position. Of course, with our model the displacement function $x(t) \to 0$ as $t \to \infty$, but there are no values $t = T$ such that $x(t) = 0$ for $t > T$ as we might expect from the physical situation. Our model only approximates the behavior

b.

Figure 5.9   (a) and (b)

of the mass. Notice also that the solution is bounded above and below by the exponential term of the solution $e^{-t/2}$ and its reflection through the horizontal axis, $-e^{-t/2}$. This is illustrated with the simultaneous display of these functions in Figure 5.9(b); the motion of the spring is illustrated in Figure 5.10.

Notice that when the system is underdamped as in Example 3, the amplitude (or **damped amplitude**) of the solution decreases as $t \to \infty$. The time interval between two successive local maxima (or minima) of $x(t)$ is called the **quasiperiod** of the solution.

 Approximate the quasiperiod of the solution in Example 3.

In Section 5.1 we developed a general formula for the displacement function. We can do the same for systems that involve damping. Assuming that the spring-mass system is underdamped, the differential equation $m\dfrac{d^2x}{dt^2} + c\dfrac{dx}{dt} + kx = 0$ has characteristic equation $mr^2 + cr + k = 0$ with roots $r = \dfrac{-c \pm i\sqrt{4mk - c^2}}{2m}$. If we let $\rho = \dfrac{c}{2m}$ and $\mu = \dfrac{\sqrt{4mk - c^2}}{2m}$ then $r = -\rho \pm \mu i$ and a general solution is $x(t) = e^{-\rho t}(c_1 \cos \mu t + c_2 \sin \mu t)$. Applying the initial conditions $x(0) = \alpha$ and $x'(0) = \beta$ yields the solution $x(t) = e^{-\rho t}\left(\alpha \cos \mu t + \dfrac{\beta + \alpha\rho}{\mu} \sin \mu t \right)$, which can be written as

$$x(t) = Ae^{-\rho t} \cos (\mu t - \phi).$$

Then, $x(t) = e^{-\rho t}(A \cos \mu t \cos \phi + A \sin \mu t \sin \phi)$. Comparing the functions

$$x(t) = e^{-\rho t}\left(\alpha \cos \mu t + \dfrac{\beta + \alpha\rho}{\mu} \sin \mu t \right)$$

and

$$x(t) = e^{-\rho t}(A \cos \mu t \cos \phi + A \sin \mu t \sin \phi),$$

we have

$$A \cos \phi = \alpha \qquad \text{and} \qquad A \sin \phi = \dfrac{\beta + \alpha\rho}{\mu}$$

which indicate that

$$\cos \phi = \dfrac{\alpha}{A} \qquad \text{and} \qquad \sin \phi = \dfrac{\beta + \alpha\rho}{A\mu}.$$

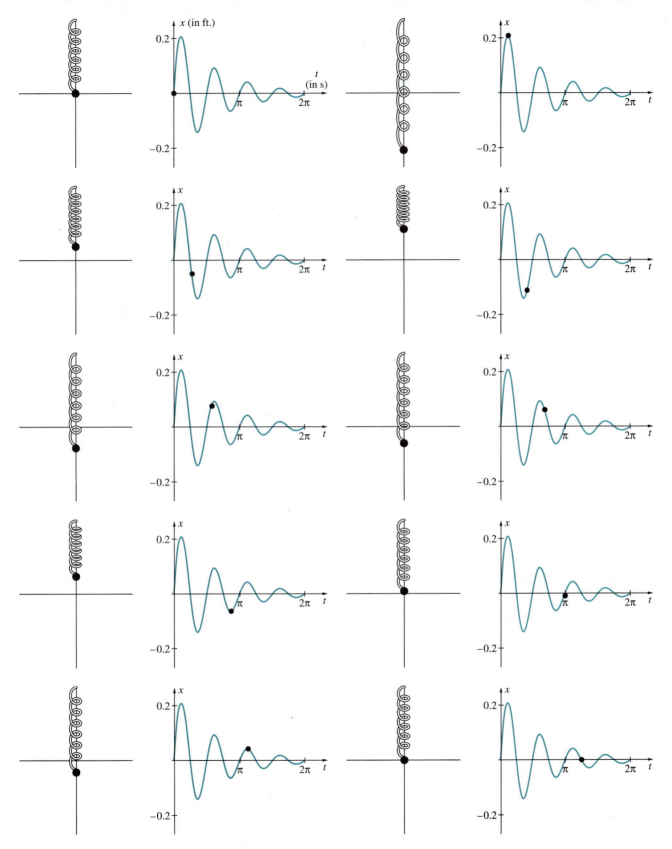

Figure 5.10

Now, $\cos^2 \phi + \sin^2 \phi = 1$, so $\dfrac{\alpha}{A^2} + \left[\dfrac{\beta + \alpha\rho}{A\mu}\right]^2 = 1$. Therefore, the decreasing amplitude of the solution is

$$A = \sqrt{\alpha^2 + \left[\dfrac{\beta + \alpha\rho}{\mu}\right]^2}$$

and

$$x(t) = e^{-\rho t}\sqrt{\alpha^2 + \left[\dfrac{\beta + \alpha\rho}{\mu}\right]^2}\, \cos(\mu t - \phi),$$

where $\phi = \cos^{-1}\left(\dfrac{\alpha}{\sqrt{\alpha^2 + \left[\dfrac{\beta + \alpha\rho}{\mu}\right]^2}}\right)$.

The quantity

$$\dfrac{2\pi}{\mu} = \dfrac{4\pi m}{\sqrt{4km - c^2}}$$

is called the **quasiperiod** of the function. (Note that functions of this type are *not* periodic.) We can also determine the times at which the mass passes through the equilibrium position from the general formula given here. We do this by setting the argument equal to odd multiples of $\pi/2$, because the cosine function is zero at these values. The mass passes through the equilibrium position at

$$t = \dfrac{\left[(2n + 1)\dfrac{\pi}{2} + \phi\right]}{\mu} = \dfrac{[m(2n + 1)\pi + 2m\phi]}{\sqrt{4mk - c^2}}, \, n = 0, \pm 1, \pm 2, \ldots$$

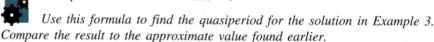 *Use this formula to find the quasiperiod for the solution in Example 3. Compare the result to the approximate value found earlier.*

Example 4

An object of mass 1 slug is attached to a spring with spring constant $k = 13$ lb/ft and is subjected to a resistive force of $F_R = 4\dfrac{dx}{dt}$ due to damping. If the initial position is $x(0) = 1$ and the initial velocity is $\dfrac{dx}{dt}(0) = 1$, determine the quasiperiod of the solution and find the values of t at which the object passes through the equilibrium position.

Solution   The initial-value problem that models this situation is

$$\begin{cases} \dfrac{d^2x}{dt^2} + 4\dfrac{dx}{dt} + 13x = 0 \\ x(0) = 1, \dfrac{dx}{dt}(0) = 1 \end{cases}$$

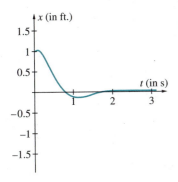

Figure 5.11 $x(t) = \sqrt{2}e^{-2t}\cos(3t - \phi)$, $\phi = \pi/4$

The characteristic equation is $r^2 + 4r + 13 = 0$ with roots $r_{1,2} = -2 \pm 3i$. A general solution is $x(t) = e^{-2t}(c_1 \cos 3t + c_2 \sin 3t)$ with derivative

$$\frac{dx}{dt}(t) = -2e^{-2t}(c_1 \cos 3t + c_2 \sin 3t) + 3e^{-2t}(-c_1 \sin 3t + c_2 \cos 3t).$$

Application of the initial conditions yields the solution $x(t) = e^{-2t}(\cos 3t + \sin 3t)$, which can be written as

$$x(t) = e^{-2t}\sqrt{\alpha^2 + \left[\frac{(\beta + \alpha\rho)}{\mu}\right]^2}\cos(\mu t - \phi) =$$

$$e^{-2t}\sqrt{1^2 + \left[\frac{1+2}{3}\right]^2}\cos(3t - \phi) = \sqrt{2}e^{-2t}\cos(3t - \phi).$$

The quasiperiod is $\dfrac{2\pi}{\mu} = \dfrac{4\pi m}{\sqrt{4km - c^2}} = \dfrac{4\pi}{\sqrt{4(13) - 4^2}} = \dfrac{2\pi}{3}$. The values of t at which the mass passes through the equilibrium are $t = \dfrac{m(2n+1)\pi + 2m\phi}{\sqrt{4mk - c^2}} = \dfrac{(2n+1)\pi + 2\phi}{6}$, $n = 0, 1, 2, \ldots$ where $\phi = \cos^{-1}\dfrac{1}{\sqrt{2}} = \dfrac{\pi}{4}$. (See Figure 5.11.)

IN TOUCH WITH TECHNOLOGY

1. Determine how the value of c affects the solution of the initial-value problem

$$\begin{cases} \dfrac{d^2x}{dt^2} + c\dfrac{dx}{dt} + 6x = 0 \\ x(0) = 0, x'(0) = 1 \end{cases},$$

where $c = 2\sqrt{6}$, $4\sqrt{6}$, and $\sqrt{6}$.

2. In Problem 1, consider the solution that results using the damping coefficient that produces critical damping with $x'(0) = -1$, 0, 1, and 2. In which case does the object pass through its equilibrium position? When does it pass through the equilibrium position?

3. Using the values of $x'(0)$ and c in Problem 2, in addition suppose that the equilibrium position is 1 unit above the floor. Does the object come into contact with the floor in any of the cases? If so, when?

4. Solve the initial-value problem $4x'' + cx' + 5x = 0$, $x(0) = 3$, $x'(0) = 0$ using $c = 1$, -4, and 0. Plot the solutions and compare them.

5. Solve the initial-value problem $x'' + cx' + x = 0$, $x(0) = -1$, $x'(0) = 3$ for values of $c = 2$ and $c = \sqrt{8}$. (Notice the effect of the coefficient of damping on the solution to the corresponding critically damped and overdamped equation.)

In Exercises 1–4, determine the mass m (slugs), spring constant k (lb/ft), and damping coefficient c in $F_R = c\dfrac{dx}{dt}$ for the given spring-mass system

$$\begin{cases} m\dfrac{d^2x}{dt^2} + c\dfrac{dx}{dt} + kx = 0 \\ x(0) = \alpha, \dfrac{dx}{dt}(0) = \beta \end{cases}$$

Describe the initial conditions. (Assume the English system.)

1. $\dfrac{d^2x}{dt^2} + 4\dfrac{dx}{dt} + 3x = 0$, $x(0) = 0$, $x'(0) = -4$

2. $\dfrac{1}{32}\dfrac{d^2x}{dt^2} + 2\dfrac{dx}{dt} + x = 0$, $x(0) = 1$, $x'(0) = 0$

*3. $\dfrac{1}{4}\dfrac{d^2x}{dt^2} + 2\dfrac{dx}{dt} + x = 0$, $x(0) = -0.5$, $x'(0) = 1$

4. $4\dfrac{d^2x}{dt^2} + 2\dfrac{dx}{dt} + 8x = 0$, $x(0) = 0$, $x'(0) = 2$

In Exercises 5–8, express the solution of the initial-value problem in the form

$$x(t) = e^{-\rho t}\sqrt{\alpha^2 + \left[\dfrac{\beta + \alpha\rho}{\mu}\right]^2}\cos(\mu t - \phi).$$

In each case, find the quasiperiod and the time at which the mass first passes through its equilibrium position.

5. $\dfrac{d^2x}{dt^2} + 4\dfrac{dx}{dt} + 13x = 0$, $x(0) = 1$, $x'(0) = -1$

6. $\dfrac{d^2x}{dt^2} + 4\dfrac{dx}{dt} + 20x = 0$, $x(0) = 1$, $x'(0) = 2$

*7. $\dfrac{d^2x}{dt^2} + 2\dfrac{dx}{dt} + 26x = 0$, $x(0) = 1$, $x'(0) = 1$

8. $\dfrac{d^2x}{dt^2} + 10\dfrac{dx}{dt} + 41x = 0$, $x(0) = 3$, $x'(0) = -2$

In Exercises 9–16, solve the initial-value problem. Classify each as overdamped or critically damped. Determine if the mass passes through its equilibrium position and if so, when.

Determine the maximum displacement of the object from the equilibrium position.

9. $\dfrac{d^2x}{dt^2} + 8\dfrac{dx}{dt} + 15x = 0$, $x(0) = 0$, $x'(0) = 1$

10. $\dfrac{d^2x}{dt^2} + 7\dfrac{dx}{dt} + 12x = 0$, $x(0) = -1$, $x'(0) = 4$

*11. $\dfrac{d^2x}{dt^2} + \dfrac{3}{2}\dfrac{dx}{dt} + \dfrac{1}{2}x = 0$, $x(0) = -1$, $x'(0) = 2$

12. $\dfrac{d^2x}{dt^2} + 5\dfrac{dx}{dt} + 4x = 0$, $x(0) = 0$, $x'(0) = 5$

13. $\dfrac{d^2x}{dt^2} + 8\dfrac{dx}{dt} + 16x = 0$, $x(0) = 4$, $x'(0) = -2$

14. $\dfrac{d^2x}{dt^2} + 6\dfrac{dx}{dt} + 9x = 0$, $x(0) = 3$, $x'(0) = -3$

*15. $\dfrac{d^2x}{dt^2} + 10\dfrac{dx}{dt} + 25x = 0$, $x(0) = -5$, $x'(0) = 1$

16. $\dfrac{d^2x}{dt^2} + \dfrac{dx}{dt} + \dfrac{1}{4}x = 0$, $x(0) = -1$, $x'(0) = 2$

17. Suppose that an object with $m = 1$ is attached to the end of a spring with spring constant $k = 1$. After reaching its equilibrium position, the object is pulled one unit above the equilibrium and released with an initial velocity v_0. If the spring-mass system is critically damped, what is the value of c?

18. A weight having mass $m = 1$ is attached to the end of a spring with $k = \dfrac{5}{4}$ and $F_R = 2\dfrac{dx}{dt}$. Determine the displacement of the mass if the object is released from the equilibrium position with an initial velocity of 3 units/s in the downward direction.

*19. A 32 lb weight is attached to the end of a spring with spring constant $k = 24$ lb/ft. If the resistive force is $F_R = 10\dfrac{dx}{dt}$, determine the displacement of the mass if it is released with no initial velocity from a position 6 in. above the equilibrium position. Determine if the mass passes through its equilibrium position and if so, when. Determine the maximum displacement of the object from the equilibrium position.

20. An object weighing 8 lb stretches a spring 6 in. beyond its natural length. If the resistive force is $F_R = 4\dfrac{dx}{dt}$, find the displacement of the mass if it is set into motion from its equilibrium position with an initial velocity of 1 ft/s in the downward direction.

21. An object of mass $m = 70$ kg is attached to the end of a spring and stretches the spring 0.25 m beyond its natural length. If the resistive force is $F_R = 280\dfrac{dx}{dt}$, find the displacement of the object if it is released from a position 3 m above its equilibrium position with no initial velocity. Does the object pass through its equilibrium position at any time?

22. Suppose that an object of mass $m = 1$ slug is attached to a spring with spring constant $k = 25$ lb/ft. If the resistive force is $F_R = 6\dfrac{dx}{dt}$, determine the displacement of the object if it is set into motion from its equilibrium position with an upward velocity of 2 ft/s. What is the quasiperiod of the motion?

*23. An object of mass $m = 4$ slugs is attached to a spring with spring constant $k = 64$ lb/ft. If the resistive force is $F_R = c\dfrac{dx}{dt}$, find the value of c so that the motion is critically damped. For what values of c is the motion underdamped?

24. An object of mass $m = 2$ slugs is attached to a spring with spring constant k lb/ft. If the resistive force is $F_R = 8\dfrac{dx}{dt}$, find the value of k so that the motion is critically damped. For what values of k is the motion underdamped? For what values of k is the motion overdamped?

25. If the quasiperiod of the underdamped motion is $\dfrac{\pi}{6}$ seconds when a $\dfrac{1}{13}$ slug mass is attached to a spring with spring constant $k = 13$ lb/ft, find the damping constant c.

26. If a mass of 0.2 kg is attached to a spring with spring constant $k = 5$ N/m that undergoes damping equivalent to $\dfrac{6}{5}\dfrac{dx}{dt}$, find the quasiperiod of the resulting motion.

27. Show that the solution $x(t)$ of the initial-value problem

$$\begin{cases} m\dfrac{d^2x}{dt^2} + c\dfrac{dx}{dt} + kx = 0 \\ x(0) = \alpha, \dfrac{dx}{dt}(0) = \beta \end{cases}$$

can be written as $x(t) = u(t) + v(t)$, where u and v satisfy the same differential equation as x, u satisfies the initial conditions $u(0) = \alpha, \dfrac{du}{dt}(0) = 0$, and v satisfies the initial conditions $v(0) = 0, \dfrac{dv}{dt}(0) = \beta$.

28. If the spring-mass system $m\dfrac{d^2x}{dt^2} + c\dfrac{dx}{dt} + kx = 0$ is either critically damped or overdamped, show that the mass can pass through its equilibrium position at most one time (independent of the initial conditions).

*29. Suppose that the spring-mass system

$$\begin{cases} m\dfrac{d^2x}{dt^2} + c\dfrac{dx}{dt} + kx = 0 \\ x(0) = \alpha, \dfrac{dx}{dt}(0) = \beta \end{cases}$$

is critically damped. If $\beta = 0$, show that $\lim_{t\to\infty} x(t) = 0$, but that there is no value $t = t_0$ such that $x(t_0) = 0$. (The mass approaches but never reaches its equilibrium position.)

30. Suppose that the spring-mass system

$$\begin{cases} m\dfrac{d^2x}{dt^2} + c\dfrac{dx}{dt} + kx = 0 \\ x(0) = \alpha, \dfrac{dx}{dt}(0) = \beta \end{cases}$$

is critically damped. If $\beta > 0$, find a condition on β so that the mass passes through its equilibrium position after it is released.

31. In the case of underdamped motion, show that the amount of time between two successive times at which the mass passes through its equilibrium is one-half of the quasiperiod.

32. In the case of underdamped motion, show that the amount of time between two successive positive maxima of the position function is $\dfrac{4\pi m}{\sqrt{4km - c^2}}$.

33. In the case of underdamped motion, show that ratio between two consecutive maxima (or minima) is $e^{2c\pi/\sqrt{4mk-c^2}}$.

34. The natural logarithm of the ratio in Exercise 33, called the **logarithmic decrement,** is

$$d = \frac{2c\pi}{\sqrt{4mk - c^2}}$$ and indicates the rate at which the

motion dies out due to damping. Notice that because m, k, and d are all quantities that can be measured in the spring-mass system, the value of d is useful in measuring the damping constant c. Compute the logarithmic decrement of the system in **(a)** Example 3 and **(b)** Example 4.

5.3 Forced Motion

In some cases, the motion of a spring is influenced by an external driving force, $f(t)$. Mathematically, this force is included in the differential equation that models the situation by

$$m\frac{d^2x}{dt^2} = -kx - c\frac{dx}{dt} + f(t).$$

The resulting initial-value problem is

$$\begin{cases} m\dfrac{d^2x}{dt^2} + c\dfrac{dx}{dt} + kx = f(t) \\[2mm] x(0) = \alpha, \dfrac{dx}{dt}(0) = \beta \end{cases}$$

Therefore, differential equations modeling forced motion are nonhomogeneous and require the method of undetermined coefficients or variation of parameters for solution. We first consider forced motion that is undamped.

Example 1

An object of mass $m = 1$ slug is attached to a spring with spring constant $k = 4$ lb/ft. Assuming there is no damping and that the object begins from rest in the equilibrium position, determine the position function of the object if it is subjected to an external force of **(a)** $f(t) = 0$; **(b)** $f(t) = 1$; **(c)** $f(t) = \cos t$.

Solution   First, we note that we must solve the initial-value problem

$$\begin{cases} \dfrac{d^2x}{dt^2} + 4x = f(t) \\[2mm] x(0) = 0, \dfrac{dx}{dt}(0) = 0 \end{cases}$$

for each of the forcing functions in (a), (b), and (c). A general solution of the homogeneous problem $\dfrac{d^2x}{dt^2} + 4x = 0$ is $x_h(t) = c_1 \cos 2t + c_2 \sin 2t$.

(a) With $f(t) = 0$, the equation is homogeneous, so we apply the initial conditions to $x(t) = c_1 \cos 2t + c_2 \sin 2t$ with derivative $\dfrac{dx}{dt}(t) = -2c_1 \sin 2t + 2c_2 \cos 2t$.

Because $x(0) = c_1 = 0$ and $\dfrac{dx}{dt}(0) = 2c_2 = 0$, $c_1 = c_2 = 0$, the solution is $x(t) = 0$.

Physically, this solution indicates that the object does not move from the equilibrium position because there is no forcing function, no initial displacement from the equilibrium position, and no initial velocity.

(b) Using the method of undetermined coefficients with a particular solution of the form $x_p(t) = A$, substitution into the differential equation $\dfrac{d^2x}{dt^2} + 4x = 1$ yields $4A = 1$, so $A = \frac{1}{4}$. Hence,

$$x(t) = x_h(t) + x_p(t) = c_1 \cos 2t + c_2 \sin 2t + \frac{1}{4}$$

with derivative $\dfrac{dx}{dt}(t) = -2c_1 \sin 2t + 2c_2 \cos 2t$. With $x(0) = c_1 + \dfrac{1}{4} = 0$ and $\dfrac{dx}{dt}(0) = 2c_2 = 0$, we have $c_1 = -\dfrac{1}{4}$ and $c_2 = 0$, so

$$x(t) = -\frac{1}{4} \cos 2t + \frac{1}{4}.$$

Notice from the graph of this function in Figure 5.12(a) that the object never moves above the equilibrium position. (Positive values of x indicate that the mass is below the equilibrium position.)

(c) In this case, we assume that $x_p(t) = A \cos t + B \sin t$. Substitution into

$$\frac{d^2x}{dt^2} + 4x = \cos t \text{ yields } 3A \cos t + 3B \sin t = \cos t,$$

so $A = \dfrac{1}{3}$ and $B = 0$. Therefore,

$$x(t) = x_h(t) + x_p(t) = c_1 \cos 2t + c_2 \sin 2t + \frac{1}{3} \cos t$$

with derivative

$$\frac{dx}{dt}(t) = -2c_1 \sin 2t + 2c_2 \cos 2t - \frac{1}{3} \sin t.$$

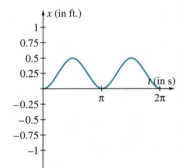

a.

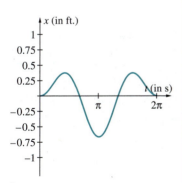

b.

Figure 5.12   (a) Graph of $x(t) = -\frac{1}{4}\cos 2t + \frac{1}{4}$ (b) Graph of $x(t) = -\frac{1}{3}\cos 2t + \frac{1}{3}\cos t$

Applying the initial conditions then gives us $x(0) = c_1 + \frac{1}{3} = 0$ and $\dfrac{dx}{dt}(0) = 2c_2 = 0$, so $c_1 = -\frac{1}{3}$ and $c_2 = 0$. Thus, $x(t) = -\frac{1}{3}\cos 2t + \frac{1}{3}\cos t$. The graph of one period of this solution is shown in Figure 5.12(b). In this case, the mass passes through the equilibrium position twice (near $t = 2$ and $t = 4$) over the period and it returns to the equilibrium without passing through it at $t = 2\pi$. *(Can you predict other values of t where this occurs?)*

How does changing the initial position to $x(0) = 1$ affect the solution in (c)? How does changing the initial velocity to $\dfrac{dx}{dt}(0) = 1$ affect the solution? Is the maximum displacement affected in either case?

When we studied nonhomogeneous equations, we considered equations in which the nonhomogeneous (right-hand side) function was a solution of the corresponding homogeneous equation. We consider this type of situation with the initial-value problem

$$\begin{cases} \dfrac{d^2x}{dt^2} + \omega^2 x = F_1 \cos \omega t + F_2 \sin \omega t + G(t) \\[2mm] x(0) = \alpha, \ \dfrac{dx}{dt}(0) = \beta \end{cases}$$

where F_1 and F_2 are constants and G is any function of t. (Note that one of the constants F_1 and F_2 can equal zero and G can be identically the zero function.) In this case, we say that ω is the **natural frequency of the system** because a general solution of the corresponding homogeneous equation is $x_h(t) = c_1 \cos \omega t + c_2 \sin \omega t$. In the case of this initial-value problem, the **forced frequency,** the frequency of the trigonometric functions in $F_1 \cos \omega t + F_2 \sin \omega t + G(t)$, equals the natural frequency.

Example 2

Investigate the effect that the forcing function $f(t) = \cos 2t$ has on the solution of the initial-value problem

$$\begin{cases} \dfrac{d^2x}{dt^2} + 4x = f(t) \\[2mm] x(0) = 0, \ \dfrac{dx}{dt}(0) = 0. \end{cases}$$

| **Solution** | As we saw in Example 1, $x_h(t) = c_1 \cos 2t + c_2 \sin 2t$. Because $f(t) = \cos 2t$ is contained in this solution, we assume that $x_p(t) = At \cos 2t + Bt \sin 2t$ with first and second derivatives

$$\frac{dx_p}{dt}(t) = A \cos 2t - 2At \sin 2t + B \sin 2t + 2Bt \cos 2t$$

and

$$\frac{d^2 x_p}{dt^2}(t) = -4A \sin 2t - 4At \cos 2t + 4B \cos 2t - 4Bt \sin 2t.$$

Substitution into $\dfrac{d^2 x}{dt^2} + 4x = \cos 2t$ yields $-4A \sin 2t + 4B \cos 2t = \cos 2t$.

Thus, $A = 0$ and $B = \dfrac{1}{4}$, so $x_p(t) = \dfrac{1}{4}t \sin 2t$ and

$$x(t) = x_h(t) + x_p(t) = c_1 \cos 2t + c_2 \sin 2t + \frac{1}{4}t \sin 2t.$$

This function has derivative

$$\frac{dx}{dt}(t) = -2c_1 \sin 2t + 2c_2 \cos 2t + \frac{1}{4} \sin 2t + \frac{1}{2}t \cos 2t,$$

so application of the initial conditions yields $x(0) = c_1 = 0$ and $\dfrac{dx}{dt}(0) = 2c_2 = 0$. Therefore, $c_1 = c_2 = 0$, so $x(t) = \dfrac{1}{4}t \sin 2t$. The graph of this solution is shown in Figure 5.13. Notice that the amplitude increases as t increases. This indicates that the spring-mass system will encounter a serious problem; either the spring will break, or the mass will eventually hit its support (like a ceiling or beam) or a lower boundary (like the ground or floor).

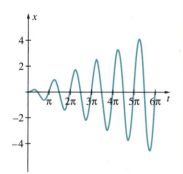

Figure 5.13   Resonance

The phenomenon illustrated in Example 2 is called **resonance** and can be extended to other situations such as vibrations in an aircraft wing, a skyscraper, a glass, or a bridge. Some of the sources of excitation that lead to the vibration of these structures include unbalanced rotating devices, vortex shedding, strong winds, rough surfaces, and moving vehicles. Many engineers must overcome problems caused when structures and machines are subjected to forced vibrations.

Over a sufficient amount of time, do changes in the initial conditions affect the motion of a spring-mass system? Experiment by changing the initial conditions in the initial-value problem in Example 2.

Let us investigate in detail initial-value problems of the form

$$\begin{cases} \dfrac{d^2x}{dt^2} + \omega^2 x = F \cos \beta t, \ \omega \neq \beta \\ \\ x(0) = 0, \ \dfrac{dx}{dt}(0) = 0 \end{cases}$$

A general solution of the corresponding homogeneous equation is $x_h(t) = c_1 \cos \omega t + c_2 \sin \omega t$. Using the method of undetermined coefficients, a particular solution is given by $x_p(t) = A \cos \beta t + B \sin \beta t$. The corresponding derivatives of this solution are

$$\frac{dx_p}{dt}(t) = -A\beta \sin \beta t + B\beta \cos \beta t \qquad \text{and} \qquad \frac{d^2x_p}{dt^2}(t) = -A\beta^2 \cos \beta t - B\beta^2 \sin \beta t.$$

Substitution into the nonhomogeneous equation $\dfrac{d^2x}{dt^2} + \omega^2 x = F \cos \beta t$ and equating the corresponding coefficients yields

$$A = \frac{F}{\omega^2 - \beta^2} \qquad \text{and} \qquad B = 0.$$

Therefore, a general solution is

$$x(t) = c_1 \cos \omega t + c_2 \sin \omega t + \frac{F}{\omega^2 - \beta^2} \cos \beta t.$$

Application of the initial conditions yields the solution

$$x(t) = \frac{F}{\omega^2 - \beta^2}(\cos \beta t - \cos \omega t).$$

Using the trigonometric identity $\frac{1}{2}[\cos (A - B) - \cos (A + B)] = \sin A \sin B$, we have

$$x(t) = \frac{2F}{\omega^2 - \beta^2} \sin \frac{(\omega + \beta)t}{2} \sin \frac{(\omega - \beta)t}{2}.$$

Notice that the solution can be represented as

$$x(t) = A(t) \sin \frac{(\omega + \beta)t}{2} \qquad \text{where } A(t) = \frac{2F}{\omega^2 - \beta^2} \sin \frac{(\omega - \beta)t}{2}.$$

Therefore, when the quantity $(\omega - \beta)$ is small, $(\omega + \beta)$ is relatively large in comparison. The function $\sin \dfrac{(\omega + \beta)t}{2}$ oscillates quite frequently because it has period $\pi/(\omega + \beta)$. Meanwhile, the function $\sin \dfrac{(\omega - \beta)t}{2}$ oscillates relatively slowly because it has period $\dfrac{\pi}{\omega - \beta}$. When we graph $x(t)$, we see that the functions $\pm \dfrac{2F}{\omega^2 - \beta^2} \sin \dfrac{(\omega - \beta)t}{2}$ form an **envelope** for the solution.

| Example 3 |

Solve the initial-value problem

$$\begin{cases} \dfrac{d^2x}{dt^2} + 4x = f(t) \\[2mm] x(0) = 0, \dfrac{dx}{dt}(0) = 0 \end{cases}$$

with (a) $f(t) = \cos 3t$ and (b) $f(t) = \cos 5t$.

| Solution | (a) A general solution of the corresponding homogeneous equation is $x_h(t) = c_1 \cos 2t + c_2 \sin 2t$. By the method of undetermined coefficients, we assume that $x_p(t) = A \cos 3t + B \sin 3t$. Substitution into $\dfrac{d^2x}{dt^2} + 4x = \cos 3t$ yields

$-5A \cos 3t - 5B \sin 3t = \cos 3t$, so $A = -\dfrac{1}{5}$ and $B = 0$. Thus, $x_p(t) = -\dfrac{1}{5} \cos 3t$

and $x(t) = x_h(t) + x_p(t) = c_1 \cos 2t + c_2 \sin 2t - \dfrac{1}{5} \cos 3t$. Because the derivative

of this function is $\dfrac{dx}{dt}(t) = -2c_1 \sin 2t + 2c_2 \cos 2t + \dfrac{3}{5} \sin 3t$, application of the

initial conditions yields $x(0) = c_1 - \dfrac{1}{5} = 0$ and $\dfrac{dx}{dt}(0) = 2c_2 = 0$. Therefore,

$c_1 = \dfrac{1}{5}$ and $c_2 = 0$, so $x(t) = \dfrac{1}{5} \cos 2t - \dfrac{1}{5} \cos 3t$. The graph of this function is

shown in Figure 5.14(a) along with the envelope functions $\dfrac{2}{5} \sin \dfrac{t}{2}$ and $-\dfrac{2}{5} \sin \dfrac{t}{2}$.

(b) In a similar manner, the solution of the initial-value problem

$$\begin{cases} \dfrac{d^2x}{dt^2} + 4x = \cos 5t \\[2mm] x(0) = 0, \dfrac{dx}{dt}(0) = 0 \end{cases}$$

is $x(t) = \dfrac{1}{21} \cos 2t - \dfrac{1}{21} \cos 5t$. The graph of the solution is shown in Figure

5.14(b) along with the envelope functions $\dfrac{2}{21} \sin \dfrac{3t}{2}$ and $-\dfrac{2}{21} \sin \dfrac{3t}{2}$.

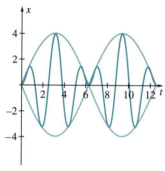

a.

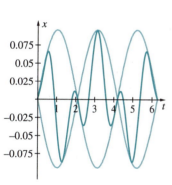

b.

Figure 5.14 (a) and (b)

Some computer algebra systems contain commands that allow us to "play" functions. If this is possible with your technology, see In Touch with Technology at the end of the section.

Oscillations like those illustrated in Example 3 are called **beats** because of the periodic variation of amplitude. This phenomenon is commonly encountered when two musicians try to simultaneously tune their instruments or when two tuning forks with almost equivalent frequencies are played at the same time.

Consider the problem

$$\begin{cases} \dfrac{d^2x}{dt^2} + 4x = \cos \beta t \\[2mm] x(0) = 0, \dfrac{dx}{dt}(0) = 0 \end{cases}$$

for β = 6, 8, and 10. What happens to the amplitude of the beats as β increases?

Example 4

Investigate the effect that the forcing function $f(t) = e^{-t} \cos 2t$ has on the initial-value problem

$$\begin{cases} \dfrac{d^2x}{dt^2} + 4x = f(t) \\[2mm] x(0) = 0, \dfrac{dx}{dt}(0) = 0 \end{cases}$$

Solution   Using the method of undetermined coefficients, a general solution of the equation is

$$x(t) = x_h(t) + x_p(t) = c_1 \cos 2t + c_2 \sin 2t + \frac{1}{17} e^{-t} \cos 2t - \frac{4}{17} e^{-t} \sin 2t.$$

Applying the initial conditions with $x(t)$ and

$$\frac{dx}{dt}(t) = -2c_1 \sin 2t + 2c_2 \cos 2t - \frac{9}{17} e^{-t} \cos 2t + \frac{2}{17} e^{-t} \sin 2t,$$

we have

$$x(0) = c_1 + \frac{1}{17} = 0 \text{ and } \frac{dx}{dt}(0) = 2c_2 - \frac{9}{17} = 0.$$

Therefore, $c_1 = -\dfrac{1}{17}$ and $c_2 = \dfrac{9}{34}$, so

$$x(t) = -\frac{1}{17} \cos 2t + \frac{9}{34} \sin 2t + \frac{1}{17} e^{-t} \cos 2t - \frac{4}{17} e^{-t} \sin 2t,$$

which is graphed in Figure 5.15. Notice that the effect of terms involving the exponential function diminishes as t increases. In this case, the forcing function $f(t) = e^{-t} \cos 2t$ approaches zero as t increases. Over time, the solution of the nonhomogeneous problem approaches that of the corresponding homogeneous problem, so we observe simple harmonic motion as $t \to \infty$.

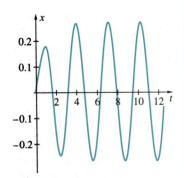

Figure 5.15

We now consider spring problems that involve forces due to damping as well as external forces. In particular, consider the initial-value problem

$$\begin{cases} m\dfrac{d^2x}{dt^2} + c\dfrac{dx}{dt} + kx = \rho \cos \lambda t \\ x(0) = \alpha, \dfrac{dx}{dt}(0) = \beta \end{cases}$$

which has a solution of the form

$$x(t) = h(t) + s(t), \text{ where } \lim_{t \to \infty} h(t) = 0 \text{ and } s(t) = c_1 \cos \lambda t + c_2 \sin \lambda t.$$

The function $h(t)$ is called the **transient** solution and $s(t)$ is known as the **steady-state** solution. Therefore, as t approaches infinity, the solution $x(t)$ approaches the steady-state solution. (Why?) Note that the steady-state solution corresponds to a particular solution obtained through the method of undetermined coefficients or variation of parameters.

Example 5

Solve the initial-value problem

$$\begin{cases} \dfrac{d^2x}{dt^2} + 4\dfrac{dx}{dt} + 13x = \cos t \\ x(0) = 0, \dfrac{dx}{dt}(0) = 1 \end{cases}$$

that models the motion of an object of mass $m = 1$ slug attached to a spring with spring constant $k = 13$ lb/ft that is subjected to a resistive force of $F_R = 4\dfrac{dx}{dt}$ and an external force of $f(t) = \cos t$. Identify the transient and steady-state solutions.

Solution A general solution of $\dfrac{d^2x}{dt^2} + 4\dfrac{dx}{dt} + 13x = 0$ is $x_h(t) = e^{-2t}(c_1 \cos 3t + c_2 \sin 3t)$. We assume that a particular solution has the form $x_p(t) = A \cos t + B \sin t$ with derivatives $\dfrac{dx_p}{dt}(t) = -A \sin t + B \cos t$ and $\dfrac{d^2x_p}{dt^2}(t) = -A \cos t - B \sin t$. After substitution into $\dfrac{d^2x}{dt^2} + 4\dfrac{dx}{dt} + 13x = \cos t$, we see that $\begin{cases} 12A + 4B = 1 \\ -4A + 12B = 0 \end{cases}$ must be satisfied. Hence, $A = \dfrac{3}{40}$ and $B = \dfrac{1}{40}$, so $x_p(t) = \dfrac{3}{40}\cos t + \dfrac{1}{40}\sin t$. Therefore,

$$x(t) = x_h(t) + x_p(t) = e^{-2t}(c_1 \cos 3t + c_2 \sin 3t) + \dfrac{3}{40}\cos t + \dfrac{1}{40}\sin t$$

with derivative

$$\frac{dx}{dt}(t) = -2e^{-2t}(c_1 \cos 3t + c_2 \sin 3t) + 3e^{-2t}(-c_1 \sin 3t + c_2 \cos 3t) -$$

$$\frac{3}{40} \sin t + \frac{1}{40} \cos t.$$

Application of the initial conditions yields

$$x(0) = c_1 + \frac{3}{40} = 0$$

and

$$\frac{dx}{dt}(0) = -2c_1 + 3c_2 + \frac{1}{40} = 1.$$

Therefore, $c_1 = -\dfrac{3}{40}$ and $c_2 = \dfrac{11}{40}$, so

$$x(t) = e^{-2t}\left(-\frac{3}{40} \cos 3t + \frac{11}{40} \sin 3t\right) + \frac{3}{40} \cos t + \frac{1}{40} \sin t.$$

This indicates that the transient solution is $e^{-2t}\left(-\dfrac{3}{40} \cos 3t + \dfrac{11}{40} \sin 3t\right)$ and the

steady-state solution is $\dfrac{3}{40} \cos t + \dfrac{1}{40} \sin t$. We graph this solution in Figure 5.16(a). The solution and the steady-state solution are graphed together in Figure 5.16(b). Notice that the two curves appear identical for $t > 2.5$. The reason for this is shown in the subsequent plot of the transient solution (in Figure 5.16(c)), which becomes quite small near $t = 2.5$.

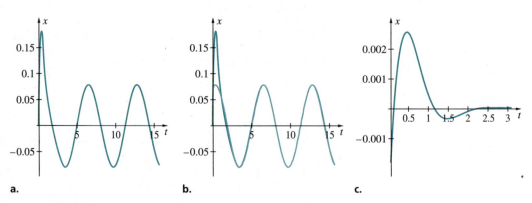

a.   b.   c.

Figure 5.16   (a)–(c)

Over a sufficient amount of time, do changes in the initial conditions in Example 5 affect the motion of the spring-mass system? Experiment by changing the initial conditions in the initial-value problem in Example 5.

IN TOUCH WITH TECHNOLOGY

1. Solve the initial-value problem $x'' + x = \cos t$, $x(0) = 0$, $x'(0) = b$ using $b = 0$ and $b = 1$. Graph the solutions simultaneously to determine the effect that the nonhomogeneous initial velocity has on the solution to the second initial-value problem as t increases.

2. Solve the initial-value problem $x'' + x = \cos \omega t$, $x(0) = 0$, $x'(0) = 1$ using values $\omega = 0.9$ and $\omega = 0.7$. Graph the solutions simultaneously to determine the effect that the value of ω has on each solution.

3. Investigate the effect that the forcing functions **(a)** $f(t) = \cos 1.9t$ and **(b)** $f(t) = \cos 2.1t$ have on the initial-value problem

$$\begin{cases} \dfrac{d^2x}{dt^2} + 4x = f(t) \\ x(0) = 0, \dfrac{dx}{dt}(0) = 0 \end{cases}$$

How do these results differ from those of Example 3? Are there more or fewer beats with these two functions?

4. Solve the initial-value problem

$$x'' + 0.1x' + x = 3\cos 2t, \; x(0) = 0, \; x'(0) = 0$$

using the method of undetermined coefficients and compare the result with the forcing function $3 \cos 2t$. Determine the phase difference between these two functions.

5. Solve the following initial-value problem involving a piecewise defined forcing function over $[0, 2]$:

$$\begin{cases} \dfrac{d^2x}{dt^2} + x = \begin{cases} t, & 0 \le t \le 1 \\ 2 - t, & 1 < t \le 2 \\ 0, & t > 2 \end{cases} \\ x(0) = a, \; x'(0) = b \end{cases}$$

Graph the solution using the initial conditions $a = 1$, $b = 1$; $a = 0$, $b = 1$; and $a = 1$, $b = 0$.

6. Consider the function $g(\gamma) = \dfrac{F}{\sqrt{(\omega^2 - \gamma^2)^2 + 4\lambda^2\gamma^2}}$ defined in Exercise 20. **(a)** Graph this function for $k = 4$, $m = 1$, $F = 2$, and the damping coefficient: $c = 2\lambda = 2, 1, 0.75, 0.50, 0.25$. **(b)** Graph the function for $k = 49$, $m = 10$, $F = 20$, and the damping coefficient: $c = 2\lambda = 2, 1, 0.75, 0.50, 0.25$. In each case, describe what happens to the maximum magnitude of $g(\gamma)$ as $c \to 0$. Also, as $c \to 0$, how does the resonance frequency relate to the natural frequency of the corresponding undamped system?

7. In order to hear beats, solve the initial-value problem

$$\begin{cases} \dfrac{d^2x}{dt^2} + \omega^2 x = F\cos\beta t, \; \omega \ne \beta \\ x(0) = 0, \dfrac{dx}{dt}(0) = 0 \end{cases}$$

using $\omega^2 = 6000$, $\beta = 5991.62$, and $F = 2$. In each case, plot and, if possible, play the solution. (*Note:* The purpose of the high frequencies is to assist in hearing the solutions when they are played.)

8. In order to hear resonance, solve the initial-value problem

$$\begin{cases} \dfrac{d^2x}{dt^2} + \omega^2 x = F\cos\beta t, \; \omega \ne \beta \\ x(0) = 0, \dfrac{dx}{dt}(0) = 0 \end{cases}$$

using $\omega^2 = 6000$, $\beta = 6000$, and $F = 2$. In each case, plot and, if possible, play the solution. (*Note:* The purpose of the high frequencies is to assist in hearing the solutions when they are played.)

1. An 8 lb weight stretches a spring 1 ft. If a 16-lb weight is then attached to the spring, it comes to rest in its equilibrium position. If it is then put into motion with a downward initial velocity of 2 ft/s, determine the displacement of the mass if there is no damping and an external force $f(t) = \cos 3t$. What is the natural frequency of the spring-mass system?

2. A 16 lb weight stretches a spring 6 in. If the mass is lowered 1 ft below its equilibrium position and released, determine the displacement of the mass if there is no damping and an external force of $f(t) = 2 \cos t$. What is the natural frequency of the spring-mass system?

*3. A 16 lb weight stretches a spring 8 in. If the mass is lowered 4 in. below its equilibrium position and released, determine the displacement of the mass if there is no damping and an external force of $f(t) = 2 \cos t$.

4. A 6 lb weight stretches a spring 6 in. The mass is raised 3 in. above its equilibrium position and released. Determine the displacement of the mass if there is no damping and an external force of $f(t) = 2 \cos 5t$.

5. An object of mass $m = 1$ kg is attached to a spring with spring constant $k = 9$ kg/m. If there is no damping and the external force is $f(t) = 4 \cos \omega t$, find the displacement of the object if $x(0) = 0$ and $x'(0) = 0$. What must the value of ω be in order for resonance to occur?

6. An object of mass $m = 2$ kg is attached to a spring with spring constant $k = 1$ kg/m. If the resistive force is $F_R = 3\dfrac{dx}{dt}$ and the external force is $f(t) = 2 \cos \omega t$, find the displacement of the object if $x(0) = 0$ and $x'(0) = 0$. Will resonance occur for any values of ω?

*7. An object of mass $m = 1$ slug is attached to a spring with spring constant $k = 25$ lb/ft. If the resistive force is $F_R = 8\dfrac{dx}{dt}$ and the external force is $f(t) = \cos t - \sin t$, find the displacement of the object if $x(0) = 0$ and $x'(0) = 0$.

8. An object of mass $m = 2$ slug is attached to a spring with spring constant $k = 6$ lb/ft. If the resistive force is $F_R = 6\dfrac{dx}{dt}$ and the external force is $f(t) = 2 \sin 2t + \cos t$, find the displacement of the object if $x(0) = 0$ and $x'(0) = 0$.

9. Suppose that an object of mass 1 slug is attached to a spring with spring constant $k = 4$ lb/ft. If the motion of the object is undamped and subjected to an external force of $f(t) = \cos t$, determine the displacement of the object if $x(0) = 0$ and $x'(0) = 0$. What functions envelope this displacement function? What is the maximum displacement of the object? If the external force is changed to $f(t) = \cos \dfrac{t}{2}$, does the maximum displacement increase or decrease?

10. An object of mass 1 slug is attached to a spring with spring constant $k = 25$ lb/ft. If the motion of the object is undamped and subjected to an external force of $f(t) = \cos \beta t$, $\beta \neq 5$, determine the displacement of the object if $x(0) = 0$ and $x'(0) = 0$. What is the value of β if the maximum displacement of the object is $\dfrac{2}{11\sqrt{2}}$ ft?

*11. An object of mass 4 slugs is attached to a spring with spring constant $k = 26$ lb/ft. It is subjected to a resistive force of $F_R = 4\dfrac{dx}{dt}$ and an external force $f(t) = 250 \sin t$. Determine the displacement of the object if $x(0) = 0$ and $x'(0) = 0$. What is the transient solution? What is the steady-state solution?

12. An object of mass 1 slug is attached to a spring with spring constant $k = 4000$ lb/ft. It is subjected to a resistive force of $F_R = 40\dfrac{dx}{dt}$ and an external force $f(t) = 600 \sin t$. Determine the displacement of the object if $x(0) = 0$ and $x'(0) = 0$. What is the transient solution? What is the steady-state solution?

13. Find the solution of the differential equation
$$m\frac{d^2x}{dt^2} + kx = F \sin \omega t, \quad \omega \neq \sqrt{\frac{k}{m}} \quad \text{that satisfies}$$
the initial conditions: **(a)** $x(0) = \alpha$, $x'(0) = 0$; **(b)** $x(0) = 0$, $x'(0) = \beta$; **(c)** $x(0) = \alpha$, $x'(0) = \beta$.

14. Find the solution of the differential equation

$$m\frac{d^2x}{dt^2} + c\frac{dx}{dt} + kx = F \sin \omega t, \quad c^2 - 4mk < 0$$

that satisfies the initial conditions: **(a)** $x(0) = \alpha$, $x'(0) = 0$; **(b)** $x(0) = 0$, $x'(0) = \beta$; **(c)** $x(0) = \alpha$, $x'(0) = \beta$.

*15. Find the solution to the initial-value problem

$$\frac{d^2x}{dt^2} + x = \begin{cases} 1, & 0 \le t \le \pi \\ 0, & t > \pi \end{cases}, \quad x(0) = 0, \ x'(0) = 0.$$

(*Hint:* Solve the initial-value problem over each interval. Choose constants appropriately so that the functions x and x' are continuous.)

16. Find the solution to the initial-value problem

$$\frac{d^2x}{dt^2} + x = \begin{cases} \cos t, & 0 \le t \le \pi \\ 0, & t > \pi \end{cases}, \quad x(0) = 0, \ x'(0) = 0.$$

17. Find the solution to the initial-value problem $\frac{d^2x}{dt^2} + x = f(t)$, $x(0) = 0$, $x'(0) = 0$ where

$$f(t) = \begin{cases} t, & 0 \le t \le 1 \\ 2 - t, & 1 < t \le 2. \\ 0, & t > 2 \end{cases}$$

(*Hint:* Solve the initial-value problem over each interval. Choose constants appropriately so that the functions x and x' are continuous.)

18. Find the solution to the initial-value problem

$$\frac{d^2x}{dt^2} + 4\frac{dx}{dt} + 13x = f(t), \quad x(0) = 0, \quad x'(0) = 0$$

where $f(t) = \begin{cases} 1, & 0 \le t \le \pi \\ 1 - t, & \pi < t \le 2\pi. \\ 0, & t > 2\pi \end{cases}$

19. Show that a general solution of

$$\frac{d^2x}{dt^2} + 2\lambda\frac{dx}{dt} + \omega^2 x = F \sin \gamma t$$

is

$$x(t) = Ae^{-\lambda t} \sin(\sqrt{\omega^2 - \lambda^2}t + \phi)$$
$$+ \frac{F}{\sqrt{(\omega^2 - \gamma^2)^2 + 4\lambda^2\gamma^2}} \sin(\gamma t + \theta),$$

where $A = \sqrt{c_1^2 + c_2^2}$, $\lambda < \omega$, and the phase angles ϕ and θ are found with $\sin \phi = \dfrac{c_1}{A}$, $\cos \phi = \dfrac{c_2}{A}$,

$$\sin \theta = \frac{-2\lambda\gamma}{\sqrt{(\omega^2 - \gamma^2)^2 + 4\lambda^2\gamma^2}},$$

and

$$\cos \theta = \frac{\omega^2 - \gamma^2}{\sqrt{(\omega^2 - \gamma^2)^2 + 4\lambda^2\gamma^2}}.$$

20. The steady-state solution (the approximation of the solution for large values of t) of the differential equation in Exercise 19 is $x_h(t) = g(\gamma) \sin(\gamma t + \theta)$ where $g(\gamma) = \dfrac{F}{\sqrt{(\omega^2 - \gamma^2)^2 + 4\lambda^2\gamma^2}}$. By differentiating g with respect to γ, show that the maximum value of $g(\gamma)$ occurs when $\gamma = \sqrt{\omega^2 - 2\lambda^2}$. The quantity $\sqrt{\omega^2 - 2\lambda^2}/2\pi$ is called the **resonance frequency** for the system. Describe the motion if the external force has frequency $\sqrt{\omega^2 - 2\lambda^2}/2\pi$.

5.4 Other Applications

L-R-C Circuits • Deflection of a Beam

L-R-C Circuits

Second-order nonhomogeneous linear ordinary differential equations arise in the study of electrical circuits after the application of *Kirchhoff's law.* Suppose that $I(t)$ is the current in the *L-R-C* series electrical circuit (shown in Figure 5.17) where *L*, *R*, and *C* represent the inductance, resistance, and capacitance of the circuit, respectively.

The voltage drops across the circuit elements in Table 5.2 have been obtained from experimental data where Q is the charge of the capacitor and $\dfrac{dQ}{dt} = I.$

Gustav Robert Kirchhoff (1824–1887). German physicist; worked in spectrum analysis, optics, and electricity. (Northwind Picture Archives)

TABLE 5.2

Circuit Element	Voltage Drop
Inductor	$L\dfrac{dI}{dt}$
Resistor	RI
Capacitor	$\dfrac{1}{C}Q$

Our goal is to model this physical situation with an initial-value problem so that we can determine the current and charge in the circuit. For convenience, the terminology used in this section is summarized in Table 5.3.

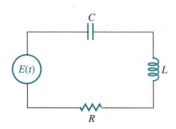

Figure 5.17   An *L-R-C* Circuit

TABLE 5.3

Electrical Quantities	Units
Inductance (L)	Henrys (H)
Resistance (R)	Ohms (Ω)
Capacitance (C)	Farads (F)
Charge (Q)	Coulombs (C)
Current (I)	Amperes (A)

The physical principle needed to derive the differential equation that models the *L-R-C* series circuit is **Kirchhoff's law.**

Kirchhoff's law:

The sum of the voltage drops across the circuit elements is equivalent to the voltage $E(t)$ impressed on the circuit.

Applying Kirchhoff's law with the voltage drops in Table 5.2 yields the differential equation $L\dfrac{dI}{dt} + RI + \dfrac{1}{C}Q = E(t)$. Using the fact that $\dfrac{dQ}{dt} = I$, we also have $\dfrac{d^2Q}{dt^2} = \dfrac{dI}{dt}$. Therefore, the equation becomes $L\dfrac{d^2Q}{dt^2} + R\dfrac{dQ}{dt} + \dfrac{1}{C}Q = E(t)$, which can be solved by the method of undetermined coefficients or the method of variation of parameters. If the initial charge and current are $Q(0) = Q_0$ and $I(0) = \dfrac{dQ}{dt}(0) = I_0$, we solve the initial-value problem

$$\begin{cases} L\dfrac{d^2Q}{dt^2} + R\dfrac{dQ}{dt} + \dfrac{1}{C}Q = E(t) \\[3mm] Q(0) = Q_0, \; I(0) = \dfrac{dQ}{dt}(0) = I_0 \end{cases}$$

for the charge $Q(t)$. The solution is differentiated to find the current $I(t)$.

Example 1

Consider the L-R-C circuit with $L = 1$ henry, $R = 40$ ohms, $C = \dfrac{1}{4000}$ farads, and $E(t) = 24$ volts. Determine the current in this circuit if there is zero initial current and zero initial charge.

Solution   Using the indicated values, the initial-value problem that we must solve is

$$\begin{cases} \dfrac{d^2Q}{dt^2} + 40\dfrac{dQ}{dt} + 4000Q = 24 \\[3mm] Q(0) = 0, \; I(0) = \dfrac{dQ}{dt}(0) = 0 \end{cases}$$

The characteristic equation of the corresponding homogeneous equation is $r^2 + 40r + 4000 = 0$ with roots $r_{1,2} = -20 \pm 60i$, so a general solution of the corresponding homogeneous equation is $Q_h(t) = e^{-20t}(c_1 \cos 60t + c_2 \sin 60t)$. Because the voltage is the constant function $E(t) = 24$, we assume that the particular solution has the form $Q_p(t) = A$. Substitution into $\dfrac{d^2Q}{dt^2} + 40\dfrac{dQ}{dt} + 4000Q = 24$ yields $4000A = 24$ or $A = \dfrac{3}{500}$. Therefore, a general solution of the nonhomogeneous equation is

$$Q(t) = Q_h(t) + Q_p(t) = e^{-20t}(c_1 \cos 60t + c_2 \sin 60t) + \frac{3}{500}$$

with derivative

$$\frac{dQ}{dt}(t) = -20e^{-20t}(c_1 \cos 60t + c_2 \sin 60t) + 60e^{-20t}(-c_1 \sin 60t + c_2 \cos 60t).$$

After application of the initial conditions, we have

$$Q(0) = c_1 + \frac{3}{500} = 0 \text{ and } \frac{dQ}{dt}(0) = -20c_1 + 60c_2 = 0.$$

Therefore, $c_1 = -\frac{3}{500}$ and $c_2 = -\frac{1}{500}$, so the charge in the circuit is

$$Q(t) = e^{-20t}\left(-\frac{3}{500} \cos 60t - \frac{1}{500} \sin 60t\right) + \frac{3}{500}$$

and the current is given by

$$\frac{dQ}{dt}(t) = -\frac{20}{500}e^{-20t}(-3 \cos 60t - \sin 60t) + \frac{60}{500}e^{-20t}(3 \sin 60t - \cos 60t)$$

$$= \frac{2}{5}e^{-20t} \sin 60t.$$

These results indicate that in time the charge approaches the constant value of $\frac{3}{500}$, which is known as the **steady-state charge.** Also, due to the exponential term, the current approaches zero as t increases. We show the graphs of $Q(t)$ and $I(t)$ in Figures 5.18(a) and (b) to verify these observations.

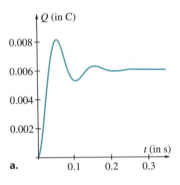

a.

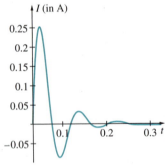

b.

Figure 5.18 (a) $Q(t)$ (b) $I(t)$

In Example 1, how is the charge $Q(t)$ affected if $E(t) = 48$ volts? What happens to $Q(t)$ if $R = 40\sqrt{10}$ ohms?

Deflection of a Beam

An important mechanical model involves the deflection of a long beam that is supported at one or both ends, as shown in Figure 5.19. Assuming that in its undeflected form the beam is horizontal, then the deflection of the beam can be expressed as a function of x. Suppose that the shape of the beam when it is deflected is given by the graph of the function $y(x) = -s(x)$, where x is the distance from the left end of the beam and s the measure of the vertical deflection from the equilibrium position. The boundary value problem that models this situation is derived as follows.

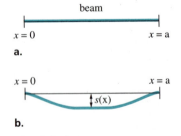

Figure 5.19 (a) and (b)

Let $m(x)$ equal the turning moment of the force relative to the point x, and $w(x)$ represent the weight distribution of the beam. These two functions are related by the equation:

$$\frac{d^2m}{dx^2} = w(x).$$

Also, the turning moment is proportional to the curvature of the beam. Hence,

$$m(x) = \frac{EI}{\left(\sqrt{1 + \left(\frac{ds}{dx}\right)^2}\right)^3} \frac{d^2s}{dx^2}$$

where E and I are constants related to the composition of the beam and the shape and size of a cross section of the beam, respectively. Notice that this equation is nonlinear. This difficulty is overcome with an approximation. For small values of s, the denominator of the right-hand side of the equation can be approximated by the constant 1. Therefore, the equation is simplified to

$$m(x) = EI\frac{d^2s}{dx^2}.$$

This equation is linear and can be differentiated twice to obtain

$$\frac{d^2m}{dx^2} = EI\frac{d^4s}{dx^4}.$$

which is then used with the equation above relating $m(x)$ and $w(x)$ to obtain the single fourth-order linear nonhomogeneous differential equation

$$EI\frac{d^4s}{dx^4} = w(x).$$

Boundary conditions for this problem may vary. In most cases, two conditions are given for each end of the beam. Some of these conditions that are specified in pairs at $x = \rho$, where $\rho = 0$ or $\rho = a$, include: $s(\rho) = 0$, $\frac{ds}{dx}(\rho) = 0$ (fixed end); $\frac{d^2s}{dx^2}(\rho) = 0$, $\frac{d^3s}{dx^3}(\rho) = 0$ (free end); $s(\rho) = 0$, $\frac{d^2s}{dx^2}(\rho) = 0$ (simple support); and $\frac{ds}{dx}(\rho) = 0$, $\frac{d^3s}{dx^3}(\rho) = 0$ (sliding clamped end).

Example 2

Solve the beam equation over the interval $0 \le x \le 1$ if $E = I = 1$, $w(x) = 48$, and the following boundary conditions are used: $s(0) = 0$, $\frac{ds}{dx}(0) = 0$ (fixed end at $x = 0$); and

(a) $s(1) = 0$, $\frac{d^2s}{dx^2}(1) = 0$ (simple support at $x = 1$)

(b) $\dfrac{d^2s}{dx^2}(1) = 0$, $\dfrac{d^3s}{dx^3}(1) = 0$ (free end at $x = 1$)

(c) $\dfrac{ds}{dx}(1) = 0$, $\dfrac{d^3s}{dx^3}(1) = 0$ (sliding clamped end at $x = 1$)

(d) $s(1) = 0$, $\dfrac{ds}{dx}(1) = 0$ (fixed end at $x = 1$).

Solution   We begin by noting that the differential equation is $\dfrac{d^4s}{dx^4} = 48$, which is a separable equation that can be solved by integrating each side four times to yield

$$s(x) = 2x^4 + c_1x^3 + c_2x^2 + c_3x + c_4$$

with derivatives

$$\frac{ds}{dx}(x) = 8x^3 + 3c_1x^2 + 2c_2x + c_3, \quad \frac{d^2s}{dx^2}(x) = 24x^2 + 6c_1x + 2c_2,$$

and

$$\frac{d^3s}{dx^3}(x) = 48x + 6c_1.$$

(Note that we could have used the method of undetermined coefficients or variation of parameters to find this solution.) We next determine the arbitrary constants for the pair of boundary conditions at $x = 0$. Because $s(0) = c_4 = 0$ and $\dfrac{ds}{dx}(0) = c_3 = 0$,

$$s(x) = 2x^4 + c_1x^3 + c_2x^2.$$

(a) Because $s(1) = 2 + c_1 + c_2 = 0$ and $\dfrac{d^2s}{dx^2}(1) = 24 + 6c_1 + 2c_2 = 0$, $c_1 = -5$ and $c_2 = 3$. Hence, $s(x) = 2x^4 - 5x^3 + 3x^2$. We can visualize the shape of the beam by graphing $y = -s(x)$ as shown in Figure 5.20(a). (b) In this case, $\dfrac{d^2s}{dx^2}(1) = 24 + 6c_1 + 2c_2 = 0$ and $\dfrac{d^3s}{dx^3}(1) = 48 + 6c_1 = 0$, so $c_1 = -8$ and $c_2 = 12$. The deflection of the beam is given by $s(x) = 2x^4 - 8x^3 + 12x^2$. We graph $y = -s(x)$ in Figure 5.20(b). We see from the graph that the end is free at $x = 1$. (c) Because $\dfrac{ds}{dx}(1) = 8 + 3c_1 + 2c_2 = 0$ and $\dfrac{d^3s}{dx^3}(1) = 48 + 6c_1 = 0$, $c_1 = -8$ and $c_2 = 8$. Therefore, $s(x) = 2x^4 - 8x^3 + 8x^2$. From the shape graph of $y = -s(x)$ in Figure 5.20(c), we see that the end at $x = 1$ is clamped as compared to the free end in (b).

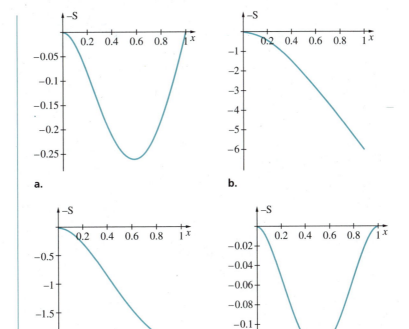

Figure 5.20 (a)–(d)

(d) Here, $s(1) = 2 + c_1 + c_2 = 0$ and $\dfrac{ds}{dx}(1) = 8 + 3c_1 + 2c_2 = 0$, so $c_1 = -4$ and $c_2 = 2$. Thus, $s(x) = 2x^4 - 4x^3 + 2x^2$. $y = -s(x)$ is graphed in Figure 5.20(d). Notice that both ends are fixed. Finally, all four graphs are shown together in Figure 5.21 to compare the different boundary conditions.

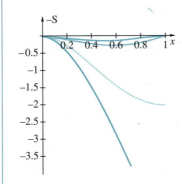

Figure 5.21

If we had used free ends at both $x = 0$ and $x = 1$ in Example 2, what is the displacement? Is this what we should expect from the physical problem?

IN TOUCH WITH TECHNOLOGY

In order to better understand the solutions to the elastic beam problem, we can use a graphics device to determine the shape of the beam under different boundary conditions.

1. Graph the solution of the beam equation if the beam has length 10, the constants E and I are such that $EI = 1$, the weight distribution is $w(x) = x^2$, and the boundary conditions are:

   (a) $s(0) = 0$, $s'(0) = 0$, (fixed end at $x = 0$); $s(10) = 0$, $s''(10) = 0$ (simple support at $x = 10$)

   (b) $s(0) = 0$, $s'(0) = 0$, (fixed end at $x = 0$); $s''(10) = 0$, $s'''(10) = 0$ (free end at $x = 10$)

   (c) $s(0) = 0$, $s'(0) = 0$, (fixed end at $x = 0$); $s'(10) = 0$, $s'''(10) = 0$ (sliding clamped end at $x = 10$)

   (d) $s(0) = 0$, $s'(0) = 0$, (fixed end at $x = 0$); $s(10) = 0$, $s'(10) = 0$ (fixed end at $x = 10$)

   Discuss the differences brought about by these conditions.

2. Repeat Problem 1 with the following boundary conditions:

   (a) $s(0) = 0$, $s''(0) = 0$, (simple support at $x = 0$); $s(10) = 0$, $s''(10) = 0$ (simple support at $x = 10$)

   (b) $s(0) = 0$, $s''(0) = 0$, (simple support at $x = 0$); $s''(10) = 0$, $s'''(10) = 0$ (free end at $x = 10$)

   (c) $s(0) = 0$, $s''(0) = 0$, (simple support at $x = 0$); $s'(10) = 0$, $s'''(10) = 0$ (sliding clamped end at $x = 10$)

   (d) $s(0) = 0$, $s''(0) = 0$, (simple support at $x = 0$); $s(10) = 0$, $s'(10) = 0$ (fixed end at $x = 10$)

3. Graph the solution of the beam equation if the beam has length 10, the constants E and I are such that $EI = 1$, the weight distribution is $w(x) = 48 \sin \dfrac{\pi x}{10}$, and the boundary conditions are the same as in Problem 2.

4. Repeat Problem 1 with $EI = 10$.

5. Repeat Problem 2 with $EI = 100$. How do these solutions compare to those in Problem 2?

6. Repeat Problem 3 with $EI = 100$. How do these solutions compare to those in Problem 3?

7. Attempt to find solutions of the beam equation in Problem 1–3 using other combinations of boundary conditions. How do the differing boundary conditions affect the solution?

EXERCISES 5.4

In Exercises 1–4, find the charge on the capacitor and the current in the L-C series circuit (in which $R = 0$) assuming that $Q(0) = 0$ and $I(0) = \dfrac{dQ}{dt}(0) = 0$.

1. $L = 2$ henry, $C = \frac{1}{32}$ farad, and $E(t) = 220$ volts.

2. $L = 2$ henry, $C = \frac{1}{50}$ farad, and $E(t) = 220$ volts.

*3. $L = \frac{1}{4}$ henry, $C = \frac{1}{64}$ farad, and $E(t) = 16t$ volts.

4. $L = \frac{1}{4}$ henry, $C = \frac{1}{64}$ farad, and $E(t) = 16 \sin 4t$ volts.

5. Find the charge $Q(t)$ on the capacitor in an L-R-C series circuit if $L = 0.2$ henry, $R = 25$ ohms, $C = 0.001$ farad, $E(t) = 0$, $Q(0) = 0$ coulombs, and $I(0) = \dfrac{dQ}{dt}(0) = 4$ amps. What is the maximum charge on the capacitor?

6. Consider the L-R-C circuit given in Exercise 5 with $E(t) = 1$. Determine the value of $Q(t)$ as t approaches infinity.

*7. Consider the solution to the *L-R-C* series circuit indicated in Exercise 5. In this case, let $E(t) = 126 \cos t + 5000 \sin t$. Determine the solution to this initial-value problem. At what time does the charge first equal zero? What are the steady-state charge and current?

8. If the resistance, R, is changed in Exercise 5 to $R = 8$ ohms, what is the resulting charge on the capacitor? What is the maximum charge attained, and when does the charge first equal zero?

9. A beam of length 10 is fixed at both ends. Determine the shape of the beam if the weight distribution is the constant function $w(x) = 8$, with constants E and I such that $EI = 100$, $EI = 10$, and $EI = 1$. What is the displacement of the beam from $s = 0$ in each case? How does the value of EI affect the solution?

10. Suppose that the beam in Exercise 9 is fixed at $x = 0$ and has simple support at $x = 10$. Determine the maximum displacement using constants E and I such that $EI = 100$, $EI = 10$, and $EI = 1$.

*11. Consider Exercise 9 with simple support at $x = 0$ and $x = 10$. How does the maximum displacement compare to that found in each case in Problems 9 and 10?

12. Determine the shape of the beam of length 10 with constants E and I such that $EI = 1$, weight distribution $w(x) = x^2$, and boundary conditions:

(a) $s(0) = 0$, $s'(0) = 0$, (fixed end at $x = 0$); $s(10) = 0$, $s''(10) = 0$ (simple support at $x = 10$)

(b) $s(0) = 0$, $s'(0) = 0$, (fixed end at $x = 0$); $s''(10) = 0$, $s'''(10) = 0$ (free end at $x = 10$)

(c) $s(0) = 0$, $s'(0) = 0$, (fixed end at $x = 0$); $s'(10) = 0$, $s'''(10) = 0$ (sliding clamped end at $x = 10$)

(d) $s(0) = 0$, $s'(0) = 0$, (fixed end at $x = 0$); $s(10) = 0$, $s'(10) = 0$ (fixed end at $x = 10$)

Discuss the differences brought about by these conditions. (See In Touch with Technology, Problem 1.)

*13. Determine the shape of the beam of length 10 with constants E and I such that $EI = 1$, weight distribution $w(x) = x^2$, and boundary conditions:

(a) $s(0) = 0$, $s''(0) = 0$, (simple support at $x = 0$); $s(10) = 0$, $s''(10) = 0$ (simple support at $x = 10$)

(b) $s(0) = 0$, $s''(0) = 0$, (simple support at $x = 0$); $s''(10) = 0$, $s'''(10) = 0$ (free end at $x = 10$)

(c) $s(0) = 0$, $s''(0) = 0$, (simple support at $x = 0$); $s'(10) = 0$, $s'''(10) = 0$ (sliding clamped end at $x = 10$)

(d) $s(0) = 0$, $s''(0) = 0$, (simple support at $x = 0$); $s(10) = 0$, $s'(10) = 0$ (fixed end at $x = 10$)

Discuss the differences brought about by these conditions. (See In Touch with Technology, Problem 2.)

14. Determine the shape of the beam of length 10 with constants E and I such that $EI = 1$, weight distribution $w(x) = 48 \sin \dfrac{\pi x}{10}$ and boundary conditions:

(a) $s(0) = 0$, $s''(0) = 0$, (simple support at $x = 0$); $s(10) = 0$, $s''(10) = 0$ (simple support at $x = 10$)

(b) $s(0) = 0$, $s''(0) = 0$, (simple support at $x = 0$); $s''(10) = 0$, $s'''(10) = 0$ (free end at $x = 10$)

(c) $s(0) = 0$, $s''(0) = 0$, (simple support at $x = 0$); $s'(10) = 0$, $s'''(10) = 0$ (sliding clamped end at $x = 10$)

(d) $s(0) = 0$, $s''(0) = 0$, (simple support at $x = 0$); $s(10) = 0$, $s'(10) = 0$ (fixed end at $x = 10$)

Discuss the differences brought about by these conditions. (See In Touch with Technology, Problem 3.)

15. Consider the *L-C* series circuit in which $E(t) = 0$ modeled by the initial-value problem

$$\begin{cases} L\dfrac{d^2Q}{dt^2} + \dfrac{1}{C}Q = 0 \\ Q(0) = Q_0, \ I(0) = \dfrac{dQ}{dt}(0) = 0 \end{cases}$$

Find the charge Q and the current I. What is the maximum charge? What is the maximum current?

16. Consider the *L-C* series circuit in which $E(t) = 0$ modeled by the initial-value problem

$$\begin{cases} L\dfrac{d^2Q}{dt^2} + \dfrac{1}{C}Q = 0 \\ Q(0) = 0, \ I(0) = \dfrac{dQ}{dt}(0) = I_0 \end{cases}$$

Find the charge Q and the current I. What is the maximum charge? What is the maximum current? How do these results compare to those of Exercise 15?

*17. Consider the *L-R-C* circuit modeled by

$$\begin{cases} L\dfrac{d^2Q}{dt^2} + R\dfrac{dQ}{dt} + \dfrac{1}{C}Q = E_0 \sin \omega t \\[2mm] Q(0) = Q_0,\ I(0) = \dfrac{dQ}{dt}(0) = I_0 \end{cases}$$

Find the **steady-state current**, $\lim_{t \to \infty} I(t)$, of this problem. Note that in this formula, $L\omega - \dfrac{1}{C\omega}$ is called the **reactance** of the circuit, and $\sqrt{\left(L\omega - \dfrac{1}{C\omega}\right)^2 + R^2}$ is called the **impedance** of the circuit, where both of these quantities are measured in ohms.

18. Compute the reactance and the impedance of the circuit in Example 1 if $E(t) = 24 \sin 4t$.

19. Show that the maximum amplitude of the steady-state current found in Exercise 17 occurs when

$\omega = \dfrac{1}{\sqrt{LC}}$. (In this case, we say that **electrical resonance** occurs.)

20. (Elastic Shaft) The differential equation that models the torsional motion of a weight suspended from an elastic shaft is $I\dfrac{d^2\theta}{dt^2} + c\dfrac{d\theta}{dt} + k\theta = T(t)$, where θ represents the amount that the weight is twisted at time t, I is the moment of inertia, c is the damping constant, k is the elastic shaft constant (similar to the spring constant), and $T(t)$ is the applied torque. Consider the differential equation with $I = 1$, $c = 4$ and $k = 13$. Find $\theta(t)$ if **(a)** $T(t) = 0$, $\theta(0) = \theta_0$, and $\dfrac{d\theta}{dt}(0) = 0$; **(b)** $T(t) = \sin \pi x$, $\theta(0) = \theta_0$, and $\dfrac{d\theta}{dt}(0) = 0$. Describe the motion that results in each case.

5.5   The Pendulum Problem

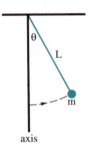

Figure 5.22   A Swinging Pendulum

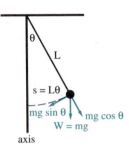

Figure 5.23   A Force Diagram for the Swinging Pendulum

Suppose that a mass m is attached to the end of a rod of length L, the weight of which is negligible. (See Figure 5.22.) We want to determine an equation that describes the motion of the mass in terms of the displacement $\theta(t)$ which is measured counterclockwise in radians from the vertical axis shown in Figure 5.22. This is possible if we are given an initial position and an initial velocity of the mass. A force diagram for this situation is shown in Figure 5.23.

Notice that the forces are determined with trigonometry using the diagram in Figure 5.23. Here, $\cos \theta = mg/x$ and $\sin \theta = mg/y$, so we obtain the forces

$$x = mg \cos \theta \text{ and } y = mg \sin \theta,$$

which are indicated in Figure 5.24.

The momentum of the mass is given by $m\dfrac{ds}{dt}$, so the rate of change of the momentum is

$$\frac{d}{dt}\left(m\frac{ds}{dt}\right) = m\frac{d^2s}{dt^2},$$

where s represents the length of the arc formed by the motion of the mass. Then, because the force $mg \sin \theta$ acts in the opposite direction of the motion of the mass, we have the equation

$$m\frac{d^2s}{dt^2} = -mg \sin \theta.$$

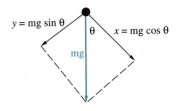

Figure 5.24

(Notice that the force $mg \cos \theta$ is offset by the force of constraint in the rod, so mg and $mg \cos \theta$ cancel each other in the sum of the forces.) Using the relationship from geometry between the length of the arc, the length of the rod, and the angle θ, $s = L\theta$, we have the relationship

$$\frac{d^2s}{dt^2} = \frac{d^2}{dt^2}(L\theta) = L\frac{d^2\theta}{dt^2}.$$

The displacement $\theta(t)$ satisfies $mL\dfrac{d^2\theta}{dt^2} = -mg \sin \theta$ or

$$mL\frac{d^2\theta}{dt^2} + mg \sin \theta = 0$$

which is a *nonlinear* equation. However, because we are only concerned with small displacements, we note from the Maclaurin series for $\sin \theta$,

$$\sin \theta = \theta - \frac{\theta^3}{3!} + \frac{\theta^5}{5!} - \cdots,$$

and that for small values of θ, $\sin \theta \approx \theta$. Therefore, we obtain the linear equation $mL\dfrac{d^2\theta}{dt^2} + mg\theta = 0$ or $\dfrac{d^2\theta}{dt^2} + \dfrac{g}{L}\theta = 0$, which approximates the original problem. If the initial displacement (position of the mass) is given by $\theta(0) = \theta_0$ and the initial velocity (the velocity with which the mass is set into motion) is given by $\dfrac{d\theta}{dt}(0) = v_0$, we have the initial-value problem

$$\begin{cases} \dfrac{d^2\theta}{dt^2} + \dfrac{g}{L}\theta = 0 \\[2mm] \theta(0) = \theta_0, \quad \dfrac{d\theta}{dt}(0) = v_0 \end{cases}$$

to find the displacement function $\theta(t)$.

Suppose that $\omega^2 = \dfrac{g}{L}$ so that the differential equation becomes $\dfrac{d^2\theta}{dt^2} + \omega^2\theta = 0$. Therefore, functions of the form

$$\theta(t) = c_1 \cos \omega t + c_2 \sin \omega t,$$

where $\omega = \sqrt{\dfrac{g}{L}}$, satisfy the equation $\dfrac{d^2\theta}{dt^2} + \dfrac{g}{L}\theta = 0$. When we use the conditions $\theta(0) = \theta_0$ and $\dfrac{d\theta}{dt}(0) = v_0$, we find that the function

$$\theta(t) = \theta_0 \cos \omega t + \frac{v_0}{\omega} \sin \omega t$$

satisfies the equation as well as the initial displacement and velocity conditions. As we did with the position function of spring-mass systems, we can write this function as a cosine function that includes a phase shift with

$$\theta(t) = \sqrt{\theta_0^2 + \frac{v_0^2}{\omega^2}}\, \cos\,(\omega t - \phi)$$

where $\phi = \cos^{-1}\left(\dfrac{\theta_0}{\sqrt{\theta_0^2 + \dfrac{v_0^2}{\omega^2}}}\right)$ and $\omega = \sqrt{\dfrac{g}{L}}$.

Note that the approximate period of $\theta(t)$ is

$$T = \frac{2\pi}{\omega} = 2\pi\sqrt{\frac{L}{g}}.$$

Example 1

Determine the displacement of a pendulum of length $L = 8$ feet if $\theta(0) = 0$ and $\dfrac{d\theta}{dt}(0) = 2$. What is the period? If the pendulum is part of a clock that ticks once for each time the pendulum makes a complete swing, how many ticks does the clock make in one minute?

Solution   Because $\dfrac{g}{L} = \dfrac{32}{8} = 4$, the initial-value problem that models this situation is

$$\begin{cases} \dfrac{d^2\theta}{dt^2} + 4\theta = 0 \\ \theta(0) = 0,\ \dfrac{d\theta}{dt}(0) = 2. \end{cases}$$

A general solution of the differential equation is $\theta(t) = c_1 \cos 2t + c_2 \sin 2t$, so application of the initial conditions yields the solution $\theta(t) = \sin 2t$. The period of this function is

$$T = 2\pi\sqrt{\frac{L}{g}} = 2\pi\sqrt{\frac{8\ \text{ft}}{32\ \text{ft/s}^2}} = \pi\ \text{sec}.$$

(Notice that we can use our knowledge of trigonometry to compare the period with $T = \dfrac{2\pi}{2} = \pi$.) Therefore, the number of ticks made by the clock per minute is calculated with the conversion $\dfrac{1\ \text{rev}}{\pi\ \text{sec}} \times \dfrac{1\ \text{tick}}{1\ \text{rev}} \times \dfrac{60\ \text{sec}}{1\ \text{min}} \approx 19.1\ \text{ticks/min}$. Hence, the clock makes approximately 19 ticks in one minute.

How is motion affected if the length of the pendulum in Example 1 is changed to L = 4?

If the pendulum undergoes a damping force that is proportional to the instantaneous velocity, the force due to damping is given by

$$F_R = b\frac{d\theta}{dt}.$$

Incorporating this force into the sum of the forces acting on the pendulum, we obtain the nonlinear equation $L\frac{d^2\theta}{dt^2} + b\frac{d\theta}{dt} + g\sin\theta = 0$. Again, using the approximation $\sin\theta \approx \theta$ for small values of θ, we use the linear equation $L\frac{d^2\theta}{dt^2} + b\frac{d\theta}{dt} + g\theta = 0$ to approximate the situation. Therefore, we solve the initial-value problem

$$\begin{cases} L\dfrac{d^2\theta}{dt^2} + b\dfrac{d\theta}{dt} + g\theta = 0 \\[2mm] \theta(0) = \theta_0, \ \dfrac{d\theta}{dt}(0) = v_0 \end{cases}$$

to find the displacement function $\theta(t)$.

Example 2

A pendulum of length $L = \dfrac{8}{5}$ ft is subjected to the resistive force $F_R = \dfrac{32}{5}\dfrac{d\theta}{dt}$ due to damping. Determine the displacement function if $\theta(0) = 1$ and $\dfrac{d\theta}{dt}(0) = 2$.

Solution The initial-value problem that models this situation is

$$\begin{cases} \dfrac{8}{5}\dfrac{d^2\theta}{dt^2} + \dfrac{32}{5}\dfrac{d\theta}{dt} + 32\theta = 0 \\[2mm] \theta(0) = 1, \ \dfrac{d\theta}{dt}(0) = 2. \end{cases}$$

Simplifying the differential equation, we obtain $\dfrac{d^2\theta}{dt^2} + 4\dfrac{d\theta}{dt} + 20\theta = 0$, which has characteristic equation $r^2 + 4r + 20 = 0$ with roots $r_{1,2} = -2 \pm 4i$. A general solution is $\theta(t) = e^{-2t}(c_1\cos 4t + c_2\sin 4t)$. Application of the initial conditions yields the solution $\theta(t) = e^{-2t}(\cos 4t + \sin 4t)$. We graph this solution in Figure 5.25. Notice that the damping causes the displacement of the pendulum to decrease over time.

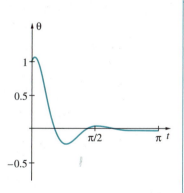

Figure 5.25

When does the object in Example 2 first pass through its equilibrium position? What is the maximum displacement from equilibrium?

In many cases, we can use computer algebra systems to obtain accurate approximations of nonlinear problems.

Example 3   Use a computer algebra system to approximate the solutions of the nonlinear problems

$$\text{(a)} \quad \begin{cases} \dfrac{d^2\theta}{dt^2} + 4 \sin\theta = 0 \\[2mm] \theta(0) = 0, \; \dfrac{d\theta}{dt}(0) = 2 \end{cases} \quad \text{and} \quad \text{(b)} \quad \begin{cases} \dfrac{8}{5}\dfrac{d^2\theta}{dt^2} + \dfrac{32}{5}\dfrac{d\theta}{dt} + 32 \sin\theta = 0 \\[2mm] \theta(0) = 1, \; \dfrac{d\theta}{dt}(0) = 2 \end{cases}$$

Compare the results to the approximations obtained in Examples 1 and 2.

Solution   We show the results obtained with a typical computer algebra system in Figure 5.26. We see that as t increases, the approximate solution obtained in Example 1 becomes less accurate. However, for small values of t, the results are nearly identical.

We show the results obtained for (b) with a typical computer algebra system in Figure 5.27. In this case, we see that the error diminishes as t increases. (Why?)

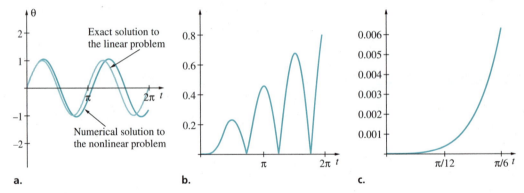

Figure 5.26   (a) The approximations are quite similar for small values of t (b) The absolute value of the difference between the two approximations   (c) The absolute value of the difference between the two approximations

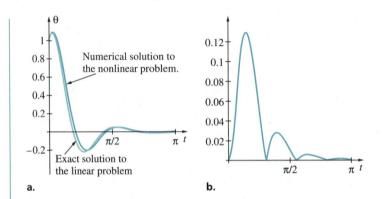

Figure 5.27   (a) The approximations are quite similar for nearly all values of t
(b) The absolute value of the difference between the two approximations

IN TOUCH WITH TECHNOLOGY

1. Use a computer algebra system to solve the initial-value problem

$$\begin{cases} \dfrac{d^2\theta}{dt^2} + \theta = 0 \\ \theta(0) = \theta_0, \ \dfrac{d\theta}{dt}(0) = v_0 \end{cases}$$

subject to the following initial conditions:

(a) $\theta(0) = 0, \ \dfrac{d\theta}{dt}(0) = 2$

(b) $\theta(0) = 2, \ \dfrac{d\theta}{dt}(0) = 0$

(c) $\theta(0) = -2, \ \dfrac{d\theta}{dt}(0) = 0$

(d) $\theta(0) = 0, \ \dfrac{d\theta}{dt}(0) = -1$

(e) $\theta(0) = 0, \ \dfrac{d\theta}{dt}(0) = -2$

(f) $\theta(0) = 1, \ \dfrac{d\theta}{dt}(0) = -1$

(g) $\theta(0) = -1, \ \dfrac{d\theta}{dt}(0) = 1.$

Plot each solution individually and plot the seven solutions simultaneously. Explain the physical interpretation of these solutions.

2. Solve the initial-value problem

$$\begin{cases} \dfrac{d^2\theta}{dt^2} + \dfrac{1}{2}\dfrac{d\theta}{dt} + \theta = 0 \\ \theta(0) = \theta_0, \ \dfrac{d\theta}{dt}(0) = v_0 \end{cases}$$

subject to the initial conditions:

(a) $\theta(0) = 1, \ \dfrac{d\theta}{dt}(0) = 0$

(b) $\theta(0) = -1, \ \dfrac{d\theta}{dt}(0) = 0$

(c) $\theta(0) = 0, \ \dfrac{d\theta}{dt}(0) = 1$

(d) $\theta(0) = 0, \ \dfrac{d\theta}{dt}(0) = -1$

(e) $\theta(0) = 1, \ \dfrac{d\theta}{dt}(0) = 1$

(f) $\theta(0) = 1, \ \dfrac{d\theta}{dt}(0) = -1$

(g) $\theta(0) = -1, \ \dfrac{d\theta}{dt}(0) = 1$

(h) $\theta(0) = -1, \ \dfrac{d\theta}{dt}(0) = -1$

(i) $\theta(0) = 1, \ \dfrac{d\theta}{dt}(0) = 2$ (j) $\theta(0) = 1, \ \dfrac{d\theta}{dt}(0) = 3$

(k) $\theta(0) = -1, \ \dfrac{d\theta}{dt}(0) = 2$

(l) $\theta(0) = -1, \ \dfrac{d\theta}{dt}(0) = 3.$

3. Plot the solutions obtained in Problem 2 individually and then plot them simultaneously. Give a physical interpretation of the results.

4. If the computer algebra system you are using has a built-in function that approximates the solution of nonlinear differential equations, solve the problem

$$\begin{cases} \dfrac{d^2\theta}{dt^2} + \dfrac{1}{2}\dfrac{d\theta}{dt} + \sin\theta = 0 \\ \theta(0) = \theta_0, \ \dfrac{d\theta}{dt}(0) = v_0 \end{cases}$$

with the initial conditions stated in Problem 2. Compare the results you obtain with each.

5. Use a built-in computer algebra system function to approximate the solution of the pendulum problem with a variable damping coefficient

$$\begin{cases} \dfrac{d^2\theta}{dt^2} + \dfrac{1}{2}(\theta^2 - 1)\dfrac{d\theta}{dt} + \theta = 0 \\ \theta(0) = \theta_0, \ \dfrac{d\theta}{dt}(0) = v_0 \end{cases}$$

using the initial conditions stated in Problem 2. Compare these results with those of Problem 4.

6. Repeat Problem 5 using

$$\begin{cases} \dfrac{d^2\theta}{dt^2} + \dfrac{1}{2}(\theta^2 - 1)\dfrac{d\theta}{dt} + \sin\theta = 0 \\ \theta(0) = \theta_0, \ \dfrac{d\theta}{dt}(0) = v_0 \end{cases}$$

1. Use the linear approximation of the model of the simple pendulum to determine the motion of a pendulum with rod length $L = 2$ ft subject to the following sets of initial conditions:

   (a) $\theta(0) = 0.05$, $\theta'(0) = 0$

   (b) $\theta(0) = 0.05$, $\theta'(0) = 1$

   (c) $\theta(0) = 0.05$, $\theta'(0) = -1$

   In each case, determine the maximum displacement (in absolute value).

2. Consider the situation indicated in Exercise 1. However, use the initial velocity $\theta'(0) = 2$. How does the maximum displacement (in absolute value) differ from that in Exercise 1(**b**)–(**c**)?

*3. Suppose that the pendulum in Exercise 1 is subjected to a resistive force with damping coefficient $b = 4\sqrt{7}$. Solve the initial-value problems given in Exercise 1, and compare the resulting motion to the undamped case.

4. Verify that $\theta(t) = C_1 \cos \omega t + C_2 \sin \omega t$ where $\omega = \sqrt{\dfrac{g}{L}}$ satisfies the equation $\dfrac{d^2\theta}{dt^2} + \dfrac{g}{L}\theta = 0$.

5. Show that $\theta(t) = \theta_0 \cos \omega t + \dfrac{v_0}{\omega} \sin \omega t$ where $\omega = \sqrt{\dfrac{g}{L}}$ is the solution of the initial-value problem

$$\begin{cases} \dfrac{d^2\theta}{dt^2} + \dfrac{g}{L}\theta = 0 \\ \theta(0) = \theta_0, \; \dfrac{d\theta}{dt}(0) = v_0 \end{cases}$$

6. Let $\theta(t) = A \cos(\omega t - \phi)$. Use $\cos(a + b) = \cos a \cos b - \sin a \sin b$ with the solution in Exercise 5 to find A so that $\theta(t)$ satisfies the initial-value problem in Exercise 5.

*7. Show that the phase angle in $\theta(t) = A \cos(\omega t - \phi)$ is

$$\phi = \cos^{-1}\left(\frac{\theta_0}{\sqrt{\theta_0{}^2 + \dfrac{v_0{}^2}{\omega^2}}} \right).$$

In Exercises 8–11, approximate the period of the motion of the pendulum using the given length.

8. $L = 1$ meter

9. $L = 2$ meters

10. $L = 2$ feet

*11. $L = 8$ feet

12. If $L = 1$ meter, how many ticks does the clock make in one minute if it ticks once for each time the pendulum makes a complete swing?

13. Assuming that a clock ticks once each time the pendulum makes a complete swing, how long (in meters) does the pendulum need to be in order for the clock to tick once per second?

14. In Exercise 13, how long (in feet) does the pendulum need to be in order for the clock to tick once per second?

*15. For the undamped problem

$$\begin{cases} \dfrac{d^2\theta}{dt^2} + \dfrac{g}{L}\theta = 0 \\ \theta(0) = \theta_0, \; \dfrac{d\theta}{dt}(0) = v_0 \end{cases},$$

what is the maximum value of $\theta(t)$? For what values of t does the maximum occur?

16. For what values of t is the pendulum vertical in Exercise 15?

17. Solve the initial-value problem

$$\begin{cases} L\dfrac{d^2\theta}{dt^2} + b\dfrac{d\theta}{dt} + g\theta = 0 \\ \theta(0) = \theta_0, \; \dfrac{d\theta}{dt}(0) = v_0 \end{cases}$$

Determine restrictions on the parameters L, b, and g that correspond to overdamping, critical damping, and underdamping. Describe the physical situation in each case.

18. Solve the initial-value problem

$$\begin{cases} L\dfrac{d^2\theta}{dt^2} + b\dfrac{d\theta}{dt} + g\theta = F \cos \gamma t \\ \theta(0) = 0, \; \dfrac{d\theta}{dt}(0) = 0 \end{cases}$$

assuming that $b^2 - 4gL < 0$. Describe the motion of the pendulum as $t \to \infty$.

*19. Consider **Van-der-Pol's equation**

$$\frac{d^2x}{dt^2} + \varepsilon(x^2 - 1)\frac{dx}{dt} + x = 0$$

where ε is a small positive number. Notice that this equation has a nonconstant damping coefficient. Because ε is small, we can approximate a solution of Van-der-Pol's equation with $x(t) = A\cos\omega t$, a solution of $\frac{d^2x}{dt^2} + x = 0$ (the equation obtained when $\varepsilon = 0$). This method of approximation is called **harmonic balance.**

(a) Substitute $x(t) = A\cos\omega t$ into the nonlinear term in Van-der-Pol's equation to obtain

$$\varepsilon(x^2 - 1)\frac{dx}{dt} = -\varepsilon A\omega\left(\frac{1}{4}A^2 - 1\right)\sin\omega t$$

$$-\frac{1}{4}A^3\sin 3\omega t.$$

(b) If we ignore the term involving the higher harmonic $\sin 3\omega t$, we have $\varepsilon(x^2 + 1)\frac{dx}{dt} \approx$

$$-\varepsilon A\omega\left(\frac{1}{4}A^2 - 1\right)\sin\omega t = \varepsilon\left(\frac{1}{4}A^2 - 1\right)\frac{dx}{dt}.$$

Substitute this expression into Van-der-Pol's equation to obtain the linear equation

$$\frac{d^2x}{dt^2} + \varepsilon\left(\frac{1}{4}A^2 - 1\right)\frac{dx}{dt} + x = 0.$$

(c) If $A = 2$ in the linear equation in **(b)**, is the approximate solution periodic?

(d) If $A \neq 2$ in the linear equation in **(b)**, is the approximate solution periodic?

20. Comment on the behavior of solutions obtained in Exercise 19 if **(a)** $A < 2$ and **(b)** $A > 2$.

Section 5.1

Hooke's law
$F = ks$

Simple harmonic motion
The initial-value problem

$$\begin{cases} m\frac{d^2x}{dt^2} + kx = 0 \\ x(0) = \alpha, \frac{dx}{dt}(0) = \beta \end{cases}$$

has solution

$$x(t) = \alpha\cos\omega t + \frac{\beta}{\omega}\sin\omega t,$$

where $\omega = \sqrt{k/m}$; the amplitude of the solution is

$$A = \sqrt{\alpha^2 + \beta^2/\omega^2}.$$

Section 5.2

Damped motion
$$m\frac{d^2x}{dt^2} + c\frac{dx}{dt} + kx = 0$$

Overdamped
$c^2 - 4mk > 0$

Critically damped
$c^2 - 4mk = 0$

Underdamped
$c^2 - 4mk < 0$

Section 5.3

Forced motion
$$m\frac{d^2x}{dt^2} + c\frac{dx}{dt} + kx = f(t)$$

Kirchhoff's law

The sum of the voltage drops across the circuit elements is equivalent to the voltage $E(t)$ impressed on the circuit.

L-R-C Circuit

$$\begin{cases} L\dfrac{d^2Q}{dt^2} + R\dfrac{dQ}{dt} + \dfrac{1}{C}Q = E(t) \\ Q(0) = Q_0, \ I(0) = \dfrac{dQ}{dt}(0) = I_0 \end{cases}$$

Deflection of a Beam

$$EI\dfrac{d^4s}{dx^4} = w(x)$$

Motion of a Pendulum

The initial-value problem

$$\begin{cases} \dfrac{d^2\theta}{dt^2} + \dfrac{g}{L}\theta = 0 \\ \theta(0) = \theta_0, \ \dfrac{d\theta}{dt}(0) = v_0 \end{cases}$$

has solution

$$\theta(t) = \theta_0 \cos \omega t + \dfrac{v_0}{\omega} \sin \omega t.$$

CHAPTER 5 REVIEW EXERCISES

1. An object weighing 32 lb stretches a spring 6 in. If the object is lowered 4 in. below the equilibrium and released from rest, determine the displacement of the object, assuming there is no damping. What is the maximum displacement of the object from equilibrium? When does the object first pass through the equilibrium position? How often does the object return to the equilibrium position?

2. If the object in Exercise 1 is released from a point 3 in. above equilibrium with a downward initial velocity of 1 ft/s, determine the displacement of the object, assuming there is no damping. What is the maximum displacement of the object from equilibrium? When does the object first pass through the equilibrium position? How often does the object return to the equilibrium position?

*3. An object of mass 5 kg is attached to the end of a spring with spring constant $k = 65$ N/m. If the object is released from the equilibrium position with an upward initial velocity of 1 m/s, determine the displacement of the object assuming the force due to damping is $F_R = 20\dfrac{dx}{dt}$. Find $\lim_{t\to\infty} x(t)$ and the quasiperiod. What is the maximum displacement of the object from equilibrium? When does the object first pass through the equilibrium position?

4. If the object in Exercise 3 is released from a point 1 m below equilibrium with zero initial velocity, determine the displacement. Find $\lim_{t\to\infty} x(t)$ and the quasiperiod. What is the maximum displacement of the object from equilibrium? When does the object first pass through the equilibrium position?

5. An object of mass 4 slugs is attached to a spring with spring constant $k = 16$ lb/ft. If there is no damping and the object is subjected to the forcing function $f(t) = 4$, determine the displacement function $x(t)$ if $x(0) = \dfrac{dx}{dt}(0) = 0$. What is the maximum displacement of the object from equilibrium? When does the object first pass through the equilibrium position?

6. If the object in Exercise 5 is subjected to the forcing function $f(t) = 4 \cos 2t$, determine the displacement. Find $\lim_{t\to\infty} x(t)$ if it exists. Describe the physical phenomenon that occurs.

*7. If the object in Exercise 5 is subjected to the forcing function $f(t) = 4 \cos t$, determine the displacement. Describe the physical phenomenon that occurs. Find the enveloping functions.

8. An object of mass 2 slugs is attached to a spring with spring constant $k = 5$ lb/ft. If the resistive force is

$F_R = 6\dfrac{dx}{dt}$ and the external force is $f(t) = 12 \cos 2t$, determine the displacement if $x(0) = \dfrac{dx}{dt}(0) = 0$. What is the steady-state solution? What is the transient solution?

9. Find the charge and current in the *L-R-C* circuit if $L = 4$ H, $R = 80\ \Omega$, $C = 1/436$ farad, and $E(t) = 100$ if $Q(0) = \dfrac{dQ}{dt}(0) = 0$. Find $\lim_{t \to \infty} Q(t)$ and $\lim_{t \to \infty} I(t)$.

10. Find the charge and current in the *L-R-C* circuit in Exercise 9 if $E(t) = 100 \sin 2t$. Find $\lim_{t \to \infty} Q(t)$ and $\lim_{t \to \infty} I(t)$. How do these limits compare to those in Exercise 9?

*11. Find the charge and current in the *L-R-C* circuit if $L = 1$ H, $R = 0\ \Omega$, $C = 10^{-4}$ farad, and $E(t) = 220$ if $Q(0) = \dfrac{dQ}{dt}(0) = 0$. Find $\lim_{t \to \infty} Q(t)$ and $\lim_{t \to \infty} I(t)$.

12. Find the charge and current in the *L-R-C* circuit in Exercise 11 if $E(t) = 100 \sin 10t$. Find $\lim_{t \to \infty} Q(t)$ and $\lim_{t \to \infty} I(t)$.

13. Determine the shape of the beam of length 10 with constants E and I such that $EI = 1$, weight distribution $w(x) = x(10 - x)$, and fixed-end boundary conditions at $x = 0$ and $x = 10$.

14. Determine the shape of the beam in Exercise 13 if there are fixed-end boundary conditions at $x = 0$ and a sliding clamped end at $x = 10$.

*15. Determine the shape of the beam in Exercise 13 if there are fixed-end boundary conditions at $x = 0$ and a free end at $x = 10$.

16. Determine the shape of the beam in Exercise 13 if there are fixed-end boundary conditions at $x = 0$ and simple support at $x = 10$.

17. Use the linear approximation of the model of the simple pendulum to determine the motion of a pendulum with rod length $L = \frac{1}{2}$ ft subject to the initial conditions $\theta(0) = 1$ and $\dfrac{d\theta}{dt}(0) = 0$. What is the maximum displacement of the pendulum from the vertical position? When does the pendulum first pass through the vertical position?

18. If the initial conditions in Exercise 17 are $\theta(0) = 0$ and $\dfrac{d\theta}{dt}(0) = -1$, what is the maximum displacement of the pendulum from the vertical position? When does the pendulum first return to the vertical position?

*19. How does the motion of the pendulum in Exercise 17 differ if it undergoes the damping force $F_R = 8\dfrac{d\theta}{dt}$?

20. Solve the model in Exercise 17 with the damping force $F_R = 8\sqrt{3}\dfrac{d\theta}{dt}$. How does the motion differ from that in Exercise 19?

21. Undamped torsional vibrations (rotations back and forth) of a wheel attached to a thin elastic rod or wire satisfy the differential equation $I_0\theta'' + k\theta = 0$ where θ is the angle measured from the state of equilibrium, I_0 is the polar moment of inertia of the wheel about its center, and k is the torsional stiffness of the rod. Solve this equation if $\dfrac{k}{I_0} = 13.69 \text{ sec}^{-2}$, the initial angle is $15° \approx 0.2168$ rad, and the initial angular velocity is $10° \text{ sec}^{-1} \approx 0.1745 \text{ rad} \cdot \text{sec}^{-1}$.

22. Determine the displacement of the spring-mass system with mass 0.250 kg, spring constant $k = 2.25 \text{ kg/sec}^2$, and driving force $f(t) = \cos t - 4 \sin t$ if there is no damping, zero initial position, and zero initial velocity. For what frequency of the driving force would there be resonance?

23. The differential equation
$$y'' + y = \begin{cases} 1 - t^2, & 0 \le t \le 1 \\ 0, & t > 1 \end{cases}, \quad y(0) = y'(0) = 0$$
can be thought of as an undamped system in which a force F acts during the interval of time $0 \le t \le 1$. This is the situation that occurs in a gun barrel when a shell is fired. The barrel is braked with heavy springs (see Figure 5.28). Solve this initial-value problem.

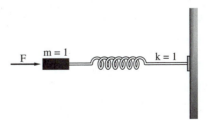

F   m = 1   k = 1

Figure 5.28

24. Consider a buoy in the shape of a cylinder of radius r, height h, and density ρ where $\rho \le 0.5$ g/cm^3 (the density of water is 1 g/cm^3). Initially, the buoy sits with its base on the surface of the water. It is then released so that it is acted upon by two forces: the force of gravity (in the downward direction) equal to the weight of the buoy, $F_1 = mg = \rho\pi r^2 hg$; and the force of buoyancy (in the upward direction) equal to the weight of the displaced water, $F_2 = \pi r^2 xg$, where $x = x(t)$ is the depth of the base of the cylinder from the surface of the water at time t. Using Newton's second law of motion with $m = \rho\pi r^2 h$, we have the differential equation $\rho\pi r^2 hx'' = -\pi r^2 xg$ or $\rho h x'' + gx = 0$. Find the displacement of the buoy if $x(0) = 1$ and $x'(0) = 0$. What is the period of the solution? What is the amplitude of the solution?

25. A cube-shaped buoy of side length ℓ and mass density ρ per unit volume is floating in a liquid of mass density ρ_0 per unit volume, where $\rho_0 > \rho$. If the buoy is slightly submerged into the liquid and released, it oscillates up and down. If there is no damping and no air resistance, the buoy is acted upon by two forces: the force of gravity (in the downward direction) which is equal to the weight of the buoy, $F_1 = mg = \rho\ell^3 g$; and the force of buoyancy (in the upward direction) which is equal to the weight of the displaced water, $F_2 = \ell^2 xg\rho_0$ where $x = x(t)$ is the depth of the base of the buoy from the surface of the water at time t. Then by Newton's second law with $m = \rho\ell^3$, we have the differential equation $\rho\ell^2 x'' + g\rho_0 x = 0$. Determine the amplitude of the motion if $\rho_0 = 1$ g/cm^3, $\rho = 0.25$ g/cm^3, $\ell = 100$ cm, $g = 980$ cm/sec^2, $x(0) = 25$ cm and $x'(0) = 0$.

26. A rabbit starts at the origin and runs with speed a due north toward a hole in a fence located at the point $(0, d)$ on the y-axis. At the same time, a dog starts at the point $(c, 0)$ on the x-axis, running at speed b in pursuit of the rabbit. (Note: The dog runs directly toward the rabbit.) The slope of the tangent line to the dog's path is $\dfrac{dy}{dx} = -\dfrac{at - y}{x}$, which can be written as $xy' = y - at$. Differentiate both sides of this equation with respect to t to obtain $xy'' = -a\dfrac{dt}{dx}$. If s is the length of the arc from $(c, 0)$ along the dog's path, then $\dfrac{ds}{dt} = b$ is the dog's speed. Also, $\dfrac{ds}{dx} = -\sqrt{1 + y'^2}$

where the negative sign indicates that s increases as x decreases. By the chain rule,

$$\frac{dt}{dx} = \frac{dt}{ds}\frac{ds}{dx} = -\frac{1}{b}\sqrt{1 + y'^2},$$

so substitution into the equation $xy'' = -a\dfrac{dt}{dx}$ yields the equation $xy'' = k\sqrt{1 + y'^2}$ where $k = \dfrac{a}{b}$. Make the substitution

$$p = y' \text{ or } y'' = \frac{dp}{dx}$$

to find the path of the dog with the initial condition $y'(c) = 0$. If $a < b$, when does the dog catch the rabbit? Does the problem make sense if $a = b$? (See Figure 5.29.)

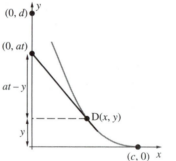

Figure 5.29

27. (Coulomb Damping) Damping that results from dry friction (like when an object slides over a dry surface) is called **Coulomb damping** or **dry-friction damping**. On the other hand, **viscous damping** is damping that can be represented by a term proportional to the velocity (like when an object such as a vibrating spring vibrates in air or when an object slides over a lubricated surface).

   Suppose that the kinetic coefficient of friction is μ. Friction forces oppose motion. Therefore, the force due to friction F is shown opposite the direction of motion in Figure 5.30. However, because F is a discontinuous function, we cannot use a single differential equation to model the motion as was done with previous damping problems. Instead, we have one equation for motion to the right and one equation for motion to the left:

   Motion to Right $x'' + \omega_n^2 x = -\dfrac{F}{m}$

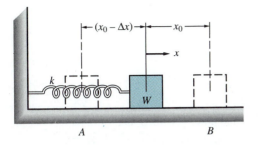

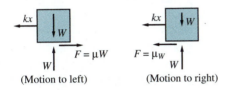

(Motion to left)   (Motion to right)

Figure 5.30   A System with Coulomb (Dry-Friction) Damping

and

Motion to Left $x'' + \omega_n^2 x = \dfrac{F}{m},$

where F represents the friction force, m the mass, and

$\omega_n^2 = \dfrac{k}{m}.$*

(a) Find a general solution of $x'' + \omega_n^2 x = \dfrac{F}{m}.$

(b) Solve the initial-value problem

$$\begin{cases} x'' + \omega_n^2 x = \dfrac{F}{m} \\ x(0) = x_0, \ x'(0) = 0 \end{cases}.$$

(c) Find the values of t for which the solution in **(b)** is valid.

(d) At the right-endpoint of the interval obtained in **(c)**, what is the displacement of the object?

(e) What must the force F be in order to guarantee that the object does not move under the given initial conditions?

28. (Self-Excited Vibrations) The differential equation that describes the motion of a spring-mass system with a single degree of freedom excited by the force Px' is

$$mx'' + cx' + kx = Px'$$

which can be rewritten as

$$x'' + \frac{c - P}{m}x' + \frac{k}{m}x = 0.†$$

(a) Show that the roots of the characteristic equations

of $x'' + \dfrac{c - P}{m}x' + \dfrac{k}{m}x = 0$ are

$$\frac{P - c}{2m} \pm \sqrt{\left(\frac{P - c}{2m}\right)^2 - \frac{k}{m}}.$$

(b) Show that if $P > c$ the motion of the system diverges **(dynamically unstable)**, if $P = c$ the solution is the solution for a free undamped system, and if $P < c$ the solution is the solution for a free damped system.

---

* M. L. James, G. M. Smith, J. C. Wolford, and P. W. Whaley, *Vibration of Mechanical and Structural Systems with Microcomputer Applications,* Harper & Row, 1989, pp. 70–72.

† Robert K. Vierck, *Vibration Analysis,* Second Edition, HarperCollins, 1979, pp. 137–139.

Differential Equations at Work:
Rack-and-Gear Systems

Consider the rack-and-gear system shown in Figure 5.31. Let T represent the kinetic energy of the system and U the change in potential energy of the system from its potential energy in the static-equilibrium position. The kinetic energy of a system is a function of the velocities of the system masses. The potential energy of a system consists of the strain energy U_e stored in elastic elements, and the energy U_g, which is a function of the vertical distances between system masses.

The rack-and-gear system consists of two identical gears of pitch radius r and centroidal mass moment of inertia $\bar{I}$, a rack of weight W, and a linear spring of stiffness k, length l, and a mass of γ per unit length. To determine the differential equation to model the motion, we differentiate the law of conservation, $T + U = $ constant, to obtain

$$\frac{d}{dt}(T + U) = 0.$$

We then use $T_{max} = U_{max}$ to determine the natural circular frequency of the system, where T_{max} represents the maximum kinetic energy and U_{max} the maximum potential energy.

Because the static displacement x_s of any point on the spring is proportional to its distance y from the spring support, we can write

$$x_s = \frac{yx}{l}.$$

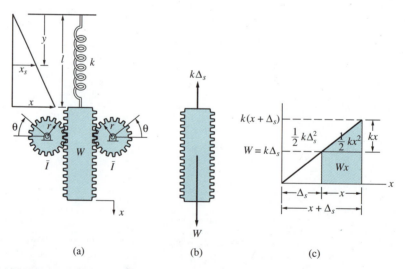

(a)   (b)   (c)

Figure 5.31   (a) A Rack-and-Gear System   (b) Static-Equilibrium Position
(c) Spring-Force Diagram

When the rack is displaced a distance x below the equilibrium position, the gears have the angular displacements θ, which are related to the displacement of the rack by

$$\theta = \frac{x}{r}.$$

Then the angular velocity of each gear is given (through differentiation of θ with respect to t) by

$$\dot{\theta} = \frac{\dot{x}}{r}.$$

(Notice that we use a "dot" to indicate differentiation with respect to t. This is a common practice in physics and engineering.) Because $x = r\theta$, we have

$$x_s = \frac{yr\theta}{l},$$

so the velocity at any point on the spring is

$$\dot{x}_s = \frac{yr\dot{\theta}}{l}.$$

The kinetic energy of a differential element of length dy of the spring is

$$dT = \frac{1}{2}\gamma(\dot{x}_s)^2 \, dy.$$

The total kinetic energy of the system in terms of θ is

$$T = \underbrace{\bar{I}\dot{\theta}^2}_{\text{Gears}} + \underbrace{\frac{1}{2}\frac{W}{g}(r\dot{\theta})^2}_{\text{Rack}} + \underbrace{\frac{\gamma}{2}\int_0^l \left(\frac{yr\dot{\theta}}{l}\right)^2 dy}_{\text{Spring}}$$

or

$$T = \left[\bar{I} + \frac{1}{2}\frac{W}{g}r^2 + \frac{m_3 r^2}{2\cdot 3}\right]\dot{\theta}^2$$

where $m_3 = \gamma l$ is the total mass of the spring.

The change in the strain energy U_e for a positive downward displacement x of the rack from the static-equilibrium position (where the spring is already displaced Δ_s from its free length (see Figure 5.30(b)) is the area of the shaded region (see Figure 5.30(c)) given by

$$U_e = \frac{1}{2}kx^2 + Wx.$$

Similarly, the change in potential energy U_g as the rack moves a distance x below the static-equilibrium position is

$$U_g = -Wx$$

and the total change in the potential energy is

$$U = U_e + U_g = \left(\frac{1}{2}kx^2 + Wx \right) + (-Wx) = \frac{1}{2}kx^2$$

or

$$U = \frac{1}{2}k(r\theta)^2.$$

Substitution of these expressions into the equation $T + U = $ constant yields

$$\left[\bar{I} + \frac{1}{2}\frac{W}{g}r^2 + \frac{m_3 r^2}{2(3)} \right]\dot{\theta}^2 + \frac{1}{2}k(r\theta)^2 = \text{constant.}$$

Differentiating with respect to t then gives us the differential equation

$$\ddot{\theta} + \left[\frac{kr^2}{2\bar{I} + \dfrac{W}{g}r^2 + \dfrac{m_3 r^2}{3}} \right]\theta = 0.*$$

1. Determine the natural frequency ω_n of the system.
2. Determine $\theta(t)$ if $\theta(0) = 0$ and $\dot{\theta}(0) = \theta_0 \omega_n$.
3. Find $T_{\max}$ and $U_{\max}$. Determine the natural frequency with these two quantities. Compare this value of ω_n with that obtained above.
4. How does the natural frequency change as r increases?

---

* M. L. James, G. M. Smith, J. C. Wolford, P. W. Whaley, *Vibration of Mechanical and Structural Systems with Microcomputer Applications,* Harper & Row, 1989, pp. 82–86.

6

ORDINARY DIFFERENTIAL EQUATIONS WITH NONCONSTANT COEFFICIENTS

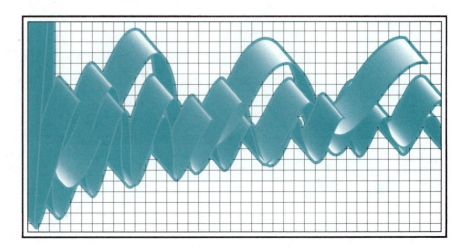

In Chapters 4 and 5, we studied techniques used to solve higher-order ordinary differential equations with constant coefficients and typical applications of these equations. In some cases, similar techniques can be used to solve differential equations with nonconstant coefficients. In other cases, different techniques must be used.

6.1   Cauchy-Euler Equations

Second-Order Cauchy-Euler Equations • Higher-Order Cauchy-Euler Equations • Variation of Parameters

In the previous two chapters, we solved linear differential equations with constant coefficients. Generally, solving an arbitrary differential equation is a formidable if not impossible task, particularly when the coefficients are not constants. However, we are able to solve certain equations with variable coefficients using techniques similar to those discussed previously. We begin by considering differential equations of the form

$$a_n x^n y^{(n)} + a_{n-1} x^{n-1} y^{(n-1)} + \cdots + a_1 xy' + a_0 y = g(x),$$

where $a_0, a_1, a_2, \ldots, a_n$ are constants, called **Cauchy-Euler** equations. (Notice that in Cauchy-Euler equations the order of each derivative in the equation equals the power of x in the corresponding coefficient.) Euler observed that equations of this type are reduced to linear equations with constant coefficients with the substitution $x = e^t$.*

Definition 6.1 Cauchy-Euler Equation

A **Cauchy-Euler differential equation** is an equation of the form

$$a_n x^n y^{(n)} + a_{n-1} x^{n-1} y^{(n-1)} + \cdots + a_1 xy' + a_0 y = g(x)$$

where $a_0, a_1, a_2, \ldots, a_n$ are constants.

Second-Order Cauchy-Euler Equations

We first consider the second-order homogeneous Cauchy-Euler equation

$$ax^2 y'' + bxy' + cy = 0.$$

Notice that the coefficient of y'' is zero if $x = 0$, so we must restrict our domain to either $x > 0$ or $x < 0$ in order to ensure that the theory of second-order equations stated in Section 4.1 holds.

Suppose that a solution to this differential equation is of the form $y = x^m$ for some constant m. Substitution of $y = x^m$ with derivatives $y' = mx^{m-1}$ and $y'' = m(m-1)x^{m-2}$ yields

$$ax^2 y'' + bxy' + cy = ax^2 m(m-1)x^{m-2} + bxmx^{m-1} + cx^m$$
$$= x^m[am(m-1) + bm + c] = 0.$$

Then $y = x^m$ is a solution of $ax^2 y'' + bxy' + cy = 0$ if m satisfies

$$am(m-1) + bm + c = 0,$$

which is called the **auxiliary** (or **characteristic**) **equation** associated with the Cauchy-Euler equation of order two. The solutions of the auxiliary equation completely deter-

---

* Carl B. Boyer, *A History of Mathematics*, Princeton University Press, 1985, p. 496.

mine the general solution of the homogeneous Cauchy-Euler equation of order two. By the quadratic formula, the solutions of the auxiliary equation

$$am(m - 1) + bm + c = am^2 + (b - a)m + c = 0$$

are

$$m_{1,2} = \frac{-(b - a) \pm \sqrt{(b - a)^2 - 4ac}}{2a}.$$

We obtain two real roots, one repeated real root, or a complex conjugate pair depending on the values of a, b, and c. We state a general solution that corresponds to the different types of roots as follows. (We derive these solutions later in the section.)

Theorem 6.1 Solving Second-Order Cauchy-Euler Equations

Let $ax^2y'' + bxy' + cy = 0$ be a homogeneous second-order Cauchy-Euler equation and let m_1 and m_2 be the solutions of the equation $am^2 + (b - a)m + c = 0$.
(a) If $m_1 \neq m_2$ are real, a general solution is

$$y = c_1x^{m_1} + c_2x^{m_2}.$$

(b) If m_1, m_2 are real and $m_1 = m_2$, a general solution is

$$y = c_1x^{m_1} + c_2x^{m_1} \ln x.$$

(c) If $m_1 = \overline{m_2} = \alpha + i\beta$, $\beta \neq 0$, a general solution is

$$y = x^{\alpha}[c_1 \cos(\beta \ln x) + c_2 \sin(\beta \ln x)].$$

Example 1   Solve $3x^2y'' - 2xy' + 2y = 0$, $x > 0$.

Solution   If $y = x^m$, $y' = mx^{m-1}$ and $y'' = m(m - 1)x^{m-2}$. Substitution into the differential equation yields

$$3x^2y'' - 2xy' + 2y = 3x^2m(m - 1)x^{m-2} - 2xmx^{m-1} + 2x^m$$
$$= x^m[3m(m - 1) - 2m + 2] = 0.$$

The auxiliary equation is

$$3m(m - 1) - 2m + 2 = 3m^2 - 5m + 2 = (3m - 2)(m - 1) = 0$$

with roots $m_1 = \frac{2}{3}$ and $m_2 = 1$. Therefore, a general solution is $y = c_1x^{2/3} + c_2x$.

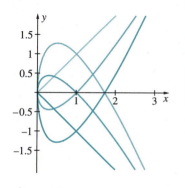

Figure 6.1

Several solutions to the equation are graphed in Figure 6.1. Identify the graph of the solution that satisfies the initial conditions $y(1) = 1$ and $y'(1) = -1$.

Example 2

Solve $x^2 y'' - xy' + y = 0$, $x > 0$.

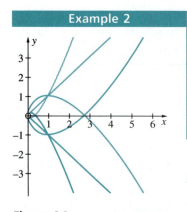

Figure 6.2

Solution

In this case, the auxiliary equation is

$$m(m - 1) - m + 1 = m^2 - 2m + 1 = (m - 1)^2 = 0$$

with root $m_1 = m_2 = 1$ of multiplicity two. A general solution is $y = c_1 x + c_2 x \ln x$.

Graphs of the solutions corresponding to various values of c_1 and c_2 are shown in Figure 6.2. Regardless of the choice of c_1 and c_2, calculate $\lim_{x \to 0^+} (c_1 x + c_2 x \ln x)$. Why is a right-hand limit necessary?

Example 3

Solve $\begin{cases} x^2 y'' - 5xy' + 10y = 0 \\ y(1) = 1, y'(1) = 0 \end{cases}$.

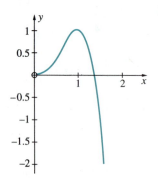

Figure 6.3

Solution

The auxiliary equation is given by

$$m(m - 1) - 5m + 10 = m^2 - 6m + 10 = 0$$

with complex conjugate roots $m = \dfrac{6 \pm \sqrt{36 - 40}}{2} = 3 \pm i$. A general solution is

$$y = x^3[c_1 \cos(\ln x) + c_2 \sin(\ln x)], \quad x > 0.$$

To find the values of c_1 and c_2 so that y satisfies the initial conditions, we first calculate

$$y' = x^2[(3c_1 + c_2) \cos(\ln x) + (3c_2 - c_1) \sin(\ln x)].$$

Applying the conditions $y(1) = 1$ and $y'(1) = 0$ yields $c_1 = 1$ and $3c_1 + c_2 = 0$ so $c_2 = -3$. The solution to the initial-value problem is $y = x^3[\cos(\ln x) - 3 \sin(\ln x)]$, which is graphed in Figure 6.3.

Show that the solution to the initial-value problem $\begin{cases} x^2 y'' - 5xy' + 10y = 0 \\ y(1) = a, y'(1) = b \end{cases}$
is $y = x^3(a \cos(\ln x) + (b - 3a) \sin(\ln x))$.

We verify the general solution given in Theorem 6.1(a) for the unequal real roots $m_1 \neq m_2$ by first noting that both $y = x^{m_1}$ and $y = x^{m_2}$ are solutions to the equation. (Why?) In addition, the set $S = \{x^{m_1}, x^{m_2}\}$ is linearly independent because the Wronskian, $W(S)$, is not zero:

$$W(S) = \begin{vmatrix} x^{m_1} & x^{m_2} \\ \dfrac{d}{dx}(x^{m_1}) & \dfrac{d}{dx}(x^{m_2}) \end{vmatrix} = \begin{vmatrix} x^{m_1} & x^{m_2} \\ m_1 x^{m_1-1} & m_2 x^{m_2-1} \end{vmatrix} = (m_2 - m_1) x^{m_1 + m_2 - 1} \neq 0.$$

So S is a fundamental set of solutions and, consequently, $y = c_1 x^{m_1} + c_2 x^{m_2}$ is a general solution of the Cauchy-Euler equation.

For the repeated real root $m_1 = m_2$ in Theorem 6.1(b), we verify the solution by recalling that if we have a solution $y_1(x)$ of the differential equation $y'' + p(x)y' + q(x)y = 0$, we construct a second linearly independent solution with the formula $y_2(x) = y_1(x) \int \dfrac{e^{-\int p(x)\,dx}}{[y_1(x)]^2}\,dx$. (See Reduction of Order.) For $m_1 = m_2$, the quantity under the radical in

$$m = \frac{-(b-a) \pm \sqrt{(b-a)^2 - 4ac}}{2a}$$

must be zero, so $m_1 = m_2 = m = \dfrac{-(b-a)}{2a}$ and a known solution is $y_1(x) = x^{-(b-a)/2a}$. To use the formula $y_2(x) = y_1(x) \int \dfrac{e^{-\int p(x)\,dx}}{[y_1(x)]^2}\,dx$, we write the equation $ax^2 y'' + bxy' + cy = 0$ as

$$y'' + \frac{b}{ax}y' + \frac{c}{ax^2}y = 0.$$

Hence, $p(x) = \dfrac{b}{ax}$. This gives us

$$y_2(x) = x^m \int \frac{e^{-\int (b/ax)\,dx}}{(x^m)^2}\,dx = x^m \int \frac{e^{-\int (b/ax)\,dx}}{(x^m)^2}\,dx = x^m \int \frac{e^{-(b/a)\ln x}}{x^{2m}}\,dx$$

$$= x^m \int \frac{x^{-(b/a)}}{x^{2[-(b-a)/2a]}}\,dx = x^m \int \frac{x^{-(b/a)}}{x^{-(b-a)/a}}\,dx = x^m \int \frac{1}{x}\,dx = x^m \ln x$$

so a general solution is $y = c_1 x^{m_1} + c_2 x^{m_1} \ln x$.

The solution associated with the complex conjugate roots is derived in a similar way. If $m_{1,2} = \alpha \pm \beta i$, a general solution is

$$y(x) = Ax^{\alpha + \beta i} + Bx^{\alpha - \beta i} = x^\alpha (Ax^{\beta i} + Bx^{-\beta i}),$$

where A and B are arbitrary constants. However, by Euler's formula,

$$x^{\beta i} = (e^{\ln x})^{\beta i} = e^{i\beta \ln x} = \cos(\beta \ln x) + i \sin(\beta \ln x)$$

and

$$x^{-\beta i} = (e^{\ln x})^{-\beta i} = e^{-i\beta \ln x} = \cos(\beta \ln x) - i \sin(\beta \ln x)$$

Adding these two solutions and dividing by 2 results in $\dfrac{1}{2}(x^{\beta i} + x^{-\beta i}) = \cos(\beta \ln x)$, while subtracting the second from the first and dividing by $2i$ results in

$$\frac{1}{2i}(x^{\beta i} - x^{-\beta i}) = \sin(\beta \ln x).$$

Because $y(x) = x^{\alpha}(Ax^{\beta i} + Bx^{-\beta i})$ is a solution for any constants A and B, the function $y_1(x) = x^{\alpha}\left(\dfrac{1}{2}x^{\beta i} + \dfrac{1}{2}x^{-\beta i}\right) = x^{\alpha}\cos(\beta \ln x)$ with $A = B = \dfrac{1}{2}$ is also a solution. Similarly, the function $y_2(x) = x^{\alpha}\left(\dfrac{1}{2i}x^{\beta i} - \dfrac{1}{2i}x^{-\beta i}\right) = x^{\alpha}\sin(\beta \ln x)$ with $A = \dfrac{1}{2i}$ and $B = -\dfrac{1}{2i}$ is a solution. Therefore, the linear combination of these two linearly independent solutions forms a *real* general solution:

$$y = c_1 y_1(x) + c_2 y_2(x) = c_1 x^{\alpha}\cos(\beta \ln x) + c_2 x^{\alpha}\sin(\beta \ln x).$$

Higher-Order Cauchy-Euler Equations

The auxiliary equation of higher-order Cauchy-Euler equations is defined in the same way, and solutions of higher-order homogeneous Cauchy-Euler equations are determined in the same manner as solutions of higher-order homogeneous differential equations with constant coefficients.

Example 4

Solve $2x^3 y''' - 4x^2 y'' - 20xy' = 0$, $x > 0$.

Solution In this case, if we assume that $y = x^m$ for $x > 0$, we have the derivatives $y' = mx^{m-1}$, $y'' = m(m-1)x^{m-2}$, and $y''' = m(m-1)(m-2)x^{m-3}$. Substitution into the differential equation and simplification yields $(2m^3 - 10m^2 - 12m)x^m = 0$. Because $x^m \neq 0$, we must solve

$$(2m^3 - 10m^2 - 12m) = m(2m + 2)(m - 6) = 0$$

for m. Hence, $m_1 = 0$, $m_2 = -1$, and $m_3 = 6$, and a general solution is $y = c_1 + c_2 x^{-1} + c_3 x^6$.

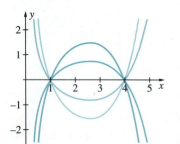

Figure 6.4 Several solutions to the $2x^3 y''' - 4x^2 y'' - 20xy' = 0$ equation that satisfy the boundary conditions $y(1) = y(4) = 0$

Find and classify all relative extrema of $y(x)$ in the interval $(1, 4)$ if y satisfies the boundary conditions $y(1) = y(4) = 0$. (See Figure 6.4.)

If a root m of the auxiliary equation is repeated r times, the r linearly independent solutions that correspond to m are x^m, $x^m \ln x$, $x^m(\ln x)^2$, ..., $x^m(\ln x)^{r-1}$.

Example 5

Solve the initial-value problem

$$\begin{cases} x^4 y^{(4)} + 4x^3 y''' + 11x^2 y'' - 9xy' + 9y = 0, x > 0 \\ y(1) = 1, y'(1) = -9, y''(1) = 27, y'''(1) = 1 \end{cases}.$$

Solution Substitution of $y(x) = x^m$, $x > 0$ into the differential equation results in

$$x^4 m(m-1)(m-2)(m-3)x^{m-4} + 4x^3 m(m-1)(m-2)x^{m-3}$$
$$+ 11x^2 m(m-1)x^{m-2} - 9xmx^{m-1} + 9x^m = 0$$

and simplification leads to the equation

$$(m^4 - 2m^3 + 10m^2 - 18m + 9)x^m = 0.$$

The auxiliary equation is

$$(m^4 - 2m^3 + 10m^2 - 18m + 9) = (m^2 + 9)(m - 1)^2 = 0.$$

The solutions are $m = \pm 3i$ and $m = 1$, which is a root of multiplicity two, so a general solution is

$$y(x) = c_1 \cos(3 \ln x) + c_2 \sin(3 \ln x) + c_3 x + c_4 x \ln x.$$

The factors of the auxiliary equation can be found with the help of a computer algebra system or through synthetic division.

The first three derivatives of this general solution are

$$y'(x) = c_3 + c_4 + 3c_2 x^{-1} \cos(3 \ln x) + c_4 \ln x - 3c_1 x^{-1} \sin(3 \ln x),$$
$$y''(x) = c_4 x^{-1} + x^{-2}[-(9c_1 + 3c_2) \cos(3 \ln x) + (3c_1 - 9c_2) \sin(3 \ln x)],$$

and

$$y'''(x) = -c_4 x^{-2} + x^{-3}[(27c_1 - 21c_2) \cos(3 \ln x) + (21c_1 + 27c_2) \sin(3 \ln x)].$$

Substitution of the initial conditions then yields the system of equations

$$\{c_1 + c_3 = 1,\ c_3 + c_4 + 3c_2 = -9,\ c_4 - 9c_1 - 3c_2 = 27,\ -c_4 + 27c_1 - 21c_2 = 1\},$$

which has the solution $\{c_1 = -\frac{12}{5},\ c_2 = -\frac{89}{30},\ c_3 = \frac{17}{5},\ c_4 = -\frac{7}{2}\}$. Therefore, the solution to the initial-value problem is $y(x) = -\frac{12}{5} \cos(3 \ln x) - \frac{89}{30} \sin(3 \ln x) + \frac{17}{5} x - \frac{7}{2} x \ln x.$

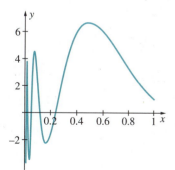

Figure 6.5

 A portion of the graph of this solution is shown in Figure 6.5. Determine $\lim_{x \to 0^+} (-\frac{12}{5} \cos(3 \ln x) - \frac{89}{30} \sin(3 \ln x) + \frac{17}{5} x - \frac{7}{2} x \ln x)$, if possible. (Hint: Graph $y(x)$ on the interval $[10^{-(n+1)}, 10^{-n}]$ for values of n like 4, 8, and 12. Is Figure 6.5 misleading? If so, describe how.)

Variation of Parameters

Of course, Cauchy-Euler equations can be nonhomogeneous, in which case the method of variation of parameters can be used to solve the problem. (Later, we show how the method of undetermined coefficients can be used.)

Example 6

Solve $x^2 y'' - xy' + 5y = x^{-1}$.

Solution We begin by finding a general solution to the corresponding homogeneous equation $x^2 y'' - xy' + 5y = 0$. The auxiliary equation is $m^2 - 2m + 5 = 0$ with roots $m_{1,2} = \dfrac{2 \pm \sqrt{4 - 20}}{2} = 1 \pm 2i$, so a general solution of the corresponding homogeneous equation is $y_h(x) = x(c_1 \cos(2 \ln x) + c_2 \sin(2 \ln x))$. A fundamental set of solutions for the corresponding homogeneous equation is $S = \{x \cos(2 \ln x), x \sin(2 \ln x)\}$, and the Wronskian is

$$W(S) = \begin{vmatrix} x \cos(2 \ln x) & x \sin(2 \ln x) \\ -x \sin(2 \ln x) \cdot \dfrac{2}{x} + \cos(2 \ln x) & x \cos(2 \ln x) \cdot \dfrac{2}{x} + \sin(2 \ln x) \end{vmatrix}$$

$$= 2x[\cos^2(2 \ln x) + \sin^2(2 \ln x)] = 2x.$$

To construct a particular solution by the method of variation of parameters, we first rewrite the equation in the form

$$y'' - \frac{1}{x}y' + \frac{5}{x^2}y = \frac{1}{x^3}.$$

Using the method of variation of parameters, we have

$$u_1(x) = -\int \frac{x \sin(2 \ln x) \cdot x^{-3}}{2x}\, dx = \frac{1}{8x^2}(\cos(2 \ln x) + \sin(2 \ln x)) \qquad \text{and}$$

$$u_2(x) = \int \frac{x \cos(2 \ln x) \cdot x^{-3}}{2x}\, dx = \frac{1}{8x^2}(\sin(2 \ln x) - \cos(2 \ln x)),$$

so a particular solution to the nonhomogeneous equation is

$$y_p(x) = u_1(x)y_1(x) + u_2(x)y_2(x)$$

$$= \frac{1}{8x^2}(\cos(2 \ln x) + \sin(2 \ln x)) \cdot x \cos(2 \ln x)$$

$$+ \frac{1}{8x^2}(\sin(2 \ln x) - \cos(2 \ln x)) \cdot x \sin(2 \ln x) = \frac{1}{8x}$$

and a general solution of the nonhomogeneous equation is

$$y(x) = y_h(x) + y_p(x) = x(c_1 \cos(2 \ln x) + c_2 \sin(2 \ln x)) + \frac{1}{8x}.$$

This general solution is plotted for various values of the constants c_1 and c_2 in Figure 6.6.

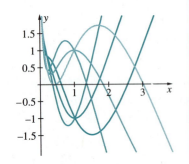

Figure 6.6

As indicated earlier, Euler solved these equations by making the substitution $x = e^t$ for $x > 0$. In doing this, the chain rule must be used to transform the derivatives with respect to x into derivatives with respect to t. If $x = e^t$, $t = \ln x$ so

$$\frac{dy}{dx} = \frac{dy}{dt}\frac{dt}{dx} = \frac{1}{x}\frac{dy}{dt}$$

and

$$\frac{d^2y}{dx^2} = \frac{d}{dx}\left(\frac{1}{x}\frac{dy}{dt}\right) = \frac{1}{x}\frac{d}{dt}\left(\frac{dy}{dt}\right)\frac{dt}{dx} + \frac{d}{dx}\left(\frac{1}{x}\right)\frac{dy}{dt} = \frac{1}{x^2}\frac{d^2y}{dt^2} - \frac{1}{x^2}\frac{dy}{dt}$$

$$= \frac{1}{x^2}\left(\frac{d^2y}{dt^2} - \frac{dy}{dt}\right).$$

We can solve Cauchy-Euler equations for negative values of x with the substitution x = −eᵗ. (See Exercise 55.)

Substitution of these derivatives into a second-order Cauchy-Euler equation yields a second-order linear differential equation with constant coefficients.

Example 7

Solve $x^2y'' - 3xy' + 13y = 4 + 3x$, $x > 0$.

Solution   After substitution of the appropriate derivatives and functions of t, we obtain the differential equation

$$x^2\frac{1}{x^2}\left(\frac{d^2y}{dt^2} - \frac{dy}{dt}\right) - 3x\frac{1}{x}\frac{dy}{dt} + 13y = 4 + 3e^t$$

which in simplified form is $\dfrac{d^2y}{dt^2} - 4\dfrac{dy}{dt} + 13y = 4 + 3e^t$. The characteristic equation of the corresponding homogeneous equation is $m^2 - 4m + 13 = 0$, which has complex conjugate roots $m_{1,2} = 2 \pm 3i$ and general solution $y_h(t) = e^{2t}(c_1 \cos 3t + c_2 \sin 3t)$. Using the method of undetermined coefficients to solve the equation

$$\frac{d^2y}{dt^2} - 4\frac{dy}{dt} + 13y = 4 + 3e^t,$$

we assume that the particular solution is $y_p(t) = A + Be^t$. The derivatives of this function are $\dfrac{dy_p}{dt}(t) = Be^t$ and $\dfrac{d^2y_p}{dt^2}(t) = Be^t$, so substitution into $\dfrac{d^2y}{dx^2} - 4\dfrac{dy}{dt} + 13y = 4 + 3e^t$ yields

$$Be^t - 4Be^t + 13A + 13Be^t = 10Be^t + 13A = 4 + 3e^t.$$

Therefore, $B = \frac{3}{10}$ and $A = \frac{4}{13}$, so $y_p(t) = \frac{4}{13} + \frac{3}{10}e^t$. A general solution in the variable t is given by

$$y(t) = e^{2t}(c_1 \cos 3t + c_2 \sin 3t) + \frac{4}{13} + \frac{3}{10}e^t.$$

Returning to the original variable $x = e^t$ (or $t = \ln x$) we have

$$y(x) = x^2(c_1 \cos(3 \ln x) + c_2 \sin(3 \ln x)) + \frac{4}{13} + \frac{3}{10}x.$$

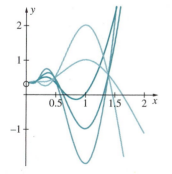

Figure 6.7

This general solution is plotted for various values of the arbitrary constants c_1 and c_2 in Figure 6.7. Identify the graph of the solution that satisfies the initial conditions $y(1) = 2$ and $y'(1) = 0$. Show that the solution that satisfies these *initial conditions is given by* $y(x) = x^2(\frac{181}{130}\cos(3\ln x) - \frac{401}{390}\sin(3\ln x)) + \frac{4}{13} + \frac{3}{10}x$ *and then approximate the first value of x greater than 10 for which (a)* $y'(x) = 0$ *and (b)* $y(x) = 0$.

*When solving nonhomogeneous Cauchy-Euler equations, the method of undetermined coefficients should be used **only** when the equation is transformed to a constant coefficient equation. On the other hand, the method of variation of parameters can be used with the original equation in standard form **or** with the transformed equation.*

IN TOUCH WITH TECHNOLOGY

1. Use a computer algebra system to assist in finding a general solution of the following Cauchy-Euler equations.

   (a) $x^3y''' + 16x^2y'' + 79xy' + 125y = 0$

   (b) $x^4y^{(4)} + 5x^3y''' - 12x^2y'' - 12xy' + 48y = 0$

   (c) $x^4y^{(4)} + 14x^3y''' + 55x^2y'' + 65xy' + 15y = 0$

   (d) $x^4y^{(4)} + 8x^3y''' + 27x^2y'' + 35xy' + 45y = 0$

   (e) $x^4y^{(4)} + 10x^3y''' + 27x^2y'' + 21xy' + 4y = 0$

2. (a) Solve the initial-value problem
   $$\begin{cases} x^3y''' + 9x^2y'' + 44xy' + 58y = 0 \\ y(1) = 2, y'(1) = 10, y''(1) = -2 \end{cases}.$$

   (b) Graph the solution on the interval [0.2, 1.8] and approximate all local minima and maxima of the solution on this interval.

3. (a) Solve the initial-value problem
   $$\begin{cases} 6x^2y'' + 5xy' - y = 0 \\ y(1) = a, y'(1) = b \end{cases}.$$

   (b) Find conditions on a and b so that $\lim_{x \to 0^+} y(x) = 0$. Graph several solutions to confirm your results.

   (c) Find conditions on a and b so that $\lim_{x \to \infty} y(x) = 0$. Graph several solutions to confirm your results.

   (d) If both a and b are not zero, is it possible to find a and b so that both $\lim_{x \to 0^+} y(x) = 0$ and $\lim_{x \to \infty} y(x) = 0$? Explain.

4. An equation of the form $f(x, y, y')y'' + g(x, y, y') = 0$ that satisfies the system
   $$\begin{cases} f_{xx} + 2pf_{xy} + p^2f_{yy} = g_{xp} + pg_{yp} - g_y \\ f_{xp} + pf_{yp} + 2f_y = g_{pp} \end{cases},$$
   where $p = y'$ is called an **exact second-order differential equation. (a)** Show that the equation
   $$x(y')^2 + yy' + xyy'' = 0$$
   is an exact second-order equation. **(b)** Show that $\phi(x, y, p) = h(x, y) + \int f(x, y, p)\, dp$ is a solution of the exact equation
   $$f(x, y, y')y'' + g(x, y, y') = 0$$
   and that a general solution of
   $$x(y')^2 + yy' + xyy'' = 0$$
   is $y^2/2 = C_1 + C_2 \ln x$.

5. (a) Find a general solution of the nonlinear equation
   $$\tfrac{4}{5}(y' - xy'')^2 = 2yy'' - (y')^2$$
   by differentiating both sides of the equation, setting the result equal to zero, and factoring. **(b)** Is the Principle of Superposition valid for this nonlinear equation? Explain.

In Exercises 1–26 find a general solution of the Cauchy-Euler equation.

1. $4x^2y'' - 8xy' + 5y = 0$

2. $3x^2y'' - 4xy' + 2y = 0$

*3. $2x^2y'' - 8xy' + 8y = 0$

4. $2x^2y'' - 7xy' + 7y = 0$

5. $4x^2y'' + 17y = 0$

6. $9x^2y'' - 9xy' + 10y = 0$

7. $2x^2y'' - 2xy' + 20y = 0$

8. $x^2y'' - 5xy' + 10y = 0$

*9. $x^2y'' + 2xy' - 6y = 0$

10. $x^2y'' + xy' - 16y = 0$

11. $x^2y'' + xy' + 4y = 0$

12. $x^2y'' + xy' + 36y = 0$

13. $x^2y'' + 5xy' + 13y = 0$

14. $x^2y'' + 2xy' + y = 0$

*15. $x^3y''' + 22x^2y'' + 124xy' + 140y = 0$

16. $x^3y''' + 12x^2y'' + 12xy' - 48y = 0$

17. $x^3y''' - 4x^2y'' - 46xy' + 100y = 0$

18. $x^3y''' + 10x^2y'' - 20xy' + 20y = 0$

19. $x^3y''' + 2x^2y'' - 4xy' + 4y = 0$

20. $x^3y''' + 2x^2y'' + 2y = 0$

*21. $x^3y''' + 4x^2y'' + 6xy' + 4y = 0$

22. $x^3y''' - 3x^2y'' - 4xy' + 34y = 0$

23. $x^3y''' + 2xy' - 2y = 0$

24. $x^4y^{(4)} - 2x^2y'' + 8xy' - 12y = 0$

25. $x^4y^{(4)} + 9x^3y''' + 11x^2y'' - 4xy' + 4y = 0$

26. $x^4y^{(4)} + 10x^3y''' + 16x^2y'' - 12xy' - 8y = 0$

In Exercises 27–40, solve the nonhomogeneous Cauchy-Euler equation.

27. $x^2y'' + 5xy' + 4y = x^{-5}$

28. $x^2y'' - 5xy' + 9y = x^3$

*29. $x^2y'' + xy' + y = x^{-2}$

30. $x^2y'' + xy' + 4y = x^{-2}$

31. $x^2y'' + 2xy' - 6y = 2x$

32. $x^2y'' + xy' - 16y = \ln x$

*33. $x^2y'' + xy' + 4y = 8$

34. $x^2y'' + xy' + 36y = x$

35. $x^2y'' + xy' + 36y = x^2$

36. $x^2y'' + 5xy' + 13y = 4$

*37. $x^3y''' + 3x^2y'' - 11xy' + 16y = x^{-3}$

38. $x^3y''' + 16x^2y'' + 65xy' + 63y = x^{-14}$

39. $x^3y''' + 16x^2y'' + 70xy' + 80y = x^{-13}$

40. $x^4y^{(4)} + 8x^3y''' + 13x^2y'' + xy' - y = x^{-7}$

In Exercises 41–52, solve the initial-value problem.

41. $\begin{cases} 3x^2y'' - 4xy' + 2y = 0 \\ y(1) = 2, y'(1) = 1 \end{cases}$

42. $\begin{cases} 2x^2y'' - 7xy' + 7y = 0 \\ y(1) = -1, y'(1) = 1 \end{cases}$

*43. $\begin{cases} x^2y'' + xy' + 4y = 0 \\ y(1) = 1, y'(1) = 0 \end{cases}$

44. $\begin{cases} x^2y'' + xy' + 2y = 0 \\ y(1) = 0, y'(1) = 2 \end{cases}$

45. $\begin{cases} x^3y''' + 10x^2y'' - 20xy' + 20y = 0 \\ y(1) = 0, y'(1) = -1, y''(1) = 1 \end{cases}$

46. $\begin{cases} x^3y''' + 15x^2y'' + 54xy' + 42y = 0 \\ y(1) = 5, y'(1) = 0, y''(1) = 0 \end{cases}$

*47. $\begin{cases} x^3y''' - 2x^2y'' + 5xy' - 5y = 0 \\ y(1) = 0, y'(1) = -1, y''(1) = 0 \end{cases}$

48. $\begin{cases} x^3y''' - 6x^2y'' + 17xy' - 17y = 0 \\ y(1) = -2, y'(1) = 0, y''(1) = 0 \end{cases}$

49. $\begin{cases} 2x^2y'' + 3xy' - y = x^{-2} \\ y(1) = 0, y'(1) = 2 \end{cases}$

50. $\begin{cases} x^2y'' + 4xy' + 2y = \ln x \\ y(1) = 2, y'(1) = 0 \end{cases}$

*51. $\begin{cases} 4x^2y'' + y = x^3 \\ y(1) = 1, y'(1) = -1 \end{cases}$

52. $\begin{cases} 9x^2y'' + 27xy' + 10y = x^{-1} \\ y(1) = 0, y'(1) = -1 \end{cases}$

53. Consider the Cauchy-Euler equation

$$ax^2y'' + bxy' + cy = f(x).$$

(a) Show that if $x = e^t$, the equation can be rewritten as $a\dfrac{d^2y}{dt^2} + (b-a)\dfrac{dy}{dt} + cy = f(e^t)$, which has constant coefficients. **(b)** Show that if $f(x) = 0$, then the three cases of **(i)** two distinct real roots, **(ii)** one repeated real root, and **(iii)** a complex conjugate pair of roots to the characteristic equation of $a\dfrac{d^2y}{dt^2} +$

$(b-a)\dfrac{dy}{dt} + cy = f(e^t)$ are the same as those found when solutions were assumed to have the form $y = x^m$.

54. Suppose that with the substitution $x = e^t$ a third-order Cauchy-Euler equation is transformed into

$$\frac{d^3y}{dt^3} - 3\frac{d^2y}{dt^2} + 3\frac{dy}{dt} - y = (D-1)^3 y = 0 \quad \text{where}$$

$D = \dfrac{d}{dt}$. **(a)** What is a general solution of the transformed equation? **(b)** What is a general solution of the original Cauchy-Euler equation in x? **(c)** If the transformed equation of a fourth-order equation is

$$\frac{d^4y}{dt^4} + 2\frac{d^2y}{dt^2} + y = 0,$$ what is a general solution of

the transformed equation and a general solution of the original fourth-order Cauchy-Euler equation in x?

55. Let $x = -e^t$. **(a)** Show that $\dfrac{dy}{dx} = \dfrac{dy}{dt}\dfrac{dt}{dx} = \dfrac{1}{x}\dfrac{dy}{dt}$

and $\dfrac{d^2y}{dx^2} = \dfrac{1}{x^2}\left(\dfrac{d^2y}{dt^2} - \dfrac{dy}{dt}\right)$. **(b)** Show that the dif-

ferential equation $ax^2y'' + bxy' + cy = f(x)$ is transformed into $a\dfrac{d^2y}{dt^2} + (b-a)\dfrac{dy}{dt} + cy = f(-e^t)$.

In Exercises 56–59, use the substitution described in Exercise 55 to solve the indicated equation or initial-value problem.

56. $x^2y'' + 4xy' + 2y = 0$, $x < 0$

57. $x^2y'' + xy' + y = x^2$, $x < 0$

58. $x^2y'' + xy' + 4y = 0$, $y(-1) = 0$, $y'(-1) = 2$

***59.** $x^2y'' - xy' + y = 0$, $y(-1) = 0$, $y'(-1) = 1$

60. Consider the second-order Cauchy-Euler equation $x^2y'' + Bxy' + y = 0$, $x > 0$, where B is a constant.

   (a) Find $\lim_{x \to \infty} y(x)$ where $y(x)$ is a general solution using the given restriction on B.

   (b) Determine if the solution is bounded as $x \to \infty$ in each case.

   (i) $B = 1$

   (ii) $B > 1$

   (iii) $B < 1$

61. Consider the second-order Cauchy-Euler equation stated in Exercise 60. Determine $\lim_{x \to 0^+} y(x)$ as well as if $y(x)$ is bounded as $x \to 0^+$ in each case.

   (a) $B = 1$

   (b) $B > 1$

   (c) $B < 1$

62. Use undetermined coefficients to solve the nonhomogeneous differential equation that was solved in Example 6 with variation of parameters. Which method is easier?

6.2   Power Series Review

Power Series: Basic Definitions and Theorems • Reindexing a Power Series

In Section 6.1, we saw that techniques similar to those learned in earlier chapters can be used to solve some linear differential equations with nonconstant coefficients. Other methods must be used if the equation is not in the form of a Cauchy-Euler equation or when some transformation cannot be performed to obtain an equation that we can solve with the techniques we have learned. In many cases, power series can be used to construct solutions of differential equations with nonconstant coefficients, as was first discovered by Newton and Leibniz.

We might observe that the series cannot contain any even-degree terms because the function $\sin x$ is an odd function and therefore must be represented by a series of odd-degree terms: even-degree terms in the series would indicate that the function is not odd. (Why?) (Note: A series of even-degree terms represents an even function.) See Exercises 27 and 28.

For example, if $y(x) = \sin x$, differentiating twice results in $y''(x) = -\sin x$, so $\dfrac{d^2y}{dx^2} = -y$. Leibniz assumed that the solution of the second-order differential equation

$$\frac{d^2y}{dx^2} = -y$$

could be written in the form $y(x) = a_1 x + a_3 x^3 + a_5 x^5 + \cdots$: according to Katz,* it was "obvious" to Leibniz that the series could not contain any even-degree terms or a constant term. (The constant term had to be zero because $\sin 0 = 0$.)

Differentiating the series $y(x) = a_1 x + a_3 x^3 + a_5 x^5 + \cdots$ twice we obtain

$$y''(x) = 3 \cdot 2a_3 x + 5 \cdot 4a_5 x^3 + 7 \cdot 6a_7 x^5 + \cdots$$

and substituting each series into the equation $\dfrac{d^2y}{dx^2} = -y$ we obtain

$$3 \cdot 2a_3 x + 5 \cdot 4a_5 x^3 + 7 \cdot 6a_7 x^5 + \cdots = -(a_1 x + a_3 x^3 + a_5 x^5 + \cdots).$$

Equating coefficients of like terms results in

$$\begin{cases} 3 \cdot 2a_3 = -a_1 \\ 5 \cdot 4a_5 = -a_3 \\ 7 \cdot 5a_7 = -a_5 \\ \quad \vdots \end{cases}$$

Setting $a_1 = 1$ (so that $y'(0) = 1$; the value of $\dfrac{d}{dx}(\sin x) = \cos x$ if $x = 0$) results in $a_3 = -\frac{1}{3!}$, $a_5 = \frac{1}{5!}$, $a_7 = -\frac{1}{7!}$, ... and thus we obtain the series for $\sin x$ about $x = 0$:

$$\sin x = x - \frac{1}{3!}x^3 + \frac{1}{5!}x^5 - \frac{1}{7!}x^7 + \cdots,$$

discovered by Leibniz prior to 1676.

Before discussing how we can use power series to construct solutions of other differential equations, we briefly review the basic properties of power series that will be used in later sections. These properties, proofs of the major theorems, and more detailed discussions are found in most calculus books.

Power Series: Basic Definitions and Theorems

Definition 6.2 Power Series

Let x_0 be a number. A **power series** in $(x - x_0)$ is a series of the form $\sum_{n=0}^{\infty} a_n(x - x_0)^n$, where a_n is a constant for all values of n.

---

* Victor J. Katz, *A History of Mathematics: An Introduction*, HarperCollins, 1993, p. 480.

Definition 6.3 Radius of Convergence

The power series $\sum_{n=0}^{\infty} a_n(x - x_0)^n$ always converges for $x = x_0$. If there is a positive number h so that the power series $\sum_{n=0}^{\infty} a_n(x - x_0)^n$ converges absolutely for all values of x in the interval $(x_0 - h, x_0 + h)$ and diverges for all values of x in the interval $(-\infty, x_0 - h) \cup (x_0 + h, +\infty)$, the power series has **radius of convergence** h. In this case, the power series may or may not converge for $x = x_0 - h$ and may or may not converge for $x = x_0 + h$. If the power series converges absolutely for all values of x, the power series has **infinite radius of convergence.** If the power series has radius of convergence $h > 0$, the power series is said to be **analytic** at $x = x_0$.

Examples of power series illustrating some of the situations that are encountered are given in Table 6.1.

TABLE 6.1

Power Series	Interval of Convergence
$\sum_{n=0}^{\infty} n!x^n$	$\{0\}$
$\sum_{n=0}^{\infty} x^n$	$(-1, 1)$
$\sum_{n=0}^{\infty} \frac{1}{n+1}x^n$	$[-1, 1)$
$\sum_{n=0}^{\infty} \frac{1}{n^2+1}x^n$	$[-1, 1]$
$\sum_{n=0}^{\infty} \frac{1}{n!}x^n$	$(-\infty, +\infty)$

In calculus, we learn that many functions can be approximated by Taylor and Maclaurin polynomials, provided that the necessary derivatives exist.

Definition 6.4 nth Degree Taylor and Maclaurin Polynomials

The **nth degree Taylor polynomial for f about $x = x_0$** is $\sum_{k=0}^{n} \frac{f^{(k)}(x_0)}{k!}(x - x_0)^k$;

the **nth degree Maclaurin polynomial for f** is the nth degree Taylor polynomial

for f about $x = 0$: $\sum_{k=0}^{n} \frac{f^{(k)}(0)}{k!}x^k$.

Taylor's Theorem tells us the conditions under which we can approximate a given function by Taylor or Maclaurin polynomials.

Theorem 6.2 Taylor's Theorem

Let f be a function with derivatives of all orders on an interval (a, b) and let $a < x_0 < b$.

If $x \in (a, b)$ and $x \neq x_0$, there is a number z between x and x_0 so that

$$f(x) = \sum_{k=0}^{n} \frac{f^{(k)}(x_0)}{k!}(x - x_0)^k + R_n(x),$$

where $R_n(x) = \frac{f^{(n+1)}(z)}{(n + 1)!}(x - x_0)^{n+1}$. Moreover, if $\lim_{n \to \infty} R_n(x) = 0$ for every x in the interval (a, b), then

$$f(x) = \sum_{k=0}^{\infty} \frac{f^{(k)}(x_0)}{k!}(x - x_0)^k.$$

If $f(x)$ can be represented by a power series at $x = x_0$,

$$f(x) = \sum_{n=0}^{\infty} a_n(x - x_0)^n,$$

with radius of convergence $h > 0$ (h may be $+\infty$), we say that $f(x)$ is **analytic*** at $x = x_0$.

Example 1

Let $f(x) = e^x$. (a) Find the Maclaurin series for $f(x)$; (b) find the nth Maclaurin polynomial for $f(x)$; and (c) show that $e^x = \sum_{k=0}^{\infty} \frac{1}{k!}x^k$ for every value of x.

Solution   For every k, $f^{(k)}(x) = e^x$, so $f^{(k)}(0) = 1$ and the Maclaurin series for $f(x)$ is $\sum_{k=0}^{\infty} \frac{f^{(k)}(0)}{k!}x^k = \sum_{k=0}^{\infty} \frac{1}{k!}x^k$. The nth Maclaurin polynomial is $\sum_{k=0}^{n} \frac{1}{k!}x^k$ and, by Taylor's Theorem,

$$R_n(x) = \frac{f^{(n+1)}(z)}{(n + 1)!}x^{n+1} = \frac{e^z}{(n + 1)!}x^{n+1}.$$

---

* Detailed discussions regarding analytic functions may be found in texts like *Functions of One Complex Variable*, Second Edition, by John B. Conway and published by Springer-Verlag, New York (1978).

If $x < 0$, then $e^z \le 1$ because z is between x and 0 and

$$|R_n(x)| = \left| \frac{e^z}{(n+1)!} x^{n+1} \right| \le \frac{|x|^{n+1}}{(n+1)!}$$

so that $\lim_{n\to\infty} |R_n(x)| \le \lim_{n\to\infty} \frac{|x|^{n+1}}{(n+1)!} = 0$. If $x > 0$, then $e^z \le e^x$ and $\lim_{n\to\infty} |R_n(x)| \le e^x \lim_{n\to\infty} \frac{x^{n+1}}{(n+1)!} = 0$. In any case, $\lim_{n\to\infty} R_n(x) = 0$, so

$$e^x = \sum_{k=0}^{\infty} \frac{1}{k!} x^k$$ for every value of x. Thus, e^x is analytic at $x = 0$; in fact, e^x is analytic for all values of x.

Frequently used Maclaurin series along with their corresponding intervals of convergence are listed in Table 6.2.

TABLE 6.2   Frequently Used Maclaurin Series

$\dfrac{1}{1-x} = \sum\limits_{n=0}^{\infty} x^n,\ x \in (-1, 1)$	$e^x = \sum\limits_{n=0}^{\infty} \dfrac{1}{n!} x^n,\ x \in (-\infty, \infty)$
$\sin x = \sum\limits_{n=0}^{\infty} \dfrac{(-1)^n x^{2n+1}}{(2n+1)!},\ x \in (-\infty, \infty)$	$\cos x = \sum\limits_{n=0}^{\infty} \dfrac{(-1)^n x^{2n}}{(2n)!},\ x \in (-\infty, \infty)$
$\ln(1+x) = \sum\limits_{n=1}^{\infty} \dfrac{(-1)^{n-1} x^n}{n},\ x \in (-1, 1]$	$\tan^{-1} x = \sum\limits_{n=0}^{\infty} \dfrac{(-1)^n x^{2n+1}}{(2n+1)!},\ x \in [-1, 1]$
$\sinh x = \sum\limits_{n=0}^{\infty} \dfrac{x^{2n+1}}{(2n+1)!},\ x \in (-\infty, \infty)$	$\cosh x = \sum\limits_{n=0}^{\infty} \dfrac{x^{2n}}{(2n)!},\ x \in (-\infty, \infty)$

Generally, constructing a Taylor series for a given function is a tedious task. However, in some cases, we can construct a Taylor series for a given function rather easily.

Definition 6.5 Geometric Series

A **geometric series** is a series of the form $\sum_{n=0}^{\infty} ar^n$. If $|r| < 1$, the geometric series converges and $\sum_{n=0}^{\infty} ar^n = \dfrac{a}{1-r}$. If $|r| \ge 1$, the geometric series diverges.

| Example 2 | Find (a) the Maclaurin series for $f(x) = \dfrac{x}{2 - 3x^2}$ and (b) the Taylor series for $g(x) = \dfrac{1}{3x - 5}$ about $x = 1$. |

Solution   For (a), we rewrite the function in order to find the Maclaurin series. Hence,

$$f(x) = \frac{x}{2 - 3x^2} = x\frac{1}{2 - 3x^2} = \frac{x}{2}\,\frac{1}{1 - \dfrac{3}{2}x^2} = \frac{x}{2}\sum_{n=0}^{\infty}\left(\frac{3}{2}x^2\right)^n = \sum_{n=0}^{\infty}\frac{3^n}{2^{n+1}}x^{2n+1}.$$

Similarly for (b) we have,

$$g(x) = \frac{1}{3x - 5} = \frac{1}{3(x - 1) + 3 - 5} = \frac{1}{3(x - 1) - 2} = -\frac{1}{2}\,\frac{1}{1 - \dfrac{3}{2}(x - 1)}$$

$$= -\frac{1}{2}\sum_{n=0}^{\infty}\left(\frac{3}{2}(x - 1)\right)^n = \sum_{n=0}^{\infty} -\frac{3^n}{2^{n+1}}(x - 1)^n.$$

For what values of x do the series in Example 2 converge?

A power series may be differentiated and integrated term-by-term on its interval of convergence, as stated in the following two theorems.

Theorem 6.3 Term-by-Term Differentiation

If the power series $\sum_{n=0}^{\infty} a_n(x - x_0)^n$ has a radius of convergence $h > 0$ (h may be $+\infty$), then the function $f(x) = \sum_{n=0}^{\infty} a_n(x - x_0)^n$ has derivatives of all orders on its interval of convergence and $f'(x) = \sum_{n=0}^{\infty} na_n(x - x_0)^{n-1}$, $f''(x) = \sum_{n=0}^{\infty} n(n - 1)a_n(x - x_0)^{n-2}$, and so on.

Theorem 6.4 Term-by-Term Integration

If the power series $\sum_{n=0}^{\infty} a_n(x - x_0)^n$ has radius of convergence $h > 0$ (h may be $+\infty$), then the series $\sum_{n=0}^{\infty} \dfrac{a_n}{n + 1}(x - x_0)^{n+1}$ has radius of convergence h. In fact, if $f(x) = \sum_{n=0}^{\infty} a_n(x - x_0)^n$, then $\int f(x)\, dx = \sum_{n=0}^{\infty} \dfrac{a_n}{n + 1}(x - x_0)^{n+1} + c$.

Example 3

(a) Use the geometric series $\dfrac{1}{1-x} = \sum_{n=0}^{\infty} x^n$, $x \in (-1, 1)$ to find the function that has as its Maclaurin series $1 + 2x + 3x^2 + 4x^3 + \cdots = \sum_{n=0}^{\infty} (n+1)x^n$. (b) Use the Maclaurin series for $f(x) = \dfrac{1}{1+x^2}$ to obtain the Maclaurin series for $f(x) = \tan^{-1} x$.

Solution   (a) First, notice that $1 + 2x + 3x^2 + 4x^3 + \cdots =$

$\dfrac{d}{dx}(1 + x + x^2 + x^3 + x^4 + \cdots)$. Therefore, $1 + 2x + 3x^2 + 4x^3 \cdots =$

$\dfrac{d}{dx}\left(\dfrac{1}{1-x}\right) = \dfrac{1}{(1-x)^2}$, so the Maclaurin series for $1/(1-x)^2$ is $\sum_{n=0}^{\infty} (n+1)x^n$.

By the ratio test, this series converges for $x \in (-1, 1)$. In Figure 6.8(a), we graph $1/(1-x)^2$ with the polynomial of degree 8 obtained from $\sum_{n=0}^{\infty} (n+1)x^n$ on the interval $[-0.75, 0.75]$. Notice that the approximation is not as accurate near the boundary of this interval.

(b) The Maclaurin series for $f(x) = \dfrac{1}{1+x^2}$ is determined as follows:

$$\frac{1}{1+x^2} = \frac{1}{1-(-x^2)} = \sum_{n=0}^{\infty} (-x^2)^n = \sum_{n=0}^{\infty} (-1)^n x^{2n}.$$

Notice that $\displaystyle\int \dfrac{1}{1+x^2}\, dx = \tan^{-1} x + c$. Because $\sum_{n=0}^{\infty} (-1)^n x^{2n}$ has interval of convergence $(-1, 1)$, we can integrate term-by-term. Therefore,

$\tan^{-1} x + c = \displaystyle\int \sum_{n=0}^{\infty} (-1)^n x^{2n}\, dx = \sum_{n=0}^{\infty} \dfrac{(-1)^n x^{2n+1}}{2n+1}$. To determine the value of the constant of integration, we substitute $x = 0$ into the equation to obtain

$$\tan^{-1} 0 + c = \sum_{n=0}^{\infty} \frac{(-1)^n 0^{2n+1}}{2n+1} = 0.$$

Thus, $c = 0$, and the Maclaurin series for $f(x) = \tan^{-1} x$ is $\sum_{n=0}^{\infty} \dfrac{(-1)^n x^{2n+1}}{2n+1}$. By using the ratio test, we determine that this series converges for $x \in [-1, 1]$. In Figure 6.8(b), we graph $f(x) = \tan^{-1} x$ along with the polynomial of degree 9 obtained from the series $\sum_{n=0}^{\infty} \dfrac{(-1)^n x^{2n+1}}{2n+1}$. Notice that the accuracy of the approximation decreases near the endpoints of the interval $[-1, 1]$.

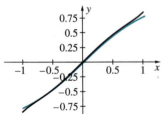

$-\!\!\!-\, p(x) = 1 + 2x + 3x^2 + 4x^3 + 5x^4 + 6x^5 + 7x^6 + 8x^7 + 9x^8$

$-\!\!\!-\, f(x) = \dfrac{1}{(1-x)^2}$

a.

$-\!\!\!-\, p(x) = x - \dfrac{1}{3}x^3 + \dfrac{1}{5}x^5 - \dfrac{1}{7}x^7 + \dfrac{1}{9}x^9$

$-\!\!\!-\, f(x) = \tan^{-1} x$

b.

Figure 6.8   (a)–(b)

Reindexing a Power Series

In the next sections, our ability to manipulate the index of a series will be important.

For example, consider the series $\sum_{n=1}^{\infty} \frac{x^n}{n}$ and $\sum_{n=0}^{\infty} \frac{x^{n+1}}{n+1}$. We write out the first few terms of each:

$$\sum_{n=1}^{\infty} \frac{x^n}{n} = x + \frac{x^2}{2} + \frac{x^3}{3} + \frac{x^4}{4} + \cdots \quad \text{and} \quad \sum_{n=0}^{\infty} \frac{x^{n+1}}{n+1} = x + \frac{x^2}{2} + \frac{x^3}{3} + \frac{x^4}{4} + \cdots.$$

Notice that although the notation differs they actually represent the same series. In fact, we obtain one from the other by changing the index. If we substitute $(n+1)$ for each occurrence of n in the series $\sum_{n=1}^{\infty} \frac{x^n}{n}$ we obtain $\sum_{n+1=1}^{\infty} \frac{x^{n+1}}{n+1} = \sum_{n=0}^{\infty} \frac{x^{n+1}}{n+1}$. (Notice that we want to make the appropriate substitution so that after simplification of the lower limit of the modified index, we obtain the desired lower limit.) We can use this technique to change the index in any series, which we will need to do in future sections.

Example 4	Write the sum $\sum_{n=0}^{\infty} 5a_n x^{n+2} + \sum_{n=2}^{\infty} a_n x^n$ as a single series.

Solution   Notice that the first term in each series contains x^2. Hence, if we change the index in one series so that it matches that of the other, we can write the expression as a single series. If we substitute $(n+2)$ for each occurrence of n in the series $\sum_{n=2}^{\infty} a_n x^n$, we have

$$\sum_{n+2=2}^{\infty} a_{n+2} x^{n+2} = \sum_{n=0}^{\infty} a_{n+2} x^{n+2}.$$

Hence,

$$\sum_{n=0}^{\infty} 5a_n x^{n+2} + \sum_{n=2}^{\infty} a_n x^n = \sum_{n=0}^{\infty} 5a_n x^{n+2} + \sum_{n=0}^{\infty} a_{n+2} x^{n+2} = \sum_{n=0}^{\infty} [5a_n + a_{n+2}] x^{n+2}.$$

An optional approach can be used for reindexing. In Example 4, we note that each index in the series $\sum_{n=2}^{\infty} a_n x^n$ must be lowered by 2, so we let $k = n - 2$ or $n = k + 2$. With this substitution, we obtain $\sum_{k+2=2}^{\infty} a_{k+2} x^{k+2} = \sum_{k=0}^{\infty} a_{k+2} x^{k+2}$. When we replace the "dummy" variable k with n, we obtain the same result as that in Example 4.

In many cases, the power series must be modified before they can be combined.

Example 5	Simplify the sum $\sum_{n=2}^{\infty} n(n-1)a_n x^{n-2} - \sum_{n=1}^{\infty} 5na_n x^n + \sum_{n=0}^{\infty} 10a_n x^n$.

Solution   We begin by determining the first term in each series. The series $\sum_{n=2}^{\infty} n(n-1)a_n x^{n-2}$ begins with x^0, $\sum_{n=1}^{\infty} 5na_n x^n$ begins with x, and $\sum_{n=0}^{\infty} 10a_n x^n$ begins with x^0. Hence, in order to write the sum as a single series, we must "pull off" enough terms in the first and third series so that they all begin with x terms. In this case, we accomplish our goal by pulling off the first term in both of these series. Thus,

$$2a_2x^0 + \sum_{n=3}^{\infty} n(n-1)a_nx^{n-2} - \sum_{n=1}^{\infty} 5na_nx^n + 10a_0x^0 + \sum_{n=1}^{\infty} 10a_nx^n.$$

Now, all three series in sigma notation begin with x terms, but we must change the index in the first sum so that it starts with $n = 1$ like the other two. Substituting $(n + 2)$ for each occurrence of n in $\sum_{n=3}^{\infty} n(n-1)a_nx^{n-2}$, we obtain

$$\sum_{n+2=3}^{\infty} (n+2)(n+2-1)a_{n+2}x^{n+2-2} = \sum_{n=1}^{\infty} (n+2)(n+1)a_{n+2}x^n.$$

Therefore,

$$2a_2x^0 + \sum_{n=1}^{\infty} (n+2)(n+1)a_{n+2}x^n - \sum_{n=1}^{\infty} 5na_nx^n + 10a_0x^0 + \sum_{n=1}^{\infty} 10a_nx^n =$$

$$2a_2 + 10a_0 + \sum_{n=1}^{\infty} [(n+2)(n+1)a_{n+2} + (10-5n)a_n]x^n.$$

IN TOUCH WITH TECHNOLOGY

One of the primary uses of Maclaurin and Taylor polynomials is to provide approximations of functions. Unfortunately, the calculation of the coefficients for these polynomials can be difficult and time-consuming. However, most computer algebra systems have built-in commands for computing Maclaurin and Taylor polynomials.

 1. Let $f(x) = (x - 1) \sin x + \cos \frac{x}{3} \sin \frac{x}{5}$. Find and graph the Maclaurin polynomial for $f(x)$ of degrees one, four, seven, and ten.

 2. Let $f(x) = \sin \frac{x}{2} + \cos 2x$. Compute and graph the Taylor polynomial of degree five for $f(x)$ about $x = \frac{\pi}{2}$.

3. In Problem 1, what do you notice about the approximation as the degree is increased? In Problem 2, near what value of x is the approximation most accurate?

4. Let $f(x) = \begin{cases} e^{-1/x^2}, & \text{if } x \neq 0 \\ 0, & \text{if } x = 0 \end{cases}$.

   (a) Compute $f^{(n)}(0)$ and the nth Maclaurin polynomial for f for $n = 0, 1, 2, \ldots, 10$.

   (b) Assume that $f^{(n)}(0) = 0$ for all values of $n \geq 11$. What is the nth Maclaurin polynomial for f for every value of n?

   (c) Calculate $g(x) = \sum_{n=0}^{\infty} \frac{f^{(n)}(0)}{n!}x^n$. Why does Taylor's Theorem fail for f at $x = 0$?

EXERCISES 6.2

In Exercises 1–5, find the Maclaurin series for the given function.

1. $f(x) = e^{-x}$

2. $g(x) = \dfrac{1}{1 - 4x}$

*3. $g(x) = \dfrac{x}{1 + x}$

4. $h(x) = \cos 5x$

*5. $h(x) = x \sin x$

In Exercises 6–9, find the Taylor series for the function about the indicated value.

6. $f(x) = \ln x$, $x_0 = 1$

7. $f(x) = e^x$, $x_0 = 1$

8. $g(x) = \sin x$, $x_0 = \frac{\pi}{4}$

*9. $g(x) = \cos x$, $x_0 = \frac{\pi}{4}$

10. Use series to show: **(a)** $\dfrac{d}{dx} \cos x = -\sin x$ and

(b) $\dfrac{d}{dx} e^x = e^x$.

11. Use the function $x \dfrac{d}{dx}\left[\dfrac{1}{1 - x}\right]$ to find the Maclaurin series for $\dfrac{x}{(1 - x)^2}$.

12. Use the Maclaurin series for $\dfrac{1}{1 - x}$ to find the Maclaurin series for $\ln\left(\dfrac{1}{1 - x}\right)$.

*13. Find the function with Maclaurin series $2 + 6x + 12x^2 + 20x^3 + \cdots$. $\left(\textit{Hint: }\text{Use the Maclaurin series for } \dfrac{1}{1 - x}.\right)$

14. Find the function with Maclaurin series

$$f(x) = \sum_{n=0}^{\infty} (-1)^n \frac{x^{n+1}}{n + 1}. \quad (\textit{Hint: Consider } f'(x).)$$

In Exercises 15–18, change the index in each of the series so that the series starts at the indicated value of n.

15. $\sum_{n=2}^{\infty} a_n x^n$, $n = 0$

16. $\sum_{n=1}^{\infty} 2a_{n-1} x^{n+1}$, $n = 2$

*17. $\sum_{n=1}^{\infty} 4a_n x^{2n+1}$, $n = 2$

18. $\sum_{n=3}^{\infty} a_{n+1} x^{2n-1}$; $n = 1$

In Exercises 19–26, rewrite the expression so that it involves a single summation symbol (Σ).

19. $\sum_{n=0}^{\infty} 5a_n x^{n+1} + \sum_{n=0}^{\infty} 3a_{n+1} x^{n+1}$

20. $\sum_{n=0}^{\infty} a_n x^{n+1} + \sum_{n=0}^{\infty} 2a_{n+1} x^{n+1} - \sum_{n=0}^{\infty} a_{n+1} x^{n+1}$

*21. $\sum_{n=0}^{\infty} 10a_n x^{n+2} + \sum_{n=2}^{\infty} a_{n+1} x^n$

22. $\sum_{n=1}^{\infty} a_n x^n + \sum_{n=0}^{\infty} 4a_{n+1} x^{n+1}$

23. $\sum_{n=2}^{\infty} a_{n-1} x^n - \sum_{n=0}^{\infty} 2a_{n+1} x^{n+1}$

24. $\sum_{n=2}^{\infty} a_n x^{n+1} + \sum_{n=1}^{\infty} a_{n-1} x^n$

*25. $\sum_{n=2}^{\infty} a_n x^{n-1} + \sum_{n=0}^{\infty} a_n x^n + \sum_{n=1}^{\infty} a_{n+1} x^n$

26. $\sum_{n=2}^{\infty} a_n x^n - \sum_{n=0}^{\infty} 2a_n x^n + \sum_{n=1}^{\infty} a_n x^{n-1}$

27. Show that if $f(x)$ is an odd function (which means that $f(-x) = -f(x)$), the coefficients of the even-degree terms in the Maclaurin series representation for f, $a_0 + a_1 x + a_2 x^2 + a_3 x^3 + \cdots$, must be zero.

28. Show that if $f(x)$ is an even function (which means that $f(-x) = f(x)$), the coefficients of the odd-degree terms in the Maclaurin series representation for f, $a_0 + a_1 x + a_2 x^2 + a_3 x^3 + \cdots$, must be zero.

29. **(Products of Power Series)** If the power series of f and g converge for $|x - h| < R$, their power series can be multiplied to find the terms of the power series representation of $f(x) \cdot g(x)$. Use this technique to find the first three nonzero terms of the power series for **(a)** $h(x) = e^x \sin x$, **(b)** $h(x) = e^{-x} \cos x$, **(c)** $h(x) = \sin x \cos x$, **(d)** $h(x) = xe^x$, and **(e)** $h(x) = x^2 \sin x$. (In **(c)**, compare your result to the first three nonzero terms in the series for $\sin 2x$.)

30. **(Quotients of Power Series)** If the power series of f and g converge for $|x - h| < R$ and if $g(x) \ne 0$ for all x in $|x - h| < R$, then $f(x)/g(x)$ has a power series expression that is valid over $|x - h| < R$. In this case, long division of the series in the numerator and denominator can be used to find the terms of the series for the quotient $f(x)/g(x)$. Use this technique to find the first three nonzero terms of the power series for

(a) $h(x) = \dfrac{\ln(1 + x)}{x - 1}$, **(b)** $h(x) = \dfrac{\sin x}{\cos x}$, **(c)** $\dfrac{1}{1 - x}$,

(d) $h(x) = \dfrac{\sin x}{x}$, **(e)** $h(x) = \dfrac{\cos x - 1}{x}$, and

(f) $h(x) = \dfrac{1}{(1 + x + x^2 + x^3 + \cdots)^2}$.

*31. Use the power series for $h(x) = \dfrac{\sin x}{x}$ to compute

$$\lim_{x \to 0} \frac{\sin x}{x}.$$

32. Use the power series for $h(x) = \dfrac{\cos x - 1}{x}$ to com-

pute $\lim_{x \to 0} \dfrac{\cos x - 1}{x}.$

*33. If $p(x)$ is a polynomial, what is its Maclaurin series? Explain.

6.3 Power Series Solutions About Ordinary Points

In Section 6.2 we saw that Maclaurin and Taylor polynomials can be used to approximate *functions*. This idea can be extended to finding or approximating the *solution of a differential equation*. We begin by introducing some terminology.

Definition 6.6 Standard Form; Ordinary and Singular Points

Given the second-order equation $a_2(x)y'' + a_1(x)y' + a_0(x)y = 0$, let $p(x) = a_1(x)/a_2(x)$ and $q(x) = a_0(x)/a_2(x)$. Then $a_2(x)y'' + a_1(x)y' + a_0(x)y = 0$ is equivalent to $y'' + p(x)y' + q(x)y = 0$, which is called the **standard form** of the equation. A number x_0 is an **ordinary point** of the differential equation means that both $p(x)$ and $q(x)$ are analytic at x_0. If x_0 is not an ordinary point, then x_0 is called a **singular point**.

Example 1

Classify $x = 0$ as an ordinary point or a singular point of the following differential equations: (a) $y'' + y' \cos x - e^x y = 0$; (b) $\dfrac{1}{x^2}y'' - \dfrac{2}{x}y' + y = 0$; and (c) $x^2 y'' + 4xy' + xy = 0$.

Solution   (a) This equation is in standard form. The functions $\cos x$ and e^x are analytic at the particular value $x = 0$ (why?), so $x = 0$ is an ordinary point of the differential equation.

(b) In standard form, the equation is $y'' - 2xy' + x^2 y = 0$. The functions $p(x) = -2x$ and $q(x) = x^2$ are polynomials so they are analytic for all values of x, including $x = 0$. Therefore, $x = 0$ is an ordinary point of the differential equation.

(c) In this case, the differential equation in standard form is $y'' + \dfrac{4}{x}y' + \dfrac{1}{x}y = 0$.

However, the functions $p(x) = \dfrac{4}{x}$ and $q(x) = \dfrac{1}{x}$ have no Maclaurin series at $x = 0$ (why?) so they are not analytic at $x = 0$. Hence, $x = 0$ is a singular point of the equation.

If x_0 is an ordinary point of the differential equation $y'' + p(x)y' + q(x)y = 0$, we can write $p(x) = \sum_{n=0}^{\infty} b_n(x - x_0)^n$ and $q(x) = \sum_{n=0}^{\infty} c_n(x - x_0)^n$, where $b_n = \dfrac{p^{(n)}(x_0)}{n!}$ and $c_n = \dfrac{q^{(n)}(x_0)}{n!}$. Substitution into the equation $y'' + p(x)y' + q(x)y = 0$ results in

$$y'' + y' \sum_{n=0}^{\infty} b_n(x - x_0)^n + y \sum_{n=0}^{\infty} c_n(x - x_0)^n = 0.$$

If we assume that $y(x)$ is analytic at x_0, we can write $y(x) = \sum_{n=0}^{\infty} a_n(x - x_0)^n$. Because a power series can be differentiated term-by-term, we can compute the first and second derivatives of y and substitute back into the equation to calculate the coefficients a_n. Thus, we obtain a power series solution of the equation. We summarize this method as follows.

Power Series Solution Method About an Ordinary Point

① Assume that $y = \sum_{n=0}^{\infty} a_n(x - x_0)^n$.

② After taking the appropriate derivatives, substitute $y = \sum_{n=0}^{\infty} a_n(x - x_0)^n$ into the differential equation.

③ Find the unknown series coefficients a_n through an equation relating the coefficients.

④ Apply any given initial conditions, if applicable.

The differentiation of power series is necessary in this method for solving differential equations, so we make a few observations about the procedure. For the Maclaurin · series $y = \sum_{n=0}^{\infty} a_n x^n$, term-by-term differentiation yields $y' = \sum_{n=0}^{\infty} na_n x^{n-1}$. Notice, however, that with the initial index value of $n = 0$, the first term of the series is 0. We rewrite the series in its equivalent form

$$y' = \sum_{n=1}^{\infty} na_n x^{n-1}.$$

Similarly,

$$y'' = \sum_{n=1}^{\infty} n(n - 1)a_n x^{n-2} = \sum_{n=2}^{\infty} n(n - 1)a_n x^{n-2}.$$

Example 2

Solve $y'' + 4x^2 y = 0$.

Solution Notice that $x = 0$ is an ordinary point of this equation. We assume that $y = \sum_{n=0}^{\infty} a_n x^n$. Substitution of this function and its derivatives

$y' = \sum_{n=1}^{\infty} na_n x^{n-1}$ and $y'' = \sum_{n=2}^{\infty} n(n-1)a_n x^{n-2}$ into $y'' + 4x^2 y = 0$ yields

$$\sum_{n=2}^{\infty} n(n-1)a_n x^{n-2} + \sum_{n=0}^{\infty} 4a_n x^{n+2} = 0.$$

The first series begins with an x^0 term while the second begins with an x^2 term. In order to have both series start with the same term, we pull off the first two terms of $\sum_{n=2}^{\infty} n(n-1)a_n x^{n-2}$. This gives us

$$2a_2 + 6a_3 x + \sum_{n=4}^{\infty} n(n-1)a_n x^{n-2} + \sum_{n=0}^{\infty} 4a_n x^{n+2} = 0.$$

Now both series begin with x^2 terms, but the indices do not match. Substituting $(n-4)$ for n in $\sum_{n=0}^{\infty} 4a_n x^{n+2}$ results in $\sum_{n-4=0}^{\infty} 4a_{n-4} x^{n-4+2} = \sum_{n=4}^{\infty} 4a_{n-4} x^{n-2}$ and the equation becomes

$$2a_2 + 6a_3 x + \sum_{n=4}^{\infty} [n(n-1)a_n + 4a_{n-4}]x^{n-2} = 0.$$

The coefficients must equal zero because the right-hand side of the equation is zero. Equating the coefficients of x^0 and x to zero yields $a_2 = a_3 = 0$. Equating the coefficient $[n(n-1)a_n + 4a_{n-4}]$ to zero yields the recurrence relation $a_n = \dfrac{-4a_{n-4}}{n(n-1)}$ for the indices in the series, $n \geq 4$. We calculate the coefficients with this formula for $n = 4, 5, \ldots, 13$ and display these values in Table 6.3. Notice that a_0 and a_1 are arbitrary.

TABLE 6.3

n	a_n	n	a_n
0	a_0	7	0
1	a_1	8	$\dfrac{4^2 a_0}{8 \cdot 7 \cdot 4 \cdot 3}$
2	0	9	$\dfrac{4^2 a_1}{9 \cdot 8 \cdot 5 \cdot 4}$
3	0	10	0
4	$-\dfrac{4a_0}{4 \cdot 3}$	11	0
5	$-\dfrac{4a_1}{5 \cdot 4}$	12	$-\dfrac{4^3 a_0}{12 \cdot 11 \cdot 8 \cdot 7 \cdot 4 \cdot 3}$
6	0	13	$-\dfrac{4^3 a_1}{13 \cdot 12 \cdot 9 \cdot 8 \cdot 5 \cdot 4}$

Hence,

$$y = a_0 + a_1 x - \frac{4a_0}{4 \cdot 3}x^4 - \frac{4a_1}{5 \cdot 4}x^5 + \frac{4^2 a_0}{8 \cdot 7 \cdot 4 \cdot 3}x^8 + \frac{4^2 a_1}{9 \cdot 8 \cdot 5 \cdot 4}x^9$$

$$- \frac{4^3 a_0}{12 \cdot 11 \cdot 8 \cdot 7 \cdot 4 \cdot 3}x^{12} - \frac{4^3 a_1}{13 \cdot 12 \cdot 9 \cdot 8 \cdot 5 \cdot 4}x^{13} + \cdots$$

$$= a_0\left(1 - \frac{4}{4 \cdot 3}x^4 + \frac{4^2}{8 \cdot 7 \cdot 4 \cdot 3}x^8 - \frac{4^3}{12 \cdot 11 \cdot 8 \cdot 7 \cdot 4 \cdot 3}x^{12} + \cdots\right)$$

$$+ a_1\left(x - \frac{4}{5 \cdot 4}x^5 + \frac{4^2}{9 \cdot 8 \cdot 5 \cdot 4}x^9 - \frac{4^3}{13 \cdot 12 \cdot 9 \cdot 8 \cdot 5 \cdot 4}x^{13} + \cdots\right).$$

Recall that when we solved second-order equations in previous sections, we needed two linearly independent solutions to form a general solution. In this example, we see that y can be expressed as the linear combination of two linearly independent functions, one multiplied by a_0 and the other by a_1. (We know that the functions are linearly independent, because they are not scalar multiples of one another.)

 In Example 2, find $y(0)$ and $y'(0)$.

Example 3

(a) Find a general solution of $(4 - x^2)\dfrac{dy}{dx} + y = 0$ and (b) solve the initial-value

problem $\begin{cases} (4 - x^2)\dfrac{dy}{dx} + y = 0 \\ y(0) = 1 \end{cases}$.

Solution (a) As in Example 2, $x = 0$ is an ordinary point of the equation, so we assume that $y = \sum_{n=0}^{\infty} a_n x^n$. Substitution of this function and its derivatives into the equation gives us

$$(4 - x^2)\frac{dy}{dx} + y = (4 - x^2)\sum_{n=1}^{\infty} na_n x^{n-1} + \sum_{n=0}^{\infty} a_n x^n$$

$$= \sum_{n=1}^{\infty} 4na_n x^{n-1} - \sum_{n=1}^{\infty} na_n x^{n+1} + \sum_{n=0}^{\infty} a_n x^n = 0.$$

Note that the first term in these three series involve x^0, x^2, and x^0, respectively. If we pull off the first two terms in the first and third series, all three series will begin with x^2. Doing so, we have

$$(4a_1 + a_0) + (8a_2 + a_1)x + \sum_{n=3}^{\infty} 4na_n x^{n-1} - \sum_{n=1}^{\infty} na_n x^{n+1} + \sum_{n=2}^{\infty} a_n x^n = 0.$$

Now the indices of these three series do not match, so we change two of the three to match the third. Substitution of $(n + 1)$ for n in $\sum_{n=3}^{\infty} 4na_n x^{n-1}$ yields

$$\sum_{n+1=3}^{\infty} 4(n + 1)a_{n+1}x^{n+1-1} = \sum_{n=2}^{\infty} 4(n + 1)a_{n+1}x^n.$$

Similarly, substitution of $(n - 1)$ for n in $\sum_{n=1}^{\infty} na_n x^{n+1}$ yields

$$\sum_{n-1=1}^{\infty} (n - 1)a_{n-1}x^{n-1+1} = \sum_{n=2}^{\infty} (n - 1)a_{n-1}x^n.$$

After combining the three series, we have the equation

$$(4a_1 + a_0) + (8a_2 + a_1)x + \sum_{n=2}^{\infty} (a_n + 4(n + 1)a_{n+1} - (n - 1)a_{n-1})x^n = 0.$$

Equating the coefficients of x^0 and x to zero yields

$$a_1 = -\frac{a_0}{4} \quad \text{and} \quad a_2 = -\frac{a_1}{8} = \frac{a_0}{32}.$$

When the coefficient of x^n, $a_n + 4(n + 1)a_{n+1} - (n - 1)a_{n-1}$, $n \geq 2$, is set to zero, we obtain the recurrence relation $a_{n+1} = \dfrac{(n - 1)a_{n-1} - a_n}{4(n + 1)}$ for the indices in the series, $n \geq 2$. We use this formula to determine the values of a_n for $n = 2, 3, \ldots, 10$, and give these values in Table 6.4.

TABLE 6.4

n	a_n	n	a_n
0	a_0	6	$\frac{69}{65536}a_0$
1	$-\frac{1}{4}a_0$	7	$-\frac{187}{262144}a_0$
2	$\frac{1}{32}a_0$	8	$\frac{1843}{8388608}a_0$
3	$-\frac{3}{128}a_0$	9	$-\frac{4859}{33554432}a_0$
4	$\frac{11}{2048}a_0$	10	$\frac{12767}{268435456}a_0$
5	$-\frac{31}{8192}a_0$		

Therefore,

$$y = a_0 - \frac{1}{4}a_0 x + \frac{1}{32}a_0 x^2 - \frac{3}{128}a_0 x^3 + \frac{11}{2048}a_0 x^4 - \frac{31}{8192}a_0 x^5 + \frac{69}{65536}a_0 x^6$$

$$- \frac{187}{262144}a_0 x^7 + \frac{1843}{8388608}a_0 x^8 - \frac{4859}{33554432}a_0 x^9 + \frac{12767}{268435456}a_0 x^{10} + \cdots$$

(b) When we apply the initial condition $y(0) = 1$, we substitute $x = 0$ into the solution obtained in (a). Hence, $a_0 = 1$, so the series solution of the initial-value problem is

$$y = 1 - \frac{1}{4}x + \frac{1}{32}x^2 - \frac{3}{128}x^3 + \frac{11}{2048}x^4 - \frac{31}{8192}x^5 + \frac{69}{65536}x^6$$

$$- \frac{187}{262144}x^7 + \frac{1843}{8388608}x^8 - \frac{4859}{33554432}x^9 + \frac{12767}{268435456}x^{10} + \cdots$$

The equation $(4 - x^2)\dfrac{dy}{dx} + y = 0$ is a separable first-order equation. Separat-

ing, integrating, and applying the initial condition yields $y = \left|\dfrac{x - 2}{x + 2}\right|^{1/4}$. We

can approximate the solution of the problem by taking a finite number of terms of the series solution. The graph of the polynomial of degree ten is shown in Figure 6.9 along with the solution obtained through separation of variables. Notice that the accuracy of the approximation decreases near $x = 2$ and $x = -2$, which are singular points of the differential equation. (Why?) The reason for this is discussed in the following theorem.

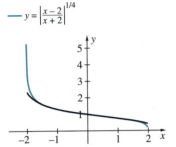

— Polynomial Approximation

— $y = \left|\dfrac{x-2}{x+2}\right|^{1/4}$

Figure 6.9 Comparison of Exact and Approximate Solutions to the Initial-Value Problem

$$\begin{cases} (4 - x^2)\dfrac{dy}{dx} + y = 0 \\ y(0) = 1 \end{cases}$$

Why does the solution in part (a) of Example 3 depend on only one unknown coefficient while that in Example 2 depends on two?

The following theorem explains where a series solution of a differential equation is valid.

Theorem 6.5 Convergence of a Power Series Solution

Let $x = x_0$ be an ordinary point of the differential equation

$$a_2(x)y''(x) + a_1(x)y'(x) + a_0(x)y(x) = 0$$

and suppose that R is the distance from $x = x_0$ to the closest singular point of the equation. Then the power series solution $y = \sum_{n=0}^{\infty} a_n(x - x_0)^n$ converges *at least* on the interval $(x_0 - R, x_0 + R)$.*

---

* A proof of this theorem may be found in more advanced texts, like *Introduction to Ordinary Differential Equations* by Albert L. Rabenstein, Academic Press, 1966, pp. 103–107.

Series solutions have been used since the days of Newton. At that time, however, mathematicians did not concern themselves with such issues as convergence, which is essential to successful application of the method.

This theorem indicates that a polynomial approximation may not be accurate near singular points of the equation. Now we understand why the approximation in Example 3 breaks down near $x = 2$ and $x = -2$; these are the closest singular points to the ordinary point $x = 0$.

Of course, $x = 0$ is not an ordinary point for every differential equation. However, because the series $y = \sum_{n=0}^{\infty} a_n(x - x_0)^n$ is easier to work with if $x_0 = 0$, we can always make a transformation so that we can use $y = \sum_{n=0}^{\infty} a_n x^n$ to solve any linear equation. For example, if $x = x_0$ is an ordinary point of a linear equation and we make the change of variable $t = x - x_0$, then $t = 0$ corresponds to $x = x_0$. Hence, $t = 0$ is an ordinary point of the transformed equation.

Example 4

Solve $xy'' + y = 0$.

Solution

In standard form the equation becomes $y'' + \frac{1}{x}y = 0$. Because $\frac{1}{x}$ is not analytic at $x = 0$, this equation has a singular point at $x = 0$. All other values of x are classified as ordinary points, so we can select one to use in our power series solution. Choosing $x = 1$, we consider the power series $y = \sum_{n=0}^{\infty} a_n(x - 1)^n$. However, with the change of variable $t = x - 1$, we have that $t = 0$ corresponds to $x = 1$. Therefore, by changing variables, we can use the series $y = \sum_{n=0}^{\infty} a_n t^n$.

Notice that
$$\frac{dy}{dx} = \frac{dy}{dt}\frac{dt}{dx} = \frac{dy}{dt} \text{ and}$$

$$\frac{d^2y}{dx^2} = \frac{d}{dx}\left(\frac{dy}{dx}\right) = \frac{d}{dt}\left(\frac{dy}{dx}\right)\frac{dt}{dx} = \frac{d}{dt}\left(\frac{dy}{dt}\right) = \frac{d^2y}{dt^2},$$

so with these substitutions into $xy'' + y = 0$, we obtain $(t + 1)\dfrac{d^2y}{dt^2} + y = 0$. Hence, we assume that $y = \sum_{n=0}^{\infty} a_n t^n$. Substitution into the transformed equation yields

$$(t + 1)\sum_{n=2}^{\infty} n(n - 1)a_n t^{n-2} + \sum_{n=0}^{\infty} a_n t^n = 0.$$

Simplification then gives us

$$\sum_{n=2}^{\infty} n(n - 1)a_n t^{n-1} + \sum_{n=2}^{\infty} n(n - 1)a_n t^{n-2} + \sum_{n=0}^{\infty} a_n t^n = 0$$

$$\sum_{n=2}^{\infty} n(n - 1)a_n t^{n-1} + 2a_2 t^0 + \sum_{n=3}^{\infty} n(n - 1)a_n t^{n-2} + a_0 t^0 + \sum_{n=1}^{\infty} a_n t^n = 0.$$

In this case, we must change the index in two of the three series. If we substitute $(n + 1)$ for n in $\sum_{n=3}^{\infty} n(n - 1)a_n t^{n-2}$ and $(n - 1)$ for n in $\sum_{n=1}^{\infty} a_n t^n$, we obtain

TABLE 6.5

n	a_n
0	a_0
1	a_1
2	$-\dfrac{a_0}{2}$
3	$\dfrac{a_0 - a_1}{6}$
4	$\dfrac{2a_1 - a_0}{24}$
5	$\dfrac{2a_0 - 5a_1}{120}$
6	$\dfrac{18a_1 - 7a_0}{720}$
7	$\dfrac{33a_0 - 85a_1}{5040}$

$$(2a_2 + a_0) + \sum_{n=2}^{\infty} n(n-1)a_n t^{n-1} + \sum_{n=2}^{\infty} (n+1)na_{n+1}t^{n-1} + \sum_{n=2}^{\infty} a_{n-1}t^{n-1} = 0$$

$$(2a_2 + a_0) + \sum_{n=2}^{\infty} [n(n-1)a_n + n(n+1)a_{n+1} + a_{n-1}]t^{n-1} = 0.$$

Equating the coefficients to zero, we determine that $a_2 = -a_0/2$ and

$$a_{n+1} = \frac{-a_{n-1} - n(n-1)a_n}{n(n+1)}$$

for the indices in the series, $n \geq 2$. We calculate the coefficients for $n = 2, 3, 4, 5,$ 6, and 7 and display the results in Table 6.5.

Therefore,

$$y = a_0 + a_1 t - \frac{a_0}{2}t^2 + \frac{a_0 - a_1}{6}t^3 + \frac{2a_1 - a_0}{24}t^4 + \frac{2a_0 - 5a_1}{120}t^5 +$$

$$\frac{18a_1 - 7a_0}{720}t^6 + \frac{33a_0 - 85a_1}{5040}t^7 + \cdots$$

Returning to the original variable, we have

$$y = a_0 + a_1(x-1) - \frac{a_0}{2}(x-1)^2 + \frac{a_0 - a_1}{6}(x-1)^3 + \frac{2a_1 - a_0}{24}(x-1)^4 +$$

$$\frac{2a_0 - 5a_1}{120}(x-1)^5 + \frac{18a_1 - 7a_0}{720}(x-1)^6 + \frac{33a_0 - 85a_1}{5040}(x-1)^7 + \cdots$$

What are two linearly independent solutions of $xy'' + y = 0$?

Power series solutions can also be used to solve linear equations that involve functions that can be expressed as a power series as well.

 Example 5   Use power series to approximate the solution of $\begin{cases} y'' + (\sin x)y = 0 \\ y(0) = 0, \, y'(0) = 1 \end{cases}$.

Solution   In order to use a power series solution, we must first note that

$$\sin x = \sum_{n=0}^{\infty} \frac{(-1)^n x^{2n+1}}{(2n+1)!} = x - \frac{x^3}{3!} + \frac{x^5}{5!} - \cdots.$$

Hence, if we assume that $y = \sum_{n=0}^{\infty} a_n x^n$, multiplying the series $(\sin x)y$ term-by-term

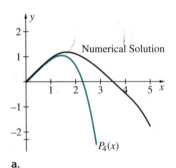

a.

b.

c.

we have

$$\sum_{n=2}^{\infty} n(n-1)a_n x^{n-2} + \left(x - \frac{x^3}{3!} + \frac{x^5}{5!} - \cdots\right)\sum_{n=0}^{\infty} a_n x^n = 0$$

$$2a_2 + 6a_3 x + 12a_4 x^2 + 20a_5 x^3 + 30a_6 x^4 + 42a_7 x^5 + \cdots$$
$$+ a_0 x + a_1 x^2 + a_2 x^3 + a_3 x^4 + a_4 x^5 + \cdots$$
$$- \frac{a_0 x^3}{3!} - \frac{a_1 x^4}{3!} - \frac{a_2 x^5}{3!} - \cdots$$
$$+ \frac{a_0 x^5}{5!} + \cdots = 0.$$

After combining like terms and equating the coefficients to zero, we have the equations:

$$2a_2 = 0, \quad 6a_3 + a_0 = 0, \quad 12a_4 + a_1 = 0, \quad 20a_5 + a_2 - \frac{a_0}{3!} = 0,$$

$$30a_6 + a_3 - \frac{a_1}{3!} = 0, \cdots.$$

Thus, $a_2 = 0$, $a_3 = -\dfrac{a_0}{3!}$, $a_4 = -\dfrac{a_1}{12}$, $a_5 = \dfrac{a_0}{120}$, and $a_6 = \dfrac{a_0 + a_1}{180}$, so the first terms of the power series solution are

$$y = a_0\left(1 - \frac{x^3}{3!} + \frac{x^5}{120} + \frac{x^6}{180} + \cdots\right) + a_1\left(x - \frac{x^4}{12} + \frac{x^6}{180} + \cdots\right).$$

Application of the initial conditions yields the solution

$$y = x - \frac{x^4}{12} + \frac{x^6}{180} + \frac{x^7}{504} - \frac{x^8}{6720} - \frac{7x^9}{25920} - \frac{x^{10}}{50400} + \frac{107x^{11}}{6652800} + \frac{31x^{12}}{6842880}$$
$$- \frac{41x^{13}}{94348800} - \frac{9293x^{14}}{21794572800} + \cdots.$$

As in Example 3, we approximate the solution of the problem by taking a finite number of terms of the series solution. In Figure 6.10, we graph various polynomial approximations along with an accurate numerical solution obtained with a computer algebra system. Notice that the polynomial approximation improves as the number of terms increases.

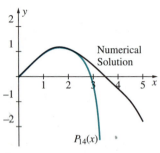

d.

Figure 6.10   (a) $P_4(x) = x - \dfrac{x^4}{12}$ (b) $P_6(x) = x - \dfrac{x^4}{12} + \dfrac{x^6}{180}$ (c) $P_{12}(x) = x - \dfrac{x^4}{12} + \cdots$
$$+ \frac{107x^{11}}{6652800} + \frac{31x^{12}}{6842880}$$ (d) $P_{14}(x) = x - \dfrac{x^4}{12} + \cdots - \dfrac{41x^{13}}{94348800} - \dfrac{9293x^{14}}{21794572800}$

IN TOUCH WITH TECHNOLOGY

1. Consider the initial-value problem $y'' + f(x)y' + y = 0$, $y(0) = 1$, $y'(0) = -1$ where

$$f(x) = \begin{cases} \dfrac{\sin x}{x}, & \text{if } x \neq 0 \\ 1, & \text{if } x = 0 \end{cases}$$. **(a)** Show that $x = 0$ is an

ordinary point of the equation. **(b)** Find a power series solution of the equation and graph an approximation of the solution on an interval. **(c)** Generate a numerical solution of the equation. Explain any unexpected results.

2. **(a)** Use power series to solve $\begin{cases} y'' - (\cos x)y = \sin x \\ y(0) = 1, \ y'(0) = 0 \end{cases}$. **(b)** Compare the polynomial approximations of degree 4, 7, 10, and 13 to the numerical solution obtained with a computer algebra system.

3. Assume that $x^2 y' - y = -1$ has a series solution of the form $y(x) = \sum_{n=0}^{\infty} a_n x^n$. **(a)** Calculate $y(x)$. **(b)** Why does Theorem 6.5 fail for this equation? **(c)** Find a general solution of the equation $x^2 y' - y = -1$.

4. The differential equation $y'' - xy = 0$ is called **Airy's equation** and arises in electromagnetic theory and quantum mechanics. Two linearly independent solutions to Airy's equation, denoted by $Ai(x)$ and $Bi(x)$, are called the **Airy functions**. The function $Ai(x) \to 0$ as $x \to \infty$ while $Bi(x) \to +\infty$ as $x \to \infty$. (Most computer algebra systems contain built-in definitions of the Airy functions.)

(a) If your computer algebra system contains built-in definitions of the Airy functions, graph each on the interval $[-15, 5]$. **(b)** Find a series solution of Airy's equation and obtain formulas for both $Ai(x)$ and $Bi(x)$. **(c)** Graph the polynomial approximation of degree n of $Ai(x)$ for $n = 6, 15, 30$, and 45 on the interval $[-15, 5]$ *Compare your results to (a), if applicable.* **(d)** Graph the polynomial approximation of degree n of $Bi(x)$ for $n = 6, 15, 30$, and 45 on the interval $[-15, 5]$. *Compare your results to (a), if applicable.*

EXERCISES 6.3

In Exercises 1–6, determine the singular points of the equations. Use these points to find an upper bound on the radius of convergence of a series solution about x_0.

1. $x^2 y'' - 2xy' + 7y = 0$, $x_0 = 1$
2. $x^2 y'' + x^2 y' + y = 0$, $x_0 = -1$
3. $(x - 2)y'' + y' - y = 0$, $x_0 = -2$
4. $(x + 5)y'' - 4y' - xy = 0$, $x_0 = 0$
5. $(x^2 - 4)y'' + 16(x + 2)y' - y = 0$, $x_0 = 1$
6. $(x^2 - 2x - 3)y'' + (x + 1)y' - (x^2 - 1)y = 0$, $x_0 = 0$

In Exercises 7–31, solve the differential equation with a power series expansion about $x = 0$. Write out at least the first five nonzero terms of each series.

7. $y'' + 3y' - 18y = 0$
8. $y'' + 3y' - 10y = 0$

*9. $y'' - 11y' + 30y = 0$
10. $y'' - 10y' + 21y = 0$
11. $y'' + 16y' + 64y = 0$
12. $y'' + 16y = 0$
*13. $y'' + y = 0$
14. $y'' + 8y' + 20y = 0$
15. $y'' + 8y' + 25y = 0$
16. $y'' - 4y' + 29y = 0$
*17. $y'' - y' - 2y = e^{-x}$
18. $y'' - 7y' - 8y = x^2$
19. $(-2 - 2x)y'' + 2y' + 4y = 0$
20. $(-1 - x)y'' - 3y' - y = 0$
*21. $(1 + 3x)y'' - 3y' - 2y = 0$
22. $(3 - 2x)y'' + 2y' - 2y = 0$
23. $y'' - xy' + 4y = 0$
24. $-y'' - xy' - 5y = 0$
*25. $(2 + 3x)y'' + 3xy' = 0$
26. $2y'' - 2xy' = 0$
27. $(2 + 2x^2)y'' + 2xy' - 3y = 0$

28. $(-4 + 5x + 3x^2)y'' + 3xy' + 4y = 0$

*29. $(2 - x^2)y'' + 2(x - 1)y' + 4y = 0$

30. $(-2 + 4x - 5x^2)y'' - (2 + 5x)y' = 0$

31. $(2 - 2x - 4x^2)y'' + (5 + 3x)y' = 0$

In Exercises 32–36, determine at least the first five nonzero terms in a power series expansion about $x = 0$ for the solution of each initial-value problem.

32. $y'' + 2xy' = 0$, $y(0) = 0$, $y'(0) = 1$

33. $y'' - 4x^2y = 0$, $y(0) = 1$, $y'(0) = 0$

34. $(3 - 2x)y'' + 2y' - 2y = 0$, $y(0) = 3$,
    $y'(0) = -2$ (See Exercise 22.)

*35. $(-1 + 2x^2)y'' + 2xy' - 3y = 0$, $y(0) = -2$,
    $y'(0) = 2$

36. $(-2 + 4x - 5x^2)y'' - (2 + 5x)y' = 0$, $y(0) = 0$,
    $y'(0) = 2$ (See Exercise 30.)

In Exercises 37–41, solve the equation with a power series expansion about $t = 0$ by making the indicated change of variable.

37. $4xy'' + y' = 0$, $x = t + 1$

38. $x(x + 2)y'' - y = 0$, $x = t - 1$

*39. $4x^2y'' + (x + 1)y' = 0$, $x = t + 2$

40. $xy'' - 2xy' + 4y = 0$, $x = t - 1$

41. $x^2y'' + (x + 4)y' + y = 0$, $x = t - 2$

In Exercises 42 and 43, determine at least the first five nonzero terms in a power series expansion about $x = 0$ of the solution to each nonhomogeneous initial-value problem.

42. $y'' + xy' = \sin x$, $y(0) = 1$, $y'(0) = 0$

43. $y'' + y' + xy = \cos x$, $y(0) = 0$, $y'(0) = 1$

44. (a) If $y = a_0 + a_1x + a_2x^2 + a_3x^3 + \cdots$, what are the first three nonzero terms of the power series for y^2? (b) Use this series to find the first three nonzero terms in the power series solution about $x = 0$ to **Van-der-Pol's equation,**

$$y'' + (y^2 - 1)y' + y = 0$$

if $y(0) = 0$ and $y'(0) = 1$.

*45. Use a method similar to that in Exercise 44 to find the first three nonzero terms of the power series solution about $x = 0$ to **Rayleigh's equation,**

$$y'' + \left(\frac{1}{3}(y')^2 - 1\right)y' + y = 0$$

if $y(0) = 1$ and $y'(0) = 0$, an equation that arises in the study of the motion of a violin string.

6.4 Series Solutions About Regular Singular Points

Regular and Irregular Singular Points • Method of Frobenius • Indicial Roots that Differ by an Integer • Equal Indicial Roots

In Section 6.3, we used a power series expansion about an ordinary point to find (or approximate) the solution of a differential equation. We noted that these series solutions may not converge near the singular points of the equation. In this section, we find that series expansions about certain singular points can be used to solve (or approximate solutions of) some differential equations.

Regular and Irregular Singular Points

We begin with the following classification of singular points.

Definition 6.7 Singular Points

> Let x_0 be a singular point of $y''(x) + p(x)y'(x) + q(x)y(x) = 0$. x_0 is a **regular singular point** of the equation if both $(x - x_0)p(x)$ and $(x - x_0)^2 q(x)$ are analytic at $x = x_0$. If x_0 is not a regular singular point, then x_0 is called an **irregular singular point** of the equation.

Sometimes this definition is difficult to apply, so we give the following equivalent definition if the coefficients $p(x)$ and $q(x)$ in $y''(x) + p(x)y'(x) + q(x)y(x) = 0$ are rational functions.

Definition 6.8 Singular Points of Equations with Rational Function Coefficients

> Suppose that $p(x)$ and $q(x)$ are rational functions. If after reducing $p(x)$ and $q(x)$ to lowest terms, the highest power of $(x - x_0)$ in the denominator of $p(x)$ is 1 and the highest power of $(x - x_0)$ in the denominator of $q(x)$ is 2, x_0 is a **regular singular point** of the equation $y''(x) + p(x)y'(x) + q(x)y(x) = 0$. Otherwise, it is an **irregular singular point.**

Example 1   Classify the singular points of each of the following equations.

(a) $y'' + \dfrac{y'}{x} + \left(1 - \dfrac{\mu^2}{x^2}\right)y = 0$ (the **Bessel equation**);

(b) $(x^2 - 16)^2 y'' + (x - 4)y' + y = 0$.

Solution   (a) $x = 0$ is a singular point of this equation because $p(x) = \dfrac{1}{x}$ is not

analytic at $x = 0$. Because $xp(x) = 1$ and $x^2 q(x) = x^2\left(1 - \dfrac{\mu^2}{x^2}\right) = x^2 - \mu^2$ are

both analytic at $x = 0$, $x = 0$ is a regular singular point.

(b) In standard form, the equation is

$$y'' + \frac{x - 4}{(x^2 - 16)^2}y' + \frac{1}{(x^2 - 16)^2}y = 0 \qquad \text{or}$$

$$y'' + \frac{1}{(x - 4)(x + 4)^2}y' + \frac{1}{(x - 4)^2(x + 4)^2}y = 0.$$

Thus, the singular points are $x = 4$ and $x = -4$. For $x = 4$, we have

$$(x - 4)p(x) = (x - 4)\frac{1}{(x - 4)(x + 4)^2} = \frac{1}{(x + 4)^2}$$

and

$$(x - 4)^2 q(x) = (x - 4)^2\frac{1}{(x - 4)^2(x + 4)^2} = \frac{1}{(x + 4)^2}.$$

Remember that to find a general solution of the second-order linear homogeneous equation $y''(x) + p(x)y'(x) + q(x)y(x) = 0$ we must find two linearly independent solutions, $y_1(x)$ and $y_2(x)$, of the equation; a general solution is given by $y = c_1 y_1(x) + c_2 y_2(x)$, where c_1 and c_2 are constants. (See Section 4.2.)

Both of these functions are analytic at $x = 4$, so $x = 4$ is a regular singular point. For $x = -4$,

$$(x + 4)p(x) = (x + 4)\frac{1}{(x - 4)(x + 4)^2} = \frac{1}{(x - 4)(x + 4)},$$

which is not analytic at $x = -4$. Thus, $x = -4$ is an irregular singular point.

Method of Frobenius for $y''(x) + p(x)y'(x) + q(x)y(x) = 0$

A series solution about a regular singular point can always be found, as stated in the following theorem. We use several examples to illustrate the method.

Theorem 6.6 Method of Frobenius

Although the Method of Frobenius was initiated by Euler, the method for finding a series expansion about a regular singular point was first published by the German mathematician Georg Frobenius (1849–1917) in 1873.

Let x_0 be a regular singular point of $y''(x) + p(x)y'(x) + q(x)y(x) = 0$. Then this differential equation has at least one solution of the form

$$y = \sum_{n=0}^{\infty} a_n(x - x_0)^{n+r}, \quad a_0 \neq 0,$$

where r is a constant that must be determined. This solution is convergent at least on some interval $|x - x_0| < R, R > 0$.*

Example 2

Find a general solution of $xy'' + (1 + x)y' - \frac{1}{16x}y = 0$.

Solution   In standard form this equation is $y'' + \frac{1 + x}{x}y' - \frac{1}{16x^2}y = 0$,

so $x = 0$ is a singular point. Moreover, because $xp(x) = x\frac{1 + x}{x} = 1 + x$ and

$x^2 q(x) = x^2\left(-\frac{1}{16x^2}\right) = -\frac{1}{16}$ are both analytic at $x = 0$, we classify $x = 0$ as a

regular singular point. According to the Method of Frobenius, there is at least one solution of the form $y = \sum_{n=0}^{\infty} a_n x^{n+r}$. Differentiating this function, we obtain

$$y' = \sum_{n=0}^{\infty} a_n(n + r)x^{n+r-1} \quad \text{and} \quad y'' = \sum_{n=0}^{\infty} a_n(n + r)(n + r - 1)x^{n+r-2}.$$

---

* A proof of this theorem may be found in more advanced texts like *Introduction to Ordinary Differential Equations* by Albert L. Rabenstein, Academic Press, 1966, pp. 113–117.

Substituting this series into the differential equation yields

$$x \sum_{n=0}^{\infty} a_n(n+r)(n+r-1)x^{n+r-2} + (1+x) \sum_{n=0}^{\infty} a_n(n+r)x^{n+r-1} -$$

$$\frac{1}{16x} \sum_{n=0}^{\infty} a_n x^{n+r} = 0$$

$$\sum_{n=0}^{\infty} a_n(n+r)(n+r-1)x^{n+r-1} + \sum_{n=0}^{\infty} a_n(n+r)x^{n+r-1} + \sum_{n=0}^{\infty} a_n(n+r)x^{n+r} -$$

$$\sum_{n=0}^{\infty} \frac{1}{16} a_n x^{n+r-1} = 0.$$

Notice that the first term in three of the four series is x^{r-1}. However, the first term in $\sum_{n=0}^{\infty} a_n(n+r)x^{n+r}$ is x^r, so we must pull off the first terms in the other three series so that they match. Hence,

$$\left[r(r-1) + r - \frac{1}{16} \right] a_0 x^{r-1} + \sum_{n=1}^{\infty} a_n(n+r)(n+r-1)x^{n+r-1} +$$

$$\sum_{n=1}^{\infty} a_n(n+r)x^{n+r-1} + \sum_{n=0}^{\infty} a_n(n+r)x^{n+r} - \sum_{n=1}^{\infty} \frac{1}{16} a_n x^{n+r-1} = 0.$$

Changing the index in the third series by substituting $(n-1)$ for each occurrence of n, we have

$$\sum_{n-1=0}^{\infty} a_{n-1}(n-1+r)x^{n-1+r} = \sum_{n=1}^{\infty} a_{n-1}(n+r-1)x^{n+r-1}.$$

After simplification, we have

$$\left[r(r-1) + r - \frac{1}{16} \right] a_0 x^{r-1} + \sum_{n=1}^{\infty} \left\{ \left[(n+r)(n+r-1) + (n+r) - \frac{1}{16} \right] a_n +$$

$$(n+r-1)a_{n-1} \right\} x^{n+r-1} = 0.$$

We equate the coefficients of like terms to zero to find the coefficients a_n and the value of r. Assuming that $a_0 \neq 0$ so that the first term of our series solution is not zero, we have from the first term the equation

$$r(r-1) + r - \frac{1}{16} = 0,$$

called the **indicial equation** because it yields the value of r. In this case,

$$r^2 - r + r - \frac{1}{16} = r^2 - \frac{1}{16} = \left(r + \frac{1}{4} \right)\left(r - \frac{1}{4} \right) = 0$$

so the roots are $r_1 = \frac{1}{4}$ and $r_2 = -\frac{1}{4}$ and the differential equation has solutions

$$y = \sum_{n=0}^{\infty} a_n x^{n+1/4} = x^{1/4} \sum_{n=0}^{\infty} a_n x^n \quad \text{and} \quad y = \sum_{n=0}^{\infty} b_n x^{n-1/4} = x^{-1/4} \sum_{n=0}^{\infty} b_n x^n.$$

Starting with the larger of the two roots, $r_1 = \frac{1}{4}$, we assume that

$$y = \sum_{n=0}^{\infty} a_n x^{n+1/4} = x^{1/4} \sum_{n=0}^{\infty} a_n x^n.$$

Equating the series coefficients to zero, we have

$$\left[\left(n + \frac{1}{4} \right) \left(n + \frac{1}{4} - 1 \right) + \left(n + \frac{1}{4} \right) - \frac{1}{16} \right] a_n + \left(n + \frac{1}{4} - 1 \right) a_{n-1} = 0,$$

so

$$a_n = \frac{-\left(n - \dfrac{3}{4} \right) a_{n-1}}{\left(n + \dfrac{1}{4} \right) \left(n - \dfrac{3}{4} \right) + \left(n + \dfrac{1}{4} \right) - \dfrac{1}{16}} = \frac{(3 - 4n) a_{n-1}}{2(2n^2 + n)}, \quad n \geq 1.$$

Several of these coefficients are calculated with this formula and given in Table 6.6. In this case, the solution is

$$y_1(x) = a_0 x^{1/4} \left(1 - \frac{1}{6} x + \frac{1}{24} x^2 - \frac{1}{112} x^3 + \frac{13}{8064} x^4 - \cdots \right).$$

If $r_2 = -\frac{1}{4}$ and we assume that $y = \sum_{n=0}^{\infty} b_n x^{n-1/4} = x^{-1/4} \sum_{n=0}^{\infty} b_n x^n$, we have

$$\left[\left(n - \frac{1}{4} \right) \left(n - \frac{1}{4} - 1 \right) + \left(n - \frac{1}{4} \right) - \frac{1}{16} \right] b_n + \left(n - \frac{1}{4} - 1 \right) b_{n-1} = 0,$$

so

$$b_n = \frac{-\left(n - \dfrac{5}{4} \right) b_{n-1}}{\left(n - \dfrac{1}{4} \right) \left(n - \dfrac{5}{4} \right) + \left(n - \dfrac{1}{4} \right) - \dfrac{1}{16}} = \frac{(5 - 4n) b_{n-1}}{2(2n^2 - n)}, \quad n \geq 1.$$

The values of several coefficients determined with this formula are given in Table 6.7. Therefore, the solution obtained with $r_2 = -\dfrac{1}{4}$ is

$$y_2(x) = b_0 x^{-1/4} \left(1 + \frac{1}{2} x - \frac{1}{8} x^2 + \frac{7}{240} x^3 - \frac{11}{1920} x^4 - \cdots \right).$$

A general solution of the differential equation is

TABLE 6.6

n	a_n
0	a_0
1	$-\dfrac{1}{6} a_0$
2	$\dfrac{1}{24} a_0$
3	$-\dfrac{1}{112} a_0$
4	$\dfrac{13}{8064} a_0$

TABLE 6.7

n	b_n
0	b_0
1	$\dfrac{1}{2} b_0$
2	$-\dfrac{1}{8} b_0$
3	$\dfrac{7}{240} b_0$
4	$-\dfrac{11}{1920} b_0$

$$y(x) = c_1 x^{1/4} \left(1 - \frac{1}{6}x + \frac{1}{24}x^2 - \frac{1}{112}x^3 + \frac{13}{8064}x^4 - \cdots \right) +$$

$$c_2 x^{-1/4} \left(1 + \frac{1}{2}x - \frac{1}{8}x^2 + \frac{7}{240}x^3 - \frac{11}{1920}x^4 - \cdots \right)$$

where c_1 and c_2 are arbitrary constants. (These two solutions are linearly independent because they are not constant multiples of one another.)

In the previous example, we found the **indicial equation** by direct substitution of
the series solution into the differential equation. In order to derive a general formula for
the indicial equation, suppose that $x = 0$ is a regular singular point of the differential
equation $y'' + p(x)y' + q(x)y = 0$. Then the functions $xp(x)$ and $x^2 q(x)$ are analytic;
both of these functions have a power series in x with a positive radius of convergence,

$$xp(x) = p_0 + p_1 x + p_2 x^2 + \cdots \qquad \text{and} \qquad x^2 q(x) = q_0 + q_1 x + q_2 x^2 + \cdots.$$

Therefore,

$$p(x) = \frac{p_0}{x} + p_1 + p_2 x + p_3 x^2 + p_4 x^3 + \cdots \qquad \text{and}$$

$$q(x) = \frac{q_0}{x^2} + \frac{q_1}{x} + q_2 + q_3 x + q_4 x^2 + q_5 x^3 + \cdots$$

Substitution of these series into the differential equation $y'' + p(x)y' + q(x)y = 0$
yields

$$\left(\sum_{n=0}^{\infty} a_n (n + r)(n + r - 1)x^{n+r-2} \right) +$$

$$\left(\frac{p_0}{x} + p_1 + p_2 x + p_3 x^2 + p_4 x^3 + \cdots \right) \left(\sum_{n=0}^{\infty} a_n (n + r)x^{n+r-1} \right) +$$

$$\left(\frac{q_0}{x^2} + \frac{q_1}{x} + q_2 + q_3 x + q_4 x^2 + q_5 x^3 + \cdots \right) \left(\sum_{n=0}^{\infty} a_n x^{n+r} \right) = 0.$$

After multiplying through by the first term in the power series for $p(x)$ and $q(x)$, we see
that the lowest term in the series is x^{n+r-2}.

$$\left(\sum_{n=0}^{\infty} a_n (n + r)(n + r - 1)x^{n+r-2} \right) + \left(\sum_{n=0}^{\infty} a_n p_0 (n + r)x^{n+r-2} \right)$$

$$+ (p_1 + p_2 x + p_3 x^2 + p_4 x^3 + \cdots) \left(\sum_{n=0}^{\infty} a_n (n + r)x^{n+r-1} \right) + \left(\sum_{n=0}^{\infty} a_n q_0 x^{n+r-2} \right)$$

$$+ \left(\frac{q_1}{x} + q_2 + q_3 x + q_4 x^2 + q_5 x^3 + \cdots \right) \left(\sum_{n=0}^{\infty} a_n x^{n+r} \right) = 0.$$

Then, with $n = 0$, we find that the coefficient of x^{r-2} is

$$-ra_0 + r^2a_0 + ra_0p_0 + a_0q_0 = a_0(r^2 + (p_0 - 1)r + q_0) = a_0(r(r - 1) + p_0r + q_0).$$

Thus, for any equation of the form $y'' + p(x)y' + q(x)y = 0$ with regular singular point $x = 0$, we have the indicial equation

$$r(r - 1) + p_0r + q_0 = 0.$$

The values of r that satisfy this equation are called the **exponents** or **indicial roots.**

Example 3

Determine the roots of the indicial equation for each of the following differential equations.

(a) $y'' - \dfrac{1}{2x}y' + \dfrac{1}{2x^2}y = 0$; (b) $xy'' + y' - 3y = 0$;

(c) $x^2y'' + x^2y' + (x - 2)y = 0$.

Solution   (a) In this case, $xp(x) = x\left(-\dfrac{1}{2x}\right) = -\dfrac{1}{2}$ and $x^2q(x) =$

$x^2\left(\dfrac{1}{2x^2}\right) = \dfrac{1}{2}$. Hence, $p_0 = -\dfrac{1}{2}$ and $q_0 = \dfrac{1}{2}$, so the indicial equation is

$$r(r - 1) + p_0r + q_0 = r(r - 1) - \dfrac{1}{2}r + \dfrac{1}{2} = r^2 - \dfrac{3}{2}r + \dfrac{1}{2} = 0.$$

Solving the equivalent equation,

$$2r^2 - 3r + 1 = (2r - 1)(r - 1) = 0,$$

we find that the roots are $r_1 = 1$ and $r_2 = \dfrac{1}{2}$.

(b) In standard form, the equation is $y'' + \dfrac{1}{x}y' - \dfrac{3}{x}y = 0$. Thus, $xp(x) = x\dfrac{1}{x} = 1$

and $x^2q(x) = x^2\left(-\dfrac{3}{x}\right) = -3x$, so $p_0 = 1$ and $q_0 = 0$ (because there is no constant

term in the expansion of $x^2q(x)$). Therefore, the indicial equation is

$$r(r - 1) + r = r^2 = 0,$$

and the equation has the equal indicial roots $r_1 = r_2 = 0$.

(c) This equation is $y'' + y' + \dfrac{x - 2}{x^2}y = 0$ in standard form. Thus, $xp(x) =$

$x(1) = x$ and $x^2q(x) = x^2\left(\dfrac{x - 2}{x^2}\right) = x - 2$, so $p_0 = 0$ and $q_0 = -2$. The indicial

equation is then

$$r(r - 1) - 2 = r^2 - r - 2 = (r - 2)(r + 1) = 0$$

with roots $r_1 = 2$ and $r_2 = -1$.

In Example 3, we see that several situations can arise when finding the roots of the indicial equation: (1) the roots are distinct and differ by a fractional value, (2) the roots are distinct and differ by an integer value, and (3) the roots are equal. We now discuss these cases in more detail.

Indicial Roots that Differ by an Integer

The following example shows that complications may arise when the roots of the indicial equation differ by an integer.

Example 4	

Find a general solution of $x^2 y'' + x^2 y' + (x - 2)y = 0$.

Solution In standard form, the equation is $y'' + y' + \dfrac{x-2}{x^2} y = 0$. Hence, $xp(x) = x$ and $x^2 q(x) = x - 2$, so $p_0 = 0$ and $q_0 = -2$. The indicial equation is

$$r(r-1) + p_0 r + q_0 = r^2 - r - 2 = (r+1)(r-2) = 0$$

with roots $r_1 = 2$ and $r_2 = -1$. For $r_1 = 2$, we have a solution of the form $y = \sum_{n=0}^{\infty} a_n x^{n+2}$. Substituting this function into the differential equation, we have

$$\sum_{n=0}^{\infty} (n+2)(n+1)a_n x^{n+2} + \sum_{n=0}^{\infty} (n+2)a_n x^{n+3} +$$

$$\sum_{n=0}^{\infty} a_n x^{n+3} - \sum_{n=0}^{\infty} 2a_n x^{n+2} = 0$$

$$(2a_0 - 2a_0)x^2 + \sum_{n=1}^{\infty} (n+2)(n+1)a_n x^{n+2} + \sum_{n=0}^{\infty} (n+2)a_n x^{n+3} +$$

$$\sum_{n=0}^{\infty} a_n x^{n+3} - \sum_{n=1}^{\infty} 2a_n x^{n+2} = 0$$

$$\sum_{n=0}^{\infty} (n+3)(n+2)a_{n+1} x^{n+3} + \sum_{n=0}^{\infty} (n+2)a_n x^{n+3} +$$

$$\sum_{n=0}^{\infty} a_n x^{n+3} - \sum_{n=0}^{\infty} 2a_{n+1} x^{n+3} = 0$$

$$\sum_{n=0}^{\infty} \{[(n+3)(n+2) - 2]a_{n+1} + [(n+2) + 1]a_n\}x^{n+3} = 0.$$

Therefore, $a_{n+1} = -\dfrac{(n+3)}{n^2 + 5n + 4} a_n = -\dfrac{(n+3)}{(n+1)(n+4)} a_n$, $n \geq 0$. We use this formula to calculate some of the coefficients and list these values in Table 6.8.

TABLE 6.8

n	a_n
0	a_0
1	$-\dfrac{3}{4} a_0$
2	$\dfrac{3}{10} a_0$
3	$-\dfrac{1}{12} a_0$
4	$\dfrac{1}{56} a_0$

Hence,

$$y_1(x) = a_0\left(x^2 - \frac{3x^3}{4} + \frac{3x^4}{10} - \frac{x^5}{12} + \frac{x^6}{56} + \cdots\right).$$

Now, for $r_2 = -1$, we assume that $y = \sum_{n=0}^{\infty} b_n x^{n-1}$. Substitution into the differential equation yields

$$\sum_{n=0}^{\infty} (n-1)(n-2)b_n x^{n-1} + \sum_{n=0}^{\infty} (n-1)b_n x^n + \sum_{n=0}^{\infty} b_n x^n - \sum_{n=0}^{\infty} 2b_n x^{n-1} = 0$$

$$(2b_0 - 2b_0)x^{-1} + \sum_{n=1}^{\infty} (n-1)(n-2)b_n x^{n-1} + $$

$$\sum_{n=0}^{\infty} (n-1)b_n x^n + \sum_{n=0}^{\infty} b_n x^n - \sum_{n=1}^{\infty} 2b_n x^{n-1} = 0$$

$$\sum_{n=1}^{\infty} (n-1)(n-2)b_n x^{n-1} + \sum_{n=1}^{\infty} (n-2)b_{n-1} x^{n-1} + \sum_{n=1}^{\infty} b_{n-1} x^{n-1} - \sum_{n=1}^{\infty} 2b_n x^{n-1} = 0$$

$$\sum_{n=1}^{\infty} \{[(n-1)(n-2) - 2]b_n + [(n-2) + 1]b_{n-1}\}x^{n-1} = 0.$$

TABLE 6.9

n	b_n
0	b_0
1	0
2	0
3	b_3
4	$-\dfrac{3}{4}b_3$
5	$\dfrac{3}{10}b_3$
6	$-\dfrac{1}{12}b_3$
7	$\dfrac{1}{56}b_3$

Therefore, $b_n = -\dfrac{(n-1)}{n^2 - 3n}b_{n-1} = -\dfrac{(n-1)}{n(n-3)}b_{n-1}$, $n \geq 1$, $n \neq 3$. The coefficients given in Table 6.9 are calculated with this formula. Notice that for $n = 3$, the recurrence relationship $n(n-3)b_n + (n-1)b_{n-1} = 0$ indicates that $2b_2 = 0$. Using these coefficients, we have the solution

$$y_2(x) = x^{-1}\left(b_0 + b_3 x^3 - \frac{3}{4}b_3 x^4 + \frac{3}{10}b_3 x^5 - \frac{1}{12}b_3 x^6 + \frac{1}{56}b_3 x^7 + \cdots\right)$$

$$= b_0 x^{-1} + b_3\left(x^2 - \frac{3}{4}x^3 + \frac{3}{10}x^4 - \frac{1}{12}x^5 + \frac{1}{56}x^6 + \cdots\right)$$

$$= b_0 x^{-1} + \frac{b_3}{a_0}y_1(x) = c_1 x^{-1} + c_2 y_1(x)$$

where c_1 and c_2 are arbitrary constants. Hence, if we had started with the smaller root, we would have obtained a general solution of the differential equation without having to carry out the procedure with the larger root. *When working with indicial roots that differ by an integer, make a note to use the smaller root first.*

In most cases, however, when the two roots of the indicial equation differ by an integer, a general solution is not obtained quite as easily as in Example 4.

Example 5

Using a series expansion about the regular singular point $x = 0$ of

$$xy'' + 3y' - y = 0, \text{ find a solution of the form } y = \sum_{n=0}^{\infty} a_n x^{n+r_2} \text{ (where } r_2 \text{ represents}$$

the smaller of the roots of the indicial equation).

Solution   In standard form, this equation is $y'' + \dfrac{3}{x}y' - \dfrac{1}{x}y = 0$. Hence, $xp(x) = x(3/x) = 3$ and $x^2 q(x) = x^2(-1/x) = -x$, so $p_0 = 3$ and $q_0 = 0$. Thus, the indicial equation is $r(r - 1) + 3r = r^2 + 2r = r(r + 2) = 0$ with roots $r_1 = 0$ and $r_2 = -2$. (Notice that we always use r_1 to denote the larger root.) Therefore, we attempt to find a solution of the form $y = \sum_{n=0}^{\infty} a_n x^{n-2}$ with derivatives $y' = \sum_{n=0}^{\infty} (n - 2)a_n x^{n-3}$ and $y'' = \sum_{n=0}^{\infty} (n - 2)(n - 3)a_n x^{n-4}$. Substitution into the differential equation yields

$$\sum_{n=0}^{\infty} (n - 2)(n - 3)a_n x^{n-3} + \sum_{n=0}^{\infty} 3(n - 2)a_n x^{n-3} - \sum_{n=0}^{\infty} a_n x^{n-2} = 0$$

$$(6a_0 - 6a_0)x^{-3} + \sum_{n=1}^{\infty} (n - 2)(n - 3)a_n x^{n-3} + \sum_{n=1}^{\infty} 3(n - 2)a_n x^{n-3} - \sum_{n=0}^{\infty} a_n x^{n-2} = 0$$

$$\sum_{n=1}^{\infty} (n - 2)(n - 3)a_n x^{n-3} + \sum_{n=1}^{\infty} 3(n - 2)a_n x^{n-3} - \sum_{n=1}^{\infty} a_{n-1} x^{n-3} = 0$$

$$\sum_{n=1}^{\infty} \{[(n - 2)(n - 3) + 3(n - 2)]a_n - a_{n-1}\}x^{n-3} = 0.$$

Equating the coefficients to zero, we have $a_n = \dfrac{a_{n-1}}{(n - 2)(n - 3 + 3)} = \dfrac{a_{n-1}}{n(n - 2)}$, $n \geq 1, n \neq 2$. Notice that from this formula, $a_1 = -a_0$. When $n = 2$, we refer to the recurrence relation $n(n - 2)a_n - a_{n-1} = 0$ obtained from the coefficient in the series solution. When $n = 2$, $2(0)a_2 - a_1 = 0$, which indicates that $a_1 = 0$. Because $a_1 = -a_0$, $a_0 = 0$. However, $a_0 \neq 0$ by assumption, so there is no solution of this form.

Due to the difficulties encountered when trying to apply the Method of Frobenius, we discuss the following situations.

Suppose that $x = 0$ is a regular singular point of $y'' + p(x)y' + q(x)y = 0$. The roots of the indicial equation $r^2 + (p_0 - 1)r + q_0 = 0$ are

$$r_1 = \frac{1 - p_0 + \sqrt{1 - 2p_0 + p_0^2 - 4q_0}}{2} \quad \text{and} \quad r_2 = \frac{1 - p_0 - \sqrt{1 - 2p_0 + p_0^2 - 4q_0}}{2}$$

where $r_1 \geq r_2$ and $r_1 - r_2 = \sqrt{1 - 2p_0 + p_0^2 - 4q_0}$.

Case 1: If $r_1 \neq r_2$ and $r_1 - r_2$ is not an integer, then there are linearly independent solutions to the equation of the form

$$y_1(x) = x^{r_1} \sum_{n=0}^{\infty} a_n x^n \quad \text{and} \quad y_2(x) = x^{r_2} \sum_{n=0}^{\infty} b_n x^n.$$

Case 2: If $r_1 \neq r_2$ and $r_1 - r_2$ is an integer, there are two linearly independent solutions to the equation of the form

$$y_1(x) = x^{r_1} \sum_{n=0}^{\infty} a_n x^n \quad \text{and} \quad y_2(x) = c y_1(x) \ln x + x^{r_2} \sum_{n=0}^{\infty} b_n x^n,\ b_0 \neq 0.$$

(In some cases, $c = 0$ in y_2.)

Case 3: If $r_1 - r_2 = 0$, there are two linearly independent solutions to the equation of the form

$$y_1(x) = x^{r_1} \sum_{n=0}^{\infty} a_n x^n \quad \text{and} \quad y_2(x) = y_1(x) \ln x + x^{r_1} \sum_{n=1}^{\infty} b_n x^n.$$

In any case, if $y_1(x)$ is a solution of the equation, a second linearly independent solution is given by $y_2(x) = y_1(x) \int \frac{e^{-\int p(x)\,dx}}{[y_1(x)]^2}\,dx$, which is obtained through reduction of order.

When solving a differential equation in Case 2, first attempt to find a general solution using $y = x^{r_2} \sum_{n=0}^{\infty} a_n x^n = \sum_{n=0}^{\infty} a_n x^{n+r_2}$, where r_2 is the smaller of the two roots. As we saw in Example 4, a general solution can sometimes be found with this procedure. However, if the contradiction $a_0 = 0$ is reached, find solutions of the form $y_1(x) = x^{r_1} \sum_{n=0}^{\infty} a_n x^n$ and $y_2(x) = c y_1(x) \ln x + x^{r_2} \sum_{n=0}^{\infty} b_n x^n$.

Example 6

Find a general solution of $xy'' + 3y' - y = 0$ using a series expansion about the regular singular point $x = 0$.

Solution In Example 3, we found that the indicial equation

$$r(r-1) + 3r = r^2 + 2r = r(r+2) = 0$$

has roots $r_1 = 0$ and $r_2 = -2$. Because we found in Example 5 that there is no series solution corresponding to $r_2 = -2$, we assume there is a solution of the form $y = \sum_{n=0}^{\infty} a_n x^n$ corresponding to $r_1 = 0$ with derivatives $y' = \sum_{n=1}^{\infty} n a_n x^{n-1}$ and $y'' = \sum_{n=2}^{\infty} n(n-1) a_n x^{n-2}$. Substitution into the differential equation yields

$$\sum_{n=2}^{\infty} n(n-1) a_n x^{n-1} + \sum_{n=1}^{\infty} 3n a_n x^{n-1} - \sum_{n=0}^{\infty} a_n x^n = 0$$

$$(3a_1 - a_0)x^0 + \sum_{n=2}^{\infty} n(n-1) a_n x^{n-1} + \sum_{n=2}^{\infty} 3n a_n x^{n-1} - \sum_{n=1}^{\infty} a_n x^n = 0$$

$$(3a_1 - a_0)x^0 + \sum_{n=2}^{\infty} \{[n(n-1) + 3n]a_n - a_{n-1}\}x^{n-1} = 0.$$

Equating the coefficients to zero, we have $a_1 = \frac{1}{3}a_0$ and

$$a_n = \frac{a_{n-1}}{n(n-1) + 3n} = \frac{a_{n-1}}{n^2 + 2n}, \quad n \geq 2.$$

We use this formula to calculate several coefficients in Table 6.10 and use them to

form $y_1(x) = a_0\left(1 + \frac{x}{3} + \frac{x^2}{24} + \frac{x^3}{360} + \cdots\right)$.

To determine a second linearly independent solution, we assume that

$$y_2(x) = cy_1(x) \ln x + \sum_{n=0}^{\infty} b_n x^{n-2}$$

and substitute this function into the differential equation to find the coefficients b_n. Because the derivatives of $y_2(x)$ are

$$y_2'(x) = \frac{cy_1(x)}{x} + cy_1'(x) \ln x + \sum_{n=0}^{\infty} (n-2)b_n x^{n-3}$$

and

$$y_2''(x) = \frac{-cy_1(x)}{x^2} + \frac{2cy_1'(x)}{x} + cy_1''(x) \ln x + \sum_{n=0}^{\infty} (n-2)(n-3)b_n x^{n-4},$$

substitution into the differential equation yields

$$x\left[\frac{-cy_1(x)}{x^2} + \frac{2cy_1'(x)}{x} + cy_1''(x) \ln x + \sum_{n=0}^{\infty} (n-2)(n-3)b_n x^{n-4}\right]$$

$$+ 3\left[\frac{cy_1(x)}{x} + cy_1'(x) \ln x + \sum_{n=0}^{\infty} (n-2)b_n x^{n-3}\right] - cy_1(x) \ln x - \sum_{n=0}^{\infty} b_n x^{n-2} = 0$$

$$\frac{-cy_1(x)}{x} + 2cy_1'(x) + cxy_1''(x) \ln x + \sum_{n=0}^{\infty} (n-2)(n-3)b_n x^{n-3}$$

$$+ \frac{3cy_1(x)}{x} + 3cy_1'(x) \ln x + \sum_{n=0}^{\infty} 3(n-2)b_n x^{n-3} - cy_1(x) \ln x - \sum_{n=0}^{\infty} b_n x^{n-2} = 0$$

$$\frac{2cy_1(x)}{x} + 2cy_1'(x) + \sum_{n=0}^{\infty} (n-2)(n-3)b_n x^{n-3} + \sum_{n=0}^{\infty} 3(n-2)b_n x^{n-3} - \sum_{n=0}^{\infty} b_n x^{n-2}$$

$$+ \underbrace{c[xy_1''(x) + 3y_1'(x) - y_1(x)]}_{\substack{\text{= 0 because } y_1 \text{ is a solution} \\ \text{of the differential equation}}} \ln x = 0.$$

TABLE 6.10

n	a_n
0	a_0
1	$\frac{1}{3}a_0$
2	$\frac{1}{24}a_0$
3	$\frac{1}{360}a_0$

Simplifying this expression gives us

$$\frac{2cy_1(x)}{x} + 2cy_1'(x) + 6b_0x^{-3} - 6b_0x^{-3} + \sum_{n=1}^{\infty}(n-2)(n-3)b_nx^{n-3} +$$

$$\sum_{n=1}^{\infty}3(n-2)b_nx^{n-3} - \sum_{n=0}^{\infty}b_nx^{n-2} = 0$$

$$\frac{2cy_1(x)}{x} + 2cy_1'(x) + \sum_{n=1}^{\infty}[(n-2)nb_n - b_{n-1}]x^{n-3} = 0.$$

Now we *choose* $a_0 = 1/c$, so

$$y_1(x) = \frac{1}{c}\left(1 + \frac{x}{3} + \frac{x^2}{24} + \frac{x^3}{360} + \cdots\right) \text{ and } y_1'(x) = \frac{1}{c}\left(\frac{1}{3} + \frac{x}{12} + \frac{x^2}{120} + \cdots\right).$$

Substitution into the previous equation then yields

$$\frac{2}{x}\left[1 + \frac{x}{3} + \frac{x^2}{24} + \frac{x^3}{360} + \cdots\right] + 2\left[\frac{1}{3} + \frac{x}{12} + \frac{x^2}{120} + \cdots\right] +$$

$$\sum_{n=1}^{\infty}[(n-2)nb_n - b_{n-1}]x^{n-3} = 0$$

$$\left(\frac{2}{x} + \frac{4}{3} + \frac{x}{4} + \frac{x^2}{45} + \cdots\right) + \sum_{n=1}^{\infty}[(n-2)nb_n - b_{n-1}]x^{n-3} = 0$$

$$\left(\frac{2}{x} + \frac{4}{3} + \frac{x}{4} + \frac{x^2}{45} + \cdots\right) + (-b_1 - b_0)x^{-2} - b_1x^{-1} +$$

$$(3b_3 - b_2)x^0 + (8b_4 - b_3)x + (15b_5 - b_4)x^2 = 0,$$

so we have the sequence of equations $-b_1 - b_0 = 0$, $-b_1 + 2 = 0$, $3b_3 - b_2 + \frac{4}{3} = 0$, $8b_4 - b_3 + \frac{1}{4} = 0$, $15b_5 - b_4 + \frac{1}{45} = 0, \ldots$. Solving these equations, we see that $b_1 = 2$ and $b_0 = -2$. However, the other coefficients depend on the value of b_2. We give these values in Table 6.11. Hence, a second linearly independent solution is given by

$$y_2(x) = cy_1(x)\ln x + x^{-2}\left(-2 + 2x + b_2x^2 + \frac{3b_2 - 4}{9}x^3 + \frac{12b_2 - 25}{288}x^4 + \frac{60b_2 - 157}{21600}x^5 + \cdots\right)$$

$$= \left(1 + \frac{x}{3} + \frac{x^2}{24} + \frac{x^3}{360} + \cdots\right)\ln x +$$

$$x^{-2}\left(-2 + 2x + b_2x^2 + \frac{3b_2 - 4}{9}x^3 + \frac{12b_2 - 25}{288}x^4 + \frac{60b_2 - 157}{21600}x^5 + \cdots\right),$$

TABLE 6.11

n	b_n
0	-2
1	2
2	b_2
3	$\dfrac{3b_2 - 4}{9}$
4	$\dfrac{12b_2 - 25}{288}$
5	$\dfrac{60b_2 - 157}{21600}$

where b_2 is arbitrary. In particular, two linearly independent solutions of the equation are $(c = 1/a_0 = 1)$

$$y_1(x) = 1 + \frac{x}{3} + \frac{x^2}{24} + \frac{x^3}{360} + \cdots$$

and $(b_2 = 0)$

$$y_2(x) = \left(1 + \frac{x}{3} + \frac{x^2}{24} + \frac{x^3}{360} + \cdots\right) \ln x +$$

$$x^{-2}\left(-2 + 2x - \frac{4}{9}x^3 - \frac{25}{288}x^4 - \frac{157}{21600}x^5 + \cdots\right).$$

A general solution is, therefore, given by $y(x) = c_1 y_1(x) + c_2 y_2(x)$, where c_1 and c_2 are arbitrary.

Explain why the choice $a_0 = 1/c$ does not affect the general solution obtained in Example 6. Hint: If $c = 0$, $b_0 = 0$, which is impossible (why?), so c cannot be zero.

Equal Indicial Roots

Several techniques can be used to solve a differential equation with equal indicial roots.

Example 7

Find a general solution of $xy'' + (2 - x)y' + \frac{1}{4x}y = 0$ by using a series about the regular singular point $x = 0$.

Solution   In standard form, the equation is $y'' + \frac{2-x}{x}y' + \frac{1}{4x^2}y = 0$. Because $xp(x) = x\left(\frac{2-x}{x}\right) = 2 - x$ and $x^2 q(x) = x^2\left(\frac{1}{4x^2}\right) = \frac{1}{4}$, $p_0 = 2$ and $q_0 = \frac{1}{4}$. The indicial equation is $r^2 + (2 - 1)r + \frac{1}{4} = 0$, which has equal roots $r_1 = r_2 = -\frac{1}{2}$, so there is a solution of the form $y_1(x) = x^{-1/2}\sum_{n=0}^{\infty} a_n x^n = \sum_{n=0}^{\infty} a_n x^{n-1/2}$. Replacing y in the equation by y_1 yields

$$x\frac{d^2}{dx^2}\left(\sum_{n=0}^{\infty} a_n x^{n-1/2}\right) + (2 - x)\frac{d}{dx}\left(\sum_{n=0}^{\infty} a_n x^{n-1/2}\right) + \frac{1}{4x}\left(\sum_{n=0}^{\infty} a_n x^{n-1/2}\right) = 0$$

$$x\sum_{n=0}^{\infty}\left(n - \frac{1}{2}\right)\left(n - \frac{3}{2}\right)a_n x^{n-5/2} + (2 - x)\sum_{n=0}^{\infty}\left(n - \frac{1}{2}\right)a_n x^{n-3/2} +$$

$$\frac{1}{4x}\sum_{n=0}^{\infty} a_n x^{n-1/2} = 0$$

$$\sum_{n=0}^{\infty}\left(n-\frac{1}{2}\right)\left(n-\frac{3}{2}\right)a_n x^{n-3/2} + \sum_{n=0}^{\infty}2\left(n-\frac{1}{2}\right)a_n x^{n-3/2} -$$

$$\sum_{n=0}^{\infty}\left(n-\frac{1}{2}\right)a_n x^{n-1/2} + \sum_{n=0}^{\infty}\frac{a_n}{4}x^{n-3/2} = 0$$

$$\sum_{n=1}^{\infty}\left[\left(n-\frac{1}{2}\right)\left(n-\frac{3}{2}\right)a_n + 2\left(n-\frac{1}{2}\right)a_n - \left(n-\frac{3}{2}\right)a_{n-1} + \frac{a_n}{4}\right]x^{n-3/2} = 0$$

$$\sum_{n=1}^{\infty}\left[a_n n^2 - \frac{2n-3}{2}a_{n-1}\right]x^{n-3/2} = 0.$$

Then, equating coefficients to zero, we find that $a_n = \dfrac{2n-3}{2n^2}a_{n-1}$. Therefore, for $a_0 \neq 0$, we obtain the coefficients given in Table 6.12. Using these coefficients,

$$y_1(x) = a_0\left(x^{-1/2} - \frac{1}{2}x^{1/2} - \frac{1}{16}x^{3/2} - \frac{1}{96}x^{5/2} - \frac{5}{3072}x^{7/2} - \frac{7}{30720}x^{9/2} - \cdots\right).$$

Choosing $a_0 = 1$, we find that

$$y_1(x) = x^{-1/2} - \frac{1}{2}x^{1/2} - \frac{1}{16}x^{3/2} - \frac{1}{96}x^{5/2} - \frac{5}{3072}x^{7/2} - \frac{7}{30720}x^{9/2} - \cdots$$

is a particular solution of the equation.

    Because the roots of the indicial equation are equal, there is a second linearly independent solution of the form $y_2(x) = y_1(x)\ln x + \sum_{n=1}^{\infty}b_n x^{n-1/2}$. Substituting y_2 into the differential equation yields

$$x\frac{d^2}{dx^2}\left(y_1(x)\ln x + \sum_{n=1}^{\infty}b_n x^{n-1/2}\right) + (2-x)\frac{d}{dx}\left(y_1(x)\ln x + \sum_{n=1}^{\infty}b_n x^{n-1/2}\right) +$$

$$\frac{1}{4x}\left(y_1(x)\ln x + \sum_{n=1}^{\infty}b_n x^{n-1/2}\right) = 0$$

$$x\left[\frac{-y_1(x)}{x^2} + \frac{2y_1'(x)}{x} + y_1''(x)\ln x + \sum_{n=1}^{\infty}\left(n-\frac{1}{2}\right)\left(n-\frac{3}{2}\right)b_n x^{n-5/2}\right] +$$

$$(2-x)\left[\frac{y_1(x)}{x} + y_1'(x)\ln x + \sum_{n=1}^{\infty}\left(n-\frac{1}{2}\right)b_n x^{n-3/2}\right] +$$

$$\frac{y_1(x)\ln x}{4x} + \frac{1}{4x}\sum_{n=1}^{\infty}b_n x^{n-1/2} = 0$$

$$\left(xy_1''(x) + (2-x)y_1'(x) + \frac{1}{4x}y_1(x)\right)\ln x +$$

TABLE 6.12

n	a_n
0	a_0
1	$-\dfrac{1}{2}a_0$
2	$-\dfrac{1}{16}a_0$
3	$-\dfrac{1}{96}a_0$
4	$-\dfrac{5}{3072}a_0$
5	$-\dfrac{7}{30720}a_0$

$$x\left[\frac{-y_1(x)}{x^2} + \frac{2y_1'(x)}{x} + \sum_{n=1}^{\infty}\left(n - \frac{1}{2}\right)\left(n - \frac{3}{2}\right)b_n x^{n-5/2}\right] +$$

$$(2 - x)\left[\frac{y_1(x)}{x} + \sum_{n=1}^{\infty}\left(n - \frac{1}{2}\right)b_n x^{n-3/2}\right] + \frac{1}{4x}\sum_{n=1}^{\infty} b_n x^{n-1/2} = 0$$

$$x\left[\frac{-y_1(x)}{x^2} + \frac{2y_1'(x)}{x} + \sum_{n=1}^{\infty}\left(n - \frac{1}{2}\right)\left(n - \frac{3}{2}\right)b_n x^{n-5/2}\right] +$$

$$(2 - x)\left[\frac{y_1(x)}{x} + \sum_{n=1}^{\infty}\left(n - \frac{1}{2}\right)b_n x^{n-3/2}\right] + \frac{1}{4x}\sum_{n=1}^{\infty} b_n x^{n-1/2} = 0$$

$$\frac{y_1(x)}{x} + 2y_1'(x) - y_1(x) + \sum_{n=1}^{\infty}\left(n - \frac{1}{2}\right)\left(n - \frac{3}{2}\right)b_n x^{n-3/2} + 2\sum_{n=1}^{\infty}\left(n - \frac{1}{2}\right)b_n x^{n-3/2} -$$

$$\sum_{n=1}^{\infty}\left(n - \frac{1}{2}\right)b_n x^{n-1/2} + \sum_{n=1}^{\infty}\frac{b_n}{4} x^{n-3/2} = 0$$

$$\frac{y_1(x)}{x} + 2y_1'(x) - y_1(x) + \sum_{n=1}^{\infty} n^2 b_n x^{n-3/2} - \sum_{n=1}^{\infty}\left(n - \frac{1}{2}\right)b_n x^{n-1/2} = 0$$

$$\frac{y_1(x)}{x} + 2y_1'(x) - y_1(x) + b_1 x^{-1/2} + \sum_{n=2}^{\infty}\left[n^2 b_n - \left(n - \frac{3}{2}\right)b_{n-1}\right]x^{n-3/2} = 0.$$

Hence, $b_1 x^{-1/2} + \displaystyle\sum_{n=2}^{\infty}\left[n^2 b_n - \left(n - \frac{3}{2}\right)b_{n-1}\right]x^{n-3/2} = y_1(x) - \frac{y_1(x)}{x} - 2y_1'(x).$

Because $y_1(x) = x^{-1/2} - \dfrac{1}{2}x^{1/2} - \dfrac{1}{16}x^{3/2} - \dfrac{1}{96}x^{5/2} - \dfrac{5}{3072}x^{7/2} -$

$\dfrac{7}{30720}x^{9/2} + \cdots$, we obtain

$$b_1 x^{-1/2} + \sum_{n=2}^{\infty}\left[n^2 b_n - \left(n - \frac{3}{2}\right)b_{n-1}\right]x^{n-3/2} =$$

$$2x^{-1/2} - \frac{1}{4}x^{1/2} + \frac{1}{384}x^{5/2} + \frac{1}{1536}x^{7/2} + \cdots.$$

Equating the coefficients of like terms on each side of this equation yields $b_1 = 2$,

$-\dfrac{b_1}{2} + 4b_2 = -\dfrac{1}{4}$, $\quad -\dfrac{3b_2}{2} + 9b_3 = 0$, $\quad -\dfrac{5b_3}{2} + 16b_4 = \dfrac{1}{384}$, $\quad -\dfrac{7b_4}{2} + 25b_5 =$

$\dfrac{a_0}{1536}, \ldots$. Thus, we have the coefficients given in Table 6.13. Therefore,

TABLE 6.13

n	b_n
1	2
2	$\dfrac{3}{16}$
3	$\dfrac{1}{32}$
4	$\dfrac{31}{6144}$
5	$\dfrac{3}{4096}$

$$y_2(x) = y_1(x) \ln x + \sum_{n=1}^{\infty} b_n x^{n-1/2}$$

$$= y_1(x) \ln x + 2x^{1/2} + \frac{3}{16}x^{3/2} + \frac{1}{32}x^{5/2} + \frac{31}{6144}x^{7/2} + \frac{3}{4096}x^{9/2} + \cdots$$

is a second linearly independent solution of the differential equation.

A general solution is $y(x) = c_1 y_1(x) + c_2 y_2(x)$, where c_1 and c_2 are arbitrary constants.

Example 8

Find a general solution of $x^2 y'' + xy' + x^2 y = 0$, $x > 0$.

Solution   In standard form, the equation is $y'' + \dfrac{1}{x}y' + y = 0$, so $x = 0$ is a regular singular point of the equation. Because $xp(x) = x(1/x) = 1$ and $x^2 q(x) = x^2$, $p_0 = 1$ and $q_0 = 0$. Therefore, the indicial equation is $r(r-1) + r = r^2 - r + r = r^2 = 0$, so $r_1 = r_2 = 0$. Hence, there is a solution of the form $y = \sum_{n=0}^{\infty} a_n x^n$. Substitution into $x^2 y'' + xy' + x^2 y = 0$ yields

$$x^2 \sum_{n=2}^{\infty} n(n-1)a_n x^{n-2} + x \sum_{n=1}^{\infty} n a_n x^{n-1} + x^2 \sum_{n=0}^{\infty} a_n x^n = 0$$

or

$$\sum_{n=2}^{\infty} n(n-1)a_n x^n + \sum_{n=1}^{\infty} n a_n x^n + \sum_{n=0}^{\infty} a_n x^{n+2} = 0.$$

After pulling off the first term of the second series and simplifying the expression, we have

$$a_1 x + \sum_{n=2}^{\infty} \{[n(n-1) + n]a_n + a_{n-2}\}x^n = 0.$$

TABLE 6.14

n	a_n
0	a_0
1	0
2	$-\dfrac{1}{2^2}a_0$
3	0
4	$\dfrac{1}{4^2 2^2}a_0$

Equating the coefficients to zero yields $a_1 = 0$ and $a_n = -\dfrac{a_{n-2}}{n^2}$, $n \geq 2$. We use this formula to calculate several of these coefficients in Table 6.14. Choosing $a_0 = 1$ we obtain that one solution to the equation is $y_1(x) = 1 - \dfrac{x^2}{4} + \dfrac{x^4}{64} - \cdots$. We use the formula $y_2(x) = y_1(x) \displaystyle\int \dfrac{e^{-\int p(x)\,dx}}{[y_1(x)]^2}\,dx$ to determine a second linearly independent solution to the differential equation as follows:

$$y_2(x) = y_1(x) \int \frac{e^{-\int 1/x\,dx}}{\left(1 - \dfrac{x^2}{4} + \dfrac{x^4}{64} - \cdots\right)^2}\,dx$$

$$= y_1(x) \int \frac{x^{-1}}{\left(1 - \frac{x^2}{2} + \frac{3x^4}{32} - \cdots\right)} \, dx \quad \text{(squaring)}$$

$$= y_1(x) \int \frac{1}{x} \frac{1}{\left(1 - \frac{x^2}{2} + \frac{3x^4}{32} - \cdots\right)} \, dx$$

$$= y_1(x) \int \frac{1}{x} \left(1 + \frac{x^2}{2} + \frac{5x^4}{32} + \cdots\right) dx \quad \text{(long division)}$$

$$= y_1(x) \int \left(\frac{1}{x} + \frac{x}{2} + \frac{5x^3}{32} + \cdots\right) dx = y_1(x)\left(\ln x + \frac{x^2}{4} + \frac{5x^4}{128} + \cdots\right)$$

$$= y_1(x) \ln x + y_1(x)\left(\frac{x^2}{4} + \frac{5x^4}{128} + \cdots\right).$$

Notice that with this alternate method for finding a second linearly independent solution, we obtain a solution of the form that was stated in Case 3. A general solution of the equation is $y = c_1 y_1(x) + c_2 y_2(x)$, where c_1 and c_2 are arbitrary constants.

We have not discussed the possibility of complex-valued roots of the indicial equation. When this occurs, the equation is solved using the procedures of Case 2. The solutions obtained are complex but they can be transformed into real solutions by taking the appropriate linear combinations, like those discussed for complex-valued roots of the characteristic equation of Cauchy-Euler differential equations.

Also, we have not mentioned if a solution can be found with a series expansion about an irregular singular point. If $x = x_0$ is an irregular point of $y'' + p(x)y' + q(x)y = 0$, then there may or may not be a solution of the form $y = \sum_{n=0}^{\infty} a_n(x - x_0)^{n+r}$.

IN TOUCH WITH TECHNOLOGY

1. The differential equation $4xy'' + 2y' + y = 0$ has the series solution

$$y(x) = c_1 x^{1/2} \sum_{n=0}^{\infty} \frac{(-1)^n}{(2n + 1)!} x^n + c_2 \sum_{n=0}^{\infty} \frac{(-1)^n}{(2n)!} x^n$$

about the regular singular point $x = 0$. **(a)** Use this formula to approximate the solution to the initial-value problem $4xy'' + 2y' + y = 0$, $y(1) = 1$, $y'(1) = 3$ by using the first ten terms in each linearly independent solution to approximate c_1 and c_2. **(b)** Use a computer algebra system to generate a numerical solution of the initial-value problem and compare these results with those obtained in **(a)** by graphing the two approximations simultaneously. (Initially use the first ten terms in the series solutions.) What happens if more terms from the series solution are used?

2. **(a)** Use the series solution found in Example 7 to approximate the solution to the initial-value problem

$$xy'' + (2 - x)y' + \frac{1}{4x}y = 0, \quad y(1) = 1, \quad y'(1) = -1.$$

(b) Use a computer algebra system to generate a numerical solution of the initial-value problem and compare these results with those obtained in **(a)** by graphing the two approximations simultaneously.

3. **Laguerre's equation** is $xy'' + (1 - x)y' + ny = 0$.
   (a) Show that $x = 0$ is a regular singular point of Laguerre's equation. **(b)** Use the Method of Frobenius to show that one solution of Laguerre's equation is

$$L_n(x) = \sum_{m=0}^{n} \frac{(-1)^m}{m!} \frac{n!}{m!(n-m)!} x^m, \quad \text{where } L_n(x) \text{ is}$$

called the **Laguerre polynomial of order n. (c)** Calcu-

late the first eight Laguerre polynomials with this formula (or find them using a built-in computer algebra system command). **(d)** Show that the Laguerre polynomials satisfy the formula $L_n(x) = \dfrac{e^x}{n!} \dfrac{d^n(x^n e^{-x})}{dx^n}$

for $n = 1,$ 2, 3, ..., 8. **(e)** Show that $\displaystyle\int_0^\infty e^{-x} L_n(x) L_m(x)\, dx = 0$ for $n \neq m$, n, $m = 1$, 2, 3, ..., 8. (This indicates that the Laguerre polynomials are **orthogonal.**) **(f)** Determine the value of $\displaystyle\int_0^\infty e^{-x} [L_n(x)]^2\, dx$ by experimenting with $n = 1$, 2, 3, ..., 8. (Note: All of the properties mentioned in this problem hold for all n. We simply chose the values $n = 1, 2, 3, \ldots, 8$ so that the calculations can be carried out with technology instead of by hand.)

EXERCISES 6.4

In Exercises 1–10, determine the singular points of each equation. In each case, classify the point as regular or irregular.

1. $x^2 y'' + 6y = 0$

2. $x^2 y'' - xy' - y = 0$

*3. $x(x + 1)y'' - \dfrac{1}{x^2} y' + 5y = 0$

4. $(x + 2)y'' - \dfrac{1}{x^2 - 4} y' + \dfrac{1}{x + 2} y = 0$

5. $(x^2 - 3x - 4)y'' - (x + 1)y' + (x^2 - 1)y = 0$

6. $(x^2 + 3x + 2)y'' - (x + 2)y' + (x^2 - 4)y = 0$

*7. $(x^2 - 25)^2 y'' - (x + 5)y' + 10y = 0$

8. $x^4 y'' - xy' + 9x^2 y = 0$

9. $x^2(x^2 - 5x + 6)y'' - x(x - 2)y' + x(x - 3)y = 0$

10. $x^2(x^2 - 16)y'' - \dfrac{x}{(x + 4)^2} y' + y = 0$

In Exercises 11–35, use the Method of Frobenius to obtain two linearly independent solutions about the regular singular point $x = 0$.

11. $4xy'' + 3y' - 2y = 0$ 12. $8xy'' + 9y' + y = 0$

*13. $2xy'' - 5y' - 3y = 0$ 14. $5xy'' + 4y' + 3y = 0$

15. $5xy'' + 8y' - xy = 0$ 16. $10xy'' + 7y' + xy = 0$

*17. $9xy'' + 14y' + (x - 1)y = 0$

18. $2xy'' + 5y' + (x + 1)y = 0$

19. $7xy'' + 10y' + (1 - x^2)y = 0$

20. $x^2 y'' - 2xy' + 4xy = 0$

*21. $x^2 y'' + xy' + (x - 1)y = 0$

22. $xy'' + 2xy' + y = 0$

23. $xy'' + 5xy' + 5y = 0$

24. $xy'' + 5y' + 5xy = 0$

*25. $y'' + \dfrac{8}{3x} y' - \left(\dfrac{2}{3x^2} - 1 \right) y = 0$

26. $y'' + \dfrac{7}{3x} y' + \left(\dfrac{1}{3x^2} + \dfrac{1}{x} \right) y = 0$

27. $y'' + \left(\dfrac{16}{3x} - 1 \right) y' - \dfrac{16}{3x^2} y = 0$

28. $y'' - \left(1 + \dfrac{3}{4x^2} \right) y = 0$

*29. $y'' + \left(\dfrac{1}{2x} - 2 \right) y' - \dfrac{35}{16x^2} y = 0$

30. $y'' + \left(\dfrac{2}{x} + 1 \right) y' - \left(\dfrac{15}{4x^2} - 1 \right) y = 0$

31. $y'' + \dfrac{7}{3x}y' + \left(\dfrac{4}{9x^2} - \dfrac{2}{x}\right)y = 0$

32. $y'' + \dfrac{3}{x}y' + \left(\dfrac{1}{x^2} + \dfrac{1}{x}\right)y = 0$

*33. $y'' - \left(\dfrac{1}{x} + 2\right)y' + \left(\dfrac{1}{x^2} + x\right)y = 0$

34. $y'' + \left(\dfrac{5}{3x} - x\right)y' + \left(\dfrac{1}{9x^2} - \dfrac{1}{x}\right)y = 0$

35. $16x^2y'' + (16x^2 - 8x)y' + 5y = 0$

In Exercises 36–41, solve the differential equation with a series expansion about $x = 0$. Compare these results with the solution obtained by solving the problem as a Cauchy-Euler equation.

36. $x^2y'' - 2y = 0$

37. $x^2y'' + 7xy' - 7y = 0$

38. $4x^2y'' + y = 0$

*39. $x^2y'' + 3xy' + y = 0$

40. $x^2y'' + 4xy' - 4y = 0$

41. $x^2y'' - 3xy' + 4y = 0$

42. The differential equation $\dfrac{d^2y}{dx^2} + p(x)\dfrac{dy}{dx} + q(x)y = 0$ has a **singular point at infinity** if after substitution of $w = \dfrac{1}{x}$ the resulting equation has a singular point at $w = 0$. Similarly, the equation has an **ordinary point at infinity** if the transformed equation has an ordinary point at $w = 0$. Use the chain rule and the substitution $w = \dfrac{1}{x}$ to show that the differential equation $\dfrac{d^2y}{dx^2} + p(x)\dfrac{dy}{dx} + q(x)y = 0$ is equivalent to

$$\frac{d^2y}{dw^2} + \left(\frac{2}{w} + \frac{p(1/w)}{w^2}\right)\frac{dy}{dw} + \frac{q(1/w)}{w^4}y = 0.$$

43. Use the definition in Exercise 42 to determine if infinity is an ordinary point or a singular point of the given differential equation.

(a) $\dfrac{d^2y}{dx^2} + xy = 0$

(b) $x^2\dfrac{d^2y}{dx^2} + x\dfrac{dy}{dx} + (x^2 - n^2)y = 0$

(c) $(1 - x^2)\dfrac{d^2y}{dx^2} - 2x\dfrac{dy}{dx} + n(n + 1)y = 0$

Some Special Equations

Legendre's Equation • The Gamma Function • Bessel's Equation

The techniques of solving differential equations through the use of power series expansions about ordinary and regular singular points can be used to solve several special ordinary differential equations. In addition to their historical significance, these equations are important because they are encountered in solving many problems in applied mathematics and physics, such as investigating the motion of a circular drumhead and finding the steady-state temperature on the surface of a sphere.

Legendre's Equation

We begin our discussion with **Legendre's equation,**

$$(1 - x^2)y'' - 2xy' + k(k + 1)y = 0,$$

where k is a constant, named after the French mathematician Adrien Marie Legendre (1752–1833). The **Legendre polynomials,** solutions of Legendre's equation, were introduced by Legendre in his three volume work *Traite des fonctions elliptiques et des integrales euleriennes* (1825–1832). Legendre encountered these polynomials while trying to determine the gravitational potential associated with a point mass.

| **Example 1** | Find a general solution of Legendre's equation $(1 - x^2)y'' - 2xy' + k(k + 1)y = 0$. |

Solution   In standard form, the equation is $y'' - \dfrac{2x}{1 - x^2}y' + \dfrac{k(k + 1)}{1 - x^2} = 0$.

Because $x = 0$ is an ordinary point, there is a solution of the form $y = \sum_{n=0}^{\infty} a_n x^n$. This solution will converge at least on the interval $(-1, 1)$, because the closest singular points to $x = 0$ are $x = \pm 1$. Substitution of this function and its derivatives into the differential equation yields

$$\sum_{n=2}^{\infty} n(n - 1)a_n x^{n-2} - \sum_{n=2}^{\infty} n(n - 1)a_n x^n - \sum_{n=1}^{\infty} 2na_n x^n + \sum_{n=0}^{\infty} k(k + 1)a_n x^n = 0$$

$$[2a_2 + k(k + 1)a_0]x^0 + [-2a_1 + k(k + 1)a_1 + 6a_3]x$$

$$+ \sum_{n=4}^{\infty} n(n - 1)a_n x^{n-2} - \sum_{n=2}^{\infty} n(n - 1)a_n x^n - \sum_{n=2}^{\infty} 2na_n x^n + \sum_{n=2}^{\infty} k(k + 1)a_n x^n = 0.$$

After substituting $(n + 2)$ for each occurrence of n in the first series and simplifying, we have

$$[2a_2 + k(k + 1)a_0]x^0 + [-2a_1 + k(k + 1)a_1 + 6a_3]x$$

$$+ \sum_{n=2}^{\infty} \{(n + 2)(n + 1)a_{n+2} + [-n(n - 1) - 2n + k(k + 1)]a_n\}x^n = 0.$$

Equating the coefficients to zero, we find that

$$a_2 = -\frac{k(k + 1)}{2}a_0, \qquad a_3 = -\frac{k(k + 1) - 2}{6}a_1 = -\frac{(k - 1)(k + 2)}{6}a_1,$$

and

$$a_{n+2} = \frac{n(n - 1) + 2n - k(k + 1)}{(n + 2)(n + 1)}a_n = \frac{(n - k)(n + k + 1)}{(n + 2)(n + 1)}a_n, \quad n \geq 2.$$

Using this formula, we find the following coefficients:

$$a_4 = \frac{(2 - k)(3 + k)}{4 \cdot 3}a_2 = -\frac{(2 - k)(3 + k)k(k + 1)}{4 \cdot 3 \cdot 2}a_0$$

$$a_5 = \frac{(3 - k)(4 + k)}{5 \cdot 4}a_3 = -\frac{(3 - k)(4 + k)(k - 1)(k + 2)}{5 \cdot 4 \cdot 3 \cdot 2}a_1$$

$$a_6 = \frac{(4 - k)(5 + k)}{6 \cdot 5}a_4 = -\frac{(4 - k)(5 + k)(2 - k)(3 + k)k(k + 1)}{6 \cdot 5 \cdot 4 \cdot 3 \cdot 2}a_0$$

$$a_7 = \frac{(5 - k)(6 + k)}{7 \cdot 6}a_5 = -\frac{(5 - k)(6 + k)(3 - k)(4 + k)(k - 1)(k + 2)}{7 \cdot 6 \cdot 5 \cdot 4 \cdot 3 \cdot 2}a_1$$

We have the two linearly independent solutions

$$y_1(x) = a_0 \left(1 - \frac{k(k+1)}{2!}x^2 - \frac{(2-k)(3+k)k(k+1)}{4!}x^4 \right.$$
$$\left. - \frac{(4-k)(5+k)(2-k)(3+k)k(k+1)}{6!}x^6 - \cdots\right)$$

and

$$y_2(x) = a_1 \left(x - \frac{(k-1)(k+2)}{6}x^3 - \frac{(3-k)(4+k)(k-1)(k+2)}{5!}x^5 \right.$$
$$\left. - \frac{(5-k)(6+k)(3-k)(4+k)(k-1)(k+2)}{7!}x^7 - \cdots\right),$$

so a general solution is

$$y = a_0 \left(1 - \frac{k(k+1)}{2!}x^2 - \frac{(2-k)(3+k)k(k+1)}{4!}x^4 - \right.$$
$$\left. \frac{(4-k)(5+k)(2-k)(3+k)k(k+1)}{6!}x^6 - \cdots\right)$$
$$+ a_1 \left(x - \frac{(k-1)(k+2)}{6}x^3 - \frac{(3-k)(4+k)(k-1)(k+2)}{5!}x^5 \right.$$
$$\left. - \frac{(5-k)(6+k)(3-k)(4+k)(k-1)(k+2)}{7!}x^7 - \cdots\right).$$

TABLE 6.15 $P_n(x)$ for $n = 0,$ $1, \ldots, 5$

$P_0(x) = 1$
$P_1(x) = x$
$P_2(x) = \frac{1}{2}(3x^2 - 1)$
$P_3(x) = \frac{1}{2}(5x^3 - 3x)$
$P_4(x) = \frac{1}{8}(35x^4 - 30x^2 + 3)$
$P_5(x) = \frac{1}{8}(63x^5 - 70x^3 + 15x)$

An interesting observation from the general solution to Legendre's equation given here is that the series solutions terminate for integer values of k. If k is an even integer, then the first series terminates, while if k is an odd integer the second series terminates. Therefore, polynomial solutions are found for integer values of k. We list several of these polynomials for suitable choices of a_0 and a_1 in Table 6.15 and graph them in Figure 6.11. Because these polynomials are useful and are encountered in numerous applications, we have a special notation for them: $P_n(x)$ is called the **Legendre polynomial of degree n** and represents the nth degree polynomial solution to Legendre's equation.

 Match each Legendre polynomial to the appropriate graph in Figure 6.11.

The Gamma Function

One of the more useful functions, which we will use shortly to solve *Bessel's equation*, is the **Gamma function**, first introduced by Euler in 1768, which is defined as follows.

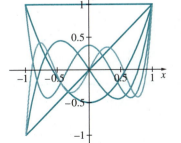

Figure 6.11 $P_0(x)$, $P_1(x)$, $P_2(x)$, $P_3(x)$, $P_4(x)$, and $P_5(x)$

Definition 6.9 Gamma Function

Notice that because integration is with respect to u, the result is a function of x.

The **Gamma function,** denoted $\Gamma(x)$, is given by

$$\Gamma(x) = \int_0^\infty e^{-u}u^{x-1}\,du, \ x > 0. \quad \text{(See Figure 6.12.)}$$

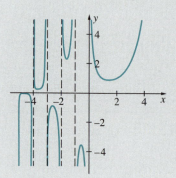

Figure 6.12   Although we have only defined $\Gamma(x)$ for $x > 0$, $\Gamma(x)$ can be defined for all real numbers *except* $x = 0, -1, -2, \ldots.$*

Example 2

Evaluate $\Gamma(1)$.

Solution

$$\Gamma(1) = \int_0^\infty e^{-u}u^{1-1}\,du = \int_0^\infty e^{-u}\,du = \lim_{b\to\infty}\ [-e^{-u}]_0^b$$
$$= \lim_{b\to\infty}\ (-e^{-b} + 1) = 1.$$

A useful property associated with the Gamma function is

$$\Gamma(x + 1) = x\Gamma(x).$$

If x is an integer, using this property we have $\Gamma(2) = \Gamma(1 + 1) = 1 \cdot \Gamma(1) = \Gamma(1)$, $\Gamma(3) = \Gamma(2 + 1) = 2\Gamma(2) = 2$, $\Gamma(4) = \Gamma(3 + 1) = 3\Gamma(3) = 3 \cdot 2$, $\Gamma(5) = \Gamma(4 + 1) = 4\Gamma(4) = 4 \cdot 3 \cdot 2, \ldots.$

and for the integer n,

$$\Gamma(n + 1) = n!$$

This property is used in solving the following equation.

Bessel's Equation

Another important equation is **Bessel's equation (of order μ),** named after the German astronomer Friedrich Wilhelm Bessel (1784–1846), who was a friend of Gauss. Bessel's equation is

$$x^2y'' + xy' + (x^2 - \mu^2)y = 0,$$

---

* This topic is discussed in most complex analysis books like *Functions of One Complex Variable,* Second Edition, by John B. Conway, Springer-Verlag, 1978, pp. 176–185.

where $\mu \geq 0$ is a constant. Bessel determined several representations of $J_\mu(x)$, the **Bessel function of order μ,** which is a solution of Bessel's equation, and noticed some of the important properties associated with the Bessel functions. The equation received its name due to Bessel's extensive work with $J_\mu(x)$, even though Euler solved the equation before Bessel.

To use a series method to solve Bessel's equation, first write the equation in standard form as

$$y'' + \frac{1}{x}y' + \frac{x^2 - \mu^2}{x^2}y = 0,$$

so $x = 0$ is a regular singular point. Using the Method of Frobenius, we assume that there is a solution of the form $y = \sum_{n=0}^{\infty} a_n x^{n+r}$. We determine the value(s) of r with the indicial equation. Because $xp(x) = x\dfrac{1}{x} = 1$ and $x^2 q(x) = x^2 \left(\dfrac{x^2 - \mu^2}{x^2}\right) = x^2 - \mu^2$, $p_0 = 1$ and $q_0 = -\mu^2$. Hence, the indicial equation is

$$r(r - 1) + p_0 r + q_0 = r(r - 1) + r - \mu^2 = r^2 - \mu^2 = 0$$

with roots $r_1 = \mu$ and $r_2 = -\mu$. We assume that $y = \sum_{n=0}^{\infty} a_n x^{n+\mu}$ with derivatives $y' = \sum_{n=0}^{\infty} (n + \mu)a_n x^{n+\mu-1}$ and $y'' = \sum_{n=0}^{\infty} (n + \mu)(n + \mu - 1)a_n x^{n+\mu-2}$. Substitution into Bessel's equation yields

$$x^2 \sum_{n=0}^{\infty} (n + \mu)(n + \mu - 1)a_n x^{n+\mu-2} + x \sum_{n=0}^{\infty} (n + \mu)a_n x^{n+\mu-1} +$$

$$(x^2 - \mu^2) \sum_{n=0}^{\infty} a_n x^{n+\mu} = 0$$

$$\sum_{n=0}^{\infty} (n + \mu)(n + \mu - 1)a_n x^{n+\mu} + \sum_{n=0}^{\infty} (n + \mu)a_n x^{n+\mu} + \sum_{n=0}^{\infty} a_n x^{n+\mu+2} -$$

$$\sum_{n=0}^{\infty} \mu^2 a_n x^{n+\mu} = 0$$

$$\mu(\mu - 1)a_0 x^\mu + (1 + \mu)\mu a_1 x^{\mu+1} + \sum_{n=2}^{\infty} (n + \mu)(n + \mu - 1)a_n x^{n+\mu}$$

$$+ \mu a_0 x^\mu + (1 + \mu)a_1 x^{\mu+1} + \sum_{n=2}^{\infty} (n + \mu)a_n x^{n+\mu} + \sum_{n=0}^{\infty} a_n x^{n+\mu+2}$$

$$- \mu^2 a_0 x^\mu - \mu^2 a_1 x^{\mu+1} - \sum_{n=2}^{\infty} \mu^2 a_n x^{n+\mu} = 0$$

$$[\mu(\mu - 1) + \mu - \mu^2]a_0 x^\mu + [(1 + \mu)\mu + (1 + \mu) - \mu^2]a_1 x^{\mu+1}$$

$$+ \sum_{n=2}^{\infty} \{[(n + \mu)(n + \mu - 1) + (n + \mu) - \mu^2]a_n + a_{n-2}\}x^{n+\mu} = 0.$$

Notice that the coefficient of $a_0 x^\mu$ is zero because $r_1 = \mu$ is a root of the indicial equation. After simplifying the other coefficients and equating them to zero, we have $(1 + 2\mu)a_1 = 0$ and

$$a_n = -\frac{a_{n-2}}{(n + \mu)(n + \mu - 1) + (n + \mu) - \mu^2} = -\frac{a_{n-2}}{n(n + 2\mu)}, \quad n \geq 2.$$

From the first equation, $a_1 = 0$. Therefore, from $a_n = -\dfrac{a_{n-2}}{n(n + 2\mu)}, n \geq 2, a_n = 0$ for all odd n. The coefficients that correspond to even indices are given by

$$a_2 = -\frac{a_0}{2(2 + 2\mu)} = -\frac{a_0}{2^2(1 + \mu)}, \quad a_4 = -\frac{a_2}{4(4 + 2\mu)} = \frac{a_0}{2^4 \cdot 2(2 + \mu)(1 + \mu)},$$

$$a_6 = -\frac{a_4}{6(6 + 2\mu)} = -\frac{a_0}{2^6 \cdot 3 \cdot 2(3 + \mu)(2 + \mu)(1 + \mu)},$$

$$a_8 = -\frac{a_6}{8(8 + 2\mu)} = \frac{a_0}{2^8 \cdot 4 \cdot 3 \cdot 2(4 + \mu)(3 + \mu)(2 + \mu)(1 + \mu)}.$$

A general formula for these coefficients is given by

$$a_{2n} = \frac{(-1)^n a_0}{2^{2n} n!(1 + \mu)(2 + \mu)(3 + \mu) \cdots (n + \mu)}, \quad n \geq 2.$$ Our solution can then be written as

$$y = \sum_{n=0}^{\infty} a_{2n} x^{2n+\mu} = \sum_{n=0}^{\infty} \frac{a_0(-1)^n x^{2n+\mu}}{2^{2n} n!(1 + \mu)(2 + \mu)(3 + \mu) \cdots (n + \mu)}$$

$$= \sum_{n=0}^{\infty} \frac{a_0(-1)^n 2^\mu}{n!(1 + \mu)(2 + \mu)(3 + \mu) \cdots (n + \mu)} \left(\frac{x}{2}\right)^{2n+\mu}.$$

Using the Gamma function, $\Gamma(x)$, we write this solution as

$$y = \sum_{n=0}^{\infty} \frac{(-1)^n}{n!\Gamma(1 + \mu + n)} \left(\frac{x}{2}\right)^{2n+\mu}, \quad \text{where } a_0 = \frac{\mu!}{2^\mu}.$$

This function, denoted $J_\mu(x)$, is called the **Bessel function of the first kind of order μ.** For the other root $r_2 = -\mu$, a similar derivation yields a second solution

$$y = \sum_{n=0}^{\infty} \frac{(-1)^n}{n!\Gamma(1 - \mu + n)} \left(\frac{x}{2}\right)^{2n-\mu},$$

which is the **Bessel function of the first kind of order $-\mu$** and is denoted $J_{-\mu}(x)$.

Now, we must determine if the functions $J_\mu(x)$ and $J_{-\mu}(x)$ are linearly independent.

Notice that if $\mu = 0$, these two functions are the same. If $\mu > 0$, $r_1 - r_2 = \mu - (-\mu) = 2\mu$. If 2μ is not an integer, by the Method of Frobenius the two solutions $J_\mu(x)$ and $J_{-\mu}(x)$ are linearly independent. Also, we can show that if 2μ is an

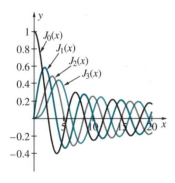

Figure 6.13

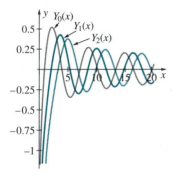

Figure 6.14

odd integer, then $J_\mu(x)$ and $J_{-\mu}(x)$ are linearly independent. In both of these cases, a general solution is given by $y = c_1 J_\mu(x) + c_2 J_{-\mu}(x)$. The graphs of the functions $J_\mu(x)$, $\mu = 0, 1, 2, 3$ are shown in Figure 6.13. Notice that these functions have numerous zeros.

 What happens to the maximum value of $J_\mu(x)$ as μ increases?

If μ is not an integer, we define the **Bessel function of the second kind of order μ** as the linear combination of the functions $J_\mu(x)$ and $J_{-\mu}(x)$. This function, denoted by $Y_\mu(x)$, is given by

$$Y_\mu(x) = \frac{\cos \mu\pi J_\mu(x) - J_{-\mu}(x)}{\sin \mu\pi}.$$

We can show that $J_\mu(x)$ and $Y_\mu(x)$ are linearly independent solutions of Bessel's equation of order μ, so a general solution of the equation is $y = c_1 J_\mu(x) + c_2 Y_\mu(x)$. We show the graphs of the functions $Y_\mu(x)$, $\mu = 0, 1, 2$ in Figure 6.14. Notice that $\lim_{x \to 0^+} Y_\mu(x) = -\infty$. We can show that if m is an integer and if $Y_m(x) = \lim_{\mu \to m} Y_\mu(x)$, then $J_m(x)$ and $Y_m(x)$ are linearly independent. Therefore, $y = c_1 J_\mu(x) + c_2 Y_\mu(x)$ is a general solution to $x^2 y'' + xy'' + (x^2 - \mu^2)y = 0$ for any value of μ.

A more general form of Bessel's equation is expressed in the form

$$x^2 y'' + xy' + (\lambda^2 x^2 - \mu^2)y = 0.$$

Through a change of variables, we can show that a general solution of this parametric Bessel equation is

$$y = c_1 J_\mu(\lambda x) + c_2 Y_\mu(\lambda x).$$

Example 3

Find a general solution of each of the following equations:

(a) $x^2 y'' + xy' + (x^2 - 16)y = 0$; (b) $x^2 y'' + xy' + \left(x^2 - \frac{1}{25}\right)y = 0$;

(c) $x^2 y'' + xy' + (9x^2 - 4)y = 0$.

Solution   (a) In this case, $\mu = 4$. Hence, $y = c_1 J_4(x) + c_2 Y_4(x)$. We graph this solution on $[1, 20]$ for various choices of the arbitrary constants in Figure 6.15(a). Notice that we must avoid graphing near $x = 0$ because of the behavior of $Y_4(x)$. (b) Because $\mu = \frac{1}{5}$, $y = c_1 J_{1/5}(x) + c_2 Y_{1/5}(x)$. This solution is graphed for several values of the arbitrary constants in Figure 6.15(b). (c) Using the parametric Bessel's equation with $\lambda = 3$ and $\mu = 2$, we have $y = c_1 J_2(3x) + c_2 Y_2(3x)$. We graph this solution for several choices of the arbitrary constants in Figure 6.15(c).

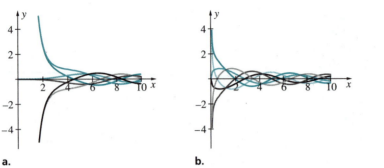

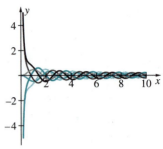

a.   b.   c.

Figure 6.15   (a) $y = c_1 J_4(x) + c_2 Y_4(x)$ (b) $y = c_1 J_{1/5}(x) + c_2 Y_{1/5}(x)$
(c) $y = c_1 J_2(3x) + c_2 Y_2(3x)$

IN TOUCH WITH TECHNOLOGY

 1. The values of x that satisfy the equation $J_0(x) = 0$ are useful in many applications in applied mathematics. **(a)** Approximate the first ten zeros (or roots) of the Bessel function of the first kind of order zero, $J_0(x)$, which is graphed in Figure 6.16. **(b)** Approximate the first nine zeros of $J_\mu(x)$ for $\mu = 1, 2, \ldots, 8$.

2. (a) Verify that the Legendre polynomials given in Table 6.15 satisfy the relationship

$$\int_{-1}^{1} P_m(x) P_n(x) \, dx = 0, \, m \neq n$$

(called an **orthogonality condition**).

(b) Evaluate $\int_{-1}^{1} [P_n(x)]^2 \, dx$ for $n = 0, 1, \ldots, 5$. How do these results compare to the value of $\dfrac{2}{2n + 1}$, for $n = 0, 1, \ldots, 5$?

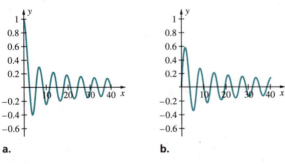

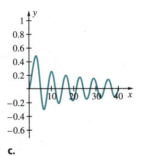

a.   b.   c.

Figure 6.16   (a) $J_0(x)$ (b) $J_1(x)$ (c) $J_2(x)$

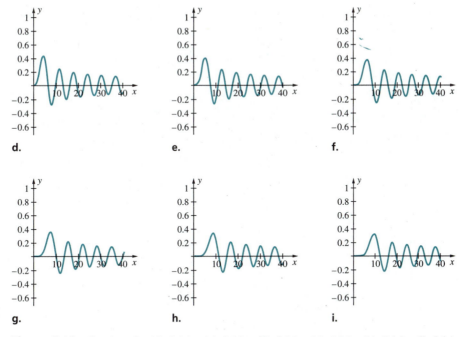

Figure 6.16 Continued (d) $J_3(x)$ (e) $J_4(x)$ (f) $J_5(x)$ (g) $J_6(x)$ (h) $J_7(x)$ (i) $J_8(x)$

EXERCISES 6.5

1. **Hermite's equation** is given by

$$y'' - 2xy' + 2ky = 0, k \geq 0.$$

Using a power series expansion about the ordinary point $x = 0$, obtain a general solution of this equation for **(a)** $k = 1$ and **(b)** $k = 3$. Show that if k is a nonnegative integer, then one of the solutions is a polynomial of degree k.

2. **Chebyshev's equation** is given by

$$(1 - x^2)y'' - xy' + k^2y = 0, n \geq 0.$$

Using a power series expansion about the ordinary point $x = 0$, obtain a general solution of this equation for **(a)** $k = 1$ and **(b)** $k = 3$. Show that if k is a nonnegative integer, then one of the solutions is a polynomial of degree k.

3. The **hypergeometric equation** is given by

$$x(1 - x)y'' + [c - (a + b + 1)x]y' - aby = 0,$$

where a, b, and c are constants. **(a)** Show that $x = 0$

and $x = 1$ are regular singular points. **(b)** Show that the roots of the indicial equation for the series $\sum_{n=0}^{\infty} a_n x^{n+r}$ are $r = 0$ and $r = 1 - c$. **(c)** Show that for $r = 0$, the solution obtained with the Method of Frobenius is

$$y_1(x) = 1 + \frac{ab}{1!c}x + \frac{a(a + 1)b(b + 1)}{2!c(c + 1)}x^2 +$$

$$\frac{a(a + 1)(a + 2)b(b + 1)(b + 2)}{3!c(c + 1)(c + 2)}x^3 + \cdots \text{ where}$$

$c \neq 0, -1, -2, \ldots$. This series is called the **hypergeometric series**. Its sum, denoted $F(a, b, c; x)$, is called the **hypergeometric function**. **(d)** Show that $F(1, b, b; x) = \dfrac{1}{1 - x}$. **(e)** Find the solution that corresponds to $r = 1 - c$.

4. **Laguerre's equation** is given by

$$xy'' + (1 - x)y' + ky = 0.$$

(a) Show that $x = 0$ is a regular singular point of

Laguerre's equation. **(b)** Use the Method of Frobenius to determine one solution of Laguerre's equation. **(c)** Show that if k is a positive integer, then the solution is a polynomial. This polynomial, denoted $L_k(x)$, is called the Laguerre polynomial of order k.

*5. Show that Legendre's equation can be written as

$$\frac{d}{dx}[(1 - x^2)y'] + k(k + 1)y = 0.$$ Use this equation to show that $P_m(x)$ and $P_n(x)$ are **orthogonal** on the interval $[-1, 1]$ by showing that

$$\int_{-1}^{1} P_m(x)P_n(x)\,dx = 0, m \neq n.$$

(*Hint:* $P_m(x)$ and $P_n(x)$ satisfy the differential equations $\dfrac{d}{dx}[(1 - x^2)P_n'(x)] + n(n + 1)P_n(x) = 0$ and

$\dfrac{d}{dx}[(1 - x^2)P_m'(x)] + m(m + 1)P_m(x) = 0,$ respectively. Multiply the first equation by $P_n(x)$ and the second by $P_m(x)$, and subtract the results. Then, integrate from -1 to 1.)

6. (Relations between Bessel functions) (a) Using

$$J_\mu(x) = \sum_{n=0}^{\infty} \frac{(-1)^n}{n!\Gamma(1 + \mu + n)}\left(\frac{x}{2}\right)^{2n+\mu} \text{ and }$$

$\Gamma(x + 1) = x\Gamma(x)$, show that $\dfrac{d}{dx}[x^\mu J_\mu(x)] =$

$x^\mu J_{\mu-1}(x).$

(b) Using $J_\mu(x) = \displaystyle\sum_{n=0}^{\infty} \frac{(-1)^n}{n!\Gamma(1 + \mu + n)}\left(\frac{x}{2}\right)^{2n+\mu},$

show that $\dfrac{d}{dx}[x^{-\mu}J_\mu(x)] = -x^{-\mu}J_{\mu+1}(x).$

(c) Using the results of parts (a) and (b), show that $J_{\mu-1}(x) - J_{\mu+1}(x) = 2J_\mu'(x).$

(d) Evaluate $\displaystyle\int x^\mu J_{\mu-1}(x)\,dx.$

7. Show that $y = J_0(kx)$ where k is a constant is a solution of the parametric Bessel equation of order zero, $xy'' + y' + k^2xy = 0.$

8. Show that the equation $xy'' + y' + k^2xy = 0$ is equivalent to $\dfrac{d}{dx}(xy') + k^2xy = 0.$ Show that

$\displaystyle\int_0^1 xJ_0(k_mx)J_0(k_nx)\,dx = 0, m \neq n.$ (This shows that $J_0(k_mx)$ and $J_0(k_nx)$ are *orthogonal* on the interval $[0, 1]$.) (*Hint:* Follow a procedure like that described in Exercise 5. Assume that $J_0(k_m) = J_0(k_n) = 0.$)

9. Find a general solution of each equation.
(a) $x^2y'' + xy' + (x^2 - \frac{1}{4})y = 0$
(b) $x^2y'' + xy' + (16x^2 - 25)y = 0$

In Exercises 10–17, solve each hypergeometric equation. Express all solutions in terms of the function $F(a, b, c; x)$. (Notice that when either a or b is a negative integer, the solution is a polynomial.) (See Exercise 3.)

10. $x(1 - x)y'' + (1 - 3x)y' - y = 0$
11. $x(1 - x)y'' + (\frac{1}{2} - 3x)y' - y = 0$
12. $x(1 - x)y'' + (-\frac{1}{2} - 3x)y' - \frac{1}{4}y = 0$
*13. $x(1 - x)y'' + y' + 2y = 0$
14. $x(1 - x)y'' + (3 - 3x)y' + 3y = 0$
15. $x(1 - x)y'' + (\frac{1}{4} - \frac{3}{4}x)y' + \frac{1}{8}y = 0$
16. $x(1 - x)y'' + (-\frac{1}{4} - x)y' + \frac{1}{4}y = 0$
*17. $x(1 - x)y'' + (1 - 2x)y' = 0$
18. Use the power series expansion of the Bessel function of the first kind of order n (n an integer),

$$J_m(x) = \sum_{n=0}^{\infty} \frac{(-1)^n}{n!(n + m)!}\left(\frac{x}{2}\right)^{2n+m}$$

to verify that $J_0'(x) = -J_1(x).$

19. Use the change of variables $y = \dfrac{v(x)}{\sqrt{x}}$ to transform Bessel's equation $x^2y'' + xy' + (x^2 - n^2)y = 0$ into the equation $v'' + \left[1 + \dfrac{\frac{1}{4} - n^2}{x^2}\right]v = 0.$ By substituting $n = \frac{1}{2}$ into the transformed equation, derive the solution to Bessel's equation with $n = \frac{1}{2}.$

20. Show that a solution of $x\dfrac{d^2y}{dx^2} + \dfrac{dy}{dx} + \dfrac{y}{v} = 0$ is

$$y = J_0\left(2\sqrt{\frac{x}{v}}\right) \text{ where } v \text{ is a constant.}$$

*21. Use integration by parts to show that $\Gamma(p + 1) = p\Gamma(p)$, $p > 0$. (Note: p is any real number.)

22. (a) Show that $\Gamma(\frac{1}{2}) = 2\displaystyle\int_0^\infty e^{-x^2}\,dx.$ (*Hint:* Let $u = x^2$.) **(b)** Use polar coordinates to evaluate $\left(\displaystyle\int_0^\infty e^{-x^2}\,dx\right)\left(\displaystyle\int_0^\infty e^{-y^2}\,dy\right) = \displaystyle\int_0^\infty\int_0^\infty e^{-(x^2+y^2)}\,dx\,dy.$ **(c)** Use the results of (a) and (b) to evaluate $\Gamma(\frac{1}{2})$ and then $\Gamma(\frac{3}{2}).$

Section 6.1

Cauchy-Euler Equation

A **Cauchy-Euler** differential equation is an equation of the form

$$a_n x^n y^{(n)} + \cdots + a_1 xy' + a_0 y = g(x),$$

where $a_0, a_1, a_2, \ldots, a_n$ are constants.

General solution of a second-order Cauchy-Euler equation

$y = c_1 x^{m_1} + c_2 x^{m_2}$, if $m_1 \neq m_2$ are real;
$y = c_1 x^{m_1} + c_2 x^{m_1} \ln x$, if $m_1 = m_2$; and
$y = x^{\alpha}[c_1 \cos(\beta \ln x) + c_2 \sin(\beta \ln x)]$ if $m_1 = \overline{m_2} = \alpha + i\beta$, $\beta \neq 0$.

Higher-Order Cauchy-Euler equation

Variation of Parameters

Section 6.2

Power series, Radius of convergence, Taylor polynomial, Maclaurin polynomial, Taylor's Theorem, Geometric series, Term-by-term integration and differentiation of a power series, Reindexing a power series

Section 6.3

Ordinary and Singular points

x_0 is an **ordinary point** of $y'' + p(x)y' + q(x)y = 0$ if both $p(x)$ and $q(x)$ are analytic at x_0. If x_0 is not an ordinary point, x_0 is called a **singular point.**

Power Series Solution Method About an Ordinary Point

1. Assume that $y = \Sigma_{n=0}^{\infty} a_n(x - x_0)^n$.

2. After taking the appropriate derivatives, substitute $y = \Sigma_{n=0}^{\infty} a_n(x - x_0)^n$ into the differential equation.
3. Find the unknown series coefficients a_n through an equation relating the coefficients.
4. Apply any given initial conditions, if applicable.

Convergence of the power series solution

Section 6.4

Regular and irregular singular points

Let x_0 be a singular point of $y'' + p(x)y' + q(x)y = 0$. x_0 is a **regular singular point** of the equation if both $(x - x_0)p(x)$ and $(x - x_0)^2 q(x)$ are analytic at x_0. If x_0 is not a regular singular point, x_0 is called an **irregular singular point** of the equation.

Method of Frobenius

Indicial equation

$r(r - 1) + p_0 r + q_0 = 0$

Indicial roots which do not differ by an integer

Indicial roots which differ by an integer

Equal indicial roots

Section 6.5

Legendre's equation

$(1 - x^2)y'' - 2xy' + k(k + 1)y = 0$

Gamma function

$\Gamma(x) = \displaystyle\int_0^{\infty} e^{-u} u^{x-1} \, du, \, x > 0$

Bessel's equation

$x^2 y'' + xy' + (x^2 - \mu^2)y = 0$

CHAPTER 6 REVIEW EXERCISES

In Exercises 1–16, solve the Cauchy-Euler equation.

1. $x^2y'' - 4xy' + 6y = 0$

2. $x^2y'' + 2xy' - 2y = 0$

*3. $x^2y'' + 7xy' + 8y = 0$

4. $6x^2y'' + 13xy' - 20y = 0$

5. $\begin{cases} 2x^2y'' + 5xy' + y = 0 \\ y(1) = 1, y'(1) = 0 \end{cases}$

6. $\begin{cases} 15x^2y'' - 7xy' - 5y = 0 \\ y(1) = 0, y'(1) = 1 \end{cases}$

*7. $x^2y'' + xy' + y = 0$

8. $x^2y'' - xy' + 5y = 0$

9. $x^2y'' + 7xy' + 25y = 0$

10. $x^2y'' + 5xy' + 20y = 0$

*11. $5x^2y'' - xy' + 2y = 0$

12. $x^3y''' + 4x^2y'' - 15xy' + 15y = 0$

13. $\begin{cases} x^3y''' + 5x^2y'' - 16xy' - 20y = 0 \\ y(1) = 2, y'(1) = 0, y''(1) = -1 \end{cases}$

14. $x^3y''' - 2x^2y'' + 13xy' - 13y = 0$

*15. $x^2y'' - 7xy' + 15y = 8x$

16. $x^2y'' + 3xy' - 8y = 7x^3$

In Exercises 17–22, solve the differential equation with a power series expansion about $x = 0$. Write out at least the first five nonzero terms of each series.

17. $y'' - 4y' + 4y = 0$

18. $y'' + 4y' + 5y = 0$

*19. $y'' + 2y' - 3y = xe^x$

20. $y'' - xy' + 4y = 0$

21. $(-1 + 2x^2)y'' + 2xy' - 3y = 0$

22. $(-1 - x)y'' - 3y' - y = 0, y(0) = 0, y'(0) = -1$

In Exercises 23–28, use the Method of Frobenius to obtain two linearly independent solutions about the regular singular point $x = 0$.

23. $3xy'' + 11y' - y = 0$

24. $4xy'' - 3y' + (x^2 + 1)y = 0$

*25. $2x^2y'' + 5xy' - 2y = 0$

26. $y'' + \left(\dfrac{3}{4x} + 1\right)y' - \dfrac{7}{2x^2}y = 0$

27. $y'' - \dfrac{7}{x}y' + \left(\dfrac{7}{x^2} - 2\right)y = 0$

28. $x^2y'' + 5xy' + (x + 3)y = 0$

In Exercises 29 and 30 find a solution to each hypergeometric equation. Express the solution in terms of the function $F(a, b, c; x)$. (See Exercise 3 in Section 6.5.)

29. $x(1 - x)y'' + (1 + 2x)y' + 10y = 0$

30. $x(1 + x)y'' + \left(\dfrac{1}{3} - 12x\right)y' - 10y = 0$

31. **(Simple Modes of a Vibrating Chain)** The equation that describes the simple modes of a vibrating chain of length ℓ is

$$x\frac{d^2y}{dx^2} + \frac{dy}{dx} + \frac{y}{\nu} = 0,$$

where y is the displacement and x is the distance from the bottom of the chain. This equation was studied extensively by Daniell Bernoulli around 1727.

(a) If the chain is fixed at the top so that $y(\ell) = 0$ and $y(0) = 1$, show that a solution to this equation is $J_0(2\sqrt{x/\nu})$.

(b) Convince yourself that for any value of ℓ, the equation $J_0(2\sqrt{\ell/\nu}) = 0$ has infinitely many solutions.

(c) If $\ell = 1$, graph $J_0(2\sqrt{\ell/\nu})$ and approximate the last ten solutions of $J_0(2\sqrt{\ell/\nu}) = 0$.

Differential Equations at Work:
The Schrödinger Equation

● The time independent **Shrödinger equation,** proposed by the Austrian physicist Erwin Schrödinger (1887–1961) in 1926, describes the relationship between particles and waves. The **Schrödinger equation in spherical coordinates** is given by

$$\frac{1}{r^2}\frac{\partial}{\partial r}\left(r^2\frac{\partial\psi}{\partial r}\right) + \frac{1}{r^2\sin\theta}\frac{\partial}{\partial\theta}\left(\sin\theta\frac{\partial\psi}{\partial\theta}\right) + \frac{1}{r^2\sin^2\theta}\frac{\partial^2\psi}{\partial\phi^2} + \frac{2\mu}{h^2}(E - V)\psi = 0.*$$

If we assume that a solution to this partial differential equation of the form

$$\psi(r, \theta, \phi) = R(r)\Theta(\theta)g(\phi)$$

exists, then by a technique called **separation of variables,** the Schrödinger equation can be rewritten as three second-order ordinary differential equations:

Azimuthal Equation $\quad\dfrac{d^2g}{d\phi^2} = -m_\ell^2 g$

Radial Equation $\quad\dfrac{1}{r^2}\dfrac{d}{dr}\left(r^2\dfrac{dR}{dr}\right) + \dfrac{2\mu}{h^2}\left(E - V - \dfrac{h^2}{2\mu}\dfrac{\ell(\ell+1)}{r^2}\right)R = 0$

Angular Equation $\quad\dfrac{1}{\sin\theta}\dfrac{d}{d\theta}\left(\sin\theta\dfrac{d\Theta}{d\theta}\right) + \left[\ell(\ell+1) - \dfrac{m_\ell^2}{\sin^2\theta}\right]\Theta = 0.$

The *orbital angular momentum quantum number* ℓ must be a nonnegative integer: $\ell = 0, \pm 1, \pm 2, \ldots$ and the *magnetic quantum number* m_ℓ depends on ℓ: $m_\ell = -\ell,$ $-\ell + 1, \ldots, -2, -1, 0, 1, 2, \ldots, \ell, \ell + 1.$

If $\ell = 0$, the radial equation is $\dfrac{1}{r^2}\dfrac{d}{dr}\left(r^2\dfrac{dR}{dr}\right) + \dfrac{2\mu}{h^2}(E - V)R = 0.$ Using

$V(r) = -\dfrac{e^2}{4\pi\epsilon_0 r}$ (the constants μ, h, e, and ϵ_0 are described later), the equation be-

comes $\dfrac{d^2R}{dr^2} + \dfrac{2}{r}\dfrac{dR}{dr} + \dfrac{2\mu}{h^2}\left(E + \dfrac{e^2}{4\pi\epsilon_0 r}\right)R = 0.$

1. Show that if $R = Ae^{-r/a_0}$ is a solution of this equation, $a_0 = \dfrac{4\pi\epsilon_0 h^2}{\mu e^2}$ and

$E = -\dfrac{h^2}{2\mu a_0^2}.$

---

* Stephen T. Thornton and Andrew Rex, *Modern Physics for Scientists and Engineers,* Saunders College Publishing, 1993, pp. 248–253.

The constants in the expression $\dfrac{4\pi\epsilon_0 h^2}{\mu e^2}$ are described as follows: $h \approx 1.054572 \times 10^{-34}$ J $\cdot$ s is the *Planck constant* ($6.6260755 \times 10^{-34}$ J $\cdot$ s divided by 2π), $\epsilon_0 \approx 8.854187817 \times 10^{-12}$ F/m is the *permittivity of vacuum*, and $e \approx 1.60217733 \times 10^{-19}$ C is the *elementary charge*. For the hydrogen atom, $\mu = 9.104434 \times 10^{-31}$ kg is the *reduced mass of the proton and electron*.

2. Calculate a_0, $2a_0$, and E for the hydrogen atom.

3. The diameter of a hydrogen atom is approximately 10^{-10} m while the energy, E, is known to be approximately 13.6 eV. How do the results you obtained in Problem 2 compare to these?

For the hydrogen atom, the radial equation can be written in the form

$$\frac{d^2R}{dr^2} + \frac{2}{r}\frac{dR}{dr} + \left[2E + \frac{2Z}{r} - \frac{\ell(\ell+1)}{r^2}\right]R = 0,$$

where E is the energy of the hydrogen atom and Z is a constant.*

When $E < 0$ and we let $p = \sqrt{-2E}$, a solution of

$$\frac{d^2R}{dr^2} + \frac{2}{r}\frac{dR}{dr} + \left[2E + \frac{2Z}{r} - \frac{\ell(\ell+1)}{r^2}\right]R = 0$$

has the form

$$R(r) = e^{-pr}u(r).$$

When this solution is substituted into the equation, we obtain the equation

$$\frac{d^2u}{dr^2} + 2\left(\frac{1}{r} - p\right)\frac{du}{dr} + \left[2\left(Z - \frac{p}{r}\right) - \frac{\ell(\ell+1)}{r^2}\right]u = 0.$$

4. Show that the solutions of the indicial equation of

$$\frac{d^2u}{dr^2} + 2\left(\frac{1}{r} - p\right)\frac{du}{dr} + \left[2\left(Z - \frac{p}{r}\right) - \frac{\ell(\ell+1)}{r^2}\right]u = 0$$

are ℓ and $-(\ell+1)$.

5. Find a power series solution to the equation of the form $u(r) = \sum_{n=0}^{\infty} a_n r^{n+\ell}$.

6. Find a series solution of

$$\frac{d^2R}{dr^2} + \frac{2}{r}\frac{dR}{dr} + \left[2E + \frac{2Z}{r} - \frac{\ell(\ell+1)}{r^2}\right]R = 0.$$

* Yi-Hsin Liu and Wai-Ning Mei, "Solution of the radial equation for hydrogen atom: series solution or Laplace transform?" *International Journal of Mathematical Education in Science and Technology,* Volume 21, Number 6, (1990) pp. 913–918.

7

INTRODUCTION TO THE LAPLACE TRANSFORM

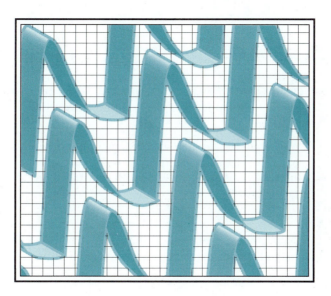

In previous chapters we investigated solving the equation

$$a_n(x)y^{(n)} + a_{n-1}(x)y^{(n-1)} + \cdots + a_1(x)y' + a_0(x)y = f(x)$$

for y. We saw that if the coefficients $a_n, a_{n-1}, \ldots, a_0$ are constants, we can find a general solution of the equation by first finding a general solution of the corresponding homogeneous equation

$$a_ny^{(n)} + a_{n-1}y^{(n-1)} + \cdots + a_1y' + a_0y = 0$$

and then finding a particular solution of

$$a_ny^{(n)} + a_{n-1}y^{(n-1)} + \cdots + a_1y' + a_0y = f(x).$$

If the coefficients $a_n, a_{n-1}, \ldots, a_0$ are not constants, the situation is more difficult. In some cases, like if the equation is a Cauchy-Euler equation, similar techniques

357

can be used. In other cases, we might be able to use power series to find a solution. In all of these cases, however, the function $f(x)$ has typically been a *smooth* function. In cases when $f(x)$ is not a smooth function, like if $f(x)$ is a piecewise-defined function, implementing methods like variation of parameters to solve the equation

$$a_n y^{(n)} + a_{n-1} y^{(n-1)} + \cdots + a_1 y' + a_0 y = f(x)$$

is usually difficult.

In this chapter, we discuss a technique that transforms the *differential* equation

$$a_n y^{(n)} + a_{n-1} y^{(n-1)} + \cdots + a_1 y' + a_0 y = f(x)$$

into an *algebraic* equation that can often be solved in order to obtain a solution of the differential equation, *even if $f(x)$ is not a smooth function.*

7.1 The Laplace Transform: Preliminary Definitions and Notation

Pierre Simon de Laplace (1749–1827) French mathematician and astronomer. (NorthWind Picture Archives)

Definition of the Laplace Transform • The Laplace Transform of sin *kt* and cos *kt* • Exponential Order • Jump Discontinuities and Piecewise Continuous Functions

We are already familiar with several operations on functions. In previous courses, we learned to add, subtract, multiply, divide, and compose functions. Another operation on functions is *differentiation,* which transforms the differentiable function $F(x)$ to its derivative $F'(x)$,

$$D_x(F(x)) = F'(x).$$

Similarly, the operation of *integration* transforms the integrable function $f(x)$ to its integral. For example, if f is integrable on an interval containing a, then

$$\int_a^x f(t)\, dt$$

is a function of x.

In this section, we introduce another operation on functions, the *Laplace transform,* and discuss several of its properties.

Definition of the Laplace Transform

Definition 7.1 Laplace Transform

Let $f(t)$ be a function defined on the interval $[0, +\infty)$. The **Laplace transform** of $f(t)$ is the function (of s)

$$\mathcal{L}\{f(t)\} = \int_0^\infty e^{-st} f(t)\, dt,$$

provided that the improper integral exists.

*The French mathematician Pierre de Laplace (1749–1827) introduced this integral transform in his work **Theorie analytique des probabilities,** published in 1812. However, Laplace is probably most famous for his contributions to astronomy and probability.*

Because the Laplace transform yields a function of s, we often use the notation $\mathcal{L}\{f(t)\} = F(s)$ to denote the Laplace transform of $f(t)$. We use the *capital* letter to denote the Laplace transform of the function named with the corresponding *small* letter.

The Laplace transform is defined as an improper integral. Recall that we evaluate improper integrals of this form by taking the limit of a definite integral:

$$\int_0^\infty f(t)\,dt = \lim_{M \to \infty} \int_0^M f(t)\,dt.$$

Example 1

Compute $\mathcal{L}\{f(t)\}$ if $f(t) = 1$.

Solution

$$F(s) = \mathcal{L}\{f(t)\} = \int_0^\infty e^{-st} \cdot 1\,dt = \lim_{M \to \infty} \int_0^M e^{-st} \cdot 1\,dt = \lim_{M \to \infty} \left[-\frac{e^{-st}}{s} \right]_{t=0}^{t=M}$$

$$= -\frac{1}{s} \lim_{M \to \infty} [e^{-sM} - 1].$$

If $s > 0$, $\lim_{M \to \infty} e^{-sM} = 0$. (Otherwise, the limit does not exist.) Therefore,

$$F(s) = \mathcal{L}\{f(t)\} = -\frac{1}{s} \lim_{M \to \infty} [e^{-sM} - 1] = -\frac{1}{s}(0 - 1) = \frac{1}{s}, \ s > 0.$$

Example 2

Compute $\mathcal{L}\{f(t)\}$ if $f(t) = e^{at}$.

Solution

$$F(s) = \mathcal{L}\{f(t)\} = \int_0^\infty e^{-st}f(t)\,dt = \int_0^\infty e^{-st}e^{at}\,dt = \int_0^\infty e^{-(s-a)t}\,dt$$

$$= \lim_{M \to \infty} \left[-\frac{e^{-(s-a)t}}{s - a} \right]_{t=0}^{t=M} = -\lim_{M \to \infty} \left(\frac{e^{-(s-a)M}}{s - a} - \frac{1}{s - a} \right).$$

If $s - a > 0$, then $\lim_{M \to \infty} e^{-(s-a)M} = 0$. Therefore,

$$F(s) = \mathcal{L}\{f(t)\} = -\lim_{M \to \infty} \left(\frac{e^{-(s-a)M}}{s - a} - \frac{1}{s - a} \right) = -\left(0 - \frac{1}{s - a} \right) = \frac{1}{s - a}, \ s > a.$$

The formula found in Example 2 can be used to avoid using the definition.

Example 3

Compute: (a) $\mathcal{L}[e^{-3t}]$ and (b) $\mathcal{L}\{e^{5t}\}$.

Solution   (a) $\mathcal{L}\{e^{-3t}\} = \dfrac{1}{s-(-3)} = \dfrac{1}{s+3}$, $s > -3$, and

(b) $\mathcal{L}\{e^{5t}\} = \dfrac{1}{s-5}$, $s > 5$.

Example 4

Compute $\mathcal{L}\{\sin t\}$.

Solution   In order to evaluate the improper integral that results, we use a table of integrals or a computer algebra system. Otherwise, we would have to use integration by parts twice.

$$\mathcal{L}\{\sin t\} = \int_0^\infty e^{-st} \sin t\, dt = \lim_{M\to\infty} \int_0^M e^{-st} \sin t\, dt$$

$$= \lim_{M\to\infty} \left[-\frac{e^{-st}}{s^2+1}(s \sin t + \cos t) \right]\Big|_{t=0}^{t=M}$$

$$= -\frac{1}{s^2+1} \lim_{M\to\infty} [e^{-sM}(s \sin M + \cos M) - 1]$$

If $s > 0$, $\lim_{M\to\infty} e^{-sM}(s \sin M + \cos M) = 0$. (Why?) Therefore,

$$\mathcal{L}\{\sin t\} = -\frac{1}{s^2+1} \lim_{M\to\infty} [e^{-sM}(s \sin M + \cos M) - 1] = \frac{1}{s^2+1}, \quad s > 0.$$

Example 5

Compute $\mathcal{L}\{f(t)\}$ if $f(t) = t$.

Solution   To compute $F(s) = \mathcal{L}\{f(t)\} = \int_0^\infty e^{-st} t\, dt$ we use integration by parts with $u = t$ and $dv = e^{-st}\, dt$. Then $du = dt$ and $v = \dfrac{-1}{s}e^{-st}$, so

$$F(s) = \mathcal{L}\{f(t)\} = \int_0^\infty e^{-st} t\, dt = \lim_{M\to\infty} \int_0^M e^{-st} t\, dt$$

$$= \lim_{M\to\infty} \left(\frac{-te^{-st}}{s}\Big|_{t=0}^{t=M} \right) + \lim_{M\to\infty} \frac{1}{s} \int_0^M e^{-st}\, dt$$

$$= 0 - \frac{1}{s^2} \lim_{M\to\infty} (e^{-st}|_{t=0}^{t=M}).$$

If $s > 0$, $\lim_{M\to\infty} e^{-sM} = 0$. Therefore,

$$F(s) = \mathcal{L}\{f(t)\} = -\frac{1}{s^2} \lim_{M\to\infty} (e^{-sM} - 1) = \frac{1}{s^2}, \quad s > 0.$$

As we can see, the definition of the Laplace transform can be difficult to apply. For example, if we wanted to calculate $\mathcal{L}\{t^n\}$ with the definition, we would have to integrate by parts n times; a difficult and time-consuming task. We will investigate other properties of the Laplace transform so that we can determine the Laplace transform of many functions more easily. We begin by discussing the *linearity property,* which enables us to use the transforms that we have already found to find the Laplace transforms of other functions.

*We are familiar with **linearity properties** through our work in calculus with derivatives and integrals. For derivatives, this property is*

$$\frac{d}{dx}[af(x) + bg(x)] = a\frac{d}{dx}[f(x)] + b\frac{d}{dx}[g(x)]$$

while for integrals, it is

$$\int [af(x) + bg(x)]\, dx = a\int f(x)\, dx + b\int g(x)\, dx.$$

Theorem 7.1 Linearity Property of the Laplace Transform

Let a and b be constants, and suppose that the Laplace transform of the functions $f(t)$ and $g(t)$ exist. Then,

$$\mathcal{L}\{af(t) + bg(t)\} = a\mathcal{L}\{f(t)\} + b\mathcal{L}\{g(t)\}.$$

PROOF OF THEOREM 7.1

$$\mathcal{L}\{af(t) + bg(t)\} = \int_0^\infty e^{-st}(af(t) + bg(t))\, dt$$

$$= a\int_0^\infty e^{-st}f(t)\, dt + b\int_0^\infty e^{-st}g(t)\, dt = a\mathcal{L}\{f(t)\} + b\mathcal{L}\{g(t)\}.$$

$\bullet\ \bullet\ \bullet$

Example 6

Calculate (a) $\mathcal{L}\{6\}$; (b) $\mathcal{L}\{5 - 2e^{-t}\}$

Solution   (a) $\mathcal{L}\{6\} = 6\mathcal{L}\{1\} = 6(1/s) = 6/s.$

(b) $\mathcal{L}\{5 - 2e^{-t}\} = 5\mathcal{L}\{1\} - 2\mathcal{L}\{e^{-t}\} = 5\left(\dfrac{1}{s}\right) - 2\left(\dfrac{1}{s - (-1)}\right) = \dfrac{5}{s} - \dfrac{2}{s + 1}.$

 For what values of s do the transforms in Example 6 exist?

The Laplace Transform of sin *kt* and cos *kt*

We can use the linearity property to calculate $\mathcal{L}\{\sin kt\}$ and $\mathcal{L}\{\cos kt\}$, where k is a constant. Recall Euler's formula $e^{i\theta} = \cos\theta + i\sin\theta$ and $e^{-i\theta} = \cos\theta - i\sin\theta$

(obtained from Euler's formula with the substitution $-\theta$) that were used in Chapter 4 to solve differential equations with complex-valued roots of the characteristic equation. Combining these expressions we obtain

$$\cos\theta = \frac{1}{2}(e^{i\theta} + e^{-i\theta}) \qquad \text{and} \qquad \sin\theta = \frac{1}{2i}(e^{i\theta} - e^{-i\theta}).$$

Then $\cos kt = \frac{1}{2}(e^{ikt} + e^{-ikt})$ and $\sin kt = \frac{1}{2i}(e^{ikt} - e^{-ikt})$. Using these representations, we have

$$\mathcal{L}\{\cos kt\} = \mathcal{L}\left\{\frac{1}{2}(e^{ikt} + e^{-ikt})\right\} = \frac{1}{2}[\mathcal{L}\{e^{ikt}\} + \mathcal{L}\{e^{-ikt}\}] = \frac{1}{2}\left[\frac{1}{s - ik} + \frac{1}{s + ik}\right]$$

$$= \frac{1}{2}\frac{s + ik + s - ik}{(s - ik)(s + ik)} = \frac{1}{2}\frac{2s}{s^2 - i^2k^2} = \frac{s}{s^2 + k^2}$$

and

$$\mathcal{L}\{\sin kt\} = \mathcal{L}\left\{\frac{1}{2i}(e^{ikt} - e^{-ikt})\right\} = \frac{1}{2i}[\mathcal{L}\{e^{ikt}\} - \mathcal{L}\{e^{-ikt}\}] = \frac{1}{2i}\left[\frac{1}{s - ik} - \frac{1}{s + ik}\right]$$

$$= \frac{1}{2i}\frac{s + ik - (s - ik)}{(s - ik)(s + ik)} = \frac{1}{2i}\frac{2ik}{s^2 - i^2k^2} = \frac{k}{s^2 + k^2}.$$

Both of these formulas are valid for $s > 0$. This can be verified if these transforms are found directly with the definition of the Laplace transform.

Exponential Order

In calculus we saw that in some cases improper integrals diverge, which means that the limit of the definite integral $\int_a^M f(x)\,dx$ does not exist as $M \to \infty$. Hence, we may believe that the Laplace transform *may not* exist for some functions. For example, $f(t) = \frac{1}{t}$ grows too rapidly near $t = 0$ for the improper integral $\int_0^\infty e^{-st}f(t)\,dt$ to exist and $f(t) = e^{t^2}$ grows too rapidly as $t \to \infty$ for the improper integral $\int_0^\infty e^{-st}f(t)\,dt$ to exist. We present the following definitions and theorems to better understand the types of functions for which the Laplace transform exists.

Definition 7.2 Exponential Order

A function $f(t)$ is of **exponential order** (of orber b) if there are numbers $b, C > 0$, and $T > 0$ such that

$$|f(t)| \le Ce^{bt}$$

for $t > T$.

We can use Definition 7.2 to show that if $\lim_{t\to\infty} f(t)e^{-bt}$ exists and is finite, $f(t)$ is of exponential order; if $\lim_{t\to\infty} f(t)e^{-bt} = +\infty$ for every value of $b > 0$, $f(t)$ is not of exponential order. (See Exercise 32.)

Example 7

(a) Show that $f(t) = t^3$ is of exponential order. (b) Show that $f(t) = e^{t^2}$ is not of exponential order.

Solution   (a) Using L'Hopital's rule several times, we have

$$\lim_{t\to\infty} f(t)e^{-bt} = \lim_{t\to\infty} \frac{f(t)}{e^{bt}} = \lim_{t\to\infty} \frac{t^3}{e^{bt}} = \lim_{t\to\infty} \frac{3t^2}{be^{bt}} = \lim_{t\to\infty} \frac{6t}{b^2 e^{bt}} = \lim_{t\to\infty} \frac{6}{b^3 e^{bt}} = 0.$$

Because the limit is finite, $f(t) = t^3$ is of exponential order.
(b) In this case,

$$\lim_{t\to\infty} f(t)e^{-bt} = \lim_{t\to\infty} \frac{f(t)}{e^{bt}} = \lim_{t\to\infty} \frac{e^{t^2}}{e^{bt}} = \lim_{t\to\infty} e^{t^2-bt} = \lim_{t\to\infty} e^{t(t-b)} = +\infty$$

for any value of b. Hence, $f(t) = e^{t^2}$ is not of exponential order.

 Is $f(t) = t^n$, where n is any positive integer, of exponential order? Why?

Example 8

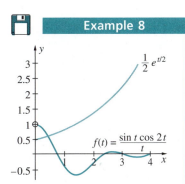

Figure 7.1

Determine values of C and T that show that $f(t) = \dfrac{\sin t \cos 2t}{t}$ is of exponential order $\frac{1}{2}$.

Solution   Suppose that we let $C = \frac{1}{2}$. Then, with $b = \frac{1}{2}$, we compare $f(t) = \dfrac{\sin t \cos 2t}{t}$ with $\frac{1}{2}e^{t/2}$ by graphing both functions in Figure 7.1. Notice that these curves intersect near $t = 0.5$. $\left(\text{This is the only point of intersection because } \lim_{t\to\infty} \dfrac{\sin t \cos 2t}{t} = 0. \text{ Why?}\right)$ We obtain a better approximation of $t \approx 0.436697$ using a numerical method. Therefore, with $T = 0.44$, $\left|\dfrac{\sin t \cos 2t}{t}\right| < \dfrac{1}{2}e^{t/2}$ for $t > T$.

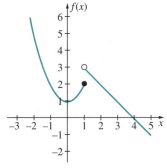

Figure 7.2   (a)

Jump Discontinuities and Piecewise Continuous Functions

In the next sections, we will see that the Laplace transform is particularly useful in solving differential equations involving piecewise or recursively defined functions. For example, the function $f(x) = \begin{cases} x^2 + 1, & \text{if } x \le 1 \\ 4 - x, & \text{if } x > 1 \end{cases}$ with graph shown in Figure 7.2(a), is a piecewise defined function. Because

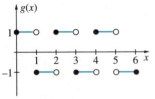

Figure 7.2   (b)

$$\lim_{x \to 1^-} f(x) = \lim_{x \to 1^-} (x^2 + 1) = 2 \ne 3 = \lim_{x \to 1^+} (4 - x) = \lim_{x \to 1^+} f(x),$$

the limit $\lim_{x \to 1} f(x)$ does not exist and $f(x)$ is not continuous if $x = 1$. Similarly, the function $g(x)$ defined by $g(x) = \begin{cases} 1, & \text{if } 0 \le x < 1 \\ -1, & \text{if } 1 \le x < 2 \end{cases}$ on the interval $[0, 2)$ and then recursively by the relationship $g(x) = g(x - 2)$ (shown in Figure 7.2(b)), is discontinuous when $x = n$, where n is any nonnegative integer. For both f and g, we say that the discontinuities are **jump discontinuities** because the left and right-hand limits both exist at the points of discontinuity, but are unequal.

Definition 7.3 Jump Discontinuity

Let $f(t)$ be defined on $[a, b]$. f has a **jump discontinuity** at $t = c$, $a < c < b$, if the one-sided limits $\lim_{t \to c^+} f(t)$ and $\lim_{t \to c^-} f(t)$ are finite, but unequal, values. $f(t)$ has a **jump discontinuity at $t = a$** if $\lim_{t \to a^+} f(t)$ is a finite value different from $f(a)$. $f(t)$ has a **jump discontinuity at $t = b$** if $\lim_{t \to b^-} f(t)$ is a finite value different from $f(b)$.

Definition 7.4 Piecewise Continuous

A function $f(t)$ is **piecewise continuous on the finite interval $[a, b]$** if $f(t)$ is continuous at every point in $[a, b]$ except at finitely many points at which $f(t)$ has a jump discontinuity. A function $f(t)$ is **piecewise continuous on $[0, \infty)$** if $f(t)$ is piecewise continuous on $[0, N]$ for all $N > 0$.

Is $f(t) = \begin{cases} \dfrac{1}{t - 1}, & 0 \le t < 1 \\ t, & 1 \le t \le 2 \end{cases}$ *piecewise continuous on [0, 2]? Is $f(t) =$*

$\begin{cases} t, & 0 \le t < 1 \\ \dfrac{1}{(t - 1)^2}, & 1 \le t \end{cases}$ *piecewise continuous on $[0, +\infty)$?*

Theorem 7.2 Sufficient Condition for Existence of $\mathcal{L}\{f(t)\}$

Suppose that $f(t)$ is a piecewise continuous function on the interval $[0, \infty)$ and that it is of exponential order b for $t > T$. Then, $\mathcal{L}\{f(t)\}$ exists for $s > b$.

PROOF OF THEOREM 7.2

We need to show that the integral $\int_0^\infty e^{-st} f(t) \, dt$ converges for $s > b$, assuming that $f(t)$

is a piecewise continuous function on the interval $[0, \infty)$ and that it is of exponential order b for $t > T$. First, we write the integral as

$$\int_0^\infty e^{-st}f(t)\, dt = \int_0^T e^{-st}f(t)\, dt + \int_T^\infty e^{-st}f(t)\, dt$$

where T is selected so that $|f(t)| \le Ce^{bt}$ for the constants b and C, $C > 0$.

Notice that because $f(t)$ is a piecewise continuous function, so is $e^{-st}f(t)$. The first of these integrals, $\int_0^T e^{-st}f(t)\, dt$, exists because it can be written as the sum of integrals over which $e^{-st}f(t)$ is continuous. The fact that $e^{-st}f(t)$ is piecewise continuous on $[T, \infty)$ is also used to show that the second integral, $\int_T^\infty e^{-st}f(t)\, dt$, converges. Because there are constants C and b such that $|f(t)| \le Ce^{bt}$, we have

$$\left| \int_T^\infty e^{-st}f(t)\, dt \right| \le \int_T^\infty |e^{-st}f(t)|\, dt \le C \int_T^\infty e^{-st}e^{bt}\, dt = C\int_T^\infty e^{-(s-b)t}\, dt$$

$$= C \lim_{M\to\infty} \int_T^M e^{-(s-b)t}\, dt = C \lim_{M\to\infty}\left[-\frac{e^{-(s-b)t}}{s-b}\right]_{t=T}^{t=M}$$

$$= -\frac{C}{s-b} \lim_{M\to\infty} (e^{-(s-b)M} - e^{-(s-b)T}).$$

Then, if $s - b > 0$, $\lim_{M\to\infty} e^{-(s-b)M} = 0$, so

$$\left| \int_T^\infty e^{-st}f(t)\, dt \right| \le \frac{Ce^{-(s-b)T}}{s-b},\; s > b.$$

Because both of the integrals $\int_0^T e^{-st}f(t)\, dt$ and $\int_T^\infty e^{-st}f(t)\, dt$ exist, $\int_0^\infty e^{-st}f(t)\, dt$ also exists for $s > b$. • • •

Example 9

Find the Laplace transform of $f(t) = \begin{cases} -1, & 0 \le t \le 4 \\ 1, & t > 4 \end{cases}$.

Solution Because $f(t)$ is a piecewise continuous function on $[0, \infty)$ and of exponential order, $F(s) = \mathcal{L}\{f(t)\}$ exists. We use the definition and evaluate the improper integral using a sum of two integrals.

$$F(s) = \mathcal{L}\{f(t)\} = \int_0^\infty f(t)e^{-st}\, dt = \int_0^4 (-1)e^{-st}\, dt + \int_4^\infty e^{-st}\, dt$$

$$= \left[\frac{e^{-st}}{s}\right]_{t=0}^{t=4} + \lim_{M\to\infty}\left[-\frac{e^{-st}}{s}\right]_{t=4}^{t=M}$$

$$= \frac{1}{s}(e^{-4s} - 1) - \frac{1}{s}\lim_{M\to\infty}(e^{-Ms} - e^{-4s}).$$

If $s > 0$, $\lim\limits_{M \to \infty} e^{-Ms} = 0$, so

$$F(s) = \mathcal{L}\{f(t)\} = \frac{1}{s}(2e^{-4s} - 1),\ s > 0.$$

Notice that Theorem 7.2 gives a sufficient condition and not a necessary condition. In other words, there are functions such as $f(t) = t^{-1/2}$ that do not satisfy the hypotheses of the previous theorem for which the Laplace transform can be found. (See Exercises 30 and 31.)

IN TOUCH WITH TECHNOLOGY

Use a computer algebra system to assist in calculating the Laplace transform of the following functions. Select particular constant values for a and k and then find a value of b so that $f(t)$ is of exponential order b. Graph $f(t)$ and e^{bt} to observe that each function is of exponential order.

1. $f(t) = t^{10}$

2. $f(t) = t^7 e^{-2t}$

3. $f(t) = e^{at} \sin kt$ (a, k constants)

4. $f(t) = e^{at} \cos kt$ (a, k constants)

5. $f(t) = t \sin kt$ (k constant)

6. $f(t) = t \cos kt$ (k constant)

7. Let $f_n(t) = \begin{cases} 1/n, & \text{if } 0 \le t \le n \\ 0, & \text{if } t > n \end{cases}$. **(a)** Graph $f_n(t)$ for $n = 10$, 1, $\frac{1}{10}$, and $\frac{1}{100}$. Is $f_n(t)$ of exponential order? Why? **(b)** Evaluate $\int_0^\infty f_n(t)\, dt$ and $\lim_{n \to 0^+} \int_0^\infty f_n(t)\, dt$. **(c)** Describe $\lim_{n \to 0^+} f_n(t)$. Is this limit a function? **(d)** Calculate $F_n(s) = \mathcal{L}\{f_n(t)\}$ and then $\lim_{n \to 0^+} F_n(s)$. Are you surprised by the result? Explain. **(e)** Is every function $F(s)$ the Laplace transform of some function $f(t)$? Why?

8. Graphically determine if the Gamma function is of exponential order. Can you prove your result analytically? Does the Gamma function have a Laplace transform? Explain.

EXERCISES 7.1

In Exercises 1–16, use the definition of the Laplace transform to compute the Laplace transform of each function.

1. $f(t) = 21t$

2. $f(t) = 7e^{-t}$

*3. $f(t) = 2e^t$

4. $f(t) = -8 \cos 3t$

5. $f(t) = 2 \sin 2t$

6. $f(t) = \begin{cases} 1 & \text{if } 0 \le t \le 2 \\ 0 & \text{if } t > 2 \end{cases}$

*7. $f(t) = \begin{cases} 0 & \text{if } 0 \le t \le 1 \\ 1 & \text{if } t > 1 \end{cases}$

8. $f(t) = \begin{cases} 1 - t & \text{if } 0 \le t \le 1 \\ 0 & \text{if } t > 1 \end{cases}$

9. $f(t) = \begin{cases} \cos t & \text{if } 0 \le t \le \pi/2 \\ 0 & \text{if } t > \pi/2 \end{cases}$

10. $f(t) = \begin{cases} \sin t & \text{if } 0 \le t \le 2\pi \\ 0 & \text{if } t > 2\pi \end{cases}$

*11. $f(t) = \begin{cases} 1 - t, & 0 < t < 3 \\ 0, & t \geq 3 \end{cases}$

12. $f(t) = \begin{cases} 3t + 2, & 0 < t < 5 \\ 0, & t \geq 5 \end{cases}$

13. $f(t) = \begin{cases} 1, & 0 \leq t < 10, \\ -1, & t \geq 10 \end{cases}$

14. $f(t) = \begin{cases} t, & 0 \leq t < 2, \\ 3, & t \geq 2 \end{cases}$

*15. $f(t) = \sin kt$

16. $f(t) = \cos kt$

In Exercises 17–28, use properties of the Laplace transform to compute the Laplace transform of each function.

17. $f(t) = 28e^t$

18. $f(t) = 15e^{-5t}$

*19. $f(t) = -18e^{3t}$

20. $f(t) = \frac{1}{4}(1 - \cos 4t)$

21. $f(t) = \frac{1}{8}(2t - \sin 2t)$

22. $f(t) = -\sin 2t$

*23. $f(t) = \cos 5t$

24. $f(t) = 1 + e^{-2t}$

25. $f(t) = 1 + \cos 2t$

26. $f(t) = e^t - \sin 5t$

*27. $f(t) = 1 + \sin 5t$

28. $f(t) = \frac{1}{5}(\cos 2t - \cos 3t)$

29. Use the identities $\sin(A + B) = \sin A \cos B + \cos A \sin B$ and $\cos(A + B) = \cos A \cos B - \sin A \sin B$ to assist in finding the Laplace transform of the following functions.

(a) $f(t) = \sin(t + \pi/4)$ (b) $f(t) = \cos(t + \pi/4)$
(c) $f(t) = \cos(t + \pi/6)$ (d) $f(t) = \sin(t + \pi/6)$

30. The **Gamma function,** $\Gamma(x)$, is defined by $\Gamma(x) = \int_0^\infty t^{x-1}e^{-t}\, dt$, $x > 0$. Show that $\mathcal{L}\{t^\alpha\} = \dfrac{\Gamma(\alpha + 1)}{s^{\alpha+1}}$, $\alpha > -1$.

*31. Use the result of Exercise 30 to find the Laplace transform of (a) $f(t) = t^{-1/2}$; (b) $f(t) = t^{1/2}$. (See Exercise 22 in Section 6.5.)

32. (a) Use Definition 7.2 to show that if $\lim_{t\to\infty} f(t)e^{-bt}$ exists and is finite, $f(t)$ is of exponential order and if $\lim_{t\to\infty} f(t)e^{-bt} = +\infty$ for every value b, $f(t)$ is not of exponential order. (b) Give an example of a function $f(t)$ of exponential order b for which $\lim_{t\to\infty} f(t)e^{-bt}$ does not exist.

33. Let $f(t)$ be a piecewise continuous function on the interval $[0, \infty)$ that is of exponential order b for $t > T$. Show that $h(t) = \int_0^t f(u)\, du$ is also of exponential order.

7.2 Properties of the Laplace Transform

As we saw in Section 7.1, the definition of the Laplace transform is not easy to apply in most cases. We now discuss several properties of the Laplace transform that let us make numerous transformations without having to use the definition. Most of the properties follow directly from our knowledge of integrals. We begin with the *shifting* (or *translation*) *property*.

Theorem 7.3 Shifting Property

If $\mathcal{L}\{f(t)\} = F(s)$ exists for $s > a$, then

$$\mathcal{L}\{e^{at}f(t)\} = F(s - a).$$

PROOF OF THEOREM 7.3

$$\mathcal{L}\{e^{at}f(t)\} = \int_0^\infty e^{-st}e^{at}f(t)\, dt = \int_0^\infty e^{-(s-a)t}f(t)\, dt = F(s - a),\ s > a.$$

Example 1

Find the Laplace transform of (a) $e^{-2t}\cos t$ and (b) $4te^{3t}$.

Solution   (a) In this case, $f(t) = \cos t$ and $a = -2$. Then $F(s) = \mathcal{L}\{\cos t\} = \dfrac{s}{s^2 + 1}$, so we replace each s with $(s - a) = (s + 2)$. Therefore,

$$\mathcal{L}\{e^{-2t}\cos t\} = \frac{s + 2}{(s + 2)^2 + 1} = \frac{s + 2}{s^2 + 4s + 5}.$$

(b) Using the linearity property, we have $\mathcal{L}\{4te^{3t}\} = 4\mathcal{L}\{te^{3t}\}$. Hence, $f(t) = t$ and $a = 3$, so we replace s in $F(s) = \mathcal{L}\{t\} = 1/s^2$ by $(s - a) = (s - 3)$. Therefore,

$$\mathcal{L}\{te^{3t}\} = \frac{1}{(s - 3)^2}, \text{ so}$$

$$\mathcal{L}\{4te^{3t}\} = \frac{4}{(s - 3)^2}.$$

In order to use the Laplace transform to solve initial-value problems, we will need to be able to compute the Laplace transform of the derivatives of an arbitrary function, provided the Laplace transform of such a function exists.

Theorem 7.4 Laplace Transform of the First Derivative

Suppose that $f(t)$ is continuous for all $t \geq 0$ and is of exponential order b for $t > T$. Also, suppose that $f'(t)$ is piecewise continuous on any closed subinterval of $[0, \infty)$. Then, for $s > b$

$$\mathcal{L}\{f'(t)\} = s\mathcal{L}\{f(t)\} - f(0).$$

PROOF OF THEOREM 7.4

Using integration by parts with $u = e^{-st}$ and $dv = f'(t)\,dt$, we have

$$\mathcal{L}\{f'(t)\} = \int_0^\infty e^{-st}f'(t)\,dt = \lim_{M\to\infty}\left(e^{-st}f(t)\Big|_{t=0}^{t=M}\right) + s\int_0^\infty e^{-st}f(t)\,dt$$

$$= -f(0) + s\mathcal{L}\{f(t)\} = s\mathcal{L}\{f(t)\} - f(0). \qquad \bullet\ \bullet\ \bullet$$

The following corollary of Theorem 7.4 is verified using induction.

Corollary 7.5 Laplace Transform of Higher Derivatives

More generally, if $f^{(i)}(t)$ is a continuous function of exponential order b on $[0, +\infty)$ for $i = 0, 1, \ldots, n - 1$ and $f^{(n)}(t)$ is piecewise continuous on $[0, +\infty)$, then for $s > b$

$$\mathcal{L}\{f^{(n)}(t)\} = s^n\mathcal{L}\{f(t)\} - s^{n-1}f(0) - \cdots - sf^{(n-2)}(0) - f^{(n-1)}(0).$$

🔩 *Use the formula in Corollary 7.5 to find an expression for $\mathscr{L}\{f''(t)\}$. Prove this result using the formula in Theorem 7.4 twice.*

We will use Theorem 7.4 and Corollary 7.5 primarily to solve initial-value problems. However, we can also use them to find the Laplace transform of the derivatives of a function if we know the Laplace transform of the function.

Example 2

Let $f(t) = t \sin t$. Compute $\mathscr{L}\{f(t)\}$, $\mathscr{L}\{f'(t)\}$, $\mathscr{L}\{f''(t)\}$, and $\mathscr{L}\{f'''(t)\}$.

Solution We have several options for finding $\mathscr{L}\{f(t)\}$, $\mathscr{L}\{f'(t)\}$, $\mathscr{L}\{f''(t)\}$, and $\mathscr{L}\{f'''(t)\}$. First, we evaluate $f'(t) = t \cos t + \sin t$, $f''(t) = 2 \cos t - t \sin t$, and $f'''(t) = -3 \sin t - t \cos t$. We then use a computer algebra system to find

$$\mathscr{L}\{f(t)\} = \frac{2s}{(s^2 + 1)^2}, \quad \mathscr{L}\{f'(t)\} = \frac{2s^2}{(s^2 + 1)^2}, \quad \mathscr{L}\{f''(t)\} = \frac{2s^3}{(s^2 + 1)^2},$$

$$\text{and} \quad \mathscr{L}\{f'''(t)\} = -\frac{2(2s^2 + 1)}{(s^2 + 1)^2}.$$

However, we can find $\mathscr{L}\{f'(t)\}$, $\mathscr{L}\{f''(t)\}$, and $\mathscr{L}\{f'''(t)\}$ using the formulas for finding the Laplace transform of a derivative. With $f(0) = 0, f'(0) = 0$, and $f''(0) = 2$, we find that

$$\mathscr{L}\{f'(t)\} = s\mathscr{L}\{f(t)\} - f(0) = s\left[\frac{2s}{(s^2 + 1)^2}\right] - 0 = \frac{2s^2}{(s^2 + 1)^2},$$

$$\mathscr{L}\{f''(t)\} = s^2\mathscr{L}\{f(t)\} - sf(0) - f'(0) = s^2\left[\frac{2s}{(s^2 + 1)^2}\right] - s(0) - 0 = \frac{2s^3}{(s^2 + 1)^2},$$

and

$$\mathscr{L}\{f'''(t)\} = s^3\mathscr{L}\{f(t)\} - s^2f(0) - sf'(0) - f''(0)$$

$$= s^3\left[\frac{2s}{(s^2 + 1)^2}\right] - s^2(0) - s(0) - 2 = -\frac{2(2s^2 + 1)}{(s^2 + 1)^2}.$$

The same results are obtained with either method.

🔩 *Use the graphs of $F(s) = \mathscr{L}\{f(t)\}$ and $\mathscr{L}\{f'(t)\}$ from Example 2 shown in Figure 7.3 to help you comment on the statement, "The derivative of the Laplace transform of f, $\dfrac{d}{ds}\mathscr{L}\{f(t)\}$, equals the Laplace transform of the derivative of f, $\mathscr{L}\left\{\dfrac{d}{dt}f(t)\right\}$." Is this statement true or false?*

Although Theorem 7.4 is useful in finding the Laplace transform of derivatives of a function f for which $\mathscr{L}\{f(t)\}$ is known, we will see in Section 7.3 that the theorem and its corollary are most useful in solving initial-value problems.

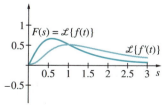

Figure 7.3

Theorem 7.6 Derivatives of the Laplace Transform

Suppose that $F(s) = \mathcal{L}\{f(t)\}$ where $f(t)$ is a piecewise continuous function on $[0, \infty)$ and of exponential order b. Then, for $s > b$,

$$\mathcal{L}\{t^n f(t)\} = (-1)^n \frac{d^n F}{ds^n}(s).$$

PROOF OF THEOREM 7.6

If $F(s) = \mathcal{L}\{f(t)\} = \int_0^\infty e^{-st} f(t)\, dt$, then

$$\frac{dF}{ds}(s) = \int_0^\infty \frac{d}{ds} e^{-st} f(t)\, dt = -\int_0^\infty t e^{-st} f(t)\, dt = -\mathcal{L}\{t f(t)\}.$$

Hence,

$$\mathcal{L}\{t f(t)\} = -\frac{dF}{ds}(s).$$

Similarly,

$$\frac{d^2 F}{ds^2}(s) = \int_0^\infty \frac{d^2}{ds^2} e^{-st} f(t)\, dt = \int_0^\infty t^2 e^{-st} f(t)\, dt = \mathcal{L}\{t^2 f(t)\}.$$

Following a similar procedure, we obtain the formula $\mathcal{L}\{t^n f(t)\} = (-1)^n \dfrac{d^n F}{ds^n}(s)$. • • •

Example 3

Find the Laplace transform of (a) $f(t) = t \cos 2t$, (b) $f(t) = t^2 e^{-3t}$.

Solution   (a) In this case, $n = 1$ and $F(s) = \mathcal{L}\{\cos 2t\} = \dfrac{s}{s^2 + 4}$. Then

$$\mathcal{L}\{t \cos 2t\} = (-1)\frac{d}{ds}\left[\frac{s}{s^2 + 4}\right] = -\frac{(s^2 + 4) - s(2s)}{(s^2 + 4)^2} = \frac{s^2 - 4}{(s^2 + 4)^2}.$$

(b) Because $n = 2$ and $F(s) = \mathcal{L}\{e^{-3t}\} = \dfrac{1}{s + 3}$, we have

$$\mathcal{L}\{t^2 e^{-3t}\} = (-1)^2 \frac{d^2}{ds^2}\left[\frac{1}{s + 3}\right] = \frac{d}{ds}\left[-\frac{1}{(s + 3)^2}\right] = \frac{2}{(s + 3)^3}.$$

Example 4

Find $\mathcal{L}\{t^n\}$.

Solution   Using Theorem 7.6 with $\mathcal{L}\{t^n\} = \mathcal{L}\{t^n \cdot 1\}$, we have $f(t) = 1$, where $F(s) = \mathcal{L}\{1\} = 1/s$. Calculating the derivatives of F, we obtain

$$\frac{dF}{ds}(s) = -\frac{1}{s^2}$$

$$\frac{d^2F}{ds^2}(s) = \frac{2}{s^3}$$

$$\frac{d^3F}{ds^3}(s) = -\frac{3 \cdot 2}{s^4}$$

$$\vdots$$

$$\frac{d^nF}{ds^n}(s) = (-1)^n \frac{n!}{s^{n+1}}.$$

Therefore,

$$\mathcal{L}\{t^n\} = \mathcal{L}\{t^n \cdot 1\} = (-1)^n(-1)^n \frac{n!}{s^{n+1}} = (-1)^{2n}\frac{n!}{s^{n+1}} = \frac{n!}{s^{n+1}}.$$

Example 5

Calculate (a) $\mathcal{L}\{t^5\}$, (b) $\mathcal{L}\{8t^3 - 5t + 10\}$, and (c) $\mathcal{L}\{t^4 e^{-5t}\}$.

Solution   (a) Because $n = 5$, $\mathcal{L}\{t^5\} = \frac{5!}{s^{5+1}} = \frac{120}{s^6}$.

(b) Using linearity, we have

$$\mathcal{L}\{8t^3 - 5t + 10\} = 8\mathcal{L}\{t^3\} - 5\mathcal{L}\{t\} + 10\mathcal{L}\{1\} = 8\frac{3!}{s^4} - 5\frac{1}{s^2} + 10\frac{1}{s}$$

$$= \frac{48}{s^4} - \frac{5}{s^2} + \frac{10}{s}.$$

(c) In this case, we use the shifting property with $f(t) = t^4$ and $a = -5$. Then, $F(s) = \mathcal{L}\{t^4\} = \frac{4!}{s^5}$. Replacing s with $(s - a) = (s + 5)$ yields $\mathcal{L}\{t^4 e^{-5t}\} = F(s + 5) = \frac{4!}{(s + 5)^5}$.

 Is the same result for $\mathcal{L}\{t^4 e^{-5t}\} = \frac{4!}{(s + 5)^5}$ *obtained by differentiating* $\mathcal{L}\{e^{-5t}\} = \frac{1}{s + 5}$ *four times?*

Using the properties of the Laplace transform discussed here, we can compute the Laplace transform of a large number of frequently encountered functions. Table 7.1 lists the Laplace transform of several of these frequently encountered functions. A more comprehensive table is found on the inside cover of this text.

TABLE 7.1   Laplace Transforms of Frequently Encountered Functions

$f(t)$	$F(s) = \mathcal{L}\{f(t)\}$	$f(t)$	$F(s) = \mathcal{L}\{f(t)\}$
1	$\dfrac{1}{s}, \quad s > 0$	$t^n, \; n = 1, 2, \ldots$	$\dfrac{n!}{s^{n+1}}, \quad s > 0$
e^{at}	$\dfrac{1}{s-a}, \quad s > a$	$t^n e^{at}, \\ n = 1, 2, \ldots$	$\dfrac{n!}{(s-a)^{n+1}}$
$\sin kt$	$\dfrac{k}{s^2 + k^2}$	$e^{at} \sin kt$	$\dfrac{k}{(s-a)^2 + k^2}$
$\cos kt$	$\dfrac{s}{s^2 + k^2}$	$e^{at} \cos kt$	$\dfrac{s-a}{(s-a)^2 + k^2}$
$\sinh kt$	$\dfrac{k}{s^2 - k^2}$	$e^{at} \sinh kt$	$\dfrac{k}{(s-a)^2 - k^2}$
$\cosh kt$	$\dfrac{s}{s^2 - k^2}$	$e^{at} \cosh kt$	$\dfrac{s-a}{(s-a)^2 - k^2}$

IN TOUCH WITH TECHNOLOGY

Some computer algebra systems contain a special package that includes commands to compute the Laplace transform of many functions. Use such a package (or an integration device) to calculate the Laplace transform of each of the following functions. Compare the results obtained to those found by using the operational properties of the Laplace transform. (Note that technology can be used to find derivatives when applying these properties.)

1. $f(t) = t^5 e^{-4t}$ **2.** $f(t) = t^4 \cos 7t$
3. $f(t) = t^{10} \sin 3t$ **4.** $f(t) = e^{-t} \sin 5t$
5. $f(t) = e^{2t} \cos 3t$ **6.** $f(t) = t^3 e^{-t} \sin 5t$
7. $f(t) = t^5 e^{2t} \cos 3t$

8. (See Problem 3, In Touch with Technology, Section 6.3.) **(a)** Show that the solution of the initial-value problem
$$\begin{cases} y'' - ty = 0 \\ y(0) = \dfrac{1}{3^{2/3}\Gamma(2/3)}, y'(0) = -\dfrac{1}{3^{1/3}\Gamma(1/3)} \end{cases}$$
is $Ai(t)$. **(b)** Explain why $Ai(t)$ has a Laplace transform $Y(s) = \mathcal{L}\{Ai(t)\}$. **(c)** Compute the Laplace transform of each side of the equation $y'' - ty = 0$, apply the initial conditions $\quad y(0) = \dfrac{1}{3^{2/3}\Gamma(2/3)} \quad$ and $\quad y'(0) =$

$-\dfrac{1}{3^{1/3}\Gamma(1/3)}$, and solve the resulting first-order linear differential equation
$$Y' + s^2 Y = \frac{1}{3^{2/3}\Gamma(2/3)}s - \frac{1}{3^{1/3}\Gamma(1/3)} \text{ to obtain}$$

$$Y(s) = e^{-s^3/3}\int_0^s \left(\frac{1}{3^{2/3}\Gamma(2/3)}u - \frac{1}{3^{1/3}\Gamma(1/3)} \right)e^{u^3/3}\,du \\ + Ce^{-s^3/3}$$
or
$$Y(s) = Ce^{-s^3/3} + \frac{1}{3}e^{-s^3/3}\left[\frac{1}{\Gamma(1/3)}\gamma(1/3, -s^3/3) + \right. \\ \left. \frac{1}{\Gamma(2/3)}s\gamma(2/3, -s^3/3) \right],$$

where $\Gamma(x)$ denotes the Gamma function and $\gamma(a, x) = \int_0^x u^{a-1}e^{-u}\,du$ denotes the **alternative incomplete Gamma function.**

Note that we must find C so that $Y(s)$ is the Laplace transform of $Ai(t)$. **(d)** To find C, show that $Y(0) = C$ and by the definition of the Laplace transform $Y(0) = \int_0^\infty Ai(t)\,dt$. Approximate C to (at least) 5 decimal places. **(e)** What is the Laplace transform of $Ai(t)$?

EXERCISES 7.2

In Exercises 1–30, compute the Laplace transform of each function.

1. $f(t) = -16e^{2t}$

2. $f(t) = 18e^{5t}$

3. $f(t) = t^7$

4. $f(t) = -6t^4 e^{-7t}$

*5. $f(t) = t^2 e^{-3t}$

6. $f(t) = t^3 e^{4t}$

7. $f(t) = t^5 e^{-4t}$

8. $f(t) = t^4 \cos t$

9. $f(t) = t \cos 3t$

10. $f(t) = 8 \cosh 4t$

11. $f(t) = 3t \sin t$

12. $f(t) = t \sin 7t$

*13. $f(t) = t \sinh 7t$

14. $f(t) = t^2 \sinh 6t$

15. $f(t) = e^{7t} \cosh t$

16. $f(t) = e^t \sinh t$

17. $f(t) = e^{-2t} \cos 4t$

18. $f(t) = -7e^{6t} \sin 2t$

19. $f(t) = e^{-2t} \sin 7t$

20. $f(t) = t \cos 2t$

*21. $f(t) = e^{5t} \cos 7t$

22. $f(t) = 20e^{-5t} \cosh 6t$

23. $f(t) = 2e^{-2t} \sinh t$

24. $f(t) = \sin 3t - 3t \cos 3t$

25. $f(t) = \sin 4t + 4t \cos 4t$

26. $f(t) = t^2 \cos 3t$

27. $f(t) = t^3 \sin 2t$

28. $f(t) = \dfrac{1}{54}(\sin 3t - 3t \cos 3t)$

*29. $f(t) = \dfrac{t}{6} \sin 3t$

30. $f(t) = \dfrac{1}{4}(\sin 2t + 2t \cos 2t)$

31. Use induction to verify Corollary 7.5.

32. Recall that the sum of the geometric series $a + ar + ar^2 + ar^3 + \cdots = \dfrac{a}{1-r}$ if $|r| < 1$. (The series diverges otherwise.) Also, recall that the Maclaurin series for $f(t) = e^t$ is $e^t = 1 + t + \dfrac{t^2}{2!} + \dfrac{t^3}{3!} + \cdots + \dfrac{t^n}{n!} + \cdots$. Take the Laplace transform of each term in the series and show that the sum of the resulting series converges to $\mathcal{L}\{e^t\}$.

33. Use the Maclaurin series $\sin t = \displaystyle\sum_{n=0}^{\infty} \dfrac{(-1)^n t^{2n+1}}{(2n+1)!}$ to verify that $\mathcal{L}\{\sin t\} = \dfrac{1}{s^2+1}$.

34. Use the Maclaurin series $\cos t = \displaystyle\sum_{n=0}^{\infty} \dfrac{(-1)^n t^{2n}}{(2n)!}$ to verify that $\mathcal{L}\{\cos t\} = \dfrac{s}{s^2+1}$.

35. Find $\mathcal{L}\{\cos^2 kt\}$.

36. Use the shifting property to verify that

(a) $\mathcal{L}\{e^{at} \sin kt\} = \dfrac{k}{(s-a)^2 + k^2}$;

(b) $\mathcal{L}\{e^{at} \cos kt\} = \dfrac{s-a}{(s-a)^2 + k^2}$.

37. Use the definitions $\cosh kt = \dfrac{e^{kt} + e^{-kt}}{2}$ and $\sinh kt = \dfrac{e^{kt} - e^{-kt}}{2}$ with the linearity property to verify that (a) $\mathcal{L}\{\cosh kt\} = \dfrac{s}{s^2 - k^2}$;

(b) $\mathcal{L}\{\sinh kt\} = \dfrac{k}{s^2 - k^2}$.

38. Use the shifting property to verify that

(a) $\mathcal{L}\{e^{at} \cosh kt\} = \dfrac{s-a}{(s-a)^2 - k^2}$;

(b) $\mathcal{L}\{e^{at} \sinh kt\} = \dfrac{k}{(s-a)^2 - k^2}$.

39. Figure 7.4(a) shows the graph of a function $f(t)$ while Figure 7.4(b) shows the graph of its Laplace transform $F(s) = \mathcal{L}\{f(t)\}$. Use Figure 7.4 to sketch the graphs of $\mathcal{L}\{e^{-2t}f(t)\}$ and $\mathcal{L}\{e^{3t}f(t)\}$.

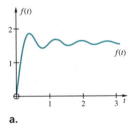

a.

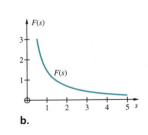

b.

Figure 7.4 (a)–(b)

7.3   The Inverse Laplace Transform

Definition of the Inverse Laplace Transform • Linear Factors (Nonrepeated) • Repeated Linear Factors • Irreducible Quadratic Factors • Laplace Transform of an Integral

Definition of the Inverse Laplace Transform

In Sections 7.1 and 7.2 we were concerned with finding the Laplace transform of a given function either through the use of the definition of the Laplace transform or with one of the properties of the Laplace transform. In this section we reverse this process: given a function $F(s)$ we want to find a function $f(t)$ such that $\mathcal{L}\{f(t)\} = F(s)$, if possible.

Definition 7.5 Inverse Laplace Transform

> The **inverse Laplace transform** of the function $F(s)$ is the function $f(t)$, if such a function exists, that satisfies $\mathcal{L}\{f(t)\} = F(s)$. We denote the inverse Laplace transform of $F(s)$ with
>
> $$f(t) = \mathcal{L}^{-1}\{F(s)\}.$$

The table of Laplace transforms given in Table 7.1 of Section 7.2 and the more comprehensive table on the inside cover of this text are useful in finding the inverse Laplace transform of a given function. In using these tables, we look for $F(s)$ in the right-hand column to find $f(t) = \mathcal{L}^{-1}\{F(s)\}$ in the corresponding left-hand column.

Example 1   Find the inverse Laplace transform of (a) $F(s) = 1/(s - 6)$; (b) $F(s) = 2/(s^2 + 4)$; (c) $F(s) = 6/s^4$; (d) $F(s) = 6/(s + 2)^4$.

Solution   (a) Because $\mathcal{L}\{e^{6t}\} = \dfrac{1}{s - 6}$, $\mathcal{L}^{-1}\left\{\dfrac{1}{s - 6}\right\} = e^{6t}$. (b) Note that $\mathcal{L}\{\sin 2t\} = \dfrac{2}{s^2 + 2^2} = \dfrac{2}{s^2 + 4}$, so $\mathcal{L}^{-1}\left\{\dfrac{2}{s^2 + 4}\right\} = \sin 2t$. (c) Because $\mathcal{L}\{t^3\} = \dfrac{3!}{s^4} = \dfrac{6}{s^4}$, $\mathcal{L}^{-1}\left\{\dfrac{6}{s^4}\right\} = t^3$. (d) Notice that $F(s) = \dfrac{6}{(s + 2)^4}$ is obtained from $F(s) = \dfrac{6}{s^4}$ by substituting $(s + 2)$ for s. Therefore by the shifting property,

$$\mathcal{L}\{e^{-2t}t^3\} = \frac{6}{(s + 2)^4}, \text{ so } \mathcal{L}^{-1}\left\{\frac{6}{(s + 2)^4}\right\} = e^{-2t}t^3.$$

Find $\mathcal{L}^{-1}\left\{\dfrac{1}{(s-4)}\right\}$ *and* $\mathcal{L}^{-1}\left\{\dfrac{s}{(s^2+4)}\right\}.$

Just as we used the linearity of the Laplace transform to find $\mathcal{L}\{f(t)\}$ for many functions f, the same is true for computing $\mathcal{L}^{-1}\{F(s)\}$.

Theorem 7.7 Linearity Property of the Inverse Laplace Transform

Suppose that $\mathcal{L}^{-1}\{F(s)\}$ and $\mathcal{L}^{-1}\{G(s)\}$ exist and are continuous on $[0, \infty)$ and that a and b are constants. Then,

$$\mathcal{L}^{-1}\{aF(s) + bG(s)\} = a\mathcal{L}^{-1}\{F(s)\} + b\mathcal{L}^{-1}\{G(s)\}.$$

Example 2 Find the inverse Laplace transform of (a) $F(s) = 1/s^3$; (b) $F(s) = -\dfrac{7}{s^2+16}$;

(c) $F(s) = \dfrac{5}{s} - \dfrac{2}{s-10}.$

Solution (a) $\mathcal{L}^{-1}\left\{\dfrac{1}{s^3}\right\} = \mathcal{L}^{-1}\left\{\dfrac{1}{2}\dfrac{2}{s^3}\right\} = \dfrac{1}{2}\mathcal{L}^{-1}\left\{\dfrac{2}{s^3}\right\} = \dfrac{1}{2}t^2.$

(b) $\mathcal{L}^{-1}\left\{-\dfrac{7}{s^2+16}\right\} = -7\mathcal{L}^{-1}\left\{\dfrac{1}{s^2+16}\right\} = -7\mathcal{L}^{-1}\left\{\dfrac{1}{4}\dfrac{4}{s^2+4^2}\right\}$

$= -\dfrac{7}{4}\mathcal{L}^{-1}\left\{\dfrac{4}{s^2+4^2}\right\} = -\dfrac{7}{4}\sin 4t.$

(c) $\mathcal{L}^{-1}\left\{\dfrac{5}{s} - \dfrac{2}{s-10}\right\} = 5\mathcal{L}^{-1}\left\{\dfrac{1}{s}\right\} - 2\mathcal{L}^{-1}\left\{\dfrac{1}{s-10}\right\} = 5 - 2e^{10t}.$

If the functions $F(s)$ are not in the forms presented in Table 7.1, we can make use of the linearity property to determine the inverse Laplace transform.

Example 3 Find the inverse Laplace transform of $F(s) = \dfrac{2s-9}{s^2+25}.$

Solution

$$\mathcal{L}^{-1}\left\{\dfrac{2s-9}{s^2+25}\right\} = 2\mathcal{L}^{-1}\left\{\dfrac{s}{s^2+25}\right\} - 9\mathcal{L}^{-1}\left\{\dfrac{1}{s^2+25}\right\}$$

$$= 2\mathcal{L}^{-1}\left\{\dfrac{s}{s^2+25}\right\} - \dfrac{9}{5}\mathcal{L}^{-1}\left\{\dfrac{5}{s^2+25}\right\}$$

$$= 2\cos 5t - \dfrac{9}{5}\sin 5t.$$

Find $\mathcal{L}^{-1}\left\{\dfrac{3s + 1}{s^2 + 16}\right\}.$

Sometimes we must complete the square in the denominator of $F(s)$ before finding $\mathcal{L}^{-1}\{F(s)\}.$

Example 4   Determine $\mathcal{L}^{-1}\left\{\dfrac{s}{s^2 + 2s + 5}\right\}.$

Solution   Notice that many of the forms of $F(s)$ in Table 7.1 involve a term of the form $s^2 + k^2$ in the denominator. Through shifting, this term is replaced by $(s - a)^2 + k^2$. We obtain a term of this form in the denominator by completing the square. This yields

$$\frac{s}{s^2 + 2s + 5} = \frac{s}{(s^2 + 2s + 1) + 4} = \frac{s}{(s + 1)^2 + 4}.$$

Because the variable s appears in the numerator, we must write it in the form $(s + 1)$ to find the inverse Laplace transform. Doing so, we find that

$$\frac{s}{s^2 + 2s + 5} = \frac{s}{(s + 1)^2 + 4} = \frac{(s + 1) - 1}{(s + 1)^2 + 4}.$$

Therefore,

$$\mathcal{L}^{-1}\left\{\frac{s}{s^2 + 2s + 5}\right\} = \mathcal{L}^{-1}\left\{\frac{(s + 1) - 1}{(s + 1)^2 + 4}\right\}$$

$$= \mathcal{L}^{-1}\left\{\frac{(s + 1)}{(s + 1)^2 + 4}\right\} - \frac{1}{2}\mathcal{L}^{-1}\left\{\frac{2}{(s + 1)^2 + 4}\right\}$$

$$= e^{-t}\cos 2t - \frac{1}{2}e^{-t}\sin 2t.$$

  Find $\mathcal{L}^{-1}\left\{\dfrac{2s}{s^2 - 4s + 8}\right\}.$

In other cases *partial fractions* must be used to obtain terms for which the inverse Laplace transform can be found. Suppose that

$$F(s) = \frac{P(s)}{Q(s)}$$

where $P(s)$ and $Q(s)$ are polynomials of degree m and n, respectively. If $n > m$, the method of partial fractions can be used to expand $F(s)$. We investigate several situations that can be solved through partial fractions: linear factors (nonrepeated), repeated linear factors, and irreducible quadratic factors (nonrepeated and repeated).

Linear Factors (Nonrepeated)

If $Q(s)$ can be written as a product of distinct linear factors

$$Q(s) = (s - q_1)(s - q_2) \cdots (s - q_n),$$

$F(s)$ can be written as

$$F(s) = \frac{A_1}{s - q_1} + \frac{A_2}{s - q_2} + \cdots + \frac{A_n}{s - q_n}$$

where $A_1, A_2, \ldots, A_n$ are constants that must be determined.

Example 5

Find $\mathcal{L}^{-1}\left\{\dfrac{3s - 4}{s(s - 4)}\right\}$.

Solution In this case we have linear factors in the denominator, so we write $F(s)$ as

$$\frac{3s - 4}{s(s - 4)} = \frac{A}{s} + \frac{B}{s - 4}.$$

Multiplying both sides of the equation by the denominator $s(s - 4)$, we have

$$3s - 4 = A(s - 4) + Bs = (A + B)s - 4A.$$

Equating the coefficients of s as well as the constant terms, we obtain the system of equations

$$\begin{cases} A + B = 3 \\ -4A = -4 \end{cases}$$

which has solution $A = 1$ and $B = 2$. Therefore,

$$\frac{3s - 4}{s(s - 4)} = \frac{A}{s} + \frac{B}{s - 4} = \frac{1}{s} + \frac{2}{s - 4},$$

so

$$\mathcal{L}^{-1}\left\{\frac{3s - 4}{s(s - 4)}\right\} = \mathcal{L}^{-1}\left\{\frac{1}{s} + \frac{2}{s - 4}\right\} = 1 + 2e^{4t}.$$

Repeated Linear Factors

Suppose that $(s - q)$ is a factor of $Q(s)$ of multiplicity k. In the partial fraction expansion of $F(s)$ that corresponds to this factor, we have

$$\frac{A_1}{s - q} + \frac{A_2}{(s - q)^2} + \cdots + \frac{A_k}{(s - q)^k},$$

where $A_1, A_2, \ldots, A_k$ are constants that must be found.

Calculate $\mathcal{L}^{-1}\left\{\dfrac{5s^2 + 20s + 6}{s^3 + 2s^2 + s}\right\}$.

Solution   Factoring the denominator, we have

$$\frac{5s^2 + 20s + 6}{s^3 + 2s^2 + s} = \frac{5s^2 + 20s + 6}{s(s^2 + 2s + 1)} = \frac{5s^2 + 20s + 6}{s(s + 1)^2}.$$

We can write this expression as

$$\frac{5s^2 + 20s + 6}{s(s + 1)^2} = \frac{A}{s} + \frac{B}{s + 1} + \frac{C}{(s + 1)^2}.$$

Multiplying both sides of the equation by the denominator $s(s + 1)^2$ yields

$$5s^2 + 20s + 6 = A(s + 1)^2 + Bs(s + 1) + Cs = (A + B)s^2 + (2A + B + C)s + A.$$

Equating like coefficients, we obtain the system

$$\begin{cases} A + B = 5 \\ 2A + B + C = 20 \\ A = 6 \end{cases}$$

which has solution $A = 6$, $B = -1$, and $C = 9$, so

$$\frac{5s^2 + 20s + 6}{s(s + 1)^2} = \frac{A}{s} + \frac{B}{s + 1} + \frac{C}{(s + 1)^2} = \frac{6}{s} - \frac{1}{s + 1} + \frac{9}{(s + 1)^2}.$$

Therefore,

$$\mathcal{L}^{-1}\left\{\frac{5s^2 + 20s + 6}{s^3 + 2s^2 + s}\right\} = \mathcal{L}^{-1}\left\{\frac{6}{s} - \frac{1}{s + 1} + \frac{9}{(s + 1)^2}\right\}$$

$$= 6\mathcal{L}^{-1}\left\{\frac{1}{s}\right\} - \mathcal{L}^{-1}\left\{\frac{1}{s + 1}\right\} + 9\mathcal{L}^{-1}\left\{\frac{1}{(s + 1)^2}\right\}$$

$$= 6 - e^{-t} + 9te^{-t}.$$

Irreducible Quadratic Factors

Suppose that $(s - a)^2 + b^2$ is a factor of $Q(s)$ that cannot be reduced to linear factors. If this irreducible quadratic is a factor of $Q(s)$ of multiplicity k, the partial fraction expansion of $F(s)$ corresponding to $(s - a)^2 + b^2$ is

$$\frac{A_1 s + B_1}{(s - a)^2 + b^2} + \frac{A_2 s + B_2}{[(s - a)^2 + b^2]^2} + \cdots + \frac{A_k s + B_k}{[(s - a)^2 + b^2]^k}.$$

Example 7

Find $\mathcal{L}^{-1}\left\{\dfrac{2s^3 - 4s - 8}{(s^2 - s)(s^2 + 4)}\right\}$.

Solution   Factoring the denominator, we obtain

$$\frac{2s^3 - 4s - 8}{(s^2 - s)(s^2 + 4)} = \frac{2s^3 - 4s - 8}{s(s - 1)(s^2 + 4)},$$

so we have the distinct linear factors s and $s - 1$ as well as the irreducible quadratic factor $s^2 + 4$. The partial fraction expansion is given by

$$\frac{2s^3 - 4s - 8}{s(s - 1)(s^2 + 4)} = \frac{A}{s} + \frac{B}{s - 1} + \frac{Cs + D}{s^2 + 4}.$$

Multiplying both sides of this equation by the denominator $s(s - 1)(s^2 + 4)$ yields

$$2s^3 - 4s - 8 = A(s - 1)(s^2 + 4) + Bs(s^2 + 4) + (Cs + D)s(s - 1)$$
$$= (A + B + C)s^3 + (-A + D - C)s^2 + (4A + 4B - D)s - 4A.$$

Equating coefficients yields the system of equations

$$\begin{cases} A + B + C = 2 \\ -A + D - C = 0 \\ 4A + 4B - D = -4 \\ \qquad\qquad -4A = -8 \end{cases}$$

with solution $A = 2$, $B = -2$, $C = 2$, and $D = 4$, so

$$\frac{2s^3 - 4s - 8}{s(s - 1)(s^2 + 4)} = \frac{A}{s} + \frac{B}{s - 1} + \frac{Cs + D}{s^2 + 4} = \frac{2}{s} - \frac{2}{s - 1} + \frac{2s + 4}{s^2 + 4}.$$

Thus,

$$\mathcal{L}^{-1}\left\{\frac{2s^3 - 4s - 8}{s(s - 1)(s^2 + 4)}\right\} = 2\mathcal{L}^{-1}\left\{\frac{1}{s}\right\} - 2\mathcal{L}^{-1}\left\{\frac{1}{s - 1}\right\} + 2\mathcal{L}^{-1}\left\{\frac{s}{s^2 + 4}\right\} +$$
$$2\mathcal{L}^{-1}\left\{\frac{2}{s^2 + 4}\right\}$$

$$= 2 - 2e^t + 2\cos 2t + 2\sin 2t.$$

Laplace Transform of an Integral

We have seen that the Laplace transform of the derivatives of a given function can be found from the Laplace transform of the function. Similarly, the Laplace transform of the integral of a given function can also be obtained from the Laplace transform of the function.

Theorem 7.8 Laplace Transform of an Integral

Suppose that $F(s) = \mathcal{L}\{f(t)\}$ where $f(t)$ is a piecewise continuous function on $[0, \infty)$ and of exponential order b. Then, for $s > b$,

$$\mathcal{L}\left\{\int_0^t f(\alpha)\, d\alpha\right\} = \frac{\mathcal{L}\{f(t)\}}{s}.$$

In other words,

$$\mathcal{L}^{-1}\left\{\frac{\mathcal{L}\{f(t)\}}{s}\right\} = \int_0^t f(\alpha)\, d\alpha.$$

Example 8

Compute $\mathcal{L}^{-1}\left\{\dfrac{1}{s(s+2)}\right\}$.

Solution   In this case, $\dfrac{1}{s(s+2)} = \dfrac{1/(s+2)}{s}$, so $\mathcal{L}\{f(t)\} = \dfrac{1}{s+2}$. Therefore,

$f(t) = \mathcal{L}^{-1}\left\{\dfrac{1}{s+2}\right\} = e^{-2t}$. Using Theorem 7.8,

$$\mathcal{L}^{-1}\left\{\frac{1}{s(s+2)}\right\} = \int_0^t e^{-2\alpha}\, d\alpha = \frac{1 - e^{-2t}}{2}.$$

Note that the same result is obtained through a partial fraction expansion of $\dfrac{1}{s(s+2)}$. Because $\dfrac{1}{s(s+2)} = \dfrac{1}{2s} - \dfrac{1}{2(s+2)}$, $\mathcal{L}^{-1}\left\{\dfrac{1}{s(s+2)}\right\} = $

$\mathcal{L}^{-1}\left\{\dfrac{1}{2s} - \dfrac{1}{2(s+2)}\right\} = \dfrac{1}{2} - \dfrac{1}{2}e^{-2t}$.

The following theorem is useful in showing that the inverse Laplace transform of a function $F(s)$ does not exist.

Theorem 7.9

Suppose that $f(t)$ is a piecewise continuous function on $[0, \infty)$ and of exponential order b. Then,

$$\lim_{s \to \infty} F(s) = \lim_{s \to \infty} \mathcal{L}\{f(t)\} = 0.$$

Example 9

Show that the inverse Laplace transform does not exist for (a) $F(s) = \dfrac{2s}{s-6}$;

(b) $F(s) = \dfrac{s^3}{s^2 + 16}$.

Solution   In both cases, we find $\lim_{s\to\infty} F(s)$. If this value is not zero, then $\mathscr{L}^{-1}\{F(s)\}$ does not exist. (a) $\lim_{s\to\infty} F(s) = \lim_{s\to\infty} \dfrac{2s}{s-6} = 2 \neq 0$, so $\mathscr{L}^{-1}\left\{\dfrac{2s}{s-6}\right\}$ does not exist. (b) $\lim_{s\to\infty} F(s) = \lim_{s\to\infty} \dfrac{s^3}{s^2+16} = \infty \neq 0$. Thus, $\mathscr{L}^{-1}\left\{\dfrac{s^3}{s^2+16}\right\}$ does not exist.

    Does $\mathscr{L}^{-1}\left\{\dfrac{s^2}{s^2+8s+16}\right\}$ *exist?*

Example 10

Suppose that $F(s) = \frac{1}{2}\tan^{-1}(3/s) + \frac{1}{2}\tan^{-1}(1/s)$. Does $\lim_{s\to\infty} F(s) = 0$ so that it may be possible to find $f(t)$ where $\mathscr{L}\{f(t)\} = F(s)$?

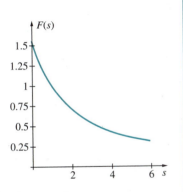

Figure 7.5

Solution   In Figure 7.5 we graph $F(s)$. From the graph we predict that $\lim_{s\to\infty} F(s) = 0$. Because $\lim_{s\to\infty} 1/s = 0$, $\lim_{s\to\infty} \tan^{-1}(3/s) = \lim_{s\to\infty} \tan^{-1}(1/s) = 0$, so we see that the limit can be evaluated directly as well. This indicates that it may be possible to find $\mathscr{L}^{-1}\{F(s)\}$ (although it is not guaranteed). With a computer algebra system, we find that

$$\mathscr{L}^{-1}\{F(s)\} = \frac{1}{2}\frac{\sin 3t}{t} + \frac{1}{2}\frac{\sin t}{t}$$

while a table of Laplace transforms shows that

$$\mathscr{L}^{-1}\{F(s)\} = \frac{\sin t \cos 2t}{t}.$$

Are these two results the same? (See Exercise 40.)

IN TOUCH WITH TECHNOLOGY

Many computer software systems contain a command for determining the partial fraction expansion of a rational function. Use this device to find the inverse Laplace transform of each of the following expressions.

1. $\dfrac{s^5 + 6s^4 + 73s^2 - 36s + 52}{s^2(s-2)^2(s^2+4s+13)}$

2. $\dfrac{1}{s^n(s^2+16)}$ for $n = 0, 1, 2, 3, 4, 5,$ and 6

3. $\dfrac{s^3 + 4s - 1}{(s^2+2s+5)(s^2-9)}$

4. $\dfrac{s^2+1}{(s^2+25)^2(s^2+6s+10)(s^2+s+1)}$

5. $\dfrac{s^3}{(s^2+25)^4}$

EXERCISES 7.3

In Exercises 1–39, compute the inverse Laplace transform of the given function.

1. $\dfrac{1}{s^2}$

2. $\dfrac{1}{s^8}$

*3. $\dfrac{1}{(s-2)^2}$

4. $\dfrac{1}{(s-5)^6}$

5. $\dfrac{1}{(s+6)^3}$

6. $\dfrac{1}{s+5}$

7. $\dfrac{4}{(s-3)^2}$

8. $\dfrac{1}{s^2+9}$

*9. $\dfrac{s}{s^2+16}$

10. $\dfrac{1}{s^2+15s+56}$

11. $\dfrac{s}{s^2-5s-14}$

12. $\dfrac{1}{s^2+12s+61}$

*13. $\dfrac{s-2}{s^2-4s}$

14. $\dfrac{s+3}{s^2+6s+5}$

15. $\dfrac{s-7}{s^2-14s+48}$

16. $\dfrac{1}{s^2-12s+35}$

*17. $\dfrac{1}{s^2-4s-12}$

18. $\dfrac{1}{s^2+2s-24}$

19. $\dfrac{1}{s^2-2s+2}$

20. $\dfrac{1}{s^2+12s+37}$

*21. $\dfrac{s-1}{s^2-2s+50}$

22. $\dfrac{s+1}{s^2+2s+37}$

23. $\dfrac{s+2}{s^2+4s+8}$

24. $\dfrac{1}{s(s^2+3)}$

*25. $\dfrac{s-7}{s^2-14s+50}$

26. $\dfrac{s}{(s^2+49)^2}$

27. $\dfrac{1}{s^2(s^2+4)}$

28. $\dfrac{s^2}{(s^2+36)^2}$

*29. $\dfrac{4-s^2}{(s^2+4)^2}$

30. $\dfrac{3s^2-1}{(s^2+1)^3}$

31. $\dfrac{s^3-75s}{(s^2+25)^3}$

32. $\dfrac{s}{(s^2-49)^2}$

*33. $\dfrac{s^3+3s}{(s^2-1)^3}$

34. $\dfrac{s^3+36s}{(s^2-36)^4}$

35. $\dfrac{s^3+147s}{(s^2-49)^3}$

36. $\dfrac{1}{s^4+13s^2+36}$

*37. $\dfrac{6-8s+s^2-2s^3}{24+10s^2+s^4}$

38. $\dfrac{4+27s+s^2+s^3}{36+13s^2+s^4}$

39. $\dfrac{-28+48s-12s^2+s^3}{380-196s+72s^2-14s^3+s^4}$

40. (a) Compute $f(t)=\mathcal{L}^{-1}\left\{\tan^{-1}\left(\dfrac{1}{s}\right)\right\}$ with

$$\mathcal{L}\{t^n f(t)\}=(-1)^n\frac{d^n}{ds^n}F(s)\text{ in the form }(n=1)$$

$$f(t)=-\frac{1}{t}\mathcal{L}^{-1}\left\{\frac{d}{ds}F(s)\right\}$$

$$=-\frac{1}{t}\mathcal{L}^{-1}\left\{\frac{d}{ds}\tan^{-1}\left(\frac{1}{s}\right)\right\}.$$

(b) Show that $\mathcal{L}^{-1}\left\{\tan^{-1}\left(\dfrac{a}{s}\right)\right\}=\dfrac{\sin at}{t}$.

(c) Show that

$$\mathcal{L}^{-1}\left\{\frac{1}{2}\tan^{-1}\left(\frac{a+b}{s}\right)+\frac{1}{2}\tan^{-1}\left(\frac{a-b}{s}\right)\right\}=$$
$$\frac{\sin at\cos bt}{t}.$$

41. (a) Use the linearity of the Laplace transform to compute $\mathcal{L}\{a\sin bt-b\sin at\}$ and $\mathcal{L}\{\cos bt-\cos at\}$.

(b) Use these results to find

$$\mathcal{L}^{-1}\left\{\frac{1}{(s^2+a^2)(s^2+b^2)}\right\}\text{ and}$$

$$\mathcal{L}^{-1}\left\{\frac{s}{(s^2+a^2)(s^2+b^2)}\right\}.\text{ (These formulas can be}$$

verified in a table of Laplace transforms or with a computer algebra system.)

42. Use the results of Exercise 41 to find

(a) $\mathcal{L}^{-1}\left\{\dfrac{10s}{(s^2+1)(s^2+16)}\right\}$;

(b) $\mathcal{L}^{-1}\left\{\dfrac{7}{(s^2+100)(s^2+1)}\right\}.$

43. Determine if the inverse Laplace transform of each of the following functions may exist by evaluating $\lim_{s\to\infty}F(s)$.

(a) $F(s)=\dfrac{s}{4-s}$

(b) $F(s)=\dfrac{3s}{s+1}$

(c) $F(s)=\dfrac{s^2}{4s+10}$

(d) $F(s)=\dfrac{5s^3}{s^2+1}$

7.4 Solving Initial-Value Problems with the Laplace Transform

Laplace transforms can be used to solve many initial-value problems. Typically, when we use Laplace transforms to solve the nth-order linear initial-value problem with constant coefficients

$$\begin{cases} a_n y^{(n)} + a_{n-1} y^{(n-1)} + \cdots + a_1 y' + a_0 y = f(x), \\ y(0) = y_0, \, y'(0) = y_0', \ldots, \, y^{(n-1)}(0) = y_0^{n-1}, \end{cases}$$

we use the following steps:

1. Compute the Laplace transform of each term in the differential equation;
2. Solve the resulting equation for $\mathcal{L}\{y(t)\} = Y(s)$; and
3. Determine $y(t)$ by computing the inverse Laplace transform of $Y(s)$.

The advantage of this method is that through the use of the property

$$\mathcal{L}\{f^{(n)}(t)\} = s^n \mathcal{L}\{f(t)\} - s^{n-1} f(0) - \cdots - s f^{(n-2)}(0) - f^{(n-1)}(0)$$

we change the *differential* equation to an *algebraic* equation that can be solved for $\mathcal{L}\{f(t)\}$.

Example 1

Solve the initial-value problem $y' - 4y = e^{4t}$, $y(0) = 0$.

Solution   We begin by taking the Laplace transform of both sides of the differential equation. Because $\mathcal{L}\{y'\} = sY(s) - y(0) = sY(s)$, we have

$$\mathcal{L}\{y' - 4y\} = \mathcal{L}\{e^{4t}\}$$

$$\mathcal{L}\{y'\} - 4\mathcal{L}\{y\} = \frac{1}{s - 4}$$

$$\underbrace{sY(s) - y(0)}_{\mathcal{L}\{y'\}} - \underbrace{4Y(s)}_{\mathcal{L}\{y\}} = \frac{1}{s - 4}$$

$$(s - 4)Y(s) = \frac{1}{s - 4}.$$

Solving for $Y(s)$ yields

$$Y(s) = \frac{1}{(s - 4)^2}.$$

By using the shifting property with $\mathcal{L}\{t\} = 1/s^2$, we have

$$y(t) = \mathcal{L}^{-1}\left\{\frac{1}{(s - 4)^2}\right\} = te^{4t}.$$

How is the solution changed if $y(0) = 1$?

As we can see from Example 1, Laplace transforms are most useful in solving nonhomogeneous equations. Problems in Chapter 4 for which the methods of undetermined coefficients or variation of parameters were difficult to apply may be more easily solved through the method of Laplace transforms.

Example 2

Use Laplace transforms to solve $y'' + 4y = e^{-t}\cos 2t$ subject to $y(0) = 0$ and $y'(0) = -1$.

Solution   Let $Y(s) = \mathcal{L}\{y(t)\}$. Then,

$$\mathcal{L}\{y'' + 4y\} = \mathcal{L}\{e^{-t}\cos 2t\}.$$

Because

$$\mathcal{L}\{y'' + 4y\} = \mathcal{L}\{y''\} + 4\mathcal{L}\{y\} = \underbrace{s^2 Y(s) - sy(0) - y'(0)}_{\mathcal{L}\{y''\}} + \underbrace{4Y(s)}_{\mathcal{L}\{y\}}$$

and

$$\mathcal{L}\{e^{-t}\cos 2t\} = \frac{s + 1}{(s + 1)^2 + 4},$$

the equation becomes

$$s^2 Y(s) - sy(0) - y'(0) + 4Y(s) = \frac{s + 1}{(s + 1)^2 + 4}.$$

Applying the initial conditions $y(0) = 0$ and $y'(0) = -1$ results in the equation

$$s^2 Y(s) + 1 + 4Y(s) = \frac{s + 1}{(s + 1)^2 + 4}$$

which we must solve for $Y(s)$:

$$s^2 Y(s) + 4Y(s) = \frac{-s^2 - s - 4}{(s + 1)^2 + 4} \quad \text{or} \quad Y(s) = \frac{-s^2 - s - 4}{(s^2 + 4)[(s + 1)^2 + 4]}.$$

With the help of a computer algebra system to perform the partial fraction decomposition, we find that

$$y(t) = \mathcal{L}^{-1}\{Y(s)\} = \mathcal{L}^{-1}\left\{\frac{-s^2 - s - 4}{(s^2 + 4)[(s + 1)^2 + 4]}\right\}$$

$$= \mathcal{L}^{-1}\left\{\frac{-8}{17(s^2 + 4)} - \frac{s}{17(s^2 + 4)} + \frac{s + 1}{17[(s + 1)^2 + 4]} - \frac{8}{17[(s + 1)^2 + 4]}\right\}$$

$$= -\frac{4}{17}\sin 2t - \frac{1}{17}\cos 2t + \frac{1}{17}e^{-t}\cos 2t - \frac{4}{17}e^{-t}\sin 2t.$$

A graph of $y(t)$ is shown in Figure 7.6(a). Notice that as $t \to \infty$, the forcing function $e^{-t}\cos 2t \to 0$. With this observation, should we expect $y(t)$ to approach the solu-

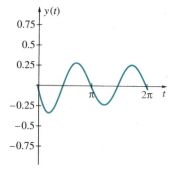

a.

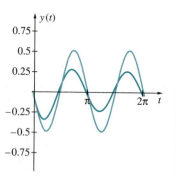

b.

Figure 7.6   (a)–(b)

tion of the homogeneous problem $\{y'' + 4y = 0,\ y(0) = 0,\ y'(0) = -1\}$? The solution of this problem is $y(t) = -\frac{1}{2}\sin 2t$, so the answer to the question is "no". We graph both functions in Figure 7.6(b) to illustrate this.

Which solution shown in Figure 7.6 has the greater amplitude? How does this relate to the original problems if we consider them as models of spring-mass systems?

Example 3

Use Laplace transforms to solve $y'' - 4y' + 4y = te^{2t}$ subject to $y(0) = 0$ and $y'(0) = 1$.

Solution Let $Y(s) = \mathcal{L}\{y(t)\}$. Then,

$$\mathcal{L}\{y'' - 4y' + 4y\} = \mathcal{L}\{te^{2t}\}$$

$$\mathcal{L}\{y''\} - 4\mathcal{L}\{y'\} + 4\mathcal{L}\{y\} = \frac{1}{(s-2)^2}$$

$$\underbrace{s^2Y(s) - sy(0) - y'(0)}_{\mathcal{L}\{y''\}} - \underbrace{4[sY(s) - y(0)]}_{\mathcal{L}\{y'\}} + \underbrace{4\,Y(s)}_{\mathcal{L}\{y\}} = \frac{1}{(s-2)^2}$$

$$(s^2 - 4s + 4)Y(s) - 1 = \frac{1}{(s-2)^2}$$

$$(s-2)^2Y(s) = \frac{1}{(s-2)^2} + 1.$$

Solving for $Y(s)$ yields

$$Y(s) = \frac{1}{(s-2)^4} + \frac{1}{(s-2)^2}.$$

Thus,

$$y(t) = \mathcal{L}^{-1}\{Y(s)\} = \mathcal{L}^{-1}\left\{\frac{1}{(s-2)^4} + \frac{1}{(s-2)^2}\right\} = \mathcal{L}^{-1}\left\{\frac{1}{(s-2)^4}\right\} + \mathcal{L}^{-1}\left\{\frac{1}{(s-2)^2}\right\}$$

$$= \frac{1}{3!}\mathcal{L}^{-1}\left\{\frac{3!}{(s-2)^4}\right\} + \mathcal{L}^{-1}\left\{\frac{1}{(s-2)^2}\right\} = \frac{1}{6}t^3e^{2t} + te^{2t}.$$

Would we expect a solution of this form if we had used the method of undetermined coefficients to solve the problem?

Some initial-value problems that involve linear differential equations with nonconstant coefficients can be solved with the method of Laplace transforms. However, Laplace transforms do not provide a general method for solving equations with nonconstant coefficients.

Example 4

Solve $y'' + 4ty' - 8y = 4$, $y(0) = 0$, $y'(0) = 0$.

Solution   First, we take the Laplace transform of both sides of the equation. Recall the property

$$\mathcal{L}\{t^n f(t)\} = (-1)^n \frac{d^n F}{ds^n}(s) = (-1)^n \frac{d^n}{ds^n} \mathcal{L}\{f(t)\}.$$

Hence, $\mathcal{L}\{ty'(t)\} = -\dfrac{d}{ds}[\mathcal{L}\{y'(t)\}]$. Then, because $\mathcal{L}\{y'(t)\} = sY(s) - y(0)$,

$$\mathcal{L}\{ty'(t)\} = -\frac{d}{ds}[\mathcal{L}\{y'(t)\}] = -\frac{d}{ds}[sY(s) - y(0)] = -sY'(s) - Y(s).$$

Thus,

$$\mathcal{L}\{y''(t)\} + 4\mathcal{L}\{ty'(t)\} - 8\mathcal{L}\{y(t)\} = 4\mathcal{L}\{1\}$$

$$s^2 Y(s) - sy(0) - y'(0) + 4[-sY'(s) - Y(s)] - 8Y(s) = \frac{4}{s}$$

$$(s^2 - 12)Y(s) - 4sY'(s) = \frac{4}{s}.$$

This is a first-order differential equation that can be solved for $Y(s)$ by using an integrating factor. Rewriting the equation as

$$Y'(s) + \left(\frac{3}{s} - \frac{s}{4}\right)Y(s) = -\frac{1}{s^2},$$

we see that an integrating factor is

$$\mu = e^{\int(3/s - s/4)\,ds} = e^{3\ln s - s^2/8} = s^3 e^{-s^2/8}.$$

Then,

$$\frac{d}{ds}[s^3 e^{-s^2/8} Y(s)] = s^3 e^{-s^2/8}\left(-\frac{1}{s^2}\right) = -se^{-s^2/8},$$

so integrating yields

$$s^3 e^{-s^2/8} Y(s) = 4e^{-s^2/8} + C$$

$$Y(s) = \frac{4}{s^3} + \frac{Ce^{s^2/8}}{s^3}.$$

Notice that in order for $y(t) = \mathcal{L}^{-1}\{Y(s)\}$ to exist, $\lim_{s\to\infty} Y(s) = 0$. This is not the case with $Y(s) = \dfrac{4}{s^3} + \dfrac{Ce^{s^2/8}}{s^3}$ unless $C = 0$. (Why?) Hence,

$$Y(s) = \frac{4}{s^3},$$

so the solution to the initial-value problem is

$$y(t) = \mathcal{L}^{-1}\left\{\frac{4}{s^3}\right\} = 2\mathcal{L}^{-1}\left\{\frac{2}{s^3}\right\} = 2t^2.$$

We can verify this solution. With $y'(t) = 4t$ and $y''(t) = 4$, $y'' + 4ty' - 8y = 4 + 4t(4t) - 8(2t^2) = 4$, as desired. The initial conditions are also satisfied because $y(0) = 0$ and $y'(0) = 0$.

Higher order initial-value problems can be solved with the method of Laplace transforms as well.

Example 5

Solve $y''' + y'' - 6y' = \sin t$, $y(0) = y'(0) = y''(0) = 0$.

Solution   Taking the Laplace transform of both sides of the equation, we find that

$$\mathcal{L}\{y'''\} + \mathcal{L}\{y''\} - 6\{y'\} = \mathcal{L}\{\sin t\}$$

$$\underbrace{s^3Y(s) - s^2y(0) - sy'(0) - y''(0)}_{\mathcal{L}\{y'''\}} + \underbrace{s^2Y(s) - sy(0) - y'(0)}_{\mathcal{L}\{y''\}}$$

$$- 6sY(s) + 6y(0) = \frac{1}{s^2 + 1}$$

$$(s^3 + s^2 - 6s)Y(s) = \frac{1}{s^2 + 1}.$$

Solving for $Y(s)$ we obtain

$$Y(s) = \frac{1}{(s^2 + 1)(s^3 + s^2 - 6s)} = \frac{1}{(s^2 + 1)s(s^2 + s - 6)} = \frac{1}{(s^2 + 1)s(s + 3)(s - 2)}.$$

Using the partial fraction expansion

$$Y(s) = \frac{1}{(s^2 + 1)s(s + 3)(s - 2)} = \frac{As + B}{s^2 + 1} + \frac{C}{s} + \frac{D}{s + 3} + \frac{E}{s - 2},$$

we find that the system of equations

$$\begin{cases} A + C + D + E = 0 \\ A + B + C - 2D + 3E = 0 \\ B - 2A - 5C + D + E = 0 \\ -6B + C - 2D + E = 0 \\ -6C = 1 \end{cases}$$

must be satisfied. Because the solution of this system is $A = \frac{7}{50}$, $B = -\frac{1}{50}$, $C = -\frac{1}{6}$, $D = \frac{1}{150}$, and $E = \frac{1}{50}$,

$$Y(s) = \frac{1}{50}\left(\frac{7s-1}{s^2+1}\right) - \frac{1}{6}\left(\frac{1}{s}\right) + \frac{1}{150}\left(\frac{1}{s+3}\right) + \frac{1}{50}\left(\frac{1}{s-2}\right).$$

Computing the inverse Laplace transform then yields

$$y(t) = \frac{1}{50}\mathcal{L}^{-1}\left\{\frac{7s-1}{s^2+1}\right\} - \frac{1}{6}\mathcal{L}^{-1}\left\{\frac{1}{s}\right\} + \frac{1}{150}\mathcal{L}^{-1}\left\{\frac{1}{s+3}\right\} + \frac{1}{50}\mathcal{L}^{-1}\left\{\frac{1}{s-2}\right\}$$

$$= \frac{7}{50}\cos t - \frac{1}{50}\sin t - \frac{1}{6} + \frac{1}{150}e^{-3t} + \frac{1}{50}e^{2t}.$$

IN TOUCH WITH TECHNOLOGY

Use the method of Laplace transforms to solve the following initial-value problems. In each case, use a computer algebra system to assist in computing the inverse Laplace transform either by using a package command for performing the task or by determining the partial fraction expansion. Graph the solution of each problem.

1. $\begin{cases} y'' + 2y' + 4y = t - e^{-t} \\ y(0) = 1, y'(0) = -1 \end{cases}$

2. $\begin{cases} y'' + 4y' + 13y = e^{-2t}\cos 3t + 1 \\ y(0) = 1, y'(0) = 1 \end{cases}$

3. $\begin{cases} y'' + 2y' + y = 2te^{-t} - e^{-t} \\ y(0) = 1, y'(0) = -1 \end{cases}$

4. In Section 6.5 we saw that **Bessel's equation** is the equation

$$x^2y'' + xy' + (x^2 - \mu^2)y = 0$$

where $\mu \geq 0$ is a constant, and a general solution is given by

$$y = c_1 J_\mu(x) + c_2 Y_\mu(x).$$

(a) Find α and β so that the solution of the initial-value problem $\begin{cases} x^2y'' + xy' + (x^2 - \mu^2)y = 0 \\ y(0) = \alpha, y'(0) = \beta \end{cases}$ is

$y = J_\mu(x)$.

(b) Compute the Laplace transform of each side of the equation $x^2y'' + xy' + (x^2 - \mu^2)y = 0$, substitute $\mathcal{L}\{y\} = Y(s)$, and solve the resulting second-order differential equation for $\mu = 1, 2, 3, 4, 5$.

(c) Use the results obtained in **(b)** to determine the Laplace transform of $J_\mu(t)$ for $\mu = 1, 2, 3, 4, 5$.

(d) Can you generalize the result you obtained in **(c)**?

EXERCISES 7.4

In Exercises 1–14, use Laplace transforms to solve the initial-value problem. In cases where the equation could be solved by other methods, describe these methods. Graph the solution on the indicated interval.

1. $y'' + 11y' + 24y = 0$, $y(0) = -1$, $y'(0) = 0$, $[-\frac{1}{2}, 1]$

2. $y'' + 8y' + 7y = 0$, $y(0) = 0$, $y'(0) = 1$, $[-\frac{1}{2}, 1]$

*3. $y'' + 3y' - 10y = 0$, $y(0) = -1$, $y'(0) = 1$, $[-\frac{1}{2}, \frac{3}{2}]$

4. $y'' - 13y' + 40y = 0$, $y(0) = 0$, $y'(0) = -2$, $[-\frac{1}{2}, \frac{1}{4}]$

5. $y'' + 4y = 0$, $y(0) = 2$, $y'(0) = 0$, $[0, 2\pi]$

6. $16y'' + 8y' + 65y = 0$, $y(0) = 0$, $y'(0) = 2$, $[0, 4\pi]$

*7. $y''' - 4y'' - 9y' + 36y = 0$, $y(0) = 1$, $y'(0) = 0$, $y''(0) = -1$, $[-1, 1]$

8. $y''' + 6y'' + 9y' + 4y = 0$, $y(0) = 0$, $y'(0) = -3$, $y''(0) = 2$, $[-1, 4]$

9. $y''' + y'' + 4y' + 4y = 0$, $y(0) = -5$, $y'(0) = 0$, $y''(0) = 0$, $[0, 3\pi]$

10. $y'' - y' - 2y = e^{-t}$, $y(0) = 2$, $y'(0) = 1$, $[-1, 1]$

*11. $y'' - 12y' + 40y = \sin(2t)$, $y(0) = 1$, $y'(0) = 0$, $[-\pi, \frac{\pi}{4}]$

12. $y'' - 2y' + 37y = e^t + \cos(3t)$, $y(0) = 0$, $y'(0) = 1$, $[-1, 1]$

*13. $y'' - y' - 12y = e^{2t} - \sin(t)$, $y(0) = -1$, $y'(0) = 1$, $[-1, 1]$

14. $y'' + 25y = 10\cos 5t$, $y(0) = 0$, $y'(0) = 0$, $[-\pi, 2\pi]$

In Exercises 15–18, solve the initial-value problem involving nonconstant coefficients.

15. $y'' + 3ty' - 6y = 3$, $y(0) = 0$, $y'(0) = 0$

16. $y'' + ty' - 2y = 4$, $y(0) = 0$, $y'(0) = 0$

*17. $y'' + 2ty' - 4y = 2$, $y(0) = 0$, $y'(0) = 0$

18. $y'' + 2ty' - 4y = 4$, $y(0) = 0$, $y'(0) = 0$

*19. Find the differential equation satisfied by $Y(s)$ for **(a)** $y'' - ty = 0$, $y(0) = 1$, $y'(0) = 0$ (**Airy's equation**) and for **(b)** $(1 - t^2)y'' - 2ty' + n(n + 1)y = 0$, $y(0) = 0$, $y'(0) = 1$ (**Legendre's equation**). What is the order of each differential equation involving Y? Is there a relationship between the power of the independent variable in the original equation and the order of the differential equation involving Y?

20. Show that application of the Laplace transform method to the initial-value problem (involving a Cauchy-Euler equation)

$$at^2 y'' + bty' + cy = 0, \; y(0) = \alpha, \; y'(0) = \beta$$

yields $as^2 Y'' + (4a - b)sY' + (2a - b + c)Y = 0$ where $Y(s) = \mathcal{L}\{y(t)\}$. Is the Laplace transform method worthwhile in the case of Cauchy-Euler equations?

21. Consider the initial-value problem (involving Bessel's equation of order 0)

$$ty'' + y' + ty = 0, \; y(0) = 1, \; y'(0) = 0$$

with solution $y = J_0(t)$, the Bessel function of order 0. Use Laplace transforms to show that $\mathcal{L}\{y(t)\} =$

$$\mathcal{L}\{J_0(t)\} = \frac{k}{\sqrt{s^2 + 1}} \text{ where } k \text{ is a constant. } (\textit{Hint:}$$

$$\mathcal{L}\{ty''\} = -\frac{d}{ds}[s^2 Y(s) - s] \text{ and}$$

$$\mathcal{L}\{ty\} = -\frac{d}{ds}[Y(s)].)$$

22. (See Exercise 21.) Use the binomial series to expand

$$\mathcal{L}\{J_0(t)\} = \frac{k}{\sqrt{s^2 + 1}} = \frac{k}{s}\left(1 + \frac{1}{s^2}\right)^{-1/2}. \text{ Take the}$$

inverse transform of each term in the expansion to

show that $J_0(t) = k\sum_{n=0}^{\infty} \frac{(-1)^n t^{2n}}{2^{2n}(n!)^2}$. Use the initial condition $J_0(0) = 1$ to show that $k = 1$. Therefore,

$$\mathcal{L}\{J_0(t)\} = \frac{1}{\sqrt{s^2 + 1}}.$$

7.5   Laplace Transforms of Several Important Functions

Piecewise Defined Functions: The Unit Step Function • Periodic Functions • Impulse Functions: The Delta Function

Piecewise Defined Functions: The Unit Step Function

An important function in modeling many physical situations is the unit step function $\mathcal{U}$, shown in Figure 7.7 and defined as follows.

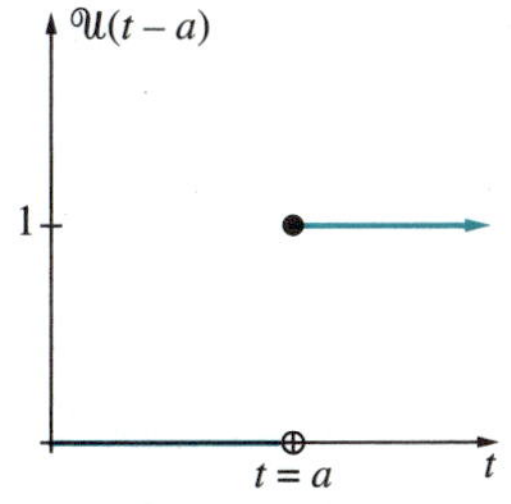

Figure 7.7

Definition 7.6 Unit Step Function

The **unit step function** $\mathcal{U}(t - a)$, where a is a given number, is defined by

$$\mathcal{U}(t - a) = \begin{cases} 0, & 0 \le t < a \\ 1, & t \ge a \end{cases}.$$

Example 1

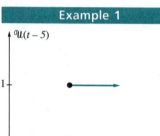

Figure 7.8

Graph (a) $\mathcal{U}(t - 5)$ and (b) $\mathcal{U}(t)$.

Solution   (a) In this case, $\mathcal{U}(t - 5) = \begin{cases} 0, & 0 \le t < 5 \\ 1, & t \ge 5 \end{cases}$, so the jump oc-

curs at $t = 5$. We graph $\mathcal{U}(t - 5)$ in Figure 7.8.
(b) Here $\mathcal{U}(t) = \mathcal{U}(t - 0)$, so $\mathcal{U}(t) = 1$ for $t \ge 0$. We graph this function in Figure 7.9.

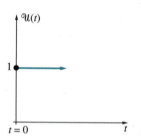

Figure 7.9

The unit step function is useful in defining functions that are piecewise continuous. Consider the function $f(t) = \mathcal{U}(t - a) - \mathcal{U}(t - b)$. If $0 \le t < a$, then $f(t) = 0 - 0 = 0$. If $a \le t < b$, then $f(t) = 1 - 0 = 1$. Finally, if $t \ge b$, then $f(t) = 1 - 1 = 0$.

Hence, $\mathcal{U}(t - a) - \mathcal{U}(t - b) = \begin{cases} 0, & 0 \le t < a \\ 1, & a \le t < b \\ 0, & t \ge b \end{cases}$, and we can define the function

$$g(t) = \begin{cases} 0, & 0 \le t < a \\ h(t), & a \le t < b \\ 0, & t \ge b \end{cases}$$

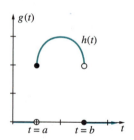

Figure 7.10

as $g(t) = h(t)[\mathcal{U}(t - a) - \mathcal{U}(t - b)]$, which is shown in Figure 7.10.
Similarly, a function like

$$f(t) = \begin{cases} g(t), & 0 \le t < a \\ h(t), & t \ge a \end{cases}$$

can be written as

$$f(t) = g(t)[\mathcal{U}(t - 0) - \mathcal{U}(t - a)] + h(t)\mathcal{U}(t - a)$$
$$= g(t)[1 - \mathcal{U}(t - a)] + h(t)\mathcal{U}(t - a).$$

The reason for writing piecewise continuous functions in terms of unit functions is because we encounter functions of this type in solving initial-value problems. Using the methods in Chapters 4 and 5, we solved the problem over each piece of the function. However, the method of Laplace transforms can be used to avoid those complicated calculations. We state the following theorem.

Theorem 7.10

Suppose that $F(s) = \mathcal{L}\{f(t)\}$ exists for $s > b \ge 0$. If a is a positive constant, then

$$\mathcal{L}\{f(t - a)\mathcal{U}(t - a)\} = e^{-as}F(s).$$

PROOF OF THEOREM 7.10

Using the definition of the Laplace transform, we obtain

$$\mathcal{L}\{f(t-a)\mathcal{U}(t-a)\} = \int_0^\infty e^{-st}f(t-a)\mathcal{U}(t-a)\,dt$$

$$= \int_0^a e^{-st}f(t-a)\underbrace{\mathcal{U}(t-a)}_{=\,0}\,dt + \int_a^\infty e^{-st}f(t-a)\underbrace{\mathcal{U}(t-a)}_{=\,1}\,dt$$

$$= \int_a^\infty e^{-st}f(t-a)\,dt$$

Changing variables with $u = t - a$ (where $du = dt$ and $t = u + a$) and changing the limits of integration, we have

$$\int_0^\infty e^{-s(u+a)}f(u)\,du = e^{-as}\int_0^\infty e^{-su}f(u)\,du = e^{-as}\mathcal{L}[f(t)] = e^{-as}F(s).$$

● ● ●

Example 2

Find (a) $\mathcal{L}\{\mathcal{U}(t-a)\}$, $a > 0$; (b) $\mathcal{L}\{(t-3)^5\mathcal{U}(t-3)\}$; (c) $\mathcal{L}\{\sin(t - \pi/6)\mathcal{U}(t - \pi/6)\}$.

Solution   (a) Because $\mathcal{L}\{\mathcal{U}(t-a)\} = \mathcal{L}\{1 \cdot \mathcal{U}(t-a)\}$, $f(t) = 1$.   Thus, $f(t-a) = 1$, and

$$\mathcal{L}\{\mathcal{U}(t-a)\} = \mathcal{L}\{1 \cdot \mathcal{U}(t-a)\} = e^{-as}\mathcal{L}\{1\} = e^{-as}\left(\frac{1}{s}\right) = \frac{e^{-as}}{s}.$$

(b) In this case $a = 3$ and $f(t) = t^5$. Thus,

$$\mathcal{L}\{(t-3)^5\mathcal{U}(t-3)\} = e^{-3s}\mathcal{L}\{t^5\} = e^{-3s}\frac{5!}{s^6} = \frac{120}{s^6}e^{-3s}.$$

(c) Here $a = \pi/6$ and $f(t) = \sin t$. Therefore,

$$\mathcal{L}\left\{\sin\left(t - \frac{\pi}{6}\right)\mathcal{U}\left(t - \frac{\pi}{6}\right)\right\} = e^{-\pi s/6}\mathcal{L}\{\sin t\} = e^{-\pi s/6}\frac{1}{s^2 + 1} = \frac{e^{-\pi s/6}}{s^2 + 1}.$$

   Find $\mathcal{L}\{\cos(t - \pi/6)\mathcal{U}(t - \pi/6)\}$.

In most cases, we must calculate

$$\mathcal{L}\{g(t)\mathcal{U}(t-a)\}$$

instead of $\mathcal{L}\{f(t-a)\mathcal{U}(t-a)\}$. To solve this problem, we let $g(t) = f(t - a)$, so $f(t) = g(t + a)$. Therefore,

$$\mathcal{L}\{g(t)\mathcal{U}(t-a)\} = e^{-as}\mathcal{L}\{g(t+a)\}.$$

Example 3

Calculate (a) $\mathcal{L}\{t^2\mathcal{U}(t - 1)\}$; (b) $\mathcal{L}\{\sin t\,\mathcal{U}(t - \pi)\}$.

Solution   (a) Because $g(t) = t^2$ and $a = 1$,

$$\mathcal{L}\{t^2\mathcal{U}(t - 1)\} = e^{-s}\mathcal{L}\{(t + 1)^2\} = e^{-s}\mathcal{L}\{t^2 + 2t + 1\} = e^{-s}\left(\frac{2}{s^3} + \frac{2}{s^2} + \frac{1}{s}\right).$$

(b) In this case, $g(t) = \sin t$ and $a = \pi$. Notice that $\sin(t + \pi) = \sin t \cos \pi + \cos t \sin \pi = -\sin t$. Thus,

$$\mathcal{L}\{\sin t\,\mathcal{U}(t - \pi)\} = e^{-\pi s}\mathcal{L}\{\sin(t + \pi)\} = e^{-\pi s}\mathcal{L}\{-\sin t\}$$

$$= -e^{-\pi s}\frac{1}{s^2 + 1} = -\frac{e^{-\pi s}}{s^2 + 1}.$$

 Find $\mathcal{L}\{\cos t\,\mathcal{U}(t - \pi)\}$.

Theorem 7.11 follows directly from Theorem 7.10.

Theorem 7.11

Suppose that $F(s) = \mathcal{L}\{f(t)\}$ exists for $s > b \geq 0$. If a is a positive constant and $f(t)$ is continuous on $[0, \infty)$, then

$$\mathcal{L}^{-1}\{e^{-as}F(s)\} = f(t - a)\mathcal{U}(t - a).$$

Example 4

Find (a) $\mathcal{L}^{-1}\left\{\dfrac{e^{-4s}}{s^3}\right\}$; (b) $\mathcal{L}^{-1}\left\{\dfrac{e^{-\pi s/2}}{s^2 + 16}\right\}$.

Solution   (a) If we write the expression e^{-4s}/s^3 in the form $e^{-as}F(s)$, we see that $a = 4$ and $F(s) = 1/s^3$. Hence, $f(t) = \mathcal{L}^{-1}\left\{\dfrac{1}{s^3}\right\} = \dfrac{1}{2}t^2$ and

$$\mathcal{L}^{-1}\left\{\frac{e^{-4s}}{s^3}\right\} = f(t - 4)\mathcal{U}(t - 4) = \frac{1}{2}(t - 4)^2\mathcal{U}(t - 4).$$

This function is shown in Figure 7.11(a). For what values of t is $f(t) > 0$?

(b) In this case, $a = \pi/2$ and $F(s) = \dfrac{1}{s^2 + 16}$. Then, $f(t) = \mathcal{L}^{-1}\left\{\dfrac{1}{s^2 + 16}\right\} = \dfrac{1}{4}\sin 4t$ and

$$\mathcal{L}^{-1}\left\{\frac{e^{-\pi s/2}}{s^2 + 16}\right\} = f\left(t - \frac{\pi}{2}\right)\mathcal{U}\left(t - \frac{\pi}{2}\right) = \frac{1}{4}\sin 4\left(t - \frac{\pi}{2}\right)\mathcal{U}\left(t - \frac{\pi}{2}\right)$$

$$= \frac{1}{4}\sin(4t - 2\pi)\mathcal{U}\left(t - \frac{\pi}{2}\right) = \frac{1}{4}\sin 4t\,\mathcal{U}\left(t - \frac{\pi}{2}\right).$$

We graph this function in Figure 7.11(b). For what values of t does $f(t) = 0$?

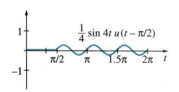

$\frac{1}{2}(t - 4)^2\,u(t - 4)$

a.

$\frac{1}{4}\sin 4t\,u(t - \pi/2)$

b.

Figure 7.11   (a)–(b)

With the unit step function, we can solve initial-value problems that involve piecewise continuous functions.

Example 5

Solve $y'' + 9y = \begin{cases} 1, & 0 \le 1 < \pi \\ 0, & t \ge \pi \end{cases}$ subject to $y(0) = y'(0) = 0$.

Solution In order to solve this initial-value problem, we must compute $\mathcal{L}\{f(t)\}$ where $f(t) = \begin{cases} 1, & 0 \le t < \pi \\ 0, & t \ge \pi \end{cases}$. Because this is a piecewise continuous function, we write it in terms of the unit step function as

$$f(t) = 1\{\mathcal{U}(t - 0) - \mathcal{U}(t - \pi)\} + 0[\mathcal{U}(t - \pi)] = \mathcal{U}(t) - \mathcal{U}(t - \pi).$$

Then,

$$\mathcal{L}\{f(t)\} = \mathcal{L}\{1 - \mathcal{U}(t - \pi)\} = \frac{1}{s} - \frac{e^{-\pi s}}{s}.$$

Hence,

$$\mathcal{L}\{y''\} + 9\mathcal{L}\{y\} = \mathcal{L}\{f(t)\}$$

$$s^2 Y(s) - sy(0) - y'(0) + 9Y(s) = \frac{1}{s} - \frac{e^{-\pi s}}{s}$$

$$(s^2 + 9)Y(s) = \frac{1}{s} - \frac{e^{-\pi s}}{s}$$

$$Y(s) = \frac{1}{s(s^2 + 9)} - \frac{e^{-\pi s}}{s(s^2 + 9)}.$$

Then,

$$y(t) = \mathcal{L}^{-1}\{Y(s)\} = \mathcal{L}^{-1}\left\{\frac{1}{s(s^2 + 9)}\right\} - \mathcal{L}^{-1}\left\{\frac{e^{-\pi s}}{s(s^2 + 9)}\right\}.$$

Consider $\mathcal{L}^{-1}\left\{\dfrac{e^{-\pi s}}{s(s^2 + 9)}\right\}$. In the form of $\mathcal{L}^{-1}\{e^{-as}F(s)\}$, $a = \pi$ and $F(s) = \dfrac{1}{s(s^2 + 9)}$. Now $f(t) = \mathcal{L}^{-1}\{F(s)\}$ can be found with either a partial fraction expansion or with the formula

$$f(t) = \mathcal{L}^{-1}\left\{\frac{1}{s(s^2 + 9)}\right\} = \int_0^t \mathcal{L}^{-1}\left\{\frac{1}{s^2 + 9}\right\} d\alpha = \int_0^t \frac{1}{3} \sin 3\alpha \, d\alpha$$

$$= -\frac{1}{3}\left[\frac{\cos 3\alpha}{3}\right]_0^t = \frac{1}{9} - \frac{1}{9}\cos 3t.$$

Then, with $\cos(3t - 3\pi) = \cos 3t \cos 3\pi + \sin 3t \sin 3\pi = -\cos 3t$, we have

$$\mathcal{L}^{-1}\left\{\frac{e^{-\pi s}}{s(s^2 + 9)}\right\} = \left[\frac{1}{9} - \frac{1}{9}\cos 3(t - \pi)\right]\mathcal{U}(t - \pi)$$

$$= \left[\frac{1}{9} - \frac{1}{9}\cos(3t - 3\pi)\right]\mathcal{U}(t - \pi) = \left[\frac{1}{9} + \frac{1}{9}\cos 3t\right]\mathcal{U}(t - \pi).$$

Combining these results yields the solution

$$y(t) = \mathscr{L}^{-1}\{Y(s)\} = \mathscr{L}^{-1}\left\{\frac{1}{s(s^2 + 9)}\right\} - \mathscr{L}^{-1}\left\{\frac{e^{-\pi s}}{s(s^2 + 9)}\right\}$$

$$= \frac{1}{9} - \frac{1}{9}\cos 3t - \left[\frac{1}{9} + \frac{1}{9}\cos 3t\right]\mathcal{U}(t - \pi).$$

Notice that we can rewrite this solution as the piecewise defined function

$$y(t) = \begin{cases} \dfrac{1}{9} - \dfrac{1}{9}\cos 3t, & 0 \le t < \pi \\[2mm] -\dfrac{2}{9}\cos 3t, & t \ge \pi \end{cases}$$

which is graphed in Figure 7.12. Can you determine where the behavior of $y(t)$ changes? Are the initial conditions satisfied?

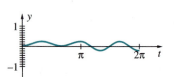

Figure 7.12

Periodic Functions

Another type of function that is encountered in many areas of applied mathematics is the *periodic* function.

Definition 7.7 Periodic Function

A function $f(t)$ is **periodic** if there is a positive number T such that

$$f(t + T) = f(t)$$

for all $t \ge 0$. The minimum value of T that satisfies this equation is called the **period** of $f(t)$.

The calculation of the Laplace transform of periodic functions is simplified through the use of the following theorem.

Theorem 7.12 Laplace Transform of Periodic Functions

Suppose that $f(t)$ is a periodic function of period T and that $f(t)$ is piecewise continuous on $[0, \infty)$. Then $\mathscr{L}\{f(t)\}$ exists for $s > 0$ and is determined with the definite integral

$$\mathscr{L}\{f(t)\} = \frac{1}{1 - e^{-sT}}\int_0^T e^{-st}f(t)\,dt.$$

PROOF OF THEOREM 7.12

We begin by writing $\mathcal{L}\{f(t)\} = \int_0^\infty e^{-st}f(t)\,dt$ as the sum

$$\mathcal{L}\{f(t)\} = \int_0^T e^{-st}f(t)\,dt + \int_T^\infty e^{-st}f(t)\,dt.$$

If we change the variable in the second integral to $u = t - T$ where $du = dt$, we obtain

$$\int_T^\infty e^{-st}f(t)\,dt = \int_0^\infty e^{-s(u+T)}\underbrace{f(u+T)}_{=f(u)}\,du = e^{-sT}\int_0^\infty e^{-su}f(u)\,du = e^{-sT}\mathcal{L}\{f(t)\}.$$

Hence,

$$\mathcal{L}\{f(t)\} = \int_0^T e^{-st}f(t)\,dt + \int_T^\infty e^{-st}f(t)\,dt = \int_0^T e^{-st}f(t)\,dt + e^{-sT}\mathcal{L}\{f(t)\}$$

which can be solved for $\mathcal{L}\{f(t)\}$ to yield

$$\mathcal{L}\{f(t)\} = \frac{1}{1-e^{-sT}}\int_0^T e^{-st}f(t)\,dt. \qquad \bullet\ \bullet\ \bullet$$

 Is the Laplace transform of a periodic function a periodic function?

Example 6

Find the Laplace transform of the periodic function $f(t) = t$, $0 \le t < 1$, $f(t + 1) = f(t)$, $t \ge 1$.

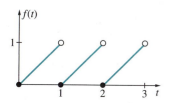

Figure 7.13

Solution   The period of f, which is graphed in Figure 7.13, is $T = 1$. Through integration by parts or a computer algebra system,

$$\mathcal{L}\{f(t)\} = \frac{1}{1-e^{-s}}\int_0^1 e^{-st}t\,dt = \frac{1}{1-e^{-s}}\left\{\left[-\frac{te^{-st}}{s}\right]_0^1 + \int_0^1 \frac{e^{-st}}{s}\,dt\right\}$$

$$= \frac{1}{1-e^{-s}}\left\{-\frac{e^{-s}}{s} - \left[\frac{e^{-st}}{s^2}\right]_0^1\right\} = \frac{1}{1-e^{-s}}\left[-\frac{e^{-s}}{s} + \frac{1-e^{-s}}{s^2}\right]$$

$$= \frac{1-(s+1)e^{-s}}{s^2(1-e^{-s})}.$$

Laplace transforms can be used more easily than other methods to solve initial-value problems with periodic forcing functions.

Example 7

Solve $y'' + y = f(t)$ subject to $y(0) = y'(0) = 0$ if $f(t) = \begin{cases} 2\sin t, & 0 \le t < \pi \\ 0, & \pi \le t < 2\pi \end{cases}$ and $f(t + 2\pi) = f(t)$. ($f(t)$ is known as the *half-wave rectification* of $2\sin t$.)

| Solution | We begin by finding $\mathcal{L}\{f(t)\}$. Because the period is $T = 2\pi$, we have |

$$\mathcal{L}\{f(t)\} = \frac{1}{1 - e^{-2\pi s}} \int_0^{2\pi} e^{-st}f(t)\, dt$$

$$= \frac{1}{1 - e^{-2\pi s}}\left[\int_0^{\pi} e^{-st} 2\sin t\, dt + \int_{\pi}^{2\pi} e^{-st}\cdot 0\, dt\right]$$

$$= \frac{2}{1 - e^{-2\pi s}} \int_0^{\pi} e^{-st}\sin t\, dt.$$

Using integration by parts, a table of integrals, or a computer algebra system yields

$$\mathcal{L}\{f(t)\} = \frac{2}{1 - e^{-2\pi s}}\left[\frac{e^{-st}(-s\sin t - \cos t)}{s^2 + 1}\right]_0^{\pi} = \frac{2}{1 - e^{-2\pi s}}\left[\frac{e^{-\pi s}}{s^2 + 1} + \frac{1}{s^2 + 1}\right]$$

$$= \frac{2(e^{-\pi s} + 1)}{(1 - e^{-2\pi s})(s^2 + 1)} = \frac{2(e^{-\pi s} + 1)}{(1 - e^{-\pi s})(1 + e^{-\pi s})(s^2 + 1)}$$

$$= \frac{2}{(1 - e^{-\pi s})(s^2 + 1)}.$$

Taking the Laplace transform of both sides of the equation and solving for $Y(s)$ gives us

$$\mathcal{L}\{y''\} + \mathcal{L}\{y\} = \mathcal{L}\{f(t)\}$$

$$s^2 Y(s) - sy(0) - y'(0) + Y(s) = \frac{2}{(1 - e^{-\pi s})(s^2 + 1)}$$

$$Y(s) = \frac{2}{(1 - e^{-\pi s})(s^2 + 1)^2}.$$

Recall from your work with the geometric series that if $|x| < 1$, then

$$\frac{1}{1 - x} = 1 + x + x^2 + x^3 + \cdots.$$

Because we do not know the inverse Laplace transform of $\dfrac{2}{(1 - e^{-\pi s})(s^2 + 1)}$, we must use a geometric series expansion of $\dfrac{1}{1 - e^{-\pi s}}$ to obtain terms for which we can calculate the inverse Laplace transform. This gives us

$$\frac{1}{1 - e^{-\pi s}} = 1 + e^{-\pi s} + e^{-2\pi s} + e^{-3\pi s} + \cdots,$$

so

$$Y(s) = (1 + e^{-\pi s} + e^{-2\pi s} + e^{-3\pi s} + \cdots)\frac{2}{(s^2 + 1)^2}$$

$$= 2\left[\frac{1}{(s^2 + 1)^2} + \frac{e^{-\pi s}}{(s^2 + 1)^2} + \frac{e^{-2\pi s}}{(s^2 + 1)^2} + \frac{e^{-3\pi s}}{(s^2 + 1)^2} + \cdots\right].$$

Then,

$$y(t) = 2\mathcal{L}^{-1}\left\{\frac{1}{(s^2+1)^2} + \frac{e^{-\pi s}}{(s^2+1)^2} + \frac{e^{-2\pi s}}{(s^2+1)^2} + \frac{e^{-3\pi s}}{(s^2+1)^2} + \cdots\right\}$$

$$= 2\mathcal{L}^{-1}\left\{\frac{1}{(s^2+1)^2}\right\} + 2\mathcal{L}^{-1}\left\{\frac{e^{-\pi s}}{(s^2+1)^2}\right\} + 2\mathcal{L}^{-1}\left\{\frac{e^{-2\pi s}}{(s^2+1)^2}\right\} +$$

$$2\mathcal{L}^{-1}\left\{\frac{e^{-3\pi s}}{(s^2+1)^2}\right\} + \cdots.$$

Notice that $\mathcal{L}^{-1}\left\{\dfrac{1}{(s^2+1)^2}\right\}$ is needed to find all of the other terms. Using a computer algebra system, we have

$$\mathcal{L}^{-1}\left\{\frac{1}{(s^2+1)^2}\right\} = \frac{1}{2}(\sin t - t\cos t).$$

Then,

$$y(t) = (\sin t - t\cos t) + [\sin(t-\pi) - (t-\pi)\cos(t-\pi)]\mathcal{U}(t-\pi) +$$

$$[\sin(t-2\pi) - (t-2\pi)\cos(t-2\pi)]\mathcal{U}(t-2\pi) +$$

$$[\sin(t-3\pi) - (t-3\pi)\cos(t-3\pi)]\mathcal{U}(t-3\pi) +$$

$$[\sin(t-4\pi) - (t-4\pi)\cos(t-4\pi)\mathcal{U}(t-4\pi) + \cdots$$

We can write y as the piecewise-defined function

$$y(t) = \begin{cases} \sin t - t\cos t, & 0 \le t < \pi \\ -\pi\cos t, & \pi \le t < 2\pi \\ \sin t - t\cos t + \pi\cos t, & 2\pi \le t < 3\pi \\ -2\pi\cos t, & 3\pi \le t < 4\pi \\ \vdots \end{cases}$$

which is graphed in Figure 7.14. Is y a smooth function? (That is, is y' continuous?) Is y periodic?

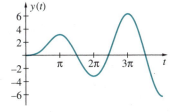

Figure 7.14

Graph the function $f(t) = \begin{cases} 2\sin t, & 0 \le t \le \pi \\ 0, & \pi \le t < 2\pi \end{cases}$ and $f(t+2\pi) = f(t)$

on the interval $[0, 6\pi]$.

Impulse Functions, The Delta Function

We now consider differential equations of the form $ax'' + bx' + cx = f(t)$ where $f(t)$ is "large" over the short interval centered at t_0, $t_0 - \alpha < t < t_0 + \alpha$, and zero otherwise. We define the **impulse** delivered by the function $f(t)$ as $I(t) = \int_{t_0-\alpha}^{t_0+\alpha} f(t)\,dt$, or because $f(t) = 0$ for t on $(-\infty, t_0 - \alpha) \cup (t_0 + \alpha, +\infty)$,

$$I(t) = \int_{-\infty}^{+\infty} f(t)\,dt.$$

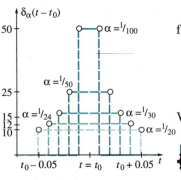

Figure 7.15

In order to better understand the impulse function, we let $f(t)$ be defined in the following manner:

$$f(t) = \delta_\alpha(t - t_0) = \begin{cases} \dfrac{1}{2\alpha}, & t_0 - \alpha < t < t_0 + \alpha \\ 0, & \text{otherwise} \end{cases}$$

We graph $\delta_\alpha(t - t_0)$ for several values of α in Figure 7.15.

 Graph $f(t) = \delta_{1/4}(t - 1)$. What is the maximum value of $f(t)$?

The impulse of $f(t) = \delta_\alpha(t - t_0)$ is given by

$$I(t) = \int_{t_0 - \alpha}^{t_0 + \alpha} f(t)\, dt = \int_{t_0 - \alpha}^{t_0 + \alpha} \frac{1}{2\alpha}\, dt = \frac{1}{2\alpha}((t_0 + \alpha) - (t_0 - \alpha)) = \frac{1}{2\alpha}(2\alpha) = 1.$$

Notice that the value of this integral does not depend on α as long as α is not zero. We now try to create the *idealized impulse function* by requiring that $\delta_\alpha(t - t_0)$ act on smaller and smaller intervals. From the integral calculation, we have

$$\lim_{\alpha \to 0} I(t) = 1.$$

We also note that

$$\lim_{\alpha \to 0} \delta_\alpha(t - t_0) = 0, \; t \neq t_0.$$

We use these properties to define the idealized unit impulse function as follows.

Definition 7.8 Unit Impulse Function

The **(idealized) unit impulse function** δ satisfies

$$\delta(t - t_0) = 0, \; t \neq t_0$$

$$\int_{-\infty}^{+\infty} \delta(t - t_0)\, dt = 1.$$

We now state the following useful theorem involving the unit impulse function.

Theorem 7.13

Suppose that $g(t)$ is a bounded and continuous function. Then,

$$\int_{-\infty}^{+\infty} \delta(t - t_0) g(t)\, dt = g(t_0).$$

The function $\delta(t - t_0)$ is known as the **Dirac delta function** and is useful in the definition of impulse forcing functions that arise in many areas of applied mathematics. The Laplace transform of $\delta(t - t_0)$ is found by using the function $\delta_\alpha(t - t_0)$ and L'Hopital's rule.

Theorem 7.14

For $t_0 \geq 0$,

$$\mathcal{L}\{\delta(t - t_0)\} = e^{-st_0}.$$

PROOF OF THEOREM 7.14

Because the Laplace transform is linear, we find $\mathcal{L}\{\delta(t - t_0)\}$ through the following calculations:

$$\mathcal{L}\{\delta(t - t_0)\} = \mathcal{L}\left\{\lim_{\alpha \to 0} \delta_\alpha(t - t_0)\right\} = \lim_{\alpha \to 0} \mathcal{L}\{\delta_\alpha(t - t_0)\}.$$

We can represent the delta function $\delta_\alpha(t - t_0)$ in terms of the unit step function as

$$\delta_\alpha(t - t_0) = \frac{1}{2\alpha}[\mathcal{U}(t - (t_0 - \alpha)) - \mathcal{U}(t - (t_0 + \alpha))].$$

Hence,

$$\mathcal{L}\{\delta_\alpha(t - t_0)\} = \mathcal{L}\left\{\frac{1}{2\alpha}[\mathcal{U}(t - (t_0 - \alpha)) - \mathcal{U}(t - (t_0 + \alpha))]\right\}$$

$$= \frac{1}{2\alpha}\left[\frac{e^{-s(t_0 - \alpha)}}{s} - \frac{e^{-s(t_0 + \alpha)}}{s}\right] = e^{-st_0}\left(\frac{e^{\alpha s} - e^{-\alpha s}}{2\alpha s}\right).$$

Therefore,

$$\mathcal{L}\{\delta(t - t_0)\} = \lim_{\alpha \to 0} L\{\delta_\alpha(t - t_0)\} = \lim_{\alpha \to 0} e^{-st_0}\left(\frac{e^{\alpha s} - e^{-\alpha s}}{2\alpha s}\right).$$

Because the limit is of the indeterminate form 0/0, we use L'Hopital's rule to find that

$$\mathcal{L}\{\delta(t - t_0)\} = \lim_{\alpha \to 0} e^{-st_0}\left(\frac{e^{\alpha s} - e^{-\alpha s}}{2\alpha s}\right) = \lim_{\alpha \to 0} e^{-st_0}\left(\frac{se^{\alpha s} + se^{-\alpha s}}{2s}\right) = e^{-st_0} \cdot 1 = e^{-st_0}.$$

• • •

Example 8

Find (a) $\mathcal{L}\{\delta(t - 1)\}$; (b) $\mathcal{L}\{\delta(t - \pi)\}$; (c) $\mathcal{L}\{\delta(t)\}$.

Solution (a) In this case $t_0 = 1$, so $\mathcal{L}\{\delta(t - 1)\} = e^{-s}$. (b) With $t_0 = \pi$, $\mathcal{L}\{\delta(t - \pi)\} = e^{-s\pi}$. (c) Because $t_0 = 0$, $\mathcal{L}\{\delta(t)\} = \mathcal{L}\{\delta(t - 0)\} = e^{-s(0)} = 1$.

Example 9

Solve $y'' + y = \delta(t - \pi)$ subject to $y(0) = y'(0) = 0$.

Solution   We solve this initial-value problem by taking the Laplace transform of both sides of the differential equation. This yields

$$\mathcal{L}\{y''\} + \mathcal{L}\{y\} = \mathcal{L}\{\delta(t - \pi)\}$$
$$s^2 Y(s) - sy(0) - y'(0) + Y(s) = e^{-\pi s}$$
$$(s^2 + 1)Y(s) = e^{-\pi s}$$
$$Y(s) = \frac{e^{-\pi s}}{s^2 + 1}$$

Hence,

$$y(t) = \mathcal{L}^{-1}\left\{\frac{e^{-\pi s}}{s^2 + 1}\right\}.$$

Because $f(t) = \mathcal{L}^{-1}\left\{\dfrac{1}{s^2 + 1}\right\} = \sin t,$

$$y(t) = \mathcal{L}^{-1}\left\{\frac{e^{-\pi s}}{s^2 + 1}\right\} = \sin(t - \pi)\mathcal{U}(t - \pi) = -\sin t\,\mathcal{U}(t - \pi).$$

The graph of $y(t)$, which can be written as $y(t) = \begin{cases} 0, & 0 \le t < \pi \\ -\sin t, & t \ge \pi \end{cases}$, is shown in Figure 7.16. Notice that $y(t) = 0$ until the impulse is applied at $t = \pi$.

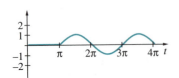

Figure 7.16

    If we interpret the problem in Example 9 as a spring-mass system with $k = m$, what is the maximum displacement of the mass from equilibrium? Does the mass come to rest?

IN TOUCH WITH TECHNOLOGY

1. Use an integration device to assist in calculating the Laplace transform of each of the following periodic functions.

(a) $f(t) = \begin{cases} 1, & 0 \le t < 1 \\ -1, & 1 \le t < 2 \end{cases}, \; f(t + 2) = f(t)$

(b) $f(t) = \begin{cases} t, & 0 \le t < 1 \\ 2 - t, & 1 \le t < 2 \end{cases}, \; f(t + 2) = f(t)$

(c) $f(t) = \begin{cases} \sin t, & 0 \le t < \pi \\ -\sin t, & \pi \le t < 2\pi \end{cases}, \; f(t + 2\pi) = f(t)$

(d) Use the results obtained above to solve the following initial-value problem for each function $f(t)$ above.

$$\begin{cases} \dfrac{d^2 x}{dt^2} + x = f(t) \\[2mm] x(0) = \dfrac{dx}{dt}(0) = 0 \end{cases}$$

Plot the solution in each case to compare the results.

2. Solve $y'' + 2y' + y = \delta(t) + 10,000\,\mathcal{U}(t - 2\pi)$ subject to $y(0) = 0$, $y'(0) = 0$. For what values of t does $y(t)$ decrease? For what values of t does $y(t)$ increase? Determine $\lim_{t \to \infty} y(t)$. Is there a relationship between this limit and the forcing function $\delta(t) + 10,000\,\mathcal{U}(t - 2\pi)$? Will the limit be affected if the forcing function is changed to $100\delta(t) + 10,000\,\mathcal{U}(t - 2\pi)$?

3. Solve $y'' + 2y' + y = \delta(t) + \delta(t - 2\pi)$ subject to $y(0) = 0$, $y'(0) = 0$. What is the maximum value of $y(t)$ and where does it occur? Determine $\lim_{t \to \infty} y(t)$. Would this limit be affected if the forcing function is changed to $\delta(t) + \delta(t - 2\pi) + \delta(t - 4\pi)$?

EXERCISES 7.5

In Exercises 1–22, find the Laplace transform of the given function.

1. $-28\,\mathcal{U}(t - 3)$

2. $-16\,\mathcal{U}(t - 6)$

*3. $3\,\mathcal{U}(t - 8) - \mathcal{U}(t - 4)$

4. $2\,\mathcal{U}(t - 5) + 7\,\mathcal{U}(t - 4)$

5. $\mathcal{U}(t - 2) - 7\,\mathcal{U}(t - 7) - 7\,\mathcal{U}(t - 6)$

6. $4\,\mathcal{U}(t - 2) + 2\,\mathcal{U}(t - 7) - 4\,\mathcal{U}(t - 8)$

*7. $-42e^{t-4}\,\mathcal{U}(t - 4)$

8. $6e^{t-2}\,\mathcal{U}(t - 2)$

9. $12 \sinh(2 - t)\,\mathcal{U}(t - 2)$

10. $\cosh(t - 2)\,\mathcal{U}(t - 2)$

*11. $-14 \sin\left(t - \dfrac{2\pi}{3}\right)\mathcal{U}\left(t - \dfrac{2\pi}{3}\right)$

12. $-3 \cos(t - 2)\,\mathcal{U}(t - 2)$

*13. $f(t) = \begin{cases} t, & 0 \le t < 1 \\ 0, & 1 \le t < 2 \end{cases}$ and $f(t) = f(t - 2)$ if $t \ge 2$ (see Figure 7.17)

Figure 7.17

14. $f(t) = \begin{cases} 1, & 0 \le t < 1 \\ 2 - t, & 1 \le t < 2 \end{cases}$ and $f(t) = f(t - 2)$ if $t \ge 2$

15. $f(t) = \begin{cases} 0, & 0 \le t < 1 \\ 1, & 1 \le t < 2 \\ 2, & 2 \le t < 3 \end{cases}$ and $f(t) = f(t - 3)$ if $t \ge 3$ (see Figure 7.18)

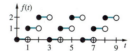

Figure 7.18

16. $f(t) = \begin{cases} t, & 0 \le t < 1 \\ 1, & 1 \le t < 2 \\ 3 - t, & 2 \le t < 3 \end{cases}$ and $f(t) = f(t - 3)$ if $t \ge 3$

17. $\delta(t - \pi)$

18. $\delta(t - 3)$

*19. $\delta(t - 1) + \delta(t - 2)$

20. $2\delta(t - 3) - \delta(t - 1)$

*21. $\sin t\,\mathcal{U}(t - 2\pi) + \delta(t - \tfrac{\pi}{2})$

22. $\cos 2t\,\mathcal{U}(t - \pi) + \delta(t - \pi)$

In Exercises 23–44, find the inverse Laplace transform of the given function.

23. $\dfrac{-3}{se^{\pi s}}$

24. $\dfrac{-10}{se^{4s}}$

*25. $\dfrac{2e^{3s} - 3}{se^{4s}}$

26. $\dfrac{3 + e^{4s}}{se^{6s}}$

27. $\dfrac{3e^{s} - 4e^{3s} - 3}{se^{6s}}$

28. $\dfrac{5 - 6e^{s} - 3e^{2s}}{se^{4s}}$

*29. $\dfrac{e^{-4s}}{s - 1}$

30. $\dfrac{e^{-3s}}{s - 1}$

31. $e^{-3s}\dfrac{s}{s^{2} + 4}$

32. $e^{-5s}\dfrac{1}{s^{2}(s^{2} + 1)}$

*33. $\dfrac{e^{-5s}}{s^2 - 7s + 10}$

34. $\dfrac{1 - e^s + se^{2s}}{s^2(e^{2s} - 1)}$

35. $\dfrac{2(1 - 2e^{2s} - e^{4s})}{s^2 e^{4s}(1 - e^{-4s})}$

36. $\dfrac{-9}{e^{4s}(s^2 - 1)}$

*37. $\dfrac{s}{e^{4s}(s^2 - 1)}$

38. $\dfrac{1}{s^3(1 + e^{-3s})}$

39. $\dfrac{1}{s^2(1 + e^{-4s})}$

40. $\dfrac{s}{(1 + e^{-2s})(s^2 + 9)}$

*41. $\dfrac{1}{(1 + e^{-3s})(s^2 + 16)}$

42. $\dfrac{1}{s(s^2 + 5)(1 + 2e^{-5s})}$

*43. $\dfrac{1}{s^2(s^2 + 4)(3e^{-2s} - 1)}$

44. $\dfrac{1}{(s^2 + 3)(4e^{-3s} - 1)}$

In Exercises 45–57, solve the initial-value problem. *Graph the solution to each problem on an appropriate interval.*

*45. $y' + 3y = f(t)$, $y(0) = 0$,
$f(t) = \begin{cases} 1, & 0 \le t < 2 \\ 0, & t \ge 2 \end{cases}$

46. $y' + 5y = f(t)$, $y(0) = 0$,
$f(t) = \begin{cases} \sin \pi t, & 0 \le t < 1 \\ 0, & t \ge 1 \end{cases}$

*47. $y'' - 4y' + 3y = f(t)$, $y(0) = 0$, $y'(0) = 1$,
$f(t) = \begin{cases} \cos\left(\dfrac{\pi}{2}t\right), & 0 \le t < 1 \\ 0, & t \ge 1 \end{cases}$

48. $y'' + 11y' + 30y = f(t)$, $y(0) = -2$, $y'(0) = 0$,
$f(t) = \begin{cases} 1, & 0 \le t < 1 \\ 2, & 1 \le t < 2 \end{cases}$ and $f(t) = f(t - 2)$ if $t \ge 2$

*49. $x'' + x = \delta(t - \pi) + 1$, $x(0) = 0$, $x'(0) = 0$

50. $x'' + x = \delta(t - \pi) + \delta(t - 2\pi)$, $x(0) = 0$, $x'(0) = 0$

51. $x'' + 3x' + 2x = \delta(t - \pi)$, $x(0) = 0$, $x'(0) = 0$

52. $x'' + 4x' + 13x = \delta(t - \pi)$, $x(0) = 0$, $x'(0) = 0$

*53. $x'' + 4x' + 13x = \delta(t - \pi) + \sin t$, $x(0) = 0$, $x'(0) = 0$

54. $x'' + 4x' + 13x = \delta(t - \pi) + \delta(t - 2\pi)$, $x(0) = 0$, $x'(0) = 0$

55. $x'' + 9x = \cos t + \delta(t - \pi)$, $x(0) = 0$, $x'(0) = 0$

56. $x'' + 9x = \cos 3t + \delta(t - 1)$, $x(0) = 0$, $x'(0) = 0$

*57. $x'' + 3x' + 2x = \delta(t - 1) + e^{-t}$, $x(0) = 0$, $x'(0) = 0$

In Exercises 58–63, solve the initial-value problem. Compare the solution to that of the corresponding homogeneous problem.

58. $x'' + 9x' = \delta(t - \pi)$, $x(0) = 0$, $x'(0) = 2$

59. $x'' + 9x' = \delta(t - \pi)$, $x(0) = 2$, $x'(0) = 0$

60. $x'' + 9x' = \delta(t - \pi)$, $x(0) = 0$, $x'(0) = 0$

*61. $x'' + 6x' + 10x = 3\delta(t - \pi)$, $x(0) = 0$, $x'(0) = 0$

62. $x'' + 6x' + 10x = 3\delta(t - \pi)$, $x(0) = 2$, $x'(0) = 0$

63. $x'' + 6x' + 10x = 3\delta(t - \pi)$, $x(0) = 0$, $x'(0) = 2$

7.6 The Convolution Theorem

The Convolution Theorem • Integral and Integrodifferential Equations

The Convolution Theorem

In many cases, we are required to determine the inverse Laplace transform of a product of two functions. Just as in integral calculus when the integral of the product of two functions did not produce the product of the integrals, neither does the inverse Laplace transform of the product yield the product of the inverse Laplace transforms. We had a preview of this when we found

$$\mathcal{L}^{-1}\left\{\frac{F(s)}{s}\right\} = \mathcal{L}^{-1}\left\{\frac{1}{s}F(s)\right\}$$

earlier in the chapter. We recall that $\mathcal{L}^{-1}\left\{\dfrac{F(s)}{s}\right\} = \displaystyle\int_0^t f(\alpha)\, d\alpha$, where

$f(t) = \mathcal{L}^{-1}\{F(s)\}$, so we realize that

$$\mathcal{L}^{-1}\left\{\frac{F(s)}{s}\right\} \ne \mathcal{L}^{-1}\left\{\frac{1}{s}\right\}\mathcal{L}^{-1}\{F(s)\} = f(t).$$

The *Convolution theorem* gives a relationship between the inverse Laplace transform of the product of two functions, $\mathcal{L}^{-1}\{F(s)G(s)\}$, and the inverse Laplace transform of each function, $\mathcal{L}^{-1}\{F(s)\}$ and $\mathcal{L}^{-1}\{G(s)\}$.

Theorem 7.15 Convolution Theorem

Suppose that $f(t)$ and $g(t)$ are piecewise continuous on $[0, \infty)$ and both of exponential order b. Further, suppose that $\mathcal{L}\{f(t)\} = F(s)$ and $\mathcal{L}\{g(t)\} = G(s)$. Then,

$$\mathcal{L}^{-1}\{F(s)G(s)\} = \mathcal{L}^{-1}\{\mathcal{L}\{(f * g)(t)\}\} = (f * g)(t) = \int_0^t f(t - v)g(v)\, dv.$$

Note that $(f * g)(t) = \int_0^t f(t - v)g(v)\, dv$ is called the **convolution integral.**

PROOF OF THE CONVOLUTION THEOREM

We prove the Convolution theorem by computing the product $F(s)G(s)$ with the definition of the Laplace transform. This yields

$$F(s)G(s) = \int_0^\infty e^{-sx}f(x)\, dx \cdot \int_0^\infty e^{-sv}g(v)\, dv$$

which can be written as the iterated integral

$$F(s)G(s) = \int_0^\infty\int_0^\infty e^{-s(x+v)}f(x)g(v)\, dx\, dv.$$

Changing variables with $x = t - v$ ($t = x + v$, $dx = dt$) then yields

$$F(s)G(s) = \int_0^\infty\int_v^\infty e^{-st}f(t - v)g(v)\, dt\, dv$$

where the region of integration R is the unbounded triangular region shown in Figure 7.19. Recall from multivariable calculus that the order of integration can be interchanged. Then we obtain

$$F(s)G(s) = \int_0^\infty\int_0^t e^{-st}f(t - v)g(v)\, dv\, dt$$

(why?) which can be written as

$$F(s)G(s) = \int_0^\infty e^{-st}\int_0^t f(t - v)g(v)\, dv\, dt = \mathcal{L}\left\{\int_0^t f(t - v)g(v)\, dv\right\} = \mathcal{L}\{(f * g)(t)\}.$$

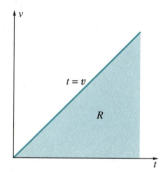

Figure 7.19

Therefore, $\mathcal{L}^{-1}\{F(s)G(s)\} = \mathcal{L}^{-1}\{\mathcal{L}\{(f * g)(t)\}\} = (f * g)(t).$ ● ● ●

Example 1

Compute $(f * g)(t)$ and $(g * f)(t)$ if $f(t) = e^{-t}$ and $g(t) = \sin t$. Verify the Convolution theorem for these functions.

Solution   We use the definition and a table of integrals (or a computer algebra system) to obtain

$$(f * g)(t) = \int_0^t f(t - v)g(v)\, dv = \int_0^t e^{-(t-v)} \sin v\, dv = e^{-t} \int_0^t e^v \sin v\, dv$$

$$= e^{-t}\left[\frac{e^v}{2}(\sin v - \cos v)\right]_0^t = e^{-t}\left[\frac{e^t}{2}(\sin t - \cos t) - \frac{1}{2}(\sin 0 - \cos 0)\right]$$

$$= \frac{1}{2}(\sin t - \cos t) + \frac{1}{2}e^{-t}.$$

With a computer algebra system, we find that

$$(g * f)(t) = \int_0^t g(t - v)f(v)\, dv = \int_0^t \sin(t - v)e^v\, dv = \frac{1}{2}(\sin t - \cos t) + \frac{1}{2}e^{-t}.$$

Now, according to the Convolution theorem, $\mathcal{L}\{f(t)\}\mathcal{L}\{g(t)\} = \mathcal{L}\{(f * g)(t)\}$. In this example, we have

$$F(s) = \mathcal{L}\{f(t)\} = \mathcal{L}\{e^{-t}\} = \frac{1}{s + 1} \quad \text{and} \quad G(s) = \mathcal{L}\{g(t)\} = \mathcal{L}\{\sin t\} = \frac{1}{s^2 + 1}.$$

Hence, $\mathcal{L}^{-1}\{F(s)G(s)\} = \mathcal{L}^{-1}\left\{\dfrac{1}{s + 1} \cdot \dfrac{1}{s^2 + 1}\right\}$ should equal $(f * g)(t)$. We can compute $\mathcal{L}^{-1}\left\{\dfrac{1}{s + 1} \cdot \dfrac{1}{s^2 + 1}\right\}$ through the partial fraction expansion

$$\frac{1}{s + 1} \cdot \frac{1}{s^2 + 1} = \frac{A}{s + 1} + \frac{Bs + C}{s^2 + 1},$$ where $A = C = \frac{1}{2}$ and $B = -\frac{1}{2}$. Therefore,

$$\mathcal{L}^{-1}\left\{\frac{1}{s + 1} \cdot \frac{1}{s^2 + 1}\right\} = \frac{1}{2}\mathcal{L}^{-1}\left\{\frac{1}{s + 1} + \frac{-s + 1}{s^2 + 1}\right\}$$

$$= \frac{1}{2}\mathcal{L}^{-1}\left\{\frac{1}{s + 1}\right\} - \frac{1}{2}\mathcal{L}^{-1}\left\{\frac{s}{s^2 + 1}\right\} + \frac{1}{2}\mathcal{L}^{-1}\left\{\frac{1}{s^2 + 1}\right\}$$

$$= \frac{1}{2}e^{-t} - \frac{1}{2}\cos t + \frac{1}{2}\sin t$$

which is the same result as that obtained for $(f * g)(t)$.

Notice that in Example 1, $(g * f)(t) = (f * g)(t)$. This is no coincidence. With a simple change of variable, we can prove this relationship in general so that the convolution integral is *commutative*. (See Exercise 29.)

Example 2

Use the Convolution theorem to find the Laplace transform of

$$h(t) = \int_0^t \cos(t - v) \sin v \, dv.$$

Solution   Notice that $h(t) = (f * g)(t)$ where $f(t) = \cos t$ and $g(t) = \sin t$. Therefore, by the Convolution theorem, $\mathcal{L}\{(f * g)(t)\} = F(s)G(s)$. Hence,

$$\mathcal{L}\{h(t)\} = \mathcal{L}\{f(t)\}\mathcal{L}\{g(t)\} = \mathcal{L}\{\cos t\}\mathcal{L}\{\sin t\} = \left(\frac{s}{s^2 + 1}\right)\left(\frac{1}{s^2 + 1}\right) = \frac{s}{(s^2 + 1)^2}.$$

 Find the Laplace transform of $h(t) = \int_0^t e^{t-v} \sin v \, dv.$

Integral and Integrodifferential Equations

The Convolution theorem is useful in solving numerous problems. In particular, this theorem can be employed to solve **integral equations,** which are equations that involve an integral of the unknown function.

Example 3

Use the Convolution theorem to solve the integral equation

$$h(t) = 4t + \int_0^t h(t - v) \sin v \, dv.$$

Solution   We need to find $h(t)$, so we first note that the integral in this equation represents $(h * g)(t)$ for $g(t) = \sin t$. Therefore, if we apply the Laplace transform to both sides of the equation, we obtain

$$\mathcal{L}\{h(t)\} = \mathcal{L}\{4t\} + \mathcal{L}\{h(t)\}\mathcal{L}\{\sin(t)\}$$

or

$$H(s) = \frac{4}{s^2} + H(s)\frac{1}{s^2 + 1}$$

where $\mathcal{L}\{h(t)\} = H(s)$. Solving for $H(s)$, we have

$$H(s)\left(1 - \frac{1}{s^2 + 1}\right) = \frac{4}{s^2},$$

so

$$H(s) = \frac{4(s^2 + 1)}{s^4} = \frac{4}{s^2} + \frac{4}{s^4}.$$

Then by computing the inverse Laplace transform, we find that

$$h(t) = \mathcal{L}^{-1}\left\{\frac{4}{s^2} + \frac{4}{s^4}\right\} = 4t + \frac{2}{3}t^3.$$

Laplace transforms are also helpful in solving **integrodifferential equations,** equations that involve a derivative as well as an integral of the unknown function.

Example 4

Solve $\dfrac{dy}{dt} + y + \displaystyle\int_0^t y(u)\,du = 1$ subject to $y(0) = 0$.

Solution   Because we must take the Laplace transform of both sides of this integrodifferential equation, we first compute

$$\mathscr{L}\left\{\int_0^t y(u)\,du\right\} = \mathscr{L}\{(1 * y)(t)\} = \mathscr{L}\{1\}\mathscr{L}\{y\} = \frac{Y(s)}{s}.$$

Hence,

$$\mathscr{L}\left\{\frac{dy}{dt}\right\} + \mathscr{L}\{y\} + \mathscr{L}\left\{\int_0^t y(u)\,du\right\} = \mathscr{L}\{1\}$$

$$sY(s) - y(0) + Y(s) + \frac{Y(s)}{s} = \frac{1}{s}$$

$$s^2Y(s) + sY(s) + Y(s) = 1$$

$$Y(s) = \frac{1}{s^2 + s + 1}.$$

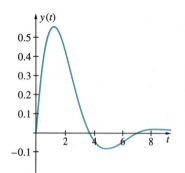

Figure 7.20   Graph of $y(t) = \dfrac{2}{\sqrt{3}}\,e^{-t/2}\sin\dfrac{\sqrt{3}}{2}t$

Because $Y(s) = \dfrac{1}{s^2 + s + 1} = \dfrac{1}{\left(s + \dfrac{1}{2}\right)^2 + \left(\dfrac{\sqrt{3}}{2}\right)^2}$, $y(t) = \dfrac{2}{\sqrt{3}}\,e^{-t/2}\sin\dfrac{\sqrt{3}}{2}t.$

The graph of y is shown in Figure 7.20. Is the initial condition satisfied?

Show that the integrodifferential equation in Example 4 is equivalent to the second-order differential equation $y'' + y' + y = 0$ by differentiating the integrodifferential equation with respect to t. Substitute $t = 0$ into the integrodifferential equation to find that $y'(0) = \dfrac{dy}{dt}\bigg|_{t=0} = 1$. Solve the initial-value problem $y'' + y' + y = 0$, $y(0) = 0$, $y'(0) = 1$. Is this problem equivalent to the initial-value problem in Example 4?

IN TOUCH WITH TECHNOLOGY

As we can see, the calculation of the convolution of two functions is not an easy task in most cases. However, with a device capable of symbolic integration, these calculations become simple; problems that are difficult by hand are completed quickly.

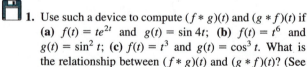

1. Use such a device to compute $(f * g)(t)$ and $(g * f)(t)$ if **(a)** $f(t) = te^{2t}$ and $g(t) = \sin 4t$; **(b)** $f(t) = t^6$ and $g(t) = \sin^2 t$; **(c)** $f(t) = t^3$ and $g(t) = \cos^3 t$. What is the relationship between $(f * g)(t)$ and $(g * f)(t)$? (See Exercise 29.)

2. Calculate (a) $\mathscr{L}^{-1}\left\{\dfrac{4}{(s-2)^2(s^2+16)}\right\}$;

(b) $\mathscr{L}^{-1}\left\{\dfrac{6!}{s^7}\left[\dfrac{1}{2s}-\dfrac{s}{2(s^2+4)}\right]\right\}$,

(c) $\mathscr{L}^{-1}\left\{\dfrac{6}{s^4}\left[\dfrac{s}{4(s^2+9)}-\dfrac{3s}{4(s^2+1)}\right]\right\}$.

3. Verify each of the following:

(a) $(\sin kt)*(\cos kt)=\dfrac{t}{2}\sin kt$;

(b) $(\sin kt)*(\sin kt)=\dfrac{1}{2k}\sin kt-\dfrac{t}{2}\cos kt$;

(c) $(\cos kt)*(\cos kt)=\dfrac{1}{2k}\sin kt+\dfrac{t}{2}\cos kt$.

(d) Use these results to verify that $\mathscr{L}^{-1}\left\{\dfrac{ks}{(s^2+k^2)^2}\right\}=$
$\dfrac{t}{2}\sin kt$, $\mathscr{L}^{-1}\left\{\dfrac{k^2}{(s^2+k^2)^2}\right\}=\dfrac{1}{2k}\sin kt-\dfrac{t}{2}\cos kt$,
and $\mathscr{L}^{-1}\left\{\dfrac{s^2}{(s^2+k^2)^2}\right\}=\dfrac{1}{2k}\sin kt+\dfrac{t}{2}\cos kt$.

EXERCISES 7.6

In Exercises 1–6, compute the convolution $(f*g)(t)$ using the indicated pair of functions.

1. $f(t)=1,\ g(t)=t^2$

2. $f(t)=e^{-3t},\ g(t)=2$

*3. $f(t)=t,\ g(t)=e^{-t}$

4. $f(t)=\sin 2t,\ g(t)=e^{-t}$

5. $f(t)=t^3,\ g(t)=\sin 4t$

6. $f(t)=\sin 2t,\ g(t)=e^{-7t}$

In Exercises 7–12, find the Laplace transform of h.

7. $h(t)=\displaystyle\int_0^t e^{-v}\,dv$

8. $h(t)=\displaystyle\int_0^t \sin v\,dv$

*9. $h(t)=\displaystyle\int_0^t (t-v)\sin v\,dv$

10. $h(t)=\displaystyle\int_0^t v\sin(t-v)\,dv$

11. $h(t)=\displaystyle\int_0^t v e^{t-v}\,dv$

12. $h(t)=\displaystyle\int_0^t e^v(t-v)^2\,dv$

In Exercises 13–18, find the inverse Laplace transform of each function using the Convolution theorem and the results of Exercise 1.

13. $\dfrac{1}{s^2(s+1)}$

14. $\dfrac{1}{s^3(s+3)}$

*15. $\dfrac{s}{(s^2+1)^2}$

16. $\dfrac{1}{(s^2+4)(s+1)}$

17. $\dfrac{1}{s^4(s^2+16)}$

18. $\dfrac{1}{(s^2+4)(s+7)}$

In Exercises 19–22, find the inverse Laplace transform of each function using the Convolution theorem.

19. $\dfrac{1}{(s^2+16)(s-10)}$

20. $\dfrac{1}{(s-4)^2(s^2+100)}$

*21. $\dfrac{1}{(s^2+100)^2}$

22. $\dfrac{s}{(s-1)^8(s^2+25)}$

In Exercises 23–26, solve the given integral equation using Laplace transforms.

23. $g(t)-t=-\displaystyle\int_0^t (t-v)g(v)\,dv$

24. $h(t)=2-\displaystyle\int_0^t h(v)\,dv$

*25. $h(t)-4e^{-2t}=\cos t-\displaystyle\int_0^t \sin(t-v)h(v)\,dv$

26. $f(t)=3-\displaystyle\int_0^t f(t-v)v\,dv$

In Exercises 27–28, solve the given integrodifferential equation using Laplace transforms.

*27. $\dfrac{dy}{dt}-4y+4\displaystyle\int_0^t y(v)\,dv=t^3 e^{2t},\ y(0)=0$

28. $\dfrac{dx}{dt}+16\displaystyle\int_0^t x(v)\,dv=\sin 4t,\ x(0)=0$

29. Show that $(f * g)(t) = (g * f)(t)$ by verifying that

$$\int_0^t f(u)g(t - u)\, du = \int_0^t g(u)f(t - u)\, dt.$$

(Therefore, the convolution integral is *commutative*.)

30. Show that convolution integral is *associative* by proving that $(f * (g * h))(t) = ((f * g) * h)(t)$.

31. Show that the convolution integral satisfies the *distributive property* $(f * (g + h))(t) = (f * g)(t) + (f * h)(t)$.

32. Show that for any constant $((kf) * g)(t) = k(f * g)(t)$.

*33. Show that $(\sin t) * (\cos kt) = (\cos t) * (\frac{1}{k} \sin kt)$.

34. Show that $t^{-1/2} * t^{-1/2} = \pi$. (See Exercise 31, Section 7.1.)

35. Express the integrodifferential equation in Exercise 27 as an equivalent second-order initial-value problem and solve this problem.

36. Express the integrodifferential equation in Exercise 28 as an equivalent second-order initial-value problem and solve this problem.

37. (Delay Equations) A **delay equation** is a differential equation that involves a delay (or shift) in the argument of the dependent variable. Consider the problem

$$y'(t) - 2y(t - 1) = t \quad \text{subject to the condition}$$
$$y(t) = y(0) \text{ for } -1 \le t \le 0$$

Laplace transforms can be used to solve equations of this type.

(a) Take the Laplace transform of both sides of the equation to obtain $sY(s) - y(0) - 2\int_0^\infty e^{-st} y(t - 1)\, dt = \frac{1}{s^2}.$

(b) Evaluate $\int_0^\infty e^{-st} y(t - 1)\, dt$ by letting $u = t - 1$ to obtain

$$\int_0^\infty e^{-st} y(t - 1)\, dt = e^{-s}\left(\int_{-1}^0 e^{-su} y(u)\, du + \int_0^\infty e^{-su} y(u)\, du\right).$$

(c) Use the condition $y(t) = y(0)$ for $-1 \le t \le 0$ to simplify the expression in **(b)**. Substitute this result into the expression in **(a)** and solve for $Y(s)$ to find that

$$Y(s) = \frac{-e^s + 2sy(0) - 2e^s sy(0) - e^s s^2 y(0)}{-2s^2 + e^s s^3}.$$

(d) Use partial fractions to find that

$$Y(s) = \frac{2(1 + 2sy(0))}{s^3(-2 + e^s s)} + \frac{1 + 2sy(0) + s^2 y(0)}{s^3}.$$

(e) Find $\mathcal{L}^{-1}\left\{\dfrac{1 + 2sy(0) + s^2 y(0)}{s^3}\right\}$.

(f) Find $\mathcal{L}^{-1}\left\{\dfrac{2(1 + 2sy(0))}{s^3(-2 + e^s s)}\right\}$ by rewriting this expression as

$$\frac{2(1 + 2sy(0))}{s^3(-2 + e^s s)} = \frac{2(1 + 2sy(0))}{s^4 e^s} \cdot \frac{1}{1 - 2s^{-1}e^{-s}}$$

$$= \frac{2(1 + 2sy(0))}{s^4 e^s} \sum_{n=0}^\infty (2s^{-1}e^{-s})^n$$

$$= 2(1 + 2sy(0)) \sum_{n=0}^\infty 2^n \frac{e^{-ns}}{s^{n+4}}$$

$$= \sum_{n=0}^\infty \left[2^{n+1}\frac{e^{-ns}}{s^{n+4}} + 2^{n+2}\, y(0)\frac{e^{-ns}}{s^{n+3}}\right].$$

(g) Show that

$$\mathcal{L}^{-1}\left\{2^{n+1}\frac{e^{-ns}}{s^{n+4}} + 2^{n+2}\, y(0)\frac{e^{-ns}}{s^{n+3}}\right\} =$$
$$\frac{2^{n+1}(t - n)^{n+2}}{(n + 3)!}\mathcal{U}(t - n)[t + 6y(0) +$$
$$(2y(0) - 1)n] \text{ so that}$$
$$\mathcal{L}^{-1}\left\{\frac{2(1 + 2sy(0))}{s^3(-2 + e^s s)}\right\} =$$
$$\sum_{n=0}^\infty \left[\frac{2^{n+1}(t - n)^{n+2}}{(n + 3)!}\mathcal{U}(t - n)[t + 6y(0) +$$
$$(2y(0) - 1)n]\right].$$

(h) Find $y(t)$. (See Figure 7.21.)

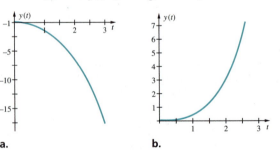

a.   b.

Figure 7.21   (a) $y(t)$ if $y(0) = -1$ (b) $y(t)$ if $y(0) = 0$

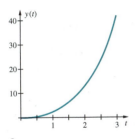

c.
Figure 7.21   (c) $y(t)$ if $y(0) = 1$

38. Solve the delay differential equation $y' = 4y(t - 2)$, $y(t) = 0$, $-2 \le t \le 0$.

7.7 Applications of Laplace Transforms

Spring-Mass Systems Revisited • L-R-C Circuits Revisited • Population Problems Revisited

Laplace transforms are useful in solving the applications that were discussed in earlier sections. However, because this method is most useful in alleviating the difficulties associated with problems involving piecewise defined functions, impulse functions, and periodic functions, we focus most of our attention on solving spring-mass systems, L-R-C circuit problems, and population problems that include functions of this type.

Spring-Mass Systems Revisited

First, we investigate the use of Laplace transforms to solve the second-order initial-value problem that models the motion of a mass attached to the end of a spring. We found in Chapter 5 that this situation is modeled by the initial-value problem

$$\begin{cases} m\dfrac{d^2x}{dt^2} + c\dfrac{dx}{dt} + kx = f(t) \\ x(0) = \alpha, \ \dfrac{dx}{dt}(0) = \beta \end{cases}$$

where m represents the mass, c the damping coefficient, k the spring constant determined by Hooke's law, and $f(t)$ the forcing function. In the following example, we consider the initial-value problem that involves a discontinuous forcing function. Instead of having to solve two initial-value problems as we did in Chapter 5, we use the Laplace transform method.

Example 1

Suppose that a mass with $m = 1$ slug is attached to a spring with spring constant $k = 1$ lb/ft. If there is no resistance due to damping, determine the displacement of the mass if it is released with zero initial velocity from its equilibrium position and is

subjected to the force $f(t) = \begin{cases} \cos t, & 0 \le t < \dfrac{\pi}{2} \\ 0, & t \ge \dfrac{\pi}{2} \end{cases}$.

Solution   The constants are $m = 1$, $c = 0$, and $k = 1$. The initial displacement is $x(0) = 0$ and the initial velocity is $\dfrac{dx}{dt}(0) = 0$. The initial-value problem that models this situation is

$$\frac{d^2x}{dt^2} + x = \begin{cases} \cos t, & 0 \le t < \dfrac{\pi}{2} \\[2mm] 0, & t \ge \dfrac{\pi}{2} \end{cases}, \quad x(0) = 0, \frac{dx}{dt}(0) = 0.$$

We must take the Laplace transform of both sides of the differential equation so we write $f(t)$ in terms of the unit step function. This gives us

$$f(t) = \cos t \left[\mathcal{U}(t - 0) - \mathcal{U}\left(t - \frac{\pi}{2}\right) \right] = \cos t \left[1 - \mathcal{U}\left(t - \frac{\pi}{2}\right) \right].$$

Then,

$$\begin{aligned} \mathcal{L}\{\cos t - \cos t\, \mathcal{U}(t - \pi/2)\} &= \mathcal{L}\{\cos t\} + \mathcal{L}\{-\cos t\, \mathcal{U}(t - \pi/2)\} \\ &= \mathcal{L}\{\cos t\} + \mathcal{L}\{\sin(t - \pi/2)\mathcal{U}(t - \pi/2)\} \\ &= \mathcal{L}\{\cos t\} + e^{-s\pi/2}\, \mathcal{L}\{\sin t\} \\ &= \frac{s}{s^2 + 1} + e^{-s\pi/2}\, \frac{1}{s^2 + 1} \end{aligned}$$

and use this result together with the initial conditions in the following calculations to find $X(s)$:

$$\mathcal{L}\left\{ \frac{d^2x}{dt^2} \right\} + \mathcal{L}\{x\} = \mathcal{L}\{\cos t - \cos t\, \mathcal{U}(t - \pi/2)\}$$

$$s^2 X(s) - sx(0) - x'(0) + X(s) = \frac{s}{s^2 + 1} + e^{-s\pi/2}\, \frac{1}{s^2 + 1}$$

$$(s^2 + 1)X(s) = \frac{s}{s^2 + 1} + e^{-s\pi/2}\, \frac{1}{s^2 + 1}$$

$$X(s) = \frac{s}{(s^2 + 1)^2} + e^{-s\pi/2}\, \frac{1}{(s^2 + 1)^2}.$$

Now, in order to compute $x(t) = \mathcal{L}^{-1}\{X(s)\}$, we must determine $\mathcal{L}^{-1}\left\{ \dfrac{1}{(s^2 + 1)^2} \right\}$. From a table of Laplace transforms or a computer algebra system, we find that

$$\mathcal{L}^{-1}\left\{ \frac{1}{(s^2 + 1)^2} \right\} = \frac{1}{2}(\sin t - t \cos t).$$

Using this formula and simplifying with trigonometric identities, we have

$$\mathcal{L}^{-1}\left\{e^{s-\pi/2}\frac{1}{(s^2+1)^2}\right\} = \frac{1}{2}\left[\sin\left(t-\frac{\pi}{2}\right)-\left(t-\frac{\pi}{2}\right)\cos\left(t-\frac{\pi}{2}\right)\right]\mathcal{U}\left(t-\frac{\pi}{2}\right)$$

$$= \frac{1}{2}\left(-\cos t - \left(t-\frac{\pi}{2}\right)\sin t\right)\mathcal{U}\left(t-\frac{\pi}{2}\right).$$

Again, with a table of Laplace transforms or computer algebra system,

$$\mathcal{L}^{-1}\left\{\frac{s}{(s^2+1)^2}\right\} = \frac{1}{2}t\sin t,$$

so

$$x(t) = \mathcal{L}^{-1}\{X(s)\} = \frac{1}{2}t\sin t + \frac{1}{2}\left(-\cos t - \left(t-\frac{\pi}{2}\right)\sin t\right)\mathcal{U}\left(t-\frac{\pi}{2}\right).$$

By eliminating the unit step function, we can write $x(t)$ in the form

$$x(t) = \begin{cases} \dfrac{1}{2}t\sin t, & 0 \le t < \dfrac{\pi}{2} \\ \dfrac{\pi}{4}\sin t - \dfrac{1}{2}\cos t, & t \ge \dfrac{\pi}{2} \end{cases}.$$

Notice that resonance occurs on the interval $0 \le t < \pi/2$. For $t \ge \pi/2$, the motion is harmonic. Therefore, although the forcing function is zero for $t \ge \pi/2$, the mass continues to follow the path defined by $x(t)$ indefinitely. We graph this function in Figure 7.22.

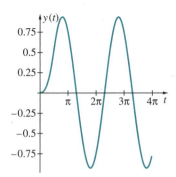

Figure 7.22

Describe the motion of the mass in Example 1 if the forcing function is changed to $f(t) = \begin{cases} \sin t, & 0 \le t < \pi \\ 0, & t \ge \pi \end{cases}$.

L-R-C Circuits Revisited

Laplace transforms can be used to solve circuit problems (see Figure 7.23) which were introduced earlier. Recall that the initial-value problems that model *L-R-C* and *L-R* circuits are

$$\begin{cases} L\dfrac{d^2Q}{dt^2} + R\dfrac{dQ}{dt} + \dfrac{1}{C}Q = E(t) \\ Q(0) = Q_0, I(0) = \dfrac{dQ}{dt}(0) = I_0 \end{cases} \quad \text{and} \quad \begin{cases} L\dfrac{dI}{dt} + RI = E(t) \\ I(0) = I_0 \end{cases},$$

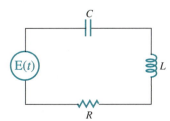

a.

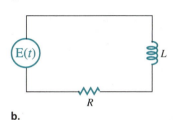

b.

Figure 7.23 (a) *L-R-C* Circuit (b) *L-R* Circuit

where L, R, and C represent the inductance, resistance, and capacitance, respectively. $Q(t)$ is the charge of the capacitor and $dQ/dt = I(t)$, where $I(t)$ is the current. $E(t)$ is the voltage supply.

Example 2

Consider the circuit with no capacitor, $R = 100\Omega$, and $L = 100\text{H}$ if
$$E(t) = \begin{cases} 100 \text{ V}, & 0 \le t < 1 \\ 0, & 1 \le t < 2 \end{cases}$$
and $E(t + 2) = E(t)$. Find the current $I(t)$ if $I(0) = 0$.

Solution   Because there is no capacitor, the differential equation that models the L-R circuit is $100\dfrac{dI}{dt} + 100I = E(t)$ and the initial-value problem is

$$\begin{cases} 100\dfrac{dI}{dt} + 100I = E(t) \\ I(0) = 0 \end{cases}$$

Notice that $E(t)$ is a periodic function, so we first compute $\mathcal{L}\{E(t)\}$.

$$\begin{aligned} \mathcal{L}\{E(t)\} &= \frac{1}{1 - e^{-2s}} \int_0^2 e^{-st}E(t)\,dt = \frac{1}{1 - e^{-2s}} \int_0^1 100e^{-st}\,dt \\ &= \frac{100}{1 - e^{-2s}}\left[-\frac{e^{-st}}{s} \right]_0^1 = \frac{100(1 - e^{-s})}{s(1 - e^{-2s})} \\ &= \frac{100(1 - e^{-s})}{s(1 - e^{-s})(1 + e^{-s})} = \frac{100}{s(1 + e^{-s})}. \end{aligned}$$

Then,

$$100\mathcal{L}\left\{\frac{dI}{dt}\right\} + 100\mathcal{L}\{I\} = \mathcal{L}\{E(t)\}$$

$$100s\mathcal{L}\{I\} - 100I(0) + 100\mathcal{L}\{I\} = \frac{100}{s(1 + e^{-s})}$$

$$\mathcal{L}\{I\} = \frac{1}{s(s + 1)(1 + e^{-s})}.$$

(Notice that we are using $\mathcal{L}\{I\}$ to represent the Laplace transform of the current $I(t)$, because a capital letter appears in the variable name.) As we did in Section 7.5, we write a power series expansion of $\dfrac{1}{1 + e^{-s}}$ as

$$\frac{1}{1 + e^{-s}} = \frac{1}{1 - (-e^{-s})} = 1 - e^{-s} + e^{-2s} - e^{-3s} + \cdots.$$

Thus,

$$\mathcal{L}\{I\} = \frac{1}{s(s+1)}[1 - e^{-s} + e^{-2s} - e^{-3s} + \cdots]$$

$$= \frac{1}{s(s+1)} - \frac{e^{-s}}{s(s+1)} + \frac{e^{-2s}}{s(s+1)} - \frac{e^{-3s}}{s(s+1)} + \cdots.$$

Because $\mathcal{L}^{-1}\left\{\dfrac{1}{s(s+1)}\right\} = 1 - e^{-t}$,

$$I(t) = (1 - e^{-t}) - (1 - e^{-(t-1)})\mathcal{U}(t - 1) + (1 - e^{-(t-2)})\mathcal{U}(t - 2) -$$
$$(1 - e^{-(t-3)})\mathcal{U}(t - 3) + \cdots.$$

We can write this function as

$$I(t) = \begin{cases} 1 - e^{-t}, & 0 \le t < 1 \\ -e^{-t} + e^{-(t-1)}, & 1 \le t < 2 \\ 1 - e^{-t} + e^{-(t-1)} - e^{-(t-2)}, & 2 \le t < 3 \\ -e^{-t} + e^{-(t-1)} - e^{-(t-2)} + e^{-(t-3)}, & 3 \le t < 4 \\ \vdots \end{cases}$$

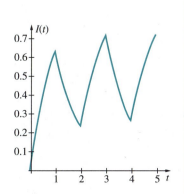

Figure 7.24

The graph of $I(t)$ is shown in Figure 7.24.

In Example 2, over what intervals does I(t) increase? Over what intervals does I(t) decrease? How do these intervals relate to the values of E(t)?

We can consider the *L-R-C* circuit in terms of the integrodifferential equation

$$L\frac{dI}{dt} + RI + \frac{1}{C}\int_0^t I(\alpha)\,d\alpha = E(t)$$

which is useful when using the method of Laplace transforms to find the current $I(t)$.

Example 3

Find the current $I(t)$ if $L = 1$ henry, $R = 6$ ohms, $C = \frac{1}{9}$ farad, $E(t) = 1$ volt, and $I(0) = 0$.

Solution   In this case, we must solve the initial-value problem

$$\begin{cases} \dfrac{dI}{dt} + 6I + 9\displaystyle\int_0^t I(\alpha)\,d\alpha = 1 \\ I(0) = 0 \end{cases}$$

Taking the Laplace transform of each side of the equation yields

$$s\mathcal{L}\{I\} - I(0) + 6\mathcal{L}\{I\} + \frac{9\mathcal{L}\{I\}}{s} = \frac{1}{s}$$

$$s^2\mathcal{L}\{I\} + 6s\mathcal{L}\{I\} + 9\mathcal{L}\{I\} = 1$$

$$\mathcal{L}\{I\} = \frac{1}{s^2 + 6s + 9} = \frac{1}{(s+3)^2}$$

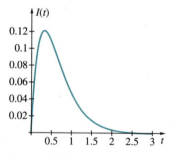

Figure 7.25

so

$$I(t) = \mathcal{L}^{-1}\left\{\frac{1}{(s + 3)^2}\right\} = te^{-3t}.$$

We graph this function in Figure 7.25. What is the maximum current and when does it occur?

Population Problems Revisited

Laplace transforms can be used to solve the population problems that were discussed as applications of first-order equations in Chapter 3. In this case, we consider a problem that involves a continuous forcing function used to describe the presence of immigration or emigration.

Example 4

Let $x(t)$ represent the population of a certain country where the rate at which the population increases depends on the growth rate of the country as well as the rate at which people are being added to or subtracted from the population due to immigration, emigration, or both. If we consider the population problem

$$x' + kx = 1000(1 + a \sin t), \quad x(0) = x_0,$$

where a is a constant, find $x(t)$.

Solution   Using the method of Laplace transforms, we have

$$\mathcal{L}\{x'\} + k\mathcal{L}\{x\} = \mathcal{L}\{1000(1 + a \sin t)\}$$

$$sX(s) - x(0) + kX(s) = 1000\left(\frac{1}{s} + \frac{a}{s^2 + 1}\right)$$

$$(s + k)X(s) = 1000\left(\frac{1}{s} + \frac{a}{s^2 + 1}\right) + x_0$$

$$X(s) = 1000\left(\frac{1}{s(s + k)} + \frac{a}{(s + k)(s^2 + 1)}\right) + \frac{x_0}{s + k}.$$

Using the partial fraction expansions

$$\frac{1}{s(s + k)} = \frac{1}{k}\left(\frac{1}{s} - \frac{1}{s + k}\right) \quad \text{and}$$

$$\frac{a}{(s + k)(s^2 + 1)} = \frac{a}{1 + k^2}\left(\frac{1}{s + k} - \frac{s}{s^2 + 1} + \frac{k}{s^2 + 1}\right)$$

and taking the inverse Laplace transform, we have

$$x(t) = \frac{1000}{k}(1 - e^{-kt}) + \frac{1000a}{1 + k^2}(e^{-kt} - \cos t + k \sin t) + x_0 e^{-kt}.$$

We can use this formula to investigate the population for particular values of a, k, and x_0. For example, if $x_0 = 2000$ and $k = 3$, we can investigate the population for various values of a. The graph of $x(t)$ is shown in Figures 7.26(a) and (b) if $a = 0.25$ and $a = 0.75$. The function $f(t) = 1000(1 + a \sin t)$ for these values describes the influx of people into and out of the country. We see that if $a = 0.25$, the influx does not vary much from a constant level of 1000; when $a = 0.75$, the change in influx is much greater. Using $x_0 = 2000$ and $k = 3$, graphs of the population for various values of a are shown in Figure 7.26(c).

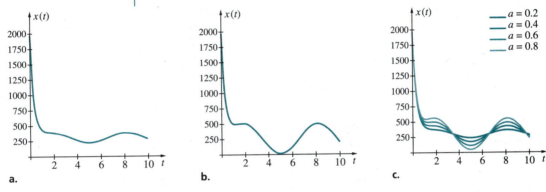

a. **b.** **c.**

Figure 7.26   (a) $a = 0.25$ (b) $a = 0.75$ (c) Various values of a

In Example 4, for what values of a is the rate at which people enter the country always positive?

IN TOUCH WITH TECHNOLOGY

In Problems 1–4, determine the motion of the object of mass m attached to a spring with spring constant k if the damping is given by $c\dfrac{dx}{dt}$ and there is a piecewise defined periodic external force $f(t)$. Use the initial conditions $x(0) = x'(0) = 0$ in each case.

1. $m = 1$, $k = 6$, $c = 5$,
$$f(t) = \begin{cases} 1, & 0 \le t < 1 \\ 0, & 1 \le t < 2 \end{cases}, f(t + 2) = f(t)$$

2. $m = 1$, $k = 6$, $c = 5$,
$$f(t) = \begin{cases} 1, & 0 \le t < 1 \\ -1, & 1 \le t < 2 \end{cases}, f(t + 2) = f(t)$$

3. $m = 1$, $k = 13$, $c = 4$,
$$f(t) = \begin{cases} \sin t, & 0 \le t < \pi \\ 0, & \pi \le t < 2\pi \end{cases}, f(t + 2\pi) = f(t)$$

4. $m = 1$, $k = 13$, $c = 4$,
$$f(t) = \begin{cases} 1, & 0 \le t < 1 \\ -1, & 1 \le t < 2 \end{cases}, f(t + 2) = f(t)$$

5. Use a graphing device to investigate the solutions obtained in Problems 1–4. How does the forcing function $f(t)$ affect the subsequent motion of the object? Does the effect of damping eventually cause the object to come to rest?

6. Consider the L-R-C circuit in which $R = 100\Omega$, $L = 0.1$ henry, $C = 10^{-3}$, and $E(t) = 155 \sin 377t$. (Notice that the frequency of the voltage source is $377/2\pi \approx 60$ Hz $= 60$ cycles/sec.)

(a) Find the current $I(t)$. **(b)** What is the maximum value of the current and where does it occur? **(c)** Find $\lim_{t\to\infty} I(t)$ (the steady-state current). **(d)** What is the frequency of the steady-state current?

7. Suppose that the L-R-C circuit in Problem 6 has a voltage source $E(t) = \begin{cases} 155 \cos 377t, & 0 \le t < 1 \\ 0, & t \ge 1 \end{cases}$.

(a) Find the current $I(t)$. **(b)** What is the maximum value of the current and where does it occur? **(c)** Find $\lim_{t\to\infty} I(t)$ (the steady-state current). **(d)** What is the frequency of the steady-state current?

8. Suppose that the L-R-C circuit in Problem 6 has the periodic voltage source $E(t) = \begin{cases} 155, & 0 < t < 1 \\ 0, & 1 \le t < 2 \end{cases}$, $E(t + 2) = E(t)$. **(a)** Find the current $I(t)$. **(b)** What is the maximum value of the current and where does it occur? **(c)** Find $\lim_{t\to\infty} I(t)$ (the steady-state current). **(d)** What is the frequency of the steady-state current?

EXERCISES 7.7

In Exercises 1 and 2, use the given initial conditions to determine the displacement of the object of mass m attached to a spring with spring constant k.

1. $m = 4$, $k = 16$, $x(0) = 1$, $x'(0) = 0$

2. $m = 1$, $k = 9$, $x(0) = 3$, $x'(0) = -2$

In Exercises 3–6, determine the displacement of the object of mass m attached to a spring with spring constant k if the damping is given by $c\dfrac{dx}{dt}$. Use the initial conditions $x(0) = 1$, $x'(0) = 0$ in each case.

3. $m = 1$, $k = 6$, $c = 5$ **4.** $m = \frac{1}{2}$, $k = 2$, $c = 2$

*5. $m = 1$, $k = 13$, $c = 4$ **6.** $m = 1$, $k = 4$, $c = 5$

In Exercises 7–10, determine the displacement of the object of mass m attached to a spring with spring constant k if the damping is given by $c\dfrac{dx}{dt}$ and there is an external force $f(t)$. In each case, use the initial conditions $x(0) = x'(0) = 0$.

7. $m = 1$, $k = 6$, $c = 5$, $f(t) = \frac{1}{4}\sin t$

8. $m = \frac{1}{2}$, $k = 2$, $c = 2$, $f(t) = te^{-t}$

*9. $m = 1$, $k = 13$, $c = 4$, $f(t) = e^{-2t}$

10. $m = 1$, $k = 4$, $c = 5$, $f(t) = e^{-4t} + 2e^{-t}$

In Exercises 11–18, determine the displacement of the object of mass m attached to a spring with spring constant k if the damping is given by $c\dfrac{dx}{dt}$ and there is a piecewise defined external force $f(t)$. In each case, use the initial conditions $x(0) = x'(0) = 0$.

*11. $m = 1$, $k = 6$, $c = 5$, $f(t) = \begin{cases} \sin 2t, & 0 \le t < \frac{\pi}{2} \\ 0, & t \ge \frac{\pi}{2} \end{cases}$

12. $m = 1$, $k = 6$, $c = 5$, $f(t) = \begin{cases} \cos \pi t, & 0 \le t < 1 \\ 0, & t \ge 1 \end{cases}$

13. $m = \frac{1}{2}$, $k = 2$, $c = 2$, $f(t) = \begin{cases} e^{-2t}, & 0 \le t < 1 \\ 0, & t \ge 1 \end{cases}$

14. $m = \frac{1}{2}$, $k = 2$, $c = 2$, $f(t) = \begin{cases} e^{-2t}, & 0 \le t < 1 \\ 1, & 1 \le t < 2 \\ 0, & t \ge 2 \end{cases}$

*15. $m = 1$, $k = 13$, $c = 4$,
$f(t) = \begin{cases} e^{-2t} \cos 3t, & 0 \le t < \pi \\ 0, & t \ge \pi \end{cases}$

16. $m = 1$, $k = 13$, $c = 4$,
$f(t) = \begin{cases} e^{-2t} \cos 3t, & 0 \le t < \pi \\ e^{-2t} \sin 3t, & \pi \le t < 2\pi \\ 0, & t \ge 2\pi \end{cases}$

17. $m = 1$, $k = 4$, $c = 5$, $f(t) = \begin{cases} e^{-t}, & 0 \le t < 1 \\ 0, & t \ge 1 \end{cases}$

18. $m = 1$, $k = 4$, $c = 5$, $f(t) = \begin{cases} \cos \pi t, & 0 \le t < 1 \\ 0, & t \ge 1 \end{cases}$

19. Show that the initial-value problem $x'' + x = \delta(t)$, $x(0) = x'(0) = 0$ is equivalent to $x'' + x = 0$, $x(0) = 0$, $x'(0) = 1$. (*Hint:* Solve each equation.)

In Exercises 20–22, assume that the mass is released with zero initial velocity from its equilibrium position.

20. Suppose that an object with mass $m = 1$ is attached to the end of a spring with spring constant 16. If there is no damping and the spring is subjected to the forcing function $f(t) = \sin t$, determine the motion of the spring if at $t = 1$, the spring is supplied with an upward shock of 4 units.

*21. An object of mass $m = 1$ is attached to a spring with $k = 13$ and is subjected to damping equivalent to $4 \frac{dx}{dt}$. Find the motion of the mass if the spring is supplied with a downward shock of 1 unit at $t = 2$.

22. An object of mass $m = 1$ is attached to a spring with $k = 13$ and is subjected to damping equivalent to $4\, dx/dt$. Find the motion if
$$f(t) = \delta(t - 1) + \delta(t - 3).$$

23. Suppose that we consider a circuit with a capacitor C, a resistor R, and a voltage supply $E(t) = \begin{cases} 100, & 0 \leq t < 1 \\ 0, & t \geq 1 \end{cases}$. If $L = 0$, find $Q(t)$ and $I(t)$ if $Q(0) = 0$, $C = 50^{-1}$ farads and $R = 50\Omega$.

24. Suppose that we consider a circuit with a capacitor C, a resistor R, and a voltage supply $E(t) = \begin{cases} 100 \sin t, & 0 \leq t < \pi \\ 0, & t \geq \pi \end{cases}$. If $L = 0$, find $Q(t)$ and $I(t)$ if $Q(0) = 0$, $C = 10^{-2}$ farads, and $R = 100\Omega$.

*25. Consider the circuit with no capacitor, $R = 100\Omega$, and $L = 100$H if $E(t) = \begin{cases} 50, & 0 \leq t < 1 \\ 0, & 1 \leq t < 2 \end{cases}$ and $E(t + 2) = E(t)$. Find the current $I(t)$ if $I(0) = 0$.

26. Consider the circuit with no capacitor, $R = 100\Omega$, and $L = 100$H if $E(t) = \begin{cases} 100 \sin t, & 0 \leq t < \pi \\ 0, & \pi \leq t < 2\pi \end{cases}$ and $E(t + 2\pi) = E(t)$. Find the current $I(t)$ if $I(0) = 0$.

27. Consider the circuit with no capacitor, $R = 100\Omega$, and $L = 100$H if $E(t) = \begin{cases} 100t, & 0 \leq t < 1 \\ 0, & 1 \leq t < 2 \end{cases}$ and $E(t + 2) = E(t)$. Find the current $I(t)$ if $I(0) = 0$.

28. Consider the circuit with no capacitor, $R = 100\Omega$, and $L = 100$H if $E(t) = \begin{cases} 100t, & 0 \leq t < 1 \\ 100(2 - t), & 1 \leq t < 2 \end{cases}$ and $E(t + 2) = E(t)$. Find the current $I(t)$ if $I(0) = 0$.

*29. Solve the L-R-C circuit integrodifferential equation that was derived before Example 3 for $I(t)$ if $L = 1$ henry, $R = 6$ ohms, $C = \frac{1}{9}$ farad, $E(t) = 100$ volts, and $I(0) = 0$.

30. Solve the L-R-C circuit integrodifferential equation for $I(t)$ if $L = 1$ henry, $R = 6$ ohms, $C = \frac{1}{9}$ farad, $E(t) = 100 \sin t$ volts, and $I(0) = 0$.

31. Solve the L-R-C circuit integrodifferential equation for $I(t)$ using the parameter values $L = C = R = 1$ and $E(t) = \begin{cases} 1, & 0 \leq t \leq \pi \\ 0, & t > \pi \end{cases}$, $I(0) = 0$.

32. Solve the L-R-C circuit integrodifferential equation for $I(t)$ using the parameter values $L = C = R = 1$ and $E(t) = \begin{cases} \cos t, & 0 \leq t \leq \pi \\ 0, & t > \pi \end{cases}$, $I(0) = 0$.

*33. Consider a circuit with $L = 1$, $R = 4$, $E(t) = \delta(t - 1)$, and no capacitor. Determine the current $I(t)$ if **(a)** $I(0) = 0$ and **(b)** $I(0) = 1$.

34. Solve the initial-value problems in Exercise 33 using $E(t) = \delta(t - 1) + \delta(t - 2)$.

35. Consider an L-R circuit in which $L = 1$ and $R = 1$, and $E(t) = \begin{cases} t, & 0 \leq t \leq 1 \\ 1, & t > 1 \end{cases}$. Find the current if $I(0) = 0$.

36. Consider the circuit in Exercise 35 with a voltage source $E(t) = \begin{cases} t, & 0 \leq t \leq 1 \\ 0, & t > 1 \end{cases}$. Find the current if $I(0) = 0$.

*37. Consider the circuit in Exercise 35 with $E(t) = \delta(t - 2) + \delta(t - 6)$. Find the current if $I(0) = 0$.

38. Consider the circuit in Exercise 35 with $E(t) = 120\delta(t - 1)$. Find the current if $I(0) = 100$.

In Exercises 39–44, solve the initial-value problem using Laplace transforms. Interpret each as a population problem. Is the population bounded?

39. $x' + 5x = 500(2 - \sin t)$, $x(0) = 10{,}000$

40. $x' + 5x = 500(2 + \cos t)$, $x(0) = 5000$

*41. $x' + 5x = 500(2 - \cos t)$, $x(0) = 5000$

42. $x' + 5x = 500(2 - \sin t)$, $x(0) = 5000$

43. $x' - 2x = 500(1 + \sin t)$, $x(0) = 5000$

44. $x' - 2x = 500(\cos t + 1)$, $x(0) = 5000$

*45. Suppose that the emigration function is
$$f(t) = \begin{cases} 5000(2 - \sin t), & 0 \leq t < 5 \\ 0, & t \geq 5 \end{cases}.$$
Solve $x' - x = f(t)$, $x(0) = 10{,}000$. Determine $\lim_{t \to \infty} f(t)$.

46. Suppose that the emigration function is
$$f(t) = \begin{cases} 5000(1 + \cos t), & 0 \le t < 10 \\ 0, & t \ge 10 \end{cases}.$$ Solve
$x' - x = f(t)$, $x(0) = 5000$. Determine $\lim_{t \to \infty} f(t)$ if it exists.

47. If the emigration function is the periodic function
$$f(t) = \begin{cases} 5000(1 + \cos t), & 0 \le t < 1 \\ 0, & 1 \le t < 2 \end{cases},$$
$f(t + 2) = f(t)$, solve $x' - x = f(t)$, $x(0) = 5000$.

48. If the emigration function is the periodic function
$$f(t) = \begin{cases} 5000(2 - \sin t), & 0 \le t < 1 \\ 0, & 1 \le t < 2 \end{cases},$$
$f(t + 2) = f(t)$, solve $x' - x = f(t)$, $x(0) = 5000$.

*49. Suppose that a patient receives glucose through an IV tube at a constant rate of c grams per minute. If at the same time the glucose is metabolized and removed from the bloodstream at a rate that is proportional to the amount of glucose present in the bloodstream, the rate at which the amount of glucose changes is mod-

eled by $dx/dt = c - kx$ where $x(t)$ is the amount of glucose in the bloodstream at time t and k is a constant. If $x(0) = x_0$, use Laplace transforms to find $x(t)$. Does $x(t)$ approach a limit as $t \to \infty$?

50. Suppose that in Exercise 49 the patient receives glucose at a variable rate $c(1 + \sin t)$. Therefore, the rate at which the amount of glucose changes is modeled by $dx/dt = c(1 + \sin t) - kx$. If $x(0) = x_0$, use Laplace transforms to find $x(t)$. How does the solution of this problem differ from that found in Exercise 49?

51. If the person in Exercise 49 receives glucose periodically according to the function
$$f(t) = \begin{cases} c, & 0 \le t < 1 \\ 0, & 1 \le t < 2 \end{cases}, \quad f(t + 2) = f(t),$$ solve the initial-value problem $dx/dt = f(t) - kx$, $x(0) = x_0$, and compare the solution with that of Exercise 50.

52. Suppose that a bacteria population satisfies the differential equation $dx/dt = x + 200\delta(t - 2)$ with $x(0) = 100$. What is the population at $t = 5$?

CHAPTER 7 SUMMARY
Concepts and Formulas

Section 7.1

Definition of the Laplace transform
$$\mathcal{L}\{f(t)\} = \int_0^\infty e^{-st}f(t)\, dt.$$

Linearity of the Laplace transform
$$\mathcal{L}\{af(t) + bg(t)\} = a\mathcal{L}\{f(t)\} + b\mathcal{L}\{g(t)\}$$

Laplace transform of $\sin kt$ and $\cos kt$

Exponential order
A function f is of **exponential order (of order b)** if there are numbers b, $C > 0$, and $T > 0$ such that
$$|f(t)| \le Ce^{bt}$$
for $t > T$.

Jump discontinuity

Piecewise continuous

Section 7.2

Shifting property
$$\mathcal{L}\{e^{at}f(t)\} = F(s - a)$$

Laplace transform of the first derivative
$$\mathcal{L}\{f'(t)\} = s\mathcal{L}\{f(t)\} - f(0)$$

Laplace transform of higher derivatives
$$\mathcal{L}\{f^{(n)}(t)\} = s^n\mathcal{L}\{f(t)\} - s^{n-1}f(0) - \cdots - f^{(n-1)}(0)$$

Derivatives of the Laplace transform
$$\mathcal{L}\{t^n f(t)\} = (-1)^n \frac{d^n F}{ds^n}(s)$$

Section 7.3

Inverse Laplace transform
$$f(t) = \mathcal{L}^{-1}\{F(s)\}$$

Linearity of the inverse Laplace transform

$\mathcal{L}^{-1}\{aF(s) + bG(s)\} = a\mathcal{L}^{-1}\{F(s)\} + b\mathcal{L}^{-1}\{G(s)\}$

Linear factors (nonrepeated)

Repeated linear factors

Irreducible quadratic factors

Laplace transform of an integral

$\mathcal{L}\left\{\int_0^t f(\alpha)\, d\alpha\right\} = \dfrac{\mathcal{L}\{f(t)\}}{s}$ or

$\mathcal{L}^{-1}\left\{\dfrac{\mathcal{L}\{f(t)\}}{s}\right\} = \displaystyle\int_0^t f(\alpha)\, d\alpha$

Section 7.4

Solving initial-value problems with the Laplace transform

Section 7.5

Unit step function

$\mathcal{U}(t - a) = \begin{cases} 0, & 0 \le t < a \\ 1, & t \ge a \end{cases}$

$\mathcal{L}\{f(t - a)\mathcal{U}(t - a)\} = e^{-as}F(s)$

$\mathcal{L}^{-1}\{e^{-as}F(s)\} = f(t - a)\mathcal{U}(t - a)$

Periodic function

$f(t + T) = f(t)$

Laplace transform of a periodic function

$\mathcal{L}\{f(t)\} = \dfrac{1}{1 - e^{-sT}} \displaystyle\int_0^T e^{-st}f(t)\, dt$

Unit impulse function

$\delta(t - t_0) = 0,\ t \ne t_0$

$\displaystyle\int_{-\infty}^{+\infty} \delta(t - t_0)\, dt = 1$

$\mathcal{L}\{\delta(t - t_0)\} = e^{-st_0}$

Section 7.6

Convolution theorem

$\mathcal{L}^{-1}\{F(s)G(s)\} = \mathcal{L}^{-1}\{\mathcal{L}\{(f * g)(t)\}\}$

$\qquad = (f * g)(t)$

$\qquad = \displaystyle\int_0^t f(t - v)g(v)\, dv.$

Integral equation

Integrodifferential equation

Section 7.7

Spring-mass system with a discontinuous forcing function

_L-R-C_ circuits in which the voltage is defined by a piecewise defined function

Population problems

CHAPTER 7 REVIEW EXERCISES

In Exercises 1–4, find the Laplace transform of each function using the definition.

1. $f(t) = 1 - t$ **2.** $f(t) = te^{-4t}$

*3. $f(t) = \begin{cases} 1, & 0 \le t < 5, \\ 0, & t \ge 5 \end{cases}$

4. $f(t) = \begin{cases} t, & 0 \le t < 1, \\ 0, & t \ge 1 \end{cases}$

In Exercises 5–26, find the Laplace transform of each function.

5. $t^5 + 5$

6. $2 \sinh 4t$

7. te^{2t}

8. t^3

9. t^3e^t

10. $2e^t \sin 5t$

11. $t \cos 3t$

12. $t \sin 2t$

13. $e^{-5t} \cos 3t$

14. $\delta(t - 2\pi)$

15. $\delta(t - 3\pi/2)$

16. $7\mathcal{U}(t - 7)$

17. $6\mathcal{U}(t - 7) - 4\mathcal{U}(t - 4)$

18. $36e^{-4t}\mathcal{U}(t - 2)$

19. $-42e^{5t}\mathcal{U}(t - 1)$

20. $5\sin(t - 5)\mathcal{U}(t - 5)$

21. $t^2\mathcal{U}(t - 2)$

22. $f(t) = \begin{cases} 1, & 0 \le t < 1 \\ -1, & 1 \le t < 2 \end{cases}$, $f(t) = f(t - 2)$ if $t \ge 2$

*23. $f(t) = \begin{cases} t, & 0 \le t < 1 \\ 2 - t, & 1 \le t < 2 \end{cases}$, $f(t) = f(t - 2)$ if $t \ge 2$

24. $f(t) = \sin \pi t$ if $0 \le t < 1$, $f(t) = f(t - 1)$ if $t \ge 1$

*25. $f(t) = \begin{cases} \sin \pi t, & \text{if } 0 \le t < 1 \\ 0, & 1 \le t < 2 \end{cases}$, $f(t) = f(t - 2)$ if $t \ge 2$

26. $f(t) = \begin{cases} t, & 0 \le t < 2 \\ 2, & 2 \le t < 4 \\ 6 - t, & 4 \le t < 6 \end{cases}$, $f(t) = f(t - 6)$ if $t \ge 6$

In Exercises 27–36, find the inverse Laplace transform of each function.

27. $\dfrac{-10}{s^2 - 25}$

28. $\dfrac{-2s}{(s^2 + 1)^2}$

*29. $\dfrac{3(40 - s^5)}{s^6}$

30. $\dfrac{2s^2 - 7s + 20}{s(s^2 - 2s + 10)}$

*31. $\dfrac{-s^2 - 15s - 52}{s(s^2 + 10s + 26)}$

32. $\dfrac{-14}{se^{2s}}$

33. $\dfrac{7 + 6e^{4s}}{se^{7s}}$

34. $\dfrac{-3e^{6-s}}{s - 6}$

35. $\dfrac{8s}{e^{3s}(s^2 + 1)}$

36. $\dfrac{-18}{(1 - e^{-3s})(s^2 + 1)}$

In Exercises 37 and 38, compute the convolution $(f * g)(t)$ using the given pair of functions.

37. $f(t) = t^2$, $g(t) = e^{-3t}$

38. $f(t) = \cos t$, $g(t) = \sin t$

In Exercises 39–48, solve the initial-value problem.

39. $y'' + 6y' + 10y = 0$, $y(0) = 0$, $y'(0) = 1$

40. $y'' - 4y' + 5y = 0$, $y(0) = 1$, $y'(0) = 0$

41. $y'' + 5ty' - 10y = 2$, $y(0) = 1$, $y'(0) = 0$

42. $y'' + 12y' + 32y = f(t)$, $y(0) = 0$, $y'(0) = -1$,

$f(t) = \begin{cases} t, & \text{if } 0 \le t < 1 \\ 2 - t, & \text{if } 1 \le t < 2 \end{cases}$ and $f(t) = f(t - 2)$

if $t \ge 2$

*43. $y'' + 6y' + 8y = f(t)$, $y(0) = 1$, $y'(0) = 0$,

$f(t) = \begin{cases} 0, & \text{if } 0 \le t < 1 \\ 1, & \text{if } t \ge 1 \end{cases}$

44. $x'' + 3x' + 2x = \delta(t - \pi) + \delta(t - 2\pi)$, $x(0) = 0$, $x'(0) = 0$

*45. $x'' + 9x = \cos t + \delta(t - \pi)$, $x(0) = 0$, $x'(0) = 0$

46. $g(t) = 5 + t - \displaystyle\int_0^t (t - v)g(v)\, dv$

*47. $g(t) = \sin t + \displaystyle\int_0^t g(t - v)e^{-v}\, dv$

48. $\dfrac{dy}{dt} - \displaystyle\int_0^t y(v)\, dv = 1 + e^{-t}$, $y(0) = 0$

49. Use the Maclaurin series expansion $\tan^{-1} x = x - \dfrac{x^3}{3} + \dfrac{x^5}{5} + \dfrac{x^7}{7} + \cdots$ to find the Maclaurin series expansion for $\tan^{-1}(1/s)$. Use this expansion to show that $\mathcal{L}^{-1}\{\tan^{-1}(1/s)\} = \dfrac{\sin t}{t}$. (*Hint:* Use the Maclaurin series expansion of $\sin x$.)

50. Use $\mathcal{L}\{t^n f(t)\} = (-1)^n \dfrac{d^n}{ds^n} F(s)$ in the form ($n = 1$)

$f(t) = -\dfrac{1}{t} \mathcal{L}^{-1}\left\{ \dfrac{d}{ds} F(s) \right\}$ to compute

(a) $\mathcal{L}^{-1}\left\{ \ln \dfrac{s - 5}{s + 2} \right\}$ and **(b)** $\mathcal{L}^{-1}\left\{ \ln \dfrac{s^2 + 4}{s^2 + 9} \right\}$.

51. Use the shifting property to evaluate **(a)** $\mathcal{L}\{\cosh t\}$; **(b)** $\mathcal{L}\{\sinh t\}$; **(c)** $\mathcal{L}\{\cos t \cosh t\}$; **(d)** $\mathcal{L}\{\sin t \cosh t\}$; **(e)** $\mathcal{L}\{\cos t \sinh t\}$; **(f)** $\mathcal{L}\{\sin t \sinh t\}$.

52. Compute $\mathcal{L}\{e^{at}t^n\}$ two ways by using: **(a)** the shifting (translation) property; **(b)** the formula for the derivatives of the Laplace transform.

In Exercises 53 and 54, solve the initial-value problem $mx'' + cx' + kx = f(t)$, $x(0) = a$, $x'(0) = b$ to determine the motion of an object attached to the end of a spring using the given parameter values.

53. $m = 4$, $c = 0$, $k = 1$, $a = 0$, $b = -1$,

$f(t) = \begin{cases} \cos 2t, & 0 \le t < \pi \\ 0, & t \ge \pi \end{cases}$

54. $m = 1$, $c = \frac{1}{2}$, $k = \frac{145}{16}$, $a = 0$, $b = 0$,

$f(t) = \delta(t - \pi)$

In Exercises 55 and 56, solve the *L-R-C* series circuit modeled
by

$$\begin{cases} L\dfrac{d^2Q}{dt^2} + R\dfrac{dQ}{dt} + \dfrac{1}{C}Q = E(t) \\ Q(0) = Q_0,\ I(0) = \dfrac{dQ}{dt}(0) = I_0 \end{cases}$$

using the given parameter values and functions. (Assume that
the units of henry, ohm, farad, and volts are used, respectively,
for *L*, *R*, *C*, and *E(t)*.)

55. $L = 1$, $R = 0$, $C = 10^{-4}$,

$$E(t) = \begin{cases} 220, & 0 \le t < 2 \\ 0, & t \ge 2 \end{cases},\ Q_0 = I_0 = 0$$

56. $L = 4$, $R = 80$, $C = 0.04$,

$$E(t) = \begin{cases} 0, & 0 \le t < 1 \\ 50, & 1 \le t < 2 \end{cases},$$
$$E(t + 2) = E(t),\ Q_0 = I_0 = 0$$

In Exercises 57 and 58, solve the *L-R-C* series circuit for *I(t)* by
interpreting the problem as an integrodifferential equation
using the given parameter values and functions. (Assume that
the units of henry, ohm, farad, and volts are used, respectively,
for *L*, *R*, *C*, and *E(t)*.)

57. $L = \frac{1}{4}$, $R = \frac{1}{2}$, $C = \frac{4}{9}$, $E(t) = 100$, $I_0 = 0$

58. $L = \frac{1}{4}$, $R = \frac{3}{2}$, $C = \frac{4}{9}$, $E(t) = 100 \sin 5t$, $I_0 = 0$

In Exercises 59 and 60, interpret the initial-value problem as a
population problem. In each case, determine if the population
approaches a limit as $t \to \infty$.

59. $x' - 2x = 100\delta(t - 1)$, $x(0) = 10{,}000$

60. $x' - 2x = \begin{cases} 100t, & 0 \le t < 1 \\ 100, & t \ge 1 \end{cases}$, $x(0) = 10{,}000$

61. Suppose that the vertical displacement of a horizontal
beam is modeled by the boundary-value problem

$$\dfrac{d^4y}{dx^4} = W,\ y(0) = 0,\ y'(0) = 0,\ y(1) = 0,\ y'(1) = 0,$$

where *W* is the constant load that is uniformly distrib-
uted along the beam. Use the two conditions given at
$x = 0$ with the method of Laplace transforms to obtain a solution that involves the arbitrary constants

$A = y''(0)$ and $B = y'''(0)$. Then apply the conditions
at $x = 1$ to find *A* and *B*.

62. Suppose that the load in Problem 61 is not constant.
Instead, it is given by $W(x) = \begin{cases} 10, & 0 \le x < \frac{1}{2} \\ 0, & \frac{1}{2} \le x < 1 \end{cases}$.

Solve

$$\dfrac{d^4y}{dx^4} = W(x),\ y(0) = 0,\ y'(0) = 0,\ y(1) = 0,\ y'(1) = 0$$

to find the displacement of the beam.

63. (Filtration Through a Burning Cigarette) Suppose
that a cigarette has original length *X* and that the tip
burns at a rate *v* so that the cigarette has length $X - vt$
at time *t*. Next, let $W(x, t)$ represent the weight per
unit length of the cigarette at position *x* and time *t*,
where *x* is measured from the burning end. Therefore,
the weight at the burning tip is $W(vt, t)$. Assume that
during steady inhalation, a constant fraction *a* of any
component of the tobacco that is burned at the tip is
drawn through the cigarette and that the weight deposited per unit length is $bw_0 e^{-bx}$ (where w_0 is the
weight at $x = 0$). During the time period Δt, a length
$v\Delta t$ is burned so that $av\Delta t W(vt, t)$ passes through the
cigarette and $abv\Delta t W(vt, t)e^{-b(x-vt)}\,\Delta x$ is deposited
between *x* and $x + \Delta x$. Therefore, the total weight
deposited by absorption in time *t* between *x* and
$x + \Delta x$ is

$$abv \int_0^t W(vu, u)e^{-b(x-vu)}\,du\,\Delta x = w(x, t)\Delta x.$$

The total weight is then $w(x, t)\Delta x + W(x, 0)\Delta x$ so
that

$$W(x, t) = W(x, 0) + abve^{-bx}\int_0^t W(vu, u)e^{bvu}\,du.{}^*$$

(a) Substitute $x = vt$ to show that

$$f(t) = W(vt, 0)e^{bvt} + abv \int_0^t f(u)\,du$$

where $f(t) = W(vt, t)e^{bvt}$.
(b) Solve the integral equation in **(a)** if $W(vt, 0) = W_0$.

* *Applications in Undergraduate Mathematics in Engineering*, The Mathematical Association of America, The Macmillan Company, New York, 1967, pp. 153–157.

Differential Equations at Work:
The Tautochrone

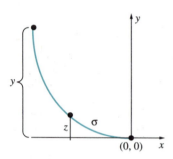

Figure 7.27

From rest, a particle slides down a frictionless curve under the force of gravity as illustrated in Figure 7.27. What must the shape of the curve be in order for the time of descent to be independent of the starting position of the particle?

The shape of the curve is found through the use of the Laplace transform. (This problem was originally solved in 1673 by Christian Huygens, a Dutch mathematician, as he studied the mathematics associated with pendulum clocks.) Suppose that the particle starts at height y and that its speed is v when it is at a height of z. If m is the mass of the particle and g is the acceleration of gravity, then the speed can be found by equating the kinetic and potential energies of the particle with

$$\frac{1}{2}mv^2 = mg(y - z)$$

which can written as

$$v = \sqrt{2g}\sqrt{y - z}.$$

To avoid confusion with the s that is usually used in the Laplace transform of functions, let σ denote the arc length along the curve from its lowest point to the particle. Therefore, the time required for the descent is

$$\text{time} = \int_0^{\sigma(y)} \frac{d\sigma}{v} = \int_0^y \frac{1}{v}\frac{d\sigma}{dz}\,dz = \int_0^y \frac{1}{v}\phi(z)\,dz$$

where $\phi(y) = d\sigma/dy$, which means that $\phi(z)$ is the value of $d\sigma/dy$ at $y = z$. Now, because the time is constant and $v = \sqrt{2g}\sqrt{y - z}$, we have

$$\int_0^y \frac{\phi(z)}{\sqrt{y - z}}\,dz = c_1$$

where c_1 is a constant. In an attempt to use a convolution, we multiply by $e^{-sy}\,dy$ and integrate. Therefore,

$$\int_0^\infty e^{-sy}\int_0^y \frac{\phi(z)}{\sqrt{y - z}}\,dz\,dy = \int_0^\infty e^{-sy}c_1\,dy$$

$$\mathscr{L}\{\phi * y^{-1/2}\} = \mathscr{L}\{c_1\}.$$

By the Convolution theorem, this simplifies to

$$\mathscr{L}\{\phi\}\mathscr{L}\{y^{-1/2}\} = \frac{c_1}{s}.$$

Then, because $\mathcal{L}\{t^{-1/2}\} = \sqrt{\dfrac{\pi}{s}}$, we have

$$\mathcal{L}\{\phi\}\sqrt{\frac{\pi}{s}} = \frac{c_1}{s} \qquad \text{or} \qquad \mathcal{L}\{\phi\} = \frac{c_1}{\sqrt{\pi}}\frac{1}{\sqrt{s}}.$$

Applying the inverse Laplace transform with $\mathcal{L}^{-1}\{s^{-1/2}\} = \dfrac{t^{-1/2}}{\sqrt{\pi}}$ then yields

$$\phi = \frac{c_1}{\sqrt{\pi}}y^{-1/2} = ky^{-1/2}.$$

Recall that $\phi(y) = d\sigma/dy$ represents arc length. Hence, $\phi(y) = \dfrac{d\sigma}{dy} = \sqrt{1 + \left(\dfrac{dx}{dy}\right)^2}$, so substitution into the previous equation gives

$$\sqrt{1 + \left(\frac{dx}{dy}\right)^2} = ky^{-1/2} \qquad \text{or} \qquad 1 + \left(\frac{dx}{dy}\right)^2 = \frac{k^2}{y}.$$

Solving for $\dfrac{dx}{dy}$ then yields $\dfrac{dx}{dy} = \sqrt{\dfrac{k^2}{y} - 1}$, which can be integrated with the substitution $y = k^2 \sin^2 \theta$.

1. Find $x(\theta)$ using this substitution.

2. Use the identity $\sin^2 \theta = \dfrac{1 - \cos 2\theta}{2}$ to obtain a similar formula for $y(\theta)$.

3. Use a device capable of parametric plotting to graph the curve $(x(\theta), y(\theta))$ on $-\dfrac{\pi}{2} \leq \theta \leq 0$ for $k = 1, 2, 3$. How does the value of k affect the curve?

4. Find the time required for an object located at $\left(x\left(-\dfrac{\pi}{2}\right), y\left(-\dfrac{\pi}{2}\right)\right)$ to move along the curve to the point $(x(0), y(0))$. Does this result depend on the value of θ?

8

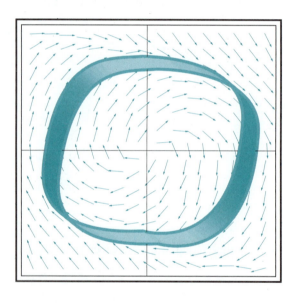

SYSTEMS OF ORDINARY DIFFERENTIAL EQUATIONS

8.1 Systems of Equations: Preliminary Theory and Methods

Preliminary Theory • Operator Notation • Solution Method Using Operator Notation

Preliminary Theory

Up to this point, we have focused our attention on solving differential equations that involve one dependent variable. However, many physical situations are modeled with more than one equation and involve more than one dependent variable. For example, if we want to determine the population of two interacting populations such as foxes and rabbits, we have two dependent variables representing the two populations; these populations depend on one independent variable representing time. Situations like this lead to systems of differential equations.

Definition 8.1 System of Ordinary Differential Equations

A **system** of ordinary differential equations is a simultaneous set of equations that involves two or more dependent variables that depend on one independent variable. A solution to the system is a set of differentiable functions that satisfies each equation on some interval I.

If the differential equations in a system are linear equations, we say that the system is a **linear system of differential equations** or a **linear system.**

Example 1

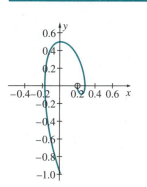

a.

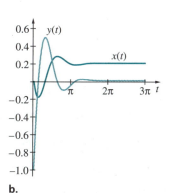

b.

Figure 8.1 (a) Graph of $\begin{cases} x(t) \\ y(t) \end{cases}$, $0 \le t \le 3\pi$ (b) Graph of $x(t)$ and $y(t)$, $0 \le t \le 3\pi$

Show that $\begin{cases} x(t) = \dfrac{1}{5}e^{-t}(e^t - \cos 2t - 3\sin 2t) \\ y(t) = -e^{-t}(\cos 2t - \sin 2t) \end{cases}$ is a solution to the system

$$\begin{cases} x' - y = 0 \\ y' + 5x + 2y = 1. \end{cases}$$

Solution The set of functions is a solution to the system of equations because

$$x' - y = -\frac{1}{5}e^{-t}(e^t - \cos 2t - 3\sin 2t) + \frac{1}{5}e^{-t}(e^t - 6\cos 2t + 2\sin 2t) +$$
$$e^{-t}(\cos 2t - \sin 2t) = 0$$

and

$$y' + 5x + 2y = 2e^{-t}(\cos 2t + \sin 2t) - e^{-t}(\cos 2t - \sin 2t) +$$
$$e^{-t}(e^t - \cos 2t - 3\sin 2t) = 1.$$

We graph this solution in Figure 8.1 in several different ways. First we graph the solution $\begin{cases} x(t) \\ y(t) \end{cases}$ parametrically. Then we graph $x(t)$ and $y(t)$ separately. Notice that $\lim_{t\to\infty} x(t) = \frac{1}{5}$ and $\lim_{t\to\infty} y(t) = 0$. (Why?) Therefore, in the parametric plot, the points on the curve approach the point $(\frac{1}{5}, 0)$.

As with other equations, under reasonable conditions a solution to a system of differential equations can always be found. The proof of the following theorem is omitted but can be found in more advanced texts.*

---

* See texts like *Principles of Differential and Integral Equations,* by C. Corduneanu and published by Chelsea Publishing Company, New York, 1977, or *Ordinary Differential Equations and Stability Theory: An Introduction,* by David A. Sánchez and published by Dover Publications, New York, 1968.

Theorem 8.1 Existence and Uniqueness

Assume that each of the functions $f_1(t, x_1, x_2, \ldots, x_n)$, $f_2(t, x_1, x_2, \ldots, x_n)$, $\ldots$, $f_n(t, x_1, x_2, \ldots, x_n)$ and the partial derivatives $\dfrac{\partial f_1}{\partial x_1}, \dfrac{\partial f_2}{\partial x_2}, \ldots, \dfrac{\partial f_n}{\partial x_n}$ are continuous in a region R containing the point $(t_0, y_1, y_2, \ldots, y_n)$. Then the initial-value problem

$$\begin{cases} x_1' = f_1(t, x_1, x_2, \ldots, x_n) \\ x_2' = f_2(t, x_1, x_2, \ldots, x_n) \\ \vdots \\ x_n' = f_n(t, x_1, x_2, \ldots, x_n) \\ x_1(t_0) = y_1, x_2(t_0) = y_2, \ldots, x_n(t_0) = y_n \end{cases}$$

has a unique solution

$$\begin{cases} x_1 = \phi_1(t) \\ x_2 = \phi_2(t) \\ \vdots \\ x_n = \phi_n(t) \end{cases}$$

on an interval I containing t_0.

Example 2   Show that the problem

$$\begin{cases} \dfrac{dx}{dt} = 2x - xy \\[2mm] \dfrac{dy}{dt} = -3y + xy \\[2mm] x(0) = 2, y(0) = \dfrac{3}{2} \end{cases}$$

has a unique solution.

Solution   We identify $f_1(t, x,y) = 2x - xy$ and $f_2(t, x, y) = -3y + xy$ with $\dfrac{\partial f_1}{\partial x} = 2 - y$ and $\dfrac{\partial f_2}{\partial y} = -3 + x$. All four of these functions are continuous on all regions containing $(0, 2, \tfrac{3}{2})$. (Why?) Thus, by the Existence and Uniqueness Theorem, a unique solution $\begin{cases} x(t) \\ y(t) \end{cases}$ to the initial-value problem exists. In this case, we used a computer algebra system to approximate the solution to this nonlinear problem. Figure 8.2(a) shows the graph of $x(t), y(t)$, and the parametric equations $\begin{cases} x(t) \\ y(t) \end{cases}$ for $0 \le t \le 10$ obtained with typical software. We can use the graph to approximate values of the solution. For example, if $t = 4$, using the graph we approximate

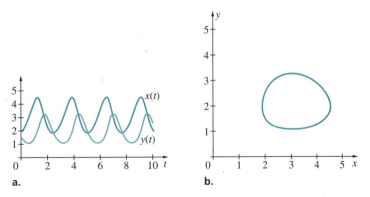

Figure 8.2 (a) Graph of $x(t)$ and $y(t)$ for $0 \leq t \leq 10$

(b) Graph of $\begin{cases} x(t) \\ y(t) \end{cases}$, for $0 \leq t \leq 10$

$x(4) \approx 4.3$ and $y(4) \approx 2.6$. The closed path in Figure 8.2(b) represents a **periodic solution** because motion that begins at a particular point on the curve eventually returns to that point and the motion repeats itself.

From the graph we see that both $x(t)$ and $y(t)$ are periodic. Use the graph to estimate the period of each.

We now turn our attention to one method for solving linear systems of differential equations.

Operator Notation

One method for solving systems of linear equations involves the differential operator

$$D = \frac{d}{dt},$$

which was first discussed in Section 4.4. We learned that the linear nth-order differential equation with constant coefficients

$$a_n \frac{d^n y}{dt^n} + a_{n-1} \frac{d^{n-1} y}{dt^{n-1}} + \cdots + a_1 \frac{dy}{dt} + a_0 y = f(t)$$

is expressed in *operator notation* as

$$(a_n D^n + a_{n-1} D^{n-1} + \cdots + a_1 D + a_0) y = f(t).$$

In a system of linear equations with constant coefficients, each equation can be written in operator notation.

Example 3

Write the following system of equations in operator notation:

$$\begin{cases} y'' + 2y' - x' + 4x = \cos t \\ \qquad x'' + y'' - y' = e^t - x + 2y \end{cases}$$

Solution   By first placing all dependent variables on the left-hand side of each equation, we have

$$\begin{cases} \quad y'' + 2y' - x' + 4x = \cos t \\ x'' + x + y'' - y' - 2y = e^t \end{cases} .$$

In operator notation, the system is

$$\begin{cases} \quad (D^2 + 2D)y - (D - 4)x = \cos t \\ (D^2 + 1)x + (D^2 - D - 2)y = e^t \end{cases} .$$

Solution Method Using Operator Notation

An advantage of operator notation is that systems of linear differential equations can be solved in much the same way as an algebraic system of equations. For example, consider the system

$$\begin{cases} Dx = y \\ Dy = -x \end{cases}$$

which can be written as

$$\begin{cases} Dx - y = 0 \\ x + Dy = 0 \end{cases} .$$

In order to eliminate one of the variables, we apply the operator $(-D)$ to the second equation to obtain the system

$$\begin{cases} \quad Dx - y = 0 \\ -Dx - D^2 y = 0 \end{cases} .$$

When these two equations are added, we have the second-order differential equation for y,

$$-D^2 y - y = 0 \qquad \text{or} \qquad (D^2 + 1)y = 0.$$

This equation has characteristic equation

$$m^2 + 1 = 0$$

with roots $m = \pm i$. Therefore,

$$y(t) = c_1 \cos t + c_2 \sin t.$$

At this point, we can repeat the elimination procedure to solve for x. On the other hand, we can use the second differential equation $x + Dy = 0$ or $x = -Dy$ to find x. Applying $(-D)$ to y then yields

$$x(t) = -Dy(t) = -D(c_1 \cos t + c_2 \sin t) = -(-c_1 \sin t + c_2 \cos t) =$$
$$c_1 \sin t - c_2 \cos t.$$

Therefore, a general solution of the system of equations is

$$\begin{cases} x(t) = c_1 \sin t - c_2 \cos t \\ y(t) = c_1 \cos t + c_2 \sin t \end{cases}$$

Of course, we can verify this result by substituting $x(t)$ and $y(t)$ into the system of equations.

If the elimination process were used to find $x(t)$, then two more arbitrary variables c_3 and c_4 would be generated. These extra coefficients must be eliminated by substituting $x(t)$ and $y(t)$ into the system and finding restrictions on these constants.

In order to better understand the number of arbitrary constants that appear in the solution, we consider the general form of the linear two-dimensional system which is written in operator notation as

$$\begin{cases} L_1 x + L_2 y = f_1(t) \\ L_3 x + L_4 y = f_2(t) \end{cases}$$

where L_1, L_2, L_3, and L_4 are *linear differential operators with constant real coefficients*. Applying the operator L_4 to the first equation and $-L_2$ to the second yields the system

$$\begin{cases} L_4 L_1 x + L_4 L_2 y = L_4 f_1(t) \\ -L_2 L_3 x - L_2 L_4 y = -L_2 f_2(t) \end{cases}$$

Now, because the operators have constant coefficients, $L_4 L_2 y = L_2 L_4 y$. When these equations are added, we have

$$(L_4 L_1 - L_2 L_3) x = g_1(t)$$

where $g_1(t) = L_4 f_1(t) - L_2 f_2(t)$. (Why?) Similarly, we can eliminate x to obtain the equation

$$(L_4 L_1 - L_2 L_3) y = g_2(t)$$

where $g_2(t) = L_1 f_2(t) - L_3 f_1(t)$. (Why?)

The left-hand side of the equations for x and y are the same (except for the variable) so the solutions x and y must depend on the same linearly independent functions, but *not* the same arbitrary constants.

The number of independent arbitrary constants in the solution will be the same as the order of the operator $L_1 L_4 - L_2 L_3$, which can be represented as the determinant

$$\begin{vmatrix} L_1 & L_2 \\ L_3 & L_4 \end{vmatrix} = L_1 L_4 - L_2 L_3.$$

Example 4

Solve the system

$$\begin{cases} x' + y' = e^t + 2x + 4y \\ x' + y' - y = e^{4t} \end{cases}.$$

Solution   We begin by writing the system in operator notation.

$$\begin{cases} (D - 2)x + (D - 4)y = e^t \\ Dx + (D - 1)y = e^{4t} \end{cases}.$$

Applying $L_4 = (D - 1)$ to the first equation and $-L_2 = -(D - 4)$ to the second yields the system

$$\begin{cases} (D - 1)(D - 2)x + (D - 1)(D - 4)y = (D - 1)e^t = e^t - e^t = 0 \\ -(D - 4)Dx - (D - 4)(D - 1)y = -(D - 4)e^{4t} = -(4e^{4t} - 4e^{4t}) = 0 \end{cases}$$

Adding then produces

$$(D - 1)(D - 2)x - (D - 4)Dx = [(D^2 - 3D + 2) - D^2 + 4D]x = (D + 2)x = 0.$$

The characteristic equation for the differential equation $(D + 2)x = 0$ is $m + 2 = 0$ with root $m = -2$ so

$$x(t) = c_1 e^{-2t}.$$

We solve for $y(t)$ by repeating the elimination procedure. Applying $L_3 = D$ to the first equation and $-L_1 = -(D - 2)$ to the second gives the system

$$\begin{cases} D(D - 2)x + D(D - 4)y = De^t = e^t \\ -(D - 2)Dx - (D - 2)(D - 1)y = -(D - 2)e^{4t} = -4e^{4t} + 2e^{4t} = -2e^{4t} \end{cases}$$

Adding the equations, we obtain the nonhomogeneous equation for y,

$$D(D - 4)y - (D - 2)(D - 1)y = e^t - 2e^{4t}.$$

A general solution of the corresponding homogeneous equation

$$D(D - 4)y - (D - 2)(D - 1)y = (D^2 - 4D - D^2 + 3D - 2)y = (-D - 2)y = 0$$

is $y_h(t) = k_1 e^{-2t}$. Using the method of undetermined coefficients, we assume there is a particular solution of the form $y_p(t) = Ae^t + Be^{4t}$. Substitution of $y_p(t)$ into the equation $(-D - 2)y = e^t - 2e^{4t}$ yields

$$(-D - 2)y_p(t) = -Dy_p(t) - 2y_p(t) = -Ae^t - 4Be^{4t} - 2Ae^t - 2Be^{4t} = \\ -3Ae^t - 6Be^{4t} = e^t - 2e^{4t},$$

so $A = -\frac{1}{3}$ and $B = \frac{1}{3}$. Therefore,

$$y(t) = y_h(t) + y_p(t) = k_1 e^{-2t} - \frac{1}{3}e^t + \frac{1}{3}e^{4t}.$$

Notice that

$$\begin{vmatrix} L_1 & L_2 \\ L_3 & L_4 \end{vmatrix} = \begin{vmatrix} D - 2 & D - 4 \\ D & D - 1 \end{vmatrix} = (D^2 - 3D + 2) - (D^2 - 4D) = D + 2.$$

There is only *one* arbitrary constant in the solution because the order of $D + 2$ is 1. Substituting $x(t)$ and $y(t)$ into the system yields

$$(D - 2)(c_1 e^{-2t}) + (D - 4)\left(k_1 e^{-2t} - \frac{1}{3}e^t + \frac{1}{3}e^{4t}\right)$$

$$= -2c_1 e^{-2t} - 2c_1 e^{-2t} - 2k_1 e^{-2t} - \frac{1}{3}e^t + \frac{4}{3}e^{4t} - 4k_1 e^{-2t} + \frac{4}{3}e^t - \frac{4}{3}e^{4t}$$

$$= (-4c_1 - 6k_1)e^{-2t} + e^t = e^t.$$

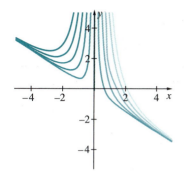

Figure 8.3

Therefore, $-4c_1 - 6k_1 = 0$, so $k_1 = -\frac{2}{3}c_1$. Making this substitution, we have

$$\begin{cases} x(t) = c_1 e^{-2t} \\ y(t) = -\dfrac{2}{3}c_1 e^{-2t} - \dfrac{1}{3}e^t + \dfrac{1}{3}e^{4t} \end{cases}.$$

Graphs of several solutions are shown in Figure 8.3.

Is there a solution to this system of equations that satisfies the initial conditions $x(0) = 1$ and $y(0) = 1$?

The determinant $\begin{vmatrix} L_1 & L_2 \\ L_3 & L_4 \end{vmatrix} = L_1 L_4 - L_2 L_3$ can be used to solve the system $\begin{cases} L_1 x + L_2 y = f_1(t) \\ L_3 x + L_4 y = f_2(t) \end{cases}$. In earlier calculations, we obtained the formulas

$$(L_1 L_4 - L_2 L_3)x = L_4 f_1(t) - L_2 f_2(t) \qquad \text{and} \qquad (L_1 L_4 - L_2 L_3)y = L_1 f_2(t) - L_3 f_1(t)$$

so

$$\begin{vmatrix} L_1 & L_2 \\ L_3 & L_4 \end{vmatrix} x = \begin{vmatrix} f_1(t) & L_2 \\ f_2(t) & L_4 \end{vmatrix} \qquad \text{and} \qquad \begin{vmatrix} L_1 & L_2 \\ L_3 & L_4 \end{vmatrix} y = \begin{vmatrix} L_1 & f_1(t) \\ L_3 & f_2(t) \end{vmatrix}.$$

Example 5

Solve the initial-value problem

$$\begin{cases} x' = x - 4y + 1 \\ y' = x + y \end{cases}, \quad x(0) = 1, \ y(0) = 0.$$

Solution In operator notation, the system is

$$\begin{cases} (D-1)x + 4y = 1 \\ -x + (D-1)y = 0 \end{cases}.$$

Then we have

$$\begin{vmatrix} D-1 & 4 \\ -1 & D-1 \end{vmatrix} x = \begin{vmatrix} 1 & 4 \\ 0 & D-1 \end{vmatrix} \qquad \text{and} \qquad \begin{vmatrix} D-1 & 4 \\ -1 & D-1 \end{vmatrix} y = \begin{vmatrix} D-1 & 1 \\ -1 & 0 \end{vmatrix}.$$

Solving for x yields $[(D-1)(D-1)+4]x = (D-1)[1] - 0 = 0 - 1 = -1$, which can be written as $(D^2 - 2D + 5)x = -1$. The corresponding homogeneous equation $(D^2 - 2D + 5)x = 0$ has characteristic equation $m^2 - 2m + 5 = 0$ with roots $m = 1 \pm 2i$, so a general solution is $x_h(t) = e^t(c_1 \cos 2t + c_2 \sin 2t)$. By the method of undetermined coefficients, we assume that a particular solution is $x_p(t) = A$. Substitution into the equation shows that $5A = -1$, so $x_p(t) = -\frac{1}{5}$ and $x(t) = e^t(c_1 \cos 2t + c_2 \sin 2t) - \frac{1}{5}$.

Next we solve for $y(t)$. Expanding $\begin{vmatrix} D-1 & 4 \\ -1 & D-1 \end{vmatrix} y = \begin{vmatrix} D-1 & 1 \\ -1 & 0 \end{vmatrix}$, we

obtain $(D^2 - 2D + 5)y = 1$. Following the same steps as those used in finding $x(t)$, we find that $y(t) = e^t(k_1 \cos 2t + k_2 \sin 2t) + \frac{1}{5}$. Because the operator

$$\begin{vmatrix} D-1 & 4 \\ -1 & D-1 \end{vmatrix} = D^2 - 2D + 5$$ is of order two, there are only two arbitrary constants in the solution. Substituting $x(t)$ and $y(t)$ into the first equation of the system yields

$$x' - x + 4y = e^t(c_1 \cos 2t + c_2 \sin 2t) + e^t(-2c_1 \sin 2t + 2c_2 \cos 2t)$$

$$-e^t(c_1 \cos 2t + c_2 \sin 2t) + \frac{1}{5} + 4e^t(k_1 \cos 2t + k_2 \sin 2t) + \frac{4}{5}$$

$$= (2c_2 + 4k_1)e^t \cos 2t + (-2c_1 + 4k_2)e^t \sin 2t + 1.$$

Therefore, $2c_2 + 4k_1 = 0$ and $-2c_1 + 4k_2 = 0$, so $k_1 = -\frac{1}{2}c_2$, $k_2 = \frac{1}{2}c_1$, and a general solution is

$$\begin{cases} x(t) = e^t(c_1 \cos 2t + c_2 \sin 2t) - \dfrac{1}{5} \\[2mm] y(t) = e^t\left(-\dfrac{1}{2}c_2 \cos 2t + \dfrac{1}{2}c_1 \sin 2t\right) + \dfrac{1}{5} \end{cases}.$$

The solution to the initial-value problem must satisfy $x(0) = 1$ and $y(0) = 0$, so $x(0) = c_1 - \frac{1}{5} = 1$ and $y(0) = -\frac{1}{2}c_2 + \frac{1}{5} = 0$, which yields $c_1 = \frac{6}{5}$ and $c_2 = \frac{2}{5}$. The solution is

$$\begin{cases} x(t) = e^t\left(\dfrac{6}{5} \cos 2t + \dfrac{2}{5} \sin 2t\right) - \dfrac{1}{5} \\[2mm] y(t) = e^t\left(-\dfrac{1}{5} \cos 2t + \dfrac{3}{5} \sin 2t\right) + \dfrac{1}{5} \end{cases},$$

which is graphed parametrically in Figure 8.4. What is the orientation of this curve?

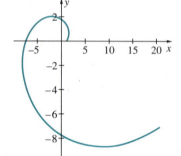

Figure 8.4

Note that if $\begin{vmatrix} L_1 & L_2 \\ L_3 & L_4 \end{vmatrix} = 0$, we say the system is **degenerate,** which means that the system either has infinitely many solutions or no solutions. If $\begin{vmatrix} L_1 & L_2 \\ L_3 & L_4 \end{vmatrix} = \begin{vmatrix} f_1(t) & L_2 \\ f_2(t) & L_4 \end{vmatrix} = \begin{vmatrix} L_1 & f_1(t) \\ L_3 & f_2(t) \end{vmatrix} = 0$, the system has **infinitely many solutions.** If $\begin{vmatrix} L_1 & L_2 \\ L_3 & L_4 \end{vmatrix} = 0$ and at least one of the other two determinants is not zero, the system has **no solutions.**

Example 6

Determine if the following systems have infinitely many or no solutions.

(a) $\begin{cases} x' - x + y' - y = -3e^{-2t} \\ x' + 2x + y' + 2y = 3e^t \end{cases}$ (b) $\begin{cases} x' + y' + y = e^t \\ x' + y' + y = e^t + 3 \end{cases}$

Solution   (a) Representing the system in operator notation, we obtain

$$\begin{cases} (D-1)x + (D-1)y = -3e^{-2t} \\ (D+2)x + (D+2)y = 3e^t \end{cases}.$$

We evaluate $\begin{vmatrix} D-1 & D-1 \\ D+2 & D+2 \end{vmatrix} = 0$, so the system either has infinitely many or no solutions. Because,

$$\begin{vmatrix} -3e^{-2t} & D-1 \\ 3e^t & D+2 \end{vmatrix} = D(-3e^{-2t}) + 2(-3e^{-2t}) - D(3e^t) + 3e^t$$

$$= 6e^{-2t} - 6e^{-2t} - 3e^t + 3e^t = 0$$

and

$$\begin{vmatrix} D-1 & -3e^{-2t} \\ D+2 & 3e^t \end{vmatrix} = D(3e^t) - 3e^t - D(-3e^{-2t}) - 2(-3e^{-2t})$$

$$= 3e^t - 3e^t - 6e^{-2t} + 6e^{-2t} = 0,$$

the system has infinitely many solutions.

(b) In operator notation, the system is $\begin{cases} Dx + (D+1)y = e^t \\ Dx + (D+1)y = e^t + 3 \end{cases}$. In this case,

$\begin{vmatrix} D & D+1 \\ D & D+1 \end{vmatrix} = 0$. However,

$$\begin{vmatrix} e^t & D+1 \\ e^t + 3 & D+1 \end{vmatrix} = D(e^t) + e^t - D(e^t + 3) - (e^t + 3) = e^t + e^t - e^t - 3$$

$$= -3 \neq 0,$$

so the system has no solutions.

Systems of three or more equations can be solved with the operator method as well.

Example 7

Solve the system of differential equations

$$\begin{cases} x' = x + 2y + z \\ y' = 6x - y \\ z' = -x - 2y - z \end{cases}.$$

Solution In operator notation, this system is

$$\begin{cases} (D-1)x - 2y - z = 0 \\ -6x + (D+1)y = 0. \\ x + 2y + (D+1)z = 0 \end{cases}$$

Because the second equation only involves x and y, we eliminate z from the first and third equations. Applying $(D+1)$ to the first equation yields

$$(D+1)(D-1)x - 2(D+1)y - (D+1)z = 0.$$

Adding this equation and the third equation then gives us

$$[(D+1)(D-1)+1]x + [2 - 2(D+1)]y = 0$$
$$D^2x - 2Dy = 0.$$

We now eliminate x by applying 6 to $D^2x - 2Dy = 0$, D^2 to $-6x + (D + 1)y = 0$ and adding to obtain

$$[D^2(D + 1) - 12D]y = (D^3 + D^2 - 12D)y = 0.$$

This homogeneous equation has characteristic equation

$$m^3 + m^2 - 12m = m(m^2 + m - 12) = m(m + 4)(m - 3) = 0$$

so

$$y(t) = c_1 + c_2e^{-4t} + c_3e^{3t}.$$

Instead of repeating the elimination process with x and z, we use the equation $y' = 6x - y$ or $x = \frac{1}{6}(y' + y)$ to find x. Therefore,

$$x(t) = \frac{1}{6}(-4c_2e^{-4t} + 3c_3e^{3t} + c_1 + c_2e^{-4t} + c_3e^{3t})$$

$$= \frac{1}{6}c_1 - \frac{1}{2}c_2e^{-4t} + \frac{2}{3}c_3e^{3t}.$$

Finally, we use $x' = x + 2y + z$ to find $z(t)$.

$$z(t) = x' - x - 2y$$

$$= 2c_2e^{-4t} + 2c_3e^{3t} - \left(\frac{1}{6}c_1 - \frac{1}{2}c_2e^{-4t} + \frac{2}{3}c_3e^{3t}\right) - 2(c_1 + c_2e^{-4t} + c_3e^{3t})$$

$$= -\frac{13}{6}c_1 + \frac{1}{2}c_2e^{-4t} - \frac{2}{3}c_3e^{3t}.$$

Another way to solve these systems is through the use of determinants. For the system of three equations

$$\begin{cases} L_1x + L_2y + L_3z = f_1(t) \\ L_4x + L_5y + L_6z = f_2(t), \\ L_7x + L_8y + L_9z = f_3(t) \end{cases}$$

we find $x(t)$, $y(t)$, and $z(t)$ with

$$\begin{vmatrix} L_1 & L_2 & L_3 \\ L_4 & L_5 & L_6 \\ L_7 & L_8 & L_9 \end{vmatrix} x = \begin{vmatrix} f_1(t) & L_2 & L_3 \\ f_2(t) & L_5 & L_6 \\ f_3(t) & L_8 & L_9 \end{vmatrix}, \qquad \begin{vmatrix} L_1 & L_2 & L_3 \\ L_4 & L_5 & L_6 \\ L_7 & L_8 & L_9 \end{vmatrix} y = \begin{vmatrix} L_1 & f_1(t) & L_3 \\ L_4 & f_2(t) & L_6 \\ L_7 & f_3(t) & L_9 \end{vmatrix},$$

and

$$\begin{vmatrix} L_1 & L_2 & L_3 \\ L_4 & L_5 & L_6 \\ L_7 & L_8 & L_9 \end{vmatrix} z = \begin{vmatrix} L_1 & L_2 & f_1(t) \\ L_4 & L_5 & f_2(t) \\ L_7 & L_8 & f_3(t) \end{vmatrix}.$$

IN TOUCH WITH TECHNOLOGY

In the same way we used computer algebra systems to solve other differential equations, we also use them to solve many linear systems.

1. Solve each of the following systems. In each case, graph $x(t)$, $y(t)$, and the parametric equations $\begin{cases} x(t) \\ y(t) \end{cases}$.

(a) $\begin{cases} x' = \dfrac{1}{3}x + \dfrac{2}{3}y \\ y' = x - \dfrac{1}{2}y \\ x(0) = 1, \, y(0) = -1 \end{cases}$

(b) $\begin{cases} x' = -5x - y \\ y' = x - 5y \\ x(0) = -2, \, y(0) = 1 \end{cases}$

(c) $\begin{cases} x' = -6x + 5y - \cos\left(\frac{1}{2}t\right) \\ y' = -\frac{5}{2}x + y + e^{-t} \\ x(0) = 2, \, y(0) = 0 \end{cases}$

2. Consider the initial-value problem

$$\begin{cases} x' = -2x - y \\ y' = \dfrac{5}{4}x \\ x(0) = 1, \, y(0) = 0 \end{cases}.$$

The solution obtained with one computer algebra system was

$$\begin{cases} x(t) = \left(\frac{1}{2} - i\right)e^{(-1-i/2)t} + \left(\frac{1}{2} + i\right)e^{(-1+i/2)t} \\ y(t) = \frac{5}{4}ie^{(-1-i/2)t} - \frac{5}{4}ie^{(-1+i/2)t} \end{cases}$$

while the solution obtained with another was

$$\begin{cases} x(t) = e^{-t}\left(\cos\left(\frac{1}{2}t\right) - 2\sin\left(\frac{1}{2}t\right)\right) \\ y(t) = \frac{5}{2}e^{-t}\sin\left(\frac{1}{2}t\right) \end{cases}.$$

Show that these solutions are equivalent.

Often, we can use a computer algebra system to generate numerical solutions to a system of equations, which is particularly useful if the system under consideration is nonlinear. (Numerical techniques for systems are discussed in more detail in Section 8.8.)

3. Figure 8.5 shows the direction field associated with the nonlinear system of equations

$$\begin{cases} \dfrac{dx}{dt} = 2x - xy \\ \dfrac{dy}{dt} = -3y + xy \end{cases}$$

for $0 \le x \le 6$ and $0 \le y \le 6$. Use the direction field to sketch the graphs of the solutions $\begin{cases} x(t) \\ y(t) \end{cases}$ that satisfy the initial conditions **(a)** $x(0) = 2$ and $y(0) = 3$; and **(b)** $x(0) = 3$ and $y(0) = 2$. **(c)** How are the solutions alike? How are they different?

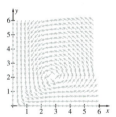

Figure 8.5

4. Under various assumptions, the nonlinear system of differential equations

$$\begin{cases} \dfrac{dX}{dt} = rX - \gamma XN - \beta XY \\ \dfrac{dI}{dt} = \beta XY - (\sigma + b + \gamma N)I \\ \dfrac{dY}{dt} = \sigma I - (\alpha + \beta + \gamma N)Y \\ \dfrac{dN}{dt} = aX - (b + \gamma N)N - \alpha Y \end{cases}$$

has been successfully used to model a fox population in which rabies is present.* Here, $X(t)$ represents the population of foxes susceptible to rabies at time t, $I(t)$ the population that has contracted the rabies virus but is not yet ill, $Y(t)$ the population that has developed rabies, and $N(t)$ the total population of the foxes. The symbols a, b, r, γ, σ, α, and β represent constants and are described in the following table.

Constant	Description	Typical Value(s)
a	a represents the *average per capita birth rate* of foxes.	1
b	$1/b$ denotes *fox life expectancy* (without resource limitations) which is typically in the range of 1.5 to 2.7 years.	0.5
r	$r = a - b$ represents the *intrinsic per capita population growth rate*.	0.5
γ	$K = \dfrac{r}{\gamma}$ represents the *fox carrying capacity* of the defined area, which is typically in the range of 0.1 to 4 foxes per km². We will compute K and r and then approximate γ.	Varies
σ	$1/\sigma$ represents the *average latent period*. This represents the average time (in years) that a fox can carry the rabies virus but not actually be ill with rabies. Typically, $1/\sigma$ is between 28 and 30 days.	12.1667
α	α represents the *death rate* of foxes with rabies. $1/\alpha$ is the life expectancy (in years) of a fox with rabies and is typically between 3 and 10 days.	73
β	β represents a *transmission coefficient*. Typically, $1/\beta$ is between 4 and 6 days.	80

(a) Generate a numerical solution to the system that satisfies the initial conditions $X(0) = 0.93$, $I(0) = 0.035$, $Y(0) = 0.035$, and $N(0) = 1.0$ valid for $0 \le t \le 40$ using the values given in the previous table if $K = 1, 2, 3, 4,$ and 8. In each case, graph $X(t)$, $I(t)$, $Y(t)$, and $N(t)$ for $0 \le t \le 40$.

(b) Repeat (a) using the initial conditions $X(0) = 0.93$, $I(0) = 0.02$, $Y(0) = 0.05$, and $N(0) = 2.0$.

(c) For both (a) and (b), estimate the smallest value of K, say K_T, so that $Y(t)$ is a periodic function.

(d) What happens to $Y(t)$ for $K < K_T$? Explain why this result does or does not make sense.

(e) Define the **basic reproductive rate** R to be

$$R = \frac{\sigma \beta K}{(\sigma + a)(\alpha + a)}$$

and

$$K_T = \frac{(\sigma + a)(\alpha + a)}{\sigma \beta}.$$

Show that if $R > 1$ then $K > K_T$ and if $R < 1$ then $K < K_T$.

(f) Use the values in the table to calculate $K_T = \dfrac{(\sigma + a)(\alpha + a)}{\sigma \beta}$. Compare the result to your approximations in (c). How do they compare?

(g) Predict how the solutions would change if the transmission coefficient β were decreased or the death rate α were increased. What if the average latent period σ were increased? Experiment with different conditions to see if you are correct.

---

* Roy M. Anderson, Helen C. Jackson, Robert N. May & Anthony M. Smith, "Population dynamics of fox rabies in Europe," *Nature*, Volume 289, February 26, 1981, pp. 765–771.

In Exercises 1–18, solve the system using the operator method.

1. $\begin{cases} x' = -2x - 2y + 4 \\ y' = -5x + y \end{cases}$

2. $\begin{cases} x' = 2x - y + t \\ y' = -4x + 2y + \sin t \end{cases}$

*3. $\begin{cases} x' = x + 2y + \cos t \\ y' = 3x + 2y \end{cases}$

4. $\begin{cases} x' = -x + 2y + 3e^t \\ y' = 3x - 6y + 4 \end{cases}$

5. $\begin{cases} x' = -3x + 2y - e^t \\ y' = -3y + 1 \end{cases}$

6. $\begin{cases} x' = 4x + y - 4\sin 2t \\ y' = 4y + \cos t \end{cases}$

*7. $\begin{cases} x'' = 2y - 5x \\ y'' = 2x - 2y \end{cases}$

8. $\begin{cases} x'' = 2y \\ y'' = -2x \end{cases}$

*9. $\begin{cases} x'' + y'' + x + y = 0 \\ y'' + 2x' - 2x - y = 0 \end{cases}$

10. $\begin{cases} x'' = 5y \\ y'' = 5x + 1 \end{cases}$

*11. $\begin{cases} x'' + y' = x - y + 1 \\ x' + y'' = x + t \end{cases}$

12. $\begin{cases} x'' = -3x + y - 1 \\ y'' = 2x - 2y + e^{3t} \end{cases}$

*13. $\begin{cases} x' = -3x + 2y + 2z \\ y' = -2x + 3y + 2z \\ z' = x + y \end{cases}$

14. $\begin{cases} x' = -3x \\ y' = -x \\ z' = 3x + z \end{cases}$

15. $\begin{cases} x' = x + 2y - 2z + \cos t \\ y' = x - y + 2z \\ z' = -x - y \end{cases}$

16. $\begin{cases} x' = -3x + 3y - 3z + e^{-2t} \\ y' = -x + 2y + z \\ z' = -3x + 2y - 3z - 1 \end{cases}$

*17. $\begin{cases} x' = x + 2y - z + 5 \\ y' = x + z - e^t \\ z' = 4x - 4y + 5z \end{cases}$

18. $\begin{cases} x' = 3x + 2y + z \\ y' = 2x - 2z - \sin 2t \\ z' = x + 3y + z \end{cases}$

In Exercises 19–24, solve the initial-value problem. For each problem, graph $x(t)$, $y(t)$, and the parametric equations $\begin{cases} x(t) \\ y(t) \end{cases}$.

19. $x' - x - 2y = 0$, $y' - 2x - y = 0$, $x(0) = 0$, $y(0) = -1$

20. $x' + 3x - 2y = 0$, $y' + 3x - 4y = 0$, $x(0) = 1$, $y(0) = 0$

*21. $x' - 3x + 2y = 0$, $y' - 2x + y = 10$, $x(0) = 0$, $y(0) = 0$

22. $x' - 5x - 9y = 0$, $y' + x - 11y = 0$, $x(0) = 0$, $y(0) = 4$

*23. $x'' = y - x$, $y'' = y - x + \sin t$, $x(0) = 0$, $x'(0) = 0$, $y(0) = 0$, $y'(0) = 0$

24. $x'' = y'' + t$, $y'' = -x''$, $x(0) = 1$, $x'(0) = 0$, $y(0) = 0$, $y'(0) = 0$

In Exercises 25 and 26, show that the system has no solution.

25. $\begin{cases} x' + x + y' = e^t \\ x' + x + y' = 2e^t \end{cases}$ **26.** $\begin{cases} x'' + y' = \cos t \\ 3x'' + 3y' = \sin t \end{cases}$

In Exercises 27 and 28, show that the system has infinitely many solutions.

27. $\begin{cases} x'' + y'' = t^2 \\ 4x'' + 4y'' = 4t^2 \end{cases}$ **28.** $\begin{cases} x'' + y'' - y' = te^t \\ 2x'' + 2y'' - 2y' = 2te^t \end{cases}$

In Exercises 29 and 30, determine the value of k so that the system has infinitely many solutions.

*29. $\begin{cases} x' + x - y' = t \\ 4y' - 4x' - 4x = kt \end{cases}$

30. $\begin{cases} y'' - x = \cos t \\ 4y'' + kx = 4\cos t \end{cases}$

In Exercises 31 and 32, determine restrictions on c so that the system has no solutions.

*31. $\begin{cases} x + x' - cy = e^{-t} \\ 2x' - 8y + 2x = 2e^{-t} \end{cases}$

32. $\begin{cases} x'' + 4y' = \cos t \\ 2x'' + 8y' = c\cos t \end{cases}$

8.2 Review of Matrix Algebra and Calculus

Basic Operations • **Determinants and Inverses** • **Eigenvalues and Eigenvectors** • **Matrix Calculus**

Because of their importance in the study of systems of linear equations, we now briefly review matrices and the basic operations associated with them. Detailed discussions of the definitions and properties discussed here are found in most introductory linear algebra texts.

Basic Operations

Definition 8.2 $n \times m$ Matrix

An $n \times m$ **matrix** is an array of the form

$$\begin{pmatrix} a_{11} & a_{12} & \cdots & a_{1m} \\ a_{21} & a_{22} & \cdots & a_{2m} \\ \vdots & \vdots & \ddots & \vdots \\ a_{n1} & a_{n2} & \cdots & a_{nm} \end{pmatrix}$$

with n rows and m columns. This matrix can be denoted $\mathbf{A} = (a_{ij})$.

We generally call an $n \times 1$ matrix $\begin{pmatrix} v_1 \\ v_2 \\ \vdots \\ v_n \end{pmatrix}$ a **column vector** and a $1 \times n$ matrix $(v_1 \quad v_2 \quad \cdots \quad v_n)$ a **row vector**.

Definition 8.3 Transpose

The **transpose** of the $n \times m$ matrix

$$\mathbf{A} = \begin{pmatrix} a_{11} & a_{12} & \cdots & a_{1m} \\ a_{21} & a_{22} & \cdots & a_{2m} \\ \vdots & \vdots & \ddots & \vdots \\ a_{n1} & a_{n2} & \cdots & a_{nm} \end{pmatrix}$$

is the $m \times n$ matrix

$$\mathbf{A}^T = \begin{pmatrix} a_{11} & a_{21} & \cdots & a_{n1} \\ a_{12} & a_{22} & \cdots & a_{n2} \\ \vdots & \vdots & \ddots & \vdots \\ a_{1m} & a_{2m} & \cdots & a_{nm} \end{pmatrix}.$$

Hence, $\mathbf{A}^T = (a_{ji})$.

Definition 8.4 Scalar Multiplication, Matrix Addition

Let $\mathbf{A} = (a_{ij})$ be an $n \times m$ matrix and c a scalar. The **scalar multiple** of $\mathbf{A}$ by c is the $n \times m$ matrix given by $c\mathbf{A} = (ca_{ij})$.

If $\mathbf{B} = (b_{ij})$ is also an $n \times m$ matrix, then the **sum** of matrices $\mathbf{A}$ and $\mathbf{B}$ is the $n \times m$ matrix $\mathbf{A} + \mathbf{B} = (a_{ij}) + (b_{ij}) = (a_{ij} + b_{ij})$.

Hence, $c\mathbf{A}$ is the matrix obtained by multiplying each element of $\mathbf{A}$ by c; $\mathbf{A} + \mathbf{B}$ is obtained by adding corresponding elements of the matrices $\mathbf{A}$ and $\mathbf{B}$ that have the same dimensions.

Example 1

Compute $3\mathbf{A} - 9\mathbf{B}$ if $\mathbf{A} = \begin{pmatrix} -1 & 4 & -2 \\ 6 & 2 & -10 \end{pmatrix}$ and $\mathbf{B} = \begin{pmatrix} 2 & -4 & 8 \\ 7 & 4 & 2 \end{pmatrix}$.

Solution Because $3\mathbf{A} = \begin{pmatrix} -3 & 12 & -6 \\ 18 & 6 & -30 \end{pmatrix}$ and

$-9\mathbf{B} = \begin{pmatrix} -18 & 36 & -72 \\ -63 & -36 & -18 \end{pmatrix}$, $3\mathbf{A} - 9\mathbf{B} = 3\mathbf{A} + (-9\mathbf{B}) = \begin{pmatrix} -21 & 48 & -78 \\ -45 & -30 & -48 \end{pmatrix}$.

Definition 8.5 Matrix Multiplication

If $\mathbf{A} = \begin{pmatrix} a_{11} & a_{12} & \cdots & a_{1j} \\ a_{21} & a_{22} & \cdots & a_{2j} \\ \vdots & \vdots & \ddots & \vdots \\ a_{n1} & a_{n2} & \cdots & a_{nj} \end{pmatrix}$ is an $n \times j$ matrix and

$\mathbf{B} = \begin{pmatrix} b_{11} & b_{12} & \cdots & b_{1m} \\ b_{21} & b_{22} & \cdots & b_{2m} \\ \vdots & \vdots & \ddots & \vdots \\ b_{j1} & b_{j2} & \cdots & b_{jm} \end{pmatrix}$ is a $j \times m$ matrix,

$$\mathbf{AB} = \begin{pmatrix} a_{11} & a_{12} & \cdots & a_{1j} \\ a_{21} & a_{22} & \cdots & a_{2j} \\ \vdots & \vdots & \ddots & \vdots \\ a_{n1} & a_{n2} & \cdots & a_{nj} \end{pmatrix} \begin{pmatrix} b_{11} & b_{12} & \cdots & b_{1m} \\ b_{21} & b_{22} & \cdots & b_{2m} \\ \vdots & \vdots & \ddots & \vdots \\ b_{j1} & b_{j2} & \cdots & b_{jm} \end{pmatrix}$$

is the unique $n \times m$ matrix

$$\mathbf{C} = \begin{pmatrix} c_{11} & c_{12} & \cdots & c_{1m} \\ c_{21} & c_{22} & \cdots & c_{2m} \\ \vdots & \vdots & \ddots & \vdots \\ c_{n1} & c_{n2} & \cdots & c_{nm} \end{pmatrix}$$

where

$$c_{11} = a_{11}b_{11} + a_{12}b_{21} + \cdots + a_{1j}b_{j1} = \sum_{k=1}^{j} a_{1k}b_{k1},$$

$$c_{12} = a_{11}b_{12} + a_{12}b_{22} + \cdots + a_{1j}b_{j2} = \sum_{k=1}^{j} a_{1k}b_{k2},$$

and

$$c_{uv} = a_{u1}b_{1v} + a_{u2}b_{2v} + \cdots + a_{uj}b_{jv} = \sum_{k=1}^{j} a_{uk}b_{kv}.$$

In other words, the element c_{uv} is obtained by multiplying each member of the uth row of $\mathbf{A}$ by the corresponding entry in the vth column of $\mathbf{B}$ and adding the result.

Example 2

Compute $\mathbf{AB}$ if $\mathbf{A} = \begin{pmatrix} 0 & 4 & 5 \\ -5 & -1 & 5 \end{pmatrix}$ and $\mathbf{B} = \begin{pmatrix} -3 & 4 \\ -5 & -4 \\ 1 & -4 \end{pmatrix}$.

Solution Because $\mathbf{A}$ is a 2×3 matrix and $\mathbf{B}$ is a 3×2 matrix, $\mathbf{AB}$ is the 2×2 matrix:

$$\mathbf{AB} = \begin{pmatrix} 0 & 4 & 5 \\ -5 & -1 & 5 \end{pmatrix} \begin{pmatrix} -3 & 4 \\ -5 & -4 \\ 1 & -4 \end{pmatrix}$$

$$= \begin{pmatrix} 0 \cdot -3 + 4 \cdot -5 + 5 \cdot 1 & 0 \cdot 4 + 4 \cdot -4 + 5 \cdot -4 \\ -5 \cdot -3 + -1 \cdot -5 + 5 \cdot 1 & -5 \cdot 4 + -1 \cdot -4 + 5 \cdot -4 \end{pmatrix}$$

$$= \begin{pmatrix} -15 & -36 \\ 25 & -36 \end{pmatrix}.$$

Definition 8.6 Identity Matrix

The $n \times n$ matrix $\begin{pmatrix} 1 & 0 & 0 & 0 \\ 0 & 1 & 0 & 0 \\ \vdots & \vdots & \ddots & \vdots \\ 0 & 0 & 0 & 1 \end{pmatrix}$ is called the $n \times n$ **identity matrix**, denoted by $\mathbf{I}$ or $\mathbf{I}_n$.

If $\mathbf{A}$ is an $n \times n$ matrix (an $n \times n$ is called a **square matrix**), then $\mathbf{IA} = \mathbf{AI} = \mathbf{A}$.

Determinants and Inverses

Definition 8.7 Determinant

If $\mathbf{A} = (a_{11})$, the **determinant** of $\mathbf{A}$, denoted by $\det(\mathbf{A})$ or $|\mathbf{A}|$, is $\det(\mathbf{A}) = a_{11}$; if

$$\mathbf{A} = \begin{pmatrix} a_{11} & a_{12} \\ a_{21} & a_{22} \end{pmatrix}, \text{ then}$$

$$\det(\mathbf{A}) = \begin{vmatrix} a_{11} & a_{12} \\ a_{21} & a_{22} \end{vmatrix} = a_{11}a_{22} - a_{12}a_{21}.$$

More generally, if $\mathbf{A} = \begin{pmatrix} a_{11} & a_{12} & \cdots & a_{1n} \\ a_{21} & a_{22} & \cdots & a_{2n} \\ \vdots & \vdots & \ddots & \vdots \\ a_{n1} & a_{n2} & \cdots & a_{nn} \end{pmatrix}$ is an $n \times n$ matrix and $\mathbf{A}_{ij}$ is

the $(n-1) \times (n-1)$ matrix obtained by deleting the ith row and jth column from $\mathbf{A}$, then

$$\det(\mathbf{A}) = \begin{vmatrix} a_{11} & a_{12} & \cdots & a_{1n} \\ a_{21} & a_{22} & \cdots & a_{2n} \\ \vdots & \vdots & \ddots & \vdots \\ a_{n1} & a_{n2} & \cdots & a_{nn} \end{vmatrix} = \sum_{j=1}^{n} (-1)^{i+j} a_{ij} \det(\mathbf{A}_{ij}) = \sum_{j=1}^{n} (-1)^{i+j} a_{ij} |\mathbf{A}ij|.$$

The number $(-1)^{i+j} a_{ij} \det(\mathbf{A}_{ij}) = (-1)^{i+j} a_{ij} |\mathbf{A}_{ij}|$ is called the **cofactor** of a_{ij}. The **cofactor matrix, $\mathbf{A}^c$,** of $\mathbf{A}$ is the matrix obtained by replacing each element of $\mathbf{A}$ by its cofactor. Hence,

$$\mathbf{A}^c = \begin{pmatrix} |\mathbf{A}_{11}| & -|\mathbf{A}_{12}| & \cdots & (-1)^{n+1}|\mathbf{A}_{1n}| \\ -|\mathbf{A}_{21}| & |\mathbf{A}_{22}| & \cdots & (-1)^{n}|\mathbf{A}_{2n}| \\ \vdots & \vdots & \ddots & \vdots \\ (-1)^{n+1}|\mathbf{A}_{n1}| & (-1)^{n}|\mathbf{A}_{n2}| & \cdots & |\mathbf{A}_{nn}| \end{pmatrix}.$$

Example 3

Calculate $|\mathbf{A}|$ if $\mathbf{A} = \begin{pmatrix} -4 & -2 & -1 \\ 5 & -4 & -3 \\ 5 & 1 & -2 \end{pmatrix}$.

Solution

$$\begin{aligned} |\mathbf{A}| &= \begin{vmatrix} -4 & -2 & -1 \\ 5 & -4 & -3 \\ 5 & 1 & -2 \end{vmatrix} \\ &= (-1)^2(-4)\begin{vmatrix} -4 & -3 \\ 1 & -2 \end{vmatrix} + (-1)^3(-2)\begin{vmatrix} 5 & -3 \\ 5 & -2 \end{vmatrix} + (-1)^4(-1)\begin{vmatrix} 5 & -4 \\ 5 & 1 \end{vmatrix} \\ &= -4((-4)(-2) - (-3)(1)) + 2((5)(-2) - (-3)(5)) - ((5)(1) - (-4)(5)) \\ &= -59. \end{aligned}$$

Definition 8.8 Adjoint and Inverse

B is an **inverse** of the $n \times n$ matrix **A** means that $\mathbf{AB} = \mathbf{BA} = \mathbf{I}$. The **adjoint, $\mathbf{A}^a$**, of an $n \times n$ matrix **A** is the transpose of the cofactor matrix: $\mathbf{A}^a = (\mathbf{A}^c)^T$. If $|\mathbf{A}| \neq 0$ and $\mathbf{B} = \dfrac{1}{|\mathbf{A}|}\mathbf{A}^a$, then $\mathbf{AB} = \mathbf{BA} = \mathbf{I}$. Therefore, if $|\mathbf{A}| \neq 0$, the inverse of **A** is given by

$$\mathbf{A}^{-1} = \frac{1}{|\mathbf{A}|}\mathbf{A}^a.$$

Example 4

Find $\mathbf{A}^{-1}$ if $\mathbf{A} = \begin{pmatrix} 2 & -1 \\ -3 & 1 \end{pmatrix}$.

Solution   In this case, $|\mathbf{A}| = \begin{vmatrix} 2 & -1 \\ -3 & 1 \end{vmatrix} = 2 - 3 = -1 \neq 0$, so $\mathbf{A}^{-1}$ exists.

Moreover, $\mathbf{A}^c = \begin{pmatrix} 1 & 3 \\ 1 & 2 \end{pmatrix}$, so $\mathbf{A}^a = \begin{pmatrix} 1 & 1 \\ 3 & 2 \end{pmatrix}$ and $\mathbf{A}^{-1} = \dfrac{1}{|\mathbf{A}|}\mathbf{A}^a = \begin{pmatrix} -1 & -1 \\ -3 & -2 \end{pmatrix}$.

Example 5

Find $\mathbf{A}^{-1}$ if $\mathbf{A} = \begin{pmatrix} -2 & -1 & 1 \\ 2 & 1 & 0 \\ 3 & 1 & -1 \end{pmatrix}$.

Solution   We begin by finding $|\mathbf{A}|$, which is given by

$$|\mathbf{A}| = (-1)^2(-2)\begin{vmatrix} 1 & 0 \\ 1 & -1 \end{vmatrix} + (-1)^3(-1)\begin{vmatrix} 2 & 0 \\ 3 & -1 \end{vmatrix} + (-1)^4(1)\begin{vmatrix} 2 & 1 \\ 3 & 1 \end{vmatrix} =$$
$$(-2)(-1) + (1)(-2) + (1)(-1) = -1.$$

We then calculate the cofactors:

$$A_{11} = (-1)^2\begin{vmatrix} 1 & 0 \\ 1 & -1 \end{vmatrix} = -1, \quad A_{12} = (-1)^3\begin{vmatrix} 2 & 0 \\ 3 & -1 \end{vmatrix} = 2,$$

$$A_{13} = (-1)^4\begin{vmatrix} 2 & 1 \\ 3 & 1 \end{vmatrix} = -1$$

$$A_{21} = (-1)^3\begin{vmatrix} -1 & 1 \\ 1 & -1 \end{vmatrix} = 0, \quad A_{22} = (-1)^4\begin{vmatrix} -2 & 1 \\ 3 & -1 \end{vmatrix} = -1,$$

$$A_{23} = (-1)^5\begin{vmatrix} -2 & -1 \\ 3 & 1 \end{vmatrix} = -1$$

$$A_{31} = (-1)^4\begin{vmatrix} -1 & 1 \\ 1 & 0 \end{vmatrix} = -1, \quad A_{32} = (-1)^5\begin{vmatrix} -2 & 1 \\ 2 & 0 \end{vmatrix} = 2,$$

$$A_{33} = (-1)^6\begin{vmatrix} -2 & -1 \\ 2 & 1 \end{vmatrix} = 0.$$

Then

$$\mathbf{A}^{-1} = \frac{1}{|\mathbf{A}|}\mathbf{A}^a = \frac{1}{-1}\begin{pmatrix} -1 & 0 & -1 \\ 2 & -1 & 2 \\ -1 & -1 & 0 \end{pmatrix} = \begin{pmatrix} 1 & 0 & 1 \\ -2 & 1 & -2 \\ 1 & 1 & 0 \end{pmatrix}.$$

For convenience, we state the following theorem. The proof is left as an exercise.

Theorem 8.2 Inverse of a 2 × 2 Matrix

Let $\mathbf{A} = \begin{pmatrix} a & b \\ c & d \end{pmatrix}$. If $|\mathbf{A}| = ad - bc \neq 0$, then

$$\mathbf{A}^{-1} = \frac{1}{ad - bc}\begin{pmatrix} d & -b \\ -c & a \end{pmatrix}.$$

Example 6

Find $\mathbf{A}^{-1}$ if $\mathbf{A} = \begin{pmatrix} 5 & -1 \\ 2 & 3 \end{pmatrix}$.

Solution   Because $|\mathbf{A}| = (5)(3) - (2)(-1) = 17$,

$$\mathbf{A}^{-1} = \frac{1}{17}\begin{pmatrix} 3 & 1 \\ -2 & 5 \end{pmatrix} = \begin{pmatrix} \frac{3}{17} & \frac{1}{17} \\ -\frac{2}{17} & \frac{5}{17} \end{pmatrix}.$$

We will almost always take advantage of a computer algebra system to perform operations on matrices. In addition, if you have taken linear algebra, you can use techniques like row reduction to find the inverse of a matrix or solve systems of equations.

The inverse $\mathbf{A}^{-1}$ can be used to solve the linear system of equations $\mathbf{Ax} = \mathbf{b}$. For example, to solve $\begin{pmatrix} 5 & -1 \\ 2 & 3 \end{pmatrix}\begin{pmatrix} x \\ y \end{pmatrix} = \begin{pmatrix} -34 \\ 17 \end{pmatrix}$ in which $\mathbf{A} = \begin{pmatrix} 5 & -1 \\ 2 & 3 \end{pmatrix}$ and $\mathbf{b} = \begin{pmatrix} -34 \\ 17 \end{pmatrix}$, we find $\mathbf{x} = \mathbf{A}^{-1}\mathbf{b} = \begin{pmatrix} \frac{3}{17} & \frac{1}{17} \\ -\frac{2}{17} & \frac{5}{17} \end{pmatrix}\begin{pmatrix} -34 \\ 17 \end{pmatrix} = \begin{pmatrix} -5 \\ 9 \end{pmatrix}$. We will find several uses for the inverse in solving systems of differential equations as well.

Eigenvalues and Eigenvectors

Definition 8.9 Eigenvalues and Eigenvectors

A *nonzero* vector $\mathbf{x}$ is an **eigenvector** of the square matrix $\mathbf{A}$ if there is a number λ, called an **eigenvalue** of $\mathbf{A}$, so that

$$\mathbf{Ax} = \lambda\mathbf{x}.$$

Note that, by definition, an eigenvector of a matrix is *never* the zero vector.

Example 7

Show that $\begin{pmatrix} -1 \\ 2 \end{pmatrix}$ and $\begin{pmatrix} 1 \\ 1 \end{pmatrix}$ are eigenvectors of $\begin{pmatrix} -1 & 2 \\ 4 & -3 \end{pmatrix}$ with corresponding eigenvalues -5 and 1, respectively.

Solution Because $\begin{pmatrix} -1 & 2 \\ 4 & -3 \end{pmatrix}\begin{pmatrix} -1 \\ 2 \end{pmatrix} = \begin{pmatrix} 5 \\ -10 \end{pmatrix} = -5\begin{pmatrix} -1 \\ 2 \end{pmatrix}$ and

$\begin{pmatrix} -1 & 2 \\ 4 & -3 \end{pmatrix}\begin{pmatrix} 1 \\ 1 \end{pmatrix} = \begin{pmatrix} 1 \\ 1 \end{pmatrix} = 1\begin{pmatrix} 1 \\ 1 \end{pmatrix}$, $\begin{pmatrix} -1 \\ 2 \end{pmatrix}$ and $\begin{pmatrix} 1 \\ 1 \end{pmatrix}$ are eigenvectors of

$\begin{pmatrix} -1 & 2 \\ 4 & -3 \end{pmatrix}$ with corresponding eigenvalues -5 and 1, respectively.

If $\mathbf{x}$ is an eigenvector of $\mathbf{A}$ with corresponding eigenvalue λ, then $\mathbf{Ax} = \lambda\mathbf{x}$. Because this equation is equivalent to the equation $(\mathbf{A} - \lambda\mathbf{I})\mathbf{x} = \mathbf{0}$, $\mathbf{x} \neq \mathbf{0}$ is an eigenvector if and only if $\det(\mathbf{A} - \lambda\mathbf{I}) = 0$.

If $\det(\mathbf{A} - \lambda\mathbf{I}) \neq 0$, what is the solution of $(\mathbf{A} - \lambda\mathbf{I})\mathbf{x} = \mathbf{0}$? Can this solution (vector) be an eigenvector of $\mathbf{A}$?

Definition 8.10 Characteristic Polynomial

The equation $\det(\mathbf{A} - \lambda\mathbf{I}) = 0$ is called the **characteristic equation** of $\mathbf{A}$; $\det(\mathbf{A} - \lambda\mathbf{I})$ is called the **characteristic polynomial** of $\mathbf{A}$

Notice that the roots of the characteristic polynomial of $\mathbf{A}$ are the eigenvalues of $\mathbf{A}$.

Example 8

Calculate the eigenvalues and corresponding eigenvectors of $\mathbf{A} = \begin{pmatrix} 4 & -6 \\ 3 & -7 \end{pmatrix}$.

Solution The characteristic polynomial of $\mathbf{A} = \begin{pmatrix} 4 & -6 \\ 3 & -7 \end{pmatrix}$ is

$$\begin{vmatrix} 4 - \lambda & -6 \\ 3 & -7 - \lambda \end{vmatrix} = \lambda^2 + 3\lambda - 10 = (\lambda + 5)(\lambda - 2).$$

Because the eigenvalues are found by solving $(\lambda + 5)(\lambda - 2) = 0$, the eigenvalues are $\lambda = -5$ and $\lambda = 2$. Let $\begin{pmatrix} x_1 \\ y_1 \end{pmatrix}$ denote the eigenvectors corresponding to the eigenvalue $\lambda = -5$. Then,

$$\left\{\begin{pmatrix} 4 & -6 \\ 3 & -7 \end{pmatrix} - (-5)\begin{pmatrix} 1 & 0 \\ 0 & 1 \end{pmatrix}\right\}\begin{pmatrix} x_1 \\ y_1 \end{pmatrix} = \mathbf{0}.$$

Simplifying yields the system of equations $\begin{cases} 9x_1 - 6y_1 = 0 \\ 3x_1 - 2y_1 = 0 \end{cases}$, so $y_1 = \frac{3}{2}x_1$. Therefore, if x_1 is any real number, then $\begin{pmatrix} x_1 \\ \frac{3}{2}x_1 \end{pmatrix}$ is an eigenvector. In particular, if

When finding an eigenvector **v** *corresponding to the eigenvalue* λ, *we see that there is actually a collection (or family) of eigenvectors corresponding to* λ. *In the study of differential equations, we will find that we only need to find one member of the collection of eigenvectors. Therefore, as we did in Example 8, we usually eliminate the arbitrary constants when we encounter them in eigenvectors by selecting particular values for the constants.*

$x_1 = 2$, then $\begin{pmatrix} 2 \\ 3 \end{pmatrix}$ is an eigenvector of $\mathbf{A} = \begin{pmatrix} 4 & -6 \\ 3 & -7 \end{pmatrix}$ with corresponding

eigenvalue $\lambda = -5$. Similarly, if we let $\begin{pmatrix} x_2 \\ y_2 \end{pmatrix}$ denote the eigenvectors corre-

sponding to $\lambda = 2$, then $\left\{ \begin{pmatrix} 4 & -6 \\ 3 & -7 \end{pmatrix} - 2 \begin{pmatrix} 1 & 0 \\ 0 & 1 \end{pmatrix} \right\} \begin{pmatrix} x_2 \\ y_2 \end{pmatrix} = \mathbf{0}$, which yields the

system $\begin{cases} 2x_2 - 6y_2 = 0 \\ 3x_2 - 9y_2 = 0 \end{cases}$, so $y_2 = \frac{1}{3} x_2$. If $x_2 = 3$, then $\begin{pmatrix} 3 \\ 1 \end{pmatrix}$ is an

eigenvector of $\mathbf{A} = \begin{pmatrix} 4 & -6 \\ 3 & -7 \end{pmatrix}$ with corresponding eigenvalue $\lambda = 2$. (Notice

that constant (scalar) multiples of the eigenvectors $\begin{pmatrix} 2 \\ 3 \end{pmatrix}$ and $\begin{pmatrix} 3 \\ 1 \end{pmatrix}$ are also

eigenvectors corresponding to $\lambda = -5$ and $\lambda = 2$, respectively.)

Example 9

Find the eigenvalues and corresponding eigenvectors of $\mathbf{A} = \begin{pmatrix} 0 & 1 \\ -1 & 0 \end{pmatrix}$.

Solution In this case, the characteristic polynomial is $\begin{vmatrix} -\lambda & 1 \\ -1 & -\lambda \end{vmatrix} = \lambda^2 + 1$, so

the eigenvalues are the roots of the equation $\lambda^2 + 1 = 0$. These are the imaginary numbers $\lambda = i$ and $\lambda = -i$ where $i = \sqrt{-1}$. The corresponding eigenvectors are found by substituting the eigenvalues into the equation $(\mathbf{A} - \lambda\mathbf{I})\mathbf{x} = \mathbf{0}$ and solving

for **x**. For $\lambda = i$, this equation is $\begin{pmatrix} -i & 1 \\ -1 & -i \end{pmatrix} \begin{pmatrix} x_1 \\ y_1 \end{pmatrix} = \begin{pmatrix} 0 \\ 0 \end{pmatrix}$, which is equivalent

to the system $\begin{cases} -ix_1 + y_1 = 0 \\ -x_1 - iy_1 = 0 \end{cases}$. Notice that the second equation of this

system is a constant multiple (i) of the first equation. Hence, an eigenvector $\begin{pmatrix} x_1 \\ y_1 \end{pmatrix}$

must satisfy $y_1 = ix_1$. Therefore, $\begin{pmatrix} x_1 \\ ix_1 \end{pmatrix} = \begin{pmatrix} 1 \\ i \end{pmatrix} x_1$ is an eigenvector for any

value of x_1. For example, if $x_1 = 1$, then $\begin{pmatrix} 1 \\ i \end{pmatrix}$ is an eigenvector. For $\lambda = -i$,

the system of equations is $\begin{pmatrix} i & 1 \\ -1 & i \end{pmatrix} \begin{pmatrix} x_2 \\ y_2 \end{pmatrix} = \begin{pmatrix} 0 \\ 0 \end{pmatrix}$, which is equivalent to

$\begin{cases} ix_2 + y_2 = 0 \\ -x_2 + iy_2 = 0 \end{cases}$. Because the second equation equals i times the first

equation, the eigenvector $\begin{pmatrix} x_2 \\ y_2 \end{pmatrix}$ must satisfy $y_2 = -ix_2$. Hence, $\begin{pmatrix} x_2 \\ -ix_2 \end{pmatrix} = \begin{pmatrix} 1 \\ -i \end{pmatrix} x_2$

is an eigenvector for any value of x_2. Therefore, if $x_2 = 1$, then $\begin{pmatrix} 1 \\ -i \end{pmatrix}$ is an

eigenvector.

Recall that the complex conjugate of the complex number $z = a + bi$ is $\bar{z} = a - bi$. Similarly, the complex conjugate of the vector $\mathbf{x} = \begin{pmatrix} a_1 + b_1 i \\ a_2 + b_2 i \\ \vdots \\ a_n + b_n i \end{pmatrix}$ is the vector $\bar{\mathbf{x}} = \begin{pmatrix} a_1 - b_1 i \\ a_2 - b_2 i \\ \vdots \\ a_n - b_n i \end{pmatrix}$. Notice that the eigenvectors corresponding to the complex conjugate eigenvalues $\lambda = i$ and $\lambda = -i$ in the previous example are $\begin{pmatrix} 1 \\ i \end{pmatrix}$ and $\begin{pmatrix} 1 \\ -i \end{pmatrix}$, which are complex conjugates. This is no coincidence. We can prove that the *eigenvectors that correspond to complex conjugate eigenvalues are themselves complex conjugates.*

Example 10   Calculate the eigenvalues and corresponding eigenvectors of the matrix

$$\mathbf{A} = \begin{pmatrix} 1 & 0 & 0 \\ 2 & 3 & 1 \\ 0 & 2 & 4 \end{pmatrix}.$$

Solution   We begin by finding the characteristic polynomial of $\mathbf{A}$ with

$$\begin{vmatrix} 1 - \lambda & 0 & 0 \\ 2 & 3 - \lambda & 1 \\ 0 & 2 & 4 - \lambda \end{vmatrix} = (1 - \lambda) \begin{vmatrix} 3 - \lambda & 1 \\ 2 & 4 - \lambda \end{vmatrix} = (1 - \lambda)(\lambda^2 - 7\lambda + 10)$$

$$= (1 - \lambda)(\lambda - 2)(\lambda - 5).$$

The eigenvalues of $\mathbf{A}$ are $\lambda = 1$, $\lambda = 2$, and $\lambda = 5$. For each eigenvalue, we find the corresponding eigenvectors by substituting the eigenvalue λ into the equation $(\mathbf{A} - \lambda \mathbf{I})\mathbf{x} = \mathbf{0}$ and solving for the vector $\mathbf{x}$. If $\lambda = 1$, we obtain

$$\begin{pmatrix} 0 & 0 & 0 \\ 2 & 2 & 1 \\ 0 & 2 & 3 \end{pmatrix} \begin{pmatrix} x_1 \\ y_1 \\ z_1 \end{pmatrix} = \begin{pmatrix} 0 \\ 0 \\ 0 \end{pmatrix}$$

which is reduced to $\begin{pmatrix} 0 & 0 & 0 \\ 1 & 0 & -1 \\ 0 & 2 & 3 \end{pmatrix} \begin{pmatrix} x_1 \\ y_1 \\ z_1 \end{pmatrix} = \begin{pmatrix} 0 \\ 0 \\ 0 \end{pmatrix}$, so $x_1 = z_1$ and $y_1 = -\frac{3}{2}z_1$.

Hence, $\begin{pmatrix} z_1 \\ -\frac{3}{2}z_1 \\ z_1 \end{pmatrix} = \begin{pmatrix} 1 \\ -\frac{3}{2} \\ 1 \end{pmatrix} z_1$ is an eigenvector for any value of z_1.

If $\lambda = 2$, we have $\begin{pmatrix} -1 & 0 & 0 \\ 2 & 1 & 1 \\ 0 & 2 & 2 \end{pmatrix} \begin{pmatrix} x_2 \\ y_2 \\ z_2 \end{pmatrix} = \begin{pmatrix} 0 \\ 0 \\ 0 \end{pmatrix}$, which is equivalent to

$\begin{pmatrix} 1 & 0 & 0 \\ 0 & 1 & 1 \\ 0 & 0 & 0 \end{pmatrix} \begin{pmatrix} x_2 \\ y_2 \\ z_2 \end{pmatrix} = \begin{pmatrix} 0 \\ 0 \\ 0 \end{pmatrix}$ in reduced form. Thus, $x_2 = 0$ and $y_2 = -z_2$, so

$\begin{pmatrix} 0 \\ -z_2 \\ z_2 \end{pmatrix} = \begin{pmatrix} 0 \\ -1 \\ 1 \end{pmatrix} z_2$ is an eigenvector for any value of z_2.

If $\lambda = 5$, then $\begin{pmatrix} -4 & 0 & 0 \\ 2 & -2 & 1 \\ 0 & 2 & -1 \end{pmatrix} \begin{pmatrix} x_3 \\ y_3 \\ z_3 \end{pmatrix} = \begin{pmatrix} 0 \\ 0 \\ 0 \end{pmatrix}$, which is equivalent to

$\begin{pmatrix} 1 & 0 & 0 \\ 0 & 2 & -1 \\ 0 & 0 & 0 \end{pmatrix} \begin{pmatrix} x_3 \\ y_3 \\ z_3 \end{pmatrix} = \begin{pmatrix} 0 \\ 0 \\ 0 \end{pmatrix}$. Therefore, $x_3 = 0$ and $z_3 = 2y_3$, so

$\begin{pmatrix} 0 \\ y_3 \\ 2y_3 \end{pmatrix} = \begin{pmatrix} 0 \\ 1 \\ 2 \end{pmatrix} y_3$ is an eigenvector for any value of y_3.

Definition 8.11 Eigenvalue of Multiplicity m

Suppose that $(\lambda - \lambda_1)^m$ where m is a positive integer is a factor of the characteristic polynomial of the $n \times n$ matrix **A**, while $(\lambda - \lambda_1)^{m+1}$ is not a factor of this polynomial. Then $\lambda = \lambda_1$ is an **eigenvalue of multiplicity m.**

We often say that the eigenvalue of an $n \times n$ matrix **A** is repeated if it is of multiplicity m where $m \geq 2$ and $m \leq n$. When trying to find the eigenvector(s) corresponding to an eigenvalue of multiplicity m, two situations may be encountered: either m or fewer than m linearly independent eigenvectors can be found that correspond to λ.

Example 11

Find the eigenvalues and corresponding eigenvectors of (a) $\mathbf{A} = \begin{pmatrix} 1 & -3 & 3 \\ 3 & -5 & 3 \\ 6 & -6 & 4 \end{pmatrix}$;

and (b) $\mathbf{B} = \begin{pmatrix} 5 & -4 & 0 \\ 1 & 0 & 2 \\ 0 & 2 & 5 \end{pmatrix}$.

Solution   (a) The eigenvalues are found with

$$\begin{vmatrix} 1-\lambda & -3 & 3 \\ 3 & -5-\lambda & 3 \\ 6 & -6 & 4-\lambda \end{vmatrix} = (1-\lambda)\begin{vmatrix} -5-\lambda & 3 \\ -6 & 4-\lambda \end{vmatrix} - (-3)\begin{vmatrix} 3 & 3 \\ 6 & 4-\lambda \end{vmatrix} +$$

$$3\begin{vmatrix} 3 & -5-\lambda \\ 6 & -6 \end{vmatrix} = 16 + 12\lambda - \lambda^3 = 0.$$

Note that $\lambda = -2$ is a root of the characteristic polynomial. Using this root with synthetic or long division, we find that $\lambda^3 - 12\lambda - 16 = (\lambda + 2)^2(\lambda - 4) = 0$. Hence, $\lambda = -2$ is an eigenvalue of multiplicity 2. When we try to find an eigenvec-

tor $\mathbf{v} = \begin{pmatrix} x_1 \\ y_1 \\ z_1 \end{pmatrix}$ corresponding to $\lambda = -2$, we see that

$$\begin{pmatrix} 3 & -3 & 3 \\ 3 & -3 & 3 \\ 6 & -6 & 6 \end{pmatrix}\begin{pmatrix} x_1 \\ y_1 \\ z_1 \end{pmatrix} = \begin{pmatrix} 0 \\ 0 \\ 0 \end{pmatrix}$$

must be satisfied. After the necessary row operations, we find that this system is equivalent to

$$\begin{pmatrix} 1 & -1 & 1 \\ 0 & 0 & 0 \\ 0 & 0 & 0 \end{pmatrix}\begin{pmatrix} x_1 \\ y_1 \\ z_1 \end{pmatrix} = \begin{pmatrix} 0 \\ 0 \\ 0 \end{pmatrix}$$

which indicates that $x_1 - y_1 + z_1 = 0$ or $x_1 = y_1 - z_1$. If $z_1 = 0$, then $x_1 = y_1$.

Hence, $\mathbf{v}_1 = \begin{pmatrix} y_1 \\ y_1 \\ 0 \end{pmatrix} = \begin{pmatrix} 1 \\ 1 \\ 0 \end{pmatrix} y_1$ is an eigenvector for any choice of y_1. On the

other hand, if in $x_1 = y_1 - z_1$, $y_1 = 0$, then $x_1 = -z_1$. Therefore,

$\mathbf{v}_2 = \begin{pmatrix} -z_1 \\ 0 \\ z_1 \end{pmatrix} = \begin{pmatrix} -1 \\ 0 \\ 1 \end{pmatrix} z_1$ (for any choice of z_1) is another linearly

independent eigenvector corresponding to $\lambda = -2$ because $\mathbf{v}_2$ is not a constant multiple of $\mathbf{v}_1$. Therefore, we have found two linearly independent eigenvectors that correspond to the eigenvalue $\lambda = -2$ of multiplicity 2. We leave it to you to find an eigenvector corresponding to $\lambda = 4$.

(b) The eigenvalues of **B** are determined by solving

$$\begin{vmatrix} 5-\lambda & -4 & 0 \\ 1 & -\lambda & 2 \\ 0 & 2 & 5-\lambda \end{vmatrix} = (5-\lambda)\begin{vmatrix} -\lambda & 2 \\ 2 & 5-\lambda \end{vmatrix} - (-4)\begin{vmatrix} 1 & 2 \\ 0 & 5-\lambda \end{vmatrix} = -(5-\lambda)^2\lambda = 0.$$

Hence, $\lambda = 5$ is an eigenvalue of multiplicity 2. In this case, when we find a corresponding eigenvector $\mathbf{v} = \begin{pmatrix} x_1 \\ y_1 \\ z_1 \end{pmatrix}$, we solve $\begin{pmatrix} 0 & -4 & 0 \\ 1 & -5 & 2 \\ 0 & 2 & 0 \end{pmatrix}\begin{pmatrix} x_1 \\ y_1 \\ z_1 \end{pmatrix} = \begin{pmatrix} 0 \\ 0 \\ 0 \end{pmatrix}$, which is

equivalent to $\begin{pmatrix} 1 & 0 & 2 \\ 0 & 1 & 0 \\ 0 & 0 & 0 \end{pmatrix}\begin{pmatrix} x_1 \\ y_1 \\ z_1 \end{pmatrix} = \begin{pmatrix} 0 \\ 0 \\ 0 \end{pmatrix}$. The components of $\mathbf{v}$ must satisfy

$x_1 = -2z_1$ and $y_1 = 0$. We find only one eigenvector $\mathbf{v} = \begin{pmatrix} -2z_1 \\ 0 \\ z_1 \end{pmatrix} = \begin{pmatrix} -2 \\ 0 \\ 1 \end{pmatrix}z_1$

that corresponds to the eigenvalue $\lambda = 5$ of multiplicity 2. We leave it to you to find an eigenvector corresponding to $\lambda = 0$.

Matrix Calculus

Definition 8.12 Derivative and Integral of a Matrix

The **derivative** of the $n \times m$ matrix $\mathbf{A}(t) = \begin{pmatrix} a_{11}(t) & a_{12}(t) & \cdots & a_{1m}(t) \\ a_{21}(t) & a_{22}(t) & \cdots & a_{2m}(t) \\ \vdots & \vdots & \ddots & \vdots \\ a_{n1}(t) & a_{n2}(t) & \cdots & a_{nm}(t) \end{pmatrix}$,

where $a_{ij}(t)$ is differentiable for all values of i and j, is

$$\frac{d}{dt}\mathbf{A}(t) = \begin{pmatrix} \frac{d}{dt}a_{11}(t) & \frac{d}{dt}a_{12}(t) & \cdots & \frac{d}{dt}a_{1m}(t) \\ \frac{d}{dt}a_{21}(t) & \frac{d}{dt}a_{22}(t) & \cdots & \frac{d}{dt}a_{2m}(t) \\ \vdots & \vdots & \ddots & \vdots \\ \frac{d}{dt}a_{n1}(t) & \frac{d}{dt}a_{n2}(t) & \cdots & \frac{d}{dt}a_{nm}(t) \end{pmatrix}.$$

The **integral** of $\mathbf{A}(t)$, where $a_{ij}(t)$ is integrable for all values of i and j, is

$$\int \mathbf{A}(t)\,dt = \begin{pmatrix} \int a_{11}(t)\,dt & \int a_{12}(t)\,dt & \cdots & \int a_{1m}(t)\,dt \\ \int a_{21}(t)\,dt & \int a_{22}(t)\,dt & \cdots & \int a_{2m}(t)\,dt \\ \vdots & \vdots & \ddots & \vdots \\ \int a_{n1}(t)\,dt & \int a_{n2}(t)\,dt & \cdots & \int a_{nm}(t)\,dt \end{pmatrix}.$$

Example 12

Find $\dfrac{d}{dt}\mathbf{A}(t)$ and $\displaystyle\int \mathbf{A}(t)\,dt$ if $\mathbf{A}(t) = \begin{pmatrix} \cos 3t & \sin 3t & e^{-t} \\ t & t\sin(t^2) & e^t \end{pmatrix}$.

Solution   We find $\dfrac{d}{dt}\mathbf{A}(t)$ by differentiating each element of $\mathbf{A}(t)$. This yields

$$\frac{d}{dt}\mathbf{A}(t) = \begin{pmatrix} -3\sin 3t & 3\cos 3t & -e^{-t} \\ 1 & \sin(t^2)+2t^2\cos(t^2) & e^t \end{pmatrix}.$$

Similarly, we find $\displaystyle\int \mathbf{A}(t)\,dt$ by integrating each element of $\mathbf{A}(t)$.

$$\int \mathbf{A}(t)\,dt = \begin{pmatrix} \dfrac{1}{3}\sin 3t + c_{11} & \dfrac{-1}{3}\cos 3t + c_{12} & -e^{-t} + c_{13} \\ \dfrac{1}{2}t^2 + c_{21} & \dfrac{-1}{2}\cos(t^2) + c_{22} & e^t + c_{23} \end{pmatrix}$$

where each c_{ij} represents an arbitrary constant.

IN TOUCH WITH TECHNOLOGY

Many computer software packages and calculators contain built-in functions for working with matrices. Use a computer or calculator to perform the following calculations in order to become familiar with these functions.

1. If $\mathbf{A} = \begin{pmatrix} 0 & 0 & -2 & 1 \\ 3 & -1 & 7 & 2 \\ -6 & 0 & 5 & -1 \\ -6 & 0 & 1 & -2 \end{pmatrix}$ and

$\mathbf{B} = \begin{pmatrix} -7 & -6 & -3 & -7 \\ 2 & -3 & 0 & 4 \\ 3 & 4 & 1 & 2 \\ 5 & 6 & 3 & 6 \end{pmatrix}$, compute **(a)** $3\mathbf{A} - 2\mathbf{B}$;

(b) $\mathbf{B}^T$; **(c)** $\mathbf{AB}$; **(d)** $|\mathbf{A}|$, $|\mathbf{B}|$, and $|\mathbf{AB}|$; **(e)** $\mathbf{A}^{-1}$.

2. Find the eigenvalues and corresponding eigenvectors of the matrices

$$\mathbf{A} = \begin{pmatrix} -4 & 4 & -4 \\ 2 & 3 & -4 \\ 5 & 0 & -1 \end{pmatrix} \text{ and } \mathbf{B} = \begin{pmatrix} 1 & -3 & 5 \\ 5 & 5 & 0 \\ -5 & -2 & 3 \end{pmatrix}.$$

3. Find $\dfrac{d}{dt}\mathbf{A}(t)$ and $\displaystyle\int \mathbf{A}(t)\,dt$ if

$$\mathbf{A}(t) = \begin{pmatrix} te^{-t} & t^2\sin 2t & \dfrac{1}{9+4t^2} \\ \cos^6 t & \sec^3 2t & \dfrac{1}{t(t-1)} \\ \dfrac{4}{\sqrt{1-t^2}} & \sin^5 t\cos t & t^3\sin^2 t \end{pmatrix}.$$

4. Calculate the eigenvalues of $\begin{pmatrix} 0 & -1-k^2 \\ 1 & 2 \end{pmatrix}$ and

$\begin{pmatrix} 0 & -1-k^2 \\ 1 & 2k \end{pmatrix}$. How do the eigenvalues change for $-\infty < k < +\infty$?

5. Nearly every computer algebra system can compute exact values of the eigenvalues and corresponding eigenvectors of an $n \times n$ matrix for $n = 2, 3,$ and 4. **(a)** Use a computer algebra system to compute the exact values of the eigenvalues and corresponding eigenvectors of $\mathbf{A} = \begin{pmatrix} 3 & 1 & 1 & 1 \\ -2 & -3 & 2 & -3 \\ -3 & 3 & -2 & 1 \\ 0 & 1 & 2 & 0 \end{pmatrix}$. Write down a portion of the result (or print it and staple it to your homework). **(b)** Compute approximations of the eigenvalues and corresponding eigenvectors of $\mathbf{A} = \begin{pmatrix} 3 & 1 & 1 & 1 \\ -2 & -3 & 2 & -3 \\ -3 & 3 & -2 & 1 \\ 0 & 1 & 2 & 0 \end{pmatrix}$. **(c)** Which results are more meaningful to you?

EXERCISES 8.2

In Exercises 1–4, perform the indicated calculation if
$\mathbf{A} = \begin{pmatrix} 2 & -5 \\ 0 & 4 \end{pmatrix}$, $\mathbf{B} = \begin{pmatrix} -1 & 0 \\ 3 & 6 \end{pmatrix}$, and $\mathbf{C} = \begin{pmatrix} 4 & 1 \\ -2 & 7 \end{pmatrix}$.

1. $\mathbf{B} - \mathbf{A}$
2. $2\mathbf{A} + \mathbf{C}$
3. $(\mathbf{B} + \mathbf{C}) - 4\mathbf{A}$
4. $(\mathbf{A} - 3\mathbf{C}) + (\mathbf{C} - 5\mathbf{B})$

In Exercises 5–8, perform the indicated calculation if
$\mathbf{A} = \begin{pmatrix} 0 & 3 & 2 \\ 5 & -1 & 4 \\ 2 & -1 & -3 \end{pmatrix}$, $\mathbf{B} = \begin{pmatrix} 4 & -2 & -3 \\ 5 & 5 & 1 \\ 4 & -5 & 2 \end{pmatrix}$, and
$\mathbf{C} = \begin{pmatrix} -2 & -4 & 0 \\ 5 & -1 & 1 \\ -1 & -3 & -3 \end{pmatrix}$.

5. $7\mathbf{A} + 3\mathbf{B}$
6. $8\mathbf{B} - 9\mathbf{A}$
7. $2\mathbf{A} - 4(\mathbf{B} + \mathbf{C})$
8. $4\mathbf{C} + 2(2\mathbf{A} - 5\mathbf{B})$

In Exercises 9–14, compute $\mathbf{AB}$ and $\mathbf{BA}$, when defined, using the given matrices.

9. $\mathbf{A} = \begin{pmatrix} 1 & -3 & -4 & 3 \\ -5 & -1 & -2 & -2 \\ 4 & -1 & 0 & -4 \end{pmatrix}$ and

$\mathbf{B} = \begin{pmatrix} -5 & 2 & 4 \\ -5 & 0 & -5 \\ 0 & 4 & 5 \\ 4 & 3 & -4 \end{pmatrix}$

10. $\mathbf{A} = \begin{pmatrix} -2 & 4 & 3 & 3 \\ -3 & 2 & 4 & -3 \end{pmatrix}$ and $\mathbf{B} = \begin{pmatrix} 0 & -2 & 2 \\ -2 & -4 & 1 \\ -2 & 0 & 4 \\ -1 & -4 & 5 \end{pmatrix}$

*11. $\mathbf{A} = \begin{pmatrix} 1 & 2 \\ 3 & 5 \end{pmatrix}$ and $\mathbf{B} = \begin{pmatrix} 1 & -1 \\ -1 & -5 \end{pmatrix}$

12. $\mathbf{A} = \begin{pmatrix} -4 & 5 \\ 1 & -5 \end{pmatrix}$ and $\mathbf{B} = \begin{pmatrix} 4 & 2 \\ 3 & 5 \end{pmatrix}$

13. $\mathbf{A} = \begin{pmatrix} 1 & -1 & -1 \\ -4 & 5 & -2 \\ 5 & -4 & -1 \end{pmatrix}$ and

$\mathbf{B} = \begin{pmatrix} -3 & -3 & -2 \\ -1 & 4 & -1 \\ -1 & 5 & 3 \end{pmatrix}$

14. $\mathbf{A} = \begin{pmatrix} 4 & 2 & 2 & 0 \\ -2 & -3 & 0 & -2 \\ -1 & -2 & -5 & 4 \\ 3 & -5 & -3 & -3 \end{pmatrix}$ and

$\mathbf{B} = \begin{pmatrix} 4 & 1 & -3 & -5 \\ 5 & -2 & -2 & 3 \\ -3 & -5 & 2 & 0 \\ 5 & 3 & -3 & -5 \end{pmatrix}$

In Exercises 15–19, find the determinant of the square matrix.

15. $A = \begin{pmatrix} -1 & -4 \\ 5 & 3 \end{pmatrix}$ **16.** $A = \begin{pmatrix} 3 & -1 \\ -5 & 4 \end{pmatrix}$

*17. $A = \begin{pmatrix} 3 & -2 & 0 \\ -1 & 3 & 0 \\ 3 & -2 & 0 \end{pmatrix}$ **18.** $A = \begin{pmatrix} 0 & 3 & 3 \\ 1 & 1 & -2 \\ -3 & 2 & -3 \end{pmatrix}$

*19. $A = \begin{pmatrix} 2 & 0 & -2 & -1 \\ -3 & -2 & 0 & 1 \\ -3 & -1 & -3 & -3 \\ 1 & 0 & 0 & 1 \end{pmatrix}$

In Exercises 20–23, find the inverse of each square matrix.

20. $A = \begin{pmatrix} 1 & -1 \\ 0 & 1 \end{pmatrix}$ ***21.** $A = \begin{pmatrix} 0 & -1 \\ 2 & 2 \end{pmatrix}$

22. $A = \begin{pmatrix} -3 & 3 & 3 \\ 3 & -2 & -2 \\ 3 & 1 & 0 \end{pmatrix}$

*23. $A = \begin{pmatrix} -1 & -1 & -2 \\ 3 & 2 & 2 \\ 0 & 0 & -2 \end{pmatrix}$

In Exercises 24–34, find the eigenvalues and corresponding eigenvectors of the matrix.

24. $A = \begin{pmatrix} -7 & -2 \\ 4 & -1 \end{pmatrix}$ ***25.** $A = \begin{pmatrix} -6 & 1 \\ -2 & -3 \end{pmatrix}$

26. $A = \begin{pmatrix} 5 & 0 \\ 0 & 5 \end{pmatrix}$ **27.** $A = \begin{pmatrix} -3 & 0 \\ 0 & -3 \end{pmatrix}$

28. $A = \begin{pmatrix} -1 & 3 \\ -3 & -7 \end{pmatrix}$ ***29.** $A = \begin{pmatrix} -7 & 4 \\ -1 & -3 \end{pmatrix}$

30. $A = \begin{pmatrix} -1 & 3 \\ -6 & -7 \end{pmatrix}$ **31.** $A = \begin{pmatrix} -1 & -1 \\ 5 & -3 \end{pmatrix}$

32. $A = \begin{pmatrix} -4 & 0 & -3 \\ 9 & 1 & 9 \\ 2 & 0 & 1 \end{pmatrix}$

*33. $A = \begin{pmatrix} -41 & -38 & 18 \\ 48 & 44 & -21 \\ 8 & 6 & -3 \end{pmatrix}$

34. $A = \begin{pmatrix} 1 & 1 & 3 \\ 1 & 0 & 1 \\ 1 & -1 & 2 \end{pmatrix}$

In Exercises 35–40, find $\dfrac{d}{dt} A(t)$ and $\displaystyle\int A(t)\,dt$.

35. $A = \begin{pmatrix} e^{2t} \\ e^{-5t} \end{pmatrix}$ **36.** $A = \begin{pmatrix} \sin 3t \\ \cos 3t \end{pmatrix}$

*37. $A = \begin{pmatrix} \cos t & t\cos t \\ \sin t & t\sin t \end{pmatrix}$ **38.** $A = \begin{pmatrix} e^{-t} & te^{-t} \\ t & t^2 \end{pmatrix}$

*39. $A = \begin{pmatrix} e^{4t} \\ \cos 3t \\ \sin 3t \end{pmatrix}$ **40.** $A = \begin{pmatrix} \dfrac{\ln t}{t} \\ t\ln t \\ \ln t \end{pmatrix}$

*41. If $A = \begin{pmatrix} a & b \\ c & d \end{pmatrix}$ and $|A| = ad - bc \neq 0$, then show that $A^{-1} = \dfrac{1}{ad - bc} \begin{pmatrix} d & -b \\ -c & a \end{pmatrix}$.

8.3   Preliminary Definitions and Notation

We first encounter systems of equations in elementary algebra courses. For example,

$$\begin{cases} 3x - 5y = -13 \\ -3x + 6y = 15 \end{cases}$$

is a system of two linear equations in two variables with solution $(x, y) = (-1, 2)$. In the same manner, we can consider a system of differential equations.

We begin our study of systems of ordinary differential equations by introducing several definitions along with some convenient notation. Let

$$\mathbf{X} = \mathbf{X}(t) = \begin{pmatrix} x_1(t) \\ x_2(t) \\ \vdots \\ x_n(t) \end{pmatrix}, \quad \mathbf{A}(t) = \begin{pmatrix} a_{11}(t) & a_{12}(t) & \cdots & a_{1n}(t) \\ a_{21}(t) & a_{22}(t) & \cdots & a_{2n}(t) \\ \vdots & \vdots & \ddots & \vdots \\ a_{n1}(t) & a_{n2}(t) & \cdots & a_{nn}(t) \end{pmatrix}, \quad \text{and} \quad \mathbf{F}(t) = \begin{pmatrix} f_1(t) \\ f_2(t) \\ \vdots \\ f_n(t) \end{pmatrix}.$$

Then, the homogeneous system of first-order linear differential equations

$$\begin{cases} x_1'(t) = a_{11}(t)x_1(t) + a_{12}(t)x_2(t) + \cdots + a_{1n}(t)x_n(t) \\ x_2'(t) = a_{21}(t)x_1(t) + a_{22}(t)x_2(t) + \cdots + a_{2n}(t)x_n(t) \\ \quad\quad\quad\quad\quad\vdots \\ x_n'(t) = a_{n1}(t)x_1(t) + a_{n2}(t)x_2(t) + \cdots + a_{nn}(t)x_n(t) \end{cases}$$

is equivalent to $\mathbf{X}'(t) = \mathbf{A}(t)\mathbf{X}(t)$, and the nonhomogeneous system

$$\begin{cases} x_1'(t) = a_{11}(t)x_1(t) + a_{12}(t)x_2(t) + \cdots + a_{1n}(t)x_n(t) + f_1(t) \\ x_2'(t) = a_{21}(t)x_1(t) + a_{22}(t)x_2(t) + \cdots + a_{2n}(t)x_n(t) + f_2(t) \\ \quad\quad\quad\quad\quad\vdots \\ x_n'(t) = a_{n1}(t)x_1(t) + a_{n2}(t)x_2(t) + \cdots + a_{nn}(t)x_n(t) + f_n(t) \end{cases}$$

is equivalent to $\mathbf{X}'(t) = \mathbf{A}(t)\mathbf{X}(t) + \mathbf{F}(t)$.

Example 1

(a) Write the homogeneous system $\begin{cases} x' = -5x + 5y \\ y' = -5x + y \end{cases}$ in matrix form. (b) Write the nonhomogeneous system $\begin{cases} x' = x + 2y - \sin t \\ y' = 4x - 3y + t^2 \end{cases}$ in matrix form.

Solution

(a) The homogeneous system $\begin{cases} x' = -5x + 5y \\ y' = -5x + y \end{cases}$ is equivalent to the system $\begin{pmatrix} x' \\ y' \end{pmatrix} = \begin{pmatrix} -5 & 5 \\ -5 & 1 \end{pmatrix} \begin{pmatrix} x \\ y \end{pmatrix}$.

(b) The nonhomogeneous system $\begin{cases} x' = x + 2y - \sin t \\ y' = 4x - 3y + t^2 \end{cases}$ is equivalent to $\begin{pmatrix} x' \\ y' \end{pmatrix} = \begin{pmatrix} 1 & 2 \\ 4 & -3 \end{pmatrix} \begin{pmatrix} x \\ y \end{pmatrix} + \begin{pmatrix} -\sin t \\ t^2 \end{pmatrix}$.

The higher order equations that we solved earlier in the text can be written as a system of first-order equations. Consider the nth-order differential equation with constant coefficients

$$y^{(n)}(t) + a_{n-1}y^{(n-1)}(t) + \cdots + a_2 y''(t) + a_1 y'(t) + a_0 y(t) = f(t).$$

Let $x_1 = y$, $x_2 = \dfrac{dx_1}{dt} = y'(t)$, $x_3 = \dfrac{dx_2}{dt} = y''(t), \ldots, x_{n-1} = \dfrac{dx_{n-2}}{dt} = y^{(n-2)}$, $x_n = \dfrac{dx_{n-1}}{dt} = y^{(n-1)}$. Then, the equation $y^{(n)}(t) + a_{n-1}y^{(n-1)}(t) + \cdots + a_2 y''(t) + a_1 y'(t) + a_0 y(t) = f(t)$ is equivalent to the system

$$\begin{cases} x_1' = x_2 \\ x_2' = x_3 \\ \vdots \\ x_{n-1}' = x_n \\ x_n' = -a_{n-1}y^{(n-1)} - \cdots - a_2 y''(t) - a_1 y'(t) - a_0 y + f(t) \\ \qquad = -a_{n-1}x_n - \cdots - a_2 x_3 - a_1 x_2 - a_0 x_1 + f(t) \end{cases}$$

which can be written in matrix form as

$$\begin{pmatrix} x_1' \\ x_2' \\ \vdots \\ x_{n-1}' \\ x_n' \end{pmatrix} = \begin{pmatrix} 0 & 1 & 0 & 0 & 0 \\ 0 & 0 & 1 & 0 & 0 \\ \vdots & \vdots & \vdots & \ddots & \vdots \\ 0 & 0 & 0 & 0 & 1 \\ -a_0 & -a_1 & -a_2 & \cdots & -a_{n-1} \end{pmatrix} \begin{pmatrix} x_1 \\ x_2 \\ \vdots \\ x_{n-1} \\ x_n \end{pmatrix} + \begin{pmatrix} 0 \\ 0 \\ 0 \\ 0 \\ f(t) \end{pmatrix}.$$

Example 2

Write the equation $y'' + 5y' + 6y = \cos t$ as a system.

Solution   We begin by letting $x_1 = y$ and $x_2 = x_1' = y'$. Then $x_2' = y'' = \cos t - 6y - 5y' = \cos t - 6x_1 - 5x_2$ and $\begin{cases} x_1' = x_2 \\ x_2' = \cos t - 6x_1 - 5x_2 \end{cases}$

can be written as $\begin{pmatrix} x_1' \\ x_2' \end{pmatrix} = \begin{pmatrix} 0 & 1 \\ -6 & -5 \end{pmatrix} \begin{pmatrix} x_1 \\ x_2 \end{pmatrix} + \begin{pmatrix} 0 \\ \cos t \end{pmatrix}.$

Nonlinear differential equations can often be written as a system of equations as well.

Example 3

The **Van-der-Pol equation** is the nonlinear ordinary differential equation

$$w'' - \mu(1 - w^2)w' + w = 0.$$

Write the Van-der-Pol equation as a system. (Notice that this equation is similar to those encountered when studying spring-mass systems, except that the coefficient of w' which corresponds to damping is not constant. Therefore, we refer to this situation as *variable damping* and will observe its effect on the system.)

Solution   Let $x_1 = w$ and $x_2 = w' = x_1'$. Then, $x_2' = w'' = \mu(1 - w^2)w' - w = \mu(1 - x_1^2)x_2 - x_1$, so the Van-der-Pol equation is equivalent to the nonlinear system $\begin{cases} x_1' = x_2 \\ x_2' = \mu(1 - x_1^2)x_2 - x_1 \end{cases}.$

In Figure 8.6, we show graphs of the solution in the initial-value problem

$$\begin{cases} x_1' = x_2 \\ x_2' = (1 - x_1^2)x_2 - x_1 \\ x_1(0) = 1, x_2(0) = 1 \end{cases}$$

on the interval $[0, 25]$. Because we let $x_2 = x_1'$, notice that $x_2(t) > 0$ when $x_1(t)$ is increasing and $x_2(t) < 0$ when $x_1(t)$ is decreasing. Note that these functions appear to become periodic as t increases.

The observation that these solutions become periodic is further confirmed by a graph of $x_1(t)$ (the horizontal axis) versus $x_2(t)$ (the vertical axis) shown in Figure 8.7, called the *phase plane*. We see that as t increases, the solution approaches a certain fixed path, called a *limit cycle*. We will find that nonlinear equations are more easily studied when they are written as a system of equations.

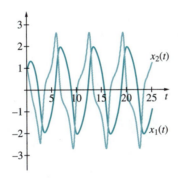

Figure 8.6   Graph of $x_1(t)$
and $x_2(t)$

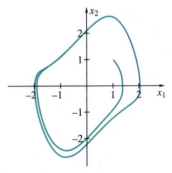

Figure 8.7   The graph of
$\begin{cases} x_1(t) \\ x_2(t) \end{cases}$ indicates that the solution approaches an isolated pe-
riodic solution, which is called
a *limit cycle.*

Graph the solution to the initial-value problem $\begin{cases} w'' + (w^2 - 1)w' + w = 0 \\ w(0) = 1, w'(0) = 1 \end{cases}$
on the interval [0, 25].

At this point, given a system of ordinary differential equations, our goal will generally be to construct an explicit, numerical, or graphical solution of the system.

We now state the theorems and terminology used in establishing the fundamentals of solving systems of differential equations. All proofs are omitted but can be found in advanced differential equations texts. In each case, we assume that the matrix $\mathbf{A}(t)$ in the systems $\mathbf{X}'(t) = \mathbf{A}(t)\mathbf{X}(t) + \mathbf{F}(t)$ and $\mathbf{X}'(t) = \mathbf{A}(t)\mathbf{X}(t)$ is an $n \times n$ matrix.

Definition 8.13 Solution Vector

A **solution vector** of the system $\mathbf{X}'(t) = \mathbf{A}(t)\mathbf{X}(t) + \mathbf{F}(t)$ on the interval I is an $n \times 1$ matrix of the form

$$\mathbf{X}(t) = \begin{pmatrix} x_1(t) \\ x_2(t) \\ \vdots \\ x_n(t) \end{pmatrix},$$

where the $x_i(t)$ are differentiable functions on I, that satisfies $\mathbf{X}'(t) = \mathbf{A}(t)\mathbf{X}(t) + \mathbf{F}(t)$.

Let $\mathbf{X}'(t) = \mathbf{A}(t)\mathbf{X}(t)$ where $\mathbf{X}(t) = \begin{pmatrix} x_1(t) \\ x_2(t) \\ \vdots \\ x_n(t) \end{pmatrix}$ and

$$\mathbf{A}(t) = \begin{pmatrix} a_{11}(t) & a_{12}(t) & \cdots & a_{1n}(t) \\ a_{21}(t) & a_{22}(t) & \cdots & a_{2n}(t) \\ \vdots & \vdots & \ddots & \vdots \\ a_{n1}(t) & a_{n2}(t) & \cdots & a_{nn}(t) \end{pmatrix} \quad \text{where} \quad a_{ij}(t) \quad \text{is} \quad \text{continuous} \quad \text{for} \quad \text{all}$$

$1 \leq j \leq n$ and $1 \leq i \leq n$. Let $\{\mathbf{\Phi}_i\}_{i=1}^{m} = \left\{ \begin{pmatrix} \Phi_{1i} \\ \Phi_{2i} \\ \vdots \\ \Phi_{ni} \end{pmatrix} \right\}_{i=1}^{m}$ be a set of m solutions of $\mathbf{X}'(t) = \mathbf{A}(t)\mathbf{X}(t)$. As with linear homogeneous equations, any linear combination of these solutions is also a solution to the homogeneous system $\mathbf{X}'(t) = \mathbf{A}(t)\mathbf{X}(t)$. (See Exercise 32.)

We define linear dependence and independence of the set of vectors $\{\mathbf{\Phi}_i\}_{i=1}^{m} = \left\{ \begin{pmatrix} \Phi_{1i} \\ \Phi_{2i} \\ \vdots \\ \Phi_{ni} \end{pmatrix} \right\}_{i=1}^{m}$ in the same way we defined linear dependence and independence of sets of functions. The set $\{\mathbf{\Phi}_i\}_{i=1}^{m} = \left\{ \begin{pmatrix} \Phi_{1i} \\ \Phi_{2i} \\ \vdots \\ \Phi_{ni} \end{pmatrix} \right\}_{i=1}^{m}$ is **linearly dependent** on an interval I if there is a set of constants $\{c_i\}_{i=1}^{m}$ not all zero such that $c_1\mathbf{\Phi}_1 + c_2\mathbf{\Phi}_2 + \cdots + c_m\mathbf{\Phi}_m = \mathbf{0}$; otherwise, the set is **linearly independent.**

Definition 8.14 Fundamental Set of Solutions

Any set $\{\mathbf{\Phi}_i\}_{i=1}^{n} = \left\{ \begin{pmatrix} \Phi_{1i} \\ \Phi_{2i} \\ \vdots \\ \Phi_{ni} \end{pmatrix} \right\}_{i=1}^{n}$ of n linearly independent solution vectors of $\mathbf{X}'(t) = \mathbf{A}(t)\mathbf{X}(t)$ on an interval I is called a **fundamental set of solutions** on I.

We can determine if a set of vectors is linearly independent or linearly dependent by computing the Wronskian.

Theorem 8.3

The set $\{\mathbf{\Phi}_i\}_{i=1}^{n} = \left\{\begin{pmatrix} \Phi_{1i} \\ \Phi_{2i} \\ \vdots \\ \Phi_{ni} \end{pmatrix}\right\}_{i=1}^{n}$ is linearly independent if and only if the **Wronskian**

$$W(\mathbf{\Phi}_1, \mathbf{\Phi}_2, \ldots, \mathbf{\Phi}_n) = \det \begin{pmatrix} \Phi_{11} & \Phi_{12} & \cdots & \Phi_{1n} \\ \Phi_{21} & \Phi_{22} & \cdots & \Phi_{2n} \\ \vdots & \vdots & \ddots & \vdots \\ \Phi_{n1} & \Phi_{n2} & \cdots & \Phi_{nn} \end{pmatrix} \neq 0.$$

Example 4

Which of the following is a fundamental set of solutions for $\begin{pmatrix} x'(t) \\ y'(t) \end{pmatrix} = \begin{pmatrix} -2 & -8 \\ 1 & 2 \end{pmatrix}\begin{pmatrix} x(t) \\ y(t) \end{pmatrix}$? (a) $\left\{\begin{pmatrix} \cos 2t \\ \sin 2t \end{pmatrix}, \begin{pmatrix} \sin 2t \\ \cos 2t \end{pmatrix}\right\}$

(b) $\left\{\begin{pmatrix} -2\sin 2t + 2\cos 2t \\ \sin 2t \end{pmatrix}, \begin{pmatrix} 4\cos 2t \\ \sin 2t - \cos 2t \end{pmatrix}\right\}$

Solution   We first remark that the equation $\begin{pmatrix} x'(t) \\ y'(t) \end{pmatrix} = \begin{pmatrix} -2 & -8 \\ 1 & 2 \end{pmatrix}\begin{pmatrix} x(t) \\ y(t) \end{pmatrix}$ is

equivalent to the system

$$\begin{cases} x' = -2x - 8y \\ y' = x + 2y \end{cases}.$$

(a) Differentiating we see that

$$\begin{pmatrix} \cos 2t \\ \sin 2t \end{pmatrix}' = \begin{pmatrix} -2\sin 2t \\ 2\cos 2t \end{pmatrix} \neq \begin{pmatrix} -2\cos 2t - 8\sin 2t \\ \cos 2t + 2\sin 2t \end{pmatrix}$$

which shows us that $\begin{pmatrix} \cos 2t \\ \sin 2t \end{pmatrix}$ is not a solution to the system.

Therefore, $\left\{\begin{pmatrix} \cos 2t \\ \sin 2t \end{pmatrix}, \begin{pmatrix} \sin 2t \\ \cos 2t \end{pmatrix}\right\}$ is not a fundamental set of solutions.

(b) You should verify that both $\begin{pmatrix} -2\sin 2t + 2\cos 2t \\ \sin 2t \end{pmatrix}$ and $\begin{pmatrix} 4\cos 2t \\ \sin 2t - \cos 2t \end{pmatrix}$

are solutions to the system. Computing the Wronskian, we have

$$\begin{vmatrix} -2\sin 2t + 2\cos 2t & 4\cos 2t \\ \sin 2t & \sin 2t - \cos 2t \end{vmatrix}$$

$$= (-2\sin 2t + 2\cos 2t)(\sin 2t - \cos 2t) - (4\cos 2t)(\sin 2t)$$

$$= -2\cos^2 2t - 2\sin^2 2t = -2.$$

Thus, the set $\left\{\begin{pmatrix} -2\sin 2t + 2\cos 2t \\ \sin 2t \end{pmatrix}, \begin{pmatrix} 4\cos 2t \\ \sin 2t - \cos 2t \end{pmatrix}\right\}$ is linearly independent and is consequently a fundamental set of solutions.

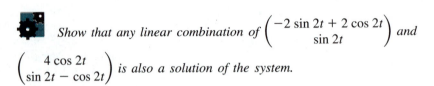

Show that any linear combination of $\begin{pmatrix} -2\sin 2t + 2\cos 2t \\ \sin 2t \end{pmatrix}$ and $\begin{pmatrix} 4\cos 2t \\ \sin 2t - \cos 2t \end{pmatrix}$ is also a solution of the system.

The following theorem states that a fundamental set of solutions cannot contain more than n vectors, because the solutions could not be linearly independent.

Theorem 8.4

Any $n + 1$ nontrivial solutions of $\mathbf{X}'(t) = \mathbf{A}(t)\mathbf{X}(t)$ are linearly dependent.

Finally, we state the theorems that indicate a fundamental set of solutions can always be found and a general solution can be constructed.

Theorem 8.5

There is a set of n nontrivial linearly independent solutions of $\mathbf{X}'(t) = \mathbf{A}(t)\mathbf{X}(t)$.

Theorem 8.6

Let $\{\mathbf{\Phi}_i\}_{i=1}^n = \left\{\begin{pmatrix} \Phi_{1i} \\ \Phi_{2i} \\ \vdots \\ \Phi_{ni} \end{pmatrix}\right\}_{i=1}^n$ be a set of n linearly independent solutions of $\mathbf{X}'(t) = \mathbf{A}(t)\mathbf{X}(t)$. Then every solution of $\mathbf{X}'(t) = \mathbf{A}(t)\mathbf{X}(t)$ is a linear combination of these solutions. Hence, a **general solution** to $\mathbf{X}'(t) = \mathbf{A}(t)\mathbf{X}(t)$ is

$$\mathbf{X}(t) = c_1\mathbf{\Phi}_1(t) + c_2\mathbf{\Phi}_2(t) + \cdots + c_n\mathbf{\Phi}_n(t).$$

Definition 8.15 Fundamental Matrix

Let $\{\Phi_i\}_{i=1}^{n} = \left\{ \begin{pmatrix} \Phi_{1i} \\ \Phi_{2i} \\ \vdots \\ \Phi_{ni} \end{pmatrix} \right\}_{i=1}^{n}$ be a set of n linearly independent solutions

of $\mathbf{X}'(t) = \mathbf{A}(t)\mathbf{X}(t)$. The matrix

$$\mathbf{\Phi}(t) = (\mathbf{\Phi}_1 \quad \mathbf{\Phi}_2 \quad \cdots \quad \mathbf{\Phi}_n) = \begin{pmatrix} \Phi_{11} & \Phi_{12} & \cdots & \Phi_{1n} \\ \Phi_{21} & \Phi_{22} & \cdots & \Phi_{2n} \\ \vdots & \vdots & \ddots & \vdots \\ \Phi_{n1} & \Phi_{n2} & \cdots & \Phi_{nn} \end{pmatrix}$$

is called a **fundamental matrix** of the system $\mathbf{X}'(t) = \mathbf{A}(t)\mathbf{X}(t)$. Thus, a **general**

solution can be written as $\mathbf{X}(t) = \mathbf{\Phi}(t)\mathbf{C}$ where $\mathbf{C} = \begin{pmatrix} c_1 \\ c_2 \\ \vdots \\ c_n \end{pmatrix}$.

Notice that the form of a general solution given above as $\mathbf{X}(t) = \mathbf{\Phi}(t)\mathbf{C}$ is equivalent to the form expressed earlier,

$$\mathbf{X}(t) = c_1\mathbf{\Phi}_1(t) + c_2\mathbf{\Phi}_2(t) + \cdots + c_n\mathbf{\Phi}_n(t).$$

Example 5

Show that $\mathbf{\Phi}(t) = \begin{pmatrix} e^{-2t} & -3e^{5t} \\ 2e^{-2t} & e^{5t} \end{pmatrix}$ is a fundamental matrix for the system

$\mathbf{X}'(t) = \begin{pmatrix} 4 & -3 \\ -2 & -1 \end{pmatrix}\mathbf{X}(t)$. Use the matrix to find a general solution of

$\mathbf{X}'(t) = \begin{pmatrix} 4 & -3 \\ -2 & -1 \end{pmatrix}\mathbf{X}(t)$.

Solution Because

$$\begin{pmatrix} e^{-2t} \\ 2e^{-2t} \end{pmatrix}' = \begin{pmatrix} -2e^{-2t} \\ -4e^{-2t} \end{pmatrix} = \begin{pmatrix} 4 & -3 \\ -2 & -1 \end{pmatrix}\begin{pmatrix} e^{-2t} \\ 2e^{-2t} \end{pmatrix} \text{ and }$$

$$\begin{pmatrix} -3e^{5t} \\ e^{5t} \end{pmatrix}' = \begin{pmatrix} -15e^{5t} \\ 5e^{5t} \end{pmatrix} = \begin{pmatrix} 4 & -3 \\ -2 & -1 \end{pmatrix}\begin{pmatrix} -3e^{5t} \\ e^{5t} \end{pmatrix},$$

both $\begin{pmatrix} e^{-2t} \\ 2e^{-2t} \end{pmatrix}$ and $\begin{pmatrix} -3e^{5t} \\ e^{5t} \end{pmatrix}$ are solutions to the system $\mathbf{X}'(t) =$

$\begin{pmatrix} 4 & -3 \\ -2 & -1 \end{pmatrix}\mathbf{X}(t)$. The solutions are linearly independent because

$$W\left(\begin{pmatrix} e^{-2t} \\ 2e^{-2t} \end{pmatrix}, \begin{pmatrix} -3e^{5t} \\ e^{5t} \end{pmatrix} \right) = \begin{vmatrix} e^{-2t} & -3e^{5t} \\ 2e^{-2t} & e^{5t} \end{vmatrix} = 7e^{3t} \neq 0.$$

The theorems and definitions introduced in the section indicate that when solving an $n \times n$ homogeneous system of linear equations $\mathbf{X}'(t) = \mathbf{A}(t)\mathbf{X}(t)$, we find n linearly independent solutions. After finding these solutions, we form a fundamental matrix that can be used to form a general solution or solve an initial-value problem.

A general solution is given by

$$\mathbf{X}(t) = \Phi(t)\mathbf{C} = \begin{pmatrix} e^{-2t} & -3e^{5t} \\ 2e^{-2t} & e^{5t} \end{pmatrix}\begin{pmatrix} c_1 \\ c_2 \end{pmatrix}$$

$$= \begin{pmatrix} c_1 e^{-2t} - 3c_2 e^{5t} \\ 2c_1 e^{-2t} + c_2 e^{5t} \end{pmatrix} = c_1\begin{pmatrix} e^{-2t} \\ 2e^{-2t} \end{pmatrix} + c_2\begin{pmatrix} -3e^{5t} \\ e^{5t} \end{pmatrix}.$$

Graphs of several solutions are shown in Figure 8.8. (a) Identify the graph of the solution that satisfies the initial conditions $x(0) = 0$ and $y(0) = 1$. (b) Find c_1 and c_2 so that $\mathbf{X}(t) = c_1\begin{pmatrix} e^{-2t} \\ 2e^{-2t} \end{pmatrix} + c_2\begin{pmatrix} -3e^{5t} \\ e^{5t} \end{pmatrix}$ satisfies $\mathbf{X}(0) = \begin{pmatrix} 0 \\ 1 \end{pmatrix}$.

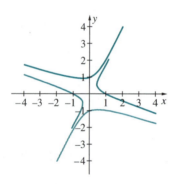

Figure 8.8

IN TOUCH WITH TECHNOLOGY

1. Show that each matrix is a fundamental matrix for the given system. In each case, use the fundamental matrix to construct a general solution to the system.

(a) $\begin{pmatrix} \cos 3t - \sin 3t & -\sin 3t - \cos 3t \\ \sin 3t & \cos 3t \end{pmatrix}$;

$$\mathbf{X}'(t) = \begin{pmatrix} -3 & -6 \\ 3 & 3 \end{pmatrix}\mathbf{X}(t)$$

(b) $\begin{pmatrix} e^t(\sin 3t - 3\cos 3t) & 5e^t\cos 3t \\ 2e^t\sin 3t & e^t(\cos 3t - 3\sin 3t) \end{pmatrix}$;

$$\mathbf{X}'(t) = \begin{pmatrix} 0 & 5 \\ -2 & 2 \end{pmatrix}\mathbf{X}(t)$$

(c) $\begin{pmatrix} e^{-t}(\sin 3t + 3\cos 3t) & 5e^{-t}\cos 3t \\ 2e^{-t}\sin 3t & e^{-t}(\cos 3t + 3\sin 3t) \end{pmatrix}$;

$$\mathbf{X}'(t) = \begin{pmatrix} 0 & -5 \\ 2 & -2 \end{pmatrix}\mathbf{X}(t)$$

(d) Use the results of (a), (b), and (c) to solve (a), (b), and (c) subject to $\mathbf{X}(0) = \begin{pmatrix} 0 \\ 1 \end{pmatrix}$. In each case, graph the resulting solution for $0 \le t \le 2\pi$. What similarities and differences between the graphs do you observe?

In Exercises 1–8, write the system of first-order equations in matrix form.

1. $\begin{cases} x' = x + 4y \\ y' = 2x - y \end{cases}$

2. $\begin{cases} x' = 2x - 3y \\ y' = x + y \end{cases}$

3. $\begin{cases} x' = y + 2x \\ y' = x - 5y \end{cases}$

4. $\begin{cases} x' = y - 4x \\ y' = y - x \end{cases}$

5. $\begin{cases} x' - x = e^t \\ y' = x + y \end{cases}$

6. $\begin{cases} x' = y \\ y' = x + \cos t \end{cases}$

7. $\begin{cases} x' - y = 0 \\ y' + x = 2\sin t \end{cases}$

8. $\begin{cases} x' + x - 3y = 0 \\ y' - x = e^{-t} \end{cases}$

In Exercises 9–14, determine if the given vectors are linearly independent.

9. $\begin{pmatrix} e^t \\ 2e^t \end{pmatrix}, \begin{pmatrix} 3e^{-t} \\ e^{-t} \end{pmatrix}$

10. $\begin{pmatrix} t \\ 1 \end{pmatrix}, \begin{pmatrix} -t \\ t \end{pmatrix}$

*11. $\begin{pmatrix} \cos 2t \\ -2\sin 2t \end{pmatrix}, \begin{pmatrix} \sin 2t \\ 2\cos 2t \end{pmatrix}$

12. $\begin{pmatrix} \cos 2t \\ -2\sin 2t \end{pmatrix}, \begin{pmatrix} 1 - 2\sin^2 t \\ -4\sin t \cos t \end{pmatrix}$

*13. $\begin{pmatrix} e^{2t} \\ e^{2t} \\ 2e^{2t} \end{pmatrix}, \begin{pmatrix} e^t \\ -e^t \\ 3e^t \end{pmatrix}, \begin{pmatrix} e^{-t} \\ e^{-t} \\ e^{-t} \end{pmatrix}$

14. $\begin{pmatrix} 6e^{4t} \\ 6e^{4t} \\ 3e^{4t} \end{pmatrix}, \begin{pmatrix} e^{-2t} \\ e^{-2t} \\ -e^{-2t} \end{pmatrix}, \begin{pmatrix} 2e^{4t} \\ 2e^{4t} \\ e^{4t} \end{pmatrix}$

In Exercises 15–20, determine whether the given matrix is a fundamental matrix for the system.

15. $\Phi(t) = \begin{pmatrix} -2e^{-8t} & 5e^{-t} \\ e^{-8t} & e^{-t} \end{pmatrix}$ and

$X(t) = \begin{pmatrix} -3 & 10 \\ 1 & -6 \end{pmatrix} X(t)$

16. $\Phi(t) = \begin{pmatrix} e^{-10t} & -3e^{5t} \\ 2e^{-10t} & e^{5t} \end{pmatrix}$ and

$X'(t) = \begin{pmatrix} 5 & 0 \\ -5 & -10 \end{pmatrix} X(t)$

*17. $\Phi(t) = \begin{pmatrix} \cos t + 2\sin t & 2\cos t - \sin t \\ \sin t & \cos t \end{pmatrix}$

and $X'(t) = \begin{pmatrix} 2 & -5 \\ 1 & -2 \end{pmatrix} X(t)$

18. $\Phi(t) = \begin{pmatrix} e^t & 3e^{3t} \\ e^t & 5e^{3t} \end{pmatrix}$ and $X'(t) = \begin{pmatrix} -2 & 3 \\ -5 & 6 \end{pmatrix} X(t)$

*19. $\Phi(t) = \begin{pmatrix} 27e^{-6t} & -e^t & 0 \\ -29e^{-6t} & -e^{-t} & e^{3t} \\ 36e^{-6t} & e^{-t} & 0 \end{pmatrix}$ and

$X'(t) = \begin{pmatrix} -2 & 0 & -3 \\ 3 & 3 & 5 \\ -4 & 0 & -3 \end{pmatrix} X(t)$

20. $\Phi(t)$

$= \begin{pmatrix} 5e^{3t} & -2e^{-t}\cos t & 2e^{-t}\sin t \\ -5e^{3t} & e^{-t}(\sin t - 2\cos t) & e^{-t}(\cos t + 2\sin t) \\ e^{3t} & 3e^{-t}\cos t & -3e^{-t}\sin t \end{pmatrix}$

and $X'(t) = \begin{pmatrix} 1 & -2 & 0 \\ -5 & -3 & -5 \\ 3 & 3 & 3 \end{pmatrix} X(t)$

In Exercises 21–26, write each equation as an equivalent system of first-order equations.

21. $x'' - 3x' + 4x = 0$

22. $x'' + 6x' + 9x = 0$

23. $x'' + 16x = t\sin t$

24. $x'' + x = e^t$

25. $y''' + 3y'' + 6y' + 3y = x$

26. $y^{(4)} + y'' = 0$

*27. **Rayleigh's equation** is

$$x'' + \mu\left[\frac{1}{3}(x')^2 - 1\right]x' + x = 0,$$

where μ is a constant.

(a) Write Rayleigh's equation as a system.

(b) Show that differentiating Rayleigh's equation and setting $x' = z$ reduces Rayleigh's equation to the Van-der-Pol equation.

In Exercises 28–30, use the given fundamental matrix to obtain a general solution to the system of first-order homogeneous equations $X' = AX$. Also, find the solution that satisfies the given initial condition.

28. $\Phi(t) = \begin{pmatrix} 2e^{4t} & e^{-t} \\ 3e^{4t} & -e^{-t} \end{pmatrix}$, $A = \begin{pmatrix} 1 & 2 \\ 3 & 2 \end{pmatrix}$, $X(0) = \begin{pmatrix} 3 \\ 2 \end{pmatrix}$

*29. $\Phi(t) = \begin{pmatrix} 3e^{4t} & e^{3t} \\ 2e^{4t} & e^{3t} \end{pmatrix}$, $A = \begin{pmatrix} 6 & -3 \\ 2 & 1 \end{pmatrix}$, $X(0) = \begin{pmatrix} 6 \\ -4 \end{pmatrix}$

30. $\Phi(t) = \begin{pmatrix} 2\cos 2t + 2\sin 2t & 2\cos 2t - 2\sin 2t \\ -\sin 2t & -\cos 2t \end{pmatrix}$,

$A = \begin{pmatrix} 2 & 8 \\ -1 & -2 \end{pmatrix}$, $X(0) = \begin{pmatrix} -8 \\ 2 \end{pmatrix}$

31. (Modeling the Motion of Spiked Volleyball) Under certain assumptions, the position of a spiked volleyball can be modeled by the system of two second-order nonlinear equations

$$\begin{cases} X'' = \dfrac{1}{m}\Big(C_M\omega^a Y'([X']^2 + [Y']^2)^{(b-1)/2} - \\ \qquad\qquad \dfrac{1}{2}C_D\rho A X'\sqrt{[X']^2 + [Y']^2}\Big) \\ \\ Y'' = -g - \dfrac{1}{m}\Big(C_M\omega^a X'([X']^2 + [Y']^2)^{(b-1)/2} + \\ \qquad\qquad \dfrac{1}{2}C_D\rho A Y'\sqrt{[X']^2 + [Y']^2}\Big) \end{cases}^{*}$$

Write this system of two second-order nonlinear equations as a system of four first-order equations.

32. (Principle of Superposition) (a) Show that any linear combination of solutions of the homogeneous system $X'(t) = A(t)X(t)$ is also a solution of the homogeneous system. **(b)** Is the Principle of Superposition ever valid for nonhomogeneous systems of equations? Explain.

8.4 Homogeneous Linear Systems with Constant Coefficients

Distinct Real Eigenvalues • Complex Conjugate Eigenvalues • Alternate Method for Solving Initial-Value Problems • Repeated Eigenvalues

Now that we have covered the necessary terminology, we turn our attention to solving linear systems with constant coefficients. Let $A = \begin{pmatrix} a_{11} & a_{12} & \cdots & a_{1n} \\ a_{22} & a_{22} & \cdots & a_{2n} \\ \vdots & \vdots & \ddots & \vdots \\ a_{n1} & a_{n2} & \cdots & a_{nn} \end{pmatrix}$ be an $n \times n$ matrix with real components. In this section, we will see that a general solution to the homogeneous system $X' = AX$ is determined by the eigenvalues and corresponding eigenvectors of A. We begin by considering the cases when the eigenvalues of A are distinct and real or the eigenvalues of A are distinct and complex, and then consider the case when A has repeated eigenvalues (eigenvalues of multiplicity greater than one).

Distinct Real Eigenvalues

In Section 8.3, we verified that a general solution of the 2×2 system $X' = \begin{pmatrix} 4 & -3 \\ -2 & -1 \end{pmatrix}X$, where the eigenvalues of $\begin{pmatrix} 4 & -3 \\ -2 & -1 \end{pmatrix}$ are $\lambda_1 = -2$ and

*Shawn S. Kao, Richard W. Sellens, and Joan M. Stevenson, "A Mathematical Model for the Trajectory of a Spiked Volleyball and Its Coaching Applications," *Journal of Applied Biomechanics*, Human Kinetics Publishers, Inc. (1994) pp. 95–109.

$\lambda_2 = 5$ and corresponding eigenvectors are $\mathbf{v}_1 = \begin{pmatrix} 1 \\ 2 \end{pmatrix}$ and $\mathbf{v}_2 = \begin{pmatrix} -3 \\ 1 \end{pmatrix}$, respectively, is

$$\mathbf{X}(t) = \Phi(t)\mathbf{C} = c_1 \begin{pmatrix} e^{-2t} \\ 2e^{-2t} \end{pmatrix} + c_2 \begin{pmatrix} -3e^{5t} \\ e^{5t} \end{pmatrix} = c_1 \begin{pmatrix} 1 \\ 2 \end{pmatrix} e^{-2t} + c_2 \begin{pmatrix} -3 \\ 1 \end{pmatrix} e^{5t}.$$

More generally, if the eigenvalues of the $n \times n$ matrix $\mathbf{A}$ are distinct, we may expect a general solution of the linear homogeneous system $\mathbf{X}' = \mathbf{AX}$ to have the form

$$\mathbf{X}(t) = c_1 \mathbf{v}_1 e^{\lambda_1 t} + c_2 \mathbf{v}_2 e^{\lambda_2 t} + \cdots + c_n \mathbf{v}_n e^{\lambda_n t},$$

where $\lambda_1, \lambda_2, \ldots, \lambda_n$ are the n distinct real eigenvalues of $\mathbf{A}$ with corresponding eigenvectors $\mathbf{v}_1, \mathbf{v}_2, \ldots, \mathbf{v}_n$, respectively. We investigate this claim by assuming that $\mathbf{X} = \mathbf{v}e^{\lambda t}$ is a solution of $\mathbf{X}' = \mathbf{AX}$. Then, $\mathbf{X}' = \lambda \mathbf{v}e^{\lambda t}$ must satisfy the system of differential equations, which implies that

$$\lambda \mathbf{v}e^{\lambda t} = \mathbf{A}\mathbf{v}e^{\lambda t}.$$

Using $\mathbf{Iv} = \mathbf{v}$ yields

$$\lambda \mathbf{Iv}e^{\lambda t} = \mathbf{A}\mathbf{v}e^{\lambda t}$$
$$\mathbf{A}\mathbf{v}e^{\lambda t} - \lambda \mathbf{Iv}e^{\lambda t} = \mathbf{0}$$
$$(\mathbf{A} - \lambda \mathbf{I})\mathbf{v}e^{\lambda t} = \mathbf{0}.$$

Then, because $e^{\lambda t} \neq 0$, we have $(\mathbf{A} - \lambda \mathbf{I})\mathbf{v} = \mathbf{0}$. In order for this system of equations to have a solution other than $\mathbf{v} = \mathbf{0}$, (remember that eigenvectors are never the zero vector), we must have

$$|\mathbf{A} - \lambda \mathbf{I}| = 0.$$

A solution λ to this equation is an eigenvalue of $\mathbf{A}$ while a nonzero vector $\mathbf{v}$ satisfying $(\mathbf{A} - \lambda \mathbf{I})\mathbf{v} = \mathbf{0}$ is an eigenvector that corresponds to λ. Hence, if $\mathbf{A}$ has n distinct eigenvalues $\{\lambda_1, \lambda_2, \ldots, \lambda_n\}$, we can find a set of n linearly independent eigenvectors $\{\mathbf{v}_1, \mathbf{v}_2, \ldots, \mathbf{v}_n\}$. From these eigenvalues and corresponding eigenvectors, we form the n linearly independent solutions

$$\mathbf{X}_1 = \mathbf{v}_1 e^{\lambda_1 t}, \ \mathbf{X}_2 = \mathbf{v}_2 e^{\lambda_2 t}, \ldots, \ \mathbf{X}_n = \mathbf{v}_n e^{\lambda_n t}.$$

 Why are these n solutions linearly independent?

Therefore, if $\mathbf{A}$ is an $n \times n$ matrix with n distinct real eigenvalues $\{\lambda_k\}_{k=1}^n$, a general solution of $\mathbf{X}' = \mathbf{AX}$ is the linear combination of the set of solutions $\{\mathbf{X}_1, \mathbf{X}_2, \ldots, \mathbf{X}_n\}$,

$$\mathbf{X}(t) = c_1 \mathbf{v}_1 e^{\lambda_1 t} + c_2 \mathbf{v}_2 e^{\lambda_2 t} + \cdots + c_n \mathbf{v}_n e^{\lambda_n t}.$$

Example 1

Find a general solution of $\mathbf{X}' = \begin{pmatrix} 5 & -1 \\ 0 & 3 \end{pmatrix}\mathbf{X}$.

Solution   The eigenvalues of $\mathbf{A} = \begin{pmatrix} 5 & -1 \\ 0 & 3 \end{pmatrix}$ are found with

$\begin{vmatrix} 5 - \lambda & -1 \\ 0 & 3 - \lambda \end{vmatrix} = (5 - \lambda)(3 - \lambda) = 0$. The eigenvalues of $\mathbf{A}$ are $\lambda_1 = 3$ and

$\lambda_2 = 5$. An eigenvector $\mathbf{v}_1 = \begin{pmatrix} x_1 \\ y_1 \end{pmatrix}$ corresponding to $\lambda_1 = 3$ satisfies the system

$\begin{pmatrix} 2 & -1 \\ 0 & 0 \end{pmatrix}\begin{pmatrix} x_1 \\ y_1 \end{pmatrix} = \begin{pmatrix} 0 \\ 0 \end{pmatrix}$, so $y_1 = 2x_1$. Choosing $x_1 = 1$, we obtain the

eigenvector $\mathbf{v}_1 = \begin{pmatrix} 1 \\ 2 \end{pmatrix}$. Similarly, an eigenvector $\mathbf{v}_2 = \begin{pmatrix} x_2 \\ y_2 \end{pmatrix}$ corresponding to

$\lambda_2 = 5$ satisfies $\begin{pmatrix} 0 & -1 \\ 0 & -2 \end{pmatrix}\begin{pmatrix} x_2 \\ y_2 \end{pmatrix} = \begin{pmatrix} 0 \\ 0 \end{pmatrix}$, which indicates that $y_2 = 0$. Hence,

$\mathbf{v}_2 = \begin{pmatrix} x_2 \\ 0 \end{pmatrix}$ is an eigenvector for any value of x_2. If we let $x_2 = 1$,

$\mathbf{v}_2 = \begin{pmatrix} 1 \\ 0 \end{pmatrix}$. Therefore, a general solution of the system $\mathbf{X}' = \begin{pmatrix} 5 & -1 \\ 0 & 3 \end{pmatrix}\mathbf{X}$ is

$$\mathbf{X}(t) = c_1\mathbf{v}_1 e^{\lambda_1 t} + c_2\mathbf{v}_2 e^{\lambda_2 t} = c_1\begin{pmatrix} 1 \\ 2 \end{pmatrix}e^{3t} + c_2\begin{pmatrix} 1 \\ 0 \end{pmatrix}e^{5t}.$$

Remember that the system $\mathbf{X}' = \begin{pmatrix} 5 & -1 \\ 0 & 3 \end{pmatrix}\mathbf{X}$ is the same as the system

$\begin{cases} x' = 5x - y \\ y' = 3y \end{cases}$. Thus, we can write the general solution obtained here as $\mathbf{X}(t) =$

$\begin{pmatrix} x(t) \\ y(t) \end{pmatrix} = \begin{pmatrix} c_1 e^{3t} + c_2 e^{5t} \\ 2c_1 e^{3t} \end{pmatrix}$ or $\begin{cases} x(t) = c_1 e^{3t} + c_2 e^{5t} \\ y(t) = 2c_1 e^{3t} \end{cases}$. Several solutions along

with the direction field for the system are shown in Figure 8.9. (Notice that each

curve corresponds to the parametric plot of the pair $\begin{cases} x(t) \\ y(t) \end{cases}$ for particular values of the

constants c_1 and c_2.) Because both eigenvalues are positive, all solutions move away from the origin as t increases. The arrows on the vectors in the direction field show this behavior.

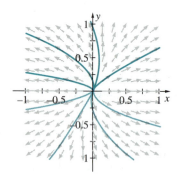

Figure 8.9

You may notice that general solutions you obtain when solving systems do not agree with those of your classmates or those given in the Exercise 8.4 Solutions. Before you become alarmed, realize that your solutions may be correct. You may have simply selected different values for the arbitrary constants in finding eigenvectors.

Complex Conjugate Eigenvalues

If $\mathbf{A}$ has complex conjugate eigenvalues $\lambda_1 = \alpha + \beta i$ and $\lambda_2 = \alpha - \beta i$ with corresponding eigenvectors $\mathbf{v}_1 = \mathbf{a} + \mathbf{b}i$ and $\mathbf{v}_2 = \mathbf{a} - \mathbf{b}i$, one solution of $\mathbf{X}' = \mathbf{A}\mathbf{X}$ is

$$\begin{aligned} \mathbf{X}(t) &= \mathbf{v}_1 e^{\lambda_1 t} = (\mathbf{a} + \mathbf{b}i)e^{(\alpha + \beta i)t} = e^{\alpha t}(\mathbf{a} + \mathbf{b}i)e^{i\beta t} = e^{\alpha t}(\mathbf{a} + \mathbf{b}i)(\cos \beta t + i \sin \beta t) \\ &= e^{\alpha t}(\mathbf{a} \cos \beta t - \mathbf{b} \sin \beta t) + ie^{\alpha t}(\mathbf{a} \sin \beta t + \mathbf{b} \cos \beta t) \\ &= \mathbf{x}_1(t) + i\mathbf{x}_2(t). \end{aligned}$$

Because $\mathbf{X}$ is a solution to the system $\mathbf{X}' = \mathbf{AX}$, we have that $\mathbf{x}_1'(t) + i\mathbf{x}_2'(t) = \mathbf{Ax}_1(t) + i\mathbf{Ax}_2(t)$. Equating the real and imaginary parts of this equation yields $\mathbf{x}_1'(t) = \mathbf{Ax}_1(t)$ and $\mathbf{x}_2'(t) = \mathbf{Ax}_2(t)$. Therefore, $\mathbf{x}_1(t)$ and $\mathbf{x}_2(t)$ are solutions to $\mathbf{X}' = \mathbf{AX}$, and any linear combination of $\mathbf{x}_1(t)$ and $\mathbf{x}_2(t)$ is also a solution. We can show that $\mathbf{x}_1(t)$ and $\mathbf{x}_2(t)$ are linearly independent (see Exercise 35), so this linear combination forms a portion of a general solution of $\mathbf{X}' = \mathbf{AX}$ where $\mathbf{A}$ is a square matrix of *any size*.

Theorem 8.7

Let $\mathbf{A}$ be a square matrix. If $\mathbf{A}$ has complex conjugate eigenvalues $\lambda_1 = \alpha + \beta i$ and $\lambda_2 = \alpha - \beta i$, $\beta \neq 0$, and corresponding eigenvectors $\mathbf{v}_1 = \mathbf{a} + \mathbf{b}i$ and $\mathbf{v}_2 = \mathbf{a} - \mathbf{b}i$, two linearly independent solutions of $\mathbf{X}' = \mathbf{AX}$ are $\mathbf{x}_1(t) = e^{\alpha t}(\mathbf{a} \cos \beta t - \mathbf{b} \sin \beta t)$ and $\mathbf{x}_2(t) = e^{\alpha t}(\mathbf{a} \sin \beta t + \mathbf{b} \cos \beta t)$.

Notice that in the case of complex conjugate eigenvalues, we are able to obtain two linearly independent solutions from knowing one of the eigenvalues and an eigenvector that corresponds to it.

If $\mathbf{A}$ is a 2×2 matrix with complex conjugate eigenvalues $\lambda_1 = \alpha + \beta i$ and $\lambda_2 = \alpha - \beta i$ and corresponding eigenvectors $\mathbf{v}_1 = \mathbf{a} + \mathbf{b}i$ and $\mathbf{v}_2 = \mathbf{a} - \mathbf{b}i$, a general solution of $\mathbf{X}' = \mathbf{AX}$ is

$$\mathbf{X}(t) = c_1\mathbf{x}_1(t) + c_2\mathbf{x}_2(t) = c_1e^{\alpha t}(\mathbf{a} \cos \beta t - \mathbf{b} \sin \beta t) + c_2e^{\alpha t}(\mathbf{a} \sin \beta t + \mathbf{b} \cos \beta t).$$

Example 2

Find a general solution to $\mathbf{X}' = \begin{pmatrix} 3 & -2 \\ 4 & -1 \end{pmatrix}\mathbf{X}$.

Solution The eigenvalues of $\mathbf{A} = \begin{pmatrix} 3 & -2 \\ 4 & -1 \end{pmatrix}$ are $\lambda_1 = 1 + 2i$ and $\lambda_2 = 1 - 2i$. An eigenvector $\mathbf{v}_1 = \begin{pmatrix} x_1 \\ y_1 \end{pmatrix}$ that corresponds to $\lambda_1 = 1 + 2i$ satisfies $\begin{pmatrix} 2 - 2i & -2 \\ 4 & -2 - 2i \end{pmatrix}\begin{pmatrix} x_1 \\ y_1 \end{pmatrix} = \begin{pmatrix} 0 \\ 0 \end{pmatrix}$. The first equation, $(2 - 2i)x_1 - 2y_1 = 0$, is equivalent to $\dfrac{2 - 2i}{4}$ times the second equation, $4x_1 + (-2 - 2i)y_1 = 0$. We can write the system as $\begin{pmatrix} 1 - i & -1 \\ 0 & 0 \end{pmatrix}\begin{pmatrix} x_1 \\ y_1 \end{pmatrix} = \begin{pmatrix} 0 \\ 0 \end{pmatrix}$, which indicates that $y_1 = (1 - i)x_1$. Choosing $x_1 = 1$, we obtain $\mathbf{v}_1 = \begin{pmatrix} x_1 \\ y_1 \end{pmatrix} = \begin{pmatrix} 1 \\ 1 - i \end{pmatrix} = \begin{pmatrix} 1 \\ 1 \end{pmatrix} + i\begin{pmatrix} 0 \\ -1 \end{pmatrix}$. Therefore, in the notation used in Theorem 8.7,

$$\mathbf{a} = \begin{pmatrix} 1 \\ 1 \end{pmatrix} \quad \text{and} \quad \mathbf{b} = \begin{pmatrix} 0 \\ -1 \end{pmatrix}.$$

With $\alpha = 1$ and $\beta = 2$ from the eigenvalues, a general solution to the system is

$$\mathbf{X}(t) = c_1e^t\left[\begin{pmatrix} 1 \\ 1 \end{pmatrix} \cos 2t - \begin{pmatrix} 0 \\ -1 \end{pmatrix} \sin 2t\right] + c_2e^t\left[\begin{pmatrix} 1 \\ 1 \end{pmatrix} \sin 2t + \begin{pmatrix} 0 \\ -1 \end{pmatrix} \cos 2t\right]$$

$$= \begin{pmatrix} c_1e^t \cos 2t + c_2e^t \sin 2t \\ c_1e^t(\cos 2t + \sin 2t) + c_2e^t(\sin 2t - \cos 2t) \end{pmatrix}.$$

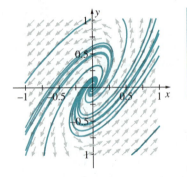

Figure 8.10 shows the graph of several solutions along with the direction field for the equation. Notice the spiraling motion of the vectors in the direction field. This is due to the product of exponential and trigonometric functions in the solution; the exponential functions cause $x(t)$ and $y(t)$ to increase while the trigonometric functions lead to rotation about the origin.

Sketch the graphs of $x(t)$, $y(t)$, and $\begin{cases} x(t) \\ y(t) \end{cases}$ in Example 2 if $x(0) = 0$ and $y(0) = 1$. Calculate $\lim_{t \to \infty} x(t)$, $\lim_{t \to \infty} y(t)$, $\lim_{t \to -\infty} x(t)$, and $\lim_{t \to -\infty} y(t)$.

Figure 8.10 We see that all (nontrivial) solutions spiral away from the origin.

Initial-value problems can be solved through the use of eigenvalues and eigenvectors as well.

Example 3

Solve $\mathbf{X}' = \begin{pmatrix} 5 & 5 & 2 \\ -6 & -6 & -5 \\ 6 & 6 & 5 \end{pmatrix} \mathbf{X}$ subject to $\mathbf{X}(0) = \begin{pmatrix} 0 \\ 0 \\ 2 \end{pmatrix}$.

Solution

The eigenvalues of $\mathbf{A} = \begin{pmatrix} 5 & 5 & 2 \\ -6 & -6 & -5 \\ 6 & 6 & 5 \end{pmatrix}$ satisfy

$$\begin{vmatrix} 5 - \lambda & 5 & 2 \\ -6 & -6 - \lambda & -5 \\ 6 & 6 & 5 - \lambda \end{vmatrix}$$

$$= (5 - \lambda) \begin{vmatrix} -6 - \lambda & -5 \\ 6 & 5 - \lambda \end{vmatrix} - 5 \begin{vmatrix} -6 & -5 \\ 6 & 5 - \lambda \end{vmatrix} + 2 \begin{vmatrix} -6 & -6 - \lambda \\ 6 & 6 \end{vmatrix}$$

$$= (5 - \lambda)(\lambda^2 + \lambda) - 5(6\lambda) + 2(6\lambda) = -\lambda(\lambda^2 - 4\lambda + 13) = 0.$$

Hence, $\lambda_1 = 0$ and $\lambda = \dfrac{4 \pm \sqrt{16 - 52}}{2} = 2 \pm 3i$, so $\lambda_2 = 2 + 3i$ and $\lambda_3 =$

$2 - 3i$. An eigenvector $\mathbf{v}_1 = \begin{pmatrix} x_1 \\ y_1 \\ z_1 \end{pmatrix}$ corresponding to $\lambda_1 = 0$ satisfies the system

$\begin{pmatrix} 5 & 5 & 2 \\ -6 & -6 & -5 \\ 6 & 6 & 5 \end{pmatrix} \begin{pmatrix} x_1 \\ y_1 \\ z_1 \end{pmatrix} = \begin{pmatrix} 0 \\ 0 \\ 0 \end{pmatrix}$ which, after row operations, is equivalent to

$\begin{pmatrix} 1 & 1 & 0 \\ 0 & 0 & 1 \\ 0 & 0 & 0 \end{pmatrix} \begin{pmatrix} x_1 \\ y_1 \\ z_1 \end{pmatrix} = \begin{pmatrix} 0 \\ 0 \\ 0 \end{pmatrix}$. Thus, $x_1 + y_1 = 0$ and $z_1 = 0$. If we choose

$y_1 = 1$, $\mathbf{v}_1 = \begin{pmatrix} -y_1 \\ y_1 \\ 0 \end{pmatrix} = \begin{pmatrix} -1 \\ 1 \\ 0 \end{pmatrix}$. One solution of the system of differential equations is $X_1 = v_1 e^{\lambda_1 t} = \begin{pmatrix} -1 \\ 1 \\ 0 \end{pmatrix} e^{(0)t} = \begin{pmatrix} -1 \\ 1 \\ 0 \end{pmatrix}$. We find two solutions that

correspond to the complex conjugate pair of eigenvalues by finding an eigenvector

$v_2 = \begin{pmatrix} x_2 \\ y_2 \\ z_2 \end{pmatrix}$ corresponding to $\lambda_2 = 2 + 3i$. This vector satisfies the system

$$\begin{pmatrix} 3-3i & 5 & 2 \\ -6 & -8-3i & -5 \\ 6 & 6 & 3-3i \end{pmatrix} \begin{pmatrix} x_2 \\ y_2 \\ z_2 \end{pmatrix} = \begin{pmatrix} 0 \\ 0 \\ 0 \end{pmatrix}, \quad \text{which can be reduced to}$$

$\begin{pmatrix} 1 & 0 & -\frac{1}{2}(1+i) \\ 0 & 1 & 1 \\ 0 & 0 & 0 \end{pmatrix} \begin{pmatrix} x_2 \\ y_2 \\ z_2 \end{pmatrix} = \begin{pmatrix} 0 \\ 0 \\ 0 \end{pmatrix}$. Therefore, the components of v_2 must

satisfy $x_2 = \dfrac{1}{2}(1+i)z_2$ and $y_2 = -z_2$. If we let $z_2 = 2$, then $\begin{pmatrix} x_2 \\ y_2 \\ z_2 \end{pmatrix} = \begin{pmatrix} 1+i \\ -2 \\ 2 \end{pmatrix} =$

$\begin{pmatrix} 1 \\ -2 \\ 2 \end{pmatrix} + \begin{pmatrix} 1 \\ 0 \\ 0 \end{pmatrix} i = a + bi$. Thus, two linearly independent solutions that

correspond to the eigenvalues $\lambda = 2 \pm 3i$ are

$$X_2 = e^{2t} \left[\begin{pmatrix} 1 \\ -2 \\ 2 \end{pmatrix} \cos 3t - \begin{pmatrix} 1 \\ 0 \\ 0 \end{pmatrix} \sin 3t \right] = \begin{pmatrix} e^{2t}(\cos 3t - \sin 3t) \\ -2e^{2t} \cos 3t \\ 2e^{2t} \cos 3t \end{pmatrix}$$

and

$$X_3 = e^{2t} \left[\begin{pmatrix} 1 \\ -2 \\ 2 \end{pmatrix} \sin 3t + \begin{pmatrix} 1 \\ 0 \\ 0 \end{pmatrix} \cos 3t \right] = \begin{pmatrix} e^{2t}(\sin 3t + \cos 3t) \\ -2e^{2t} \sin 3t \\ 2e^{2t} \sin 3t \end{pmatrix},$$

so a general solution is

$$X(t) = c_1 X_1 + c_2 X_2 + c_3 X_3$$

$$= c_1 \begin{pmatrix} -1 \\ 1 \\ 0 \end{pmatrix} + c_2 \begin{pmatrix} e^{2t}(\cos 3t - \sin 3t) \\ -2e^{2t} \cos 3t \\ 2e^{2t} \cos 3t \end{pmatrix} + c_3 \begin{pmatrix} e^{2t}(\sin 3t + \cos 3t) \\ -2e^{2t} \sin 3t \\ 2e^{2t} \sin 3t \end{pmatrix}$$

$$= \begin{pmatrix} -c_1 + c_2 e^{2t}(\cos 3t - \sin 3t) + c_3 e^{2t}(\sin 3t + \cos 3t) \\ c_1 - 2c_2 e^{2t} \cos 3t - 2c_3 e^{2t} \sin 3t \\ 3c_2 e^{2t} \cos 3t + 2c_3 e^{2t} \sin 3t \end{pmatrix}.$$

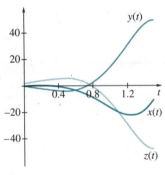

a.

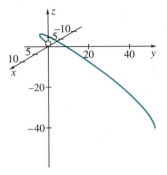

b.

Figure 8.11 (a) Graph of $x(t)$, $y(t)$, and $z(t)$. Identify each graph. (b) Graph of $\begin{cases} x(t) \\ y(t), \\ z(t) \end{cases}$

$0 \le t \le \frac{\pi}{2}$

Application of the initial condition $\mathbf{X}(0) = \begin{pmatrix} 0 \\ 0 \\ 2 \end{pmatrix}$ yields

$$\mathbf{X}(0) = \begin{pmatrix} -c_1 + c_2 + c_3 \\ c_1 - 2c_2 \\ 2c_2 \end{pmatrix} = \begin{pmatrix} 0 \\ 0 \\ 2 \end{pmatrix},$$ which gives us the system of equations

$$\begin{cases} -c_1 + c_2 + c_3 = 0 \\ c_1 - 2c_2 = 0 \\ 2c_2 = 2 \end{cases}$$ with solution $c_2 = 1$, $c_1 = 2$, and $c_3 = 1$. Therefore, the solution of the initial-value problem is

$$\mathbf{X}(t) = \begin{pmatrix} -2 + e^{2t}(\cos 3t - \sin 3t) + e^{2t}(\sin 3t + \cos 3t) \\ 2 - 2e^{2t} \cos 3t - 2e^{2t} \sin 3t \\ 2e^{2t} \cos 3t + 2e^{2t} \sin 3t \end{pmatrix}$$

$$= \begin{pmatrix} -2 + 2e^{2t} \cos 3t \\ 2 - 2e^{2t} \cos 3t - 2e^{2t} \sin 3t \\ 2e^{2t} \cos 3t + 2e^{2t} \sin 3t \end{pmatrix}$$

In Figure 8.11, we graph $x(t)$, $y(t)$, and $z(t)$ and we show the parametric plot of $\{x(t), y(t), z(t)\}$ in three dimensions. Notice that $\lim_{t\to\infty} x(t) = \infty$, $\lim_{t\to\infty} y(t) = \infty$, and $\lim_{t\to\infty} z(t) = \infty$, so the solution is directed away from the initial point $(0, 0, 2)$.

 Verify that $\mathbf{X}(t) = \begin{pmatrix} -2 + 2e^{2t} \cos 3t \\ 2 - 2e^{2t} \cos 3t - 2e^{2t} \sin 3t \\ 2e^{2t} \cos 3t + 2e^{2t} \sin 3t \end{pmatrix}$ *satisfies the initial conditions in Example 3.*

Alternate Method for Solving Initial-Value Problems

We can also use a fundamental matrix to help us solve homogeneous initial-value problems. If $\mathbf{\Phi}(t)$ is a fundamental matrix for the homogeneous system $\mathbf{X}' = \mathbf{A}\mathbf{X}$, a general solution is $\mathbf{X}(t) = \mathbf{\Phi}(t)\mathbf{C}$, where $\mathbf{C}$ is a constant vector. Given the initial condition $\mathbf{X}(0) = \mathbf{X}_0$, then through substitution into $\mathbf{X}(t) = \mathbf{\Phi}(t)\mathbf{C}$,

$$\mathbf{X}(0) = \mathbf{\Phi}(0)\mathbf{C}$$
$$\mathbf{X}_0 = \mathbf{\Phi}(0)\mathbf{C}$$
$$\mathbf{C} = \mathbf{\Phi}^{-1}(0)\mathbf{X}_0.$$

Therefore, the solution to the initial-value problem $\mathbf{X}' = \mathbf{A}\mathbf{X}$, $\mathbf{X}(0) = \mathbf{X}_0$ is $\mathbf{X}(t) = \mathbf{\Phi}(t)\mathbf{\Phi}^{-1}(0)\mathbf{X}_0$.

Example 4

Use a fundamental matrix to solve the initial-value problem $\mathbf{X}' = \begin{pmatrix} 1 & 1 \\ 4 & -2 \end{pmatrix} \mathbf{X}$ subject to $\mathbf{X}(0) = \begin{pmatrix} 1 \\ -2 \end{pmatrix}$.

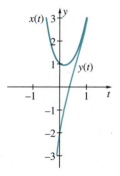

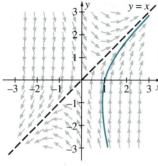

a.

b.

Figure 8.12 (a) Graph of $x(t)$ and $y(t)$. Identify each graph.

(b) Graph of $\begin{cases} x(t) \\ y(t) \end{cases}$ along with the direction field associated with the system of equations.

> **Solution** The eigenvalues of $\mathbf{A} = \begin{pmatrix} 1 & 1 \\ 4 & -2 \end{pmatrix}$ are $\lambda_1 = 2$ and $\lambda_2 = -3$ with corresponding eigenvectors $\mathbf{v}_1 = \begin{pmatrix} 1 \\ 1 \end{pmatrix}$ and $\mathbf{v}_2 = \begin{pmatrix} 1 \\ -4 \end{pmatrix}$, respectively. A fundamental matrix is then given by $\Phi(t) = \begin{pmatrix} e^{2t} & e^{-3t} \\ e^{2t} & -4e^{-3t} \end{pmatrix}$. We calculate $\Phi^{-1}(0)$ by observing that $\Phi(0) = \begin{pmatrix} 1 & 1 \\ 1 & -4 \end{pmatrix}$, so $\Phi^{-1}(0) = \frac{1}{-4-1} \begin{pmatrix} -4 & -1 \\ -1 & 1 \end{pmatrix} = \left(-\frac{1}{5}\right) \begin{pmatrix} -4 & -1 \\ -1 & 1 \end{pmatrix}$. Hence,
>
> $$\mathbf{X}(t) = \Phi(t)\Phi^{-1}(0)\mathbf{X}_0 = \begin{pmatrix} e^{2t} & e^{-3t} \\ e^{2t} & -4e^{-3t} \end{pmatrix} \left(-\frac{1}{5}\right) \begin{pmatrix} -4 & -1 \\ -1 & 1 \end{pmatrix} \begin{pmatrix} 1 \\ -2 \end{pmatrix}$$
>
> $$= \left(-\frac{1}{5}\right) \begin{pmatrix} e^{2t} & e^{-3t} \\ e^{2t} & -4e^{-3t} \end{pmatrix} \begin{pmatrix} -2 \\ -3 \end{pmatrix} = \begin{pmatrix} \dfrac{2}{5}e^{2t} + \dfrac{3}{5}e^{-3t} \\ \dfrac{2}{5}e^{2t} - \dfrac{12}{5}e^{-3t} \end{pmatrix}.$$
>
> Figure 8.12 shows a graph of the solution. Notice that as $t \to \infty$, the values of $x(t)$ and $y(t)$, where $\mathbf{X}(t) = \begin{pmatrix} x(t) \\ y(t) \end{pmatrix}$, are both close to $\frac{2}{5}e^{2t}$ because $\lim_{t\to\infty} e^{-3t} = 0$. This means that the solution approaches the line $y = x$ because for large values of t, $x(t)$ and $y(t)$ are approximately the same.

 Find the solution that satisfies the initial condition $\mathbf{X}(0) = \begin{pmatrix} 0 \\ 0 \end{pmatrix}$.

Repeated Eigenvalues

We now consider the case of repeated eigenvalues. This is more complicated than the other cases because two situations can arise. As we discovered in Section 8.2, an eigenvalue of multiplicity m can either have m corresponding linearly independent eigenvectors or it can have fewer than m corresponding linearly independent eigenvectors. In the case of m linearly independent eigenvectors, a general solution is found in the same manner as the case of m distinct eigenvalues.

Example 5

Solve $\mathbf{X}' = \begin{pmatrix} 1 & -3 & 3 \\ 3 & -5 & 3 \\ 6 & -6 & 4 \end{pmatrix} \mathbf{X}$.

> **Solution** We found the eigenvalues $\lambda_1 = \lambda_2 = -2$ and $\lambda_3 = 4$ of $\mathbf{A} = \begin{pmatrix} 1 & -3 & 3 \\ 3 & -5 & 3 \\ 6 & -6 & 4 \end{pmatrix}$ in Example 11 in Section 8.2. We also found that the

eigenvalue $\lambda_1 = \lambda_2 = -2$ of multiplicity 2 has two corresponding linearly indepen-

dent eigenvectors, $\mathbf{v}_1 = \begin{pmatrix} 1 \\ 1 \\ 0 \end{pmatrix}$ and $\mathbf{v}_2 = \begin{pmatrix} -1 \\ 0 \\ 1 \end{pmatrix}$. An eigenvector $\mathbf{v}_3 = \begin{pmatrix} x_3 \\ y_3 \\ z_3 \end{pmatrix}$ that

corresponds to $\lambda_3 = 4$ satisfies the system $\begin{pmatrix} -3 & -3 & 3 \\ 3 & -9 & 3 \\ 6 & -6 & 0 \end{pmatrix} \begin{pmatrix} x_3 \\ y_3 \\ z_3 \end{pmatrix} = \begin{pmatrix} 0 \\ 0 \\ 0 \end{pmatrix}$, which can

be reduced with row operations to the system $\begin{pmatrix} 1 & 0 & -\frac{1}{2} \\ 0 & 1 & -\frac{1}{2} \\ 0 & 0 & 0 \end{pmatrix} \begin{pmatrix} x_3 \\ y_3 \\ z_3 \end{pmatrix} = \begin{pmatrix} 0 \\ 0 \\ 0 \end{pmatrix}$. Hence,

$x_3 = \dfrac{z_3}{2}$ and $y_3 = \dfrac{z_3}{2}$. Choosing $z_3 = 2$, we obtain $\mathbf{v}_3 = \begin{pmatrix} 1 \\ 1 \\ 2 \end{pmatrix}$. A general solution

is

$$\mathbf{X}(t) = c_1 \mathbf{v}_1 e^{\lambda_1 t} + c_2 \mathbf{v}_2 e^{\lambda_2 t} + c_3 \mathbf{v}_3 e^{\lambda_3 t}$$

$$= c_1 \begin{pmatrix} 1 \\ 1 \\ 0 \end{pmatrix} e^{-2t} + c_2 \begin{pmatrix} -1 \\ 0 \\ 1 \end{pmatrix} e^{-2t} + c_3 \begin{pmatrix} 1 \\ 1 \\ 2 \end{pmatrix} e^{4t} = \begin{pmatrix} (c_1 - c_2)e^{-2t} + c_3 e^{4t} \\ c_1 e^{-2t} + c_3 e^{4t} \\ c_2 e^{-2t} + 2c_3 e^{4t} \end{pmatrix}.$$

Figure 8.13 shows graphs of several solutions to the system along with the direc-
tion field. Notice that if $c_3 \neq 0$, then $\lim_{t \to \infty} x(t) = \infty$, $\lim_{t \to \infty} y(t) = \infty$, and
$\lim_{t \to \infty} z(t) = \infty$. Therefore, many of the vectors in the direction field are directed
away from the origin.

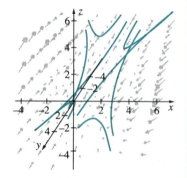

Figure 8.13

*Find conditions on x_0, y_0, and z_0, if possible, so that the limit as $t \to \infty$ of
the solution $\mathbf{X}(t)$ that satisfies the initial condition $\mathbf{X}(0) = \begin{pmatrix} x_0 \\ y_0 \\ z_0 \end{pmatrix}$ is $\mathbf{0}$.*

If an eigenvalue of multiplicity m has fewer than m linearly independent eigenvectors, we proceed in a manner that is similar to the situation that arose in Chapter 4 when
we encountered repeated roots of characteristic equations. Consider a system with the
repeated eigenvalue $\lambda_1 = \lambda_2$ and corresponding eigenvector $\mathbf{v}_1$. (Assume that there is
not a second linearly independent eigenvector corresponding to $\lambda_1 = \lambda_2$.) With the
eigenvalue λ_1 and corresponding eigenvector $\mathbf{v}_1$, we obtain the solution to the system
$\mathbf{X}_1 = \mathbf{v}_1 e^{\lambda_1 t}$. To find a second linearly independent solution corresponding to λ_1, instead of multiplying $\mathbf{X}_1$ by t as we did in Chapter 4, we suppose that a second linearly
independent solution corresponding to λ_1 is of the form

$$\mathbf{X}_2 = (\mathbf{v}_2 t + \mathbf{w}_2)e^{\lambda_1 t}.$$

In order to find the vectors $\mathbf{v}_2$ and $\mathbf{w}_2$, we substitute $\mathbf{X}_2$ into $\mathbf{X}' = \mathbf{AX}$. Because $\mathbf{X}_2' = \lambda_1(\mathbf{v}_2 t + \mathbf{w}_2)e^{\lambda_1 t} + \mathbf{v}_2 e^{\lambda_1 t}$, we have

$$\mathbf{X}_2' = \mathbf{AX}_2$$
$$\lambda_1(\mathbf{v}_2 t + \mathbf{w}_2)e^{\lambda_1 t} + \mathbf{v}_2 e^{\lambda_1 t} = \mathbf{A}(\mathbf{v}_2 t + \mathbf{w}_2)e^{\lambda_1 t}$$
$$\lambda_1 \mathbf{v}_2 t + (\lambda_1 \mathbf{w}_2 + \mathbf{v}_2) = \mathbf{A}\mathbf{v}_2 t + \mathbf{A}\mathbf{w}_2.$$

Equating coefficients yields $\lambda_1 \mathbf{v}_2 = \mathbf{A}\mathbf{v}_2$ and $\lambda_1 \mathbf{w}_2 + \mathbf{v}_2 = \mathbf{A}\mathbf{w}_2$. The equation $\lambda_1 \mathbf{v}_2 = \mathbf{A}\mathbf{v}_2$ indicates that $\mathbf{v}_2$ is an eigenvector that corresponds to λ_1, so $\mathbf{v}_2 = \mathbf{v}_1$. Simplifying $\lambda_1 \mathbf{w}_2 + \mathbf{v}_2 = \mathbf{A}\mathbf{w}_2$, we find that

$$\lambda_1 \mathbf{w}_2 + \mathbf{v}_2 = \mathbf{A}\mathbf{w}_2$$
$$\mathbf{v}_2 = \mathbf{A}\mathbf{w}_2 - \lambda_1 \mathbf{w}_2$$
$$\mathbf{v}_2 = (\mathbf{A} - \lambda_1 \mathbf{I})\mathbf{w}_2.$$

Hence, $\mathbf{w}_2$ satisfies the equation

$$(\mathbf{A} - \lambda_1 \mathbf{I})\mathbf{w}_2 = \mathbf{v}_1.$$

Therefore, a *second linearly independent solution* corresponding to the eigenvalue λ_1 has the form

$$\mathbf{X}_2 = (\mathbf{v}_1 t + \mathbf{w}_2)e^{\lambda_1 t}$$

where $\mathbf{w}_2$ satisfies $(\mathbf{A} - \lambda_1 \mathbf{I})\mathbf{w}_2 = \mathbf{v}_1$.

Theorem 8.8 Repeated Eigenvalues with One Eigenvector

> Let $\mathbf{A}$ be a square matrix. If $\mathbf{A}$ has a repeated eigenvalue $\lambda_1 = \lambda_2$ with only one corresponding (linearly independent) eigenvector $\mathbf{v}_1$, two linearly independent solutions of $\mathbf{X}' = \mathbf{AX}$ are $\mathbf{X}_1 = \mathbf{v}_1 e^{\lambda_1 t}$ and $\mathbf{X}_2 = (\mathbf{v}_1 t + \mathbf{w}_2)e^{\lambda_1 t}$, where $\mathbf{w}_2$ satisfies $(\mathbf{A} - \lambda_1 \mathbf{I})\mathbf{w}_2 = \mathbf{v}_1$.

If $\mathbf{A}$ is a 2×2 matrix with the repeated eigenvalue $\lambda_1 = \lambda_2$ with only one corresponding (linearly independent) eigenvector $\mathbf{v}_1$, a general solution to $\mathbf{X}' = \mathbf{AX}$ is

$$\mathbf{X}(t) = c_1 \mathbf{v}_1 e^{\lambda_1 t} + c_2(\mathbf{v}_1 t + \mathbf{w}_2)e^{\lambda_1 t}$$

where $\mathbf{w}_2$ is found by solving $(\mathbf{A} - \lambda_1 \mathbf{I})\mathbf{w}_2 = \mathbf{v}_1$.

Suppose that $\mathbf{A}$ has a repeated eigenvalue $\lambda_1 = \lambda_2$ and we can find only one corresponding (linearly independent) eigenvector $\mathbf{v}_1$. What happens if you try to find a second linearly independent solution of the form $\mathbf{v}_2 t e^{\lambda_1 t}$ as we did in solving higher order equations with repeated roots of the characteristic equation in Section 4.3?

Example 6

Find a general solution to $\mathbf{X}'(t) = \begin{pmatrix} -8 & -1 \\ 16 & 0 \end{pmatrix} \mathbf{X}(t)$.

Solution   The eigenvalues of $\mathbf{A} = \begin{pmatrix} -8 & -1 \\ 16 & 0 \end{pmatrix}$ are $\lambda_1 = \lambda_2 = -4$. An eigenvector $\mathbf{v}_1 = \begin{pmatrix} x_1 \\ y_1 \end{pmatrix}$ that corresponds to $\lambda_1 = -4$ satisfies the system $\begin{pmatrix} -4 & -1 \\ 16 & 4 \end{pmatrix}\begin{pmatrix} x_1 \\ y_1 \end{pmatrix} = \begin{pmatrix} 0 \\ 0 \end{pmatrix}$, which is equivalent to $\begin{pmatrix} 4 & 1 \\ 0 & 0 \end{pmatrix}\begin{pmatrix} x_1 \\ y_1 \end{pmatrix} = \begin{pmatrix} 0 \\ 0 \end{pmatrix}$. Hence $y_1 = -4x_1$, so if we choose $x_1 = 1$, $\mathbf{v}_1 = \begin{pmatrix} 1 \\ -4 \end{pmatrix}$ and one solution to the system is $\mathbf{X}_1 = \begin{pmatrix} 1 \\ -4 \end{pmatrix} e^{-4t}$. To find $\mathbf{w}_2 = \begin{pmatrix} x_2 \\ y_2 \end{pmatrix}$ in a second linearly independent solution $\mathbf{X}_2 = (\mathbf{v}_2 t + \mathbf{w}_2)e^{\lambda_1 t}$, we solve $(\mathbf{A} - \lambda_1 \mathbf{I})\mathbf{w}_2 = \mathbf{v}_1$ which in this case is

$$\begin{pmatrix} -4 & -1 \\ 16 & 4 \end{pmatrix}\begin{pmatrix} x_2 \\ y_2 \end{pmatrix} = \begin{pmatrix} 1 \\ -4 \end{pmatrix}.$$

This system is equivalent to

$$\begin{pmatrix} 4 & 1 \\ 0 & 0 \end{pmatrix}\begin{pmatrix} x_2 \\ y_2 \end{pmatrix} = \begin{pmatrix} -1 \\ 0 \end{pmatrix},$$

which indicates that $4x_2 + y_2 = -1$. If we let $x_2 = 0$, then $y_2 = -1$, $\mathbf{w}_2 = \begin{pmatrix} 0 \\ -1 \end{pmatrix}$, and a second linearly independent solution is

$$\mathbf{X}_2 = \left(\begin{pmatrix} 1 \\ -4 \end{pmatrix} t + \begin{pmatrix} 0 \\ -1 \end{pmatrix} \right) e^{-4t}.$$

Hence, a general solution is

$$\mathbf{X}(t) = c_1 \begin{pmatrix} 1 \\ -4 \end{pmatrix} e^{-4t} + c_2 \left(\begin{pmatrix} 1 \\ -4 \end{pmatrix} t + \begin{pmatrix} 0 \\ -1 \end{pmatrix} \right) e^{-4t} = \begin{pmatrix} c_1 + c_2 t \\ (-4c_1 - c_2) - 4tc_2 \end{pmatrix} e^{-4t}.$$

Figure 8.14 shows the graph of several solutions to the system along with the direction field. Notice that the behavior of these solutions differs from those of the systems solved earlier in the section due to the repeated eigenvalues. We see from the formula for $\mathbf{X}(t)$ that $\lim_{t \to \infty} x(t) = 0$ and $\lim_{t \to \infty} y(t) = 0$. (Why?) In addition, these solutions approach $(0, 0)$ tangent to $y = -4x$, the line through the origin that is parallel to the vector $\mathbf{v}_1 = \begin{pmatrix} 1 \\ -4 \end{pmatrix}$. (We will discuss why this occurs in Section 8.7.)

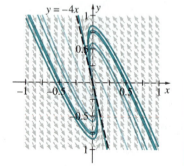

Figure 8.14

Example 7

Solve $\mathbf{X}' = \begin{pmatrix} 5 & 3 & -3 \\ 2 & 4 & -5 \\ -4 & 2 & -3 \end{pmatrix} \mathbf{X}$.

Solution

The eigenvalues of $\mathbf{A} = \begin{pmatrix} 5 & 3 & -3 \\ 2 & 4 & -5 \\ -4 & 2 & -3 \end{pmatrix}$ are determined with

$$\begin{vmatrix} 5-\lambda & 3 & -3 \\ 2 & 4-\lambda & -5 \\ -4 & 2 & -3-\lambda \end{vmatrix} = -\lambda^3 + 6\lambda^2 + 15\lambda + 8 = -(\lambda + 1)(\lambda + 1)(\lambda - 8) = 0$$

to be $\lambda_1 = \lambda_2 = -1$ and $\lambda_3 = 8$. An eigenvector $\mathbf{v}_1 = \begin{pmatrix} x_1 \\ y_1 \\ z_1 \end{pmatrix}$ that corresponds to

$\lambda_1 = -1$ satisfies $\begin{pmatrix} 6 & 3 & -3 \\ 2 & 5 & -5 \\ -4 & 2 & -2 \end{pmatrix}\begin{pmatrix} x_1 \\ y_1 \\ z_1 \end{pmatrix} = \begin{pmatrix} 0 \\ 0 \\ 0 \end{pmatrix}$. This system is equivalent to

$\begin{pmatrix} 1 & 0 & 0 \\ 0 & 1 & -1 \\ 0 & 0 & 0 \end{pmatrix}\begin{pmatrix} x_1 \\ y_1 \\ z_1 \end{pmatrix} = \begin{pmatrix} 0 \\ 0 \\ 0 \end{pmatrix}$. Hence, $x_1 = 0$ and $y_1 - z_1 = 0$. If we let $y_1 = 1$,

then $\mathbf{v}_1 = \begin{pmatrix} 0 \\ 1 \\ 1 \end{pmatrix}$. Therefore, one solution to the system is $\mathbf{X}_1 = \begin{pmatrix} 0 \\ 1 \\ 1 \end{pmatrix}e^{-t}$. A

second linearly independent solution $\mathbf{X}_2 = (\mathbf{v}_1 t + \mathbf{w}_2)e^{\lambda_1 t}$ is found by solving

$(\mathbf{A} - \lambda_1\mathbf{I})\mathbf{w}_2 = \mathbf{v}_1$ given by $\begin{pmatrix} 6 & 3 & -3 \\ 2 & 5 & -5 \\ -4 & 2 & -2 \end{pmatrix}\begin{pmatrix} x_2 \\ y_2 \\ z_2 \end{pmatrix} = \begin{pmatrix} 0 \\ 1 \\ 1 \end{pmatrix}$ for the vector $\mathbf{w}_2 = \begin{pmatrix} x_2 \\ y_2 \\ z_2 \end{pmatrix}$.

This system is equivalent to $\begin{pmatrix} 1 & 0 & 0 \\ 0 & 1 & -1 \\ 0 & 0 & 0 \end{pmatrix}\begin{pmatrix} x_2 \\ y_2 \\ z_2 \end{pmatrix} = \begin{pmatrix} -\frac{1}{8} \\ \frac{1}{4} \\ 0 \end{pmatrix}$, which indicates

that $x_2 = -\frac{1}{8}$ and $y_2 - z_2 = \frac{1}{4}$. Hence, if $z_2 = 0$, $y_2 = \frac{1}{4}$, so $\mathbf{w}_2 = \begin{pmatrix} -\frac{1}{8} \\ \frac{1}{4} \\ 0 \end{pmatrix}$.

A second linearly independent solution is then $\mathbf{X}_2 = \left(\begin{pmatrix} 0 \\ 1 \\ 1 \end{pmatrix}t + \begin{pmatrix} -\frac{1}{8} \\ \frac{1}{4} \\ 0 \end{pmatrix}\right)e^{-t}$. A

third linearly independent solution is found using an eigenvector $\mathbf{v}_3 = \begin{pmatrix} x_3 \\ y_3 \\ z_3 \end{pmatrix}$ corre-

sponding to $\lambda_3 = 8$, which satisfies $\begin{pmatrix} -3 & 3 & -3 \\ 2 & -4 & -5 \\ -4 & 2 & -11 \end{pmatrix}\begin{pmatrix} x_3 \\ y_3 \\ z_3 \end{pmatrix} = \begin{pmatrix} 0 \\ 0 \\ 0 \end{pmatrix}$. Because

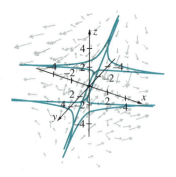

Figure 8.15

this system is equivalent to $\begin{pmatrix} 1 & 0 & \frac{9}{2} \\ 0 & 1 & \frac{7}{2} \\ 0 & 0 & 0 \end{pmatrix} \begin{pmatrix} x_3 \\ y_3 \\ z_3 \end{pmatrix} = \begin{pmatrix} 0 \\ 0 \\ 0 \end{pmatrix}$, the components of $\mathbf{v}_3$ must satisfy $x_3 + \frac{9}{2}z_3 = 0$ and $y_3 + \frac{7}{2}z_3 = 0$. If we let $z_3 = -2$, then $x_3 = 9$ and $y_3 = 7$. Hence, $\mathbf{v}_3 = \begin{pmatrix} 9 \\ 7 \\ -2 \end{pmatrix}$ and $\mathbf{X}_3 = \begin{pmatrix} 9 \\ 7 \\ -2 \end{pmatrix} e^{8t}$. A general solution is then given by

$$\mathbf{X}(t) = c_1\mathbf{X}_1 + c_2\mathbf{X}_2 + c_3\mathbf{X}_3 = c_1 \begin{pmatrix} 0 \\ 1 \\ 1 \end{pmatrix} e^{-t} + c_2 \left(\begin{pmatrix} 0 \\ 1 \\ 1 \end{pmatrix} t + \begin{pmatrix} -\frac{1}{8} \\ \frac{1}{4} \\ 0 \end{pmatrix} \right) e^{-t} + c_3 \begin{pmatrix} 9 \\ 7 \\ -2 \end{pmatrix} e^{8t}$$

$$= \begin{pmatrix} -\frac{1}{8}c_2 e^{-t} + 9c_3 e^{8t} \\ c_1 e^{-t} + c_2(t + \frac{1}{4})e^{-t} + 7c_3 e^{8t} \\ c_1 e^{-t} + c_2 t e^{-t} - 2c_3 e^{8t} \end{pmatrix}.$$

Figure 8.15 shows the graph of several solutions to the system along with the direction field.

Using the formula for $\mathbf{X}$ in Example 7 and the direction field in Figure 8.15, determine if the solutions approach the origin at t increases if $c_3 \neq 0$. What happens to the solutions if $c_3 = 0$ and either $c_1 \neq 0$ or $c_2 \neq 0$?

A similar method is carried out in the case of three equal eigenvalues $\lambda_1 = \lambda_2 = \lambda_3$ where we can find only one (linearly independent) eigenvector $\mathbf{v}_1$. When we encounter this situation, we assume that

$$\mathbf{X}_1 = \mathbf{v}_1 e^{\lambda_1 t}, \quad \mathbf{X}_2 = (\mathbf{v}_2 t + \mathbf{w}_2)e^{\lambda_1 t}, \quad \text{and} \quad \mathbf{X}_3 = \left(\mathbf{v}_3 \frac{t^2}{2} + \mathbf{w}_3 t + \mathbf{u}_3 \right) e^{\lambda_1 t}.$$

Substitution of these solutions into the system of differential equations yields the following system of equations, which is solved for the unknown vectors $\mathbf{v}_2, \mathbf{w}_2, \mathbf{v}_3, \mathbf{w}_3$, and $\mathbf{u}_3$:

$$\lambda_1\mathbf{v}_2 = \mathbf{A}\mathbf{v}_2, \quad (\mathbf{A} - \lambda_1\mathbf{I})\mathbf{w}_2 = \mathbf{v}_2, \quad \lambda_1\mathbf{v}_3 = \mathbf{A}\mathbf{v}_3, \quad (\mathbf{A} - \lambda_1\mathbf{I})\mathbf{w}_3 = \mathbf{v}_3,$$
$$\text{and} \quad (\mathbf{A} - \lambda_1\mathbf{I})\mathbf{u}_3 = \mathbf{w}_3.$$

Similar to the previous case, $\mathbf{v}_3 = \mathbf{v}_2 = \mathbf{v}_1$, $\mathbf{w}_2 = \mathbf{w}_3$, and the vector $\mathbf{u}_3$ is found by solving the system

$$(\mathbf{A} - \lambda_1 I)\mathbf{u}_3 = \mathbf{w}_2.$$

The three linearly independent solutions have the form

$$\mathbf{X}_1 = \mathbf{v}_1 e^{\lambda_1 t}, \quad \mathbf{X}_2 = (\mathbf{v}_1 t + \mathbf{w}_2)e^{\lambda_1 t}, \quad \text{and} \quad \mathbf{X}_3 = \left(\mathbf{v}_1 \frac{t^2}{2} + \mathbf{w}_2 t + \mathbf{u}_3 \right) e^{\lambda_1 t}.$$

Notice that this method is generalized for instances when the multiplicity of the repeated eigenvalue is greater than three.

Example 8

Solve $\mathbf{X}' = \begin{pmatrix} 1 & 1 & 1 \\ 2 & 1 & -1 \\ -3 & 2 & 4 \end{pmatrix} \mathbf{X}$.

Solution The eigenvalues are found by solving

$$\begin{vmatrix} 1 - \lambda & 1 & 1 \\ 2 & 1 - \lambda & -1 \\ -3 & 2 & 4 - \lambda \end{vmatrix}$$

$$= (1 - \lambda)\begin{vmatrix} 1 - \lambda & -1 \\ 2 & 4 - \lambda \end{vmatrix} - (1)\begin{vmatrix} 2 & -1 \\ -3 & 4 - \lambda \end{vmatrix} + (1)\begin{vmatrix} 2 & 1 - \lambda \\ -3 & 2 \end{vmatrix}$$

$$= -\lambda^3 + 6\lambda^2 - 12\lambda + 8 = -(\lambda - 2)(\lambda^2 - 4\lambda + 4) = -(\lambda - 2)^3 = 0.$$

Hence $\lambda_1 = \lambda_2 = \lambda_3 = 2$. We find eigenvectors $\mathbf{v}_1 = \begin{pmatrix} x_1 \\ y_1 \\ z_1 \end{pmatrix}$ corresponding to

$\lambda = 2$ with the system $\begin{pmatrix} -1 & 1 & 1 \\ 2 & -1 & -1 \\ -3 & 2 & 2 \end{pmatrix}\begin{pmatrix} x_1 \\ y_1 \\ z_1 \end{pmatrix} = \begin{pmatrix} 0 \\ 0 \\ 0 \end{pmatrix}$. With elementary row

operations, we reduce this system to $\begin{pmatrix} 1 & 0 & 0 \\ 0 & 1 & 1 \\ 0 & 0 & 0 \end{pmatrix}\begin{pmatrix} x_1 \\ y_1 \\ z_1 \end{pmatrix} = \begin{pmatrix} 0 \\ 0 \\ 0 \end{pmatrix}$. Thus,

$x_1 = 0$ and $y_1 + z_1 = 0$. If we select $z_1 = 1$, then $y_1 = -1$ and $\mathbf{v}_1 = \begin{pmatrix} 0 \\ -1 \\ 1 \end{pmatrix}$.

Therefore we can find only one (linearly independent) eigenvector corresponding to $\lambda = 2$. Using this eigenvalue and eigenvector, we find that one solution to the

system is $\mathbf{X}_1 = \mathbf{v}_1 e^{2t} = \begin{pmatrix} 0 \\ -1 \\ 1 \end{pmatrix} e^{2t}$. The vector $\mathbf{w}_2 = \begin{pmatrix} x_2 \\ y_2 \\ z_2 \end{pmatrix}$ in a second

linearly independent solution of the form $\mathbf{X}_2 = (\mathbf{v}_1 t + \mathbf{w}_2)e^{2t}$ is found by solving

the system $(\mathbf{A} - \lambda \mathbf{I})\mathbf{w}_2 = \mathbf{v}_1$. This system is $\begin{pmatrix} -1 & 1 & 1 \\ 2 & -1 & -1 \\ -3 & 2 & 2 \end{pmatrix}\begin{pmatrix} x_2 \\ y_2 \\ z_2 \end{pmatrix} = \begin{pmatrix} 0 \\ -1 \\ 1 \end{pmatrix}$

which is equivalent to $\begin{pmatrix} 1 & 0 & 0 \\ 0 & 1 & 1 \\ 0 & 0 & 0 \end{pmatrix}\begin{pmatrix} x_2 \\ y_2 \\ z_2 \end{pmatrix} = \begin{pmatrix} -1 \\ -1 \\ 0 \end{pmatrix}$. Hence $x_2 = -1$ and $y_2 +$

$z_2 = -1$, so if we choose $z_2 = 0$, then $y_2 = -1$. Therefore $\mathbf{w}_2 = \begin{pmatrix} -1 \\ -1 \\ 0 \end{pmatrix}$, so $\mathbf{X}_2 =$

$\left(\begin{pmatrix} 0 \\ -1 \\ 1 \end{pmatrix} t + \begin{pmatrix} -1 \\ -1 \\ 0 \end{pmatrix} \right) e^{2t}$. Finally, we must determine the vector $\mathbf{u}_3 = \begin{pmatrix} x_3 \\ y_3 \\ z_3 \end{pmatrix}$

in the third linearly independent solution $\mathbf{X}_3 = \left(\mathbf{v}_1 \dfrac{t^2}{2} + \mathbf{w}_2 t + \mathbf{u}_3 \right) e^{\lambda_1 t}$ by solving the system $(\mathbf{A} - \lambda\mathbf{I})\mathbf{u}_3 = \mathbf{w}_2$. This yields $\begin{pmatrix} -1 & 1 & 1 \\ 2 & -1 & -1 \\ -3 & 2 & 2 \end{pmatrix}\begin{pmatrix} x_3 \\ y_3 \\ z_3 \end{pmatrix} = \begin{pmatrix} -1 \\ -1 \\ 0 \end{pmatrix}$

which is equivalent to $\begin{pmatrix} 1 & 0 & 0 \\ 0 & 1 & 1 \\ 0 & 0 & 0 \end{pmatrix}\begin{pmatrix} x_3 \\ y_3 \\ z_3 \end{pmatrix} = \begin{pmatrix} -2 \\ -3 \\ 0 \end{pmatrix}$. Therefore, $x_3 = -2$ and

$y_3 + z_3 = -3$. If we select $z_3 = 0$, then $y_3 = -3$. Hence $\mathbf{u}_3 = \begin{pmatrix} -2 \\ -3 \\ 0 \end{pmatrix}$, so a third

linearly independent solution is $\mathbf{X}_3 = \left(\begin{pmatrix} 0 \\ -1 \\ 1 \end{pmatrix}\dfrac{t^2}{2} + \begin{pmatrix} -1 \\ -1 \\ 0 \end{pmatrix}t + \begin{pmatrix} -2 \\ -3 \\ 0 \end{pmatrix}\right)e^{2t}$. A

general solution is then given by

$\mathbf{X}(t)$

$= c_1\mathbf{X}_1 + c_2\mathbf{X}_2 + c_3\mathbf{X}_3$

$= c_1\begin{pmatrix} 0 \\ -1 \\ 1 \end{pmatrix}e^{2t} + c_2\left(\begin{pmatrix} 0 \\ -1 \\ 1 \end{pmatrix}t + \begin{pmatrix} -1 \\ -1 \\ 0 \end{pmatrix}\right)e^{2t} + c_3\left(\begin{pmatrix} 0 \\ -1 \\ 1 \end{pmatrix}\dfrac{t^2}{2} + \begin{pmatrix} -1 \\ -1 \\ 0 \end{pmatrix}t + \begin{pmatrix} -2 \\ -3 \\ 0 \end{pmatrix}\right)e^{2t}$

$= \begin{pmatrix} -c_2 e^{2t} + c_3(-t-2)e^{2t} \\ -c_1 e^{2t} + c_2(-t-1)e^{2t} + c_3\left(-\dfrac{t^2}{2} - t - 3\right)e^{2t} \\ c_1 e^{2t} + c_2 t e^{2t} + c_3\left(\dfrac{t^2}{2}\right)e^{2t} \end{pmatrix}.$

IN TOUCH WITH TECHNOLOGY

1. Solve the systems **(a)** $\begin{cases} x' = 2x - y \\ y' = -x + 3y \end{cases}$;

(b) $\begin{cases} x' = 2x \\ y' = 3x + 2y \end{cases}$; **(c)** $\begin{cases} x' = x + 4y \\ y' = -2x - y \end{cases}$ subject

to $x(0) = 1$ and $y(0) = 1$. In each case, graph the solution parametrically and individually.

2. Find a general solution of the system $\mathbf{X}' = \begin{pmatrix} 0 & 0 & 3 \\ 1 & -4 & 2 \\ 0 & -4 & 1 \end{pmatrix}\mathbf{X}$. Graph the solution for various values of the constants.

3. How do the general solutions and direction field of the system $\mathbf{X}' = \begin{pmatrix} 2 & \lambda \\ 1 & 0 \end{pmatrix}\mathbf{X}$ change as λ goes from -2 to 0? Solve the system for values of λ between -2 and 0 and note how the solution changes.

4. Consider the initial-value problem

$$\begin{cases} \mathbf{X}' = \begin{pmatrix} 1 & -\frac{7}{3} \\ -2 & -\frac{8}{3} \end{pmatrix}\mathbf{X} \\ x(0) = x_0, y(0) = y_0 \end{cases}.$$

(a) Graph the direction field associated with the system

$\mathbf{X}' = \begin{pmatrix} 1 & -\frac{7}{3} \\ -2 & -\frac{8}{3} \end{pmatrix}\mathbf{X}$. **(b)** Find conditions on x_0 and y_0 so that at least one of $x(t)$ or $y(t)$ approaches zero as t approaches infinity. **(c)** Is it possible to choose x_0 and y_0 so that both $x(t)$ and $y(t)$ approach zero as t approaches infinity?

5. (a) Find a general solution of $\mathbf{X}' = \mathbf{AX}$ if

$$\mathbf{X}(t) = \begin{pmatrix} x_1(t) \\ x_2(t) \\ x_3(t) \\ x_4(t) \end{pmatrix} \quad \text{and} \quad \text{(i)} \quad \mathbf{A} = \begin{pmatrix} \lambda & 0 & 0 & 0 \\ 0 & \lambda & 0 & 0 \\ 0 & 0 & \lambda & 0 \\ 0 & 0 & 0 & \lambda \end{pmatrix},$$

$$\text{(ii)} \ \mathbf{A} = \begin{pmatrix} \lambda & 1 & 0 & 0 \\ 0 & \lambda & 0 & 0 \\ 0 & 0 & \lambda & 0 \\ 0 & 0 & 0 & \lambda \end{pmatrix}, \ \text{(iii)} \ \mathbf{A} = \begin{pmatrix} \lambda & 1 & 0 & 0 \\ 0 & \lambda & 1 & 0 \\ 0 & 0 & \lambda & 0 \\ 0 & 0 & 0 & \lambda \end{pmatrix},$$

$$\text{and (iv)} \ \mathbf{A} = \begin{pmatrix} \lambda & 1 & 0 & 0 \\ 0 & \lambda & 1 & 0 \\ 0 & 0 & \lambda & 1 \\ 0 & 0 & 0 & \lambda \end{pmatrix}. \ \textbf{(b)} \ \text{For each sys-}$$

tem in (a), find the solution that satisfies the initial

condition $\mathbf{X}(0) = \begin{pmatrix} -1 \\ 0 \\ 1 \\ 2 \end{pmatrix}$ if $\lambda = -\frac{1}{2}$ and then

graph $x_1(t)$, $x_2(t)$, $x_3(t)$, and $x_4(t)$ for $0 \le t \le 10$. How are the solutions similar? How are they different? **(c)** Indicate how to generalize the results obtained in (a). Explain how you would find a general solution of

$$\mathbf{X}' = \mathbf{AX} \text{ if } \mathbf{A} = \begin{pmatrix} \lambda & 1 & 0 & 0 & 0 \\ 0 & \lambda & 1 & 0 & 0 \\ 0 & 0 & \lambda & 1 & 0 \\ 0 & 0 & 0 & \lambda & 1 \\ 0 & 0 & 0 & 0 & \lambda \end{pmatrix}. \text{ How would}$$

you find a general solution of $\mathbf{X}' = \mathbf{AX}$ for the 5×5

$$\text{matrix } \mathbf{A} = \begin{pmatrix} \lambda & 1 & 0 & 0 & 0 \\ 0 & \lambda & 1 & 0 & 0 \\ 0 & 0 & \lambda & 1 & 0 \\ 0 & 0 & 0 & \lambda & 1 \\ 0 & 0 & 0 & 0 & \lambda \end{pmatrix}? \text{ How would you find}$$

a general solution of $\mathbf{X}' = \mathbf{AX}$ for the $n \times n$ matrix

$$\mathbf{A} = \begin{pmatrix} \lambda & 1 & 0 & \cdots & 0 & 0 \\ 0 & \lambda & 1 & \cdots & 0 & 0 \\ \vdots & \vdots & & \ddots & \vdots & \vdots \\ 0 & 0 & 0 & \cdots & \lambda & 1 \\ 0 & 0 & 0 & \cdots & 0 & \lambda \end{pmatrix}?$$

EXERCISES 8.4

In Exercises 1–26, find a general solution of the system.

1. $\mathbf{X}' = \begin{pmatrix} 1 & -10 \\ -7 & 10 \end{pmatrix}\mathbf{X}$

2. $\begin{cases} x' = x - 2y \\ y' = 2x + 6y \end{cases}$

*3. $\begin{cases} \dfrac{dx}{dt} = 6x - y \\ \dfrac{dy}{dt} = 5x \end{cases}$

4. $\mathbf{X}' = \begin{pmatrix} 4 & 3 \\ -5 & -4 \end{pmatrix}\mathbf{X}$

5. $\begin{cases} x' = 7x \\ y' = 5x - 8y \end{cases}$

6. $\begin{cases} \dfrac{dx}{dt} = 8x + 9y \\ \dfrac{dy}{dt} = -2x - 3y \end{cases}$

*7. $\mathbf{X}' = \begin{pmatrix} -5 & 3 \\ 2 & -10 \end{pmatrix}\mathbf{X}$

8. $\begin{cases} x' = 8x + 5y \\ y' = -10x - 6y \end{cases}$

9. $\begin{cases} \dfrac{dx}{dt} = -6x - 4y \\ \dfrac{dy}{dt} = -3x - 10y \end{cases}$

10. $X' = \begin{pmatrix} 1 & 8 \\ -2 & -7 \end{pmatrix} X$

*11. $\begin{cases} x' = -6x + 2y \\ y' = -2x - 10y \end{cases}$

12. $X' = \begin{pmatrix} 0 & 8 \\ 2 & 0 \end{pmatrix} X$

13. $X' = \begin{pmatrix} 0 & 8 \\ -2 & 0 \end{pmatrix} X$

14. $\begin{cases} \dfrac{dx}{dt} = -3y \\ \dfrac{dy}{dt} = 3x \end{cases}$

*15. $\begin{cases} x' = y \\ y' = -13x - 4y \end{cases}$

16. $X' = \begin{pmatrix} -1 & 2 \\ -2 & -1 \end{pmatrix} X$

17. $X' = \begin{pmatrix} 4 & 0 & 1 \\ 0 & -2 & 0 \\ 0 & 0 & -1 \end{pmatrix} X$

18. $X' = \begin{pmatrix} 4 & -4 & 2 \\ 5 & -4 & 4 \\ 2 & -1 & 1 \end{pmatrix} X$

*19. $X' = \begin{pmatrix} -3 & 3 & -4 \\ 0 & -3 & 0 \\ 0 & -5 & -4 \end{pmatrix} X$

20. $X' = \begin{pmatrix} -3 & -2 & -3 \\ 2 & 1 & 3 \\ 5 & 3 & -3 \end{pmatrix} X$

21. $X' = \begin{pmatrix} 5 & -1 & 3 \\ -4 & -1 & -2 \\ -4 & 2 & -3 \end{pmatrix} X$

22. $X' = \begin{pmatrix} -3 & 4 & 5 \\ 4 & 3 & 5 \\ -4 & 2 & -3 \end{pmatrix} X$

*23. $X' = \begin{pmatrix} -5 & 4 & -5 \\ 0 & -1 & 0 \\ 5 & 1 & 1 \end{pmatrix} X$

24. $X' = \begin{pmatrix} -2 & 1 & -4 \\ 0 & -3 & 3 \\ 1 & 1 & 2 \end{pmatrix} X$

25. $\begin{cases} x' = x + 2y + 3z \\ y' = y + 2z \\ z' = -2y + z \end{cases}$

26. $\begin{cases} x' = y \\ y' = z \\ z' = x - y + z \end{cases}$

In Exercises 27–34, solve the initial-value problem.

27. $X' = \begin{pmatrix} 1 & 1 \\ 0 & 2 \end{pmatrix} X$, $X(0) = \begin{pmatrix} 0 \\ 4 \end{pmatrix}$

28. $X' = \begin{pmatrix} -4 & 0 \\ 2 & 4 \end{pmatrix} X$, $X(0) = \begin{pmatrix} 8 \\ 0 \end{pmatrix}$

*29. $X' = \begin{pmatrix} 4 & 0 \\ 2 & 4 \end{pmatrix} X$, $X(0) = \begin{pmatrix} 8 \\ 0 \end{pmatrix}$

30. $X' = \begin{pmatrix} 4 & 8 \\ 0 & 4 \end{pmatrix} X$, $X(0) = \begin{pmatrix} 8 \\ 8 \end{pmatrix}$

31. $X' = \begin{pmatrix} 0 & -4 \\ 4 & 0 \end{pmatrix} X$, $X(0) = \begin{pmatrix} 0 \\ 8 \end{pmatrix}$

32. $X' = \begin{pmatrix} 0 & 13 \\ -1 & -4 \end{pmatrix} X$, $X(0) = \begin{pmatrix} 0 \\ 4 \end{pmatrix}$

*33. $X' = \begin{pmatrix} 4 & 0 & 1 \\ -2 & 1 & 0 \\ -2 & 0 & 1 \end{pmatrix} X$, $X(0) = \begin{pmatrix} -1 \\ 2 \\ 0 \end{pmatrix}$

34. $X' = \begin{pmatrix} 1 & 0 & 2 \\ -1 & -1 & -1 \\ -1 & 0 & -2 \end{pmatrix} X$, $X(0) = \begin{pmatrix} 4 \\ 0 \\ 8 \end{pmatrix}$

35. Show that $x_1(t) = e^{\alpha t}(\mathbf{a} \cos \beta t - \mathbf{b} \sin \beta t)$ and $x_2(t) = e^{\alpha t}(\mathbf{a} \sin \beta t + \mathbf{b} \cos \beta t)$ are linearly independent functions.

36. Show that in the 3×3 system $X' = AX$ with the eigenvalue λ of multiplicity 3 with one corresponding eigenvector $\mathbf{v}_1$, the three linearly independent solutions are $X_1 = \mathbf{v}_1 e^{\lambda_1 t}$, $X_2 = (\mathbf{v}_1 t + \mathbf{w}_2) e^{\lambda_1 t}$, and $X_3 = \left(\mathbf{v}_1 \dfrac{t^2}{2} + \mathbf{w}_2 t + \mathbf{u}_3 \right) e^{\lambda_1 t}$ where $\mathbf{u}_3$ satisifies $(A - \lambda I)\mathbf{u}_3 = \mathbf{w}_2$ and $\mathbf{w}_2$ satisfies $(A - \lambda I)\mathbf{w}_2 = \mathbf{v}_1$.

In Exercises 37–42, without solving each system, match each system in Group A with the graph of its direction field in Group B.

Group A

Group B

37. $\begin{cases} x' = x \\ y' = 2y \end{cases}$ **(a)**

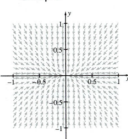

41. $\begin{cases} x' = -x - y \\ y' = 2x \end{cases}$ **(e)**

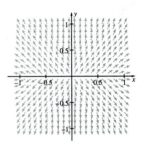

38. $\begin{cases} x' = -x \\ y' = 2y \end{cases}$ **(b)**

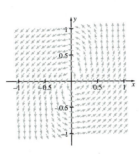

42. $\begin{cases} x' = x - y \\ y' = 2x \end{cases}$ **(f)**

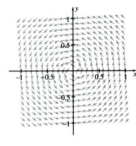

39. $\begin{cases} x' = -x \\ y' = -2y \end{cases}$ **(c)**

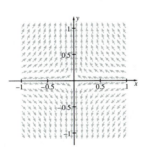

43. Show that both $\begin{cases} x(t) = e^{2t}(c_1 \cos t - c_2 \sin t) \\ y(t) = e^{2t}(c_1 \sin t + c_2 \cos t) \end{cases}$ and

$$\begin{cases} x(t) = \dfrac{1}{2}e^{(2-i)t}[c_1(1 + e^{2it}) + c_2 i(e^{2it} - 1)] \\ y(t) = \dfrac{1}{2}e^{(2-i)t}[c_1 i(1 - e^{2it}) + c_2(1 + e^{2it})] \end{cases}$$

are general solutions of the system
$$\begin{cases} x'(t) = 2x(t) - y(t) \\ y'(t) = x(t) + 2y(t) \end{cases}.$$

40. $\begin{cases} x' = -y \\ y' = 2x \end{cases}$ **(d)**

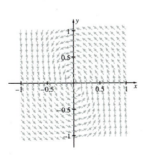

8.5 Nonhomogeneous First-Order Systems: Undetermined Coefficients and Variation of Parameters

Undetermined Coefficients • Variation of Parameters

In Chapter 4, we learned how to solve nonhomogeneous differential equations through the use of undetermined coefficients and variation of parameters. Here we approach the solution of systems of nonhomogeneous equations using these methods.

$$\text{Let}\quad \mathbf{X} = \mathbf{X}(t) = \begin{pmatrix} x_1(t) \\ x_2(t) \\ \vdots \\ x_n(t) \end{pmatrix},\quad \mathbf{A} = \begin{pmatrix} a_{11} & a_{12} & \cdots & a_{1n} \\ a_{21} & a_{22} & \cdots & a_{2n} \\ \vdots & \vdots & \ddots & \vdots \\ a_{n1} & a_{n2} & \cdots & a_{nn} \end{pmatrix},\quad \mathbf{F}(t) = \begin{pmatrix} f_1(t) \\ f_2(t) \\ \vdots \\ f_n(t) \end{pmatrix},\quad \text{and}$$

$\mathbf{\Phi}(t)$ be a fundamental matrix of the system $\mathbf{X}' = \mathbf{AX}$. Then a general solution to the homogeneous system $\mathbf{X}' = \mathbf{AX}$ is $\mathbf{X} = \mathbf{\Phi}(t)\mathbf{C}$ where $\mathbf{C} = \begin{pmatrix} c_1 \\ c_2 \\ \vdots \\ c_n \end{pmatrix}$ is an $n \times 1$

constant matrix. To find a general solution to the linear nonhomogeneous system $\mathbf{X}' = \mathbf{AX} + \mathbf{F}(t)$, we proceed in the same way we did with linear nonhomogeneous equations in Chapter 4: if $\mathbf{X}_P$ is a particular solution of the nonhomogeneous system, then all other solutions $\mathbf{X}$ of the system can be written in the form $\mathbf{X} = \mathbf{\Phi}(t)\mathbf{C} + \mathbf{X}_P$ (see Exercise 36).

Undetermined Coefficients

We use the method of undetermined coefficients to find a particular solution $\mathbf{X}_P$ to a nonhomogeneous system in much the same way as we approached nonhomogeneous higher order equations with constant coefficients in Chapter 4. The main difference is that the coefficients are *constant vectors* when we work with systems. For example, if we consider $\mathbf{X}' = \mathbf{AX} + \mathbf{F}(t)$ where $\mathbf{F}(t) = \begin{pmatrix} e^{-2t} \\ 4 \end{pmatrix} = \begin{pmatrix} 1 \\ 0 \end{pmatrix}e^{-2t} + \begin{pmatrix} 0 \\ 4 \end{pmatrix}$ and none of the terms in $\mathbf{F}(t)$ satisfy the corresponding homogeneous system $\mathbf{X}' = \mathbf{AX}$, we assume that a particular solution has the form $\mathbf{X}_P(t) = \mathbf{a}e^{-2t} + \mathbf{b}$ where $\mathbf{a}$ and $\mathbf{b}$ are constant vectors. On the other hand if $\lambda = -2$ is an eigenvalue of $\mathbf{A}$, we assume that $\mathbf{X}_P(t) = \mathbf{a}te^{-2t} + \mathbf{b}e^{-2t} + \mathbf{c}.$

Example 1

Solve $\mathbf{X}' = \begin{pmatrix} 0 & 8 \\ 2 & 0 \end{pmatrix}\mathbf{X} + \begin{pmatrix} e^{3t} \\ t \end{pmatrix}.$

Solution   In this case, $\mathbf{F}(t) = \begin{pmatrix} e^{3t} \\ t \end{pmatrix} = \begin{pmatrix} 1 \\ 0 \end{pmatrix}e^{3t} + \begin{pmatrix} 0 \\ 1 \end{pmatrix}t$ and a general

solution to the corresponding homogeneous system $\mathbf{X}' = \begin{pmatrix} 0 & 8 \\ 2 & 0 \end{pmatrix}\mathbf{X}$ is $\mathbf{X}_h(t) =$

$c_1 \begin{pmatrix} 2 \\ 1 \end{pmatrix} e^{4t} + c_2 \begin{pmatrix} -2 \\ 1 \end{pmatrix} e^{-4t}$. Notice that none of the components of $\mathbf{F}(t)$ are in $\mathbf{X}_h(t)$, so we assume that there is a particular solution of the form $\mathbf{X}_P(t) = \mathbf{a}e^{3t} + \mathbf{b}t + \mathbf{c}$. Then, $\mathbf{X}_P'(t) = 3\mathbf{a}e^{3t} + \mathbf{b}$ and substitution into the nonhomogeneous system

$$\mathbf{X}' = \mathbf{AX} + \begin{pmatrix} 1 \\ 0 \end{pmatrix} e^{3t} + \begin{pmatrix} 0 \\ 1 \end{pmatrix} t, \text{ where } \mathbf{A} = \begin{pmatrix} 0 & 8 \\ 2 & 0 \end{pmatrix}, \text{ yields}$$

$$3\mathbf{a}e^{3t} + \mathbf{b} = \mathbf{Aa}e^{3t} + \mathbf{Ab}t + \mathbf{Ac} + \begin{pmatrix} 1 \\ 0 \end{pmatrix} e^{3t} + \begin{pmatrix} 0 \\ 1 \end{pmatrix} t.$$

Collecting like terms, we obtain the system of equations

$$\begin{cases} 3\mathbf{a} = \mathbf{Aa} + \begin{pmatrix} 1 \\ 0 \end{pmatrix} & \text{(Coefficients of } e^{3t}) \\ \mathbf{b} = \mathbf{Ac} & \text{(Constant terms)} \\ \mathbf{Ab} + \begin{pmatrix} 0 \\ 1 \end{pmatrix} = \mathbf{0} & \text{(Coefficients of } t) \end{cases}$$

From the coefficients of e^{3t}, we find that $(\mathbf{A} - 3\mathbf{I})\mathbf{a} = \begin{pmatrix} -1 \\ 0 \end{pmatrix}$ or $\begin{pmatrix} -3 & 8 \\ 2 & -3 \end{pmatrix} \begin{pmatrix} a_1 \\ a_2 \end{pmatrix} = \begin{pmatrix} -1 \\ 0 \end{pmatrix}$ where $\mathbf{a} = \begin{pmatrix} a_1 \\ a_2 \end{pmatrix}$. This system has the unique solution $\mathbf{a} = \begin{pmatrix} -\frac{3}{7} \\ -\frac{2}{7} \end{pmatrix}$. Next, we solve the system $\mathbf{Ab} + \begin{pmatrix} 0 \\ 1 \end{pmatrix} = \mathbf{0}$ or $\begin{pmatrix} 0 & 8 \\ 2 & 0 \end{pmatrix} \begin{pmatrix} b_1 \\ b_2 \end{pmatrix} = \begin{pmatrix} 0 \\ -1 \end{pmatrix}$ for $\mathbf{b}$. This yields the unique solution $\mathbf{b} = \begin{pmatrix} -\frac{1}{2} \\ 0 \end{pmatrix}$. Finally, we solve $\mathbf{b} = \mathbf{Ac}$ or $\begin{pmatrix} 0 & 8 \\ 2 & 0 \end{pmatrix} \begin{pmatrix} c_1 \\ c_2 \end{pmatrix} = \begin{pmatrix} -\frac{1}{2} \\ 0 \end{pmatrix}$ for $\mathbf{c}$ which gives us $\mathbf{c} = \begin{pmatrix} 0 \\ -\frac{1}{16} \end{pmatrix}$. A particular solution to the nonhomogeneous system is then

$$\mathbf{X}_P(t) = \mathbf{a}e^{3t} + \mathbf{b}t + \mathbf{c} = \begin{pmatrix} -\frac{3}{7} \\ -\frac{2}{7} \end{pmatrix} e^{3t} + \begin{pmatrix} -\frac{1}{2} \\ 0 \end{pmatrix} t + \begin{pmatrix} 0 \\ -\frac{1}{16} \end{pmatrix} = \begin{pmatrix} -\frac{3}{7}e^{3t} - \frac{1}{2}t \\ -\frac{2}{7}e^{3t} - \frac{1}{16} \end{pmatrix},$$

so a general solution to $\mathbf{X}' = \mathbf{AX} + \begin{pmatrix} 1 \\ 0 \end{pmatrix} e^{3t} + \begin{pmatrix} 0 \\ 1 \end{pmatrix} t$ is

$$\mathbf{X}(t) = \mathbf{X}_h(t) + \mathbf{X}_P(t) = \begin{pmatrix} 2c_1e^{4t} - 2c_2e^{-4t} - \frac{3}{7}e^{3t} - \frac{1}{2}t \\ c_1e^{4t} + c_2e^{-4t} - \frac{2}{7}e^{3t} - \frac{1}{16} \end{pmatrix}.$$

To give another illustration of how the form of a particular solution is selected, suppose that $\mathbf{F}(t) = \begin{pmatrix} 4\sin 2t \\ e^{-t} \end{pmatrix} = \begin{pmatrix} 4 \\ 0 \end{pmatrix} \sin 2t + \begin{pmatrix} 0 \\ 1 \end{pmatrix} e^{-t}$ in Example 1. In this case, we assume that $\mathbf{X}_P(t) = \mathbf{a}\cos 2t + \mathbf{b}\sin 2t + \mathbf{c}e^{-t}$ and find the vectors $\mathbf{a}$, $\mathbf{b}$, and $\mathbf{c}$ through substitution into the nonhomogeneous system.

Find a particular solution to $X' = \begin{pmatrix} 0 & 8 \\ 2 & 0 \end{pmatrix} X + \begin{pmatrix} 4\sin 2t \\ e^{-t} \end{pmatrix}$.

Variation of Parameters

Variation of parameters can be used to solve linear nonhomogeneous systems as well. In much the same way that we derived the method of variation of parameters for solving higher order differential equations, we assume that a particular solution of the nonhomogeneous system can be expressed in the form

$$X_p(t) = \mathbf{\Phi}(t)V(t), \text{ where } V(t) = \begin{pmatrix} v_1(t) \\ v_2(t) \\ \vdots \\ v_n(t) \end{pmatrix}.$$

Notice that $X_p' = \mathbf{\Phi}(t)V'(t) + \mathbf{\Phi}'(t)V(t)$. Then if X_p satisfies $X' = AX + F(t)$, we have

$$\mathbf{\Phi}(t)V'(t) + \mathbf{\Phi}'(t)V(t) = A\mathbf{\Phi}(t)V(t) + F(t).$$

However, the fundamental matrix $\mathbf{\Phi}(t)$ satisfies $X' = AX$, so $\mathbf{\Phi}'(t) = A\mathbf{\Phi}(t)$. Hence,

$$\mathbf{\Phi}(t)V'(t) + A\mathbf{\Phi}(t)V(t) = A\mathbf{\Phi}(t)V(t) + F(t)$$
$$\mathbf{\Phi}(t)V'(t) = F(t).$$

Multiplying both sides of this equation by $\mathbf{\Phi}^{-1}(t)$ yields

$$\mathbf{\Phi}^{-1}(t)\mathbf{\Phi}(t)V'(t) = \mathbf{\Phi}^{-1}(t)F(t)$$
$$V'(t) = \mathbf{\Phi}^{-1}(t)F(t).$$

Therefore, $V(t) = \int \mathbf{\Phi}^{-1}(t)F(t)\,dt$, so a particular solution is

$$X_p(t) = \mathbf{\Phi}(t)\int \mathbf{\Phi}^{-1}(t)F(t)\,dt,$$

and a general solution to the system is

$$X(t) = \mathbf{\Phi}(t)C + X_p(t) = \mathbf{\Phi}(t)C + \mathbf{\Phi}(t)\int \mathbf{\Phi}^{-1}(t)F(t)\,dt.$$

Example 2

Solve $X' = \begin{pmatrix} -5 & 3 \\ 2 & -10 \end{pmatrix} X + \begin{pmatrix} e^{-2t} \\ 1 \end{pmatrix}$ using variation of parameters.

Solution   In order to apply variation of parameters, we first calculate a fundamental matrix for the corresponding homogeneous system $X' = \begin{pmatrix} -5 & 3 \\ 2 & -10 \end{pmatrix} X$. The eigenvalues of $A = \begin{pmatrix} -5 & 3 \\ 2 & -10 \end{pmatrix}$ are $\lambda_1 = -4$ and $\lambda_2 = -11$ with corresponding eigenvectors $v_1 = \begin{pmatrix} 3 \\ 1 \end{pmatrix}$ and $v_2 = \begin{pmatrix} -\frac{1}{2} \\ 1 \end{pmatrix}$, respectively. A general

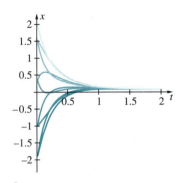

a.

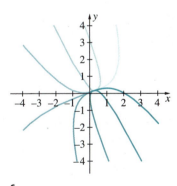

b.

c.

Figure 8.16 (a) Graph of $x(t)$ for various initial conditions (b) Graph of $y(t)$ for various initial conditions (c) Graph of $\begin{cases} x(t) \\ y(t) \end{cases}$ for various initial conditions

solution to the homogeneous system is $\mathbf{X}_h(t) = c_1 \begin{pmatrix} -\frac{1}{2} \\ 1 \end{pmatrix} e^{-11t} + c_2 \begin{pmatrix} 3 \\ 1 \end{pmatrix} e^{-4t}$,

so a fundamental matrix is given by $\mathbf{\Phi}(t) = \begin{pmatrix} -\frac{1}{2}e^{-11t} & 3e^{-4t} \\ e^{-11t} & e^{-4t} \end{pmatrix}$.

Because $\mathbf{\Phi}^{-1}(t) = \dfrac{1}{-\frac{7}{2}e^{-15t}} \begin{pmatrix} e^{-4t} & -3e^{-4t} \\ -e^{-11t} & -\frac{1}{2}e^{-11t} \end{pmatrix} = \begin{pmatrix} -\frac{2}{7}e^{11t} & \frac{6}{7}e^{11t} \\ \frac{2}{7}e^{4t} & \frac{1}{7}e^{4t} \end{pmatrix}$,

$$\mathbf{\Phi}^{-1}(t)\mathbf{F}(t) = \begin{pmatrix} \frac{-2}{7}e^{11t} & \frac{6}{7}e^{11t} \\ \frac{2}{7}e^{4t} & \frac{1}{7}e^{4t} \end{pmatrix} \begin{pmatrix} e^{-2t} \\ 1 \end{pmatrix} = \begin{pmatrix} \frac{-2}{7}e^{9t} + \frac{6}{7}e^{11t} \\ \frac{2}{7}e^{2t} + \frac{1}{7}e^{4t} \end{pmatrix}.$$

Therefore,

$$\mathbf{V}(t) = \int \mathbf{\Phi}^{-1}(t)\mathbf{F}(t)\,dt = \int \begin{pmatrix} \frac{-2}{7}e^{9t} + \frac{6}{7}e^{11t} \\ \frac{2}{7}e^{2t} + \frac{1}{7}e^{4t} \end{pmatrix} dt = \begin{pmatrix} \frac{-2}{63}e^{9t} + \frac{6}{77}e^{11t} \\ \frac{1}{7}e^{2t} + \frac{1}{28}e^{4t} \end{pmatrix}.$$

By variation of parameters, we have the particular solution

$$\mathbf{X}_p(t) = \mathbf{\Phi}(t) \int \mathbf{\Phi}^{-1}(t)\mathbf{F}(t)\,dt =$$

$$\begin{pmatrix} \frac{-1}{2}e^{-11t} & 3e^{-4t} \\ e^{-11t} & e^{-4t} \end{pmatrix} \begin{pmatrix} \frac{-2}{63}e^{9t} + \frac{6}{77}e^{11t} \\ \frac{1}{7}e^{2t} + \frac{1}{28}e^{4t} \end{pmatrix} = \begin{pmatrix} \frac{3}{44} + \frac{4}{9}e^{-2t} \\ \frac{5}{44} + \frac{1}{9}e^{-2t} \end{pmatrix}.$$

Therefore, a general solution is given by

$$\mathbf{X}(t) = \mathbf{\Phi}(t)\mathbf{C} + \mathbf{X}_p(t) = \begin{pmatrix} \frac{-1}{2}e^{-11t} & 3e^{-4t} \\ e^{-11t} & e^{-4t} \end{pmatrix} \begin{pmatrix} c_1 \\ c_2 \end{pmatrix} + \begin{pmatrix} \frac{3}{44} + \frac{4}{9}e^{-2t} \\ \frac{5}{44} + \frac{1}{9}e^{-2t} \end{pmatrix}.$$

In Figure 8.16, we graph $x(t)$, $y(t)$, and $\begin{cases} x(t) \\ y(t) \end{cases}$ for several values of c_1 and c_2.

Evaluate $\lim_{t\to\infty} x(t)$ and $\lim_{t\to\infty} y(t)$. Does the choice of c_1 or c_2 affect the limit?

Variation of parameters allows us to solve systems that we cannot solve using the method of undetermined coefficients.

Example 3

Solve the initial-value problem $\begin{cases} x' = 2x - 5y + \csc t \\ y' = x - 2y + \sec t \\ x(\pi/4) = 5, \ y(\pi/4) = 1 \end{cases}$, $0 < t < \pi/2$.

Solution   First, we write the system as $\mathbf{X}' = \begin{pmatrix} 2 & -5 \\ 1 & -2 \end{pmatrix} \mathbf{X} + \begin{pmatrix} \csc t \\ \sec t \end{pmatrix}$ and

solve the corresponding homogeneous system $\mathbf{X}' = \begin{pmatrix} 2 & -5 \\ 1 & -2 \end{pmatrix} \mathbf{X}$ to obtain

$$\mathbf{X}_h(t) = c_1 \begin{pmatrix} \cos t + 2 \sin t \\ \sin t \end{pmatrix} + c_2 \begin{pmatrix} -5 \sin t \\ \cos t - 2 \sin t \end{pmatrix}.$$ A fundamental matrix is

$$\mathbf{\Phi}(t) = \begin{pmatrix} \cos t + 2 \sin t & -5 \sin t \\ \sin t & \cos t - 2 \sin t \end{pmatrix} \text{ with inverse}$$

$$\mathbf{\Phi}^{-1}(t) = \begin{pmatrix} \cos t - 2 \sin t & 5 \sin t \\ -\sin t & \cos t + 2 \sin t \end{pmatrix}. \text{ Then,}$$

$$\mathbf{V}(t) = \int \mathbf{\Phi}^{-1}(t) \mathbf{F}(t) \, dt = \int \begin{pmatrix} \dfrac{\cos t}{\sin t} - 2 + \dfrac{5 \sin t}{\cos t} \\ \dfrac{2 \sin t}{\cos t} \end{pmatrix} dt =$$

$$\begin{pmatrix} \ln (\sin t) - 2t - 5 \ln (\cos t) \\ -2 \ln (\cos t) \end{pmatrix},$$

so a particular solution is

$$\mathbf{X}_p(t) = \mathbf{\Phi}(t) \int \mathbf{\Phi}^{-1}(t) \mathbf{F}(t) \, dt$$

$$= \begin{pmatrix} \cos t + 2 \sin t & -5 \sin t \\ \sin t & \cos t - 2 \sin t \end{pmatrix} \begin{pmatrix} \ln (\sin t) - 2t - 5 \ln (\cos t) \\ -2 \ln (\cos t) \end{pmatrix}$$

$$= \begin{pmatrix} \cos t \ln (\sin t) - 2t \cos t - 5 \cos t \ln (\cos t) + 2 \sin t \ln (\sin t) - 4t \sin t \\ \sin t \ln (\sin t) - 2t \sin t - \sin t \ln (\cos t) - 2 \cos t \ln (\cos t) \end{pmatrix}.$$

A general solution to the nonhomogeneous system is

$$\mathbf{X}(t) = \mathbf{X}_h(t) + \mathbf{X}_p(t) =$$
$$\begin{pmatrix} \cos t \, (c_1 + \ln (\sin t) - 2t - 5 \ln (\cos t)) + \sin t \, (2c_1 - 5c_2 + 2 \ln (\sin t) - 4t) \\ \cos t \, (c_2 - 2 \ln (\cos t)) + \sin t \, (c_1 - 2c_2 + \ln (\sin t) - 2t - \ln (\cos t)) \end{pmatrix}.$$

To find the solution that satisfies the initial conditions $x(\pi/4) = 5$ and $y(\pi/4) = 1$, we solve the system of equations

$$\begin{cases} x\left(\dfrac{\pi}{4}\right) = \dfrac{3}{2} c_1 \sqrt{2} - \dfrac{5}{2} c_2 \sqrt{2} + \dfrac{1}{2} \sqrt{2} \ln 2 - \dfrac{3}{4} \pi \sqrt{2} = 5 \\ y\left(\dfrac{\pi}{4}\right) = \dfrac{1}{2} c_1 \sqrt{2} - \dfrac{1}{2} c_2 \sqrt{2} + \dfrac{1}{2} \sqrt{2} \ln 2 - \dfrac{1}{4} \pi \sqrt{2} = 1 \end{cases}$$

for c_1 and c_2, which yields $c_1 = \frac{1}{2}\pi - 2\ln 2 \approx 0.184501966$ and $c_2 = -\sqrt{2} - \ln 2 \approx -2.107360743$. We use the values of c_1 and c_2 to graph $x(t)$ and $y(t)$ for $0 < t < \pi/2$ in Figure 8.17.

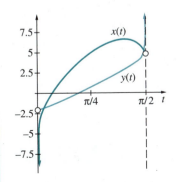

Figure 8.17

 Find $\lim_{t\to 0^+} x(t)$, $\lim_{t\to 0^+} y(t)$, $\lim_{t\to \pi/2^-} x(t)$, and $\lim_{t\to \pi/2^-} y(t)$ for the solution to the initial-value problem in Example 3.

IN TOUCH WITH TECHNOLOGY

1. Use a computer algebra system to solve each of the following initial-value problems. In each case, graph $x(t)$, $y(t)$, and the parametric equations $\begin{cases} x(t) \\ y(t) \end{cases}$ for the indicated values of t.

   (a) $\begin{cases} \begin{pmatrix} x' \\ y' \end{pmatrix} = \begin{pmatrix} -7 & -3 \\ -2 & -2 \end{pmatrix}\begin{pmatrix} x \\ y \end{pmatrix} + \begin{pmatrix} 1 \\ te^{-t} \end{pmatrix};\ 0 \le t \le 10 \\ x(0) = 0, \quad y(0) = 1 \end{cases}$

   (b) $\begin{cases} \mathbf{X}' = \begin{pmatrix} 8 & 10 \\ -7 & -9 \end{pmatrix}\mathbf{X} + \begin{pmatrix} t^2 e^{-2t} \\ -te^t \end{pmatrix};\ 0 \le t \le 3 \\ x(0) = 1, \quad y(0) = 0 \end{cases}$

   (c) $\begin{cases} \dfrac{dx}{dt} = 2x - 5y + \sin 4t \\ \dfrac{dy}{dt} = 4x - 2y - te^{-t};\ 0 \le t \le 2\pi \\ x(0) = 1,\ y(0) = 1 \end{cases}$

2. (a) Show that
   $$\begin{cases} x = -\frac{6}{13}e^{-4t} + \frac{6}{13}e^{9t} - \frac{3}{13}e^{9t}\int_0^t e^{-9s}f(s)\,ds + \\ \qquad \frac{16}{13}e^{-4t}\int_0^t e^{4s}f(s)\,ds \\ \\ y = -\frac{3}{13}e^{-4t} + \frac{16}{13}e^{9t} - \frac{8}{13}e^{9t}\int_0^t e^{-9s}f(s)\,ds + \\ \qquad \frac{8}{13}e^{-4t}\int_0^t e^{4s}f(s)\,ds \end{cases}$$

   is the solution to the initial-value problem

   $$\begin{cases} \dfrac{dx}{dt} = -7x + 6y + f(t) \\ \dfrac{dy}{dt} = -8x + 12y \\ x(0) = 0,\ y(0) = 1 \end{cases}.$$

(b) Is it possible to choose $f(t)$ so that $\lim_{t\to\infty} x(t) = 0$? So that $\lim_{t\to\infty} y(t) = 0$? So that both $\lim_{t\to\infty} x(t) = 0$ and $\lim_{t\to\infty} y(t) = 0$? Why or why not? Provide evidence of your results by graphing various solutions.

In Problems 3–5, graphically compare solutions to each non-homogeneous system and the corresponding homogeneous system.

3. $\mathbf{X}'(t) = \begin{pmatrix} -5 & 6 \\ 0 & -7 \end{pmatrix}\mathbf{X}(t) + \begin{pmatrix} 1 \\ t \end{pmatrix}$, $\mathbf{X}(0) = \begin{pmatrix} 1 \\ -1 \end{pmatrix}$

4. $\begin{cases} x'(t) = -6x(t) + t \\ y'(t) = -x(t) - 6y(t) + t^2 \\ x(0) = 0,\ y(0) = 1 \end{cases}$

5. $\mathbf{X}'(t) = \begin{pmatrix} 9 & 5 \\ -8 & -3 \end{pmatrix}\mathbf{X}(t) + \begin{pmatrix} e^{3t}\sin 2t \\ e^{3t} \end{pmatrix}$,

$\mathbf{X}(0) = \begin{pmatrix} 1 \\ 0 \end{pmatrix}$

EXERCISES 8.5

In Exercises 1–24, solve the system by undetermined coefficients or variation of parameters.

1. $\mathbf{X}' = \begin{pmatrix} -5 & 6 \\ 0 & -7 \end{pmatrix}\mathbf{X} + \begin{pmatrix} 1 \\ t \end{pmatrix}$

2. $\mathbf{X}' = \begin{pmatrix} 3 & 8 \\ 2 & 9 \end{pmatrix}\mathbf{X} + \begin{pmatrix} t^2 \\ 1 \end{pmatrix}$

*3. $\begin{cases} x' = -6x + t \\ y' = -x - 6y + t^2 \end{cases}$

4. $\mathbf{X}' = \begin{pmatrix} 2 & 2 \\ 3 & 1 \end{pmatrix}\mathbf{X} + \begin{pmatrix} e^{-4t} \\ e^t \end{pmatrix}$

5. $\mathbf{X}' = \begin{pmatrix} -3 & -4 \\ -2 & -10 \end{pmatrix}\mathbf{X} + \begin{pmatrix} e^{2t} \\ e^{11t} \end{pmatrix}$

6. $\mathbf{X}' = \begin{pmatrix} -9 & -9 \\ 4 & 3 \end{pmatrix}\mathbf{X} + \begin{pmatrix} e^{3t} \\ te^{3t} \end{pmatrix}$

*7. $\begin{cases} x' = -3x + 7y + \cos t \\ y' = -x + 5y + e^{-4t} \end{cases}$

8. $\mathbf{X}' = \begin{pmatrix} -6 & -3 \\ 4 & 1 \end{pmatrix}\mathbf{X} + \begin{pmatrix} te^{2t} \\ \sin 2t \end{pmatrix}$

9. $\mathbf{X}' = \begin{pmatrix} -6 & -4 \\ 4 & -6 \end{pmatrix}\mathbf{X} + \begin{pmatrix} e^{-6t} \\ e^{-6t} \end{pmatrix}$

10. $\begin{cases} x' = -3x + 2y + e^{-t}\sec 2t \\ y' = -10x + 5y + e^{-t}\csc 2t \end{cases}$

*11. $\mathbf{X}' = \begin{pmatrix} 9 & 5 \\ -8 & -3 \end{pmatrix}\mathbf{X} + \begin{pmatrix} e^{3t}\sin 2t \\ e^{3t} \end{pmatrix}$

12. $\begin{cases} x' = 6x - 5y + e^{5t} \\ y' = x + 4y + e^{5t}\cos 2t \end{cases}$

13. $\begin{cases} x' = -y + \sec t \\ y' = x \end{cases}$

14. $\begin{cases} x' = -y + 4\tan t \\ y' = x \end{cases}$

*15. $\begin{cases} x' = y \\ y' = -x - 2\cot t \end{cases}$

16. $\begin{cases} x' = -y \\ y' = x - 2\csc t \end{cases}$

17. $\mathbf{X}' = \begin{pmatrix} 0 & 0 & 4 \\ 2 & -2 & 5 \\ -3 & 4 & -1 \end{pmatrix}\mathbf{X} + \begin{pmatrix} t \\ 1 \\ t^2 \end{pmatrix}$

18. $\begin{cases} x' = -2x - 2y - 2z + 1 \\ y' = -2y + z + t^2 \\ z' = -2y - 5z + t \end{cases}$

*19. $\begin{cases} x' = -x + y - z + e^{-4t} \\ y' = 4x + y - 5z + te^{-2t} \\ z' = -x - 2y - 2z + t \end{cases}$

20. $\mathbf{X}' = \begin{pmatrix} 1 & 5 & -4 \\ -2 & -6 & 2 \\ 0 & 0 & 6 \end{pmatrix}\mathbf{X} + \begin{pmatrix} te^{4t} \\ 1 \\ e^{6t} \end{pmatrix}$

21. $\mathbf{X}' = \begin{pmatrix} 2 & 1 & 2 \\ 3 & 0 & 2 \\ 1 & 3 & 1 \end{pmatrix}\mathbf{X} + \begin{pmatrix} 0 \\ t \\ e^t \end{pmatrix}$

22. $\mathbf{X}' = \begin{pmatrix} 6 & 0 & -5 \\ 3 & 2 & 0 \\ 0 & 0 & 6 \end{pmatrix}\mathbf{X} + \begin{pmatrix} 1 \\ e^{-t} \\ t \end{pmatrix}$

*23. $\mathbf{X}' = \begin{pmatrix} 3 & 2 & -3 \\ 1 & 1 & 1 \\ 0 & -4 & -4 \end{pmatrix} \mathbf{X} + \begin{pmatrix} e^{4t} \\ 0 \\ t \end{pmatrix}$

24. $\mathbf{X}' = \begin{pmatrix} -4 & 5 & -1 \\ -1 & 2 & 0 \\ -3 & -2 & 0 \end{pmatrix} \mathbf{X} + \begin{pmatrix} \sin t \\ e^{-4t} \\ 0 \end{pmatrix}$

In Exercises 25–30, solve the initial-value problem.

25. $\mathbf{X}' = \begin{pmatrix} -1 & 3 \\ 0 & 2 \end{pmatrix} \mathbf{X} + \begin{pmatrix} t \\ 0 \end{pmatrix}, \ \mathbf{X}(0) = \begin{pmatrix} 0 \\ 0 \end{pmatrix}$

26. $\mathbf{X}' = \begin{pmatrix} 0 & 1 \\ -3 & -4 \end{pmatrix} \mathbf{X} + \begin{pmatrix} 0 \\ \sin t \end{pmatrix}, \ \mathbf{X}(0) = \begin{pmatrix} 0 \\ 1 \end{pmatrix}$

*27. $\mathbf{X}' = \begin{pmatrix} 3 & 4 \\ 2 & 1 \end{pmatrix} \mathbf{X} + \begin{pmatrix} 0 \\ e^t \end{pmatrix}, \ \mathbf{X}(0) = \begin{pmatrix} -1 \\ 1 \end{pmatrix}$

28. $\mathbf{X}' = \begin{pmatrix} 6 & -3 \\ 2 & 1 \end{pmatrix} \mathbf{X} + \begin{pmatrix} t \\ t^2 - 1 \end{pmatrix}, \ \mathbf{X}(0) = \begin{pmatrix} 0 \\ 1 \end{pmatrix}$

29. $\mathbf{X}' = \begin{pmatrix} 3 & 2 \\ -5 & 1 \end{pmatrix} \mathbf{X} + \begin{pmatrix} 0 \\ 10 \end{pmatrix}, \ \mathbf{X}(0) = \begin{pmatrix} 1 \\ -2 \end{pmatrix}$

30. $\mathbf{X}' = \begin{pmatrix} 3 & -1 \\ 4 & -1 \end{pmatrix} \mathbf{X} + \begin{pmatrix} \cos t \\ \sin t \end{pmatrix}, \ \mathbf{X}(0) = \begin{pmatrix} 0 \\ 0 \end{pmatrix}$

In Exercises 31–35, without solving each of the following initial-value problems, match each problem with the graph of its solution.

Initial-value problem	Graph of $x(t)$ (in dark blue) and $y(t)$ (in light blue)	Graph of $\begin{cases} x(t) \\ y(t) \end{cases}$
31. $\begin{cases} x' = 2x + y \\ y' = -8x - 2y \\ x(0) = 0, y(0) = 1 \end{cases}$	(a) 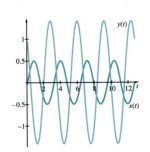	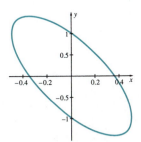
32. $\begin{cases} x' = 2x + y + 1 \\ y' = -8x - 2y + t \\ x(0) = 0, y(0) = 1 \end{cases}$	(b) 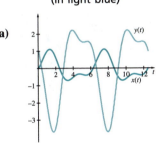	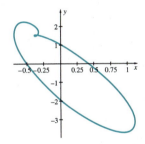
33. $\begin{cases} x' = 2x + y + \sin t \\ y' = -8x - 2y \\ x(0) = 0, y(0) = 1 \end{cases}$	(c) 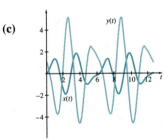	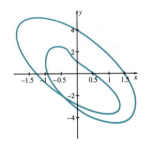

34. $\begin{cases} x' = 2x + y + \sin 2t \\ y' = -8x - 2y \\ x(0) = 0,\ y(0) = 1 \end{cases}$ **(d)**

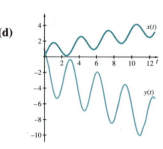

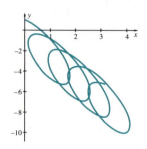

35. $\begin{cases} x' = 2x + y + \sin 3t \\ y' = -8x - 2y \\ x(0) = 0,\ y(0) = 1 \end{cases}$ **(e)**

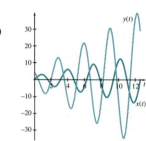

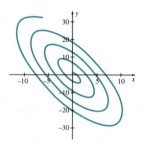

*36. Let $\mathbf{X} = \mathbf{X}(t) = \begin{pmatrix} x_1(t) \\ x_2(t) \\ \vdots \\ x_n(t) \end{pmatrix}$,

$\mathbf{A} = \begin{pmatrix} a_{11} & a_{12} & \cdots & a_{1n} \\ a_{21} & a_{22} & \cdots & a_{2n} \\ \vdots & \vdots & \ddots & \vdots \\ a_{n1} & a_{n2} & \cdots & a_{nn} \end{pmatrix}$, $\mathbf{F}(t) = \begin{pmatrix} f_1(t) \\ f_2(t) \\ \vdots \\ f_n(t) \end{pmatrix}$, and

$\mathbf{\Phi}(t)$ be a fundamental matrix of the system $\mathbf{X}' = \mathbf{AX}$.

(a) Show that if $\mathbf{X}_1$ and $\mathbf{X}_2$ are any two solutions of $\mathbf{X}' = \mathbf{AX} + \mathbf{F}(t)$, then $\mathbf{X}_1 - \mathbf{X}_2$ is a solution to $\mathbf{X}' = \mathbf{AX}$. **(b)** Show that if $\mathbf{X}_p$ is a particular solution to $\mathbf{X}' = \mathbf{AX} + \mathbf{F}(t)$ and $\mathbf{X}_{\text{any}}$ is any solution to $\mathbf{X}' = \mathbf{AX} + \mathbf{F}(t)$, then there is an $n \times 1$ constant vector

$\mathbf{C} = \begin{pmatrix} c_1 \\ c_2 \\ \vdots \\ c_n \end{pmatrix}$ so that $\mathbf{X}_{\text{any}} = \mathbf{\Phi}(t)\mathbf{C} + \mathbf{X}_P$.

8.6 Laplace Transforms

In many cases, Laplace transforms can be used to solve initial-value problems that involve a system of linear differential equations. This method is applied in much the same way that it was in solving initial-value problems involving higher order differential equations except that a system of algebraic equations is obtained after taking the Laplace transform of each equation. After solving for the Laplace transform of each of the unknown functions, the inverse Laplace transform is used to find each unknown function in the solution of the system.

Example 1

Solve $\mathbf{X}' = \begin{pmatrix} 0 & 1 \\ 1 & 0 \end{pmatrix}\mathbf{X} + \begin{pmatrix} \sin t \\ 2 \cos t \end{pmatrix}$ subject to $\mathbf{X}(0) = \begin{pmatrix} 2 \\ 0 \end{pmatrix}$.

Solution

Let $\mathbf{X}(t) = \begin{pmatrix} x(t) \\ y(t) \end{pmatrix}$. Then we can rewrite this problem as

$$\begin{cases} x' = y + \sin t \\ y' = x + 2 \cos t \end{cases}, \quad x(0) = 2,\ y(0) = 0.$$

Taking the Laplace transform of both sides of each equation yields the system

$$\begin{cases} sX(s) - x(0) = Y(s) + \dfrac{1}{s^2 + 1} \\ sY(s) - y(0) = X(s) + \dfrac{2s}{s^2 + 1} \end{cases}$$

which is equivalent to

$$\begin{cases} sX(s) - Y(s) = \dfrac{1}{s^2 + 1} + 2 \\ -X(s) + sY(s) = \dfrac{2s}{s^2 + 1} \end{cases}$$

Multiplying the second equation by s and adding the two equations, we have

$$(s^2 - 1)Y(s) = \frac{2s^2}{s^2 + 1} + \frac{2s^2 + 1}{s^2 + 1}$$

$$Y(s) = \frac{2s^2 + 1}{(s^2 + 1)(s^2 - 1)} + \frac{2}{s^2 - 1} = \frac{4s^2 + 3}{(s^2 + 1)(s^2 - 1)}$$

$$= \frac{7}{4(s - 1)} - \frac{7}{4(s + 1)} + \frac{1}{2(s^2 + 1)}.$$

Taking the inverse Laplace transform then yields

$$y(t) = \frac{7}{4}e^t - \frac{7}{4}e^{-t} + \frac{1}{2}\sin t.$$

At this point, we can solve the system for $X(s)$ and compute $x(t)$ with the inverse Laplace transform. However, we can find $x(t)$ more easily by substituting $y(t)$ into the second differential equation $y' = x + 2\cos t$. Because $y'(t) = \frac{7}{4}e^t + \frac{7}{4}e^{-t} + \frac{1}{2}\cos t$ and $x = y' - 2\cos t$, we have

$$x(t) = y'(t) - 2\cos t = \frac{7}{4}e^t + \frac{7}{4}e^{-t} + \frac{1}{2}\cos t - 2\cos t = \frac{7}{4}e^t + \frac{7}{4}e^{-t} - \frac{3}{2}\cos t.$$

In Figure 8.18, we graph $x(t)$, $y(t)$, and $\begin{cases} x(t) \\ y(t) \end{cases}$. What is the orientation of $\begin{cases} x(t) \\ y(t) \end{cases}$?

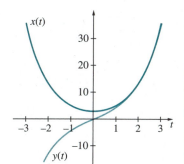

a.

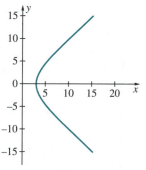

b.

Figure 8.18 (a) Graph of $x(t)$ and $y(t)$ for $-3 \le t \le 3$
(b) Graph of $\begin{cases} x(t) \\ y(t) \end{cases}$ for $-3 \le t \le 3$

In Section 8.1, we solved systems that involve higher order differential equations with differential operators. In many cases these systems can be solved with Laplace transforms as well.

Example 2

Solve $\begin{cases} x'' = 3x' - y' - 2x + y \\ x' + y' = 2x - y \end{cases}$, $x(0) = 0$, $x'(0) = 0$, $y(0) = -1$.

Solution We begin by taking the Laplace transform of both equations. For the first equation, this yields

$$\mathcal{L}\{x''\} = \mathcal{L}\{3x' - y' - 2x + y\}$$
$$s^2X(s) - sx(0) - x'(0) = 3(sX(s) - x(0)) - (sY(s) - y(0)) - 2X(s) + Y(s)$$
$$s^2X(s) = 3sX(s) - sY(s) - 1 - 2X(s) + Y(s)$$
$$(s^2 - 3s + 2)X(s) + (s - 1)Y(s) = -1$$

and for the second equation,

$$\mathcal{L}\{x' + y'\} = \mathcal{L}\{2x - y\}$$
$$sX(s) - x(0) + sY(s) - y(0) = 2X(s) - Y(s)$$
$$(s - 2)X(s) + (s + 1)Y(s) = -1.$$

After factoring the coefficient of $X(s)$ in the first equation, we have the system of equations

$$\begin{cases} (s - 2)(s - 1)X(s) + (s - 1)Y(s) = -1 \\ (s - 2)X(s) + (s + 1)Y(s) = -1. \end{cases}$$

Multiplying the second equation by the factor $-(s - 1)$ yields

$$-(s - 1)(s - 2)X(s) - (s - 1)(s + 1)Y(s) = s - 1.$$

Adding this result to the first equation then eliminates $X(s)$, so we can solve for $Y(s)$ with

$$[(s - 1) - (s - 1)(s + 1)]Y(s) = s - 2$$
$$-s(s - 1)Y(s) = s - 2$$
$$Y(s) = -\frac{s - 2}{s(s - 1)}.$$

Using the partial fraction expansion $Y(s) = -\frac{s - 2}{s(s - 1)} = -\left[\frac{A}{s} + \frac{B}{s - 1}\right]$ shows that $A = 2$ and $B = -1$, so

$$Y(s) = -\left[\frac{2}{s} - \frac{1}{s - 1}\right] \quad \text{and} \quad y(t) = \mathcal{L}^{-1}\left\{-\left[\frac{2}{s} - \frac{1}{s - 1}\right]\right\} = -2 + e^t.$$

In a similar manner, we eliminate $Y(s)$ from the original system of equations by multiplying the first equation by the factor $(s + 1)$ and the second equation by $-(s - 1)$. This yields

$$\begin{cases} (s - 2)(s - 1)(s + 1)X(s) + (s - 1)(s + 1)Y(s) = -(s + 1) \\ -(s - 2)(s - 1)X(s) - (s - 1)(s + 1)Y(s) = s - 1. \end{cases}$$

Adding these equations results in

$$(s - 2)(s - 1)[s + 1 - 1]X(s) = -2$$
$$s(s - 2)(s - 1)X(s) = -2$$
$$X(s) = \frac{-2}{s(s - 2)(s - 1)}.$$

With the partial fraction expansion $X(s) = \dfrac{-2}{s(s-2)(s-1)} = \dfrac{A}{s} + \dfrac{B}{s-2} + \dfrac{C}{s-1}$, we find that $A = -1$, $B = -1$, and $C = 2$. Therefore,

$$X(s) = -\frac{1}{s} - \frac{1}{s-2} + \frac{2}{s-1}, \text{ so } x(t) = \mathcal{L}^{-1}\left\{-\frac{1}{s} - \frac{1}{s-2} + \frac{2}{s-1}\right\}$$
$$= -1 - e^{2t} + 2e^{t}.$$

In Figure 8.19 we graph $x(t)$, $y(t)$, and $\begin{cases} x(t) \\ y(t) \end{cases}$. What is the orientation of $\begin{cases} x(t) \\ y(t) \end{cases}$?

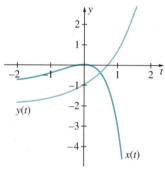
a.

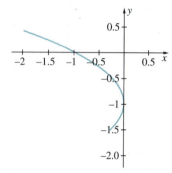
b.

Figure 8.19 (a) Graph of $x(t)$ and $y(t)$ for $-2 \le t \le 2$ (b) Graph of $\begin{cases} x(t) \\ y(t) \end{cases}$ for $-2 \le t \le 2$

IN TOUCH WITH TECHNOLOGY

1. Use Laplace transforms to solve the system $\begin{cases} x' - y = e^{-t} \\ y' + 5x + 2y = \sin 3t \end{cases}$. Graph the solution for different initial conditions.

2. Use Laplace transforms to solve $\begin{cases} 2x'' - 5y' = 0 \\ -y'' - 3y - x' = \sin t \end{cases}$ subject to the conditions $x(0) = 1$, $x'(0) = 0$, $y(0) = 1$, and $y'(0) = 1$. Graph the solution.

In Exercises 1–26, use Laplace transforms to solve each initial-value problem. Graph $x(t)$, $y(t)$, and $\begin{cases} x(t) \\ y(t) \end{cases}$. Describe other methods that could be used to solve the system.

1. $\begin{cases} x' - 2x + 3y = 0 \\ y' + 9x + 4y = 0 \end{cases}$, $x(0) = 0$, $y(0) = 4$

2. $\begin{cases} x' + 9x - 2y = 0 \\ y' + 10x - 3y = 0 \end{cases}$, $x(0) = -2$, $y(0) = 0$

*3. $\begin{cases} x' - 5x - 5y = 0 \\ y' + 4x + 3y = 0 \end{cases}$, $x(0) = 2$, $y(0) = 0$

4. $\begin{cases} x' + 2x - 4y = 0 \\ y' + 2x - 2y = 0 \end{cases}$, $x(0) = 0$, $y(0) = 6$

*5. $\begin{cases} x' - 5x + 4y - 2z = 0 \\ y' + 2x + 2y + 2z = 0, \\ z' - z = 0 \\ x(0) = 0,\ y(0) = 0,\ z(0) = 15 \end{cases}$

6. $\begin{cases} x' + 5x - 4y + 2z = 0 \\ y' - 6x - 2y - 2z = 0, \\ z' - 5x - 7y + 6z = 0 \\ x(0) = 2,\ y(0) = 0,\ z(0) = 2 \end{cases}$

*7. $\begin{cases} x' - x - 3y = e^{4t} \\ y' - 5x + y = 0 \end{cases}$, $x(0) = 0$, $y(0) = 0$

8. $\begin{cases} x' + 4x + 4y = 0 \\ y' + 5x + 3y = e^{t} \end{cases}$, $x(0) = 0$, $y(0) = 1$

9. $\begin{cases} x' + 2x - 3y = 0 \\ y' - 2x + y = 0 \end{cases}$, $x(0) = 1$, $y(0) = 0$

10. $\begin{cases} x' - 3x - 6y = 0 \\ y' - x + 2y = 0 \end{cases}$, $x(0) = 0$, $y(0) = -1$

11. $\begin{cases} x' - 5x - 4y = e^{-t} \\ y' + 2x + 4y = e^{2t} \end{cases}$, $x(0) = -1$, $y(0) = 1$

12. $\begin{cases} x' + 6x + y = t \\ y' + 2x + 7y = e^{4t} \\ x(0) = 0,\ y(0) = 2 \end{cases}$

*13. $\begin{cases} x' + 2x - 4y = \cos 2t \\ y' + 5x - 2y = \sin 2t \\ x(0) = 0,\ y(0) = -1 \end{cases}$

14. $\begin{cases} x' - 4x + 5y = e^{-t} \\ y' - 5x + 2y = \sin 2t \\ x(0) = -2,\ y(0) = 0 \end{cases}$

*15. $\begin{cases} x' - y = 0 \\ y' + x = f(t), \text{ where} \\ x(0) = 0,\ y(0) = 0 \end{cases}$

$f(t) = \begin{cases} \sin t, & \text{if } 0 \le t < \pi \\ 0, & \text{if } t \ge \pi \end{cases}$

16. $\begin{cases} x' + 7x + 4y = f(t) \\ y' + 6x - 3y = 0, \text{ where} \\ x(0) = 1,\ y(0) = 0 \end{cases}$

$f(t) = \begin{cases} 1, & \text{if } 0 \le t < 1 \\ 0, & \text{if } t \ge 1 \end{cases}$

17. $\begin{cases} x' - 2x + 4y = 0 \\ y' - x - 2y = f(t), \text{ where} \\ x(0) = -1,\ y(0) = 0 \end{cases}$

$f(t) = \begin{cases} 1, & \text{if } 0 \le t < 1 \\ -1, & \text{if } 1 \le t < 2 \\ 0, & \text{if } t \ge 2 \end{cases}$

18. $\begin{cases} x' - 2x + 7y = f(t) \\ y' - x + 2y = 0, \text{ where} \\ x(0) = 0,\ y(0) = -1 \end{cases}$

$f(t) = \begin{cases} t, & \text{if } 0 \le t < 1 \\ 2 - t, & \text{if } 1 \le t < 2 \end{cases}$ and
$f(t) = f(t - 2)$ if $t \ge 2$

*19. $\begin{cases} x' + 2x + 3y = 0 \\ y' - x + 6y = f(t), \text{ where} \\ x(0) = 1,\ y(0) = 0 \end{cases}$

$f(t) = \begin{cases} 0, & \text{if } 0 \le t < 1 \\ 1, & \text{if } 1 \le t < 2 \\ 2, & \text{if } 2 \le t < 3 \end{cases}$ and
$f(t) = f(t - 3)$ if $t \ge 3$

20. $\begin{cases} x' + 5x - 7y = f(t) \\ y' + 3x - 5y = 0, \text{ where} \\ x(0) = 0,\ y(0) = -1 \end{cases}$

$f(t) = \begin{cases} t, & \text{if } 0 \le t < 1 \\ 1, & \text{if } 1 \le t < 2 \\ 3 - t, & \text{if } 2 \le t < 3 \end{cases}$ and
$f(t) = f(t - 3)$ if $t \ge 3$

*21. $\begin{cases} -2\dfrac{d^2x}{dt^2} - 2\dfrac{dy}{dt} = 0 \quad x(0) = 2,\ x'(0) = 1, \\[2mm] \dfrac{d^2y}{dt^2} + y - \dfrac{dx}{dt} = \cos t \quad y(0) = 1,\ y'(0) = 2 \end{cases}$

22. $\begin{cases} -2x'' - x = 0 \quad x(0) = 1,\ x'(0) = 1,\ y(0) = 0, \\ 2y'' - 2y + x' = e^{t} \quad y'(0) = 1 \end{cases}$

23. $\begin{cases} -2x'' - 2y' = 0 \qquad x(0) = 2,\ x'(0) = 1, \\ -2y'' + 2y - 6x' = \sin t \quad y(0) = -2,\ y'(0) = 0 \end{cases}$

24. $\begin{cases} 2\dfrac{d^{2}x}{dt^{2}} - 2x - 2y = 1 \qquad x(0) = -2,\ x'(0) = -1, \\[2mm] \qquad \dfrac{d^{2}y}{dt^{2}} + y = \sin t \quad y(0) = -1,\ y'(0) = 0 \end{cases}$

*25. $\begin{cases} -x'' - 2x + 2y' = e^{t} \quad x(0) = 1,\ x'(0) = -1, \\ -2y'' - 2y = t \quad y(0) = -1,\ y'(0) = -2 \end{cases}$

26. $\begin{cases} -2x + 2y - y' = t \qquad x(0) = -2,\ x'(0) = 1, \\ -y'' + y - x + 2x' = \sin t \quad y(0) = -1,\ y'(0) = 2 \end{cases}$

*27. (Vibration Absorbers) Vibration absorbers can be used to virtually eliminate vibration in systems where it is particularly undesirable and to reduce excessive amplitudes of vibration in others. A typical type of vibration absorber consists of a spring-mass system constructed so that its natural frequency is easily varied. This system is then attached to the principal system that is to have its amplitude of vibration reduced, and the frequency of the absorber system is then adjusted until the desired result is achieved. (See Figure 8.20.)

If the frequency ω of the disturbing force $F_0 \sin \omega t$ is near or equal to the natural frequency $\omega_n = \sqrt{\dfrac{k_1}{m_1}}$

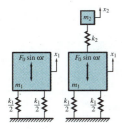

Figure 8.20   The Principal system and the Vibration Absorber Attached to the Principal System

of the system, the amplitude of vibration of the system could become very large due to resonance. If the absorber spring-mass system, made up of components with k_2 and m_2, is attached to the principal system, the amplitude of the mass can be reduced to almost zero if the natural frequency of the absorber is adjusted until it equals that of the disturbing force

$$\omega = \sqrt{\frac{k_2}{m_2}}.$$

These types of absorbers are designed to have little damping and are "tuned" by varying k_2, m_2, or both. This problem is modeled by the system of differential equations

$$\begin{cases} m_1 \dfrac{d^{2}x_1}{dt^{2}} + (k_1 + k_2)x_1 - k_2x_2 = F_0 \sin \omega t \\[3mm] m_2 \dfrac{d^{2}x_2}{dt^{2}} - k_2x_1 + k_2x_2 = 0 \end{cases}$$

(a) Solve the system for x_1 and x_2.

(b) When is the amplitude of x_1 equal to zero? What does this represent?

The mass ratio m_2/m_1 is an important parameter in the design of the absorber. To see the effect on the response of the system, transform these parameters to the nondimensional form

$$\omega_{22}^{2} = \frac{k_2}{m_2} = \frac{k_1}{m_1} \quad \text{and} \quad \mu = \frac{m_2}{m_1} = \frac{k_2}{k_1}.$$

(c) If A_1 is the amplitude of x_1, express $\dfrac{A_1}{F_0/k_1}$ as a function of $\dfrac{\omega}{\omega_{22}}$.

(d) Plot the absolute value of $\dfrac{A_1}{F_0/k_1}$ for $\mu = 0.2$.

(e) When does $\dfrac{A_1}{F_0/k_1}$ become infinite? Do these values correspond to resonance?

(f) In designing the vibration absorber, what frequencies should the absorber not be "tuned" to in order to avoid resonance?

8.7   Nonlinear Systems, Linearization, and Classification of Equilibrium Points

Real Distinct Eigenvalues • Repeated Eigenvalue • Complex Conjugate Eigenvalues • Nonlinear Systems

We now turn our attention to systems of equations of the form

$$\begin{cases} \dfrac{dx}{dt} = f(x, y) \\ \dfrac{dy}{dt} = g(x, y) \end{cases},$$

which are called **autonomous systems** because f and g do not depend explicitly on the independent variable t.

Example 1

Determine if the following systems are autonomous. (In each case, $'$ denotes differentiation with respect to t.)

(a) $\begin{cases} x' = x - ty \\ y' = x + y \end{cases}$; (b) $\begin{cases} x' = y + 2xy \\ y' = x^2 y \end{cases}$.

Solution   (a) The independent variable t appears explicitly in the first differential equation in the system; this system is not classified as autonomous. (b) In this case, t does not appear in either equation (other than in the derivative). Therefore, this system is autonomous.

In order to better understand autonomous systems, we introduce some terminology.

Definition 8.16 Equilibrium Point

A point (x_0, y_0) is an **equilibrium point** of the system $\begin{cases} x' = f(x, y) \\ y' = g(x, y) \end{cases}$ if $f(x_0, y_0) = 0$ and $g(x_0, y_0) = 0$.

Example 2

Find the equilibrium points of the system $\begin{cases} x' = 2x - y \\ y' = -x + 3y \end{cases}$.

Solution   The equilibrium points of the system $\begin{cases} x' = 2x - y \\ y' = -x + 3y \end{cases}$ are the solutions to the system of equations $\begin{cases} 2x - y = 0 \\ -x + 3y = 0 \end{cases}$. The only solution to this system is $(0, 0)$, so the only equilibrium point is $(0, 0)$.

Before we move on to nonlinear systems, we first investigate properties of systems of the form

$$\begin{cases} x' = ax + by \\ y' = cx + dy \end{cases}$$

where $\begin{vmatrix} a & b \\ c & d \end{vmatrix} = ad - bc \neq 0$, which has only the one equilibrium point $(0, 0)$. We have solved many systems of this type using the eigenvalues and corresponding eigenvectors of $\mathbf{A} = \begin{pmatrix} a & b \\ c & d \end{pmatrix}$ and have seen that the solutions vary greatly.

The behavior of the solutions of this system and the classification of the equilibrium point depend on the eigenvalues and corresponding eigenvectors of the system. We more thoroughly investigate the cases that can arise in solving this system by considering the classification of the equilibrium point $(0, 0)$ based on the eigenvalues and corresponding eigenvectors of $\begin{pmatrix} a & b \\ c & d \end{pmatrix}$.

Real Distinct Eigenvalues

Suppose that λ_1 and λ_2 are real eigenvalues of $\mathbf{A} = \begin{pmatrix} a & b \\ c & d \end{pmatrix}$ where $\lambda_2 < \lambda_1$ with corresponding eigenvectors $\mathbf{v}_1$ and $\mathbf{v}_2$, respectively. Then, a general solution of $\mathbf{X}' = \mathbf{A}\mathbf{X}$ is

$$\mathbf{X}(t) = \begin{pmatrix} x(t) \\ y(t) \end{pmatrix} = c_1 \mathbf{v}_1 e^{\lambda_1 t} + c_2 \mathbf{v}_2 e^{\lambda_2 t} = e^{\lambda_1 t}[c_1 \mathbf{v}_1 + c_2 \mathbf{v}_2 e^{(\lambda_2 - \lambda_1)t}].$$

(1) If both eigenvalues are negative, suppose that $\lambda_2 < \lambda_1 < 0$ so $\lambda_2 - \lambda_1 < 0$. This means that $e^{(\lambda_2 - \lambda_1)t}$ is very small for large values of t, so $\mathbf{X}(t) \approx c_1 \mathbf{v}_1 e^{\lambda_1 t}$ is small for large values of t. If $c_1 \neq 0$, then $\lim_{t \to \infty} \mathbf{X}(t) = \mathbf{0}$ along the line through $(0, 0)$ in the direction of $\mathbf{v}_1$. If $c_1 = 0$, then $\mathbf{X}(t) = c_2 \mathbf{v}_2 e^{\lambda_2 t}$. Again, because $\lambda_2 < 0$, $\lim_{t \to \infty} \mathbf{X}(t) = \mathbf{0}$ along the line through $(0, 0)$ in the direction of $\mathbf{v}_2$. In this case, $(0, 0)$ is a **stable node.**

(2) If both eigenvalues are positive, suppose that $0 < \lambda_2 < \lambda_1$. Then $e^{\lambda_1 t}$ and $e^{\lambda_2 t}$ both become unbounded as t increases. If $c_1 \neq 0$, then $\mathbf{X}(t)$ becomes unbounded along the line through $(0, 0)$ in the direction of $\mathbf{v}_1$. If $c_1 = 0$, then $\mathbf{X}(t)$ becomes unbounded along the line through $(0, 0)$ in the direction given by $\mathbf{v}_2$. In this case, $(0, 0)$ is an **unstable node.**

(3) If the eigenvalues have opposite signs, suppose that $\lambda_2 < 0 < \lambda_1$ and $c_1 \neq 0$. Then, $\mathbf{X}(t)$ becomes unbounded along the line through $(0, 0)$ in the direction of $\mathbf{v}_1$ as it did in (2). However, if $c_1 = 0$, then due to the fact that $\lambda_2 < 0$, $\lim_{t \to \infty} \mathbf{X}(t) = \mathbf{0}$ along the line through $(0, 0)$ determined by $\mathbf{v}_2$. If the initial point $\mathbf{X}(0)$ is not on the line through $(0, 0)$ determined by $\mathbf{v}_2$, then the line given by $\mathbf{v}_1$ is an asymptote for the solution. We say that $(0, 0)$ is a **saddle point** in this case.

Example 3

Classify the equilibrium point $(0, 0)$ of the systems: (a) $\begin{cases} x' = 5x + 3y \\ y' = -4x - 3y \end{cases}$;

(b) $\begin{cases} x' = x - 2y \\ y' = 3x - 4y \end{cases}$; (c) $\begin{cases} x' = -x - 2y \\ y' = 3x + 4y \end{cases}$.

Solution   (a) The eigenvalues are found by solving

$$\begin{vmatrix} 5 - \lambda & 3 \\ -4 & -3 - \lambda \end{vmatrix} = \lambda^2 - 2\lambda - 3 = (\lambda - 3)(\lambda + 1) = 0.$$

Because these eigenvalues $\lambda_1 = 3$ and $\lambda_2 = -1$ have opposite signs, $(0, 0)$ is a saddle point. Note that eigenvectors corresponding to λ_1 and λ_2 are $\mathbf{v}_1 = \begin{pmatrix} -3 \\ 2 \end{pmatrix}$ and $\mathbf{v}_2 = \begin{pmatrix} 1 \\ -2 \end{pmatrix}$, respectively. Therefore, the solution becomes unbounded along $y = -\frac{2}{3}x$, the line through $(0, 0)$ in the direction associated with the positive eigenvalue, $\mathbf{v}_1 = \begin{pmatrix} -3 \\ 2 \end{pmatrix}$, because $\lim_{t \to \infty} e^{3t} = \infty$. Along $y = -2x$, the line through $(0, 0)$ in the direction associated with the negative eigenvalue $\mathbf{v}_2 = \begin{pmatrix} 1 \\ -2 \end{pmatrix}$, the solution approaches $(0, 0)$ because $\lim_{t \to \infty} e^{-t} = 0$. The direction field and graphs of several solutions to the system are shown in Figure 8.21(a). Notice that the behavior of the solutions agrees with what we observed from the formula for $\mathbf{X}$ in our earlier discussion.

(b) Because the characteristic equation is

$$\begin{vmatrix} 1 - \lambda & -2 \\ 3 & -4 - \lambda \end{vmatrix} = \lambda^2 + 3\lambda + 2 = (\lambda + 1)(\lambda + 2) = 0,$$

the eigenvalues $\lambda_1 = -1$ and $\lambda_2 = -2$ are both negative; $(0, 0)$ is a stable node. In this case, corresponding eigenvectors are $\mathbf{v}_1 = \begin{pmatrix} 1 \\ 1 \end{pmatrix}$ and $\mathbf{v}_2 = \begin{pmatrix} 2 \\ 3 \end{pmatrix}$, so the solutions approach $(0, 0)$ along the lines through $(0, 0)$ in the direction of these vectors, $y = x$ and $y = \frac{3}{2}x$. The direction field and graphs of several solutions to the system are shown in Figure 8.21(b).

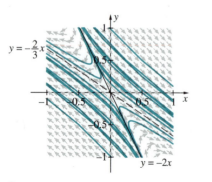

a.

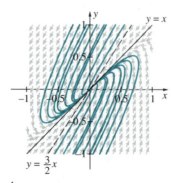

b.

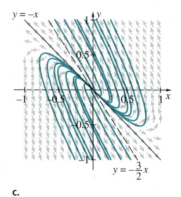

c.

Figure 8.21   (a)–(c)

(c) Because

$$\begin{vmatrix} -1 - \lambda & -2 \\ 3 & 4 - \lambda \end{vmatrix} = \lambda^2 - 3\lambda + 2 = (\lambda - 2)(\lambda - 1) = 0,$$

the eigenvalues $\lambda_1 = 2$ and $\lambda_2 = 1$ are both positive. Therefore, $(0, 0)$ is an unstable node. Note that corresponding eigenvectors are $\mathbf{v}_1 = \begin{pmatrix} 2 \\ -3 \end{pmatrix}$ and $\mathbf{v}_2 = \begin{pmatrix} 1 \\ -1 \end{pmatrix}$, respectively. The solutions become unbounded along the lines through $(0, 0)$ determined by these vectors, $y = -\frac{3}{2}x$ and $y = -x$. The direction field and graphs of several solutions to the system are shown in Figure 8.21(c).

Repeated Eigenvalue

We recall from our previous experience with repeated eigenvalues of a 2×2 system that the eigenvalue can have two linearly independent eigenvectors associated with it or only one (linearly independent) eigenvector associated with it. We investigate the behavior of solutions in the case of repeated eigenvalues by considering both of these possibilities.

(1) If the eigenvalue $\lambda = \lambda_1 = \lambda_2$ has two corresponding linearly independent eigenvectors $\mathbf{v}_1$ and $\mathbf{v}_2$, a general solution is

$$\mathbf{X}(t) = c_1 \mathbf{v}_1 e^{\lambda t} + c_2 \mathbf{v}_2 e^{\lambda t} = (c_1 \mathbf{v}_1 + c_2 \mathbf{v}_2) e^{\lambda t}.$$

If $\lambda > 0$, then $\mathbf{X}(t)$ becomes unbounded along the lines through $(0, 0)$ determined by the vectors $c_1 \mathbf{v}_1 + c_2 \mathbf{v}_2$, where c_1 and c_2 are arbitrary constants. In this case, we call the equilibrium point a **degenerate unstable node** (or an **unstable star**). On the other hand, if $\lambda < 0$, then $\mathbf{X}(t)$ approaches $(0, 0)$ along these lines, and we call $(0, 0)$ a **degenerate stable node** (or **stable star**). Note that the name "star" was selected due to the shape of the solutions.

(2) If $\lambda = \lambda_1 = \lambda_2$ has only one corresponding (linearly independent) eigenvector $\mathbf{v}_1$, a general solution is

$$\mathbf{X}(t) = c_1 \mathbf{v}_1 e^{\lambda t} + c_2 [\mathbf{v}_1 t + \mathbf{w}_2] e^{\lambda t} = (c_1 \mathbf{v}_1 + c_2 \mathbf{w}_2) e^{\lambda t} + c_2 \mathbf{v}_1 t e^{\lambda t}$$

where $(\mathbf{A} - \lambda \mathbf{I}) \mathbf{w}_2 = \mathbf{v}_1$. If we write this solution as

$\mathbf{X}(t) = t e^{\lambda t} \left[\dfrac{1}{t} (c_1 \mathbf{v}_1 + c_2 \mathbf{w}_2) + c_2 \mathbf{v}_1 \right]$, we can more easily investigate the behavior

of this solution. If $\lambda < 0$, then $\lim_{t \to \infty} t e^{\lambda t} = 0$ and

$\lim_{t \to \infty} \left[\dfrac{1}{t} (c_1 \mathbf{v}_1 + c_2 \mathbf{w}_2) + c_2 \mathbf{v}_1 \right] = c_2 \mathbf{v}_1$. The solutions approach $(0, 0)$ along the

line through $(0, 0)$ determined by $\mathbf{v}_1$, and we call $(0, 0)$ a **degenerate stable node.** If $\lambda > 0$, the solutions become unbounded along this line, and we say that $(0, 0)$ is a **degenerate unstable node.**

Example 4

Classify the equilibrium point $(0, 0)$ in the systems: (a) $\begin{cases} x' = x + 9y \\ y' = -x - 5y \end{cases}$;

(b) $\begin{cases} x' = 2x \\ y' = 2y \end{cases}$.

Solution	(a) The eigenvalues are found by solving

$$\begin{vmatrix} 1 - \lambda & 9 \\ -1 & -5 - \lambda \end{vmatrix} = \lambda^2 + 4\lambda + 4 = (\lambda + 2)^2 = 0.$$

Hence, $\lambda_1 = \lambda_2 = -2$. In this case, an eigenvector $\mathbf{v}_1 = \begin{pmatrix} x_1 \\ y_1 \end{pmatrix}$ satisfies

$\begin{pmatrix} 3 & 9 \\ -1 & -3 \end{pmatrix}\begin{pmatrix} x_1 \\ y_1 \end{pmatrix} = \begin{pmatrix} 0 \\ 0 \end{pmatrix}$, which is equivalent to $\begin{pmatrix} 1 & 3 \\ 0 & 0 \end{pmatrix}\begin{pmatrix} x_1 \\ y_1 \end{pmatrix} = \begin{pmatrix} 0 \\ 0 \end{pmatrix}$, so there is only one corresponding (linearly independent) eigenvector $\mathbf{v}_1 = \begin{pmatrix} -3y_1 \\ y_1 \end{pmatrix} = \begin{pmatrix} -3 \\ 1 \end{pmatrix}y_1$. Because $\lambda = -2 < 0$, $(0, 0)$ is a degenerate stable node. Notice that in the graph of several members of the family of solutions of this system along with the direction field in Figure 8.22(a), the solutions approach $(0, 0)$ along the line in the direction of $\mathbf{v}_1 = \begin{pmatrix} -3 \\ 1 \end{pmatrix}$ given by $y = -\dfrac{x}{3}$.

(b) Solving the characteristic equation

$$\begin{vmatrix} 2 - \lambda & 0 \\ 0 & 2 - \lambda \end{vmatrix} = (2 - \lambda)^2 = 0,$$

we have $\lambda_1 = \lambda_2 = 2$. However, because an eigenvector $\mathbf{v}_1 = \begin{pmatrix} x_1 \\ y_1 \end{pmatrix}$ satisfies the system $\begin{pmatrix} 0 & 0 \\ 0 & 0 \end{pmatrix}\begin{pmatrix} x_1 \\ y_1 \end{pmatrix} = \begin{pmatrix} 0 \\ 0 \end{pmatrix}$, any nonzero choice of $\mathbf{v}_1$ is an eigenvector. If we select two linearly independent vectors such as $\mathbf{v}_1 = \begin{pmatrix} 1 \\ 0 \end{pmatrix}$ and $\mathbf{v}_2 = \begin{pmatrix} 0 \\ 1 \end{pmatrix}$, we obtain two linearly independent eigenvectors corresponding to $\lambda_1 = \lambda_2 = 2$. (Note: The choice of these two vectors does not change the value of the solution, because of the form of the general solution in this case.) Because $\lambda = 2 > 0$, we classify $(0, 0)$ as a degenerate unstable node (or star). Some of these solutions along

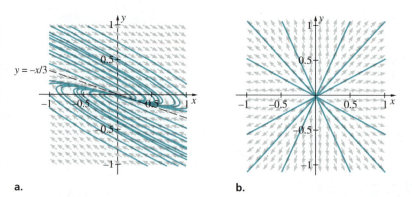

a.   **b.**

Figure 8.22   (a)–(b)

with the direction field are graphed in Figure 8.22(b). Notice that they become unbounded in the direction of any vector in the xy-plane because $\mathbf{v}_1 = \begin{pmatrix} 1 \\ 0 \end{pmatrix}$ and $\mathbf{v}_2 = \begin{pmatrix} 0 \\ 1 \end{pmatrix}$ and every vector in the xy-plane can be written as a linear combination of these vectors.

Complex Conjugate Eigenvalues

We have seen that if the eigenvalues of the system of differential equations are $\lambda_1 = \alpha + \beta i$ and $\lambda_2 = \alpha - \beta i$ with corresponding eigenvectors $\mathbf{v}_1 = \mathbf{a} + \mathbf{b}i$ and $\mathbf{v}_2 = \mathbf{a} - \mathbf{b}i$, then two linearly independent solutions of the system are

$$\mathbf{X}_1(t) = e^{\alpha t}(\cos \beta t \, \mathbf{a} - \sin \beta t \, \mathbf{b}) \quad \text{and} \quad \mathbf{X}_2(t) = e^{\alpha t}(\cos \beta t \, \mathbf{b} + \sin \beta t \, \mathbf{a}).$$

A general solution is $\mathbf{X}(t) = c_1 \mathbf{X}_1(t) + c_2 \mathbf{X}_2(t)$, so there are constants A_1, A_2, B_1, and B_2 such that x and y are given by

$$\mathbf{X}(t) = \begin{pmatrix} x(t) \\ y(t) \end{pmatrix} = \begin{pmatrix} A_1 e^{\alpha t} \cos \beta t + A_2 e^{\alpha t} \sin \beta t \\ B_1 e^{\alpha t} \cos \beta t + B_2 e^{\alpha t} \sin \beta t \end{pmatrix}.$$

(1) If $\alpha = 0$, a general solution is

$$\mathbf{X}(t) = \begin{pmatrix} x(t) \\ y(t) \end{pmatrix} = \begin{pmatrix} A_1 \cos \beta t + A_2 \sin \beta t \\ B_1 \cos \beta t + B_2 \sin \beta t \end{pmatrix}.$$

Both x and y are periodic. In fact, if $A_2 = B_1 = 0$, then

$$\mathbf{X}(t) = \begin{pmatrix} x(t) \\ y(t) \end{pmatrix} = \begin{pmatrix} A_1 \cos \beta t \\ B_2 \sin \beta t \end{pmatrix}.$$

In rectangular coordinates this solution is

$$\frac{x^2}{A_1^2} + \frac{y^2}{B_2^2} = 1$$

where the graph is either a circle or an ellipse centered at $(0, 0)$ depending on the value of A_1 and B_2. Hence, $(0, 0)$ is classified as a **center.** Note that the motion around these circles or ellipses is either clockwise or counterclockwise for all solutions.

(2) If $\alpha \neq 0$, $e^{\alpha t}$ is present in the solution. The $e^{\alpha t}$ term causes the solution to spiral around the equilibrium point. If $\alpha > 0$, the solution spirals away from $(0, 0)$, so we classify $(0, 0)$ as an **unstable spiral.** If $\alpha < 0$, the solution spirals toward $(0, 0)$, so we say that $(0, 0)$ is a **stable spiral.**

| Example 5 |

Classify the equilibrium point $(0, 0)$ in each of the following systems: (a) $\begin{cases} x' = -y; \\ y' = x \end{cases}$

(b) $\begin{cases} x' = x - 5y \\ y' = x - 3y \end{cases}$.

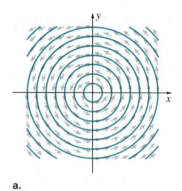

a.

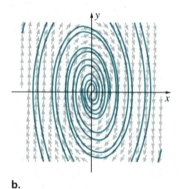

b.

Figure 8.23   (a)–(b)

| Solution | (a) The eigenvalues are found by solving

$$\begin{vmatrix} -\lambda & -1 \\ 1 & -\lambda \end{vmatrix} = \lambda^2 + 1 = 0.$$

Hence, $\lambda = \pm i$. Because these eigenvalues have zero real part (i.e., are purely imaginary), $(0, 0)$ is a center. Several solutions along with the direction field are graphed in Figure 8.23(a). The vectors indicate that the solutions move counterclockwise around $(0, 0)$. We also see this by observing the equations in the system, $x' = -y$ and $y' = x$. In the first quadrant where $x > 0$ and $y > 0$, these equations indicate that $x' < 0$ and $y' > 0$. Therefore, on solutions in the first quadrant, x decreases while y increases. Similarly, in the second quadrant where $x < 0$ and $y > 0$, $x' < 0$ and $y' < 0$, so x and y both decrease in this quadrant. (Similar observations are made in the other two quadrants.)

(b) Because the characteristic equation is

$$\begin{vmatrix} 1 - \lambda & -5 \\ 1 & -3 - \lambda \end{vmatrix} = \lambda^2 + 2\lambda + 2 = 0,$$

the eigenvalues are $\lambda = \dfrac{-2 \pm \sqrt{4 - 8}}{2} = -1 \pm i$. Thus, $(0, 0)$ is a stable spiral, because $\alpha = -1 < 0$. Several solutions along with the direction field are graphed in Figure 8.23(b). Notice that solutions spiral towards the origin because the eigenvalues have negative real part.

In Table 8.1, we summarize the criteria for classifying the equilibrium point $(0, 0)$ of the system $\begin{cases} x' = ax + by \\ y' = cx + dy \end{cases}$, which will be useful in classifying equilibrium points of nonlinear systems. (Degenerate cases are not included in the table.)

TABLE 8.1

Eigenvalues	Classification
λ_1, λ_2 Real; $\lambda_1 > \lambda_2 > 0$	Unstable Node
λ_1, λ_2 Real; $\lambda_2 < \lambda_1 < 0$	Stable Node
λ_1, λ_2 Real; $\lambda_2 < 0 < \lambda_1$	Saddle
$\lambda_1 = \alpha + \beta i, \lambda_2 = \alpha - \beta i; \beta \neq 0; \alpha > 0$	Unstable Spiral
$\lambda_1 = \alpha + \beta i, \lambda_2 = \alpha - \beta i; \beta \neq 0; \alpha < 0$	Stable Spiral
$\lambda_1 = \alpha + \beta i, \lambda_2 = \alpha - \beta i; \beta \neq 0; \alpha = 0$	Center

Nonlinear Systems

When working with nonlinear systems, we can often gain a great deal of information concerning the system by making a *linear approximation* near each equilibrium point of the nonlinear system and solving the linear system. Although the solution to the linearized system only approximates the solution to the nonlinear system, the general behavior of solutions to the nonlinear system near each equilibrium is the same as that of the corresponding linear system *in most cases*. The first step towards approximating a nonlinear system near each equilibrium point is to find the equilibrium points of the system and to linearize the system at each of these points.

Recall from multivariable calculus that if $z = F(x, y)$ is a differentiable function, the tangent plane to the surface S given by the graph of $z = F(x, y)$ at the point (x_0, y_0) is

$$z = F_x(x_0, y_0)(x - x_0) + F_y(x_0, y_0)(y - y_0) + F(x_0, y_0).$$

Near each equilibrium point (x_0, y_0) of the nonlinear system

$$\begin{cases} x' = f(x, y) \\ y' = g(x, y) \end{cases}$$

the system can be *approximated* with

$$\begin{cases} x' = f_x(x_0, y_0)(x - x_0) + f_y(x_0, y_0)(y - y_0) + f(x_0, y_0) \\ y' = g_x(x_0, y_0)(x - x_0) + g_y(x_0, y_0)(y - y_0) + g(x_0, y_0) \end{cases}$$

where we have used the tangent plane to approximate f and g in the two differential equations. Because $f(x_0, y_0) = 0$ and $g(x_0, y_0) = 0$ (why?), the *approximate* system is

$$\begin{cases} x' = f_x(x_0, y_0)(x - x_0) + f_y(x_0, y_0)(y - y_0) \\ y' = g_x(x_0, y_0)(x - x_0) + g_y(x_0, y_0)(y - y_0) \end{cases}$$

which can be written in matrix form as

$$\begin{pmatrix} x' \\ y' \end{pmatrix} = \begin{pmatrix} f_x(x_0, y_0) & f_y(x_0, y_0) \\ g_x(x_0, y_0) & g_y(x_0, y_0) \end{pmatrix} \begin{pmatrix} x - x_0 \\ y - y_0 \end{pmatrix}.$$

Note that we often call this system the **linearized system corresponding to the nonlinear system** or the **associated linearized system** due to the fact that we have removed the nonlinear terms from the original system.

The equilibrium point (x_0, y_0) of the system $\begin{cases} x' = f(x, y) \\ y' = g(x, y) \end{cases}$ is classified by the eigenvalues of the matrix

$$J(x_0, y_0) = \begin{pmatrix} f_x(x_0, y_0) & f_y(x_0, y_0) \\ g_x(x_0, y_0) & g_y(x_0, y_0) \end{pmatrix},$$

*Notice that the linearization must be carried out for **each** equilibrium point.*

which is called the **Jacobian matrix.** After determining the Jacobian matrix for each equilibrium point, we find the eigenvalues of the matrix in order to classify the corresponding equilibrium point according to the following criteria.

> ### Classification of Equilibrium Points of a Nonlinear System

Let (x_0, y_0) be an equilibrium point of the system $\begin{cases} x' = f(x, y) \\ y' = g(x, y) \end{cases}$ and let λ_1 and λ_2 be the eigenvalues of the matrix

$$\begin{pmatrix} f_x(x_0, y_0) & f_y(x_0, y_0) \\ g_x(x_0, y_0) & g_y(x_0, y_0) \end{pmatrix}$$

of the associated linearized system about the equilibrium point.

(a) If (x_0, y_0) is classified as a node (stable or unstable), saddle, or spiral (stable or unstable) in the associated linear system, (x_0, y_0) has the same classification in the nonlinear system.

(b) If (x_0, y_0) is classified as a center in the associated linear system, (x_0, y_0) may be a center, unstable spiral, or stable spiral in the nonlinear system, so we cannot classify (x_0, y_0) in this situation. (See Exercise 28.)

These findings are summarized in Table 8.2 where classification of (x_0, y_0) in the associated linearized system is given in the left-hand column and the corresponding classification in the nonlinear system appears in the right-hand column.

TABLE 8.2

Associated Linearized System	Nonlinear System
Stable Node	Stable Node
Unstable Node	Unstable Node
Stable Spiral	Stable Spiral
Unstable Spiral	Unstable Spiral
Saddle	Saddle
Center	No Conclusion

Example 6

Find and classify the equilibrium points of $\begin{cases} x' = 1 - y \\ y' = x^2 - y^2 \end{cases}$.

Solution   We begin by finding the equilibrium points of this nonlinear system by solving $\begin{cases} 1 - y = 0 \\ x^2 - y^2 = 0 \end{cases}$. Because $y = 1$ from the first equation, substitution into the second equation yields $x^2 - 1 = 0$. Therefore, $x = \pm 1$, so the two equilibrium points are $(1, 1)$ and $(-1, 1)$. Because $f(x, y) = 1 - y$ and $g(x, y) = x^2 - y^2$, $f_x(x, y) = 0$, $f_y(x, y) = -1$, $g_x(x, y) = 2x$, and $g_y(x, y) = -2y$, so the Jacobian matrix is $J(x, y) = \begin{pmatrix} 0 & -1 \\ 2x & -2y \end{pmatrix}$. Next, we classify each equilibrium point by finding the eigenvalues of the Jacobian matrix of each linearized system.

For $(1, 1)$, we obtain the Jacobian matrix $J(1, 1) = \begin{pmatrix} 0 & -1 \\ 2 & -2 \end{pmatrix}$ with eigenvalues that satisfy $\begin{vmatrix} -\lambda & -1 \\ 2 & -2-\lambda \end{vmatrix} = \lambda^2 + 2\lambda + 2 = 0$. Hence, $\lambda_1 = -1 + i$ and $\lambda_2 = -1 - i$. Because these eigenvalues are complex-valued with negative real part, we classify $(1, 1)$ as a stable spiral in the associated linearized system. Therefore, $(1, 1)$ is a stable spiral in the nonlinear system.

For $(-1, 1)$, we obtain $J(-1, 1) = \begin{pmatrix} 0 & -1 \\ -2 & -2 \end{pmatrix}$. In this case, the eigenvalues are solutions of $\begin{vmatrix} -\lambda & -1 \\ -2 & -2-\lambda \end{vmatrix} = \lambda^2 + 2\lambda - 2 = 0$. Thus, $\lambda_1 = \dfrac{-2 + 2\sqrt{3}}{2} = -1 + \sqrt{3} > 0$ and $\lambda_2 = \dfrac{-2 - 2\sqrt{3}}{2} = -1 - \sqrt{3} < 0$, so $(-1, 1)$ is a saddle in the associated linearized system and this classification carries over to the nonlinear system. In Figure 8.24, we graph solutions to this nonlinear system approximated with the use of a computer algebra system. We can see how the solutions move toward and away from the equilibrium points by observing the arrows on the vectors in the direction field.

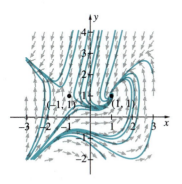

Figure 8.24

Example 7

Find and classify the equilibrium points of $\begin{cases} x' = x(7 - x - 2y) \\ y' = y(5 - x - y) \end{cases}$.

Solution The equilibrium points of this system satisfy $\begin{cases} x(7 - x - 2y) = 0 \\ y(5 - x - y) = 0 \end{cases}$. Thus, $\{x = 0 \text{ or } 7 - x - 2y = 0\}$ and $\{y = 0 \text{ or } 5 - x - y = 0\}$. If $x = 0$, then $y(5 - y) = 0$ so $y = 0$ or $y = 5$, and we obtain the equilibrium points $(0, 0)$ and $(0, 5)$. If $y = 0$, then $x(7 - x) = 0$, which indicates that $x = 0$ or $x = 7$. The corresponding equilibrium points are $(0, 0)$ (which we found earlier) and $(7, 0)$. The other possibility that leads to an equilibrium point is the solution to $\begin{cases} 7 - x - 2y = 0 \\ 5 - x - y = 0 \end{cases}$ which is $x = 3$ and $y = 2$. The equilibrium point associated with this possibility is $(3, 2)$.

The Jacobian matrix is

$$J(x, y) = \begin{pmatrix} 7 - 2x - 2y & -2x \\ -y & 5 - x - 2y \end{pmatrix}.$$

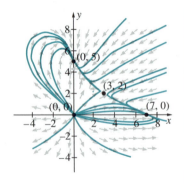

Figure 8.25

We classify each of the equilibrium points (x_0, y_0) of the associated linearized system using the eigenvalues of $J(x_0, y_0)$.

$$J(0, 0) = \begin{pmatrix} 7 & 0 \\ 0 & 5 \end{pmatrix}; \ \lambda_1 = 7, \ \lambda_2 = 5; \ (0, 0) \text{ is an unstable node.}$$

$$J(0, 5) = \begin{pmatrix} -3 & 0 \\ -5 & -5 \end{pmatrix}; \ \lambda_1 = -3, \ \lambda_2 = -5; \ (0, 5) \text{ is a stable node.}$$

$$J(7, 0) = \begin{pmatrix} -7 & -14 \\ 0 & -2 \end{pmatrix}; \ \lambda_1 = -2, \ \lambda_2 = -7; \ (7, 0) \text{ is a stable node.}$$

$$J(3, 2) = \begin{pmatrix} -3 & -6 \\ -2 & -2 \end{pmatrix}; \ \lambda_1 = 1, \ \lambda_2 = -6; \ (3, 2) \text{ is a saddle.}$$

In each case, the classification carries over to the nonlinear system. In Figure 8.25, we graph several approximate solutions and the direction field to this nonlinear system through the use of a computer algebra system. Notice the behavior near each equilibrium point.

Identify the equilibrium points in Figure 8.25.

IN TOUCH WITH TECHNOLOGY

 1. Classify the equilibrium point $(0, 0)$ for the system $X' = AX$ if

(a) $A = \begin{pmatrix} -6 & 3 \\ 5 & -4 \end{pmatrix}$;   (b) $A = \begin{pmatrix} -3 & -1 \\ 5 & 3 \end{pmatrix}$;

(c) $A = \begin{pmatrix} 0 & -1 \\ 2 & 3 \end{pmatrix}$;   (d) $A = \begin{pmatrix} -1 & -1 \\ 5 & -3 \end{pmatrix}$;

(e) $A = \begin{pmatrix} 10 & -1 \\ 6 & 8 \end{pmatrix}$; and (f) $A = \begin{pmatrix} -\frac{3}{7} & \frac{1}{3} \\ -1 & \frac{3}{7} \end{pmatrix}$.

For each system, graph the associated direction field.

2. An exciting system to explore is the system

$$\begin{cases} x' = \mu x + y - x(x^2 + y^2) \\ y' = \mu y - x - y(x^2 + y^2) \end{cases}$$

for fixed values of μ.

(a) Show that the only equilibrium point of the system is $(0, 0)$.

(b) Show that the eigenvalues of the Jacobian matrix $\begin{pmatrix} \mu & 1 \\ -1 & \mu \end{pmatrix}$ are $\mu \pm i$ and classify the equilibrium point $(0, 0)$.

(c) How do you *think* the direction field of the system changes as μ goes from -1 to 1?

(d) Graph the direction field and various solutions of the system for several values of μ between -1 and 1.

(e) How do your results in (d) differ from your predictions in (c)?

In Exercises 1–12, classify the equilibrium point $(0, 0)$ of the system.

1. $\begin{cases} x' = -5x - 4y \\ y' = -9x - 5y \end{cases}$ 2. $\begin{cases} x' = -x - 7y \\ y' = -3x + 3y \end{cases}$

*3. $\begin{cases} x' = 8x - 4y \\ y' = 9x - 4y \end{cases}$ 4. $\begin{cases} x' = 4x - 2y \\ y' = -6x + 8y \end{cases}$

5. $\begin{cases} x' = -10x - 2y \\ y' = -2x - 10y \end{cases}$ 6. $\begin{cases} x' = 6y \\ y' = -3x - 9y \end{cases}$

*7. $\begin{cases} x' = -x + 10y \\ y' = 3x + 7y \end{cases}$ 8. $\begin{cases} x' = -3x - y \\ y' = x + 4y \end{cases}$

9. $\begin{cases} x' = -9y \\ y' = 6x - 5y \end{cases}$ 10. $\begin{cases} x' = -8x + 5y \\ y' = -2x - 2y \end{cases}$

*11. $\begin{cases} x' = -3x + 5y \\ y' = -10x + 3y \end{cases}$ 12. $\begin{cases} x' = -4x - 4y \\ y' = 8x + 4y \end{cases}$

In Exercises 13–20, classify the equilibrium points of the nonlinear system.

13. $\begin{cases} x' = x^2 - 10x + 16 \\ y' = y + 1 \end{cases}$ 14. $\begin{cases} x' = x(3 - y) \\ y' = x + y + 1 \end{cases}$

*15. $\begin{cases} x' = y - x \\ y' = x + y - 2xy \end{cases}$ 16. $\begin{cases} x' = 1 - x^2 \\ y' = y \end{cases}$

17. $\begin{cases} x' = xy \\ y' = x^2 + y^2 - 4 \end{cases}$ 18. $\begin{cases} x' = 2x + y^2 \\ y' = -2x + 4y \end{cases}$

*19. $\begin{cases} x' = 1 - xy \\ y' = y - x^3 \end{cases}$ 20. $\begin{cases} x' = y - \dfrac{x}{2} \\ y' = 1 - xy \end{cases}$

21. **(Population)** Suppose that we consider the relationship between a host population and a parasite population that is modeled by the nonlinear system

$$\begin{cases} x' = (a_1 - b_1 x - c_1 y)x \\ y' = (-a_2 + c_2 x)y \end{cases},$$

where $x(t)$ represents the number in the host population at time t, $y(t)$ the number in the parasite population at time t, a_1 the growth rate of the host population, a_2 the starvation rate of the parasite population, and b_1 the interference when the host population overpopulates. The constants c_1 and c_2 relate the interactions between the two populations leading to

decay in the host population and growth in the parasite population, respectively. Find and classify the equilibrium point(s) of the system if all constants in the system are positive.

22. **(Economics)** Let $x(t)$ represent the income of a company and $y(t)$ the amount of consumer spending. Also suppose that z represents the rate of company expenditures. The system of nonlinear equations that models this situation is

$$\begin{cases} x' = x - ay \\ y' = b(x - y - z) \end{cases} \text{ where } 1 < a < \infty, \text{ and } 1 \le b.$$

(a) If $z = z_0$ is constant, find and classify the equilibrium point of the system. Also consider the special case if $b = 1$.

(b) If the expenditure depends on income according to the relationship $z = z_0 + cx$ $(c > 0)$, find and classify the equilibrium points (if they exist) of the system.

In Exercises 23–27, match each of the following systems in Group A with its direction field in Group B.

Group A	Group B

23. $\begin{cases} x' = 4y^2 + 3x + 2 \\ y' = 4x^2 - 4 \end{cases}$ (a)

24. $\begin{cases} x' = y^2 + 2xy \\ y' = x^2 - 4x - 5 \end{cases}$ (b)

*25. $\begin{cases} x' = -5xy - y^3 \\ y' = 3xy + y + 2x \end{cases}$

(c)

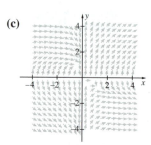

26. $\begin{cases} x' = -3y - 3xy \\ y' = 5y^2 - x + 5 \end{cases}$

(d)

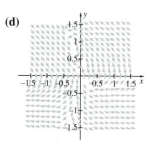

27. $\begin{cases} x' = x^2 y^2 \\ y' = xy^2 + yx^2 \end{cases}$

(e)

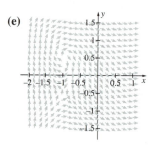

28. Consider the nonlinear system

$$\begin{cases} \dfrac{dx}{dt} = y + \mu x(x^2 + y^2) \\[2mm] \dfrac{dy}{dt} = -x + \mu y(x^2 + y^2) \end{cases}$$

(a) Show that this system has the equilibrium point $(0, 0)$ and the linearized system about this point is

$$\begin{cases} \dfrac{dx}{dt} = y \\[2mm] \dfrac{dy}{dt} = -x \end{cases}$$. Also, show that $(0, 0)$ is classified as

a center of the linearized system.

(b) Consider the change of variables to polar coordinates $x = r \cos \theta$ and $y = r \sin \theta$. Use these equations with the chain rule to show that $\dfrac{dx}{dt} =$

$\dfrac{dr}{dt} \cos \theta - r \dfrac{d\theta}{dt} \sin \theta$ and $\dfrac{dy}{dt} = \dfrac{dr}{dt} \sin \theta +$

$r \dfrac{d\theta}{dt} \cos \theta$. Use elimination with these equations

to show that $\dfrac{dr}{dt} = \dfrac{dx}{dt} \cos \theta + \dfrac{dy}{dt} \sin \theta$ and

$r \dfrac{d\theta}{dt} = -\dfrac{dx}{dt} \sin \theta + \dfrac{dy}{dt} \cos \theta$. Transform the

original system to polar coordinates and substitute the equations that result in the equations for $\dfrac{dr}{dt}$ and $r \dfrac{d\theta}{dt}$ to obtain the system in polar coordinates

$$\begin{cases} \dfrac{dr}{dt} = \mu r^3 \\[3mm] \dfrac{d\theta}{dt} = -1 \end{cases}$$

(c) What does the equation $\dfrac{d\theta}{dt} = -1$ indicate about the rotation of solutions to the system? According to $\dfrac{dr}{dt} = \mu r^3$, does r increase or decrease for $r > 0$? Using these observations, is the equilibrium point $(0, 0)$ which was classified as a center for the linearized system also classified as a center for the nonlinear system? If not, how would you classify it?

*29. Consider the first-order equation $\dfrac{dy}{dx} = f(y)$. If $f(y_0) = 0$, then $y = y_0$ is called an **equilibrium solution** of the equation. The equilibrium solution $y = y_0$ is **stable** if solutions converge to $y = y_0$ as x increases. Otherwise, $y = y_0$ is **unstable.** (Thus, $y = y_0$ is stable if a solution that starts near $y = y_0$ remains near $y = y_0$.) (a) Show that the equilibrium solutions of $\dfrac{dy}{dx} = 3y - y^2$ are $y = 0$ and $y = 3$. (b) Show that $\dfrac{dy}{dx} < 0$ if $y < 0$ or $y > 3$; $\dfrac{dy}{dx} > 0$ if $0 < y < 3$. (c) Show that $y = 0$ is unstable and $y = 3$ is stable by considering the sign on $\dfrac{dy}{dx}$ on these intervals as well as the graph of the direction field in Figure 8.26.

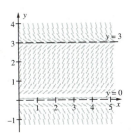

Figure 8.26

(d) Solve the initial-value problem $\left\{\dfrac{dy}{dx} = 3y - y^2,\right.$

$\left.y(0) = -1\right\}.$ Calculate $\lim_{x \to \infty} y(x).$

(e) Repeat part (d) with $\left\{\dfrac{dy}{dx} = 3y - y^2,\ y(0) = 1\right\}.$

(f) Repeat part (d) with $\left\{\dfrac{dy}{dx} = 3y - y^2,\ y(0) = 4\right\}.$

In Exercises 30–35, find and classify (as stable or unstable) the equilibrium solutions of the first-order equations.

30. $\dfrac{dy}{dx} = 3y$

31. $\dfrac{dy}{dx} = -y$

32. $\dfrac{dy}{dx} = y^2 - y$

33. $\dfrac{dy}{dx} = 16y - 8y^2$

34. $\dfrac{dy}{dx} = 12 + 4y - y^2$

35. $\dfrac{dy}{dx} = y^2 - 5y + 4$

8.8 Numerical Methods

Euler's Method • Runge-Kutta Method • Computer Algebra Systems

Because it may be difficult or even impossible to construct an explicit solution to some systems of differential equations, we now turn our attention to some numerical methods that are used to construct solutions to systems of differential equations.

Euler's Method

Euler's method for approximation which was discussed for first-order equations may be extended to include systems of first-order equations. The initial-value problem

$$\begin{cases} \dfrac{dx}{dt} = f(t, x, y) \\[2mm] \dfrac{dy}{dt} = g(t, x, y) \\[2mm] x(t_0) = x_0,\ y(t_0) = y_0 \end{cases}$$

is approximated at each step by the recursive relationship based on the Taylor expansion of x and y up to order h,

$$\begin{cases} x_{n+1} = x_n + hf(t_n, x_n, y_n) \\ y_{n+1} = y_n + hg(t_n, x_n, y_n) \end{cases},$$

where $t_n = t_0 + nh,\ n = 0, 1, 2, \ldots .$

| Example 1 |

Use Euler's method with $h = 0.1$ and $h = 0.05$ to approximate the solution to the initial-value problem:

$$\begin{cases} \dfrac{dx}{dt} = x - y + 1 \\[2mm] \dfrac{dy}{dt} = x + 3y + e^{-t}. \\[2mm] x(0) = 0, \; y(0) = 1 \end{cases}$$

Compare these results with the exact solution to the system of equations.

| Solution | In this case, $f(x, y) = x - y + 1$, $g(x, y) = x + 3y + e^{-t}$, $t_0 = 0$, $x_0 = 0$, and $y_0 = 1$, so we use the formulas

$$\begin{cases} x_{n+1} = x_n + h(x_n - y_n + 1) \\ y_{n+1} = y_n + h(x_n + 3y_n + e^{-t_n}) \end{cases}$$

where $t_n = (0.1)n$, $n = 0, 1, 2, \ldots$.
  If $n = 0$, then

$$\begin{cases} x_1 = x_0 + h(x_0 - y_0 + 1) = 0 \\ y_1 = y_0 + h(x_0 + 3y_0 + e^{-t_0}) = 1.4 \end{cases}$$

The exact solution of this problem, which can be determined using the method of variation of parameters, is

$$\begin{cases} x(t) = -\dfrac{3}{4} - \dfrac{e^{-t}}{9} + \dfrac{31e^{2t}}{36} - \dfrac{11te^{2t}}{6} . \\[3mm] y(t) = \dfrac{1}{4} - \dfrac{2e^{-t}}{9} + \dfrac{35e^{2t}}{36} + \dfrac{11te^{2t}}{6} \end{cases}$$

TABLE 8.3

t_n	x_n (approx)	x_n (exact)	y_n (approx)	y_n (exact)
0.0	0.0	0.0	1.0	1.0
0.1	0.0	−0.02270	1.4	1.46032
0.2	−0.04	−0.10335	1.91048	2.06545
0.3	−0.13505	−0.26543	2.5615	2.85904
0.4	−0.30470	−0.54011	3.39053	3.89682
0.5	−0.57423	−0.96841	4.44423	5.24975
0.6	−0.97607	−1.60412	5.78076	7.00806
0.7	−1.55176	−2.51737	7.47226	9.28638
0.8	−2.35416	−3.79926	9.60842	12.23
0.9	−3.45042	−5.56767	12.3005	16.0232
1.0	−4.9255	−7.97468	15.6862	20.8987

In Table 8.3, we display the results obtained with this method and compare them to the actual function values.

Because the accuracy of this approximation diminishes as t increases, we attempt to improve the approximation by decreasing the increment size. We do this by entering the value $h = 0.05$ and repeating the above procedure. We show the results of this method in Table 8.4. Notice that the approximations are more accurate with the smaller value of h.

TABLE 8.4

t_n	x_n (approx)	x_n (exact)	y_n (approx)	y_n (exact)
0.0	0.0	0.0	1.0	1.0
0.05	0.0	−0.00532	1.2	1.21439
0.10	−0.01	−0.02270	1.42756	1.46032
0.15	−0.03188	−0.05447	1.68644	1.74231
0.20	−0.06779	−0.10335	1.98084	2.06545
0.25	−0.12023	−0.17247	2.31552	2.43552
0.30	−0.192013	−0.26543	2.69577	2.85904
0.35	−0.28640	−0.38639	3.12758	3.34338
0.40	−0.40710	−0.54011	3.61763	3.89682
0.45	−0.55834	−0.73203	4.17344	4.52876
0.50	−0.74493	−0.96841	4.80342	5.24975
0.55	−0.97234	−1.25639	5.51701	6.07171
0.60	−1.24681	−1.60412	6.32479	7.00806
0.65	−1.57529	−2.02091	7.23861	8.07394
0.70	−1.96609	−2.51737	8.27174	9.28638
0.75	−2.42798	−3.10558	9.43902	10.6645
0.80	−2.97133	−3.79926	10.7571	12.23
0.85	−3.60776	−4.61405	12.2446	14.0071
0.90	−4.35037	−5.56767	13.9222	16.0232
0.95	−5.214	−6.68027	15.8134	18.3088
1.00	−6.21537	−7.97468	17.944	20.8987

Runge-Kutta Method

Because we would like to be able to improve the approximation without using such a small value for h, we seek to improve the method. As with first-order equations, the Runge-Kutta method can be extended to systems. In this case, the recursive formula at each step is

$$\begin{cases} x_{n+1} = x_n + \dfrac{h}{6}(k_1 + 2k_2 + 2k_3 + k_4) \\ y_{n+1} = y_n + \dfrac{h}{6}(m_1 + 2m_2 + 2m_3 + m_4) \end{cases}$$

where

$$k_1 = f(t_n, x_n, y_n) \qquad\qquad m_1 = g(t_n, x_n, y_n)$$

$$k_2 = f\left(t_n + \frac{h}{2}, x_n + \frac{hk_1}{2}, y_n + \frac{hm_1}{2}\right) \quad m_2 = g\left(t_n + \frac{h}{2}, x_n + \frac{hk_1}{2}, y_n + \frac{hm_1}{2}\right)$$

$$k_3 = f\left(t_n + \frac{h}{2}, x_n + \frac{hk_2}{2}, y_n + \frac{hm_2}{2}\right) \quad m_3 = g\left(t_n + \frac{h}{2}, x_n + \frac{hk_2}{2}, y_n + \frac{hm_2}{2}\right)$$

$$k_4 = f(t_n + h, x_n + hk_3, y_n + hm_3) \qquad m_4 = g(t_n + h, x_n + hk_3, y_n + hm_3).$$

Example 2

Use the Runge-Kutta method to approximate the solution to the initial-value problem

$$\begin{cases} \dfrac{dx}{dt} = x - y + 1 \\ \dfrac{dy}{dt} = x + 3y + e^{-t} \\ x(0) = 0,\ y(0) = 1 \end{cases}$$

for $h = 0.1$. Compare these results with the exact solution to the system of equations as well as those obtained with Euler's method.

Solution   Because $f(x, y) = x - y + 1$, $g(x, y) = x + 3y + e^{-t}$, $t_0 = 0$, $x_0 = 0$, and $y_0 = 1$, we use the formulas

$$\begin{cases} x_{n+1} = x_n + \dfrac{h}{6}(k_1 + 2k_2 + 2k_3 + k_4) \\ y_{n+1} = y_n + \dfrac{h}{6}(m_1 + 2m_2 + 2m_3 + m_4) \end{cases}$$

where

$$k_1 = f(t_n, x_n, y_n) = x_n - y_n + 1 \qquad m_1 = g(t_n, x_n, y_n) = x_n + 3y_n + e^{-t_n}$$

$$k_2 = \left(x_n + \frac{hk_1}{2}\right) - \left(y_n + \frac{hm_1}{2}\right) + 1 \quad m_2 = \left(x_n + \frac{hk_1}{2}\right) + 3\left(y_n + \frac{hm_1}{2}\right) + e^{-(t_n+h/2)}$$

$$k_3 = \left(x_n + \frac{hk_2}{2}\right) - \left(y_n + \frac{hm_2}{2}\right) + 1 \quad m_3 = \left(x_n + \frac{hk_2}{2}\right) + 3\left(y_n + \frac{hm_2}{2}\right) + e^{-(t_n+h/2)}$$

$$k_4 = (x_n + hk_3) - (y_n + hm_3) + 1 \qquad m_4 = (x_n + hk_3) + 3(y_n + hm_3) + e^{-(t_n+h)}$$

For example, if $n = 0$, then

$$k_1 = x_0 - y_0 + 1 = 0 - 1 + 1 = 0$$
$$m_1 = x_0 + 3y_0 + e^{-t_0} = 0 + 3 + 1 = 4$$

$$k_2 = \left(x_0 + \frac{hk_1}{2}\right) - \left(y_0 + \frac{hm_1}{2}\right) + 1 = -1 - \frac{4(0.1)}{2} + 1 = -0.2$$

$$m_2 = \left(x_0 + \frac{hk_1}{2}\right) + 3\left(y_0 + \frac{hm_1}{2}\right) + e^{-(t_0+h/2)} = 3\left(1 + \frac{4(0.1)}{2}\right) + e^{-0.05} \approx 4.55123$$

$$k_3 = \left(x_0 + \frac{hk_2}{2}\right) - \left(y_0 + \frac{hm_2}{2}\right) + 1 = \frac{(0.1)(0.2)}{2} - 1 - \frac{(0.1)(4.55123)}{2} + 1$$
$$\approx -0.23756$$

$$m_3 = \left(x_0 + \frac{hk_2}{2}\right) + 3\left(y_0 + \frac{hm_2}{2}\right) + e^{-(t_0+h/2)}$$

$$= \frac{(0.1)(0.2)}{2} + 3\left(1 + \frac{(0.1)(4.55123)}{2}\right) + e^{-0.05} \approx 4.62391$$

$$k_4 = (x_0 + hk_3) - (y_0 + hm_3) + 1 = (0.1)(-0.23756) - 1 + (0.1)(4.62391) + 1$$
$$\approx -0.48615$$
$$m_4 = (x_0 + hk_3) + 3(y_0 + hm_3) + e^{-(t_0+h)}$$
$$= (0.1)(-0.23756) + 3(1 + (0.1)(4.62391)) + e^{-0.1} \approx 5.26826.$$

Therefore,

$$x_1 = x_0 + \frac{0.1}{6}(k_1 + 2k_2 + 2k_3 + k_4)$$

$$= 0 + \frac{0.1}{6}[0 + 2(-0.2) + 2(-0.23756) + -0.48615] \approx -0.0226878$$

and

$$y_1 = y_0 + \frac{0.1}{6}(m_1 + 2m_2 + 2m_3 + m_4)$$

$$= 1 + \frac{0.1}{6}[4 + 2(4.55123) + 2(4.62391) + 5.26826] \approx 1.46031.$$

In Table 8.5, we show the results obtained with this method and compare them to the exact values. Notice that the Runge-Kutta method is much more accurate than Euler's method. In fact, the Runge-Kutta with $h = 0.1$ is more accurate than Euler's method with $h = 0.05$. (Compare the results here to those given in Table 8.4.)

TABLE 8.5

t_n	x_n (approx)	x_n (exact)	y_n (approx)	y_n (exact)
0.0	0.0	0.0	1.0	1.0
0.1	−0.02269	−0.02270	1.46031	1.46032
0.2	−0.10332	−0.10335	2.06541	2.06545
0.3	−0.26538	−0.26543	2.85897	2.85904
0.4	−0.54002	−0.54011	3.8967	3.89682
0.5	−0.96827	−0.96841	5.24956	5.24975
0.6	−1.60391	−1.60412	7.00778	7.00806
0.7	−2.51707	−2.51737	9.28596	9.28638
0.8	−3.79882	−3.79926	12.2294	12.23
0.9	−5.56704	−5.56767	16.0223	16.0232
1.0	−7.97379	−7.97468	20.8975	20.8987

The Runge-Kutta method can be extended to systems of first-order equations so it can be used to solve higher order differential equations. This is accomplished by transforming the higher order equation into a system of first-order equations. We illustrate this with the pendulum equation that we have solved in several situations (by using the approximation $\sin x \approx x$).

Example 3

Use the Runge-Kutta method with $h = 0.1$ to approximate the solution to the nonlinear initial-value problem $x'' + \sin x = 0$, $x(0) = 0$, $x'(0) = 1$.

Solution   We begin by transforming the second-order equation into a system of first-order equations. We do this by letting $x' = y$, so $y' = x'' = -\sin x$. Hence, $f(t, x, y) = y$, and $g(t, x, y) = -\sin x$. With the Runge-Kutta method, we obtain the approximate values given in Table 8.6 under the heading RK. Also in Table 8.6 under the heading "linear," we give the corresponding values of the initial-value problem $x'' + x = 0$, $x(0) = 0$, $x'(0) = 1$. We approximate the nonlinear equation $x'' + \sin x = 0$ with $x'' + x = 0$ because $\sin x \approx x$ for small values of x. The solution of $x'' + x = 0$, $x(0) = 0$, $x'(0) = 1$ is found to be $x = \sin t$, so $y = x' = \cos t$. Because the use of the approximation $\sin x \approx x$ is linear, we expect the approximations obtained with the Runge-Kutta method (which is a fourth-order method) to be more accurate than those obtained by solving $x'' + x = 0$, $x(0) = 0$, $x'(0) = 1$.

TABLE 8.6

t_n	x_n (RK)	x_n (linear)	y_n (RK)	y_n (linear)
0.0	0.0	0.0	1.0	1.0
0.1	0.09983	0.09983	0.99500	0.99500
0.2	0.19867	0.198669	0.98013	0.98007
0.3	0.29553	0.29552	0.95566	0.95534
0.4	0.38950	0.389418	0.922061	0.92106
0.5	0.47966	0.47943	0.87994	0.87758
0.6	0.56523	0.56464	0.83002	0.82534
0.7	0.64544	0.64422	0.77309	0.764842
0.8	0.71964	0.71736	0.70999	0.69671
0.9	0.78726	0.78333	0.641545	0.62161
1.0	0.84780	0.84147	0.568569	0.54030

Detailed discussions regarding the error involved in using Euler's method or the Runge-Kutta method to approximate solutions to systems of equations can be found in advanced numerical analysis texts.

Computer Algebra Systems

Numerical and graphical solutions generated by computer algebra systems are useful in helping us observe and understand behavior of the solution(s) to a differential equation, especially when we do not wish to utilize numerical methods like Euler's method or the Runge-Kutta method.

Example 4

Rayleigh's equation is the nonlinear equation $x'' + \left(\frac{1}{3}(x')^2 - 1\right)x' + x = 0$ and arises in the study of the motion of a violin string. We write Rayleigh's equation as a system by letting $y = x'$. Then,

$$y' = x'' = -\left(\frac{1}{3}(x')^2 - 1\right)x' - x = -\left(\frac{1}{3}y^2 - 1\right)y - x$$

so Rayleigh's equation is equivalent to the system

$$\begin{cases} x' = y \\ y' = -\left(\frac{1}{3}y^2 - 1\right)y - x \end{cases}$$

We see that the only equilibrium point of this system is $(0, 0)$. (a) Classify the equilibrium point $(0, 0)$. (b) Is it possible to find $y_0 \neq 0$ so that the solution to the initial-value problem

$$
\begin{cases}
x' = y \\
y' = -\left(\dfrac{1}{3}y^2 - 1\right)y - x \\
x(0) = 0, \ y(0) = y_0
\end{cases}
$$

is periodic?

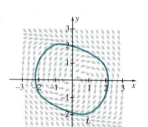

Figure 8.27

Solution   The associated linearized system about the point $(0, 0)$ is $\mathbf{X}' = \begin{pmatrix} 0 & 1 \\ -1 & 1 \end{pmatrix}\mathbf{X}$ and the eigenvalues of $\begin{pmatrix} 0 & 1 \\ -1 & 1 \end{pmatrix}$ are $\lambda_{1,2} = \dfrac{1}{2} \pm \dfrac{\sqrt{3}}{2}i$ so $(0, 0)$ is an unstable spiral. This result is confirmed by the direction field of the system for $-3 \leq x \leq 3$ and $-3 \leq y \leq 3$, shown in Figure 8.27.

In the direction field, we see that all solutions appear to tend to a closed curve, L. Indeed, we see that the graph of the solution that satisfies the initial conditions $x(0) = 0.1$ and $y(0) = 0$ in Figure 8.28(a) tends toward L. Choosing initial conditions outside of L yields the same result; Figure 8.28(b) shows the graph of the solution that satisfies $x(0) = 0$ and $y(0) = 3$.

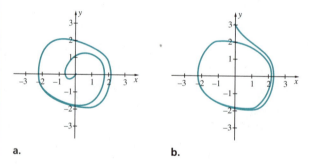

a.   b.

Figure 8.28   (a)–(b) All solutions tend to L

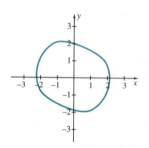

Figure 8.29   An isolated periodic solution like this is called a *limit cycle*. This limit cycle is *stable* because all solutions spiral into it.*

Can we find L? We use Figure 8.27 to approximate a point at which L intersects the y-axis. We obtain $y = 1.9$. We find a numerical solution that satisfies the initial conditions $x(0) = 0$ and $y(0) = 1.9$ and graph the result in Figure 8.29. The solution does appear to be periodic.

Approximate the period of the periodic solution in Example 4. (Hint: Graph $x(t)$ and $y(t)$ together for $0 \leq t \leq T$ for various values of T.)

---

* See texts like *Nonlinear Ordinary Differential Equations* (Second edition) by D. W. Jordan and P. Smith, which is published by Oxford University Press, 1987, for detailed discussions about limit cycles and their significance.

IN TOUCH WITH TECHNOLOGY

Find the exact solution of the following initial-value problems and then approximate the solution with the Runge-Kutta method using $h = 0.1$. Compare the results by graphing the two solutions.

1. $\begin{cases} x' = y \\ y' = 2x - y \end{cases}$, $x(0) = 0$, $y(0) = 1$

2. $\begin{cases} x' = y \\ y' = -4x - 13y \end{cases}$, $x(0) = -1$, $y(0) = 1$

3. $\begin{cases} x' = y \\ y' = -2y - x \end{cases}$, $x(0) = 2$, $y(0) = 0$

In addition to being able to implement algorithms like Euler's method and the Runge-Kutta method, most computer algebra systems contain built-in commands that allow you to generate numerical solutions of initial-value problems.

4. (FitzHugh-Nagumo Model) Under certain assumptions, the **FitzHugh-Nagumo equation,** which arises in the study of the impulses in a nerve fiber, can be written as the system of ordinary differential equations

$$\begin{cases} \dfrac{dV}{d\xi} = W \\[2mm] \dfrac{dW}{d\xi} = F(V) + R - uW, \\[2mm] \dfrac{dR}{d\xi} = \dfrac{\epsilon}{u}(bR - V - a) \end{cases}$$

where $F(V) = \frac{1}{3}V^3 - V$.* **(a)** Graph the solution to the FitzHugh-Nagumo equation that satisfies the initial conditions $V(0) = 1$, $W(0) = 0$, $R(0) = 1$ if $\epsilon = 0.08$, $a = 0.7$, $b = 0$, and $u = 1$. **(b)** Graph the solution that satisfies the initial conditions $V(0) = 1$, $W(0) = 0.5$, $R(0) = 0.5$ if $\epsilon = 0.08$, $a = 0.7$, $b = 0.8$, and $u = 0.6$. **(c)** Approximate the maximum and minimum values, if they exist, of V, W, and R in (a) and (b).

5. (a) Graph the direction field associated with the nonlinear system $\begin{cases} x' = y \\ y' = -\sin x \end{cases}$ for $-7 \leq x \leq 7$ and $-4 \leq y \leq 4$.

(b) (i) Approximate the solution to the initial-value problem $\begin{cases} x_1' = y_1 \\ y_1' = -\sin x_1 \\ x_1(0) = 0, \ y_1(0) = 1 \end{cases}$.

(ii) Graph $\begin{cases} x_1(t) \\ y_1(t) \end{cases}$ for $0 \leq t \leq 7$ and display the graph together with the direction field. Does it appear as though the vectors in the vector field are tangent to the solution curve?

(iii) Approximate the solution to the initial-value problem $\begin{cases} x_2' = y_2 \\ y_2' = -\sin x_2 \\ x_2(0) = 0, \ y_2(0) = 2 \end{cases}$.

(iv) Graph $\begin{cases} x_2(t) \\ y_2(t) \end{cases}$ for $0 \leq t \leq 7$ and display the graph together with the direction field. Does it appear as though the vectors in the vector field are tangent to the solution curve?

(v) Graph $\begin{cases} x_1(t) + x_2(t) \\ y_1(t) + y_2(t) \end{cases}$ for $0 \leq t \leq 7$ and display the graph together with the direction field. Does it appear as though the vectors in the vector field are tangent to this curve? Is $\begin{cases} x_1(t) + x_2(t) \\ y_1(t) + y_2(t) \end{cases}$ a solution to the system $\begin{cases} x' = y \\ y' = -\sin x \end{cases}$? Explain.

(c) Solve the initial-value problem $\begin{cases} x' = y \\ y' = -\sin x \\ x(0) = 0, \ y(0) = 3 \end{cases}$ and graph the solution parametrically for $0 \leq t \leq 7$. Is this the graph of $\begin{cases} x_1(t) + x_2(t), \\ y_1(t) + y_2(t) \end{cases}$?

(d) Is the Principle of Superposition valid for nonlinear systems? Explain.

---

* J. D. Murray, *Mathematical Biology,* Springer-Verlag, 1990, pp. 161–166. Alwyn C. Scott, ''The electrophysics of a nerve fiber,'' *Reviews of Modern Physics,* Vol. 47, No. 2 (April 1975) pp. 487–533.

In Exercises 1–6, use Euler's method with $h = 0.1$ to approximate the solution of the initial-value problem at the given value of t.

1. $\begin{cases} x' = -x + 2y + 5 \\ y' = 2x - y + 4 \end{cases}$, $x(0) = 1$, $y(0) = 0$, $t = 1$

2. $\begin{cases} x' = x + y + t \\ y' = x - y \end{cases}$, $x(0) = 1$, $y(0) = -1$, $t = 1$

*3. $\begin{cases} x' = 3x - 5y \\ y' = x - 2y + t^2 \end{cases}$, $x(0) = -1$, $y(0) = 0$, $t = 1$

4. $\begin{cases} x' = x + 2y + \cos t \\ y' = 5x - 2y \end{cases}$, $x(0) = 0$, $y(0) = 1$, $t = 1$

5. $\begin{cases} x' = x^2 - y \\ y' = x + y \end{cases}$, $x(0) = -1$, $y(0) = 0$, $t = 1$

6. $\begin{cases} x' = xy \\ y' = x - y \end{cases}$, $x(0) = 1$, $y(0) = 1$, $t = 1$

In Exercises 7–12, use the Runge-Kutta method with $h = 0.1$ to approximate the solution of the initial-value problem at the given value of t.

7. $\begin{cases} x' = x - 3y + e^t \\ y' = -x + 6y \end{cases}$, $x(0) = 0$, $y(0) = 1$, $t = 1$

8. $\begin{cases} x' = 4x - y \\ y' = -x + 5y + 6\sin t \end{cases}$, $x(0) = -1$, $y(0) = 0$, $t = 1$

*9. $\begin{cases} x' = x - 8y \\ y' = 3x - y + e^t \cos 2t \end{cases}$, $x(0) = 0$, $y(0) = 0$, $t = 1$

10. $\begin{cases} x' = 5x + y + \sqrt{t+1} \\ y' = x - 2y \end{cases}$, $x(0) = 0$, $y(0) = 0$, $t = 2$

11. $\begin{cases} x' = 2y \\ y' = -xy \end{cases}$, $x(0) = 0$, $y(0) = 1$, $t = 1$

12. $\begin{cases} x' = x\sqrt{y} \\ y' = x - y \end{cases}$, $x(1) = 1$, $y(1) = 1$, $t = 2$

In Exercises 13–18, use Euler's method with $h = 0.1$ to approximate the solution of the initial-value problem by transforming the second-order equation to a system of first-order equations. Compare the approximation with the exact solution at the given value of t.

13. $x'' + 3x' + 2x = 0$, $x(0) = 0$, $x'(0) = -3$, $t = 1$

14. $x'' + 4x' + 4x = 0$, $x(0) = 4$, $x'(0) = 0$, $t = 1$

*15. $x'' + 9x = 0$, $x(0) = 0$, $x'(0) = 3$, $t = 1$

16. $x'' + 4x' + 13x = 0$, $x(0) = 0$, $x'(0) = 12$, $t = 1$

17. $t^2 x'' + tx' + 16x = 0$, $x(1) = 0$, $x'(1) = 4$, $t = 2$

18. $t^2 x'' + 3tx' + x = 0$, $x(1) = 0$, $x'(1) = 2$, $t = 2$

In Exercises 19–24, use the Runge-Kutta method with $h = 0.1$ to approximate the solution of the initial-value problem in the earlier exercise. Compare the results obtained to those in Exercises 13–18.

19. Exercise 13   **20.** Exercise 14
*21. Exercise 15   **22.** Exercise 16
23. Exercise 17   **24.** Exercise 18

CHAPTER 8 SUMMARY
Concepts and Formulas

Section 8.1

System of Ordinary Differential Equations
A **system** of ordinary differential equations is a simultaneous set of equations that involves two or more dependent variables that depend on one independent variable. A **solution** to the system is a set of functions that satisfies each equation on some interval I.

Existence and Uniqueness Theorem
Operator Notation
$D = \dfrac{d}{dt}$ can be used to express

$$a_n \frac{d^n y}{dt^n} + a_{n-1}\frac{d^{n-1}y}{dt^{n-1}} + \cdots + a_1 \frac{dy}{dt} + a_0 y = f(t)$$

as $(a_n D^n + a_{n-1}D^{n-1} + \cdots + a_1 D + a_0)y = f(t)$.

Section 8.2

Matrix, Transpose of a Matrix

Scalar Multiplication, Matrix Addition, Matrix Multiplication

Identity Matrix (2 × 2)

$$\mathbf{I} = \begin{pmatrix} 1 & 0 \\ 0 & 1 \end{pmatrix}$$

Determinant of a 2 × 2 Matrix

$$\begin{vmatrix} a_{11} & a_{12} \\ a_{21} & a_{22} \end{vmatrix} = a_{11}a_{22} - a_{12}a_{21}$$

Inverse of a 2 × 2 Matrix

If $\mathbf{A} = \begin{pmatrix} a & b \\ c & d \end{pmatrix}$ and $|\mathbf{A}| = ad - bc \neq 0$, then

$$\mathbf{A}^{-1} = \frac{1}{ad - bc} \begin{pmatrix} d & -b \\ -c & a \end{pmatrix}.$$

Eigenvalues and Eigenvectors

A nonzero vector $\mathbf{x}$ is an **eigenvector** of the square matrix $\mathbf{A}$ if there is a number λ, called an **eigenvalue** of $\mathbf{A}$, so that $\mathbf{Ax} = \lambda\mathbf{x}$.

Characteristic Polynomial

The equation $\det(\mathbf{A} - \lambda\mathbf{I}) = 0$ is called the **characteristic equation** of $\mathbf{A}$; $\det(\mathbf{A} - \lambda\mathbf{I})$ is called the **characteristic polynomial** of $\mathbf{A}$; the roots of the characteristic polynomial of $\mathbf{A}$ are the eigenvalues of $\mathbf{A}$.

Eigenvalue of Multiplicity m

Suppose that $(\lambda - \lambda_1)^m$ where m is a positive integer is a factor of the characteristic polynomial of the $n \times n$ matrix $\mathbf{A}$ while $(\lambda - \lambda_1)^{m+1}$ is not a factor of this polynomial. Then $\lambda = \lambda_1$ is an **eigenvalue of multiplicity m.**

Derivative and Integral of a Matrix

The **derivative** of the $n \times m$ matrix

$$\mathbf{A}(t) = \begin{pmatrix} a_{11}(t) & a_{12}(t) & \cdots & a_{1m}(t) \\ a_{21}(t) & a_{22}(t) & \cdots & a_{2m}(t) \\ \vdots & \vdots & \ddots & \vdots \\ a_{n1}(t) & a_{n2}(t) & \cdots & a_{nm}(t) \end{pmatrix},$$

where $a_{ij}(t)$ is differentiable for all values of i and j, is

$$\frac{d}{dt}\mathbf{A}(t) = \begin{pmatrix} \dfrac{d}{dt}a_{11}(t) & \dfrac{d}{dt}a_{12}(t) & \cdots & \dfrac{d}{dt}a_{1m}(t) \\ \dfrac{d}{dt}a_{21}(t) & \dfrac{d}{dt}a_{22}(t) & \cdots & \dfrac{d}{dt}a_{2m}(t) \\ \vdots & \vdots & \ddots & \vdots \\ \dfrac{d}{dt}a_{n1}(t) & \dfrac{d}{dt}a_{n2}(t) & \cdots & \dfrac{d}{dt}a_{nm}(t) \end{pmatrix}.$$

The **integral** of $\mathbf{A}(t)$, where $a_{ij}(t)$ is integrable for all values of i and j, is

$$\int \mathbf{A}(t)dt = \begin{pmatrix} \int a_{11}(t)dt & \int a_{12}(t)dt & \cdots & \int a_{1m}(t)dt \\ \int a_{21}(t)dt & \int a_{22}(t)dt & \cdots & \int a_{2m}(t)dt \\ \vdots & \vdots & \ddots & \vdots \\ \int a_{n1}(t)dt & \int a_{n2}(t)dt & \cdots & \int a_{nm}(t)dt \end{pmatrix}.$$

Section 8.3

Solution Vector

A **solution vector** of the system $\mathbf{X}'(t) = \mathbf{A}(t)\mathbf{X}(t) + \mathbf{F}(t)$ on the interval I is an $n \times 1$ matrix of the form

$$\mathbf{X}(t) = \begin{pmatrix} x_1(t) \\ x_2(t) \\ \vdots \\ x_n(t) \end{pmatrix}$$

where the $x_i(t)$ are differentiable functions that satisfy $\mathbf{X}'(t) = \mathbf{A}(t)\mathbf{X}(t) + \mathbf{F}(t)$ on I.

Fundamental Set of Solutions

A set $\{\boldsymbol{\Phi}_i\}_{i=1}^n = \left\{ \begin{pmatrix} \Phi_{1i} \\ \Phi_{2i} \\ \vdots \\ \Phi_{ni} \end{pmatrix} \right\}_{i=1}^n$ of n linearly independent solution vectors of $\mathbf{X}'(t) = \mathbf{A}(t)\mathbf{X}(t)$ on an interval I is called a **fundamental set of solutions** on I.

Wronskian

$$W(\boldsymbol{\Phi}_1, \boldsymbol{\Phi}_2, \ldots \boldsymbol{\Phi}_n) = \det \begin{pmatrix} \Phi_{11} & \Phi_{12} & \cdots & \Phi_{1n} \\ \Phi_{21} & \Phi_{22} & \cdots & \Phi_{2n} \\ \vdots & \vdots & \ddots & \vdots \\ \Phi_{n1} & \Phi_{n2} & \cdots & \Phi_{nn} \end{pmatrix}$$

General Solution

A **general solution** of $\mathbf{X}'(t) = \mathbf{A}(t)\mathbf{X}(t)$ is $\mathbf{X}(t) = c_1\mathbf{\Phi}_1(t) + c_2\mathbf{\Phi}_2(t) + \cdots + c_n\mathbf{\Phi}_n(t)$ where $\{\mathbf{\Phi}_i\}_{i=1}^n$ is a set of n linearly independent solution vectors of the system.

Fundamental Matrix

$$\mathbf{\Phi}(t) = (\mathbf{\Phi}_1 \quad \mathbf{\Phi}_2 \quad \cdots \quad \mathbf{\Phi}_n) = \begin{pmatrix} \Phi_{11} & \Phi_{12} & \cdots & \Phi_{1n} \\ \Phi_{21} & \Phi_{22} & \cdots & \Phi_{2n} \\ \vdots & \vdots & \ddots & \vdots \\ \Phi_{n1} & \Phi_{n2} & \cdots & \Phi_{nn} \end{pmatrix}$$

is called a **fundamental matrix** of the system $\mathbf{X}'(t) = \mathbf{A}(t)\mathbf{X}(t)$ where $\{\mathbf{\Phi}_i\}_{i=1}^n$ is a set of n linearly independent solution vectors of the system.

Section 8.4

Distinct Real Eigenvalues

If $\mathbf{A}$ is an $n \times n$ matrix with n distinct real eigenvalues $\{\lambda_k\}_{k=1}^n$, a general solution to $\mathbf{X}' = \mathbf{AX}$ is the linear combination of the set of solutions $\{\mathbf{X}_1, \mathbf{X}_2, \ldots, \mathbf{X}_n\}$,

$$\mathbf{X}(t) = c_1\mathbf{v}_1 e^{\lambda_1 t} + c_2\mathbf{v}_2 e^{\lambda_2 t} + \cdots + c_n\mathbf{v}_n e^{\lambda_n t}.$$

Complex Conjugate Eigenvalues

If $\mathbf{A}$ has complex conjugate eigenvalues $\lambda_1 = \alpha + \beta i$ and $\lambda_2 = \alpha - \beta i$ and corresponding eigenvectors $\mathbf{v}_1 = \mathbf{a} + \mathbf{b}i$ and $\mathbf{v}_2 = \mathbf{a} - \mathbf{b}i$, two linearly independent solutions of $\mathbf{X}' = \mathbf{AX}$ are $\mathbf{X}_1(t) = e^{\alpha t}(\mathbf{a}\cos\beta t - \mathbf{b}\sin\beta t)$ and $\mathbf{X}_2(t) = e^{\alpha t}(\mathbf{a}\sin\beta t + \mathbf{b}\cos\beta t)$.

Repeated Eigenvalues

If the system $\mathbf{X}' = \mathbf{AX}$ has the repeated eigenvalue $\lambda_1 = \lambda_2$ with only one corresponding eigenvector $\mathbf{v}_1$, two linearly independent solutions corresponding to $\lambda_1 = \lambda_2$ are $\mathbf{X}_1 = \mathbf{v}_1 e^{\lambda_1 t}$ and $\mathbf{X}_2 = (\mathbf{v}_1 t + \mathbf{w}_2)e^{\lambda_1 t}$ where $\mathbf{w}_2$ satisfies $(\mathbf{A} - \lambda_1\mathbf{I})\mathbf{w}_2 = \mathbf{v}_1$.

Section 8.5

Variation of Parameters

A general solution to $\mathbf{X}' = \mathbf{AX} + \mathbf{F}(t)$ is

$$\mathbf{X}(t) = \mathbf{\Phi}(t)\mathbf{C} + \mathbf{X}_p(t) = \mathbf{\Phi}(t)\mathbf{C} + \mathbf{\Phi}(t)\int \mathbf{\Phi}^{-1}(t)\mathbf{F}(t)\,dt$$

where $\mathbf{\Phi}(t)$ is a fundamental matrix of the system $\mathbf{X}' = \mathbf{AX}$.

Section 8.7

Eigenvalues	Classification
Real: $\lambda_1 \leq \lambda_2 < 0$	Stable Node
Real: $\lambda_1 < 0 < \lambda_2$	Saddle
Real: $0 < \lambda_1 \leq \lambda_2$	Unstable Node
Complex: $\lambda_1 = \overline{\lambda_2} = \alpha + \beta i,\ \beta \neq 0,\ \alpha < 0$	Stable Spiral
Complex: $\lambda_1 = \overline{\lambda_2} = \alpha + \beta i,\ \beta \neq 0,\ \alpha > 0$	Unstable Spiral
Complex: $\lambda_1 = \overline{\lambda_2} = \alpha + \beta i,\ \beta \neq 0,\ \alpha = 0$	Center

Section 8.8

Euler's Method

The initial-value problem

$$\begin{cases} \dfrac{dx}{dt} = f(t, x, y) \\[2mm] \dfrac{dy}{dt} = g(t, x, y) \\[2mm] x(t_0) = x_0,\ y(t_0) = y_0 \end{cases}$$

is approximated at each step by the recursive relationship based on the Taylor expansion of x and y up to order h:

$$\begin{cases} x_{n+1} = x_n + hf(t_n, x_n, y_n) \\ y_{n+1} = y_n + hg(t_n, x_n, y_n) \end{cases}$$

where $t_n = t_0 + nh$, $n = 0, 1, 2, \ldots$.

Runge-Kutta Method

The Runge-Kutta method for systems uses the recursive formula at each step

$$\begin{cases} x_{n+1} = x_n + \dfrac{h}{6}(k_1 + 2k_2 + 2k_3 + k_4) \\[3mm] y_{n+1} = y_n + \dfrac{h}{6}(m_1 + 2m_2 + 2m_3 + m_4) \end{cases}$$

where

$k_1 = f(t_n, x_n, y_n)$

$k_2 = f\left(t_n + \dfrac{h}{2}, x_n + \dfrac{hk_1}{2}, y_n + \dfrac{hm_1}{2}\right)$

$k_3 = f\left(t_n + \dfrac{h}{2}, x_n + \dfrac{hk_2}{2}, y_n + \dfrac{hm_2}{2}\right)$

$k_4 = f(t_n + h, x_n + hk_3, y_n + hm_3)$

and

$m_1 = g(t_n, x_n, y_n)$

$m_2 = g\left(t_n + \dfrac{h}{2}, x_n + \dfrac{hk_1}{2}, y_n + \dfrac{hm_1}{2}\right)$

$m_3 = g\left(t_n + \dfrac{h}{2}, x_n + \dfrac{hk_2}{2}, y_n + \dfrac{hm_2}{2}\right)$

$m_4 = g(t_n + h, x_n + hk_3, y_n + hm_3).$

CHAPTER 8 REVIEW EXERCISES

In Exercises 1–9, find the eigenvalues and corresponding eigenvectors of A.

1. $A = \begin{pmatrix} -1 & 6 \\ 6 & 8 \end{pmatrix}$

2. $A = \begin{pmatrix} -3 & 6 \\ 4 & -1 \end{pmatrix}$

*3. $A = \begin{pmatrix} -1 & 2 \\ -1 & -4 \end{pmatrix}$

4. $A = \begin{pmatrix} 3 & -5 \\ 2 & 5 \end{pmatrix}$

5. $A = \begin{pmatrix} 1 & 0 & 1 \\ 0 & 1 & 0 \\ -1 & -2 & 3 \end{pmatrix}$

6. $A = \begin{pmatrix} 3 & -2 & -3 \\ -2 & 0 & 1 \\ 0 & 0 & -3 \end{pmatrix}$

*7. $A = \begin{pmatrix} 3 & 1 & 1 \\ -1 & 3 & 0 \\ 2 & 0 & 3 \end{pmatrix}$

8. $A = \begin{pmatrix} 0 & -3 & -2 \\ -4 & 3 & 4 \\ -3 & 0 & -3 \end{pmatrix}$

9. $A = \begin{pmatrix} -4 & 3 & 0 \\ 0 & -2 & 3 \\ 0 & -3 & -2 \end{pmatrix}$

In Exercises 10–34, find a general solution to each system or solve the initial-value problem.

10. $X' = \begin{pmatrix} -1 & -5 \\ 2 & 6 \end{pmatrix} X$

11. $X' = \begin{pmatrix} 2 & -4 \\ -1 & 5 \end{pmatrix} X$

12. $X' = \begin{pmatrix} -1 & -6 \\ 0 & -7 \end{pmatrix} X, \; X(0) = \begin{pmatrix} -1 \\ 2 \end{pmatrix}$

*13. $X' = \begin{pmatrix} 2 & 0 \\ 1 & -1 \end{pmatrix} X, \; X(0) = \begin{pmatrix} 3 \\ 3 \end{pmatrix}$

14. $X' = \begin{pmatrix} -4 & 5 \\ -5 & 4 \end{pmatrix} X$

15. $X' = \begin{pmatrix} -\frac{3}{2} & 5 \\ -\frac{1}{2} & \frac{3}{2} \end{pmatrix} X$

16. $X' = \begin{pmatrix} 4 & -5 \\ 4 & -4 \end{pmatrix} X$

*17. $X' = \begin{pmatrix} -4 & -1 \\ 5 & -2 \end{pmatrix} X$

18. $X' = \begin{pmatrix} 4 & 2 \\ -1 & 2 \end{pmatrix} X$

19. $X' = \begin{pmatrix} 0 & 8 \\ -2 & 0 \end{pmatrix} X$

20. $X' = \begin{pmatrix} 1 & 1 \\ -5 & -1 \end{pmatrix} X, \; X(0) = \begin{pmatrix} 1 \\ -3 \end{pmatrix}$

*21. $X' = \begin{pmatrix} 2 & -1 \\ 18 & -4 \end{pmatrix} X, \; X(0) = \begin{pmatrix} 2 \\ 3 \end{pmatrix}$

22. $X' = \begin{pmatrix} 0 & 2 \\ -2 & -4 \end{pmatrix} X$

23. $X' = \begin{pmatrix} 8 & -1 \\ 1 & 6 \end{pmatrix} X$

24. $X' = \begin{pmatrix} 1 & -3 \\ 1 & 5 \end{pmatrix} X + \begin{pmatrix} -2e^{4t} \\ 2e^{4t} \end{pmatrix}$

*25. $X' = \begin{pmatrix} 8 & -9 \\ 1 & -2 \end{pmatrix} X - 8 \begin{pmatrix} 1 \\ -1 \end{pmatrix} e^{7t}$

26. $\begin{cases} x' = 3x + 3y - 1 \\ y' = -4x - 10y + 1 \end{cases}$

27. $\begin{cases} x' = 3x + 4y + 1 \\ y' = 2x + y - 1 \end{cases}$

28.
$$\begin{cases} x' = -3x - \dfrac{5}{2}y + \dfrac{5}{2}\sin 2t \\ y' = 4x + 3y + 2\cos 2t - 3\sin 2t \\ x(0) = -1,\ y(0) = 1 \end{cases}$$

*29. $\mathbf{X}' = \begin{pmatrix} 0 & -1 \\ 1 & 0 \end{pmatrix}\mathbf{X} + \begin{pmatrix} -\sin t \\ \cos t \end{pmatrix},\ \mathbf{X}(0) = \begin{pmatrix} 1 \\ 0 \end{pmatrix}$

30.
$$\begin{cases} y'' = -x - 2y & x(0) = 0,\ x'(0) = 1, \\ x'' = -2x - 4y' & y(0) = 0,\ y'(0) = 0 \end{cases}$$

31.
$$\begin{cases} y'' - 3y' = -x' - 2y + x & x(0) = 0,\ x'(0) = 1, \\ y' + x' = 2y - x & y(0) = 0,\ y'(0) = -1 \end{cases}$$

32.
$$\begin{cases} x'' = -3x + y + \cos 4t & x(0) = 0,\ x'(0) = 0, \\ y'' = 2x - 2y & y(0) = 0,\ y'(0) = 0 \end{cases}$$

*33.
$$\begin{cases} x' = x - 2y \\ y' = 3y + f(t) \\ x(0) = 0,\ y(0) = 0 \end{cases},\ f(t) = \begin{cases} 1, & 0 \le t < 2 \\ 0, & t \ge 2 \end{cases}$$

34.
$$\begin{cases} x' = -x + \dfrac{1}{2}y + f(t) \\ y' = -\dfrac{3}{4}x + \dfrac{1}{4}y \\ x(0) = 0,\ y(0) = 0 \end{cases},\ f(t) = \begin{cases} -1, & 0 \le t < 1 \\ 1, & 1 \le t < 2 \\ 0, & t \ge 2 \end{cases}$$

*35. (Modeling Testosterone Production) The level of testosterone in men can be modeled by the system of

delay equations

$$\begin{cases} \dfrac{dR}{dt} = f(T) - b_1 R \\[1mm] \dfrac{dL}{dt} = g_1 R - b_2 L \\[1mm] \dfrac{dT}{dt} = g_2 L(t - \tau) - b_3 T \end{cases} \qquad .*$$

Show that if $f(0) > 0$ and $f(T)$ is a one-to-one decreasing function, then the equilibrium point of this system is (R_0, L_0, T_0), where

$$L_0 = \frac{b_3 T_0}{g_2}, \quad R_0 = \frac{b_3 b_2 T_0}{g_1 g_2}, \quad \text{and}$$

$$f(T_0) - \frac{b_1 b_2 b_3 T_0}{g_1 g_2} = 0$$

and that the associated linearized system is

$$\begin{cases} \dfrac{dx}{dt} = f'(T_0)z - b_1 x \\[1mm] \dfrac{dy}{dt} = g_1 x - b_2 y \\[1mm] \dfrac{dz}{dt} = g_2 y(t - \tau) - b_3 z \end{cases}$$

Differential Equations at Work:
Controlling the Spread of a Disease

● See the section **Differential Equations at Work: Modeling the Spread of a Disease** for an introduction to the terminology used in this section.

If a person becomes immune to a disease after recovering from it, and births and deaths in the population are not taken into account, then the percent of persons susceptible to becoming infected with the disease, $S(t)$, the percent of people in the population infected with the disease, $I(t)$, and the percent of the population recovered and immune to the disease, $R(t)$, can be modeled by the system

$$\begin{cases} S'(t) = -\lambda SI \\ I'(t) = \lambda SI - \gamma I \\ R'(t) = \gamma I \\ S(0) = S_0,\ I(0) = I_0,\ R(0) = 0 \end{cases}.$$

---

* J. D. Murray, *Mathematical Biology,* Springer-Verlag, 1990, pp. 166–175.

Because $S(t) + I(t) + R(t) = 1$, once we know S and I, we can compute R with

$$R(t) = 1 - S(t) - I(t).$$

This model is called an **SIR model without vital dynamics** because once a person has had the disease he becomes immune to it, and because births and deaths are not taken into consideration. This model might be used to model diseases that are **epidemic** to a population: those diseases that persist in a population for short periods of time (less than one year). Such diseases typically include influenza, measles, rubella, and chicken pox.

1. Show that if $S_0 < \dfrac{\gamma}{\lambda}$, the disease dies out, while an epidemic results if $S_0 > \dfrac{\gamma}{\lambda}$.

2. Show that $\dfrac{dI}{dS} = -\dfrac{(\lambda S - \gamma)I}{\lambda SI} = -1 + \dfrac{\rho}{S}$, where $\rho = \dfrac{\gamma}{\lambda}$, has solution

$$I + S - \rho \ln S = I_0 + S_0 - \rho \ln S_0.$$

3. What is the maximum value of I?

   When diseases persist in a population for long periods of time, births and deaths must be taken into consideration. If a person becomes immune to a disease after recovering from it and births and deaths in the population are taken into account, then the percent of persons susceptible to becoming infected with the disease, $S(t)$, and the percent of people in the population infected with the disease, $I(t)$, can be modeled by the system

$$\begin{cases} S'(t) = -\lambda SI + \mu - \mu S \\ I'(t) = \lambda SI - \gamma I - \mu I \quad .^* \\ S(0) = S_0, I(0) = I_0 \end{cases}$$

This model is called an **SIR model with vital dynamics** because once a person has had the disease he becomes immune to it, and because births and deaths are taken into consideration. This model might be used to model diseases that are **endemic** to a population: those diseases that persist in a population for long periods of time (ten or twenty years). Smallpox is an example of a disease that was endemic until it was eliminated in 1977.

4. Show that the equilibrium points of the system $\begin{cases} S'(t) = -\lambda SI + \mu - \mu S \\ I'(t) = \lambda SI - \gamma I - \mu I \end{cases}$ are

   $S = 1, I = 0$ and $S = \dfrac{\gamma + \mu}{\lambda}$, $I = \dfrac{\mu[\lambda - (\gamma + \mu)]}{\lambda(\gamma + \mu)}$.

   Because $S(t) + I(t) + R(t) = 1$, it follows that $S(t) + I(t) \leq 1$.

---

* Herbert W. Hethcote, "Three Basic Epidemiological Models," *Applied Mathematical Ecology,* edited by Simon A. Levin, Thomas G. Hallan, and Louis J. Gross, Springer-Verlag (1989) pp. 119–143. Roy M. Anderson and Robert M. May, "Directly Transmitted Infectious Diseases: Control by Vaccination," *Science,* Volume 215 (February 26, 1982) pp. 1053–1060. J. D. Murray, *Mathematical Biology,* Springer-Verlag, 1990, pp. 611–618.

5. Use the fact that $S(t) + I(t) \le 1$ to determine conditions on γ, μ, and λ so that the system $\begin{cases} S'(t) = -\lambda SI + \mu - \mu S \\ I'(t) = \lambda SI - \gamma I - \mu I \end{cases}$ has the equilibrium point $S = \dfrac{\gamma + \mu}{\lambda}$, $I = \dfrac{\mu[\lambda - (\gamma + \mu)]}{\lambda(\gamma + \mu)}$. In this case, classify the equilibrium point.

6. Use the fact that $S(t) + I(t) \le 1$ to determine conditions on γ, μ, and λ so that the system $\begin{cases} S'(t) = -\lambda SI + \mu - \mu S \\ I'(t) = \lambda SI - \gamma I - \mu I \end{cases}$ does *not* have the equilibrium point $S = \dfrac{\gamma + \mu}{\lambda}$, $I = \dfrac{\mu[\lambda - (\gamma + \mu)]}{\lambda(\gamma + \mu)}$.

The following table shows the average infectious period and typical contact numbers for several diseases during certain epidemics.

Disease	Infectious Period (Average) $\dfrac{1}{\gamma}$	γ	Typical contact number σ
Measles	6.5	0.153846	14.9667
Chicken pox	10.5	0.0952381	11.3
Mumps	19	0.0526316	8.1
Scarlet fever	17.5	0.0571429	8.5

Let us assume that the average lifetime, $\dfrac{1}{\mu}$, is 70 so that $\mu = 0.0142857$.

7. For each of the diseases listed in the following table, use the formula $\sigma = \dfrac{\lambda}{\gamma + \mu}$ to calculate the daily contact rate λ.

Disease	λ
Measles	
Chicken pox	
Mumps	
Scarlet fever	

Diseases like those listed above can be controlled once an effective and inexpensive vaccine has been developed. It is virtually impossible to vaccinate everybody against a disease; we would like to determine the percentage of a population that needs to be vaccinated to eliminate a disease from the population under consideration. A population of people has **herd immunity** to a disease when enough people are immune to the disease so that if it is introduced into the population, it will not spread throughout the population. In order to have herd immunity, an infected person must infect less than one uninfected person during the time the person is infectious. Thus, we must have

$$\sigma S < 1.$$

Because $I + S + R = 1$, when $I = 0$ we have that $S = 1 - R$; consequently, herd immunity is achieved when

$$\sigma(1 - R) < 1$$
$$\sigma - \sigma R < 1$$
$$-\sigma R < 1 - \sigma$$
$$R > \frac{\sigma - 1}{\sigma} = 1 - \frac{1}{\sigma}.$$

8. For each of the diseases listed in the following table, estimate the minimum percentage of a population that needs to be vaccinated to achieve herd immunity.

Disease	Minimum Value of R to Achieve Herd Immunity
Measles	
Chicken pox	
Mumps	
Scarlet fever	

9. Using the values obtained in the previous exercises, for each disease in the tables graph the direction field and several solutions $\left(I(t),\ S(t),\ R(t),\ \text{and}\ \begin{Bmatrix} S(t) \\ I(t) \end{Bmatrix} \right)$ using both models. Discuss scenarios in which each model is valid and note any significant differences between the two models.

10. What are some possible ways that an epidemic can be controlled?

9

APPLICATIONS OF SYSTEMS OF ORDINARY DIFFERENTIAL EQUATIONS

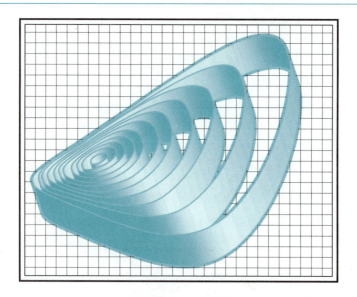

9.1 Mechanical and Electrical Problems with First-Order Linear Systems

L-R-C Circuits with Loops • *L-R-C* Circuits with One Loop • *L-R-C* Circuits with the Two Loops • Spring-Mass Systems

L-R-C Circuits with Loops

As indicated in Chapter 5, an electrical circuit can be modeled with a linear ordinary differential equation with constant coefficients. In this section, we illustrate how a circuit involving loops can be described as a system of linear ordinary differential equations with constant coefficients. This derivation is based on the following principles.

As was the case in Chapter 5, we use the following standard symbols for the components of the circuit:

$$I(t) = \text{current where } I(t) = \frac{dQ}{dt}(t), \ Q(t) = \text{charge}, \ R = \text{resistance}, \ C = \text{capacitance},$$

$$E = \text{voltage, and } L = \text{inductance}.$$

The relationships corresponding to the drops in voltage in the various components of the circuit are restated in Table 9.1.

TABLE 9.1

Circuit Element	Voltage Drop
Inductor	$L\dfrac{dI}{dt}$
Resistor	RI
Capacitor	$\dfrac{1}{C}Q$
Voltage Source	$-E(t)$

L-R-C Circuit with One Loop

In determining the drops in voltage around the circuit, we consistently add the voltages in the clockwise direction. The positive direction is from the negative symbol toward the positive symbol associated with the voltage source. In summing the voltage drops encountered in the circuit, a drop across a component is added to the sum if the positive direction through the component agrees with the clockwise direction. Otherwise this drop is subtracted. In the case of the *L-R-C* circuit with one loop involving each type of component shown in Figure 9.1, the current is equal around the circuit by Kirchhoff's current law.

Also, by Kirchhoff's voltage law, we have

a.

b.

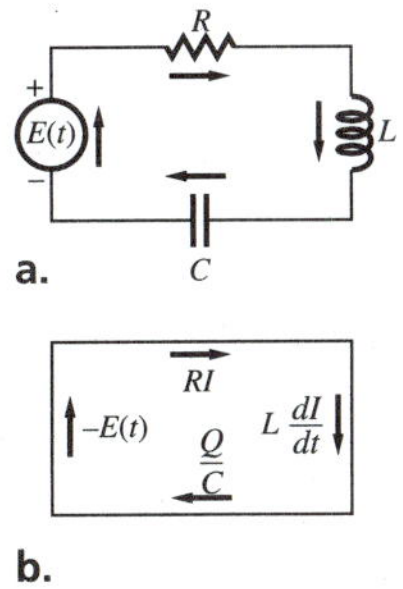

Figure 9.1 (a)–(b) A Simple *L-R-C* Circuit

$$RI + L\frac{dI}{dt} + \frac{1}{C}Q - E(t) = 0.$$

Solving this equation for dI/dt and using the relationship $dQ/dt = I$, we have the system of differential equations

$$\begin{cases} \dfrac{dQ}{dt} = I \\[2ex] \dfrac{dI}{dt} = -\dfrac{1}{LC}Q - \dfrac{R}{L}I + \dfrac{E(t)}{L} \end{cases}$$

with initial conditions $Q(0) = Q_0$ and $I(0) = I_0$ on charge and current, respectively. The method of variation of parameters (for systems) can be used to solve problems of this type.

Example 1

Determine the charge and current in the *L-R-C* circuit with $L = 1$ henry, $R = 2$ ohms, $C = \frac{4}{3}$ farads and $E(t) = e^{-t}$ if $Q(0) = 1$ and $I(0) = 1$.

Solution   We begin by modeling the circuit with a system of differential equations. In this case, we have

$$\begin{cases} \dfrac{dQ}{dt} = I \\[2ex] \dfrac{dI}{dt} = -\dfrac{3}{4}Q - 2I + e^{-t} \end{cases}$$

with initial conditions $Q(0) = 1$ and $I(0) = 1$. We can write this nonhomogeneous system in matrix form as

$$\begin{pmatrix} \dfrac{dQ}{dt} \\[2ex] \dfrac{dI}{dt} \end{pmatrix} = \begin{pmatrix} 0 & 1 \\[1ex] -\dfrac{3}{4} & -2 \end{pmatrix}\begin{pmatrix} Q \\ I \end{pmatrix} + \begin{pmatrix} 0 \\ e^{-t} \end{pmatrix}.$$

The eigenvalues of the corresponding homogeneous system are $\lambda_1 = -\frac{1}{2}$ and $\lambda_2 = -\frac{3}{2}$ with corresponding eigenvectors $\mathbf{v}_1 = \begin{pmatrix} -2 \\ 1 \end{pmatrix}$ and $\mathbf{v}_2 = \begin{pmatrix} 2 \\ -3 \end{pmatrix}$, respectively. Thus, a fundamental matrix is

$$\mathbf{\Phi}(t) = \begin{pmatrix} -2e^{-t/2} & 2e^{-3t/2} \\[1ex] e^{-t/2} & -3e^{-3t/2} \end{pmatrix}$$

and

$$\mathbf{\Phi}^{-1}(t) = \begin{pmatrix} -\dfrac{3}{4}e^{t/2} & -\dfrac{1}{2}e^{t/2} \\[2ex] -\dfrac{1}{4}e^{3t/2} & -\dfrac{1}{2}e^{3t/2} \end{pmatrix}.$$

Then, by variation of parameters, if we let $\mathbf{X}(t) = \begin{pmatrix} Q(t) \\ I(t) \end{pmatrix}$ we have

$$\mathbf{X}(t) = \mathbf{\Phi}(t)\mathbf{\Phi}^{-1}(0)\mathbf{X}(0) + \mathbf{\Phi}(t)\int_0^t \mathbf{\Phi}^{-1}(u)\mathbf{F}(u)\,du$$

$$= \begin{pmatrix} -2e^{-t/2} & 2e^{-3t/2} \\ e^{-t/2} & -3e^{-3t/2} \end{pmatrix} \begin{pmatrix} -\dfrac{3}{4} & -\dfrac{1}{2} \\ -\dfrac{1}{4} & -\dfrac{1}{2} \end{pmatrix} \begin{pmatrix} 1 \\ 1 \end{pmatrix}$$

$$+ \begin{pmatrix} -2e^{-t/2} & 2e^{-3t/2} \\ e^{-t/2} & -3e^{-3t/2} \end{pmatrix} \int_0^t \begin{pmatrix} -\dfrac{3}{4}e^{u/2} & -\dfrac{1}{2}e^{u/2} \\ -\dfrac{1}{4}e^{3u/2} & -\dfrac{1}{2}e^{3u/2} \end{pmatrix} \begin{pmatrix} 0 \\ e^{-u} \end{pmatrix} du$$

$$= \begin{pmatrix} -2e^{-t/2} & 2e^{-3t/2} \\ e^{-t/2} & -3e^{-3t/2} \end{pmatrix} \begin{pmatrix} -\dfrac{5}{4} \\ -\dfrac{3}{4} \end{pmatrix} + \begin{pmatrix} -2e^{-t/2} & 2e^{-3t/2} \\ e^{-t/2} & -3e^{-3t/2} \end{pmatrix} \int_0^t \begin{pmatrix} -\dfrac{1}{2}e^{-u/2} \\ -\dfrac{1}{2}e^{u/2} \end{pmatrix} du$$

$$= \begin{pmatrix} \dfrac{5}{2}e^{-t/2} - \dfrac{3}{2}e^{-3t/2} \\ -\dfrac{5}{4}e^{-t/2} + \dfrac{9}{4}e^{-3t/2} \end{pmatrix} + \begin{pmatrix} -4e^{-t} + 2e^{-t/2} + 2e^{-3t/2} \\ 4e^{-t} - e^{-t/2} - 3e^{-3t/2} \end{pmatrix}$$

$$= \begin{pmatrix} \dfrac{9}{2}e^{-t/2} + \dfrac{1}{2}e^{-3t/2} - 4e^{-t} \\ -\dfrac{9}{4}e^{-t/2} - \dfrac{3}{4}e^{-3t/2} + 4e^{-t} \end{pmatrix}.$$

We plot the solution $\mathbf{X}(t) = \begin{pmatrix} Q(t) \\ I(t) \end{pmatrix}$ parametrically in Figure 9.2(a). Notice that $\lim_{t\to\infty} Q(t) = \lim_{t\to\infty} I(t) = 0$, so the solution approaches $(0, 0)$ as t increases.

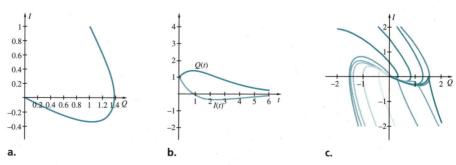

a.   b.   c.

Figure 9.2   (a)–(c)

We also plot $Q(t)$ and $I(t)$ simultaneously in Figure 9.2(b). Finally, in Figure 9.2(c), we graph $\mathbf{X}(t) = \begin{pmatrix} Q(t) \\ I(t) \end{pmatrix}$ for other initial conditions.

In Example 1, do the limits $\lim_{t \to \infty} Q(t) = \lim_{t \to \infty} I(t) = 0$ hold for all choices of the initial conditions?

L-R-C Circuit with Two Loops

The differential equations that model the circuit become more difficult to derive as the number of loops in the circuit increase. For example, consider the circuit that contains two loops as shown in Figure 9.3.

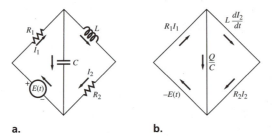

a.   **b.**

Figure 9.3   (a)–(b) A Two-Loop Circuit

In this case, the current through the capacitor is equivalent to $I_1 - I_2$. Summing the voltage drops around each loop, we obtain the system of equations

$$\begin{cases} R_1 I_1 + \dfrac{1}{C}Q - E(t) = 0 \\[2mm] L\dfrac{dI_2}{dt} + R_2 I_2 - \dfrac{1}{C}Q = 0 \end{cases}$$

Solving the first equation for I_1 yields $I_1 = \dfrac{1}{R_1}E(t) - \dfrac{1}{R_1 C}Q$. Using the relationship $\dfrac{dQ}{dt} = I = I_1 - I_2$ gives us the system

$$\begin{cases} \dfrac{dQ}{dt} = -\dfrac{1}{R_1 C}Q - I_2 + \dfrac{E(t)}{R_1} \\[4mm] \dfrac{dI_2}{dt} = \dfrac{1}{LC}Q - \dfrac{R_2}{L}I_2 \end{cases}$$

Example 2

Find $Q(t)$, $I(t)$, $I_1(t)$, and $I_2(t)$ in the L-R-C circuit with two loops given that $R_1 = R_2 = 1$ ohm, $C = 1$ farad, $L = 1$ henry, and $E(t) = e^{-t}$ if $Q(0) = 1$ and $I_2(0) = 3$.

Solution The nonhomogeneous system that models this circuit is

$$\begin{cases} \dfrac{dQ}{dt} = -Q - I_2 + e^{-t} \\[2mm] \dfrac{dI_2}{dt} = Q - I_2 \end{cases}$$

with initial conditions $Q(0) = 1$ and $I_2(0) = 3$. As in Example 1, we use the method of variation of parameters to solve the problem. In matrix form this system is

$$\begin{pmatrix} \dfrac{dQ}{dt} \\[2mm] \dfrac{dI_2}{dt} \end{pmatrix} = \begin{pmatrix} -1 & -1 \\ 1 & -1 \end{pmatrix} \begin{pmatrix} Q \\ I_2 \end{pmatrix} + \begin{pmatrix} e^{-t} \\ 0 \end{pmatrix}.$$

The eigenvalues of the corresponding homogeneous system are $\lambda_{1,2} = -1 \pm i$ and an eigenvector corresponding to $\lambda_1 = -1 + i$ is $\mathbf{v}_1 = \begin{pmatrix} i \\ 1 \end{pmatrix} = \begin{pmatrix} 0 \\ 1 \end{pmatrix} + i\begin{pmatrix} 1 \\ 0 \end{pmatrix}$. Two linearly independent solutions of the corresponding homogeneous system are

$$\mathbf{X}_1(t) = \begin{pmatrix} Q(t) \\ I_2(t) \end{pmatrix} = e^{-t} \cos t \begin{pmatrix} 0 \\ 1 \end{pmatrix} - e^{-t} \sin t \begin{pmatrix} 1 \\ 0 \end{pmatrix} = \begin{pmatrix} -e^{-t} \sin t \\ e^{-t} \cos t \end{pmatrix}$$

and

$$\mathbf{X}_2(t) = \begin{pmatrix} Q(t) \\ I_2(t) \end{pmatrix} = e^{-t} \sin t \begin{pmatrix} 0 \\ 1 \end{pmatrix} + e^{-t} \cos t \begin{pmatrix} 1 \\ 0 \end{pmatrix} = \begin{pmatrix} e^{-t} \cos t \\ e^{-t} \sin t \end{pmatrix},$$

so a fundamental matrix is

$$\mathbf{\Phi}(t) = \begin{pmatrix} -e^{-t} \sin t & e^{-t} \cos t \\ e^{-t} \cos t & e^{-t} \sin t \end{pmatrix}$$

and

$$\mathbf{\Phi}^{-1}(t) = \begin{pmatrix} -e^{t} \sin t & e^{t} \cos t \\ e^{t} \cos t & e^{t} \sin t \end{pmatrix}.$$

Therefore,

$$\mathbf{X}(t) = \mathbf{\Phi}(t)\mathbf{\Phi}^{-1}(0)\mathbf{X}(0) + \mathbf{\Phi}(t) \int_0^t \mathbf{\Phi}^{-1}(u)\mathbf{F}(u)\, du$$

$$= \begin{pmatrix} -e^{-t} \sin t & e^{-t} \cos t \\ e^{-t} \cos t & e^{-t} \sin t \end{pmatrix} \begin{pmatrix} 0 & 1 \\ 1 & 0 \end{pmatrix} \begin{pmatrix} 1 \\ 3 \end{pmatrix}$$

$$+ \begin{pmatrix} -e^{-t} \sin t & e^{-t} \cos t \\ e^{-t} \cos t & e^{-t} \sin t \end{pmatrix} \int_0^t \begin{pmatrix} -e^{u} \sin u & e^{u} \cos u \\ e^{u} \cos u & e^{u} \sin u \end{pmatrix} \begin{pmatrix} e^{-u} \\ 0 \end{pmatrix} du$$

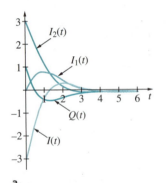

a.

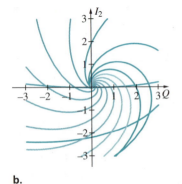

b.

Figure 9.4   (a)–(b)

$$= \begin{pmatrix} -e^{-t}\sin t & e^{-t}\cos t \\ e^{-t}\cos t & e^{-t}\sin t \end{pmatrix}\begin{pmatrix} 3 \\ 1 \end{pmatrix} + \begin{pmatrix} -e^{-t}\sin t & e^{-t}\cos t \\ e^{-t}\cos t & e^{-t}\sin t \end{pmatrix}\int_0^t \begin{pmatrix} -\sin u \\ \cos u \end{pmatrix} du$$

$$= \begin{pmatrix} e^{-t}\cos t - 3e^{-t}\sin t \\ 3e^{-t}\cos t + e^{-t}\sin t \end{pmatrix} + \begin{pmatrix} -e^{-t}\sin t & e^{-t}\cos t \\ e^{-t}\cos t & e^{-t}\sin t \end{pmatrix}\begin{pmatrix} \cos t - 1 \\ \sin t \end{pmatrix}$$

$$= \begin{pmatrix} e^{-t}\cos t - 3e^{-t}\sin t \\ 3e^{-t}\cos t + e^{-t}\sin t \end{pmatrix} + \begin{pmatrix} e^{-t}\sin t \\ e^{-t} - e^{-t}\cos t \end{pmatrix}$$

$$= \begin{pmatrix} e^{-t}\cos t - 2e^{-t}\sin t \\ 2e^{-t}\cos t + e^{-t} + e^{-t}\sin t \end{pmatrix}.$$

Because $dQ/dt = I$ and $Q(t) = e^{-t}\cos t - 2e^{-t}\sin t$, differentiation yields

$$I(t) = -e^{-t}\cos t - e^{-t}\sin t + 2e^{-t}\sin t - 2e^{-t}\cos t = -3e^{-t}\cos t + e^{-t}\sin t.$$

Also, because $I_1(t) = I(t) + I_2(t)$,

$$I_1(t) = I(t) + I_2(t) = -3e^{-t}\cos t + e^{-t}\sin t + e^{-t}\sin t + 2e^{-t}\cos t + e^{-t}$$
$$= -e^{-t}\cos t + 2e^{-t}\sin t + e^{-t}.$$

We graph $Q(t)$, $I_1(t)$, $I_2(t)$, and $I(t)$ together in Figure 9.4(a). In Figure 9.4(b), we graph $\begin{cases} Q(t) \\ I_2(t) \end{cases}$ parametrically to show the phase plane for the system of non-homogeneous equations using several different initial conditions. Notice that some of the solutions overlap, which does not occur if a system is homogeneous.

🔧   *Find the limit of $Q(t)$, $I_1(t)$, $I_2(t)$, and $I(t)$ as $t \to \infty$. Does a change in initial conditions affect these limits?*

Spring-Mass Systems

The displacement of a mass attached to the end of a spring was modeled with a second-order linear differential equation with constant coefficients in Chapter 5. This situation can be expressed as a system of first-order ordinary differential equations as well. Recall that if there is no external forcing function, the second-order differential equation that models the situation is

$$m\frac{d^2x}{dt^2} + c\frac{dx}{dt} + kx = 0,$$

where m is the mass of the object attached to the end of the spring, c is the damping coefficient, and k is the spring constant found with Hooke's law. This equation is transformed into a system of equations with the substitution $\dfrac{dx}{dt} = y$. Then, solving the differential equation for $\dfrac{d^2x}{dt^2}$, we have $\dfrac{dy}{dt} = \dfrac{d^2x}{dt^2} = -\dfrac{k}{m}x - \dfrac{c}{m}\dfrac{dx}{dt}$, which yields the system

$$\begin{cases} \dfrac{dx}{dt} = y \\[2mm] \dfrac{dy}{dt} = -\dfrac{k}{m}x - \dfrac{c}{m}y \end{cases}$$

In previous chapters, the displacement of the spring was illustrated as a function of time. Problems of this type may also be investigated using the phase plane. In the following example, the phase plane corresponding to the various situations encountered by spring-mass systems discussed in previous sections (undamped, damped, over-damped, and critically damped) are determined.

Example 3

Solve the system of differential equations to find the displacement of the mass if $m = 1$, $c = 0$, and $k = 1$.

Solution   In this case, the system is

$$\begin{cases} \dfrac{dx}{dt} = y \\[2mm] \dfrac{dy}{dt} = -x \end{cases}$$

The eigenvalues are solutions of $\begin{vmatrix} -\lambda & 1 \\ -1 & -\lambda \end{vmatrix} = \lambda^2 + 1 = 0$, so $\lambda_{1,2} = \pm i$. An eigen-

vector corresponding to $\lambda_1 = i$ is $\mathbf{v}_1 = \begin{pmatrix} 1 \\ i \end{pmatrix} = \begin{pmatrix} 1 \\ 0 \end{pmatrix} + i\begin{pmatrix} 0 \\ 1 \end{pmatrix}$, so two linearly

independent solutions are $\mathbf{X}_1(t) = \begin{pmatrix} 1 \\ 0 \end{pmatrix}\cos t - \begin{pmatrix} 0 \\ 1 \end{pmatrix}\sin t = \begin{pmatrix} \cos t \\ -\sin t \end{pmatrix}$ and

$\mathbf{X}_2(t) = \begin{pmatrix} 1 \\ 0 \end{pmatrix}\sin t + \begin{pmatrix} 0 \\ 1 \end{pmatrix}\cos t = \begin{pmatrix} \sin t \\ \cos t \end{pmatrix}$. A general solution is

$$\mathbf{X}(t) = \begin{pmatrix} x(t) \\ y(t) \end{pmatrix} = c_1\mathbf{X}_1(t) + c_2\mathbf{X}_2(t) = \begin{pmatrix} c_1\cos t + c_2\sin t \\ -c_1\sin t + c_2\cos t \end{pmatrix}.$$

Notice that this system is equivalent to the second-order differential equation $\dfrac{d^2x}{dt^2} + x = 0$, which we solved in Chapters 4 and 5. At that time, we found a general solution to be $x(t) = c_1\cos t + c_2\sin t$, which is the same as the first component of $\mathbf{X}(t) = \begin{pmatrix} x(t) \\ y(t) \end{pmatrix}$ obtained here. We graph this function for several values of the arbitrary constants in Figure 9.5(a) to illustrate the periodic motion of the mass. Also notice that $(0, 0)$ is the equilibrium point of the system. Because the eigenvalues are $\lambda = \pm i$, we classify the origin as a center. We graph the phase plane of this system in Figure 9.5(b).

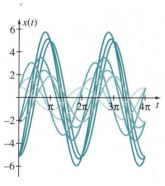

a.

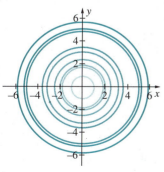

b.

Figure 9.5 (a)–(b)

By observing the phase plane in Figure 9.5 and the corresponding system of differential equations in Example 3, describe the motion of the object in each quadrant and determine the sign on the velocity $\dfrac{dx}{dt}$ of the object in each quadrant.

IN TOUCH WITH TECHNOLOGY

Use a graphing device to graph the solutions to Exercises 1–4 simultaneously and parametrically. Also determine the limit of these solutions as $t \to \infty$.

1. Solve the one-loop L-R-C circuit with $L = 1$, $R = 40$, $C = 0.004$, and $E(t) = 120 \sin t$. ($Q(0) = 0$ coulombs, $I(0) = 0$ amps)

2. Solve the one-loop L-R-C circuit with $L = 4$, $R = 80$, $C = 0.08$, and $E(t) = 120e^{-t} \sin t$. ($Q(0) = 10^{-6}$ coulomb, $I(0) = 0$ amps)

3. Solve the two-loop L-R-C circuit with $L = 1$, $R_1 = R_2 = 40$, $C = 0.004$, and $E(t) = 220 \cos t$. ($Q(0) = 0$ coulombs, $I_2(0) = 0$ amps)

4. Solve the two-loop L-R-C circuit with $L = 1$, $R_1 = 40$, $R_2 = 80$, $C = 0.004$, and $E(t) = 150e^{-t} \cos t$. ($Q(0) = 10^{-6}$ coulomb, $I_2(0) = 0$ amps)

5. Use the system derived in Exercise 6 to solve the three-loop circuit shown in Figure 9.6 if $R_1 = R_2 = 2$, $L_1 = L_2 = L_3 = 1$, $C = 1$, $E(t) = 90$, and the initial conditions are $Q(0) = 0$, $I_2(0) = I_3(0) = 0$. Find $I_1(t)$ and determine $\lim_{t\to\infty} Q(t)$, $\lim_{t\to\infty} I_1(t)$, $\lim_{t\to\infty} I_2(t)$, and $\lim_{t\to\infty} I_3(t)$.

6. Solve the system of differential equations to find the displacement of the spring-mass system if $m = 1$, $c = 1$, and $k = \frac{1}{2}$. Graph several solutions in the phase plane for this system. How is the equilibrium point $(0, 0)$ classified?

7. Solve the system of differential equations to find the displacement of the spring-mass system given that $m = 1$, $c = 2$, and $k = \frac{3}{4}$. Graph several solutions in the phase plane for this system. How is the equilibrium point $(0, 0)$ classified?

EXERCISES 9.1

Solve each of the following systems for charge and current using the procedures discussed in the example problems.

*1. Solve the one-loop L-R-C circuit with $L = 3$ henrys, $R = 10$ ohms, $C = 0.1$ farad, and: (a) $E(t) = 0$; (b) $E(t) = e^{-t}$. ($Q(0) = 0$ coulombs, $I(0) = 1$ amp)

2. Solve the one-loop L-R-C circuit with $L = 1$ henry, $R = 20$ ohms, $C = 0.01$ farad, and: (a) $E(t) = 0$; (b) $E(t) = 1200$. ($Q(0) = 0$ coulombs, $I(0) = 1$ amp)

3. Find $Q(t)$, $I(t)$, $I_1(t)$, and $I_2(t)$ in the two-loop L-R-C circuit with $L = 1$, $R_1 = 2$, $R_2 = 1$, $C = \frac{1}{2}$, and: (a) $E(t) = 0$; (b) $E(t) = 2e^{-t/2}$. ($Q(0) = 10^{-6}$ coulomb, $I_2(0) = 0$ amps)

4. Find $Q(t)$, $I(t)$, $I_1(t)$, and $I_2(t)$ in the two-loop L-R-C circuit with $L = 1$, $R_1 = 2$, $R_2 = 1$, $C = \frac{1}{2}$, and: (a) $E(t) = 90$; (b) $E(t) = 90 \sin t$. ($Q(0) = 0$ coulombs, $I_2(0) = 0$ amps)

*5. Find $Q(t)$, $I(t)$, $I_1(t)$, and $I_2(t)$ in the two-loop L-R-C circuit with $L = 1$, $R_1 = 1$, $R_2 = 3$, $C = 1$, and: **(a)** $E(t) = 0$, $Q(0) = 10^{-6}$, and $I_2(0) = 0$; **(b)** $E(t) = 90$, $Q(0) = 0$, $I_2(0) = 0$.

6. Consider the circuit made up of three loops illustrated in Figure 9.6. In this circuit, the current through the resistor R_2 is $I_2 - I_3$, and the current through the capacitor is $I_1 - I_2$. Using these quantities in the voltage drop sum equations, model this circuit with the three-dimensional system:

$$\begin{cases} -E(t) + R_1 I_1 + \dfrac{1}{C}Q = 0 \\ -\dfrac{1}{C}Q + L_2\dfrac{dI_2}{dt} + R_2(I_2 - I_3) = 0 \\ E(t) - R_2(I_2 - I_1) + L_3\dfrac{dI_3}{dt} = 0 \end{cases}$$

Using the relationship $dQ/dt = I_1 - I_2$ and solving the first equation for I_1, show that we obtain the system

$$\begin{cases} \dfrac{dQ}{dt} = -\dfrac{1}{R_1 C}Q - I_2 + \dfrac{E(t)}{R_1} \\ \dfrac{dI_2}{dt} = \dfrac{1}{L_2 C}Q - \dfrac{R_2}{L_2}I_2 + \dfrac{R_2}{L_2}I_3 \\ \dfrac{dI_3}{dt} = \dfrac{R_2}{L_3}I_2 - \dfrac{R_2}{L_3}I_3 - \dfrac{E(t)}{L_3} \end{cases}$$

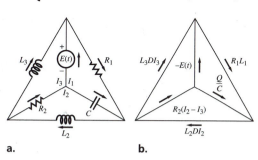

a.   b.

Figure 9.6 (a)–(b) A Three-Loop Circuit

In Exercises 7–11, solve the three-loop circuit using the given values and initial conditions.

7. $L_2 = L_3 = 1$ henry, $R_1 = 1$ ohm, $R_2 = 1$ ohm, $C = 1$ farad, and: **(a)** $E(t) = 0$ volts; **(b)** $E(t) = e^{-t}$ volts. ($Q(0) = 10^{-6}$ coulomb, $I_2(0) = I_3(0) = 0$ amps)

8. $L_2 = L_3 = 1$ henry, $R_1 = R_2 = 1$ ohms, $C = 1$ farad, and: **(a)** $E(t) = 90$ volts; **(b)** $E(t) = 90 \sin t$ volts. ($Q(0) = 0$ coulomb, $I_2(0) = I_3(0) = 0$ amps)

*9. $L_2 = 1$ henry, $L_3 = 1$ henry, $R_1 = R_2 = 1$ ohm, $C = 1$ farad, and: **(a)** $E(t) = 90$ volts; **(b)** $E(t) = 90 \sin t$ volts. ($Q(0) = 0$ coulombs, $I_2(0) = 1$ amp, $I_3(0) = 0$ amps)

10. $L_2 = 3$ henrys, $L_3 = 1$ henry, $R_1 = R_2 = 1$ ohm, $C = 1$ farad, and: **(a)** $E(t) = 0$ volts; **(b)** $E(t) = 90 \sin t$ volts. ($Q(0) = 0$ coulombs, $I_2(0) = 1$ amp, $I_3(0) = 0$ amps)

11. $L_2 = 4$ henrys, $L_3 = 1$ henry, $R_1 = R_2 = 1$ ohm, $C = 1$ farad, and: **(a)** $E(t) = 90$ volts; **(b)** $E(t) = 90 \sin t$ volts. ($Q(0) = 0$ coulombs, $I_2(0) = 1$ amp, $I_3(0) = 0$ amps)

12. Show that the system of differential equations that models the four-loop circuit shown in Figure 9.7 is

$$\begin{cases} L_1\dfrac{dI_1}{dt} = -(R_1 + R_2)I_1 + R_2 I_2 + R_1 I_4 + E(t) \\ L_2\dfrac{dI_2}{dt} = R_2 I_1 - (R_2 + R_3)I_2 + R_3 I_3 \\ L_3\dfrac{dI_3}{dt} = R_3 I_2 - (R_3 + R_4)I_3 + R_4 I_4 \\ L_4\dfrac{dI_4}{dt} = R_1 I_1 + R_4 I_3 - (R_1 + R_4)I_4 \end{cases}$$

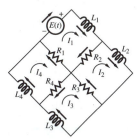

Figure 9.7 A Four-Loop Circuit

In Exercises 13–20, transform the second-order equation to a system of first-order equations and classify the system as undamped, overdamped, underdamped, or critically damped by finding the eigenvalues of the corresponding system. Also classify the equilibrium point $(0, 0)$.

*13. $\frac{1}{2}x'' + \frac{9}{2}x = 0$

14. $\frac{1}{4}x'' + 4x = 0$

15. $\frac{1}{2}x'' + 5x' + \frac{9}{2}x = 0$

16. $\frac{1}{4}x'' + x' + \frac{5}{4}x = 0$

***17.** $\frac{1}{2}x'' + 5x' + 25x = 0$

18. $\frac{1}{3}x'' + \frac{4}{3}x' + \frac{13}{3}x = 0$

19. $\frac{1}{2}x'' + 5x' + \frac{25}{2}x = 0$

20. $x'' + 6x' + 9x = 0$

***21.** Solve Exercises 13, 17, and 19 with the initial conditions $x(0) = 1, \dfrac{dx}{dt}(0) = y(0) = 0$.

22. Solve Exercises 17–20 with the initial conditions $x(0) = 0, \dfrac{dx}{dt}(0) = y(0) = 1$.

23. Find the eigenvalues for the spring-mass system

$$\begin{cases} \dfrac{dx}{dt} = y \\[2mm] \dfrac{dy}{dt} = -\dfrac{k}{m}x - \dfrac{c}{m}y \end{cases}$$

How do these values relate to overdamping, critical damping, and underdamping?

24. (a) Find the equilibrium point of the spring-mass system

$$\begin{cases} \dfrac{dx}{dt} = y \\[2mm] \dfrac{dy}{dt} = -\dfrac{k}{m}x - \dfrac{c}{m}y \end{cases}$$

(b) Find restrictions on m, c, and k to classify this point as a center, stable node, or stable spiral.

(c) Can the equilibrium point be unstable for any choice of the positive constants m, c, and k? Is a saddle possible?

9.2 Diffusion and Population Problems with First-Order Linear Systems

Diffusion Through a Membrane • Mixture Problems • Population Problems

Diffusion Through a Membrane

Solving problems to determine the diffusion of a substance (such as glucose or salt) in a medium (like a blood cell) also leads to first-order systems of linear ordinary differential equations. For example, consider the situation shown in Figure 9.8 in which two solutions of a substance are separated by a membrane. The amount of substance that passes through the membrane at any particular time is proportional to the difference in the concentrations of the two solutions. The constant of proportionality, P, is called the **permeability** of the membrane and describes the ability of the substance to permeate the membrane (where $P > 0$). If we let $x(t)$ and $y(t)$ represent the amount of substance at time t on each side of the membrane, and V_1 and V_2 represent the (constant) volume of each solution, respectively, then the system of differential equations is given by

$$\begin{cases} \dfrac{dx}{dt} = P\left(\dfrac{y}{V_2} - \dfrac{x}{V_1}\right) \\[3mm] \dfrac{dy}{dt} = P\left(\dfrac{x}{V_1} - \dfrac{y}{V_2}\right) \end{cases}$$

where the initial amounts of x and y are given with the initial conditions $x(0) = x_0$ and $y(0) = y_0$. (Notice that the amount of the substance divided by the volume is the *concentration* of the solution.)

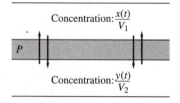

Concentration: $\dfrac{x(t)}{V_1}$

P

Concentration: $\dfrac{y(t)}{V_2}$

Figure 9.8

In this system, if $\dfrac{y(t)}{V_2} > \dfrac{x(t)}{V_1}$, is $\dfrac{dx}{dt} > 0$ or is $\dfrac{dx}{dt} < 0$? Also, is $\dfrac{dy}{dt} > 0$ or

is $\dfrac{dy}{dt} < 0$? Using these results, does the substance move from the side with a

lower concentration to that with a higher concentration or is the opposite true?

Example 1

Suppose that two salt concentrations of equal volume V are separated by a membrane of permeability P. Given that $P = V$, determine the amount of salt in each concentration at time t if $x(0) = 2$ and $y(0) = 10$.

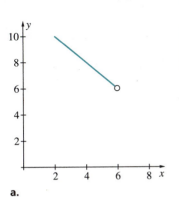

a.

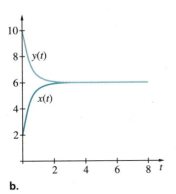

b.

Figure 9.9   (a)–(b)

Solution   In this case, the initial-value problem that models the situation is

$$\begin{cases} \dfrac{dx}{dt} = y - x \\[2mm] \dfrac{dy}{dt} = x - y \end{cases}, \quad x(0) = 2, \ y(0) = 10.$$

The eigenvalues of $\mathbf{A} = \begin{pmatrix} -1 & 1 \\ 1 & -1 \end{pmatrix}$ are $\lambda_1 = 0$ and $\lambda_2 = -2$. Corresponding

eigenvectors are found to be $\mathbf{v}_1 = \begin{pmatrix} 1 \\ 1 \end{pmatrix}$ and $\mathbf{v}_2 = \begin{pmatrix} -1 \\ 1 \end{pmatrix}$, so a general solution is

$$\mathbf{X}(t) = \begin{pmatrix} x(t) \\ y(t) \end{pmatrix} = c_1 \begin{pmatrix} 1 \\ 1 \end{pmatrix} + c_2 \begin{pmatrix} -1 \\ 1 \end{pmatrix} e^{-2t} = \begin{pmatrix} c_1 - c_2 e^{-2t} \\ c_1 + c_2 e^{-2t} \end{pmatrix}.$$

Because $\mathbf{X}(0) = \begin{pmatrix} c_1 - c_2 \\ c_1 + c_2 \end{pmatrix} = \begin{pmatrix} 2 \\ 10 \end{pmatrix}$, we have that $c_1 = 6$ and $c_2 = 4$, so the solution is

$$\mathbf{X}(t) = \begin{pmatrix} x(t) \\ y(t) \end{pmatrix} = \begin{pmatrix} 6 - 4e^{-2t} \\ 6 + 4^{-2t} \end{pmatrix}.$$

We graph this solution parametrically in Figure 9.9(a). We then graph $x(t)$ and $y(t)$ together in Figure 9.9(b). Notice that the amount of salt in each concentration approaches 6, which is the average value of the two initial amounts.

 In Example 1, if $x(0) = x_0$ and $y(0) = y_0$, does $\lim_{t \to \infty} x(t) = \lim_{t \to \infty} y(t) =$

$\dfrac{1}{2}(x_0 + y_0)?$

Mixture Problems

Consider the interconnected tanks that are shown in Figure 9.10 in which a salt solution is allowed to flow according to the given information. Let $x(t)$ and $y(t)$ represent the amount of salt in Tank 1 and Tank 2, respectively. Using this information, we set up

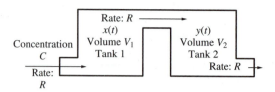

Figure 9.10

two differential equations to describe the rate at which x and y change with respect to time. Notice that the rate at which *liquid* flows into each tank equals the rate at which it flows out, so the volume of liquid in each tank remains constant. If we consider Tank 1, we can determine a first-order differential equation for dx/dt with

$$\frac{dx}{dt} = (\text{Rate at which salt enters Tank 1}) - (\text{Rate at which salt leaves Tank 1})$$

where the rate at which salt enters Tank 1 is R gal/min $\times$ C lb/gal $= RC$ lb/min and the rate at which it leaves is R gal/min $\times \dfrac{x}{V_1}$ lb/gal $= \dfrac{Rx}{V_1}$ lb/min $\left(\text{where } \dfrac{x}{V_1} \text{ is the salt concentration in Tank 1}\right)$. Therefore,

$$\frac{dx}{dt} = RC - \frac{Rx}{V_1}.$$

Similarly, we find $\dfrac{dy}{dt}$ to be

$$\frac{dy}{dt} = \frac{Rx}{V_1} - \frac{Ry}{V_2}.$$

We use the initial conditions $x(0) = x_0$ and $y(0) = y_0$ to solve the nonhomogeneous system

In deriving this system of equations, we used rates in gal/min and concentrations in lb/gal. In general, rates are given by (volume of liquid)/time and concentrations are given by (amount of salt)/(volume of liquid).

$$\begin{cases} \dfrac{dx}{dt} = RC - \dfrac{Rx}{V_1} \\[2mm] \dfrac{dy}{dt} = \dfrac{Rx}{V_1} - \dfrac{Ry}{V_2} \end{cases}$$

for $x(t)$ and $y(t)$.

Example 2

Determine the amount of salt in each tank in Figure 9.10 if $V_1 = V_2 = 500$ gallons, $R = 5$ gal/min, $C = 3$ lb/gal, $x_0 = 50$ lb, and $y_0 = 100$ lb.

Solution   In this case, the initial-value problem is

$$\begin{cases} \dfrac{dx}{dt} = (5)(3) - \dfrac{5x}{500} = 15 - \dfrac{x}{100} \\[3mm] \dfrac{dy}{dt} = \dfrac{5x}{500} - \dfrac{5y}{500} = \dfrac{x}{100} - \dfrac{y}{100} \end{cases}, \quad x(0) = 50, \ y(0) = 100$$

which in matrix form is $\mathbf{X}' = \begin{pmatrix} -\frac{1}{100} & 0 \\ \frac{1}{100} & -\frac{1}{100} \end{pmatrix}\mathbf{X} + \begin{pmatrix} 15 \\ 0 \end{pmatrix} = \mathbf{AX} + \mathbf{F}(t), \ \mathbf{X}(0) =$

$\begin{pmatrix} 50 \\ 100 \end{pmatrix}$. The matrix $\mathbf{A}$ has the repeated eigenvalue $\lambda_1 = \lambda_2 = -\frac{1}{100}$ for

which we can find one (linearly independent) eigenvector $\mathbf{v}_1 = \begin{pmatrix} 0 \\ 1 \end{pmatrix}$. Therefore,

one solution to the corresponding homogeneous system $\mathbf{X}' = \mathbf{AX}$ is $\mathbf{X}_1(t) =$

$\begin{pmatrix} 0 \\ 1 \end{pmatrix}e^{-t/100}$ and a second solution is $\mathbf{X}_2(t) = \left[\begin{pmatrix} 0 \\ 1 \end{pmatrix}t + \begin{pmatrix} 100 \\ 0 \end{pmatrix} \right]e^{-t/100}$, so

$$\mathbf{X}_h(t) = c_1\begin{pmatrix} 0 \\ 1 \end{pmatrix}e^{-t/100} + c_2\left[\begin{pmatrix} 0 \\ 1 \end{pmatrix}t + \begin{pmatrix} 100 \\ 0 \end{pmatrix} \right]e^{-t/100} = \begin{pmatrix} 100c_2e^{-t/100} \\ c_1e^{-t/100} + c_2te^{-t/100} \end{pmatrix}.$$

Notice that $\mathbf{F}(t) = \begin{pmatrix} 15 \\ 0 \end{pmatrix}$ is not contained in $\mathbf{X}_h(t)$, so with the method of

undetermined coefficients we assume a particular solution has the form $\mathbf{X}_p(t) = \mathbf{a} =$

$\begin{pmatrix} a_1 \\ a_2 \end{pmatrix}$ and substitute into the nonhomogeneous system $\mathbf{X}' = \mathbf{AX} + \mathbf{F}(t)$.

This yields $\begin{pmatrix} 0 \\ 0 \end{pmatrix} = \begin{pmatrix} -\frac{1}{100} & 0 \\ \frac{1}{100} & -\frac{1}{100} \end{pmatrix}\begin{pmatrix} a_1 \\ a_2 \end{pmatrix} + \begin{pmatrix} 15 \\ 0 \end{pmatrix} = \begin{pmatrix} -\frac{1}{100}a_1 + 15 \\ \frac{1}{100}a_1 - \frac{1}{100}a_2 \end{pmatrix}$ with solu-

tion $a_1 = 1500$ and $a_2 = 1500$. Therefore, $\mathbf{X}_p(t) = \begin{pmatrix} 1500 \\ 1500 \end{pmatrix}$ and

$$\mathbf{X}(t) = \mathbf{X}_h(t) + \mathbf{X}_p(t) = \begin{pmatrix} 100c_2e^{-t/100} + 1500 \\ c_1e^{-t/100} + c_2te^{-t/100} + 1500 \end{pmatrix}.$$

Application of the initial conditions then gives us $\mathbf{X}(0) = \begin{pmatrix} 100c_2 + 1500 \\ c_1 + 1500 \end{pmatrix} =$

$\begin{pmatrix} 50 \\ 100 \end{pmatrix}$, so $c_1 = -1400$ and $c_2 = -1450/100 = -29/2$. The solution to

the initial-value problem is

$$\mathbf{X}(t) = \begin{pmatrix} x(t) \\ y(t) \end{pmatrix} = \begin{pmatrix} -1450e^{-t/100} + 1500 \\ -1400e^{-t/100} - \dfrac{29}{2}te^{-t/100} + 1500 \end{pmatrix}.$$

In Figure 9.11 we graph $x(t)$ and $y(t)$ together. Notice that each function approaches a limit of 1500, which means that the amount of salt in each tank tends towards a value of 1500 lb.

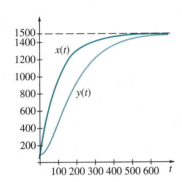

Figure 9.11

⚙ *In Example 2, is there a value of t for which $x(t) = y(t)$? If so, what is this value? Which function increases most rapidly for smaller values of t?*

Population Problems

In Chapter 3 we discussed population problems that were based on the simple principle that the rate at which a population grows (or decays) is proportional to the number present in the population at any time t. This idea can be extended to problems involving more than one population and leads to systems of ordinary differential equations. We illustrate several situations through the following examples. Note that in each problem we determine the rate at which a population P changes with the equation

$$\frac{dP}{dt} = \text{(rate entering)} - \text{(rate leaving)}.$$

We begin by determining the population in two neighboring territories where the populations x and y of the territories depend on several factors. The birth rate of x is a_1 while that of y is b_1. The rate at which citizens of x move to y is a_2 while that at which citizens move from y to x is b_2. After assuming that the mortality rate of each territory is disregarded, we determine the respective populations of these two territories for any time t.

Using the simple principles of previous examples, the rate at which population x changes is

$$\frac{dx}{dt} = a_1 x - a_2 x + b_2 y = (a_1 - a_2)x + b_2 y$$

while the rate at which population y changes is

$$\frac{dy}{dt} = b_1 y - b_2 y + a_2 x = (b_1 - b_2)y + a_2 x.$$

Therefore, the system of equations that must be solved is

$$\begin{cases} \dfrac{dx}{dt} = (a_1 - a_2)x + b_2 y \\[2mm] \dfrac{dy}{dt} = a_2 x + (b_1 - b_2)y \end{cases}$$

where the initial populations of the two territories $x(0) = x_0$ and $y(0) = y_0$ are given.

Example 3

Determine the populations $x(t)$ and $y(t)$ in each territory if $a_1 = 5$, $a_2 = 4$, $b_1 = 5$, and $b_2 = 3$ given that $x(0) = 14$ and $y(0) = 7$.

Solution   In this case, the initial-value problem that models the situation is

$$\begin{cases} \dfrac{dx}{dt} = x + 3y \\[2mm] \dfrac{dy}{dt} = 4x + 2y \end{cases} , \quad x(0) = 14 \text{ and } y(0) = 7.$$

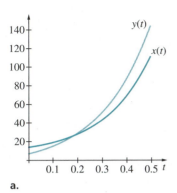

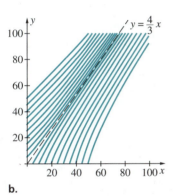

a.

b.

Figure 9.12 (a)–(b)

The eigenvalues of $A = \begin{pmatrix} 1 & 3 \\ 4 & 2 \end{pmatrix}$ and $\lambda_1 = -2$ and $\lambda_2 = 5$. Corresponding eigenvectors are found to be $\mathbf{v}_1 = \begin{pmatrix} 1 \\ -1 \end{pmatrix}$ and $\mathbf{v}_2 = \begin{pmatrix} 3 \\ 4 \end{pmatrix}$, so a general solution is

$$\mathbf{X}(t) = \begin{pmatrix} x(t) \\ y(t) \end{pmatrix} = c_1 \begin{pmatrix} 1 \\ -1 \end{pmatrix} e^{-2t} + c_2 \begin{pmatrix} 3 \\ 4 \end{pmatrix} e^{5t} = \begin{pmatrix} c_1 e^{-2t} + 3c_2 e^{5t} \\ -c_1 e^{-2t} + 4c_2 e^{5t} \end{pmatrix}.$$

Application of the initial condition $\mathbf{X}(0) = \begin{pmatrix} x(0) \\ y(0) \end{pmatrix} = \begin{pmatrix} 14 \\ 7 \end{pmatrix}$ yields the system $\begin{cases} c_1 + 3c_2 = 14 \\ -c_1 + 4c_2 = 7 \end{cases}$, so $c_1 = 5$ and $c_2 = 3$. Therefore, the solution is $\mathbf{X}(t) = \begin{pmatrix} x(t) \\ y(t) \end{pmatrix} = \begin{pmatrix} 5e^{-2t} + 9e^{5t} \\ -5e^{-2t} + 12e^{5t} \end{pmatrix}$. We graph these two population functions in Figure 9.12(a). In Figure 9.12(b), we graph several solutions to the system of differential equations for various initial conditions in the phase plane. As we can see, all solutions move away from the origin in the direction of the eigenvector $\mathbf{v}_2 = \begin{pmatrix} 3 \\ 4 \end{pmatrix}$, which corresponds to the positive eigenvalue $\lambda_2 = 5$.

Population problems that involve more than two neighboring populations can be solved with a system of differential equations as well. Suppose that the population of three neighboring territories, x, y, and z depends on several factors. The birth rates of x, y, and z are a_1, b_1, and c_1, respectively. The rate at which citizens of x move to y is a_2 while that at which citizens move from x to z is a_3. Similarly, the rate at which citizens of y move to x is b_2 while that at which citizens move from y to z is b_3. Also, the rate at which citizens of z move to x is c_2 while that at which citizens move from z to y is c_3. (This information is summarized in Table 9.2.) If the mortality rate of each territory is ignored in the model, we can determine the respective populations of the three territories for any time t.

The system of equations to determine $x(t)$, $y(t)$, and $z(t)$ is similar to that derived in the previous example. The rate at which population x changes is

$$\frac{dx}{dt} = a_1 x - a_2 x - a_3 x + b_2 y + c_2 z = (a_1 - a_2 - a_3)x + b_2 y + c_2 z,$$

TABLE 9.2

From	To			Birth rate
	x	y	z	
x	—	a_2	a_3	a_1
y	b_2	—	b_3	b_1
z	c_2	c_3	—	c_1

while the rate at which population y changes is

$$\frac{dy}{dt} = b_1 y - b_2 y - b_3 y + a_2 x + c_3 z = (b_1 - b_2 - b_3)y + a_2 x + c_3 z,$$

and that of z is

$$\frac{dz}{dt} = c_1 z - c_2 z - c_3 z + a_3 x + b_3 y = (c_1 - c_2 - c_3)z + a_3 x + b_3 y.$$

We must solve the 3×3 system

$$\begin{cases} \dfrac{dx}{dt} = (a_1 - a_2 - a_3)x + b_2 y + c_2 z \\[2mm] \dfrac{dy}{dt} = a_2 x + (b_1 - b_2 - b_3)y + c_3 z, \\[2mm] \dfrac{dz}{dt} = a_3 x + b_3 y + (c_1 - c_2 - c_3)z \end{cases}$$

where the initial populations $x(0) = x_0$, $y(0) = y_0$, and $z(0) = z_0$ are given.

Example 4

Determine the population of the three territories if $a_1 = 3$, $a_2 = 0$, $a_3 = 2$, $b_1 = 4$, $b_2 = 2$, $b_3 = 1$, $c_1 = 5$, $c_2 = 3$, and $c_3 = 0$ if $x(0) = 50$, $y(0) = 60$, and $z(0) = 25$.

Solution   In this case, the system of differential equations is

$$\begin{cases} \dfrac{dx}{dt} = x + 2y + 3z \\[2mm] \dfrac{dy}{dt} = y \\[2mm] \dfrac{dz}{dt} = 2x + y + 2z \end{cases}.$$

Because the characteristic polynomial is

$$\begin{vmatrix} 1 - \lambda & 2 & 3 \\ 0 & 1 - \lambda & 0 \\ 2 & 1 & 2 - \lambda \end{vmatrix} = (1 - \lambda)^2(2 - \lambda) + 3(-2)(1 - \lambda) =$$

$$(1 - \lambda)(\lambda + 1)(\lambda - 4) = 0,$$

the eigenvalues are $\lambda_1 = 4$, $\lambda_2 = 1$, and $\lambda_3 = -1$, with corresponding eigenvectors

$$\mathbf{v}_1 = \begin{pmatrix} 1 \\ 0 \\ 1 \end{pmatrix}, \mathbf{v}_2 = \begin{pmatrix} 1 \\ -6 \\ 4 \end{pmatrix}, \text{ and } \mathbf{v}_3 = \begin{pmatrix} -3 \\ 0 \\ 2 \end{pmatrix}, \text{ respectively. A general solution is}$$

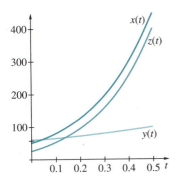

Figure 9.13

$$\mathbf{X}(t) = \begin{pmatrix} x(t) \\ y(t) \\ z(t) \end{pmatrix} = c_1 \begin{pmatrix} 1 \\ 0 \\ 1 \end{pmatrix} e^{4t} + c_2 \begin{pmatrix} 1 \\ -6 \\ 4 \end{pmatrix} e^t + c_3 \begin{pmatrix} -3 \\ 0 \\ 2 \end{pmatrix} e^{-t} =$$

$$\begin{pmatrix} c_1 e^{4t} + c_2 e^t - 3c_3 e^{-t} \\ -6c_2 e^t \\ c_1 e^{4t} + 4c_2 e^t + 2c_3 e^{-t} \end{pmatrix}.$$

Using the initial conditions, we find that
$$\begin{cases} c_1 + c_2 - 3c_3 = 50 \\ -6c_2 = 60, \\ c_1 + 4c_2 + 2c_3 = 25 \end{cases} \text{so } c_1 = 63,$$

$c_2 = -10$, and $c_3 = 1$. Therefore, the solution is $\mathbf{X}(t) =$

$$\begin{pmatrix} x(t) \\ y(t) \\ z(t) \end{pmatrix} = \begin{pmatrix} 63e^{4t} - 10e^t - 3e^{-t} \\ 60e^t \\ 63e^{4t} - 40e^t + 2e^{-t} \end{pmatrix}.$$ We graph these three population func-

tions in Figure 9.13. Notice that although population y is initially greater than populations x and z, these populations increase at a much higher rate than does y.

 In Example 4, does population y approach a limit or do all three populations increase exponentially?

IN TOUCH WITH TECHNOLOGY

 1. Solve the initial-value problem to find the concentration of a substance on each side of a permeable membrane modeled by the system

$$\begin{cases} \dfrac{dx}{dt} = P\left(\dfrac{y}{V_2} - \dfrac{x}{V_1}\right) \\ \dfrac{dy}{dt} = P\left(\dfrac{x}{V_1} - \dfrac{y}{V_2}\right) \end{cases}, \quad x(0) = a, \ y(0) = b.$$

(a) Find $\lim_{t\to\infty} x(t)$ and $\lim_{t\to\infty} y(t)$.

(b) Determine a condition so that $x(t) > y(t)$ as $t \to \infty$. When does $\lim_{t\to\infty} x(t) = \lim_{t\to\infty} y(t)$ as $t \to \infty$?

(c) Find a condition so that $x(t)$ is an increasing function. Find a condition so that $y(t)$ is an increasing function. Can these functions increase simultaneously?

(d) If $V_1 = V_2$ and $a = b$, describe what eventually happens to $x(t)$ and $y(t)$.

2. Investigate the effect that the membrane permeability has on the diffusion of a substance. Suppose that $V_1 = V_2 = 1$, $x(0) = 1$, $y(0) = 2$, and **(a)** $P = 0.25$, **(b)** $P = 0.5$, **(c)** $P = 1.0$, and **(d)** $P = 2.0$. Graph the solution in each case both parametrically and simultaneously. Describe the effect that the value of P has on the corresponding solution.

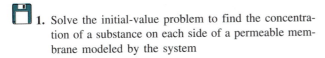 **3.** Solve the initial-value problem

$$\begin{cases} \dfrac{dx}{dt} = (a_1 - a_2)x + b_1 y \\ \dfrac{dy}{dt} = a_2 x + (b_1 - b_2)y \end{cases}, \quad x(0) = x_0, \ y(0) = y_0.$$

Are there possible parameter values so that the functions $x(t)$ and $y(t)$ are periodic? Are there possible parameter values so that the functions $x(t)$ and $y(t)$ experience exponential decay? For what parameter values do the two populations experience exponential growth?

EXERCISES 9.2

In Exercises 1–6, solve the diffusion problem with one perme-able membrane with the indicated initial conditions and param-eter values. Find the limiting concentration of each solution.

1. $P = 0.5$, $V_1 = V_2 = 1$, $x(0) = 1$, $y(0) = 2$

2. $P = 0.5$, $V_1 = V_2 = 1$, $x(0) = 2$, $y(0) = 1$

*3. $P = 2$, $V_1 = 4$, $V_2 = 2$, $x(0) = 0$, $y(0) = 4$

4. $P = 2$, $V_1 = \frac{1}{2}$, $V_2 = \frac{1}{4}$, $x(0) = 8$, $y(0) = 0$

5. $P = 6$, $V_1 = 2$, $V_2 = 8$, $x(0) = 4$, $y(0) = 1$

6. $P = 6$, $V_1 = 2$, $V_2 = 8$, $x(0) = 1$, $y(0) = 4$

In Exercises 7–8, use the tanks shown in Figure 9.10 with $R = 4$ gal/min, $C = \frac{1}{2}$ lb/gal, $V_1 = V_2 = 20$ gallons, and the given initial conditions. **(a)** Determine $x(t)$ and $y(t)$. **(b)** Find $\lim_{t\to\infty} x(t)$ and $\lim_{t\to\infty} y(t)$. **(c)** Does one of the tanks contain more salt than the other tank for all values of t?

*7. $x(0) = y(0) = 0$ **8.** $x(0) = 0$, $y(0) = 2$

9. Suppose that pure water is pumped into Tank 1 (in Figure 9.10) at a rate of 4 gal/min (that is, $R = 4$, $C = 0$) and that $V_1 = V_2 = 20$ gallons. Determine the amount of salt in each tank at time t if $x(0) = y(0) = 4$. Calculate $\lim_{t\to\infty} x(t)$ and $\lim_{t\to\infty} y(t)$. Which function decreases more rapidly?

10. If $x(0) = y(0) = 0$ in Exercise 9, how much salt is in each tank at any time t?

*11. Use the tanks in Figure 9.10 with $R = 5$ gal/min, $C = 3$ lb/gal, $V_1 = 100$ gallons, $V_2 = 50$ gallons, and the initial conditions $x(0) = y(0) = 0$. **(a)** Determine $x(t)$ and $y(t)$. **(b)** Find $\lim_{t\to\infty} x(t)$ and $\lim_{t\to\infty} y(t)$.

12. Solve Exercise 11 with $V_1 = 50$ and $V_2 = 100$. How do the limiting values of $x(t)$ and $y(t)$ compare to those in Exercise 11?

13. Consider the tanks shown in Figure 9.14 where $R_1 = 3$ liters/min, $R_2 = 4$ liters/min, $R_3 = 1$ liter/min, $C = 1$ kg/liter, $V_1 = V_2 = 50$ liters. If $x(0) = y(0) = 0$,

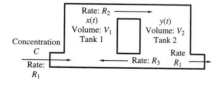

Figure 9.14

determine the amount of salt in each tank at time t. Find $\lim_{t\to\infty} x(t)$ and $\lim_{t\to\infty} y(t)$. Is there a time (other than $t = 0$) at which each tank contains the same amount of salt?

14. Solve the tank problem described in Exercise 13 using the initial conditions $x(0) = 0$ and $y(0) = 6$. For how many values of t do the functions $x(t)$ and $y(t)$ agree?

*15. Find the amount of salt in each tank in Figure 9.15 if $R = 1$ gal/min, $C = 1$ lb/gal, $V_1 = 1$ gal, $V_2 = \frac{1}{2}$ gal, $x(0) = 2$, and $y(0) = 4$. What is the maximum amount of salt (at any value of t) in each tank? Find $\lim_{t\to\infty} x(t)$ and $\lim_{t\to\infty} y(t)$.

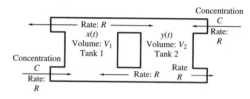

Figure 9.15

16. Determine the amount of salt in each tank described in Exercise 15 using the initial conditions $x(0) = 4$ and $y(0) = 0$. How do these functions differ from those in Exercise 15?

17. Consider the three tanks in Figure 9.16 in which the amount of salt in Tanks 1, 2, and 3 is given by $x(t)$, $y(t)$, and $z(t)$, respectively. Find $x(t)$, $y(t)$, and $z(t)$ if $R = 5$ gal/min, $C = 2$ lb/gal, $V_1 = V_2 = V_3 = 50$ gallons, and $x(0) = y(0) = z(0) = 0$. Find $\lim_{t\to\infty} x(t)$, $\lim_{t\to\infty} y(t)$, and $\lim_{t\to\infty} z(t)$.

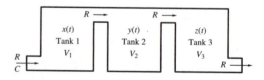

Figure 9.16

18. Solve the problem described in Exercise 17 using the unequal volumes $V_1 = 100$, $V_2 = 50$, and $V_3 = 25$. How do the limiting values of $x(t)$, $y(t)$, and $z(t)$ differ from those in Exercise 17?

In Exercises 19–22, solve the initial-value problem using the given parameters to find the population in two neighboring territories. Do either of the populations approach a finite limit? If so, what is the limit?

19. $a_1 = 10$, $a_2 = 9$, $b_1 = 2$, $b_2 = 1$; $x(0) = 10$, $y(0) = 20$

20. $a_1 = 4$, $a_2 = 4$, $b_1 = 1$, $b_2 = 1$; $x(0) = 4$, $y(0) = 4$

*21. $a_1 = 2$, $a_2 = 0$, $b_1 = 2$, $b_2 = 3$; $x(0) = 5$, $y(0) = 10$

22. $a_1 = 1$, $a_2 = 1$, $b_1 = 1$, $b_2 = 1$; $x(0) = 5$, $y(0) = 10$

In Exercises 23–26, solve the initial-value problem using given parameters to find the population in three neighboring territories. Which population is largest at $t = 1$?

23. $a_1 = 10$, $a_2 = 6$, $a_3 = 7$, $b_1 = 6$, $b_2 = 3$, $b_3 = 3$, $c_1 = 5$, $c_2 = 7$, $c_3 = 1$; $x(0) = 17$, $y(0) = 0$, $z(0) = 34$

24. $a_1 = 2$, $a_2 = 1$, $a_3 = 4$, $b_1 = 6$, $b_2 = 4$, $b_3 = 5$, $c_1 = 2$, $c_2 = 8$, $c_3 = 4$; $x(0) = 0$, $y(0) = 4$, $z(0) = 2$

*25. $a_1 = 7$, $a_2 = 2$, $a_3 = 4$, $b_1 = 7$, $b_2 = 5$, $b_3 = 8$, $c_1 = 7$, $c_2 = 1$, $c_3 = 2$; $x(0) = 8$, $y(0) = 2$, $z(0) = 0$

26. $a_1 = 7$, $a_2 = 2$, $a_3 = 4$, $b_1 = 7$, $b_2 = 5$, $b_3 = 8$, $c_1 = 7$, $c_2 = 1$, $c_3 = 2$; $x(0) = 0$, $y(0) = 0$, $z(0) = 16$

27. Suppose that a radioactive substance X decays into another unstable substance Y which in turn decays into a stable substance Z. Show that we can model this situation through the system of differential equations

$$\begin{cases} \dfrac{dx}{dt} = -ax \\[2mm] \dfrac{dy}{dt} = ax - by, \text{ where } a \text{ and } b \text{ are positive constants.} \\[2mm] \dfrac{dz}{dt} = by \end{cases}$$

(Assume that one unit of X decomposes into one unit of Y, and one unit of Y decomposes into one unit of Z.)

28. Solve the system of differential equations in Exercise 27 if $a \neq b$, $x(0) = x_0$, $y(0) = y_0$, and $z(0) = z_0$. Find $\lim_{t\to\infty} x(t)$, $\lim_{t\to\infty} y(t)$, and $\lim_{t\to\infty} z(t)$.

*29. Solve the system in Exercise 27 with $x(0) = x_0$, $y(0) = y_0$, and $z(0) = z_0$, if $a = b$. Find $\lim_{t\to\infty} x(t)$, $\lim_{t\to\infty} y(t)$, and $\lim_{t\to\infty} z(t)$.

30. In Exercise 27, what is the half-life of substance X?

31. In the reaction described in Exercise 27, show that if k units of X are added per year and h units of Z are removed, then the situation is described with the system

$$\begin{cases} \dfrac{dx}{dt} = -ax + k \\[2mm] \dfrac{dy}{dt} = ax - by, \\[2mm] \dfrac{dz}{dt} = by - h \end{cases}$$

where a, b, k, and h are positive constants.

32. Solve the system described in Exercise 31 if $a \neq b$, $x(0) = x_0$, $y(0) = y_0$, and $z(0) = z_0$. Find $\lim_{t\to\infty} x(t)$ and $\lim_{t\to\infty} y(t)$. How do these limits differ from those in Exercise 27? If $k = h$, then find $\lim_{t\to\infty} z(t)$. If $k > h$, then find $\lim_{t\to\infty} z(t)$. Describe the corresponding physical situation.

In Exercises 33–36, solve the radioactive decay model in Exercise 27 with the given parameter values and initial conditions. If $z(t)$ represents the amount (grams) of substance Z after t hours, how many grams of Z is eventually produced?

33. $a = 6$, $b = 1$, $x_0 = 7$, $y_0 = 1$, $z_0 = 8$

34. $a = 4$, $b = 2$, $x_0 = 10$, $y_0 = 2$, $z_0 = 4$

35. $a = 4$, $b = 2$, $x_0 = 1$, $y_0 = 1$, $z_0 = 1$

36. $a = 1$, $b = 4$, $x_0 = 2$, $y_0 = 2$, $z_0 = 2$

In Exercises 37–40, solve the radioactive decay model in Exercise 31 with $a = 1$, $b = 1$, and the given parameter values and initial conditions. Describe what eventually happens to the amount of each substance.

37. $k = 2$, $h = 1$, $x_0 = 2$, $y_0 = 1$, $z_0 = 2$

38. $k = 0$, $h = 10$, $x_0 = 4$, $y_0 = 2$, $z_0 = 1$

*39. $k = 10$, $h = 0$, $x_0 = 8$, $y_0 = 2$, $z_0 = 2$

40. $k = 0$, $h = 5$, $x_0 = 1$, $y_0 = 10$, $z_0 = 5$

Coupled Spring-Mass Systems   The Double Pendulum

Coupled Spring-Mass Systems

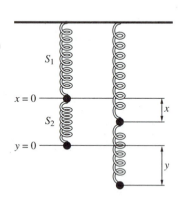

Figure 9.17   A Coupled Spring-Mass System

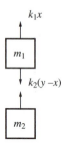

Figure 9.18   Force Diagram for a Coupled Spring-Mass System

The displacement of a mass attached to the end of a spring was modeled with a second-order linear differential equation with constant coefficients in Chapter 5. Similarly, if a second spring and mass are attached to the end of the first mass as shown in Figure 9.17, the model becomes that of a system of second-order equations. To more precisely state the problem, let masses m_1 and m_2 be attached to the ends of springs S_1 and S_2 having spring constants k_1 and k_2, respectively. Spring S_2 is then attached to the base of mass m_1. Suppose that $x(t)$ and $y(t)$ represent the vertical displacement from the equilibrium position of springs S_1 and S_2, respectively. Because spring S_2 undergoes both elongation and compression when the system is in motion (due to the spring S_1 and the mass m_2), according to Hooke's law, S_2 exerts the force $k_2(y - x)$ on m_2 while S_1 exerts the force $-k_1x$ on m_1. Therefore, the force acting on mass m_1 is the sum $-k_1x + k_2(y - x)$ and the force acting on m_2 is $-k_2(y - x)$. (In Figure 9.18 we show the forces acting on the two masses where up is the negative direction and down is positive.) Using Newton's second law ($F = ma$) with each mass, we have the system

$$\begin{cases} m_1\dfrac{d^2x}{dt^2} = -k_1x + k_2(y - x) \\[2mm] m_2\dfrac{d^2y}{dt^2} = -k_2(y - x) \end{cases}$$

The initial displacement and velocity of the two masses m_1 and m_2 are given by $x(0)$, $x'(0)$, $y(0)$, and $y'(0)$, respectively. Because this system involves second-order equations, Laplace transforms can be used to solve problems of this type. Recall the following property of the Laplace transform: $\mathcal{L}\{f''(t)\} = s^2F(s) - sf(0) - f'(0)$ where $F(s)$ is the Laplace transform of $f(t)$. This property is of great use in solving this problem because both equations involve second derivatives. We will see that the method is similar to that used in solving initial-value problems involving a single equation. With systems, however, we end up solving a system of algebraic equations after taking the Laplace transform of each equation.

Example 1

Consider the spring-mass system with $m_1 = m_2 = 1$, $k_1 = 3$, and $k_2 = 2$. Find the position functions $x(t)$ and $y(t)$ if $x(0) = 0$, $x'(0) = 1$, $y(0) = 1$, and $y'(0) = 0$.

Solution   In order to find $x(t)$ and $y(t)$, we must solve the initial-value problem

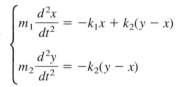

$$\begin{cases} \dfrac{d^2x}{dt^2} = -5x + 2y \\[2mm] \dfrac{d^2y}{dt^2} = 2x - 2y \end{cases}, \quad x(0) = 0,\ x'(0) = 1,\ y(0) = 1,\ y'(0) = 0.$$

Taking the Laplace transform of both sides of each equation, we have

$$\begin{cases} s^2 X(s) - sx(0) - x'(0) = -5X(s) + 2Y(s) \\ s^2 Y(s) - sy(0) - y'(0) = 2X(s) - 2Y(s) \end{cases}$$

which is simplified to obtain the system

$$\begin{cases} (s^2 + 5)X(s) - 2Y(s) = 1 \\ -2X(s) + (s^2 + 2)Y(s) = s \end{cases}$$

Solving for $X(s)$, we have

$$X(s) = \frac{s^2 + 2s + 2}{(s^2 + 5)(s^2 + 2) - 4} = \frac{s^2 + 2s + 2}{s^4 + 7s^2 + 6} = \frac{s^2 + 2s + 2}{(s^2 + 1)(s^2 + 6)}.$$

Then, with the partial fraction expansion

$$X(s) = \frac{As + B}{s^2 + 1} + \frac{Cs + D}{s^2 + 6},$$

we find that $A = \dfrac{2}{5}$, $B = \dfrac{1}{5}$, $C = -\dfrac{2}{5}$, and $D = \dfrac{4}{5}$ so

$$X(s) = \frac{2s + 1}{5(s^2 + 1)} + \frac{2(2 - s)}{5(s^2 + 6)}.$$

Taking the inverse Laplace transform then yields

$$x(t) = \frac{1}{5} \sin t + \frac{2}{5} \cos t - \frac{2}{5} \cos \sqrt{6}t + \frac{4}{5\sqrt{6}} \sin \sqrt{6}t.$$

Instead of solving the system to find $Y(s)$, we use the differential equation $\dfrac{d^2 x}{dt^2} = -5x + 2y$ to find $y(t)$. Because $y = \dfrac{1}{2}(x'' + 5x)$, and

$$x'(t) = \frac{1}{5} \cos t - \frac{2}{5} \sin t + \frac{2\sqrt{6}}{5} \sin \sqrt{6}t + \frac{4}{5} \cos \sqrt{6}t$$

and

$$x''(t) = -\frac{1}{5} \sin t - \frac{2}{5} \cos t + \frac{12}{5} \cos \sqrt{6}t - \frac{4\sqrt{6}}{5} \sin \sqrt{6}t,$$

$$y(t) = \frac{1}{2}\left(-\frac{1}{5} \sin t - \frac{2}{5} \cos t + \frac{12}{5} \cos \sqrt{6}t - \frac{4\sqrt{6}}{5} \sin \sqrt{6}t + \sin t + \right.$$

$$\left. 2 \cos t - 2 \cos \sqrt{6}t + \frac{4}{\sqrt{6}} \sin \sqrt{6}t\right)$$

$$= \frac{2}{5} \sin t + \frac{4}{5} \cos t - \frac{\sqrt{6}}{15} \sin \sqrt{6}t + \frac{1}{5} \cos \sqrt{6}t.$$

In Figure 9.19(a), we graph $x(t)$ and $y(t)$ simultaneously. Note that the initial point of $y(t)$ is $(0, 1)$ while that of $x(t)$ is $(0, 0)$. Of course, these functions can be graphed

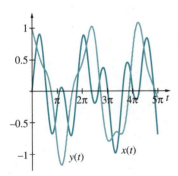

a.

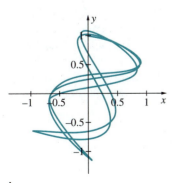

b.

Figure 9.19 (a)–(b)

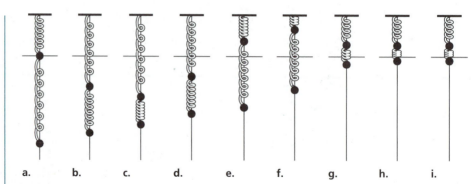

Figure 9.20
(a) $t = 0$ (b) $t = \frac{1}{2}$ (c) $t = 1$ (d) $t = \frac{3}{2}$ (e) $t = 2$ (f) $t = \frac{5}{2}$ (g) $t = 3$
(h) $t = \frac{7}{2}$ (i) $t = 4$

parametrically in the xy-plane as shown in Figure 9.19(b). Notice that this phase plane is different from those discussed in previous sections. One of the reasons for this is that the equations in the system of differential equations are second-order instead of first-order. Finally, in Figure 9.20, we illustrate the motion of the spring-mass system by graphing the springs for several values of t.

In Example 1, what is the maximum displacement of each spring? Does one of the springs always attain its maximum displacement before the other?

If external forces $F_1(t)$ and $F_2(t)$ are applied to the masses, the system of equations becomes

$$\begin{cases} m_1 \dfrac{d^2x}{dt^2} = -k_1 x + k_2(y - x) + F_1(t) \\[2mm] m_2 \dfrac{d^2y}{dt^2} = -k_2(y - x) + F_2(t) \end{cases}$$

We investigate the effects of these external forcing functions in Problem 1 of **In Touch with Technology,** which is again solved through the method of Laplace transforms.

The previous situation can be modified to include a third spring with spring constant k_3 between the base of the mass m_2 and a lower support as shown in Figure 9.21. The motion of the spring-mass system is affected by the third spring. Using the techniques of the earlier case, this model becomes

$$\begin{cases} m_1 \dfrac{d^2x}{dt^2} = -k_1 x + k_2(y - x) + F_1(t) \\[2mm] m_2 \dfrac{d^2y}{dt^2} = -k_3 y - k_2(y - x) + F_2(t) \end{cases}$$

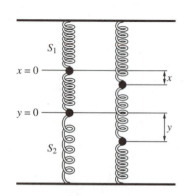

Figure 9.21

Example 2	Consider the problem with $m_1 = m_2 = 1$ and $k_1 = k_2 = k_3 = 1$. Determine $x(t)$ and $y(t)$ if $x(0) = 0$, $x'(0) = -1$, $y(0) = 0$, and $y'(0) = 1$. (Assume there are no external forces.)

Solution Using these values, we obtain the following initial-value problem:

$$\begin{cases} \dfrac{d^2x}{dt^2} = -2x + y \\[2mm] \dfrac{d^2y}{dt^2} = x - 2y \end{cases}, \quad x(0) = 0,\ x'(0) = -1,\ y(0) = 0,\ y'(0) = 1.$$

Taking the Laplace transform then yields

$$\begin{cases} s^2X(s) - sx(0) - x'(0) = -2X(s) + Y(s) \\ s^2Y(s) - sy(0) - y'(0) = X(s) - 2Y(s) \end{cases}$$

which is equivalent to

$$\begin{cases} (s^2 + 2)X(s) - Y(s) = -1 \\ -X(s) + (s^2 + 2)Y(s) = 1 \end{cases}.$$

Solving for X, we have

$$X(s) = \frac{-(s^2 + 1)}{(s^2 + 2)^2 - 1} = \frac{-(s^2 + 1)}{s^4 + 4s^2 + 3} = \frac{-(s^2 + 1)}{(s^2 + 1)(s^2 + 3)} = -\frac{1}{s^2 + 3},$$

and taking the inverse Laplace transform yields

$$x(t) = -\frac{1}{\sqrt{3}}\sin\sqrt{3}t.$$

Then $\dfrac{dx}{dt} = -\cos\sqrt{3}t$ and $\dfrac{d^2x}{dt^2} = \sqrt{3}\sin\sqrt{3}t$. Solving the first differential equation for y, we have $y = x'' + 2x$. Therefore,

$$y(t) = \frac{d^2x}{dt^2} + 2x = \sqrt{3}\sin\sqrt{3}t - \frac{2}{\sqrt{3}}\sin\sqrt{3}t = \left(\sqrt{3} - \frac{2}{\sqrt{3}}\right)\sin\sqrt{3}t$$

$$= \frac{1}{\sqrt{3}}\sin\sqrt{3}t.$$

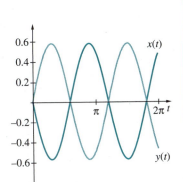

Figure 9.22

We graph these two functions in Figure 9.22. These curves (along with the formulas) indicate that when the mass m_1 is s units above (below) its equilibrium position, then m_2 is s units below (above) its equilibrium.

The Double Pendulum

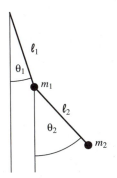

Figure 9.23 A Double Pendulum

In a method similar to that of the simple pendulum in Chapter 5 and that of the coupled spring system, the motion of a double pendulum (see Figure 9.23) is modeled by the following system of equations using the approximation $\sin\theta \approx \theta$ for small displacements:

$$\begin{cases} (m_1 + m_2)\ell_1^2\theta_1'' + m_2\ell_1\ell_2\theta_2'' + (m_1 + m_2)\ell_1 g\theta_1 = 0 \\ m_2\ell_2^2\theta_2'' + m_2\ell_1\ell_2\theta_1'' + m_2\ell_2 g\theta_2 = 0 \end{cases}$$

where θ_1 represents the displacement of the upper pendulum, and θ_2 that of the lower pendulum. Also, m_1 and m_2 represent the mass attached to the upper and lower pendulums, respectively, while the length of each is given by ℓ_1 and ℓ_2.

Example 3

Suppose that $m_1 = 3$, $m_2 = 1$, and each pendulum has length 16. If $g = 32$, determine $\theta_1(t)$ and $\theta_2(t)$ if $\theta_1(0) = 1$, $\theta_1'(0) = 0$, $\theta_2(0) = 0$, and $\theta_2'(0) = -1$.

Solution   In this case, the system is

$$\begin{cases} 4(16)^2\theta_1'' + 16^2\theta_2'' + 4(16)(32)\theta_1 = 0 \\ 16^2\theta_1'' + 16^2\theta_2'' + (16)(32)\theta_2 = 0 \end{cases}$$

which can be simplified to obtain

$$\begin{cases} 4\theta_1'' + \theta_2'' + 8\theta_1 = 0 \\ \theta_1'' + \theta_2'' + 2\theta_2 = 0 \end{cases}.$$

If we let $\mathcal{L}\{\theta_1(t)\} = X(s)$ and $\mathcal{L}\{\theta_2(t)\} = Y(s)$, we have

$$\begin{cases} 4[s^2X(s) - s\theta_1(0) - \theta_1'(0)] + [s^2Y(s) - s\theta_2(0) - \theta_2'(0)] + 8X(s) = 0 \\ [s^2X(s) - s\theta_1(0) - \theta_1'(0)] + [s^2Y(s) - s\theta_2(0) - \theta_2'(0)] + 2Y(s) = 0 \end{cases}$$

or

$$\begin{cases} 4(s^2 + 2)X(s) + s^2Y(s) = 4s - 1 \\ s^2X(s) + (s^2 + 2)Y(s) = s - 1 \end{cases}.$$

Solving this system for $X(s)$, we obtain

$$X(s) = \frac{3s^3 + 8s - 2}{3s^4 + 16s^2 + 16} = \frac{3s^3 + 8s - 2}{(3s^2 + 4)(s^2 + 4)}.$$

With the partial fraction expansion $X(s) = \dfrac{As + B}{3s^2 + 4} + \dfrac{Cs + D}{s^2 + 4}$, we find that

$$X(s) = \frac{3}{4}\frac{2s - 1}{3s^2 + 4} + \frac{1}{4}\frac{2s + 1}{s^2 + 4}. \text{ Then,}$$

$$\theta_1(t) = \frac{1}{2}\cos\frac{2t}{\sqrt{3}} - \frac{\sqrt{3}}{8}\sin\frac{2t}{\sqrt{3}} + \frac{1}{2}\cos 2t + \frac{1}{8}\sin 2t.$$

Differentiating, we have

$$\theta_1'(t) = -\frac{1}{\sqrt{3}}\sin\frac{2t}{\sqrt{3}} - \frac{1}{4}\cos\frac{2t}{\sqrt{3}} - \sin 2t + \frac{1}{4}\cos 2t$$

and

$$\theta_1''(t) = -\frac{2}{3}\cos\frac{2t}{\sqrt{3}} + \frac{1}{2\sqrt{3}}\sin\frac{2t}{\sqrt{3}} - 2\cos 2t - \frac{1}{2}\sin 2t.$$

Using the differential equation $4\theta_1'' + \theta_2'' + 8\theta_1 = 0$ yields $\theta_2'' = -4\theta_1'' - 8\theta_1$. Therefore,

$$\theta_2''(t) = -4\theta_1''(t) - 8\theta_1(t) = \frac{1}{\sqrt{3}}\sin\frac{2t}{\sqrt{3}} - \frac{4}{3}\cos\frac{2t}{\sqrt{3}} + \sin 2t + 4\cos 2t.$$

Integrating, we have

$$\theta_2'(t) = -\frac{1}{2}\cos\frac{2t}{\sqrt{3}} - \frac{2}{\sqrt{3}}\sin\frac{2t}{\sqrt{3}} - \frac{1}{2}\cos 2t + 2\sin 2t + c_1.$$

Application of the initial condition $\theta_2'(0) = -1$ yields $\theta_2'(0) = -\frac{1}{2} - \frac{1}{2} + c_1 = -1$, so $c_1 = 0$. Again, by integrating, we obtain

$$\theta_2(t) = -\frac{\sqrt{3}}{4}\sin\frac{2t}{\sqrt{3}} + \cos\frac{2t}{\sqrt{3}} - \frac{1}{4}\sin 2t - \cos 2t + c_2.$$

Then, because $\theta_2(0) = 0$, $\theta_2(0) = 1 - 1 + c_2 = 0$, so $c_2 = 0$, which indicates that

$$\theta_2(t) = -\frac{\sqrt{3}}{4}\sin\frac{2t}{\sqrt{3}} + \cos\frac{2t}{\sqrt{3}} - \frac{1}{4}\sin 2t - \cos 2t.$$

These two functions are graphed together in Figure 9.24(a) and parametrically in Figure 9.24(b) to show the solution in the phase plane. We also show the motion of the double pendulum for several values of t in Figure 9.25.

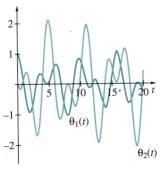

a.

b.

Figure 9.24 (a)–(b)

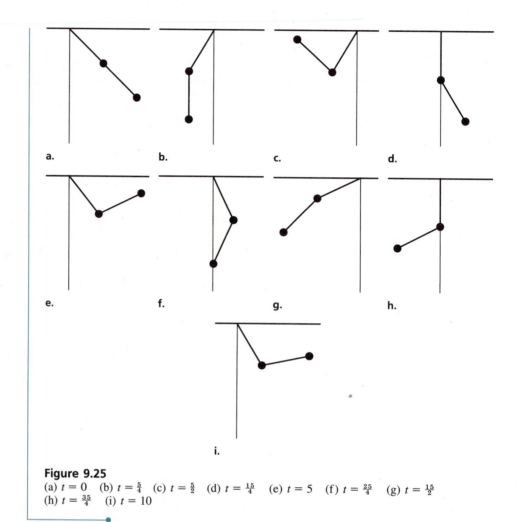

Figure 9.25
(a) $t = 0$ (b) $t = \frac{5}{4}$ (c) $t = \frac{5}{2}$ (d) $t = \frac{15}{4}$ (e) $t = 5$ (f) $t = \frac{25}{4}$ (g) $t = \frac{15}{2}$
(h) $t = \frac{35}{4}$ (i) $t = 10$

Does the system in Example 3 come to rest? Which pendulum experiences a greater displacement from equilibrium?

IN TOUCH WITH TECHNOLOGY

1. Solve the problem of the forced coupled spring-mass system with $m_1 = m_2 = 1$, $k_1 = 3$, and $k_2 = 2$ if the forcing functions are $F_1(t) = 1$ and $F_2(t) = \sin t$ and the initial conditions are $x(0) = 0$, $x'(0) = 1$, $y(0) = 1$, $y'(0) = 0$. Graph the solution parametrically as well as simultaneously. How does the motion differ from that of Example 1? What eventually happens to this system? Will the objects eventually come to rest?

2. Consider the three-spring problem shown in Figure 9.21 with $m_1 = m_2 = 1$ and $k_1 = k_2 = k_3 = 1$. Determine $x(t)$ and $y(t)$ if $x(0) = 0$, $x'(0) = -1$, $y(0) = 0$, and $y'(0) = 1$. Graph the solution both simultaneously and parametrically. When does the object attached to the top spring first pass through its equilibrium position? When does the object attached to the second spring first pass through its equilibrium position?

3. Solve Problem 2 with the forcing functions $F_1(t) = 1$ and $F_2(t) = \cos t$. Graph the solution both simultaneously and parametrically. When does the object attached to the top spring first pass through its equilibrium position? When does the object attached to the second spring first pass through its equilibrium position? How does the motion differ from that in Problem 2?

4. Consider the double pendulum system with $m_1 = 3$ slugs, $m_2 = 1$ slug, and $\ell_1 = \ell_2 = 16$ feet. If $g = 32$ ft/s^2, determine $\theta_1(t)$ and $\theta_2(t)$ if $\theta_1(0) = 1$, $\theta_1'(0) = 0$, $\theta_2(0) = 0$, $\theta_2'(0) = -1$. Graph the solution parametrically and simultaneously.

5. Suppose that in the double pendulum $m_1 = 3$ slugs, $m_2 = 1$ slug, and each pendulum has length 4 feet. If $g = 32$ ft/s^2, determine $\theta_1(t)$ and $\theta_2(t)$ if $\theta_1(0) = 1$, $\theta_1'(0) = 0$, $\theta_2(0) = 0$, $\theta_2'(0) = -1$. Graph the solution parametrically and simultaneously. How does the length of each pendulum affect the motion as compared to that of Problem 4?

6. Suppose that $m_1 = m_2 = 1$ slug, and each pendulum has length 4 feet in the double pendulum system. If $g = 32$ ft/s^2, determine $\theta_1(t)$ and $\theta_2(t)$ if $\theta_1(0) = 1$, $\theta_1'(0) = 0$, $\theta_2(0) = 0$, $\theta_2'(0) = -1$. Graph the solution parametrically and simultaneously. How does the mass of the first object affect the motion as compared to that of Problem 5?

7. Consider the physical situation shown in Figure 9.26 where the uniform bar has mass m_1, a centroidal moment of inertia $\bar{I}$, and is supported by two springs each with spring constant k. A mass m_3 is attached at the center of gravity (Point G) of the bar by another spring with spring constant k. Using the coordinates q_1, q_2, and q_3 as shown in Figure 9.26, we find that the motion is modeled with

$$\begin{cases} m_1 \dfrac{d^2 q_1}{dt^2} + 3kq_1 - kq_3 = 0 \\[2mm] \bar{I} \dfrac{d^2 q_2}{dt^2} + 2k\ell^2 q_2 = 0. \\[2mm] m_3 \dfrac{d^2 q_3}{dt^2} - kq_1 + kq_3 = 0 \end{cases}$$

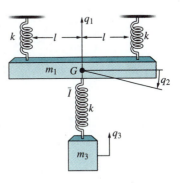

Figure 9.26

(a) Determine the motion of the system if $k = 2$, $m_1 = 2$, $m_3 = 1$, $\bar{I} = 1$, $\ell = 1$ and the initial conditions are $q_1(0) = 0$, $q_1'(0) = 1$, $q_2(0) = 0$, $q_2'(0) = 0$, $q_3(0) = 0$, and $q_3'(0) = 0$. What is the maximum displacement of q_1, q_2, and q_3? (b) If the initial conditions are $q_1(0) = 0$, $q_1'(0) = 0$, $q_2(0) = 0$, $q_2'(0) = 0$, $q_3(0) = 1$, and $q_3'(0) = 0$, determine the motion of the system. What is the maximum displacement of q_1, q_2, and q_3? (c) How do the changes in the initial conditions affect the maximum displacement of each component of the system?

8. Consider the system of three springs shown in Figure 9.27 where springs S_1, S_2, and S_3 have spring constants k_1, k_2, and k_3, respectively, and have objects of mass m_1, m_2, and m_3 attached to them. In this case, summing the forces acting on each mass and applying Newton's second law of motion yields

$$\begin{cases} m_1 \dfrac{d^2 x_1}{dt^2} = -k_1 x_1 + k_2(x_2 - x_1) \\[2mm] m_2 \dfrac{d^2 x_2}{dt^2} = -k_2(x_2 - x_1) + k_3(x_3 - x_2). \\[2mm] m_3 \dfrac{d^2 x_3}{dt^2} = -k_3(x_3 - x_2) \end{cases}$$

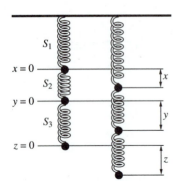

Figure 9.27

Solve this system if $m_1 = 2$, $m_2 = m_3 = 1$, $k_1 = \frac{1}{2}$, $k_2 = 1$, and $k_3 = 2$, using the initial conditions $x_1(0) = x_2(0) = x_3(0) = 0$, $\dfrac{dx_1}{dt}(0) = -2$, $\dfrac{dx_2}{dt}(0) = 1$, and $\dfrac{dx_3}{dt}(0) = 2$. When does each mass first pass through its equilibrium position?

EXERCISES 9.3

In Exercises 1–6, solve the coupled spring-mass system modeled by

$$\begin{cases} m_1 \dfrac{d^2x}{dt^2} = -k_1 x + k_2(y - x) \\[2mm] m_2 \dfrac{d^2y}{dt^2} = -k_2(y - x) \end{cases}$$

using the indicated parameter values and initial conditions.

1. $m_1 = 2$, $m_2 = 2$, $k_1 = 6$, $k_2 = 4$, $x(0) = 0$, $x'(0) = 0$, $y(0) = 1$, $y'(0) = 0$

2. $m_1 = 2$, $m_2 = 2$, $k_1 = 6$, $k_2 = 4$, $x(0) = 0$, $x'(0) = -1$, $y(0) = 0$, $y'(0) = 0$

*3. $m_1 = 2$, $m_2 = 1$, $k_1 = 4$, $k_2 = 2$, $x(0) = 1$, $x'(0) = 0$, $y(0) = 0$, $y'(0) = 0$

4. $m_1 = 2$, $m_2 = 1$, $k_1 = 4$, $k_2 = 2$, $x(0) = 0$, $x'(0) = 0$, $y(0) = 0$, $y'(0) = -1$

5. $m_1 = 2$, $m_2 = 2$, $k_1 = 3$, $k_2 = 2$, $x(0) = 1$, $x'(0) = 0$, $y(0) = 0$, $y'(0) = 0$

6. $m_1 = 2$, $m_2 = 2$, $k_1 = 3$, $k_2 = 2$, $x(0) = 0$, $x'(0) = 1$, $y(0) = 0$, $y'(0) = 0$

In Exercises 7–10, solve the coupled spring-mass system modeled with

$$\begin{cases} m_1 \dfrac{d^2x}{dt^2} = -k_1 x + k_2(y - x) + F_1(t) \\[2mm] m_2 \dfrac{d^2y}{dt^2} = -k_2(y - x) + F_2(t) \end{cases}$$

where $m_1 = 2$, $m_2 = 1$, $k_1 = 4$, $k_2 = 2$, $x(0) = 1$, $x'(0) = 0$, $y(0) = 0$, $y'(0) = 0$ using the given forcing functions.

7. $F_1(t) = 1$, $F_2(t) = \sin t$
8. $F_1(t) = 1$, $F_2(t) = 1$
*9. $F_1(t) = \cos 2t$, $F_2(t) = 0$
10. $F_1(t) = \cos t$, $F_2(t) = \sin t$

In Exercises 11–18, solve the coupled pendulum problem using the parameter values $m_1 = 3$ slugs, $m_2 = 1$ slug, $\ell_1 = \ell_2 = 16$ feet, and $g = 32$ ft/s^2 and the indicated initial conditions.

11. $\theta_1(0) = 0$, $\theta_1'(0) = 0$, $\theta_2(0) = 0$, $\theta_2'(0) = -1$
12. $\theta_1(0) = 1$, $\theta_1'(0) = 0$, $\theta_2(0) = 0$, $\theta_2'(0) = 1$
*13. $\theta_1(0) = 1$, $\theta_1'(0) = 0$, $\theta_2(0) = 0$, $\theta_2'(0) = 0$
14. $\theta_1(0) = 1$, $\theta_1'(0) = 1$, $\theta_2(0) = 0$, $\theta_2'(0) = 0$
15. $\theta_1(0) = 0$, $\theta_1'(0) = 1$, $\theta_2(0) = 0$, $\theta_2'(0) = 0$
16. $\theta_1(0) = 0$, $\theta_1'(0) = 1$, $\theta_2(0) = -1$, $\theta_2'(0) = 0$
*17. $\theta_1(0) = 0$, $\theta_1'(0) = 0$, $\theta_2(0) = -1$, $\theta_2'(0) = 0$
18. $\theta_1(0) = 0$, $\theta_1'(0) = 0$, $\theta_2(0) = -1$, $\theta_2'(0) = -1$

19. Consider the physical situation of two pendulums coupled with a spring as shown in Figure 9.28. The motion of this pendulum-spring system is approximated by solving the second-order system

$$\begin{cases} mx'' + m\omega_0^2 x = -k(x - y) \\ my'' + m\omega_0^2 y = -k(y - x) \end{cases}$$

where L is the length of each pendulum, g is the gravitational constant, and $\omega_0^2 = g/L$. Use the method of

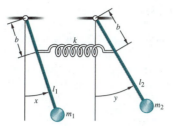

Figure 9.28

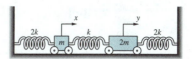

Figure 9.29

Laplace transforms to solve this system if the initial conditions are $x(0) = a$, $x'(0) = b$, $y(0) = c$, $y'(0) = d$.

20. Solve the initial-value problem

$$\begin{cases} mx'' + m\omega_0^2 x = -k(x - y) \\ my'' + m\omega_0^2 y = -k(y - x) \\ x(0) = -1,\ x'(0) = 0 \\ y(0) = 1,\ y'(0) = 0. \end{cases}$$

What is the maximum displacement of $x(t)$ and $y(t)$?

*21. Solve the initial-value problem

$$\begin{cases} mx'' + m\omega_0^2 x = -k(x - y) \\ my'' + m\omega_0^2 y = -k(y - x), \end{cases}$$
$$x(0) = 0,\ x'(0) = 1,\ y(0) = 0,\ y'(0) = 0.$$

Verify that the initial conditions are satisfied.

22. The physical situation shown in Figure 9.29 is modeled by the system of differential equations

$$\begin{cases} m\dfrac{d^2x}{dt^2} + 3kx - ky = 0 \\[2mm] 2m\dfrac{d^2y}{dt^2} + 3ky - kx = 0 \end{cases}$$

If $x(0) = 0$, $x'(0) = 1$, $y(0) = 0$, and $y'(0) = 0$, find x and y. Compare these results to those found if $x(0) = 0$, $x'(0) = 0$, $y(0) = 0$, and $y'(0) = 1$. Does this model resemble one that was introduced earlier in the section?

23. Write the system
$$\begin{cases} m_1\dfrac{d^2x}{dt^2} = -k_1x + k_2(y - x) \\[2mm] m_2\dfrac{d^2y}{dt^2} = -k_3y - k_2(y - x) \end{cases}$$
as a system of four first-order equations with the substitutions $x = x_1$, $x_1' = x_2$, $y = y_1$, $y_1' = y_2$.

24. Show that the eigenvalues of the 4×4 coefficient matrix in the system of equations obtained in Exercise 23 are two complex conjugate pairs of the form $\pm q_1 i$ and $\pm q_2 i$ where q_1 and q_2 are real numbers. What does this tell you about the solutions to this system? Does this agree with the solution obtained with Laplace transforms?

9.4 Nonlinear Systems of Equations

Biological Systems: Predator-Prey Interaction • Physical Systems: Variable Damping

Several special equations and systems that arise in the study of many areas of applied mathematics can be solved using the techniques of Chapter 8. These include the predator-prey population dynamics problem, the Van-der-Pol equation that models variable damping in a spring-mass system, and the Bonhoeffer-Van-der-Pol (BVP) oscillator.

Biological Systems: Predator-Prey Interaction

Let $x(t)$ and $y(t)$ represent the number of members at time t of the prey and predator populations, respectively. (Examples of such populations include fox/rabbit and shark/seal.) Suppose that the positive constant a is the birth rate of $x(t)$ so that in the absence of the predator

$$\frac{dx}{dt} = ax$$

and that $c > 0$ is the death rate of y which indicates that

$$\frac{dy}{dt} = -cy$$

in the absence of the prey population. In addition to these factors, the number of interactions between predator and prey affects the number of members in the two populations. Note that an interaction increases the growth of the predator population and decreases the growth of the prey population, because an interaction between the two populations indicates that a predator overtakes a member of the prey population. In order to include these interactions in the model, we assume that the number of interactions is directly proportional to the product of $x(t)$ and $y(t)$. Therefore, the rate at which $x(t)$ changes with respect to time is

$$\frac{dx}{dt} = ax - bxy,$$

where $b > 0$. Similarly, the rate at which $y(t)$ changes with respect to time is

$$\frac{dy}{dt} = -cy + dxy,$$

where $d > 0$. These equations and initial conditions form the **Lotka-Volterra problem**

$$\begin{cases} \dfrac{dx}{dt} = ax - bxy \\[2mm] \dfrac{dy}{dt} = -cy + dxy \end{cases}, \quad x(0) = x_0, \; y(0) = y_0.$$

Alfred James Lotka (1880–1949) American biophysicist, born in the Ukraine; wrote first text on mathematical biology. (UPI/Bettman)

| **Example 1** | Find and classify the equilibrium points of the Lotka-Volterra system. |

Solution   Solving $\begin{cases} ax - bxy = x(a - by) = 0 \\ -cy + dxy = y(-c + dx) = 0 \end{cases}$, we have $x = 0$ or $y = a/b$ and $y = 0$ or $x = c/d$. The equilibrium points are $(0, 0)$ and $(c/d, a/b)$. The Jacobian matrix of the nonlinear system is $J(x, y) = \begin{pmatrix} a - by & -bx \\ dy & -c + dx \end{pmatrix}$. Near $(0, 0)$, we have $J(0, 0) = \begin{pmatrix} a & 0 \\ 0 & -c \end{pmatrix}$ with eigenvalues $\lambda_1 = -c$ and $\lambda_2 = a$. Because

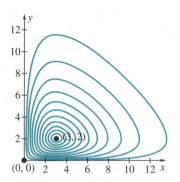

Figure 9.30 Typical Solutions of the Lotka-Volterra System

these eigenvalues are real with opposite signs, we classify $(0, 0)$ as a saddle. Similarly, near $(c/d, a/b)$, we have $J(c/d, a/b) = \begin{pmatrix} 0 & -bc/d \\ ad/b & 0 \end{pmatrix}$ with eigenvalues $\lambda_{1,2} = \pm i\sqrt{ac}$, so the point $\left(\frac{c}{d}, \frac{a}{b}\right)$ is classified as a center.

In Figure 9.30, we show several solutions (which were found using different initial populations) parametrically in the phase plane of this system with $a = 2$, $b = 1$, $c = 3$ and $d = 1$. Notice that all of the solutions oscillate about the center. These solutions reveal the relationship between the two populations: prey, $x(t)$, and predator, $y(t)$. As we follow one cycle counterclockwise beginning, for example, near the point $(\frac{3}{2}, 1)$ (see Figure 9.31), we notice that as the prey population, $x(t)$, increases, the predator population, $y(t)$, first slightly decreases (is that really possible?) and then increases until the predator becomes overpopulated. Then, because the prey population is too small to supply the predator population, the predator population decreases, which leads to an increase in the population of the prey. At this point, because the number of predators becomes too small to control the prey population, the number in the prey population becomes overpopulated and the cycle repeats itself.

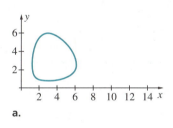
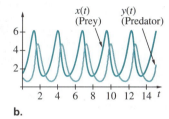

a.   b.

Figure 9.31 (a)–(b)

How does the period of a solution with an initial point near the equilibrium point $(\frac{c}{d}, \frac{a}{b})$ compare to that of a solution with an initial point that is not located near this point?

Physical Systems: Variable Damping

In some physical systems, energy is fed into the system when there are small oscillations while energy is taken from the system when there are large oscillations. This indicates that the system undergoes ''negative damping'' for small oscillations and ''positive damping'' for large oscillations. A differential equation that models this situation is **Van-der-Pol's equation**

$$x'' + \mu(x^2 - 1)x' + x = 0$$

where μ is a positive constant. We can transform this second-order differential equation into a system of first-order differential equations with the substitution $x' = y$. Hence,

Balthasar Van-der-Pol
Dutch applied mathematician and engineer (UPI/Bettman)

$y' = x'' = -x - \mu(x^2 - 1)x' = -x - \mu(x^2 - 1)y$, so the corresponding system of equations is

$$\begin{cases} x' = y \\ y' = -x - \mu(x^2 - 1)y \end{cases}$$

which is solved using an initial position $x(0) = x_0$ and an initial velocity $y(0) = x'(0) = y_0$. Notice that $\mu(x^2 - 1)$ represents the damping coefficient. This system models variable damping because $\mu(x^2 - 1) < 0$ when $-1 < x < 1$ and $\mu(x^2 - 1) > 0$ when $|x| > 1$. Therefore, damping is negative for the small oscillations, $-1 < x < 1$, and positive for the large oscillations, $|x| > 1$.

Example 2

Find and classify the equilibrium points of the system of differential equations that is equivalent to Van-der-Pol's equation.

Solution   We find the equilibrium points by solving

$$\begin{cases} y = 0 \\ -x - \mu(x^2 - 1)y = 0 \end{cases}$$

From the first equation, we see that $y = 0$. Substitution of $y = 0$ into the second equation yields $x = 0$ as well. Therefore, the equilibrium point is $(0, 0)$.

The Jacobian matrix for this system is

$$J(x, y) = \begin{pmatrix} 0 & 1 \\ -1 - 2\mu xy & -\mu(x^2 - 1) \end{pmatrix}.$$

At $(0, 0)$, we have the matrix

$$J(0, 0) = \begin{pmatrix} 0 & 1 \\ -1 & \mu \end{pmatrix}.$$

We find the eigenvalues of $J(0, 0)$ by solving

$$\begin{vmatrix} -\lambda & 1 \\ -1 & \mu - \lambda \end{vmatrix} = \lambda^2 - \mu\lambda + 1 = 0$$

which has roots $\lambda_{1,2} = \dfrac{\mu \pm \sqrt{\mu^2 - 4}}{2}$. Notice that if $\mu > 2$, then both eigenvalues are positive and real, so we classify $(0, 0)$ as an **unstable node.** On the other hand, if $0 < \mu < 2$, the eigenvalues are a complex conjugate pair with a positive real part. Hence, $(0, 0)$ is an **unstable spiral.** (We omit the case when $\mu = 2$ because the eigenvalues are repeated.)

In Figure 9.32, we show several curves in the phase plane that begin at different points for various values of μ. In each figure, we see that all of the curves approach a curve called a **limit cycle.** Physically, the fact that the system has a limit cycle indicates that for all oscillations the motion eventually becomes periodic, which is represented by a closed curve in the phase plane.

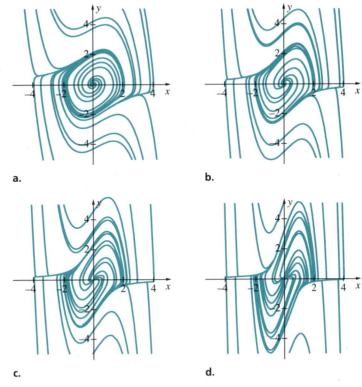

Figure 9.32 (a) $\mu = \frac{1}{2}$ (b) $\mu = 1$ (c) $\mu = \frac{3}{2}$ (d) $\mu = 3$

On the other hand, in Figure 9.33 we graph the solution that satisfies the initial conditions $x(0) = 1$ and $y(0) = 0$ parametrically and individually for various values of μ. Notice that for small values of μ the system more closely approximates that of the harmonic oscillator because the damping coefficient is small. The curves are more circular than those for larger values of μ.

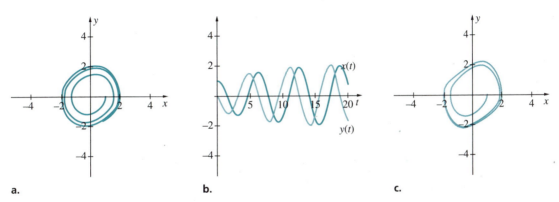

Figure 9.33 (a)–(b) $\mu = \frac{1}{4}$ (c)–(d) $\mu = \frac{1}{2}$ (e)–(f) $\mu = 1$ (g)–(h) $\mu = \frac{3}{2}$
(i)–(j) $\mu = 2$ (k)–(l) $\mu = 3$

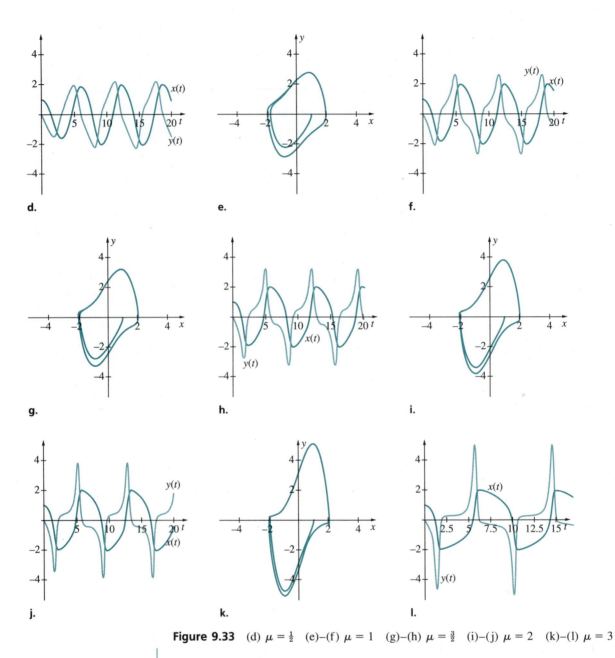

d.   **e.**   **f.**

g.   **h.**   **i.**

j.   **k.**   **l.**

Figure 9.33   (d) $\mu = \frac{1}{2}$ (e)–(f) $\mu = 1$ (g)–(h) $\mu = \frac{3}{2}$ (i)–(j) $\mu = 2$ (k)–(l) $\mu = 3$

Graph several solution curves in the phase plane for the Van-der-Pol equation if $\mu = 0.1$ and $\mu = 0.001$. Compare these graphs to the corresponding solution curves for the equation $x'' + x = 0$. Are they similar? Why?

IN TOUCH WITH TECHNOLOGY

1. The **Bonhoeffer-Van-der-Pol (BVP) oscillator** is the system of ordinary differential equations

$$\begin{cases} \dfrac{dx}{dt} = x - \dfrac{x^3}{3} - y + I(t) \\[2mm] \dfrac{dy}{dt} = c(x + a - by) \end{cases}$$
. **(a)** Find and classify
the equilibrium points of this system if $I(t) = 0$, $a = c = 1$ and $b = 1$. **(b)** Graph the direction field associated with the system and then approximate the phase plane by graphing several solutions near each equilibrium point.

2. Find $x_0 > 0$ so that the solution to the initial-value problem

$$\begin{cases} x' = y \\ y' = -x - (x^2 - 1)y \\ x(0) = x_0, \ y(0) = 0 \end{cases}$$

is periodic. Confirm your result graphically.

3. **(Competing Species)** (See Problem 1 of **In Touch with Technology,** Section 1.5.) The system of equations

$$\begin{cases} \dfrac{dx}{dt} = x(a - b_1x - b_2y) \\[2mm] \dfrac{dy}{dt} = y(c - d_1x - d_2y) \end{cases}$$
,

where a, b_1, b_2, c, d_1, and d_2 represent positive constants, can be used to model the population of two species, represented by $x(t)$ and $y(t)$, competing for a common food supply.
   (a) (i) Find and classify the equilibrium points of the system if $a = 1$, $b_1 = 2$, $b_2 = 1$, $c = 1$, $d_1 = 0.75$, and $d_2 = 2$. **(ii)** Graph several solutions by using different initial populations parametrically in the phase plane. **(iii)** Find $\lim_{t \to \infty} x(t)$ and $\lim_{t \to \infty} y(t)$ if both $x(0)$ and $y(0)$ are not zero. Compare your result to **(ii)**.
   (b) (i) Find and classify the equilibrium points of the system if $a = 1$, $b_1 = 1$, $b_2 = 1$, $c = 0.67$, $d_1 = 0.75$, and $d_2 = 1$. **(ii)** Graph several solutions by using different initial populations parametrically in

the phase plane. **(iii)** Determine the fate of the species with population $y(t)$. What happens to the species with population $x(t)$? What happens to the species with population $y(t)$ if the species with population $x(t)$ is suddenly removed? (*Hint:* How does the term $-d_1xy$ affect the equation $dy/dt = y(c - d_1x - d_2y)$?) **(c)** Find conditions on the positive constants a, b_1, b_2, c, d_1, and d_2 so that **(i)** the system has exactly one equilibrium point in the first quadrant and **(ii)** the system has no equilibrium point in the first quadrant. **(iii)** Is it possible to choose positive constants a, b_1, b_2, c, d_1, and d_2 so that each population grows without bound? If so, illustrate graphically.

4. **(Chemical Reactor)** Consider the continuous-flow stirred tank reactor shown in Figure 9.34. In this reaction, a stream of chemical C flows into the tank of volume V, and products as well as residue flow out of the tank at a constant rate q. Because the reaction is continuous, the composition and temperature of the contents of the tank are constant and are the same as the composition and temperature of the stream flowing out of the tank. Suppose that a concentration c_{in} of the chemical C flows into the tank while a concentration c flows out of the tank. In addition, suppose that the chemical C changes into products at a rate proportional to the concentration c of C, where the constant of proportionality $k(T)$ depends on the temperature T,

$$k(T) = Ae^{-B/T}.$$

Because

$$\left(\begin{matrix} \text{Rate of change} \\ \text{of amount of } C \end{matrix}\right) = (\text{Rate in of } C) -$$

$$(\text{Rate out of } C) - \left(\begin{matrix} \text{Rate that} \\ C \text{ disappears} \\ \text{by the reaction} \end{matrix}\right),$$

Flow Rate q
Concentration c_{in} → | Volume V | → Flow Rate q
Temperature T_{in} Concentration c
      Tank   Temperature T

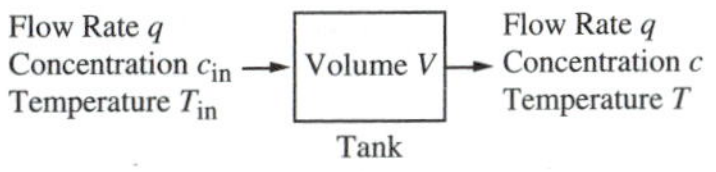

Figure 9.34
Continuous-Flow Stirred Tank Reactor

we have the differential equation

$$\frac{d}{dt}(Vc) = qc_{in} - qc - Vk(T)c.*$$

(a) Show that if V is constant, then this equation is equivalent to $\dfrac{dc}{dt} = \dfrac{q}{V}(c_{in} - c) - k(T)c.$

(b) In a similar manner, we balance the heat of the reaction with

$$\begin{pmatrix} \text{Rate of change} \\ \text{of heat content} \end{pmatrix}$$
$$= (\text{Rate in of Heat}) - (\text{Rate out of Heat})$$
$$- \begin{pmatrix} \text{Heat removed} \\ \text{by cooling} \end{pmatrix} + \begin{pmatrix} \text{Heat produced} \\ \text{by reaction} \end{pmatrix}.$$

Let C_p be the specific heat so that the heat content per unit volume of the reaction mixture at temperature T is $C_p T$. If H is the rate at which heat is generated by the reaction and $VS(T)$ is the rate at which heat is removed from the system by a cooling system, then we have the differential equation

$$VC_p \frac{dT}{dt} = qC_p T_{in} - qC_p T - VS(T) + HVk(T)c$$

where T_{in} is the temperature at which the chemical flows into the tank. Solve this equation for dT/dt.

(c) The equations in (a) and (b) form a system of non-linear ordinary differential equations. Show that the equilibrium point of this system satisfies the equations

$$c_{in} - c = \frac{V}{q}ck(T) \quad \text{and} \quad T - T_{in} = \frac{HV}{qC_p}ck(T)$$

if we assume there is no cooling system. Solve the first equation for c and substitute into the

second equation to find that

$$T - T_{in} = \frac{HVc_{in}}{qC_p + VC_p k(T)}k(T).$$ We call values of T that satisfy this equation **steady state temperatures.**

(d) If $T_{in} = 1$, $HVc_{in} = VC_p = qC_p = 1$, and $k(T) = e^{-1/T}$, graph $y = T - T_{in}$ and

$$y = \frac{HVc_{in}}{qC_p + VC_p k(T)}k(T)$$ to determine the number of roots of the equation in (c).

(e) If $T_{in} = 0.15$, $HVc_{in} = 1$, $VC_p = 0.25$, $qC_p = 0.015$, and $k(T) = e^{-3/T}$, graph $y = T - T_{in}$ and $y = \dfrac{HVc_{in}}{qC_p + VC_p k(T)}k(T)$ to determine the number of roots of the equation in (c). How does this situation differ from that in (d)?

(f) Notice that $y = T - T_{in}$ describes heat removal while $y = \dfrac{HVc_{in}}{qC_p + VC_p k(T)}k(T)$ describes heat production. Therefore, if the slope of the heat production curve is greater than that of the heat removal curve, the steady state is *unstable*. On the other hand, if the slope of the heat production curve is less than or equal to that of the heat removal curve, then the steady state is *stable*. Use this information to determine which of the temperatures found in (d) and (e) are stable and which are unstable. What do the slopes of these curves represent?

(g) Approximate the solution to the system of differential equations using the parameter values given in (d) and (e). Does the stability of the system correspond to those found in (f)?

---

* *Applications in Undergraduate Mathematics in Engineering,* The Mathematical Association of America, The Macmillan Company, New York, 1967, pp. 122–125. *Computational Methods for Process Simulation,* W. Fred Ramirez, Butterworths Series in Chemical Engineering, Boston, MA, 1989, pp. 175–182.

1. The phase paths of the Lotka-Volterra model

$$
\begin{cases}
\dfrac{dx}{dt} = ax - bxy \\[2mm]
\dfrac{dy}{dt} = -cy + dxy
\end{cases}
$$

are given by $\dfrac{dy}{dx} = \dfrac{dy/dt}{dx/dt} = \dfrac{-cy + dxy}{ax - bxy}$. Solve this

separable equation to find an implicit equation of the phase paths.

2. Consider the predator-prey model

$$
\begin{cases}
\dfrac{dx}{dt} = (a_1 - b_1 x - c_1 y)x \\[2mm]
\dfrac{dy}{dt} = (-a_2 + c_2 x)y
\end{cases}
$$

where $x(t)$ represents the prey population, $y(t)$ the predator population, and the constants a_1, a_2, b_1, c_1, c_2 are all positive. In this model, the term $-b_1 x^2$ represents the interference that occurs when the prey population becomes too large. **(a)** If $a_1 = 2, a_2 = b_1 = c_1 = c_2 = 1$, locate and classify the three equilibrium points. **(b)** If $a_2 = 2, a_1 = b_1 = c_1 = c_2 = 1$, classify the two equilibrium points.

*3. The national economy can be modeled with a nonlinear system of differential equations. If I represents the national income, C the rate of consumer spending, and G the rate of government spending, a simple model of the economy is

$$
\begin{cases}
\dfrac{dI}{dt} = I - aC \\[2mm]
\dfrac{dC}{dt} = b(I - C - G)
\end{cases},
$$

where a and b are constants. **(a)** Suppose that $a = b = 2$ and $G = k$, ($k =$ constant). Find and classify the equilibrium point. **(b)** If $a = 2, b = 1$, and $G = k$, find and classify the equilibrium point.

4. If $a = b = 2$ and $G = k_1 + k_2 I$ ($k_2 > 0$), show that there is no equilibrium point if $k_2 \geq \frac{1}{2}$ for the national economy model in Exercise 3. Describe the economy under these conditions.

5. The differential equation $x'' + \sin x = 0$ can be used to describe the motion of a pendulum. In this case,

$x(t)$ represents the displacement from the position $x = 0$. Represent this second-order equation as a system of first-order equations. Find and classify the equilibrium points of this system. Describe the physical significance of these points. Graph several paths in the phase plane of the system.

6. Show that the paths in the phase plane of $x'' + \sin x = 0$ satisfy the first-order equation $\dfrac{dy}{dx} = -\dfrac{\sin x}{y}$. (See Exercise 5.) Use separation of variables to show that paths are $\frac{1}{2}y^2 - \cos x = C$ where C is a constant. Graph several paths. Do these graphs agree with your result in Exercise 5?

*7. (a) Write the equation $\dfrac{d^2 x}{dt^2} + k^2 x = 0$, which models the **simple harmonic oscillator,** as a system of first-order equations. **(b)** Show that paths in the phase plane satisfy $\dfrac{dy}{dx} = -k^2 \dfrac{x}{y}$. (See Exercise 6.) **(c)** Solve the equation in (b) to find the paths. **(d)** What is the equilibrium point of this system and how is it classified? How do the paths in (c) compare with what you expect to see in the phase plane?

8. Repeat Exercise 7 for the equation $\dfrac{d^2 x}{dt^2} - k^2 x = 0$.

9. Suppose that a satellite is in flight on the line between a planet of mass M_1 and its moon of mass M_2 which are a constant distance R apart. The distance x between the satellite and the planet satisfies the nonlinear second-order equation

$$
x'' = -\frac{gM_1}{x^2} + \frac{gM_2}{(R - x)^2}
$$

where g is the gravitational constant. Transform this equation into a system of first-order equations. Find and classify the equilibrium point of the linearized system.

10. Consider the nonlinear autonomous system

$$
\begin{cases}
\dfrac{dx}{dt} = -y + x(1 - x^2 - y^2) \\[2mm]
\dfrac{dy}{dt} = x + y(1 - x^2 - y^2)
\end{cases}.
$$

(a) Show that $(0, 0)$ is an equilibrium point of this system. Classify $(0, 0)$.

(b) Show that $x = r \cos \theta$, $y = r \sin \theta$ transforms the system to the (uncoupled) system
$$\begin{cases} \dfrac{dr}{dt} = r(1 - r^2) \\ \dfrac{d\theta}{dt} = 1 \end{cases}$$

(c) Show that this system has solution
$$r(t) = \frac{1}{\sqrt{1 + ae^{-2t}}}, \quad \theta(t) = t + b \text{ so that}$$
$$x(t) = \frac{1}{\sqrt{1 + ae^{-2t}}} \cos(t + b),$$
$$y(t) = \frac{1}{\sqrt{1 + ae^{-2t}}} \sin(t + b).$$

(d) Calculate $\lim_{t \to \infty} x(t)$ and $\lim_{t \to \infty} y(t)$ to show that all solutions approach the circle $x^2 + y^2 = 1$ as $t \to \infty$. Thus, $x^2 + y^2 = 1$ is a limit cycle.

*11. Consider Liénard's equation $\dfrac{d^2x}{dt^2} + f(x)\dfrac{dx}{dt} + g(x) = 0$ where $f(x)$ and $g(x)$ are continuous. Show that this equation can be written as the system
$$\begin{cases} \dfrac{dx}{dt} = y - F(x) \\ \dfrac{dy}{dt} = -g(x) \end{cases} \quad \text{where } F(x) = \int_0^x f(u)\,du.$$

12. (See Exercise 11.) Liénard's theorem states that if **(i)** $F(x)$ is an odd function, **(ii)** $F(x)$ is zero only at $x = 0$, $x = a$, $x = -a$ (for some $a > 0$), **(iii)** $F(x) \to \pm\infty$ monotonically for $x > a$, and **(iv)** $g(x)$ is an odd function where $g(x) > 0$ for all $x > 0$; then Liénard's equation has a unique limit cycle. Use Liénard's theorem to determine which of the following equations has a unique limit cycle:

(a) $\dfrac{d^2x}{dt^2} + \varepsilon(1 - x^2)\dfrac{dx}{dt} + x = 0$, $\varepsilon > 0$;

(b) $\dfrac{d^2x}{dt^2} + 3x^2\dfrac{dx}{dt} + x^3 = 0$.

13. Consider the system of autonomous equations
$$\begin{cases} \dfrac{dx}{dt} = f(x, y) \\ \dfrac{dy}{dt} = g(x, y) \end{cases}$$
. **Bendixson's theorem** (or Negative Criterion) states that if $f_x(x, y) + g_y(x, y)$ is a continuous function that is either always positive or always negative in a particular region R of the phase plane, then the system has no limit cycle in R. Use this the-

orem to determine if the given system has no limit cycle in the phase plane.

(a) $\begin{cases} \dfrac{dx}{dt} = x^3 + x + 7y \\ \dfrac{dy}{dt} = x^2y \end{cases}$; **(b)** $\begin{cases} \dfrac{dx}{dt} = y - x \\ \dfrac{dy}{dt} = 3x - y \end{cases}$;

(c) $\begin{cases} \dfrac{dx}{dt} = xy^2 \\ \dfrac{dy}{dt} = x^2 + 8y \end{cases}$

14. Let E and s be positive constants and suppose that f is a continuous odd function that approaches a finite limit as $x \to \infty$, is increasing, and is concave down for $x > 0$. The voltages over the deflection plate in a sweeping circuit for an oscilloscope (Androkov and Chaiken, 1949) are determined by solving
$$\begin{cases} \dfrac{dV_1}{dt} = -sV_1 + f(E - V_2) \\ \dfrac{dV_2}{dt} = -sV_2 + f(E - V_1) \end{cases}$$
. Use Bendixson's theorem (see Exercise 13) to show that this system has no limit cycles.

*15. Consider the relativistic equation for the central orbit of a planet, $\dfrac{d^2u}{d\theta^2} + u - ku^2 = \alpha$, where k and α are positive constants (k is very small), $u = 1/r$, and r and θ are polar coordinates. **(a)** Write this second-order equation as a system of first-order equations. **(b)** Show that $\left(\dfrac{1 - \sqrt{1 - 4k\alpha}}{2k}, 0\right)$ is a center in the linearized system.

16. Consider the system of equations $\begin{cases} \dfrac{dx}{dt} = P(x, y) \\ \dfrac{dy}{dt} = Q(x, y) \end{cases}$ where P and Q are polynomials of degree n. Finding the maximum number of limit cycles of this system, called the **Hilbert number** H_n, has been investigated as part of Hilbert's 16th Problem. Several of these numbers are known: $H_0 = 0$, $H_1 = 0$, $H_2 \geq 4$, $H_n \geq \dfrac{n - 1}{2}$ (if n is odd), and $H_n < \infty$. Determine the Hilbert number (or a restriction on the Hilbert number) for each system: **(a)** $\begin{cases} \dfrac{dx}{dt} = 10 \\[2mm] \dfrac{dy}{dt} = -5 \end{cases}$; **(b)** $\begin{cases} \dfrac{dx}{dt} = x - y \\[2mm] \dfrac{dy}{dt} = y + 1 \end{cases}$;

(c) $\begin{cases} \dfrac{dx}{dt} = y^2 - 2xy \\[2mm] \dfrac{dy}{dt} = x^2 + y^2 \end{cases}$; and **(d)** $\begin{cases} \dfrac{dx}{dt} = x + 5y^7 \\[2mm] \dfrac{dy}{dt} = 10x^7 + y^4 \end{cases}$.

17. In a mechanical system, suppose that $x(t)$ represents position at time t, K kinetic energy, and V potential energy where $K = \dfrac{1}{2}m(x)\left(\dfrac{dx}{dt}\right)^2$ (m is a positive function) and $V = V(x)$. If the system is **conservative**, then the total energy E of the system remains constant during motion, which indicates that $\dfrac{1}{2}m(x)\left(\dfrac{dx}{dt}\right)^2 + V(x) = E$. Show that with the change of variable $u = \displaystyle\int \sqrt{m(x)}\,dx$ this equation becomes $\dfrac{d^2u}{dt^2} + Q'(u) = 0$ where $Q'(u) = \dfrac{V'(x)}{\sqrt{m(x)}}$. This means that an equation of the form $\dfrac{d^2x}{dt^2} = f(x)$ is a **conservative system** where f is a function representing force per unit mass that does not depend on dx/dt.

18. Consider the conservative system $\dfrac{d^2x}{dt^2} = f(x)$ in which f is a continuous function. Suppose that $V(x) = -\displaystyle\int f(x)\,dx$ so that $V'(x) = -f(x)$. **(a)** Use the substitution $dx/dt = y$ to transform $\dfrac{d^2x}{dt^2} = f(x)$ into the system of first-order equations $\begin{cases} \dfrac{dx}{dt} = y \\[2mm] \dfrac{dy}{dt} = -V'(x) \end{cases}$. **(b)** Find the equilibrium points of the system in (a). What is the physical significance of these points? **(c)** Use the chain rule and $dx/dt = y$ to show that $\dfrac{dy}{dt} = y\dfrac{dy}{dx}$. **(d)** Show that the paths in the phase plane are $\tfrac{1}{2}y^2 + V(x) = C$.

*19. (See Exercise 18.) What are the paths in the phase plane if $V(x) = \tfrac{1}{2}x^2$? How does this compare with the

classification of the equilibrium point of the corresponding system of first-order equations? Notice that if V has a local minimum at $x = a$, the system has a center at the corresponding point in the phase plane.

20. (See Exercise 18.) What are the paths in the phase plane if $V(x) = -\tfrac{1}{2}x^2$? How does this compare with the classification of the equilibrium point of the corresponding system of first-order equations? Notice that if V has a local maximum at $x = a$, the system has a saddle at the corresponding point in the phase plane.

21. Use the observations made in Exercises 19–20 concerning the relationship between the local extrema of V and the classification of equilibrium points in the phase plane to classify the equilibrium points of a conservative system with **(a)** $V'(x) = x^2 - 1$; **(b)** $V'(x) = x - x^3$.

22. Which of the following physical systems are conservative? **(a)** The motion of a pendulum modeled by $\dfrac{d^2x}{dt^2} + \sin x = 0$. **(b)** A spring-mass system that disregards damping and external forces. **(c)** A spring-mass system that includes damping.

*23. Consider a brake that acts on a wheel. Assuming that the force due to friction depends only on the angular velocity of the wheel, $\dfrac{d\theta}{dt}$, we have $I\dfrac{d^2\theta}{dt^2} = -FR\,\text{sgn}\left(\dfrac{d\theta}{dt}\right)$ to describe the spinning motion of the wheel where R is the radius of the brake drum, F is the frictional force, I is the moment of inertia of the wheel, and $\text{sgn}(x) = \begin{cases} 1, & x > 0 \\ 0, & x = 0 \\ -1, & x < 0 \end{cases}$. **(a)** Let $d\theta/dt = y$ and transform this second-order equation into a system of first-order equations. **(b)** What are the equilibrium points of this system? **(c)** Show that $\dfrac{d^2\theta}{dt^2} = \dfrac{d\theta}{dt}\dfrac{d}{d\theta}\left(\dfrac{d\theta}{dt}\right)$. **(d)** Use the relationship in **(c)** to show that the paths in the phase plane are $\dfrac{1}{2}I\left(\dfrac{d\theta}{dt}\right)^2 = -FR\theta + C$, $\dfrac{d\theta}{dt} > 0$ and $\dfrac{1}{2}I\left(\dfrac{d\theta}{dt}\right)^2 = FR\theta + C$, $\dfrac{d\theta}{dt} < 0$. What are these paths?

24. The equation $\dfrac{d^2x}{dt^2} + k^2 \sin x = 0$ that models the motion of a pendulum is approximated with $\dfrac{d^2x}{dt^2} + k^2\left(x - \dfrac{1}{6}x^3\right) = 0$. Why? Write this equation as a

system, then locate and classify its equilibrium points. How do these findings differ from those of $\dfrac{d^2x}{dt^2} + k^2 \sin x = 0$?

CHAPTER 9 SUMMARY
Concepts and Formulas

L-R-C Circuit with One Loop

$$\begin{cases} \dfrac{dQ}{dt} = I \\[2mm] \dfrac{dI}{dt} = -\dfrac{1}{LC}Q - \dfrac{R}{L}I + \dfrac{E(t)}{L} \\[2mm] Q(0) = Q_0,\ I(0) = I_0 \end{cases}$$

L-R-C Circuit with Two Loops

$$\begin{cases} \dfrac{dQ}{dt} = -\dfrac{1}{R_1 C}Q - I_2 + \dfrac{E(t)}{R_1} \\[2mm] \dfrac{dI_2}{dt} = \dfrac{1}{LC}Q - \dfrac{R_2}{L}I_2 \\[2mm] Q(0) = Q_0,\ I_2(0) = I_0 \end{cases}$$

Spring-Mass System

$$\begin{cases} \dfrac{dx}{dt} = y \\[2mm] \dfrac{dy}{dt} = -\dfrac{k}{m}x - \dfrac{c}{m}y \end{cases}$$

Mixture Problem with Two Tanks

$$\begin{cases} \dfrac{dx}{dt} = RC - \dfrac{Rx}{V_1} \\[2mm] \dfrac{dy}{dt} = \dfrac{Rx}{V_1} - \dfrac{Ry}{V_2} \end{cases}$$

Population of Two Neighboring Territories

$$\begin{cases} \dfrac{dx}{dt} = (a_1 - a_2)x + b_1 y \\[2mm] \dfrac{dy}{dt} = a_2 x + (b_1 - b_2)y \end{cases}$$

Population of Three Neighboring Territories

$$\begin{cases} \dfrac{dx}{dt} = (a_1 - a_2 - a_3)x + b_2 y + c_2 z \\[2mm] \dfrac{dy}{dt} = a_2 x + (b_1 - b_2 - b_3)y + c_3 z \\[2mm] \dfrac{dz}{dt} = a_3 x + b_3 y + (c_1 - c_2 - c_3)z \end{cases}$$

Diffusion through a Membrane

$$\begin{cases} \dfrac{dx_1}{dt} = P\left(\dfrac{x_2}{V_2} - \dfrac{x_1}{V_1}\right) \\[2mm] \dfrac{dx_2}{dt} = P\left(\dfrac{x_1}{V_1} - \dfrac{x_2}{V_2}\right) \end{cases}$$

Coupled Spring-Mass System: Two Springs

$$\begin{cases} m_1 \dfrac{d^2x}{dt^2} = -k_1 x + k_2(y - x) + F_1(t) \\[2mm] m_2 \dfrac{d^2y}{dt^2} = -k_2(y - x) + F_2(t) \end{cases}$$

Coupled Spring-Mass System: Three Springs

$$\begin{cases} m_1 \dfrac{d^2x}{dt^2} = -k_1x + k_2(y - x) + F_1(t) \\[3mm] m_2 \dfrac{d^2y}{dt^2} = -k_3y - k_2(y - x) + F_2(t) \end{cases}$$

The Double Pendulum

$$\begin{cases} (m_1 + m_2)\ell_1^2\theta_1'' + m_2\ell_1\ell_2\theta_2'' + (m_1 + m_2)\ell_1 g\theta_1 = 0 \\ m_2\ell_2^2\theta_2'' + m_2\ell_1\ell_2\theta_1'' + m_2\ell_2 g\theta_2 = 0 \end{cases}$$

Section 9.4

Predator-Prey (Lotka-Volterra system)

$$\begin{cases} \dfrac{dx}{dt} = ax - bxy \\[3mm] \dfrac{dy}{dt} = -cy + dxy \end{cases}$$

Van-der-Pol's Equation (System)

$$x'' + \mu(x^2 - 1)x' + x = 0; \quad \begin{cases} x' = y \\ y' = -x - \mu(x^2 - 1)y \end{cases}$$

CHAPTER 9 REVIEW EXERCISES

1. Solve the one-loop L-R-C circuit with $L = 1$ henry, $R = \frac{5}{3}$ ohms, $C = \frac{3}{2}$ farads, $Q(0) = 10^{-6}$ coulomb, $I(0) = 0$ amperes, and: **(a)** $E(t) = 0$ volts; **(b)** $E(t) = e^{-t}$ volts.

2. Solve the one-loop L-R-C circuit with $L = 3$ henrys, $R = 10$ ohms, $C = 0.1$ farad, $Q(0) = 0$ coulomb, $I(0) = 0$ amperes, and: **(a)** $E(t) = 120$ volts; **(b)** $E(t) = 120 \sin t$ volts.

*3. Solve the two-loop L-R-C circuit with $L = 1$ henry, $R_1 = R_2 = 1$ ohm, $C = 1$ farad, $Q(0) = 10^{-6}$ coulomb, $I_2(0) = 0$ amperes, and: **(a)** $E(t) = 0$ volts; **(b)** $E(t) = 120$ volts.

4. Solve the two-loop L-R-C circuit with $L = 1$ henry, $R_1 = R_2 = 1$ ohm, $C = 1$ farad, $Q(0) = 0$ coulombs, $I_2(0) = 0$ amperes, and: **(a)** $E(t) = 120e^{-t/2}$ volts; **(b)** $E(t) = 120 \cos t$ volts.

5. Transform the second-order equation to a system of first-order equations and classify the system as undamped, overdamped, underdamped, or critically damped by finding the eigenvalues of the corresponding system. Solve with the initial conditions $x(0) = 1$, $\dfrac{dx}{dt}(0) = y(0) = 0$. **(a)** $x'' + 2x' + x = 0$; **(b)** $x'' + 4x = 0$

6. Transform the second-order equation to a system of first-order equations and classify the system as undamped, overdamped, underdamped, or critically damped by finding the eigenvalues of the correspond-

ing system. Solve with the initial conditions $x(0) = 0$, $\dfrac{dx}{dt}(0) = y(0) = 1$. **(a)** $x'' + 4x = 0$;

(b) $x'' + 4x' + \dfrac{7}{4}x = 0$

*7. Solve the diffusion problem with one permeable membrane with $P = 0.5$, $V_1 = V_2 = 1$, $x(0) = 5$, and $y(0) = 10$.

8. Solve the mixture problem with two tanks (see Figure 9.10) with $R = 4$ gal/min, $V_1 = V_2 = 80$, $C = 3$ lb/gal, $x(0) = 10$, and $y(0) = 20$.

9. Investigate the behavior of solutions of the two-dimensional population problem with

$a_1 = 3$, $a_2 = 1$, $b_1 = 3$, $b_2 = 1$, $x(0) = 20$, and $y(0) = 10$.

10. Investigate the behavior of solutions of the two-dimensional population problem with

$a_1 = 1$, $a_2 = 2$, $b_1 = 1$, $b_2 = 1$, $x(0) = 4$, and $y(0) = 8$.

*11. Solve the three-dimensional population problem with

$a_1 = 2$, $a_2 = 1$, $a_3 = 4$, $b_1 = 6$, $b_2 = 4$, $b_3 = 5$, $c_1 = 2$, $c_2 = 8$, $c_3 = 4$, $x(0) = 4$, $y(0) = 4$, and $z(0) = 8$.

12. Solve the spring-mass system with two springs, as shown in Figure 9.17, using $m_1 = m_2 = k_1 = 6$,

$k_2 = 4$ and the initial conditions $x(0) = 0$, $x'(0) = 2$, $y(0) = 0$, and $y'(0) = 0$. (Assume no external forcing functions.)

13. Solve the spring-mass system with two springs, as shown in Figure 9.17, if the forcing functions $F_1(t) = 2 \cos t$ and $F_2(t) = 0$ are included with the parameters $m_1 = m_2 = 1$, $k_1 = 3$, $k_2 = 2$ and initial conditions $x(0) = 0$, $x'(0) = 0$, $y(0) = 0$, and $y'(0) = 0$. What physical phenomenon occurs in $y(t)$?

14. Solve the spring-mass system with three springs (as shown in Figure 9.21) if the forcing functions $F_1(t) = 2 \cos t$ and $F_2(t) = 0$ are included with the parameters $m_1 = m_2 = 1$, $k_1 = 1$, $k_2 = 1$, $k_3 = 1$ and initial conditions $x(0) = 0$, $x'(0) = 0$, $y(0) = 0$, and $y'(0) = 0$.

*15. Solve the double pendulum problem with $m_1 = 3$, $m_2 = 1$, $\ell_1 = \ell_2 = 16$ ft, $g = 32$ ft/s^2, and the initial conditions $\theta_1(0) = 0$, $\theta_1'(0) = 0$, $\theta_2(0) = 1$, $\theta_2'(0) = -1$.

16. Solve the double pendulum problem with $m_1 = 3$, $m_2 = 1$, $\ell_1 = \ell_2 = 16$ ft, $g = 32$ ft/s^2, and the initial conditions $\theta_1(0) = 0$, $\theta_1'(0) = -1$, $\theta_2(0) = 0$, $\theta_2'(0) = 1$.

17. The nonlinear second-order equation that describes the motion of a damped pendulum is $\dfrac{d^2x}{dt^2} + b\dfrac{dx}{dt} + \dfrac{g}{L}\sin x = 0$. **(a)** Make the substitution $dx/dt = y$ to write the equation as the system

$$\begin{cases} \dfrac{dx}{dt} = y \\ \dfrac{dy}{dt} = -by - \dfrac{g}{L}\sin x \end{cases}$$

(b) Show that the equilibrium points of the system are $(n\pi, 0)$ where n is an integer. **(c)** Show that if n is odd, then $(n\pi, 0)$ is a saddle. **(d)** Show that if n is even, then $(n\pi, 0)$ is either a stable spiral or a stable node.

18. Suppose that the predator-prey model is altered so that the prey population $x(t)$ follows the logistic equation when the predator population $y(t)$ is not present. With this assumption, the system is

$$\begin{cases} \dfrac{dx}{dt} = ax - kx - bxy \\ \dfrac{dy}{dt} = -cy + dxy \end{cases}$$

where a, b, c, d, and k are positive constants, $a \neq k$, and the ratio a/k is much larger than c/d. Find and classify the equilibrium points of this system.

19. **(Production of Monochlorobenzene)** The Ajax Pharmaceutical Company has often thought of making its own monochlorobenzene C_6H_5Cl from benzene C_6H_6 and chloride Cl_2 (with a small amount of ferric chloride as a catalyst) instead of purchasing it from another company. The chemical reactions are

$$C_6H_6 + Cl_2 \longrightarrow C_6H_5Cl + HCl$$
$$C_6H_5Cl + Cl_2 \longrightarrow C_6H_4Cl_2 + HCl$$
$$C_6H_4Cl_2 + Cl_2 \longrightarrow C_6H_4Cl_3 + HCl$$

where HCl is hydrogen chloride, $C_6H_4Cl_2$ is dichlorobenzene, and $C_6H_4Cl_3$ is trichlorobenzene. Experimental data indicate that the formation of trichlorobenzene is small. Therefore we will neglect it, so the rate equations are given by

$$\begin{cases} -\dfrac{dx_A}{dt} = k_1 x_A \\ \dfrac{dx_B}{dt} = k_1 x_A - k_2 x_B \end{cases}$$

where x_A is the mole fraction (dimensionless) of benzene and x_B is the mole fraction of monochlorobenzene. The formation of trichlorobenzene is neglected so we assume that x_C, the mole fraction of dichlorobenzene, is found with $x_C = 1 - x_A - x_B$.

The rate constants k_1 and k_2 have been determined experimentally. We give these constants in the following table.* If $x_A(0) = 1$ and $x_B(0) = 0$, solve the system for each of these three temperatures. Graph $x_A(t)$ and $x_B(t)$ simultaneously for each temperature. Is there a relationship between the temperature and the time at which $x_B(t) > x_A(t)$?

	40°C	55°C	70°C
k_1 (hr^{-1})	0.0965	0.412	1.55
k_2 (hr^{-1})	0.0045	0.055	0.45

---

* Samuel W. Bodman, *The Industrial Practice of Chemical Process Engineering*, The M.I.T. Press, 1968, pp. 18–24.

20. (Airplane Wing) A small airplane is modeled using three lumped masses as shown in Figure 9.35. We assume in this simplified model of the airplane that the wings are uniform beams of length ℓ and stiffness factor EI where E and I are constants. (E depends on the material from which the beam is made and I depends on the shape and size of the beam.) We also assume that $m_1 = m_3$ and $m_2 = 4m_1$, where m_1 and m_3 represent the mass of each wing and m_2 represents the mass of the body of the airplane.

We find the displacements x_1, x_2, and x_3 by solving the system

$$M\begin{pmatrix} x_1'' \\ x_2'' \\ x_3'' \end{pmatrix} + K\begin{pmatrix} x_1 \\ x_2 \\ x_3 \end{pmatrix} = \begin{pmatrix} 0 \\ 0 \\ 0 \end{pmatrix}$$

where

$$K = \frac{EI}{\ell^3}\begin{pmatrix} 3 & -3 & 0 \\ -3 & 6 & -3 \\ 0 & -3 & 3 \end{pmatrix} \quad \text{and}$$

$$M = m_1\begin{pmatrix} 1 & 0 & 0 \\ 0 & 4 & 0 \\ 0 & 0 & 1 \end{pmatrix}.$$

We can then write the system as

$$\begin{pmatrix} 1 & 0 & 0 \\ 0 & 4 & 0 \\ 0 & 0 & 1 \end{pmatrix}\begin{pmatrix} x_1'' \\ x_2'' \\ x_3'' \end{pmatrix} + \frac{EI}{m_1\ell^3}\begin{pmatrix} 3 & -3 & 0 \\ -3 & 6 & -3 \\ 0 & -3 & 3 \end{pmatrix}\begin{pmatrix} x_1 \\ x_2 \\ x_3 \end{pmatrix} = \begin{pmatrix} 0 \\ 0 \\ 0 \end{pmatrix}.$$

(a) Solve the system $\begin{pmatrix} 1 & 0 & 0 \\ 0 & 4 & 0 \\ 0 & 0 & 1 \end{pmatrix}\begin{pmatrix} x_1'' \\ x_2'' \\ x_3'' \end{pmatrix} +$

$\frac{EI}{m_1\ell^3}\begin{pmatrix} 3 & -3 & 0 \\ -3 & 6 & -3 \\ 0 & -3 & 3 \end{pmatrix}\begin{pmatrix} x_1 \\ x_2 \\ x_3 \end{pmatrix} = \begin{pmatrix} 0 \\ 0 \\ 0 \end{pmatrix}$ subject to the

initial conditions $x_1(0) = 0$, $x_1'(0) = 1$, $x_2(0) = 0$, $x_2'(0) = -\frac{1}{2}$, $x_3(0) = 0$, $x_3'(0) = 1$.*

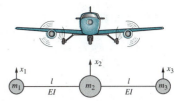

Figure 9.35 A Simple Model of an Airplane

(b) Determine the period of x_1, x_2, and x_3 for $c = 0.0001$, 0.01, 0.1, 1, and 2 where $c = \frac{EI}{m_1\ell^3}$.

(c) Graph x_1, x_2, and x_3 over several periods for $c = 0.0001$, 0.01, 0.1, 1, and 2.

(d) Illustrate the motion of the components of the airplane under these conditions.

21. (Trailer) A 9000-lb trailer is connected to a 4000 lb car by way of a flexible hitch with a modulus of 750 lb/in. (See Figure 9.36.) While traveling on a level highway, find the frequency of oscillatory movement of the trailer **(a)** assuming that the car is coasting so that both vehicles move freely on the roadway and **(b)** considering that the car is operating under power so that its drive train is firmly engaged and the rear wheels do not slip on the paving.†

Figure 9.36 A Car and a Trailer

(*Hint:* Consider the system shown in Figure 9.37.)

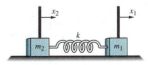

Figure 9.37

* M. L. James, G. M. Smith, J. C. Wolford, P. W. Whaley, *Vibration of Mechanical and Structural Systems,* Harper and·Row, 1989, pp. 346–349.

† Robert K. Vierck, *Vibration Analysis,* Harper Collins, 1979, p. 256.

Differential Equations at Work:
Free Vibration of a Three-Story Building

The tuned mass damper in the John Hancock Building, Boston. (Courtesy of MTS Systems Corp.)

If you have ever gone to the top of a tall building, like the Sears Tower, World Trade Center, or Empire State Building on a windy day you may have been acutely aware of the *sway* of the building. In fact, all buildings sway, or vibrate, naturally. Usually, we are only aware of the sway of a building when we are in a very tall building or in a building during an event like an earthquake. In some tall buildings, like the John Hancock Building in Boston, the sway of the building during high winds is reduced by installing a tuned mass damper at the top of the building that oscillates at the same frequency as the building but out of phase. We will investigate the sway of a three-story building and then try to determine how we would investigate the sway of a building with more stories.

We make two assumptions to solve this problem. First, we assume that the mass distribution of the building can be represented by the lumped masses at the different levels. Second, we assume that the girders of the structure are infinitely rigid compared with the supporting columns. With these assumptions, we can determine the motion of the building by interpreting the columns as springs in parallel.

Assume that the coordinates x_1, x_2, and x_3 as well as the velocities and accelerations are positive to the right. If we assume that $x_3 > x_2 > x_1$, the forces that the columns exert on the masses are shown in Figure 9.38.

In applying Newton's second law of motion, recall that we assumed that acceleration is in the *positive* direction. Therefore, we sum forces in the same direction as the acceleration positively, and others negatively. With this configuration, Newton's second law on each of the three masses yields the following system of differential equations:

$$
\begin{cases}
-k_1 x_1 + k_2(x_2 - x_1) = m_1 \dfrac{d^2 x_1}{dt^2} \\[2mm]
-k_2(x_2 - x_1) + k_3(x_3 - x_2) = m_2 \dfrac{d^2 x_2}{dt^2} \\[2mm]
-k_3(x_3 - x_2) = m_3 \dfrac{d^2 x_3}{dt^2}
\end{cases}
$$

which we write as

$$
\begin{cases}
m_1 \dfrac{d^2 x_1}{dt^2} + (k_1 + k_2)x_1 - k_2 x_2 = 0 \\[2mm]
m_2 \dfrac{d^2 x_2}{dt^2} - k_2 x_1 + (k_2 + k_3)x_2 - k_3 x_3 = 0, \\[2mm]
m_3 \dfrac{d^2 x_3}{dt^2} - k_3 x_2 + k_3 x_3 = 0
\end{cases}
$$

where m_1, m_2, and m_3 represent the mass of the building on the first, second, and third levels, and k_1, k_2, and k_3, corresponding to the spring constants, represent the total stiffness of the columns supporting a given floor.*

1. Show that this system can be written in matrix form as

$$\begin{pmatrix} m_1 & 0 & 0 \\ 0 & m_2 & 0 \\ 0 & 0 & m_3 \end{pmatrix}\begin{pmatrix} x_1'' \\ x_2'' \\ x_3'' \end{pmatrix} + \begin{pmatrix} k_1 + k_2 & -k_2 & 0 \\ -k_2 & k_2 + k_3 & -k_3 \\ 0 & -k_3 & k_3 \end{pmatrix}\begin{pmatrix} x_1 \\ x_2 \\ x_3 \end{pmatrix} = 0.$$

The matrix $\begin{pmatrix} k_1 + k_2 & -k_2 & 0 \\ -k_2 & k_2 + k_3 & -k_3 \\ 0 & -k_3 & k_3 \end{pmatrix}$ is called the **stiffness matrix** of the system.

2. Find a general solution to the system. What can you conclude from your results?

   Suppose that $m_2 = 2m_1$ and $m_3 = 3m_1$. Then we can rewrite the system in the form

$$\begin{pmatrix} 1 & 0 & 0 \\ 0 & 2 & 0 \\ 0 & 0 & 3 \end{pmatrix}\begin{pmatrix} x_1'' \\ x_2'' \\ x_3'' \end{pmatrix} + \frac{1}{m_1}\begin{pmatrix} k_1 + k_2 & -k_2 & 0 \\ -k_2 & k_2 + k_3 & -k_3 \\ 0 & -k_3 & k_3 \end{pmatrix}\begin{pmatrix} x_1 \\ x_2 \\ x_3 \end{pmatrix} = 0.$$

3. Find exact and numerical solutions to the system subject to the initial conditions $x_1(0) = 0$, $x_1'(0) = \frac{1}{4}$, $x_2(0) = 0$, $x_2'(0) = -\frac{1}{2}$, $x_3(0) = 0$, and $x_3'(0) = 1$ if $k_1 = 3$, $k_2 = 2$, and $k_3 = 1$ for $m_1 = 1$, 10, 100, and 1000.

4. Determine the period of x_1, x_2, and x_3 for $m_1 = 1$, 10, 100, and 1000.

5. Graph x_1, x_2, and x_3 over several periods for $m_1 = 1$, 10, 100, and 1000.

6. Illustrate the motion of the building under these conditions.

7. How does the system change if we consider a five-story building? a fifty-story building? a one-hundred-story building?

---

* M. L. James, G. M. Smith, J. C. Wolford, P. W. Whaley, *Vibration of Mechanical and Structural Systems with Microcomputer Applications,* Harper & Row, 1989, pp. 282–286. Robert K. Vierck, *Vibration Analysis,* Second Edition, Harper Collins, 1979, pp. 266–290.

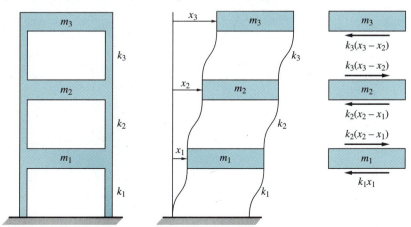

Figure 9.38 Diagram used to Model the Sway of a Three-Story Building

Answers to Selected Exercises

CHAPTER 1

Exercises 1.1

1. (a) ordinary; (b) second-order; (c) linear

3. (a) partial; (c) linear

5. (a) ordinary; (b) first-order; (c) nonlinear

7. (a) partial; (c) linear

9. (a) ordinary; (b) second-order; (c) nonlinear

11. (a) partial; (c) nonlinear

13. (a) ordinary; (b) first-order; and (c) linear in
$$y\left(\frac{dy}{dx} = 2x - y\right); \text{ nonlinear in } x\left(\frac{dx}{dy} = \frac{1}{2x - y}\right)$$

15. (a) ordinary; (b) first-order; (c) nonlinear in
$$y\left(\frac{dy}{dx} = \frac{2x - y}{y}\right), \text{ nonlinear in } x\left(\frac{dx}{dy} = \frac{y}{2x - y}\right)$$

17. linear; first-order

19. nonlinear

21. (a) $\begin{cases} x' = y \\ y' = y + 6x \end{cases}$

(b) $\begin{cases} x' = y \\ y' = \dfrac{1}{4}(-4y - 37x) \end{cases}$ (c) $\begin{cases} x' = y \\ y' = -\dfrac{g}{L}\sin x \end{cases}$

(d) $\begin{cases} x' = y \\ y' = \mu(1 - x^2)y - x \end{cases}$

(e) $\begin{cases} x' = y \\ y' = \dfrac{1}{t}[(t - b)y + ax] \end{cases}$

Exercises 1.2

1. $\dfrac{dy}{dx} + 2y = (-2e^{-2x}) + 2e^{-2x} = 0$

3. $\dfrac{dy}{dx} + y = \left(-e^{-x} + \dfrac{1}{2}\sin x + \dfrac{1}{2}\cos x\right) +$
$\left(e^{-x} - \dfrac{1}{2}\cos x + \dfrac{1}{2}\sin x\right) = \sin x$

5. $\dfrac{d^2y}{dx^2} + 9\dfrac{dy}{dx} = 81Be^{-9x} + 9(-9Be^{-9x}) = 0$

7. $\dfrac{dx}{dt} = \left(-A + \dfrac{t}{4} - \dfrac{1}{2}\right)\sin t +$
$\left(B + \dfrac{t^2}{4} - \dfrac{t}{2} + \dfrac{1}{4}\right)\cos t;$
$\dfrac{d^2x}{dt^2} = \left(-A + \dfrac{3t}{4} - 1\right)\cos t +$
$\left(-B - \dfrac{t^2}{4} + \dfrac{t}{2}\right)\sin t$

9. $\dfrac{dy}{dx} = 2Be^{2x} - 2Ce^{-2x}; \dfrac{d^2y}{dx^2} = 4Be^{2x} + 4Ce^{-2x}; \dfrac{d^3y}{dx^3} = 8Be^{2x} - 8Ce^{-2x}$

11. $x^2(30Ax^4 + 42Bx^5) - 12x(6Ax^5 + 7Bx^6) + 42(Ax^6 + Bx^7) = 0$

13. $2x + 2y\dfrac{dy}{dx} = 0 \Rightarrow \dfrac{dy}{dx} = -\dfrac{x}{y}; 0^2 + y^2 = 16 \Rightarrow y = \pm 4;$
$(0, \pm 4)$

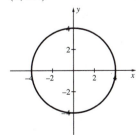

15. $3x^2 + 2xy + x^2 \dfrac{dy}{dx} = 0 \Longrightarrow \dfrac{dy}{dx} = -\dfrac{3x^2 + 2xy}{x^2}$;

$1^3 + y = 100 \Longrightarrow y = 99$; $(1, 99)$

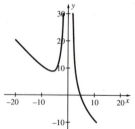

17. $\dfrac{y}{x} + \dfrac{dy}{dx} \ln x + \cos y - x \sin y \dfrac{dy}{dx} = 0 \Longrightarrow$

$(\ln x - x \sin y) \dfrac{dy}{dx} = -\cos y - \dfrac{y}{x} \Longrightarrow$

$\dfrac{dy}{dx} = \dfrac{\cos y + \dfrac{y}{x}}{x \sin y - \ln x}$; $y \ln 1 + \cos y = 0 \Longrightarrow$

$\cos y = 0 \Longrightarrow y = \dfrac{(2n+1)\pi}{2}$, $n = 0, \pm 1, \pm 2, \ldots$

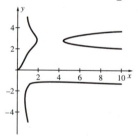

19. $y(x) = -\dfrac{1}{2} \cos(x^2) + C$

21. $y(x) = \ln|\ln x| + C$

23. $y(x) = -xe^{-x} - e^{-x} + C$

25. $y(x) = \displaystyle\int \dfrac{x - x^2}{(x+1)(x^2+1)} \, dx$

$= \displaystyle\int \left[-\dfrac{1}{x+1} + \dfrac{1}{x^2+1} \right] dx$

$= -\ln|x+1| + \tan^{-1} x + C$

(partial fractions)

Exercises 1.3

1. $y(x) = 2e^{-2x}$

3. $y(x) = \dfrac{1}{7}e^{4x} - \dfrac{1}{7}e^{-3x}$

27. $y(x) = \displaystyle\int (4 - x^2)^{3/2} \, dx = \dfrac{x}{4}(4 - x^2)^{3/2} +$

$\dfrac{3x}{2}\sqrt{4 - x^2} + 6 \sin^{-1}\left(\dfrac{x}{2}\right) + C$ (trig. substitution;

$x = 2 \sin \theta$)

29. $x(0) = 3$; $\dfrac{dx}{dt} = -12 \sin 4t + 9 \cos 4t$, $\dfrac{dx}{dt}(0) = 9$

31. $u_t = 16ke^{-16t} \cos 4x$, $u_{xx} = 16e^{-16t} \cos 4x$; $u(\pi, 0) = 2$;

$\lim_{t \to \infty} u(x, t) = 3$

33. $m = 1$, $m = 2$

35. $y(x) = e^{-x} + Ce^{-2x}$

37. Because $\dfrac{d\psi(x)}{dx} = \dfrac{An\pi}{L} \cos\left(\dfrac{n\pi x}{L}\right)$ and $\dfrac{d^2\psi(x)}{dx^2} =$

$-\dfrac{An^2\pi^2}{L^2} \sin\left(\dfrac{n\pi x}{L}\right)$,

$-\dfrac{h^2}{2m} \dfrac{d^2\psi(x)}{dx^2} + U(x)\psi(x) = \dfrac{h^2}{2m} \dfrac{An^2\pi^2}{L^2} \sin\left(\dfrac{n\pi x}{L}\right) =$

$EA \sin\left(\dfrac{n\pi x}{L}\right)$.

Therefore, $E = \dfrac{h^2 n^2 \pi^2}{2mL^2}$.

39. $1 + \dfrac{2x}{y} - \dfrac{x^2}{y^2} \dfrac{dy}{dx} = 0 \Longrightarrow \dfrac{dy}{dx} =$

$\dfrac{y^2 + 2xy}{x^2}$; $y = \dfrac{x^2}{C - x}$

so $y = 0$ is singular

41. $a[C_1 y_1''(x) + C_2 y_2''(x)] + b[C_1 y_1'(x) + C_2 y_2'(x)] +$

$c[C_1 y_1(x) + C_2 y_2(x)]$

$= C_1[ay_1''(x) + by_1'(x) + cy_1(x)] +$

$C_2[ay_2''(x) + by_2'(x) + cy_2(x)]$

$= C_1(0) + C_2(0) = 0$

5. $y(x) = \dfrac{1}{3} \sin 3x$

7. $y(x) = 8e^{3x} - 8e^{-3x}$

9. $y(x) = -\dfrac{3}{4} - \dfrac{1}{2}x + \dfrac{3}{4}e^{2x}$

11. $y(x) = 8x^6 - 7x^7$

13. $y(x) = x^4 - \dfrac{1}{2}x^2 + 2x + 1$

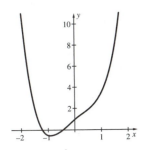

15. $y(x) = -\sin\left(\dfrac{1}{x}\right) + 2$

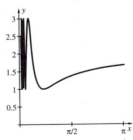

17. $y(x) = B\sin 2x$

19. No solution; $y(\pi/2) = B\sin 3\pi = 0 \neq -2$

21. $v(t) = \dfrac{mg}{c} + e^{-ct/m}\left(-\dfrac{mg}{c} + v_0\right)$, $\displaystyle\lim_{t\to\infty} v(t) = \dfrac{mg}{c}$

Exercises 1.4

1.

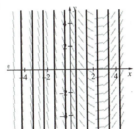

3.

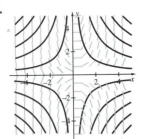

5.

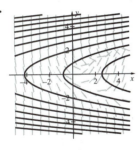

7.

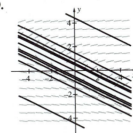

9.

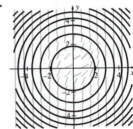

11. $y = \tan^{-1}x + C$

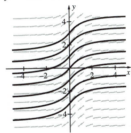

13. $y = \ln|x + 1| + C$

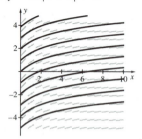

15. $\dfrac{dy}{dx} - (x^2 \cos x + 2y/x) = 2Cx +$

$2x \sin x - \dfrac{2(Cx^2 + x^2 \sin x)}{x} = 0$

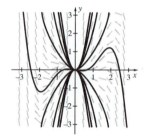

17. $x^2y - \tan x + y^2 = C \Rightarrow x^2\dfrac{dy}{dx} + 2y\dfrac{dy}{dx} + 2xy -$

$\sec^2 x = 0 \Rightarrow \dfrac{dy}{dx} = \dfrac{\sec^2 x - 2xy}{x^2 + 2y}$

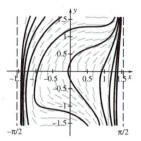

19. $\tan^{-1}(y/x) + x = C \Rightarrow 1 + \dfrac{y'/x - y/x^2}{1 + y^2/x^2} = 0 \Rightarrow 1 +$

$\dfrac{xy' - y}{x^2 + y^2} = 0 \Rightarrow \dfrac{dy}{dx} = y' = \dfrac{y - y^2 - x^2}{x}$

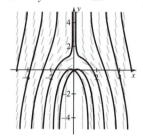

21.

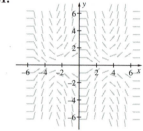 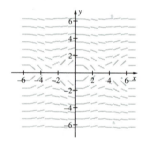

Exercises 1.5

1. $k = \dfrac{\ln 3}{6}$

3. $P(t) = 10\left(\dfrac{3}{2}\right)^{t/3}$

5. $P(t) = 1000\left(\dfrac{25}{23}\right)^{-t/50}$

7. $k = -\dfrac{1}{5000}\ln\left(\dfrac{99}{1249}\right) \approx 0.00051$

9. $v(t) = 32 - 32e^{-t}$, $v(1) \approx 20.23$ ft/s

11. $0.2I'' + 300I' + 10^5I = 0$

13. $4x'' + 16x = 0$

15. $u_t(x, t) = -\pi^2 e^{-\pi^2 t}\cos \pi x$; $u_{xx}(x, t) = -\pi^2 e^{-\pi^2 t}\cos \pi x$

Chapter 1 Review Exercises

1. (a) ordinary; **(b)** first-order; **(c)** linear

3. (a) ordinary; **(b)** second-order; **(c)** linear

5. (a) partial; **(c)** nonlinear

7. $\dfrac{dy}{dx} = \dfrac{1}{2}(e^x + \sin x - \cos x)$ so

$\dfrac{dy}{dx} - y - \sin x = \dfrac{1}{2}(e^x + \sin x - \cos x) -$

$\dfrac{1}{2}(e^x - \cos x - \sin x) - \sin x = 0$

9. $\dfrac{dy}{dx} = -3e^{3x}(\cos 6x + 3 \sin 6x)$ and $\dfrac{d^2y}{dx^2} =$

$-9e^{3x}(7\cos 6x + \sin 6x)$ so

$y'' - 6y' + 45y = -9e^{3x}(7 \cos 6x + \sin 6x) +$
$\qquad\qquad 18e^{3x}(\cos 6x + 3 \sin 6x) +$
$\qquad\qquad 45e^{3x}(\cos 6x - \sin 6x)$
$\qquad\qquad = 0.$

10. $\dfrac{dy}{dx} = 5Ax^4 - \dfrac{3B}{x^4};\ \dfrac{d^2y}{dx^2} = 20Ax^3 + \dfrac{12B}{x^5}$

11. $x^2 y'' + 3xy' + 2y = x^2 \cdot \dfrac{2[2 \cos(\ln x) + \sin(\ln x)]}{x^3} +$

$3x \cdot \dfrac{-2 \cos(\ln x)}{x^2} + 2 \cdot \dfrac{\cos(\ln x) - \sin(\ln x)}{x} = 0$

13. $y' = 3Ae^{3x} + 4Be^{4x}$ and $y'' = 9Ae^{3x} + 16Be^{4x}$ so

$y'' - 7y' + 12y = 9Ae^{3x} + 16Be^{4x} - 21Ae^{3x} - 28Be^{4x}$
$\qquad\qquad + 12Ae^{3x} + 12Be^{4x} + 2 = 2$

15. $(y \cos xy + \sin x)\, dx + x \cos xy\, dy = 0 \Rightarrow$

$\dfrac{dy}{dx} = -\dfrac{\sin x + y \cos xy}{x \cos xy}$

$\cos xy\left(x \dfrac{dy}{dx} + y\right) + \sin x = 0 \Rightarrow \dfrac{dy}{dx} =$

$-\dfrac{\sin x + y \cos xy}{x \cos xy}$

17. $y = -x^2 \cos x + 2 \cos x + 2x \sin x + C$

19. $y = \dfrac{x}{2}\sqrt{x^2 - 1} + \dfrac{1}{2}\ln(x + \sqrt{x^2 - 1}) + C$

21. $y = \dfrac{x}{4} + \cos 2x$

23. $y = -\dfrac{1}{3}\cos^3 x + \dfrac{1}{3}$

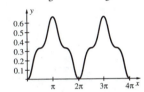

25. $\dfrac{du}{dt} = -70ke^{-kt};\ u(0) = 100;\ \lim\limits_{t \to \infty} u(t) = 30$

27. $u_t(x, t) = -\pi^2 ke^{-\pi^2 kt} \sin \pi x + 4\pi^2 ke^{-4\pi^2 kt} \sin 2\pi x$
$u_{xx}(x, t) = -\pi^2 ke^{-\pi^2 kt} \sin \pi x + 4\pi^2 ke^{-4\pi^2 kt} \sin 2\pi x$
$u(1, 0) = 0;\ \lim\limits_{t \to \infty} u(x, t) = 0$

29. $u_{xx}(x, y) = \dfrac{2xy}{(x^2 + y^2)^2};\ u_{yy}(x, y) = \dfrac{-2xy}{(x^2 + y^2)^2}$

C H A P T E R   2

Exercises 2.1

1. $y^4 = \dfrac{8}{7}x^3 + C$

3. $\dfrac{1}{y^6} = \dfrac{18}{7x^7} + C$

5. $5y - \dfrac{9}{7y^7} + 6x + x^4 = C$

7. $\cosh 4y = 2 \sinh 3x + C$

9. $-\dfrac{2}{3y^{3/2}} + \dfrac{2}{3}x^3 + \dfrac{8}{3}x^{3/2} = C$

11. $-3 \cos x = 4 \sin y + C$

13. $\dfrac{5x^6}{6} - 4 \sin x = -\dfrac{2}{9}\sin 9y + \dfrac{2}{7}\cos 7y + C$

15. $20 \cosh y = -\dfrac{1}{6}\sinh 6x - \dfrac{5}{4}\cosh 4x + C$

17. $10x - \dfrac{7}{3}e^{-3x} = e^y - 2y^4 + C$

19. $-3 \cos x + \dfrac{1}{3}\cos 3x = \dfrac{1}{4}\sin 4y - 4 \sin y + C$

21. $2y + \dfrac{5}{y} + \sin 2x + 2x = C$

23. $\dfrac{1}{2}\tan^2 y = \dfrac{1}{8}\cos^4 2x + C$

25. $-\dfrac{1}{2}\cos(x^2) = 2\sin\sqrt{y} + C$

27. $\dfrac{1}{1-\sin y} = \dfrac{1}{4}\sin^4 x + C$

29. $\dfrac{2}{3}(\ln x)^{3/2} = -\dfrac{1}{3}e^{3/y} + C$

31. $y\ln(3y) - y = \dfrac{1}{18}x^6 + C$

33. $\dfrac{5}{2}\ln|y+1| - \dfrac{1}{2}\ln|7y+1| - \ln|x+3| + \dfrac{8}{3}\ln|3x+1| = C$

35. $-\dfrac{2}{11}\sin^{11/2} y + \dfrac{2}{7}\sin^{7/2} y = 2x - \dfrac{1}{2}\sin 4x + C$

37. $y + 2\ln(y^2+1) - \tan^{-1} y = \dfrac{1}{2}\sin^{-1}\left(\dfrac{2x}{3}\right) + C$

39. $\dfrac{x}{\sqrt{1-x^2}} = -\dfrac{4}{25}e^{3y}\cos 4y + \dfrac{3}{25}e^{3y}\sin 4y + C$

41. $y^2 e^y - 2ye^y + 2e^y = -\dfrac{1}{5}e^{-2x}\cos 4x - \dfrac{1}{10}e^{-2x}\sin 4x + C$

43. $y(x) = \dfrac{1}{4}x^4 + 4$

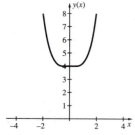

45. $y(x) = \ln|\sec x + \tan x| + 2$

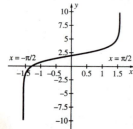

47. $\dfrac{1}{2}y^2 = \dfrac{2}{3}x^{3/2} + 2$ so $y = \sqrt{\dfrac{4}{3}x^{3/2} + 4}$

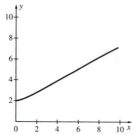

49. $\dfrac{1}{2}y^2 + y = e^x - 1$ so $y = -\sqrt{2e^x - 1} - 1$

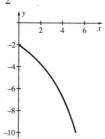

51. $\dfrac{1}{2}(\ln y)^2 = x + \dfrac{1}{2}$ so $y = e^{\sqrt{2x+1}}$

53. $y(x) = \tan^{-1} x + 1$

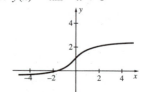

55. $y(x) = \dfrac{2x}{x-2}$

59. (c)

61. (b)

63. $N(0) = N_0 \Rightarrow N(t) = N_0 e^{-kt}$; $D = \ln 10/k \approx 2.30259k^{-1}$

64. (b) $C = \dfrac{e^{k_d t_0}}{N_0}\left(1 - N_0\dfrac{k_c}{k_d}\right)$

Exercises 2.2

1. Yes, $n = 1$

3. No

5. Yes, $n = 0$

7. Yes, $n = 0$

9. No

11. $-\ln\left|\dfrac{2x}{y} - 1\right| + 2\ln\left|\dfrac{x}{y} - 1\right| = -\ln|y| + C$

13. $y(x) = x + Cx^{-2}$

15. $\dfrac{1}{2}x^2y^2 - xy^3 = C$

17. $x + y = Ce^{y/x}$

19. $y^3 = 3x^3\ln|x| + Cx^3$

21. $(y - x)^{5/2} = C(y + x)^{3/2}$

23. $y = -x\ln|x| + Cx$

25. $y(x) = -\ln|x + 1| + C$, $y = 0$

27. $x^{5/6} = \dfrac{C\sqrt{2y - x}}{(3y - 2x)^{2/3}}$

29. $y = \dfrac{x^4 - C^2}{2Cx}$

31. $y = Ce^{y/(4x)}$

33. $\dfrac{2}{3}\sqrt{\dfrac{x}{y}} - \ln|y| = C$

35. $xy - \dfrac{1}{2}y^2 = C$

37. $x = vy \Rightarrow dx = v\,dy + y\,dv \Rightarrow \dfrac{dx}{dy} = v + y\dfrac{dv}{dy}$

$v + y\dfrac{dv}{dy} = \dfrac{2}{v}e^{-v} + v \Rightarrow ve^v\,dv = \dfrac{2dy}{y} \Rightarrow ve^v - e^v =$

$2\ln|y| + C$

$e^{x/y}\left(\dfrac{x}{y} - 1\right) = 2\ln|y| + C$

39. $x = yv \Rightarrow dx = v\,dy + y\,dv \Rightarrow \dfrac{dx}{dy} = v + y\dfrac{dv}{dy}$

$x\ln\dfrac{x}{y}\,dy = y\,dx \Rightarrow \dfrac{dx}{dy} = \dfrac{x}{y}\ln\dfrac{x}{y} \Rightarrow v + y\dfrac{dv}{dy} = v\ln v \Rightarrow$

$\dfrac{dv}{v(\ln v - 1)} = \dfrac{dy}{y}$

$\ln|\ln v - 1| = \ln|y| + C \Rightarrow \ln\left|\ln\dfrac{x}{y} - 1\right| = \ln|y| + C$

41. $y^2 = \dfrac{1}{2}x^2(1 + x^2)$ or $y = \dfrac{x\sqrt{1 + x^2}}{\sqrt{2}}$

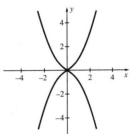

43. $y = xv \Rightarrow dy = v\,dx + x\,dv \Rightarrow \dfrac{dy}{dx} = v + x\dfrac{dv}{dx}$

$\dfrac{dy}{dx} = \dfrac{y}{x} + \sqrt{1 + \left(\dfrac{y}{x}\right)^2} \Rightarrow v + x\dfrac{dv}{dx} = v + \sqrt{1 + v^2} \Rightarrow$

$x\dfrac{dv}{dx} = \sqrt{1 + v^2}$

$\dfrac{dv}{\sqrt{1 + v^2}} = \dfrac{dx}{x} \Rightarrow \ln|v + \sqrt{1 + v^2}| = \ln|x| + C \Rightarrow$

$\ln\left|\dfrac{y}{x} + \sqrt{1 + \left(\dfrac{y}{x}\right)^2}\right| = \ln|x| + C$

$y(1) = 0;\ \ln|1| = \ln|1| + C \Rightarrow C = 0$

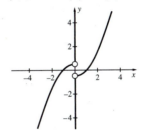

45. $\dfrac{y^3}{3x^3} = -\ln|x| + 9$ or $y = \sqrt[3]{3}x(9 - \ln|x|)^{1/3}$

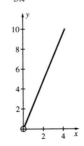

47. $y = ux \Rightarrow y^4 \, dx + (x^4 - xy^3) \, dx = 0 \Rightarrow$
$(ux)^4 \, dx + (x^4 - xu^3x^3)(u \, dx + x \, du) = 0$

$\Rightarrow x^4 u \, dx + x^5(1 - u^3) \, du = 0 \Rightarrow \dfrac{1}{x} \, dx = \dfrac{u^3 - 1}{u} \, du \Rightarrow$

$\ln|x| = \dfrac{1}{3} u^3 - \ln|u| + C$

$\Rightarrow \ln|x| = \dfrac{1}{3}(y/x)^3 - \ln|y/x| + C$

$y(1) = 2 \Rightarrow 0 = \dfrac{1}{3} \cdot 8 - \ln 2 + C \Rightarrow C = \dfrac{1}{3}(3 \ln 2 - 8) \approx -1.07352$

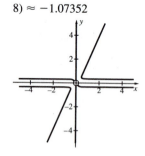

49. $y^2 = 2x^2 \ln|x|$

57. (c) $y = \dfrac{2}{3\sqrt{3}} x^{3/2}$

59. $f(x) = x - 1; \; g(x) = x^2 - x$

general solution: $(xc - y) - 1 = c^2 - c \Rightarrow y = cx - 1 + c - c^2$

singular solution: $\dfrac{d}{dx}[xy' - y - 1] = \dfrac{d}{dx}[(y')^2 - y'] \Rightarrow$
$xy'' + y' - y' = 2y'y'' - y''$

$(x - 2y' + 1)y'' = 0 \Rightarrow y' = \dfrac{x+1}{2} \Rightarrow y = \dfrac{1}{4}x^2 + \dfrac{x}{2} - \dfrac{3}{4}$

61. $f(x) = 1 - 2x; \; g(x) = x^{-2}$; general solution:

$1 - 2(xc - y) = c^{-2} \Rightarrow y = \dfrac{1}{2}(c^{-2} + 2cx - 1)$

singular solution: $y = \dfrac{1}{2}(3x^{2/3} - 1)$

Exercises 2.3

1. Exact

3. Exact

5. Exact

7. Not exact

9. Exact

11. $y = x^3 + C$

13. $xy^2 = C$

15. $x^2 + xy^3 + 4y = C$

17. $x^2y + \dfrac{1}{3}y^3 = C$

19. $x \sin^2 y = C$

21. $\dfrac{1}{y} = C + Cx^{-1} + x^{-1}$

23. $y^2 \cos(x^2) = C$

25. $x + y \sin(xy) = C$

27. $(3 + x) \sin(x + y) = C$

29. $y = -x \ln|x| + Cx$

31. $x^2y^2 = 1$

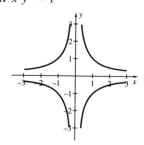

33. $y = -\dfrac{x^3 + 1}{x^2 - 1}$

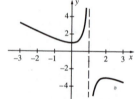

35. $xe^y - x^2y = 0$

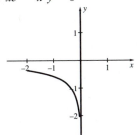

$\tan^{-1} x - xy^2 = C;\ \tan^{-1} 0 - 0 \cdot 0^2 = C \Rightarrow C = 0;$
$\tan^{-1} x - xy^2 = 0$

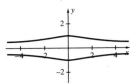

47. $xy = C$

49. $\mu(y) = \exp\left(\int \dfrac{1}{y}\,dy\right) = \exp(\ln|y|) = y,\ y > 0$

$y^2\,dx + (2xy - y^2e^y)\,dy = 0 \Rightarrow f_x(x, y) = y^2 \Rightarrow$
$f(x, y) = xy^2 + g(y) \Rightarrow f_y(x, y) = 2xy + g'(y) \Rightarrow$
$g'(y) = -y^2e^y \Rightarrow g(y) = -y^2e^y + 2ye^y - 2e^y \Rightarrow$
$xy^2 - y^2e^y + 2ye^y - 2e^y = C$

37. $xy^2 + \cos 2x + y = 2$

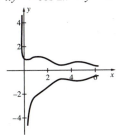

51. $-\dfrac{y}{x} + 2x + \dfrac{1}{2}y^2 = C$

39. $f_x(x, y) = \dfrac{1}{1 + x^2} - y^2 \Rightarrow f(x, y) = \tan^{-1} x - xy^2 +$
$g(y) \Rightarrow f_y(x, y) = -2xy + g'(y)$

53. $\dfrac{1}{y} = \dfrac{C\sqrt{2x^2 - 1} - 1}{x}$

Exercises 2.4

1. $y = \dfrac{1}{3}x^2 + Cx^{-1}$

3. $y = e^x - x^{-1}e^x + Cx^{-1}$

5. $y = \ln(1 + x^2) + C + x^2\ln(1 + x^2) + Cx^2$

7. $y = \sqrt{x^2 - 1}(2\sqrt{x^2 - 1} + C)$

9. $y = -\dfrac{2}{5}e^{-2x} + \dfrac{1}{5}e^{-2x}\tan x + C\sec x$

11. $y = -x^3 + 4x + (x^2 - 4)^{3/2}\ln|x + \sqrt{x^2 - 4}| +$
$C(x^2 - 4)^{3/2}$

13. $y = \dfrac{1}{3}x^4 + \dfrac{3}{4}x^2 - \dfrac{27}{8} + C\sqrt{4x^2 - 9}$

15. $y = x^2 + \dfrac{49}{9} + C\sqrt{9x^2 + 49}$

17. $y = (e^{x^2/2}x^2 - 2e^{x^2/2} + C)e^{-x^2/2} = x^2 - 2 + Ce^{-x^2/2}$

19. $x = -y^2 - 2y - 2 + Ce^y$

21. $x = 3y^2 + Cy$

23. $p = \dfrac{1}{3}t^4 + Ct$

25. $y = 4e^x(x + 1)$

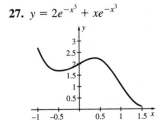

27. $y = 2e^{-x^3} + xe^{-x^3}$

29. $y = x^{-1}(1 + \sin x)$

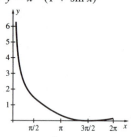

31. $y = \dfrac{x^2}{2(e^x + 1)} + \dfrac{2}{e^x + 1}$

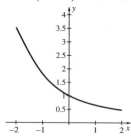

33. $x = -2 + 4e^t - t$

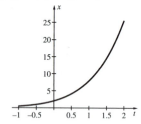

37. $y = e^{-x} + \dfrac{2}{5}\cos x + \dfrac{1}{5}\sin x + Ce^{-2x}$

39. (a)

41. (a) $N(t) = Ae^{\mu t} - A$,
 (b) $\mu = 0.693147$, and
 (c) $3e^{0.693147 \cdot 24} - 3 \approx 5.03314 \times 10^7$ and
 $3e^{0.693147 \cdot 36} - 3 \approx 2.06157 \times 10^{11}$

43. $p(x) \equiv 0$

45. $y = \begin{cases} 2e^{-x} + x - 1, & 0 \le x \le 1. \\ 2e^{-x}, & x > 1 \end{cases}$

47. $y = \begin{cases} e^{-2x}, & 0 \le x \le 1 \\ e^2 e^{-4x}, & x > 1 \end{cases}$

51. $y^2 = -2x - 2 + Ce^x$ or $y = \sqrt{Ce^x - 2x - 2}$

53. $\dfrac{1}{y^2} = -\dfrac{2\cos x + 2x\sin x + C}{x}$

55. $y^{3/2} = -\dfrac{9}{20}\cos x + \dfrac{3}{20}\sin x + Ce^{3x}$

59. $y = (Ce^{-p} - 2)(p + 1) + (2p + 1)$

61. $z = \dfrac{1}{v}\left(-\dfrac{1}{2} + Ce^{2v}\right)$

Exercises 2.5

1. (a) yes; (b) no; (c) no

3. In this case, $f(x, y) = y^{1/5}$, so $\dfrac{\partial f}{\partial y}(x, y) = \dfrac{1}{5}y^{-4/5}$ is not continuous at $(0, 0)$; uniqueness is not guaranteed. Solutions: $y = \left(\dfrac{4}{5}x\right)^{5/4}$, $y = 0$.

5. Because $f(x, y) = 2\sqrt{|y|} = \begin{cases} 2\sqrt{y}, & y \ge 0 \\ 2\sqrt{-y}, & y < 0 \end{cases}$,
$\dfrac{\partial f}{\partial y}(x, y) = \begin{cases} y^{-1/2}, & y > 0 \\ -(-y)^{-1/2}, & y < 0 \end{cases}$ is not continuous at $(0, 0)$. Therefore, the hypotheses of the Existence and Uniqueness Theorem are not satisfied.

7. Yes. $y = e^{2(x^{3/2} - 1)/3}$

9. Yes. $f(x, y) = \sin y - \cos x$ and $\dfrac{\partial f}{\partial y}(x, y) = \cos y$ are continuous in a region containing $(\pi, 0)$.

11. $y = \sec x \Longrightarrow y' = \sec x \tan x = y \tan x$ and $y(0) = \sec 0 = 1$; $\dfrac{\partial f}{\partial y}(x, y) = x$ is continuous on $-\dfrac{\pi}{2} < x < \dfrac{\pi}{2}$ and $f(x, y) = \sec x$ is continuous on $-\dfrac{\pi}{2} < x < \dfrac{\pi}{2}$ so the largest interval on which the solution is valid is $-\dfrac{\pi}{2} < x < \dfrac{\pi}{2}$.

13. $f(x, y) = \sqrt{y^2 - 1}$ and $\dfrac{\partial f}{\partial y}(x, y) = \dfrac{1}{2}(y^2 - 1)^{-1/2} \cdot 2y = \dfrac{y}{\sqrt{y^2 - 1}}$; unique solution guaranteed for (a) only.

Exercises 2.6

1. 47.3742, 63.2572
3. 1.8857, 2.09847
5. 79.8458, 123.048
7. 1.95109, 1.95388
9. 83.6491, 88.6035
11. 2.37754, 2.41897

13. 185.34, 206.981
15. 1.95547, 1.95609
17. 90.6405, 90.6927
19. 2.43501, 2.43514
21. 216.582, 216.992
23. 1.95629, 1.95629

Chapter 2 Review Exercises

1. $y^3 = \dfrac{1}{5}x^6 + C$

3. $\sinh y - \dfrac{1}{2}\cosh x = C$

5. $y^5 = e^{5x} + C$

7. $\dfrac{1}{2}(\ln y)^2 + \dfrac{1}{2}e^{-2x} = C$

9. $\dfrac{1}{37}e^y \cos 6y + \dfrac{6}{37}e^y \sin 6y = -\dfrac{1}{20}\cos 10x +$
$\dfrac{1}{8}\cos 4x + C$

11. $-\dfrac{1}{2}x^2 + xy + \dfrac{1}{2}y^2 = C$

13. $y^2 + x = C$
15. $y = C(y^2 - 4x)^{5/2}$

17. $\dfrac{1}{3}x^3 y - \cos x - \sin y = C$

19. $x^2 \ln y + 2y = C$
21. $y = 1 + Ce^{-x^2/2}$

23. $y = \dfrac{\cos x + x \sin x + C}{x}$

25. $y^3 = \dfrac{3}{4}e^x + Ce^{-3x}$

27. $y = cx + 2\ln c$; Sing: $y = 2\ln\left(\dfrac{-2}{x}\right) - 2,\ x < 0$

29. $x = \dfrac{8}{3}p + Cp^{-2},\ y = \dfrac{4}{3}p^2 + 2Cp^{-1}$

31. $\sin(x - y) + y = \pi$
33. $x \sin y - y \sin x = 0$
35. $y \ln x + x \ln y = 0$

37.

n	x_n	y_n (Euler's)	y_n (Improved Euler's)	y_n (Runge-Kutta of Order 4)
0	1	1	1	1
1	1.05	1	1.00559	1.00678
2	1.1	1.01118	1.0181	1.01921
3	1.15	1.02608	1.03383	1.03486
4	1.2	1.04368	1.052	1.05296
5	1.25	1.06345	1.07219	1.07309
6	1.3	1.08505	1.0941	1.09494
7	1.35	1.10823	1.11752	1.11831
8	1.4	1.13281	1.14228	1.14303
9	1.45	1.15866	1.16825	1.16896
10	1.5	1.18565	1.19533	1.19601
11	1.55	1.21368	1.22343	1.22407
12	1.6	1.24268	1.25247	1.25307
13	1.65	1.27256	1.28237	1.28295
14	1.7	1.30328	1.31309	1.31364
15	1.75	1.33478	1.34456	1.34509
16	1.8	1.36699	1.37675	1.37726
17	1.85	1.3999	1.40961	1.4101
18	1.9	1.43344	1.44311	1.44357
19	1.95	1.46759	1.4772	1.47764
20	2.	1.50232	1.51186	1.51229

39.

n	x_n	y_n (Euler's)	y_n (Improved Euler's)	y_n (Runge-Kutta of Order 4)
0	0	1	1	1
1	0.05	1	1.00125	1.00125
2	0.1	1.0025	1.00501	1.00501
3	0.15	1.0075	1.01129	1.01129
4	0.2	1.01503	1.02012	1.02013
5	0.25	1.02511	1.03156	1.03157
6	0.3	1.03779	1.04564	1.04565
7	0.35	1.0531	1.06242	1.06245
8	0.4	1.07112	1.08198	1.08201
9	0.45	1.09189	1.10437	1.10441
10	0.5	1.11548	1.12966	1.12971
11	0.55	1.14194	1.15789	1.15795
12	0.6	1.17132	1.1891	1.18918
13	0.65	1.20364	1.22329	1.2234
14	0.7	1.23889	1.26043	1.26056
15	0.75	1.27702	1.30042	1.30058
16	0.8	1.31791	1.34309	1.34329
17	0.85	1.36139	1.38818	1.38842
18	0.9	1.40717	1.43532	1.43561
19	0.95	1.45488	1.48403	1.48438
20	1.	1.50399	1.53369	1.5341

C H A P T E R   3

Exercises 3.1

1. $y = \dfrac{x}{2} + k$

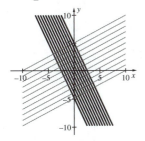

3. $y^2 \ln y - \dfrac{1}{2}y^2 + x^2 = k$

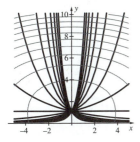

5. $xy = k$

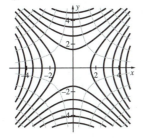

7. $x^2y + \dfrac{1}{3}y^3 = k$

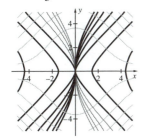

9. $\dfrac{1}{2}x^2e^{2y} + \dfrac{1}{2}e^{2y}y - \dfrac{1}{4}e^{2y} = k$

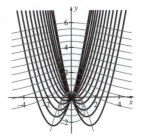

11. $\dfrac{4}{3}y^{3/2} + x = k$

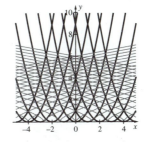

13. $y = -\dfrac{1}{k - \ln|x|} = \dfrac{1}{\ln|x/k|}$

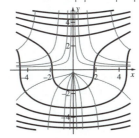

15. $y = e^{-x} + k$

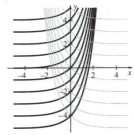

17. $x^3 + y^3 = k$

19. $e^y + e^{-y} = ke^{-x^2/2}$

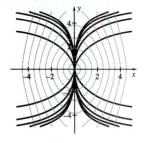

21. Yes. The orthogonal trajectories of $y^2 - 2cx = c^2$ satisfy
$\dfrac{dy}{dx} = \dfrac{y}{x \pm \sqrt{x^2 + y^2}}$. One solution of this
homogeneous equation can be written as $y^2 + 2Cx = C^2$. Now, replace C by $-c$.

23. Equipotential lines: $x^2 + y^2 = c$; orthogonal traj.: $y = kx$

25. (c) $\dfrac{1}{2}x^2 + xy - \dfrac{1}{2}y^2 = k_1$ and $-\dfrac{1}{2}x^2 + xy + \dfrac{1}{2}y^2 = k_2$

Exercises 3.2

1. 102400, $t = 16.23$ days

3. $t \approx 2.71$ days

5. $y(1) = 99.93$ g, $y(500) = 70.71$ g

7. $y(24) = 1036.8$ g, $t \approx 22.81$ hours

9. $t \approx 12.92$ hours, $y(15) = \dfrac{1}{8}y_0$

13. 9.72 days

15. $y(100) \approx 0.96y_0$

17. $t_{\text{tool}} \approx 3561.13$ years old, $t_{\text{fossil}} \approx 4222.81$ years old,
$t_{\text{fossil}} - t_{\text{tool}} \approx 661.68$ years. No.

21. $y = \dfrac{1,000,000}{1 + 9e^{-t/100}}$; $y(25) \approx 124,856$;
$\lim\limits_{t \to \infty} y(t) = 1,000,000$

23. $y = \dfrac{200}{1 + 199\left(\dfrac{3}{199}\right)^t}$; $y(2) \approx 191$; because there is no t
so that $y = 200$, all students do not learn of the rumor.

25. $y(t) = 1 + 499e^{-5t}$; $y(20) \approx 1$; quickly

27. $\lim\limits_{m \to \infty} S_0\left(1 + \dfrac{k}{m}\right)^{mt}$ is found with L'Hopital's rule.
Because
$$\lim\limits_{m \to \infty} \ln\left(1 + \dfrac{k}{m}\right)^{mt} = \lim\limits_{m \to \infty} mt \ln\left(1 + \dfrac{k}{m}\right) =$$
$$\lim\limits_{m \to \infty} \dfrac{\ln\left(1 + \dfrac{k}{m}\right)}{\dfrac{1}{mt}} = \lim\limits_{m \to \infty} \dfrac{\dfrac{-k/m^2}{1 + \dfrac{k}{m}}}{-\dfrac{1}{tm^2}} = \lim\limits_{m \to \infty} \dfrac{kt}{1 + \dfrac{k}{m}} = kt,$$
$$\lim\limits_{m \to \infty} S_0\left(1 + \dfrac{k}{m}\right)^{mt} = S_0 \lim\limits_{m \to \infty}\left(1 + \dfrac{k}{m}\right)^{mt} = S_0 e^{kt}.$$

Exercises 3.3

1. $t \approx 55.85$ min

3. 12.59 min

5. $t \approx -2.45$ hr, 12:30 P.M.

7. $T(5) \approx 76.3°F$

9. $75°F$

11. $t \approx 4.7$ min

13. $u(t) = \dfrac{-5}{9 + \pi^2}\left(-8\pi^2 - (2\pi^2 + 27)e^{-t/4}\right.$

$\left. + 3\pi \sin\left(\dfrac{\pi t}{12}\right) + 9\cos\left(\dfrac{\pi t}{12}\right) - 72\right)$

15. $u(t) = \dfrac{-5}{9 + \pi^2}\left(-14\pi^2 + \pi^2 e^{-t/4} + 3\pi \sin\left(\dfrac{\pi t}{12}\right)\right.$

$\left. + 9\cos\left(\dfrac{\pi t}{12}\right) - 126\right)$

17. If $R_1 = R_2$, the volume remains constant, so $V(t) = V_0$. If $R_1 > R_2$, V increases. If $R_1 < R_2$, V decreases.

19. $R_1 = 4\dfrac{\text{gal}}{\text{min}}$, $R_2 = 3\dfrac{\text{gal}}{\text{min}}$

Exercises 3.4

1. $v(t) = 32 - 32e^{-t}$; $v(2) \approx 27.67$ ft/s

3. $v(t) = \dfrac{1}{2} + \dfrac{15}{2}e^{-64t}$; $v(1) \approx 0.5$ ft/s

5. $v(t) = 4 - 68e^{-8t}$; $v(0) = 0$ when

$t = -\dfrac{1}{8}\ln\left(\dfrac{1}{17}\right) \approx 0.354$ s

7. $\left\{\begin{array}{l} \dfrac{dv}{dt} = 32 - v, v(0) = 0 \end{array}\right\} \Rightarrow v(t) = 32 - 32e^{-t}$

$\left\{\begin{array}{l} \dfrac{ds}{dt} = v(t), s(0) = 0 \end{array}\right\} \Rightarrow s(t) = 32t + 32e^{-t} - 32;$

$s(4) \approx 96.59$ ft $< 300 \Rightarrow 203.41$ ft above the ground

9. $v(t) = \dfrac{49}{5} - \dfrac{49}{5}e^{-t}$, $\lim\limits_{t\to\infty} v(t) = \dfrac{49}{5}$

11. $v(t) = -9800 + 9900e^{-t/1000}$; $v(t) = 0 \Rightarrow t \approx 10.152$ s; $s(t) = -9800t + 9900000(-e^{-t/1000} + 1)$; $s(10.152) \approx 506.76$ m

13. $\left\{\begin{array}{l} \dfrac{dv}{dt} = -g, v(0) = v_0 \end{array}\right\} \Rightarrow v(t) = -gt + v_0;$

$\left\{\begin{array}{l} \dfrac{ds}{dt} = v(t), s(0) = s_0 \end{array}\right\} \Rightarrow s(t) = -\dfrac{1}{2}gt^2 + v_0 t + s_0$

Chapter 3 Review Exercises

1. $y = -\dfrac{x}{4} + C$

3. $e^y(x - y + 1) = C$

$\dfrac{dV}{dt} = 4 - 3 = 1$, $V(0) = 200 \Rightarrow V(t) = t + 200$

$\dfrac{dy}{dt} = \left(2\dfrac{\text{lb}}{\text{gal}}\right)\left(4\dfrac{\text{gal}}{\text{min}}\right) - \left(\dfrac{y}{t + 200}\right)\left(3\dfrac{\text{gal}}{\text{min}}\right) =$

$8 - \dfrac{3y}{t + 200}$; $y(0) = 10$

$\dfrac{dy}{dt} + \dfrac{3y}{t + 200} = 8 \Rightarrow \mu = e^{\int 3/(t+200)dt} = e^{3\ln(t+200)} = (t + 200)^3$

$\dfrac{d}{dt}[(t + 200)^3 y] = 8(t + 200)^3 \Rightarrow (t + 200)^3 y = 2(t + 200)^4 + C \Rightarrow y = 2(t + 200) + C(t + 200)^{-3}$

$10 = 400 + C \cdot 200^{-3} \Rightarrow C = -390 \cdot 200^3 \Rightarrow y = 2t + 400 - 390 \cdot 200^3(t + 200)^{-3}$

15. Because the object reaches its max. height when $v = -gt + v_0 = 0$ or $t = \dfrac{v_0}{g}$ and the air resistance is ignored, the object hits the ground when $t = \dfrac{2v_0}{g}$. Therefore, the velocity at this time is

$v\left(\dfrac{2v_0}{g}\right) = -g\left(\dfrac{2v_0}{g}\right) + v_0 = v_0$.

17. $c = 5$

19. The velocity of the parachutist after the parachute is opened is given by $v(t) = 8\dfrac{17e^{8t} + 13}{17e^{8t} - 13}$; the limiting velocity is $\lim\limits_{t\to\infty} v(t) = 8$.

21. **(b)** $\dfrac{dv}{dt} = \dfrac{dv}{dr}\dfrac{dr}{dt} = v\dfrac{dv}{dr}$; **(c)** $\lim\limits_{r\to\infty} v^2 = v_0^2 - 2gR$

23. $g \approx 32$ ft/s$^2 \approx 0.006$ mi/s^2; $v_0 = \sqrt{2gR} = \sqrt{2(.165)(0.006)(1080)} \approx 1.46$ mi/s

25. $Q(t) = E_0 C + e^{-t/(RC)}(-E_0 C + Q_0)$;

$I(t) = -\dfrac{1}{RC}e^{-t/(RC)}(-E_0 C + Q_0)$

27. $v(t) = 12(4 - \sqrt{3}) - 12(4 - \sqrt{3})e^{-t/3}$; $x(t) = 12(4 - \sqrt{3})t + 36(4 - \sqrt{3})e^{-t/3} - 36(4 - \sqrt{3})$

5. $y = P_0 3^{t/4}$; $t = \dfrac{4\ln 5}{\ln 3} \approx 5.86$ days

7. $y = y_0 \left(\dfrac{1}{2}\right)^{t/1700}$; $y(50) \approx 0.9798 y_0$ (97.98% of y_0)

9. $y = \dfrac{1000}{1 + 3\left(\dfrac{1}{3}\right)^t}$; $y = 750 \Rightarrow t = \dfrac{\ln 9}{\ln 3} \approx 2$ days

11. $T(t) = 90 - 50\left(\dfrac{1}{2}\right)^{t/20}$; $T(30) = 90 - 50\left(\dfrac{1}{2}\right)^{3/2} \approx$ 72.3°F

13. $\dfrac{dT}{dt} = k(T - 325)$, $T(0) = 100$, $T(45) = 150$, $T(-1) = $?

$T(t) = Ce^{kt} + 325$; $T(0) = 100 \Rightarrow C = -225 \Rightarrow T(t) = -225e^{kt} + 325$

$T(45) = 150 \Rightarrow -225e^{45k} + 325 = 150 \Rightarrow$

$e^{45k} = \dfrac{175}{225} = \dfrac{7}{9} \Rightarrow e^k = \left(\dfrac{7}{9}\right)^{1/45}$

$T(t) = -225\left(\dfrac{7}{9}\right)^{t/45} + 325 \Rightarrow T(-1) \approx 98.74°F$

15. $v(t) = 4 - 4e^{-8t}$; $v(3) \approx 4$ft/s; $s(t) = 4t + \dfrac{1}{2}e^{-8t} - \dfrac{1}{2}$; $s(3) \approx 11.5$ ft

17. $\dfrac{dv}{dt} = -9.8 - v$, $v(0) = 40 \Rightarrow v(t) = -\dfrac{99}{5} + \dfrac{299}{5}e^{-t}$;

$v(t) = 0 \Rightarrow t = -\ln\dfrac{99}{299} \approx 1.11$ s

$\dfrac{ds}{dt} = v$, $s(0) = 0 \Rightarrow s(t) = -\dfrac{99}{5}t - \dfrac{299}{5}e^{-t} + \dfrac{299}{5}$;

$s\left(-\ln\dfrac{99}{299}\right) \approx 18.11$ ft

19. $\dfrac{dv}{dt} = 32 - \dfrac{1}{2}v^2$, $v(0) = 30 \Rightarrow v(t) = \dfrac{8(19e^{8t} + 11)}{19e^{8t} - 11}$; $\lim_{t \to \infty} v(t) = 8$ ft/s

21. $m = 230$ kg $\Rightarrow \dfrac{dv}{dt} = \dfrac{82}{115} - \dfrac{637}{230000}v$, $v(0) = 0$;

$v(t) = \dfrac{164000}{637}(1 - e^{-637t/230000})$

$v(t) = 12 \Rightarrow t = -\dfrac{230000}{637}\ln\dfrac{39089}{41000} \approx 17.23$ s

$\dfrac{dy}{dt} = v$, $y(0) = 0 \Rightarrow y = \dfrac{164000}{637}t +$

$\dfrac{37720000000}{405769}(e^{-637t/230000} - 1)$

$H = y(17.23) \approx 104.17798$

23. **(a)** $4r^2 = 32\cos^2\theta - 16$; $r = 2\sec\theta + 2$;

(c) $r = -6\cos^2\dfrac{\theta}{2} + 6$

C H A P T E R　4

Exercises 4.1

1. $W(S) = 1 \neq 0$ for any x; lin. indep.

3. $W(S) = 2e^{-10x} \neq 0$ for any x; lin. indep.

5. $W(S) = 3e^{-6x} \neq 0$ for any x; lin. indep.

7. $W(S) = 0$ for all x; lin. dep.

9. $W(S) = 16e^{5x} \neq 0$ for any x; lin. indep.

11. $W(S) = -37e^{-9x} \neq 0$ for any x; lin. indep.

13. $W(S) = 1080e^{-12x} \neq 0$ for any x; lin. indep.

15. $y' = c_1e^x - c_2e^{-x}$; $y'' = c_1e^x + c_2e^{-x}$

17. $y' = e^{-x}(-c_1 + c_2 - c_2x)$; $y'' = e^{-x}(c_1 - 2c_2 + c_2x)$

19. $y' = -2c_1\sin 2x + 2c_2\cos 2x$; $y'' = -4c_1\cos 2x - 4c_2\sin 2x$

21. $y = e^{-x} - 2e^{2x}$

23. $y = \sin 2x - \cos 2x$

25. $y = e^{4x}(1 - x)$

27. $y = 3x^{-7} - x$

29. $y = \cos x + \sin x + x\sin x$

31. $\begin{vmatrix} y_1 & y_2 \\ y_1' & y_2' \end{vmatrix} = \begin{vmatrix} y_1 & y_2 \\ -p(x)y_1 & -p(x)y_2 \end{vmatrix} = -p(x)(y_1y_2 - y_1y_2) = 0 \Rightarrow y_1, y_2$ are lin. dep.

32. **(d)** $\begin{vmatrix} \cosh x & \sinh x \\ \sinh x & \cosh x \end{vmatrix} = \cosh^2 x - \sinh^2 x = 1$

Exercises 4.2

1–10. See the calculation of $W(S)$ in Section 4.1. Also, verify that each function is a solution of the differential equation.

13. $p(x) = 6$; $y_1(x) = e^{-2x}$

$$y_2(x) = e^{-2x} \int \frac{e^{-\int 6dx}}{[e^{-2x}]^2}\, dx = e^{-2x} \int \frac{e^{-6x}}{e^{-4x}}\, dx$$

$$= e^{-2x} \int e^{-2x}\, dx = e^{-2x}\left(-\frac{1}{2}e^{-2x}\right)$$

$$= -\frac{1}{2}e^{-4x}$$

15. $y_2(x) = xe^{-5x}$

17. $y_2(x) = \cos 7x$

19. $y_2(x) = x^{-3}$

21. $y_2(x) = x \ln x$

23. $a = 2$, $b = -1$, $c = -2$

Exercises 4.3

1. $y = c_1 e^{-2x} + c_2 e^{-6x}$

3. $y = c_1 e^{-4x} + c_2 e^{x}$

5. $y = c_1 \cos 4x + c_2 \sin 4x$

7. $y = c_1 \cos \sqrt{7}x + c_2 \sin \sqrt{7}x$

9. $y = c_1 e^{3x} \cos 4x + c_2 e^{3x} \sin 4x$

11. $y = c_1 e^{-3x} \cos 3x + c_2 e^{-3x} \sin 3x$

13. $y = c_1 e^{3x/7} + c_2 e^{-x}$

15. $y = c_1 e^{3x} + c_2 x e^{3x}$

17. $y = c_1 + c_2 e^{5x} + c_3 x e^{5x}$

19. $y = c_1 + c_2 x + c_3 e^{-x/8}$

21. $y = c_1 e^{x} + c_2 e^{-4x} \cos 4x + c_3 e^{-4x} \sin 4x$

23. $y = c_1 e^{5x} + c_2 e^{-3x} + c_3 e^{4x}$

25. $y = c_1 e^{-8x} + c_2 e^{-2x} + c_3 x e^{-2x}$

27. $y = c_1 e^{4x} + c_2 e^{-3x} + c_3 x e^{-3x}$

29. $y = c_1 + c_2 e^{-x/2} + c_3 e^{x/4} \cos \dfrac{x\sqrt{3}}{4} + c_4 e^{x/4} \sin \dfrac{x\sqrt{3}}{4}$

31. $y = c_1 + c_2 x + c_3 x^2 + c_4 e^{-4x/9}$

33. $y = c_1 e^{-4x} + c_2 e^{2x} + c_3 e^{3x} + c_4 e^{4x}$

35. $y = c_1 e^{-x} + c_2 e^{2x} + c_3 e^{-4x} + c_4 x e^{-4x}$

37. $y = c_1 e^{x} + c_2 e^{2x} + c_3 e^{-2x} \cos 4x + c_4 e^{-2x} \sin 4x$

39. $y = c_1 e^{-2x} + c_2 x e^{-2x} + c_3 e^{x} \cos x + c_4 e^{x} \sin x$

41. $y = c_1 \cos 4x + c_2 \sin 4x + c_3 x \cos 4x + c_4 x \sin 4x$

43. $y = 1 + 2e^{x}$

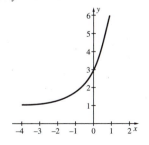

45. $y = e^{3x} - e^{-4x}$

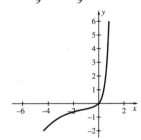

47. $y = \dfrac{2}{9}e^{4x} - \dfrac{2}{9}e^{-x/2}$

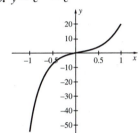

49. $y = 2 \cos 6x - \sin 6x$

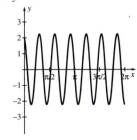

51. $y = 4e^x - 4xe^x$

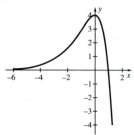

53. $y = 2e^{-x} \sin 2x + e^{-x} \cos 2x$

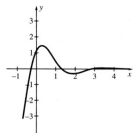

55. $y = 1 + 2x$

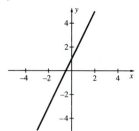

59. $a(x) = \dfrac{1}{x^2 + 1}$; $b(x) = -\dfrac{2(-6 + x - 6x^2)}{x^2 + 1}$;

$c(x) = 45(x^2 + 1)$

$w'' - \left[\dfrac{-2x/(x^2 + 1)^2}{1/(x^2 + 1)} + \dfrac{2(-6 + x - 6x^2)}{x^2 + 1} \right] w' +$
$45w = 0$

$w'' - \left[\dfrac{-2x}{x^2 + 1} + \dfrac{2(-6 + x - 6x^2)}{x^2 + 1} \right] w' + 45w = 0$

$w'' + 12w' + 45w = 0 \Rightarrow w(x) = c_1 e^{-6x} \cos 3x +$
$c_2 e^{-6x} \sin 3x$

$y(x) = \dfrac{(-6c_1 + 3c_2) \cos 3x + (-6c_2 - 3c_1) \sin 3x}{c_1 \cos 3x + c_2 \sin 3x}(x^2 + 1)$

61. $a(x) = x^2 \cos x$; $b(x) = -\dfrac{x \tan x + 2x - 2}{x}$;

$c(x) = \dfrac{2 \sec x}{x^2}$

$w'' - \left[\dfrac{2x \cos x - x^2 \sin x}{x^2 \cos x} + \dfrac{x \tan x + 2x - 2}{x} \right] w' +$
$2w = 0$

$w'' - 2w' + 2w = 0 \Rightarrow w(x) = c_1 e^x \cos x + c_2 e^x \sin x$

$y(x) = \dfrac{(c_1 + 3c_2) \cos x + (c_2 - 3c_1) \sin x}{(x^2 \cos x)(c_1 \cos x + c_2 \sin x)}$

63. $a(x) = \dfrac{x}{x^2 + 4}$; $b(x) = -\dfrac{x^2 - 4}{x}$; $c(x) = \dfrac{4(x^2 + 4)}{x}$

$w'' - \left[\dfrac{4 - x^2}{(x^2 + 4)} \dfrac{x^2 + 4}{x} + \dfrac{x^2 - 4}{x} \right] w' + 4w = 0$

$w'' + 4w = 0 \Rightarrow w(x) = c_1 \cos 2x + c_2 \sin 2x$

$y(x) = \dfrac{2(-c_1 \sin 2x + c_2 \cos 2x)}{c_1 \cos 2x + c_2 \sin 2x} \dfrac{x^2 + 4}{x}$

65. $y^{(8)} - 9y^{(7)} + 32y^{(6)} + 50y^{(5)} - 939y^{(4)} + 4207y''' -$
$10130y'' + 12948y' - 6760y = 0$

71. (a) $y = e^{-3x}(c_1 \cosh \sqrt{7}x + c_2 \sinh \sqrt{7}x)$;

(b) $y = e^{5x/2}\left(c_1 \cosh \dfrac{x}{2} + c_2 \sinh \dfrac{x}{2}\right)$;

(c) $y = e^{3x}(c_1 \cosh 5x + c_2 \sinh 5x)$;

(d) $y = c_1 \cosh 4x + c_2 \sinh 4x$

73. (a) $y = c_1 e^{-x} + c_2 e^x$ or $y = c_3 e^{-2x} + c_4 e^{2x}$; the
Principle of Superposition is not valid: $c_1 e^{-x} + c_2 e^x +$
$c_3 e^{-2x} + c_4 e^{2x}$ is not a solution of the equation unless
$c_1 = c_2 = 0$ or $c_3 = c_4 = 0$. **(b)** $y = c_1 e^{-x} + c_2 e^x$; the
Principle of Superposition is valid.

Exercises 4.4

1. $y = c_1 e^{-2x} + c_2 e^{4x/3} + \dfrac{5}{8}$

3. $y = c_1 e^{-3x/5} + c_2 e^{3x/2} + \dfrac{x}{3} - \dfrac{1}{3}$

5. $y = c_1 e^{-5x} \cos 4x + c_2 e^{-5x} \sin 4x - \dfrac{1}{80} e^{3x}$

7. $y = c_1 e^x + c_2 x e^x - \dfrac{3}{2} \sin x$

9. D

11. D^3

13. $D - 4$

15. $D^2 + 4$

17. $D(D^2 + 9)$

19. $(D - 8)^3$

21. $(D - 3)^4$

23. $D^2 + 2D + 10$

25. $(D^2 - 16D + 145)^3$

27. $(D + 4)^5(D^2 + 4D + 8)^2$

31. The set of functions $\{e^{-2x}, e^{2x}\}$ is a fundamental set of solutions to $y'' - 4y = 0$. Therefore, if $y(x) - y_p(x)$ is a solution to $y'' - 4y = 0$, there are constants c_1 and c_2 so that $y(x) - y_p(x) = c_1 e^{-2x} + c_2 e^{2x}$.

33. $3x^2 y'' - xy' + 8y = 0$

35. $xy'' + 6y' - 6y = 0$

37. $4xy'' + 3y' - 3y = 0$

39. $x^3 y''' + 2x^2 y'' + xy'' + 10xy = 0$

41. $y' = 4c_1 x^3 - c_2 x^{-2};\ y'' = 12c_1 x^2 + 2c_2 x^{-3}$

43. $y' = c_1 \cos(2 \ln x) + c_2 \sin(2 \ln x) +$
$2(-c_1 \sin(2 \ln x) + c_2 \cos(2 \ln x))$
$= (c_1 + 2c_2) \cos(2 \ln x) + (c_2 - 2c_1) \sin(2 \ln x)$

$y'' = -\dfrac{2}{x}(c_1 + 2c_2) \sin(2 \ln x) +$

$\dfrac{2}{x}(c_2 - 2c_1) \cos(2 \ln x)$

45. $y' = 3c_1 x^2 + 2c_2 x + 2;\ y'' = 6c_1 x + 2c_2$

47. No. If $y = 1$, $xD(1) = 0$. However, $Dx(1) = 1$.

Exercises 4.5

1. $y = c_1 e^{2x} + c_2 e^{3x} + \dfrac{1}{2} e^x$

3. $y = c_1 e^{-7x} + c_2 x e^{-7x} + \dfrac{1}{2} x^2 e^{-7x}$

5. $y = c_1 e^{3x} \cos x + c_2 e^{3x} \sin x + 1$

7. $y = c_1 e^{-4x} + c_2 - \dfrac{2}{9} e^{-x} - \dfrac{1}{3} x e^{-x}$

9. $y = c_1 \cos\dfrac{x}{3} + c_2 \sin\dfrac{x}{3} + x$

11. $y = c_1 e^{-x} \cos 4x + c_2 e^{-x} \sin 4x + \dfrac{x}{8} e^{-x} \sin 4x$

13. $y = c_1 e^{5x} + c_2 e^{-3x} + 8x e^{5x}$

15. $y = c_1 e^x \cos 5x + c_2 e^x \sin 5x + 3x + e^{-x}$

17. $y = c_1 e^{2x} \cos x + c_2 e^{2x} \sin x + 5x + 4$

19. $y = e^{-3x}(c_1 \cos 5x + c_2 \sin 5x) + \dfrac{1}{10} x e^{-3x} \sin 5x +$

$e^{5x}\left(-\dfrac{3}{544} \cos 3x + \dfrac{5}{544} \sin 3x\right)$

21. $y = e^{-4x}(c_1 \cos 2x + c_2 \sin 2x)$
$+ x e^{-4x}\left(2 \cos 2x + \dfrac{1}{2} \sin 2x\right)$

23. $y = c_1 + c_2 \cos 4x + c_3 \sin 4x + \dfrac{5}{16} x^2$

25. $y = c_1 e^{-4x} + c_2 e^{-x} + c_3 e^{4x} + x e^{3x}$

27. $y = c_1 e^{2x} + c_2 e^x \cos 4x + c_3 e^x \sin 4x - \dfrac{1}{50} e^{-2x}$

29. $y = c_1 e^{-3x} + c_2 e^{2x} \cos 4x + c_3 e^{2x} \sin 4x + x^2$

31. $y = c_1 e^{-3x} + c_2 e^{-2x} + c_3 e^{2x} + c_4 e^{3x} + \dfrac{1}{50} \cos x$

33. $y = c_1 e^{2x} + c_2 e^{3x} + c_3 \cos 4x + c_4 \sin 4x + 10$

35. $y = c_1 e^{-3x} + c_2 x e^{-3x} + c_3 e^{2x} \cos 2x + c_4 e^{2x} \sin 2x + 2x$

37. $y = c_1 e^{-4x} + c_2 x e^{-4x} + c_3 e^{4x} + c_4 x e^{4x} + x^2$

39. $y = c_1 e^{-x} \cos x + c_2 e^{-x} \sin x + c_3 e^{2x} \cos x +$

$c_4 e^{2x} \sin x - \dfrac{1}{10} x + \dfrac{1}{50}$

41. $y = c_1 e^x \cos x + c_2 e^x \sin x + c_3 x e^x \cos x + c_4 x e^x \sin x + 1$

43. $y = e^{3x} - e^{3x/5} - \dfrac{5}{9}$

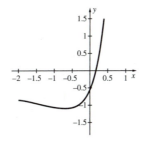

45. $y = 9e^{-x/2} - 7e^{-2x/3} - \dfrac{1}{2}$

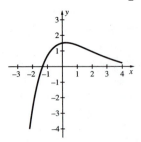

47. $y = -7e^{-x} + 7e^{x} - 6x - x^{3}$

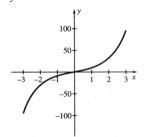

49. $y = \dfrac{2}{3}\sin 2x - \dfrac{7}{3}\sin x$

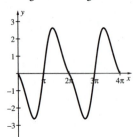

51. $y = -\dfrac{3}{4}e^{-x} + 3e^{x} + e^{-3x}$

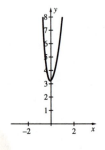

53. $y = x - 2 + 3e^{-x} + 2xe^{-x}$

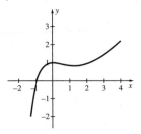

55. $y = \dfrac{1}{3}e^{-4x} + \dfrac{2}{3}e^{2x}$

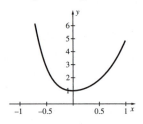

57. $y = 4e^{-3x}\sin x + e^{-3x}\cos x - \dfrac{2}{5}$

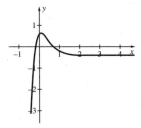

Exercises 4.6

1. $y = c_1 e^x + c_2 e^{-2x} + \dfrac{1}{2}$

3. $y = c_1 e^{4x/5} + c_2 e^{-x} + \dfrac{3}{4}$

5. $y = c_1 e^{4x} + c_2 e^{-2x} - 4x + 1$

7. $y = c_1 e^{5x/4} + c_2 e^{-3x/4} - 5x + \dfrac{8}{3}$

9. $y = c_1 e^{-x} \cos 5x + c_2 e^{-x} \sin 5x - 13x + 1$

11. $y = c_1 e^x + c_2 e^{-4x} + 8x^2 + 12x + 13$

13. $y = c_1 e^{-x/4} + c_2 e^{-x/2} + 5x^2 - 60x + 280$

15. $y = c_1 e^{4x} + c_2 e^{2x} - 32x^3 - 72x^2 - 84x - 45$

17. $y = c_1 + c_2 e^{2x} - 4 \sin 3x + \dfrac{8}{3} \cos 3x$

19. $y = c_1 e^{3x} \cos 2x + c_2 e^{3x} \sin 2x + \dfrac{4}{3} \cos 2x + \sin 2x$

21. $y = c_1 e^{3x} + c_2 e^{-3x} - \cos 3x - 3x \sin 3x$

23. $y = c_1 e^{3x} + c_2 e^{2x} + \cos 3x + 5 \sin 3x$

25. $y = c_1 e^{-2x} + c_2 x e^{-2x} - 4x \cos 2x + 3 \cos 2x - 4x^2 \sin 2x + 4x \sin 2x$

27. $y = c_1 e^{-4x} + c_2 e^{5x} + \dfrac{1}{10} e^x$

29. $y = c_1 e^{-x} + c_2 e^{5x} - 36x^3 e^{5x} + 18x^2 e^{5x} - 6x e^{5x}$

31. $y = c_1 e^{4x} + c_2 e^{3x} - \dfrac{1}{2} x^4 e^{4x} + 2x^3 e^{4x} - 6x^2 e^{4x} + 12x e^{4x}$

33. $y = c_1 e^{-3x} + c_2 e^{-2x} + c_3 e^{-x} + x e^{-3x} + \dfrac{3}{4} x e^{-x} - \dfrac{1}{4} x^2 e^{-x}$

35. $y = c_1 e^{-4x} + e^{-3x}(c_2 \cos x + c_3 \sin x) + \left(\dfrac{1}{4} x^2 + \dfrac{1}{2} x\right) e^{-4x} + x e^{-3x}\left(-\dfrac{1}{2} \cos x + \dfrac{1}{2} \sin x\right)$

37. $y = c_1 e^{4x} + c_2 e^{-5x} \cos x + c_3 e^{-5x} \sin x + e^x$

39. $y = c_1 e^{3x} + c_2 x e^{3x} + c_3 \cos 2x + c_4 \sin 2x + 3x + 2$

41. $y = c_1 e^{2x} + c_2 x e^{2x} + c_3 e^{3x} \cos x + c_4 e^{3x} \sin x + e^{-x}$

43. $y = 6x - 1 + e^{-3x}$

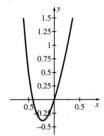

45. $y = \dfrac{7}{2} 2 e^{2x} - \dfrac{7}{2} e^{-2x} - 8x$

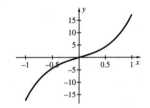

47. $y = e^{2x} + e^{-3x} - \dfrac{1}{2} x - \dfrac{1}{12}$

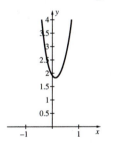

49. $y = \dfrac{3}{2} - e^{-4x} - \dfrac{1}{4} \cos x + \sin x$

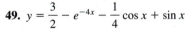

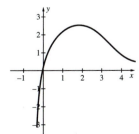

51. $y = \dfrac{7}{4}e^{-3x}\sin 4x - \dfrac{1}{25}$

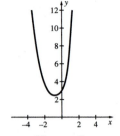

53. $y = \dfrac{1}{3}e^{3x} - \dfrac{x}{3}e^{3x} + \dfrac{1}{3}x^2 + \dfrac{2}{9}x - \dfrac{1}{3}$

55. $y = 5e^x - 7e^{2x} + 6xe^{2x} - 2x^2e^{2x} + 3x + \dfrac{3}{2}x^2 - \dfrac{3}{2}$

57. $y = 2 + e^{2x} - \dfrac{1}{2}x - \dfrac{1}{2}x^2 - \dfrac{1}{3}x^3$

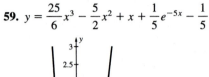

59. $y = \dfrac{25}{6}x^3 - \dfrac{5}{2}x^2 + x + \dfrac{1}{5}e^{-5x} - \dfrac{1}{5}$

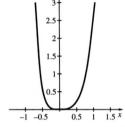

61. $y = 13 - \dfrac{25}{2}e^{-x} - \dfrac{1}{2}\cos 5x - \dfrac{5}{2}\sin 5x$

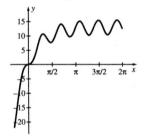

63. (a)

65. (a) $y = c_1 e^{4x} - \dfrac{1}{32} - \dfrac{1}{8}x - \dfrac{1}{4}x^2;$

(b) $y = c_1 e^{-x} + \dfrac{1}{5}\cos 2x + \dfrac{2}{5}\sin 2x;$

(c) $y = c_1 e^x + \dfrac{1}{3}e^{4x}$

Exercises 4.7

1. $y = c_1 e^{2x} + c_2 e^{5x} - \dfrac{1}{2}e^{3x}$

3. $y = c_1 \cos 4x + c_2 \sin 4x + \dfrac{x}{4}\sin 4x$

5. $y = c_1 e^{-2x} + c_2 x e^{-2x} + \dfrac{1}{2}x^2 e^{-2x}$

7. $y = c_1 e^{2x}\cos 3x + c_2 e^{2x}\sin 3x + \dfrac{4x}{3}e^{2x}\sin 3x$

9. $y = c_1 e^{-2x} \cos 4x + c_2 e^{-2x} \sin 4x + \dfrac{1}{8} x e^{-2x}$

11. $y = c_1 \cos 4x + c_2 \sin 4x - \dfrac{1}{4} x \cos 4x +$

$\dfrac{1}{16} \sin 4x \ln|\sin 4x|$

13. $y = c_1 e^{-x} \cos 7x + c_2 e^{-x} \sin 7x - \dfrac{1}{7} x e^{-x} \cos 7x +$

$\dfrac{1}{49} e^{-x} \sin 7x \ln|\sin 7x|$

15. $y = c_1 e^x \cos 5x + c_2 e^x \sin 5x +$

$e^x \cos 5x \left(\dfrac{1}{25} \ln|\cos 5x| - \dfrac{x}{5} \right) +$

$e^x \sin 5x \left(\dfrac{1}{25} \ln|\sin 5x| + \dfrac{x}{5} \right)$

17. $y = c_1 e^{3x} \cos 5x + c_2 e^{3x} \sin 5x +$

$e^{3x} \cos 5x \left(\dfrac{1}{25} \sin 5x - \dfrac{1}{25} \ln|\sec 5x + \tan 5x| \right) -$

$\dfrac{1}{25} e^{3x} \sin 5x \cos 5x$

19. $y = c_1 e^{6x} \cos x + c_2 e^{6x} \sin x + x e^{6x} \sin x +$
$e^{6x} \cos x \ln|\cos x|$

21. $y = c_1 e^{3x} + c_2 e^{-3x} - \dfrac{1}{18} + \dfrac{1}{18} e^{3x} \ln(1 + e^{-3x}) -$

$\dfrac{1}{18} e^{-3x} \ln(1 + e^{3x})$

23. $y = c_1 e^{-x} + c_2 e^x + \left(\dfrac{1}{4} + \dfrac{1}{2} x \right) e^{-x} + \left(\dfrac{1}{2} x - \dfrac{1}{4} \right) e^x$ or

$y = c_1 e^{-x} + c_2 e^x + \dfrac{1}{2} x e^{-x} + \dfrac{1}{2} x e^x$

25. $y = c_1 e^x + c_2 x e^x + x e^x \ln x$

27. $y = c_1 e^{-4x} + c_2 x e^{-4x} + \dfrac{1}{6} x^{-2} e^{-4x}$

29. $y = c_1 e^{-3x} + c_2 x e^{-3x} + \dfrac{1}{4} x e^{-5x} - x e^{-3x} + \dfrac{1}{2} x^2 e^{-3x} +$
$x e^{-3x} \ln x$ or

$y = c_1 e^{-3x} + c_2 x e^{-3x} + \dfrac{1}{4} x e^{-5x} + \dfrac{1}{2} x^2 e^{-3x} + x e^{-3x} \ln x$

31. $y = c_1 e^{-2x} + c_2 x e^{-2x} +$
$\sqrt{1 - x^2} \left(\dfrac{1}{3} e^{-2x} + \dfrac{1}{6} x^2 e^{-2x} \right) + \dfrac{1}{2} x e^{-2x} \sin^{-1} x$

33. $y = c_1 e^{5x} + c_2 x e^{5x} - \dfrac{3}{4} x^2 e^{5x} + \dfrac{1}{2} x^2 e^{5x} \ln|2x|$

35. $y = c_1 e^{-4x} + c_2 x e^{-4x} + x e^{-4x} \tan^{-1} x - \dfrac{1}{2} e^{-4x} \ln(1 + x^2)$

36. $y = c_1 \cos \dfrac{x}{2} + c_2 \sin \dfrac{x}{2} +$

$\cos \dfrac{x}{2} \left[-2x + 4 \ln \left| \cos \dfrac{x}{2} \right| \right] + \sin \dfrac{x}{2} \left[2x + 4 \ln \left| \sin \dfrac{x}{2} \right| \right]$

37. $y = c_1 + c_2 \cos x + c_3 \sin x - \cos x \ln|\cos x|$

39. $y = c_1 + c_2 x + c_3 e^{2x} - x + x \ln x - \dfrac{1}{2}$

41. $y = c_1 e^{2x} + c_2 e^{5x} + c_3 e^{-3x} - \dfrac{1}{14} e^{4x}$

43. $y = c_1 e^x + c_2 e^{3x} + c_3 e^{-4x} + \dfrac{3}{85} \cos x - \dfrac{7}{170} \sin x$

45. $y = -\cos x \ln(1 + \sin x) + \cos x \ln(\cos x) + 2 \sin x$
$+ 2 \cos x$

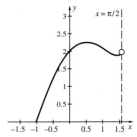

47. $y = -\dfrac{1}{2} + \dfrac{1}{25} \cos 4x + \dfrac{1}{50} \sin 4x + e^{-2x} - e^{-3x}$

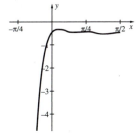

49. $y = e^{-4x} \left(-\dfrac{1}{32} - \dfrac{1}{4} x - x^2 \right) + \dfrac{1}{32} e^{4x}$

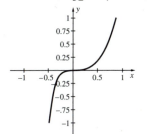

51. $y = -\dfrac{18}{13}e^x + \dfrac{35}{78}xe^x + \dfrac{1}{18}e^{-2x}\sin 3x$

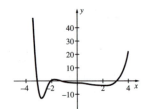

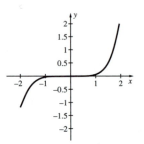

55. $y = c_1 e^{-3x} + c_2 e^{-x} - \cos 2x + 8\sin 2x$

61. $y = x^2 e^x + e^x - xe^x$

63. $y = 3x + 4 + \dfrac{3}{2}e^{3x} - \dfrac{11}{2}e^x$

53. $y = -\dfrac{1}{4}x^4 - x^3 - 3x^2 - 6x - 6 + 6e^x$

65. $y = c_1\sin(2\ln x) + c_2\cos(2\ln x) + \dfrac{x}{5}$

Chapter 4 Review Exercises

1. Lin. indep.

3. Lin. indep.

5. Lin. indep.

7. Lin. dep.

9. Lin. indep.

11. Lin. indep.

19. $y_2(x) = x^2 + x$

21. $y = c_1 e^{-2x} + c_2 e^{-5x}$

23. $y = c_1 e^{-4x/3} + c_2 e^{x/2}$

25. $y = c_1 e^{-x} + c_2 xe^{-x}$

27. $y = c_1 e^{-x} + c_2 e^{-2x}$

29. $y = c_1 e^{5x}\cos 3x + c_2 e^{5x}\sin 3x$

31. $y = c_1 e^{x/2} + c_2 e^{2x}$

33. $y = c_1 e^{x/3} + c_2 e^{2x/5}$

35. $y = c_1 e^{-x/4} + c_2 e^{x/5}$

37. $y = c_1 e^{-x/2} + c_2 e^{-x/6}$

39. $y = c_1 + c_2 e^{-x} + c_3 e^{-x/2}$

41. $y = e^{-2x}\left(c_1\sin\dfrac{2x}{3} + c_2\cos\dfrac{2x}{3}\right)$

43. $y = e^{-2x/3}(c_1\sin x + c_2\cos x)$

45. $y = c_1 e^{-2x} + c_2 e^{4x} + \dfrac{x}{8} - \dfrac{1}{32}$

47. $y = c_1 + c_2 e^{-5x} + \dfrac{2x}{25} - \dfrac{1}{5}x^2 + \dfrac{1}{3}x^3$

49. $y = c_1 + c_2 e^{4x} - \dfrac{12}{17}\cos x + \dfrac{3}{17}\sin x$

51. $y = c_1 e^{-x}\cos 2x + c_2 e^{-x}\sin 2x - \dfrac{12}{17}\cos 2x + \dfrac{3}{17}\sin 2x$

53. $y = c_1 + c_2 e^{2x} - \dfrac{1}{10}\cos 4x - \dfrac{1}{20}\sin 4x$

55. $y = c_1 e^{3x}\cos 2x + c_2 e^{3x}\sin 2x + \dfrac{3}{29}e^{-2x}$

57. $y = c_1 e^{-4x} + c_2 e^{-3x} - 6xe^{-4x} - 3x^2 e^{-4x} - x^3 e^{-4x}$

59. $y = c_1 e^{-2x} + c_2 xe^{-2x} + c_3 e^{4x} + \dfrac{x}{36}e^{4x} + \dfrac{1}{12}x^2 e^{-2x}$

61. $y = c_1 e^{-x}\cos 2x + c_2 e^{-x}\sin 2x + c_3 xe^{-x}\cos 2x +$

$\quad c_4 xe^{-x}\sin 2x + \dfrac{4}{625} - \dfrac{8x}{125} + \dfrac{1}{25}x^2$

63. $y = -\dfrac{2}{3}e^{-8x} + \dfrac{2}{3}e^{-2x}$

65. $y = \cos 5x$

67. $y = -\dfrac{19}{25}e^{-4x} + \dfrac{19}{25}e^x + \dfrac{x}{5}e^x$

69. $y = \dfrac{1}{2}x\sin x$

71. $y = c_1\cos x + c_2\sin x + \sin x\ln|\sin x| - x\cos x$

73. $y = \dfrac{1}{2x}e^{4x} - e^{4x} + \dfrac{x}{2}e^{4x}$

75. $y = -\dfrac{1}{4}e^x + xe^x + \dfrac{1}{2}x^2e^x \ln x - \dfrac{3}{4}x^2e^x$

81. (a) $W(y_1, y_2) = Ce^{-\int 3dx} = Ce^{-3x};$

$$W(e^{-4x}, e^x) = \begin{vmatrix} e^{-4x} & e^x \\ -4e^{-4x} & e^x \end{vmatrix} = e^{-3x} + 4e^{-3x} = 5e^{-3x}$$

(b) $W(y_1, y_2) = Ce^{-\int 4dx} = Ce^{-4x};$
$W(e^{-2x}\cos 3x, e^{-2x}\sin 3x) = 3e^{-4x}$

(c) $W(y_1, y_2) = Ce^{-\int 4dx} = Ce^{-4x};\ W(e^{-2x}, xe^{-2x}) = e^{-4x}$

(d) $W(y_1, y_2) = Ce^{-\int 0dx} = C;\ W(\cos 3x, \sin 3x) = 3$

83. (a) $y = \dfrac{1}{2}e^{x-1} - e^{-1-x}$

(b) $\lambda = 0 \Rightarrow y = 0;\ \lambda < 0 \Rightarrow y = 0;\ \lambda > 0 \Rightarrow \lambda_n = (n\pi/p)^2 \Rightarrow y_n = \sin(n\pi x/p),\ n = 1, 2, \ldots$

(c) $\lambda = 0 \Rightarrow y = 1;\ \lambda < 0 \Rightarrow y = 0;\ \lambda > 0 \Rightarrow \lambda_n = (n\pi/p)^2 \Rightarrow y_n = \cos(n\pi x/p),\ n = 1, 2, \ldots$

(d) $\lambda = 0 \Rightarrow y = 0;\ \lambda > 0 \Rightarrow y = 0;\ \lambda < 0 \Rightarrow \lambda_n = -(n\pi/2)^2 \Rightarrow y_n = e^{-x}\sin(n\pi x/2),\ n = 1, 2, \ldots$

C H A P T E R   5

Exercises 5.1

1. $m = 4$ slugs, $k = 9$ lb/ft; released 1 ft above eq. with zero init. vel.

3. $m = \dfrac{1}{4}$ slugs, $k = 16$ lb/ft; released 0.75 ft (8 inches) below eq. with an upward initial vel. of 2 ft/s.

5. $x(t) = 5\cos(t - \phi),\ \phi = -\cos^{-1}\dfrac{3}{5} \approx 0.93$ rads; per. $= 2\pi$; amp $= 5$

7. $x(t) = \dfrac{\sqrt{65}}{4}\cos(4t - \phi),\ \phi = \cos^{-1}\left(-\dfrac{8}{\sqrt{65}}\right) \approx$

3.02 rads; per. $= \dfrac{\pi}{2}$; amp $= \dfrac{\sqrt{65}}{4}$

9. $x(t) = \dfrac{\sqrt{2}}{3}\cos(3t - \phi),\ \phi = -\cos^{-1}\left(\dfrac{1}{\sqrt{2}}\right) =$

$\dfrac{-\pi}{4}$ rads; per. $= \dfrac{2\pi}{3}$; amp $= \dfrac{\sqrt{2}}{3}$

11. $x(t) = \cos 8t$, max. dis. $= 1$ ft when $-8\sin 8t = 0$ or

$$8t = n\pi \Rightarrow t = \dfrac{n\pi}{8},\ n = 0, 1, 2,\ \ldots$$

13. $x(t) = -\dfrac{1}{4}\cos 8t;\ t = \dfrac{\pi}{16}$ s; $x(5) \approx 0.167$ ft;

$x(t) = \dfrac{1}{8}\sin 8t;\ t = \dfrac{\pi}{8}$ s

15. As k increases, the frequency at which the spring-mass system passes through equilibrium increases.

17. $b = \dfrac{\sqrt{1023}}{16}$

19. $m = 2$ slugs

21. $\omega\sqrt{\alpha^2 + \dfrac{\beta^2}{\omega^2}}$

23. $\dfrac{d^2y}{dt^2} + \dfrac{\pi r^2 \rho}{m}y = 0$

25. $y(t) = 1.61\cos 3.5t - 0.856379\sin 3.5t$; max. displacement $= \sqrt{(1.61)^2 + (0.856379)^2} \approx 1.8236$ ft.

Exercises 5.2

1. $m = 1,\ c = 4,\ k = 3$; released from equilibrium with an upward init. vel. of 4 ft/s.

3. $m = \dfrac{1}{4},\ c = 2,\ k = 1$; released 6 inches above equil. with a downward init. vel. of 1 ft/s.

5. $x(t) = \dfrac{\sqrt{10}}{3}e^{-2t}\cos(3t - \phi),$

$\phi = \cos^{-1}\dfrac{3}{\sqrt{10}} \approx 0.32$ rads, Q.P.: $\dfrac{2\pi}{3},\ t \approx 0.63$

7. $x(t) = \dfrac{\sqrt{29}}{5}e^{-t}\cos(5t - \phi),$

$\phi = \cos^{-1}\dfrac{5}{\sqrt{29}} \approx 0.38$ rads, Q.P.: $\dfrac{2\pi}{5},\ t \approx 0.39$

9. $x(t) = -\dfrac{1}{2}e^{-5t} + \dfrac{1}{2}e^{-3t}$; overdamped; does not pass through equilibrium; $x\left(\dfrac{1}{2}\ln\left(\dfrac{5}{3}\right)\right) \approx 0.093$

11. $x(t) = -3e^{-t} + 2e^{-t/2}$; overdamped; $x(t) = 0 \Rightarrow$

$t = 2\ln\left(\dfrac{3}{2}\right) \approx 0.811$; max. dis. $= 1$ at $t = 0$

13. $x(t) = 4e^{-4t} + 14te^{-4t}$; critically damped; does not pass through equilibrium; max. dis. $= 4$ at $t = 0$

15. $x(t) = -5e^{-5t} - 24te^{-5t}$; critically damped; does not pass through equilibrium; max. dis. $= 5$ at $t = 0$

17. $c = 2$

19. $x(t) = -\dfrac{3}{2}e^{-4t} + e^{-6t}$; does not pass through equilibrium; max. dis. $= \frac{1}{2}$ at $t = 0$

21. $x(t) = e^{-2t}\left(-3\cos\left(4\sqrt{\dfrac{11}{5}}t\right) - \dfrac{3}{2}\sqrt{\dfrac{5}{11}}\sin\left(4\sqrt{\dfrac{11}{5}}t\right)\right)$ or $x(t) = $

$\dfrac{21}{2\sqrt{11}}\cos\left(4\sqrt{\dfrac{11}{5}}t - \phi\right)$, where $\phi =$

$\cos^{-1}\dfrac{-2\sqrt{11}}{7} \approx 2.81646$; $x(t) = 0 \Rightarrow$

$t = \dfrac{1}{4}\sqrt{\dfrac{5}{11}}\left(\dfrac{(2n+1)\pi}{2} + \phi\right)$, n any integer or

$t \approx 0.739471, 1.26899, 1.7985, 2.32802, 2.85753 \ldots$

23. $c = 32$, $0 < c < 32$

25. $c = \dfrac{10}{13}$

27. $m\dfrac{d^2}{dt^2}(u+v) + c\dfrac{d}{dt}(u+v) + k(u+v) = 0 \Rightarrow$

$m\dfrac{d^2u}{dt^2} + c\dfrac{du}{dt} + ku = 0$ and $m\dfrac{d^2v}{dt^2} + c\dfrac{dv}{dt} + kv = 0$;

$x(0) = u(0) + v(0) = \alpha \Rightarrow u(0) = \alpha$ and $v(0) = 0$;

$\dfrac{dx}{dt}(0) = \dfrac{du}{dt}(0) + \dfrac{dv}{dt}(0) = \beta \Rightarrow \dfrac{du}{dt}(0) = 0$ and

$\dfrac{dv}{dt}(0) = \beta$

29. $x(t) = c_1e^{-\rho t} + c_2te^{-\rho t}$, $\rho = \dfrac{c}{2m}$; $x(0) = \alpha$,

$x'(0) = 0 \Rightarrow x(t) = \alpha e^{-\rho t}(1 + \rho t)$; $x(t) = 0 \Rightarrow$

$t = -\dfrac{1}{\rho} < 0$

Exercises 5.3

1. $x(t) = -\dfrac{2}{7}\cos 4t + \dfrac{1}{2}\sin 4t + \dfrac{2}{7}\cos 3t$, $\omega = 4$

3. $x(t) = \dfrac{35}{141}\cos(4\sqrt{3}t) + \dfrac{4}{47}\cos t$

5. $x(t) = \begin{cases} \dfrac{4}{\omega^2 - 9}(\cos 3t - \cos \omega t), & \omega \neq 3 \\[2ex] \dfrac{2}{3}t\sin 3t, & \omega = 3 \end{cases}$;

resonance occurs if $\omega = 3$

7. $x(t) = -e^{-4t}\left(\dfrac{1}{20}\cos 3t + \dfrac{7}{120}\sin 3t\right) + \dfrac{1}{40}(2\cos t - \sin t)$

9. $x(t) = \dfrac{1}{3}\cos t - \dfrac{1}{3}\cos 2t$; $\pm\dfrac{2}{3}\sin\dfrac{t}{2}$; decreases

11. $x(t) = e^{-t/2}\left(2\cos\dfrac{5t}{2} - 4\sin\dfrac{5t}{2}\right) - 2\cos t + 11\sin t$;

trans: $e^{-t/2}\left(2\cos\dfrac{5t}{2} - 4\sin\dfrac{5t}{2}\right)$; steady-state:

$-2\cos t + 11\sin t$

13. **(a)** $x(t) = \alpha\cos\left(\sqrt{\dfrac{k}{m}}t\right) - \dfrac{F\omega}{k - m\omega^2}\sqrt{\dfrac{m}{k}}\sin\left(\sqrt{\dfrac{k}{m}}t\right) + \dfrac{F}{k - m\omega^2}\sin \omega t$

(b) $x(t) = \left[\beta - \dfrac{F\omega}{k - m\omega^2}\right]\sqrt{\dfrac{m}{k}}\sin\left(\sqrt{\dfrac{k}{m}}t\right) + \dfrac{F}{k - m\omega^2}\sin \omega t$

(c) $x(t) = \alpha\cos\left(\sqrt{\dfrac{k}{m}}t\right) + \left[\beta - \dfrac{F\omega}{k - m\omega^2}\right]\sqrt{\dfrac{m}{k}}\sin\left(\sqrt{\dfrac{k}{m}}t\right) + \dfrac{F}{k - m\omega^2}\sin \omega t$

15. $x(t) = \begin{cases} 1 - \cos t, & 0 \le t \le \pi \\ -2\cos t, & t > \pi \end{cases}$

17. $x(t) = \begin{cases} t - \sin t, \, 0 \le t \le 1 \\ -2\sin 1\cos t + (2\cos 1 - 1)\sin t + 2 - t, \, 1 < t \le 2 \\ (\sin 2 - 2\sin 1)\cos t + 4\cos 1\sin^2\dfrac{1}{2}\sin t, \, t > 2 \end{cases}$

Exercises 5.4

1. $Q(t) = \dfrac{55}{8} - \dfrac{55}{8}\cos 4t$

3. $Q(t) = \dfrac{t}{4} - \dfrac{1}{64}\sin 16t$

5. $Q(t) = \dfrac{8\sqrt{7}}{175}e^{-(125/2)t}\sin\left(\dfrac{25\sqrt{7}}{2}t\right)$; max.: 0.0225 at $t \approx 0.0147$

7. $Q(t) = e^{-125t/2}\left(-\dfrac{12185}{12502813}\cos\left(\dfrac{25\sqrt{7}}{2}t\right) - \right.$
$\left. \dfrac{26554371}{312570325\sqrt{7}}\sin\left(\dfrac{25\sqrt{7}}{2}t\right)\right) + \dfrac{12185}{12502813}\cos t +$
$\dfrac{62526875}{12502813}\sin t$. steady-state charge: $\lim\limits_{t\to\infty} Q(t) =$
$\dfrac{12185}{12502813}\cos t + \dfrac{62526875}{12502813}\sin t$; steady-state current:
$-\dfrac{12185}{12502813}\sin t + \dfrac{62526875}{12502813}\cos t$

9. **(a)** $s(x) = \dfrac{1}{300}x^4 + \dfrac{1}{3}x^2 - \dfrac{1}{15}x^3$; **(b)** $s(x) = \dfrac{1}{30}x^4 +$
$\dfrac{10}{3}x^2 - \dfrac{2}{3}x^3$; **(c)** $s(x) = \dfrac{1}{3}x^4 + \dfrac{100}{3}x^2 - \dfrac{20}{3}x^3$

11. **(a)** $s(x) = \dfrac{1}{300}x^4 + \dfrac{10}{3}x - \dfrac{1}{15}x^3$;

(b) $s(x) = \dfrac{1}{30}x^4 + \dfrac{100}{3}x - \dfrac{2}{3}x^3$;

(c) $s(x) = \dfrac{1}{3}x^4 + \dfrac{1000}{3}x - \dfrac{20}{3}x^3$; simple support leads to larger max. displacement.

13. **(a)** $s(x) = \dfrac{1}{360}x^6 + \dfrac{10000}{9}x - \dfrac{125}{9}x^3$;

(d) $s(x) = \dfrac{1}{360}x^6 + \dfrac{1250}{3}x - \dfrac{125}{18}x^3$

15. $Q(t) = Q_0\cos\dfrac{t}{\sqrt{LC}}$; $I(t) = -\dfrac{Q_0}{\sqrt{LC}}\sin\dfrac{t}{\sqrt{LC}}$; Max.
$Q = Q_0$; Max. $I = \dfrac{Q_0}{\sqrt{LC}}$

17. $\dfrac{E_0}{\sqrt{\left(L\omega - \dfrac{1}{C\omega}\right)^2 + R^2}}\left[\dfrac{R\sin\omega t}{\sqrt{\left(L\omega - \dfrac{1}{C\omega}\right)^2 + R^2}} - \right.$
$\left. \dfrac{\left(L\omega - \dfrac{1}{C\omega}\right)\cos\omega t}{\sqrt{\left(L\omega - \dfrac{1}{C\omega}\right)^2 + R^2}}\right]$

Exercises 5.5

1. **(a)** $\theta(t) = \dfrac{1}{20}\cos 4t$; **(b)** $\theta(t) = \dfrac{1}{20}\cos 4t + \dfrac{1}{4}\sin 4t$;

(c) $\theta(t) = \dfrac{1}{20}\cos 4t - \dfrac{1}{4}\sin 4t$; max. displacement:

(a) $\dfrac{1}{20}$; **(b)** and **(c)**: $\dfrac{\sqrt{26}}{20} \approx 0.255$

3. **(a)** $\theta(t) = \dfrac{\sqrt{7}}{60}e^{-t\sqrt{7}}\sin 3t + \dfrac{1}{20}e^{-t\sqrt{7}}\cos 3t$;

(b) $\theta(t) = \left(\dfrac{\sqrt{7}}{60} + \dfrac{1}{3}\right)e^{-t\sqrt{7}}\sin 3t + \dfrac{1}{20}e^{-t\sqrt{7}}\cos 3t$;

(c) $\theta(t) = \left(\dfrac{\sqrt{7}}{60} - \dfrac{1}{3}\right)e^{-t\sqrt{7}}\sin 3t + \dfrac{1}{20}e^{-t\sqrt{7}}\cos 3t$

9. $T = 2\pi\sqrt{\dfrac{2}{9.8}} \approx 2.83$ s

11. $T = 2\pi\sqrt{\dfrac{8}{32}} \approx 3.14$ s

13. $2\pi\sqrt{\dfrac{L}{9.8}} = 1 \Rightarrow L = 0.248$ m

17. $b^2 - 4Lg > 0$, overdamped; $b^2 - 4Lg = 0$, critically damped; $b^2 - 4Lg < 0$, underdamped

19. **(c)** yes; **(d)** no

Chapter 5 Review Exercises

1. $x(t) = \dfrac{1}{3}\cos 8t$; max. dis. $= \dfrac{1}{3}$; $t = \dfrac{\pi}{16}, \dfrac{\pi}{8}$

3. $x(t) = -\dfrac{1}{3}e^{-2t}\sin 3t$, $\lim\limits_{t\to\infty} x(t) = 0$; quasiper. $= \dfrac{2\pi}{3}$;
max. dis. $= \left|x\left(\dfrac{1}{3}\tan^{-1}\dfrac{3}{2}\right)\right| \approx 0.144$; $t = \dfrac{\pi}{3}$

5. $x(t) = \dfrac{1}{4} - \dfrac{1}{4}\cos 2t$; maximum displacement is $\dfrac{1}{2}$ and occurs when $t = \dfrac{\pi}{2}$.

7. $x(t) = \dfrac{1}{3}\cos t - \dfrac{1}{3}\cos 2t$; beats; $\pm\dfrac{2}{3}\sin\dfrac{t}{2}$

9. $Q(t) = \dfrac{25}{109} - \dfrac{25}{109}e^{-10t}\cos 3t - \dfrac{250}{327}e^{-10t}\sin 3t$;

$I(t) = \dfrac{25}{3}e^{-10t}\sin 3t$; $\lim\limits_{t\to\infty} Q(t) = \dfrac{25}{109}$; $\lim\limits_{t\to\infty} I(t) = 0$

11. $Q(t) = \dfrac{11}{500} - \dfrac{11}{500}\cos 100t$; $I(t) = \dfrac{11}{5}\sin 100t$;

$\lim\limits_{t\to\infty} Q(t)$ and $\lim\limits_{t\to\infty} I(t)\,DNE$

13. $s(x) = \dfrac{250}{3}x^2 - \dfrac{125}{9}x^3 + \dfrac{1}{12}x^5 - \dfrac{1}{360}x^6$

15. $s(x) = \dfrac{1250}{3}x^2 - \dfrac{250}{9}x^3 + \dfrac{1}{12}x^5 - \dfrac{1}{360}x^6$

17. $\theta(t) = \cos 8t$; max. dis. $= 1$; $t = \dfrac{\pi}{16}$

19. $\theta(t) = e^{-8t} + 8te^{-8t}$; the motion is not periodic.

21. $\theta(t) = 0.2168\cos 3.7t + 0.0471622\sin 3.7t$

23. $y(t) = \begin{cases} 3 - t^2 - 3\cos t, & 0 \le t \le 1 \\ (-2 + 2\cos 1 - \cos 2 + 2\sin 1)\cos t - \\ \quad (2\cos 1 - 2\sin 1 + \sin 2)\sin t, & t > 1 \end{cases}$

25. $x(t) = 25\cos\left(\dfrac{7\sqrt{5}}{25}t\right)$

27. $x(t) = A\sin\omega_n t + B\cos\omega_n t + \dfrac{F}{k}$;

$x(t) = \left(x_0 - \dfrac{F}{k}\right)\cos\omega_n t + \dfrac{F}{k}$

C H A P T E R   6

Exercises 6.1

1. $y = c_1\sqrt{x} + c_2 x^{5/2}$

3. $y = c_1 x + c_2 x^4$

5. $y = c_1\sqrt{x}\cos(2\ln x) + c_2\sqrt{x}\sin(2\ln x)$

7. $y = c_1 x\cos(3\ln x) + c_2 x\sin(3\ln x)$

9. $y = c_1 x^2 + c_2 x^{-3}$

11. $y = c_1\sin(2\ln x) + c_2\cos(2\ln x)$

13. $y = c_1 x^{-2}\sin(3\ln x) + c_2 x^{-2}\cos(3\ln x)$

15. $y = c_1 x^{-10} + c_2 x^{-7} + c_3 x^{-2}$

17. $y = c_1 x^2 + c_2 x^{-5} + c_3 x^{10}$

19. $y = c_1 x + c_2 x^2 + c_3 x^{-2}$

21. $y = c_1 x^{-1} + c_2\sin(2\ln x) + c_3\cos(2\ln x)$

23. $y = c_1 x + c_2 x\sin(\ln x) + c_3 x\cos(\ln x)$

25. $y = c_1 x^{-4} + c_2 x^{-1} + c_3 x + c_4 x\ln x$

27. $y = \dfrac{1}{9x^5}(9c_1 x^3 + 9c_2 x^3\ln x + 1)$

29. $y = \dfrac{1}{5x^2}(1 - 5c_1 x^2\sin(\ln x) + 5c_2 x^2\cos(\ln x))$

31. $y = -\dfrac{1}{2}x + c_1 x^2 + c_2 x^{-3}$

33. $y = 2 + c_1\sin(2\ln x) + c_2\cos(2\ln x)$

35. $y = \dfrac{1}{40}x^2 + c_1\sin(6\ln x) + c_2\cos(6\ln x)$

37. $y = \dfrac{1}{25x^4}(25c_1 + 25c_2 x^6 + 25c_3 x^6\ln x + x)$

39. $y = \dfrac{1}{648x^{13}}(648c_1 x^8 + 648c_2 x^9 + 648c_3 x^9\ln x - 1)$

41. $y = \dfrac{9}{5}x^{1/3} + \dfrac{1}{5}x^2$

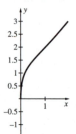

43. $y = \cos(2\ln x)$

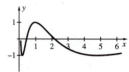

45. $y = \dfrac{1}{44}x^{-10} + \dfrac{8}{11}x - \dfrac{3}{4}x^2$

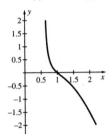

47. $y = \dfrac{3}{2}x - \dfrac{3}{2}x^2 \cos(\ln x) + \dfrac{1}{2}x^2 \sin(\ln x)$

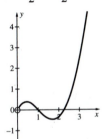

49. $y = \dfrac{1}{5x^2} + \dfrac{22}{15}\sqrt{x} - \dfrac{5}{3x}$

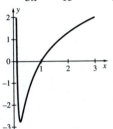

51. $y = \dfrac{1}{25}x^3 + \dfrac{24}{25}\sqrt{x} - \dfrac{8}{5}\sqrt{x}\ln x$

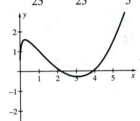

57. $y = \dfrac{1}{5}x^2 + c_1 \sin(\ln(-x)) + c_2 \cos(\ln(-x))$

59. $y = x\ln(-x)$

Exercises 6.2

1. $e^{-x} = 1 - x + \dfrac{x^2}{2!} - \dfrac{x^3}{3!} + \cdots = \displaystyle\sum_{k=0}^{\infty} \dfrac{(-1)^k x^k}{k!}$

3. $\dfrac{x}{1+x} = x - x^2 + x^3 - x^4 + \cdots = \displaystyle\sum_{k=0}^{\infty} (-1)^k x^{k+1}$

5. $x \sin x = x\left(x - \dfrac{x^3}{3!} + \dfrac{x^5}{5!} - \dfrac{x^7}{7!} + \cdots\right)$

$\qquad = x^2 - \dfrac{x^4}{3!} + \dfrac{x^6}{5!} - \dfrac{x^8}{7!} + \cdots = \displaystyle\sum_{k=0}^{\infty} \dfrac{(-1)^k x^{2k+2}}{(2k+1)!}$

7. $e^x = e + e(x - 1) + \dfrac{e}{2!}(x - 1)^2 + \dfrac{e}{3!}(x - 1)^3 + \cdots$

$\qquad = e \displaystyle\sum_{k=0}^{\infty} \dfrac{(x - 1)^k}{k!}$

9. $\cos x = \dfrac{\sqrt{2}}{2} - \dfrac{\sqrt{2}}{2}\left(x - \dfrac{\pi}{4}\right) - \dfrac{\sqrt{2}}{2}\cdot\dfrac{1}{2!}\left(x - \dfrac{\pi}{4}\right)^2 +$

$\qquad \dfrac{\sqrt{2}}{2}\cdot\dfrac{1}{3!}\left(x - \dfrac{\pi}{4}\right)^3 + \cdots$

$\qquad = \displaystyle\sum_{k=0}^{\infty} \cos\left(\dfrac{(2k + 1)\pi}{4}\right)\dfrac{\left(x - \dfrac{\pi}{4}\right)^k}{k!}$

11. $x + 2x^2 + 3x^3 + 4x^4 + \cdots = \displaystyle\sum_{k=0}^{\infty} (k + 1)x^{k+1}$

13. $f(x) = \dfrac{2}{(1 - x)^3}$

15. $\displaystyle\sum_{n=0}^{\infty} a_{n+2} x^{n+2}$

17. $\displaystyle\sum_{n=2}^{\infty} 4a_{n-1}x^{2n-1}$

19. $\displaystyle\sum_{n=0}^{\infty} [5a_n + 3a_{n+1}]x^{n+1}$

21. $\displaystyle\sum_{n=0}^{\infty} [10a_n + a_{n+3}]x^{n+2}$

23. $\displaystyle\sum_{n=0}^{\infty} a_{n+1}[x^{n+2} - 2x^{n+1}] = -2a_1x + \sum_{n=1}^{\infty} (a_{n-1} - 2a_n)x^n$

25. $\displaystyle\sum_{n=0}^{\infty} [2a_{n+2}x^{n+1} + a_nx^n] = a_0 + \sum_{n=1}^{\infty} (a_n + 2a_{n+1})x^n$

29. (a) $e^x \sin x = x + x^2 + \dfrac{1}{3}x^3 + \cdots$;

(b) $e^{-x} \cos x = 1 - x + \dfrac{1}{3}x^3 + \cdots$;

(c) $\sin x \cos x = x - \dfrac{2}{3}x^3 + \dfrac{2}{15}x^5 + \cdots$;

(d) $xe^x = x + x^2 + \dfrac{1}{2}x^3 + \cdots$;

(e) $x^2 \sin x = x^3 - \dfrac{1}{6}x^5 + \dfrac{1}{120}x^7 + \cdots$

31. $\displaystyle\lim_{x\to 0} \frac{\sin x}{x} = \lim_{x\to 0} \frac{1}{x}\left(x - \frac{x^3}{3!} + \frac{x^5}{5!} - \cdots\right)$

$\displaystyle = \lim_{x\to 0}\left(1 - \frac{x^2}{3!} + \frac{x^4}{5!} - \cdots\right) = 1$

Exercises 6.3

1. $x = 0$, $R \geq 1$

3. $x = 2$, $R \geq 4$

5. $x = \pm 2$, $R \geq 1$

7. $y = c_1 + c_2x + \left(9c_1 - \dfrac{3}{2}c_2\right)x^2 + \left(-9c_1 + \dfrac{9}{2}c_2\right)x^3 +$

$\left(\dfrac{81}{4}c_1 - \dfrac{45}{8}c_2\right)x^4 + \cdots$

9. $y = c_1 + c_2x + \left(-15c_1 + \dfrac{11}{2}c_2\right)x^2 +$

$\left(-55c_1 + \dfrac{91}{6}c_2\right)x^3 + \left(-\dfrac{455}{4}c_1 + \dfrac{671}{24}c_2\right)x^4 + \cdots$

11. $y = c_1 + c_2x + (-32c_1 - 8c_2)x^2 +$

$\left(\dfrac{512}{3}c_1 + 32c_2\right)x^3 + \left(-512c_1 - \dfrac{256}{3}c_2\right)x^4 + \cdots$

13. $y = c_1 + c_2x - \dfrac{1}{2}c_1x^2 - \dfrac{1}{6}c_2x^3 + \dfrac{1}{24}c_1x^4 + \cdots$

15. $y = c_1 + c_2x + \left(-\dfrac{25}{2}c_1 - 4c_2\right)x^2 +$

$\left(\dfrac{100}{3}c_1 + \dfrac{13}{2}c_2\right)x^3 + \left(-\dfrac{325}{8}c_1 - \dfrac{14}{3}c_2\right)x^4 + \cdots$

17. $y = c_1 + c_2x + \left(c_1 + \dfrac{1}{2}c_2 + \dfrac{1}{2}\right)x^2 +$

$\left(\dfrac{1}{3}c_1 + \dfrac{1}{2}c_2\right)x^3 + \left(\dfrac{1}{8} + \dfrac{1}{4}c_1 + \dfrac{5}{24}c_2\right)x^4 + \cdots$

19. $y = c_1 + c_2x + \left(c_1 + \dfrac{1}{2}c_2\right)x^2 + \dfrac{1}{3}c_2x^3 + \dfrac{1}{6}c_1x^4 + \cdots$

21. $y = c_1 + c_2x + \left(c_1 + \dfrac{3}{2}c_2\right)x^2 + \dfrac{1}{3}c_2x^3 + \dfrac{1}{6}c_1x^4 + \cdots$

23. $y = c_1 + c_2x - 2c_1x^2 - \dfrac{1}{2}c_1x^3 + \dfrac{1}{3}c_1x^4 + \cdots$

25. $y = c_1 + c_2x - \dfrac{1}{4}c_2x^3 + \dfrac{3}{16}c_2x^4 - \dfrac{9}{80}c_2x^5 + \cdots$

27. $y = c_1 + c_2x + \dfrac{3}{4}c_1x^2 + \dfrac{1}{12}c_2x^3 - \dfrac{5}{32}c_1x^4 + \cdots$

29. $y = c_1 + c_2x + \left(\dfrac{1}{2}c_2 - c_1\right)x^2 + \left(-\dfrac{1}{3}c_1 - \dfrac{1}{3}c_2\right)x^3$

$+ \left(\dfrac{1}{6}c_1 - \dfrac{5}{24}c_2\right)x^4 + \cdots$

31. $y = c_1 + c_2x - \dfrac{5}{4}c_2x^2 + \dfrac{3}{8}c_2x^3 - \dfrac{29}{192}c_2x^4 + \cdots$

33. $y = 1 + \dfrac{1}{3}x^4 + \dfrac{1}{42}x^8 + \dfrac{1}{1386}x^{12} + \dfrac{1}{83160}x^{16} + \cdots$

35. $y = -2 + 2x + 3x^2 - \dfrac{1}{3}x^3 + \dfrac{5}{4}x^4 + \cdots$

37. $y = c_1 + c_2(x - 1) - \dfrac{c_2}{8}(x - 1)^2 + \dfrac{5c_2}{96}(x - 1)^3 + \cdots$

39. $y = c_1 + c_2(x - 2) - \dfrac{3c_2}{32}(x - 2)^2 + \dfrac{41c_2}{1536}(x - 2)^3 + \cdots$

41. $y = c_1 + c_2(x + 2) + \left(-\dfrac{1}{8}c_1 - \dfrac{1}{4}c_2\right)(x + 2)^2$

$+ \left(-\dfrac{1}{48}c_1 - \dfrac{1}{8}c_2\right)(x + 2)^3 + \cdots$

43. $y = x - \dfrac{1}{8}x^4 + \dfrac{1}{40}x^5 - \dfrac{1}{360}x^6 + \dfrac{17}{5040}x^7 -$
$\dfrac{1}{1120}x^8 + \cdots$

45. $y = 1 - \dfrac{1}{2}x^2 - \dfrac{1}{6}x^3 + \dfrac{1}{40}x^5 + \cdots$

Exercises 6.4

1. $x = 0$, regular
3. $x = 0$, irregular; $x = -1$, regular
5. $x = 4$, regular
7. $x = 5$, irregular; $x = -5$, regular
9. $x = 0$, regular; $x = 2$, regular; $x = 3$, regular

11. $y = c_1 x^{1/4}\left(1 + \dfrac{2}{5}x + \dfrac{2}{45}x^2 + \dfrac{4}{1755}x^3 + \cdots\right) +$
$c_2\left(1 + \dfrac{2}{3}x + \dfrac{2}{21}x^2 + \dfrac{4}{693}x^3 + \cdots\right)$

13. $y = c_1 x^{7/2}\left(1 + \dfrac{1}{3}x + \dfrac{1}{22}x^2 + \dfrac{1}{286}x^3 + \cdots\right) +$
$c_2\left(1 - \dfrac{3}{5}x + \dfrac{3}{10}x^2 - \dfrac{3}{10}x^3 + \cdots\right)$

15. $y = c_1 x^{-3/5}\left(1 + \dfrac{1}{14}x^2 + \dfrac{1}{952}x^4 + \cdots\right) +$
$c_2\left(1 + \dfrac{1}{26}x^2 + \dfrac{1}{2392}x^4 + \cdots\right)$

17. $y = c_1 x^{-5/9}\left(1 + \dfrac{1}{4}x - \dfrac{3}{104}x^2 - \dfrac{29}{6864}x^3 + \cdots\right) +$
$c_2\left(1 + \dfrac{1}{14}x - \dfrac{13}{644}x^2 - \dfrac{59}{61824}x^3 + \cdots\right)$

19. $y = c_1 x^{-3/7}\left(1 - \dfrac{1}{4}x + \dfrac{1}{88}x^2 + \dfrac{29}{1584}x^3 + \cdots\right) +$
$c_2\left(1 - \dfrac{1}{10}x + \dfrac{1}{340}x^2 + \dfrac{113}{8160}x^3 + \cdots\right)$

21. $y = c_1 x\left(1 - \dfrac{1}{3}x + \dfrac{1}{24}x^2 - \dfrac{1}{360}x^3 + \dfrac{1}{8640}x^4 - \cdots\right) +$
$c_2 x^{-1}\ln x\left(x^2 - \dfrac{1}{3}x^3 + \dfrac{1}{24}x^4 - \dfrac{1}{360}x^5 + \cdots\right) +$
$c_2 x^{-1}\left(-2 - 2x + \dfrac{4}{9}x^3 - \dfrac{25}{288}x^4 + \dfrac{157}{21600}x^5 + \cdots\right)$

23. $y = c_1 x\left(1 - 5x + \dfrac{25}{2}x^2 - \dfrac{125}{6}x^3 + \cdots\right) +$
$c_2 \ln x\left(-5x + 25x^2 - \dfrac{125}{2}x^3 + \dfrac{625}{6}x^4 + \cdots\right) +$
$c_2\left(1 - 5x + \dfrac{125}{4}x^3 + \cdots\right)$

25. $y = c_1 x^{-2}\left(1 + \dfrac{3}{2}x^2 - \dfrac{9}{40}x^4 + \cdots\right) +$
$c_2 x^{1/3}\left(1 - \dfrac{3}{26}x^2 + \dfrac{9}{1976}x^4 + \cdots\right)$

27. $y = c_1 x^{-16/3}\left(1 + x + \dfrac{1}{2}x^2 + \dfrac{1}{6}x^3 + \cdots\right) +$
$c_2 x\left(1 + \dfrac{3}{22}x + \dfrac{9}{550}x^2 + \dfrac{27}{15400}x^3 + \cdots\right)$

29. $y = c_1 x^{7/4}\left(1 + \dfrac{7}{8}x + \dfrac{77}{160}x^2 + \dfrac{77}{384}x^3 + \cdots\right) +$
$c_2 x^{-5/4}\ln x\left(\dfrac{15}{8}x^3 + \dfrac{105}{64}x^4 + \cdots\right)$
$+ c_2 x^{-5/4}\left(12 + 15x + \dfrac{15}{4}x^2 - \dfrac{13}{2}x^3 + \cdots\right)$

31. $y = c_1 x^{-2/3}\left(1 + 2x + x^2 + \dfrac{2}{9}x^3 + \cdots\right) +$
$c_2 x^{-2/3}\ln x\left(1 + 2x + x^2 + \dfrac{2}{9}x^3 + \cdots\right)$
$+ c_2 x^{-2/3}\left(-4x - 3x^2 - \dfrac{22}{27}x^3 + \cdots\right)$

33. $y = c_1 x\left(1 + 2x + 2x^2 + \dfrac{11}{9}x^3 + \cdots\right) +$
$c_2 x \ln x\left(1 + 2x + 2x^2 + \dfrac{11}{9}x^3 + \cdots\right)$
$+ c_2 x\left(-2x - 3x^2 - \dfrac{64}{27}x^3 + \cdots\right)$

35. $y = c_1 x^{5/4}\left(1 - \dfrac{5}{8}x + \dfrac{15}{64}x^2 - \dfrac{65}{1024}x^3 + \dfrac{221}{16384}x^4 - \cdots\right) +$
$c_2 x^{1/4}\ln x\left(-\dfrac{1}{4}x + \dfrac{5}{32}x^2 - \dfrac{15}{256}x^3 + \dfrac{65}{4096}x^4 - \dfrac{221}{65536}x^5 + \cdots\right) +$
$c_2 x^{1/4}\left(1 - x + \dfrac{33}{64}x^2 - \dfrac{131}{768}x^3 + \dfrac{6167}{147456}x^4 - \dfrac{48149}{5898240}x^5 + \cdots\right)$

37. $y = c_1 x + c_2 x^{-7}$

39. $y = c_1 x^{-1} + c_2 x^{-1} \ln x$

41. $y = c_1 x^2 + c_2 x^2 \ln x$

Exercises 6.5

1. (a) $y = c_1\left(1 - x^2 - \dfrac{1}{6}x^4 - \dfrac{1}{30}x^6 + \cdots\right) + c_2 x;$

   (b) $y = c_1\left(1 - 3x^2 + \dfrac{1}{2}x^4 + \cdots\right) + c_2\left(x - \dfrac{2}{3}x^3\right)$

3. $F(1, b, b; x) = 1 + x + x^2 + x^3 + \cdots = \dfrac{1}{1 - x}$

9. (a) $y = c_1 J_{1/4}(x) + c_2 Y_{1/4}(x);$

   (b) $y = c_1 J_5(3x) + c_2 Y_5(3x)$

11. $y = c_1 F\left(1, 1, \dfrac{1}{2}; x\right) + c_2 x^{1/2} F\left(\dfrac{3}{2}, \dfrac{3}{2}, \dfrac{3}{2}; x\right)$

13. $y_1(x) = F(2, -1, 1; x) = 1 - 2x;$ Second lin. indep. soln.:

$$y_2(x) = (1 - 2x)\left[\ln\left|\dfrac{x - 1}{2x - 1}\right| + \dfrac{1}{2(2x - 1)}\right]$$

15. $y = c_1 F\left(-\dfrac{1}{2}, \dfrac{1}{4}, -\dfrac{1}{4}; x\right) + c_2 x^{3/4} F\left(\dfrac{1}{4}, 1, \dfrac{7}{4}; x\right)$

17. $y_1(x) = F(1, 0, 1; x) = 1;$ Second lin. indep. soln.:
$y_2(x) = x + \ln|x - 1|$

Chapter 6 Review Exercises

1. $y = c_1 x^2 + c_2 x^3$

3. $y = c_1 x^{-4} + c_2 x^{-2}$

5. $y = c_1 x^{-1} + c_2 x^{-1/2},\ y = -x^{-1} + 2x^{-1/2}$

7. $y = c_1 \sin(\ln x) + c_2 \cos(\ln x)$

9. $y = x^{-3}[c_1 \sin(4 \ln x) + c_2 \cos(4 \ln x)]$

11. $y = x^{3/5}\left[c_1 \sin\left(\dfrac{1}{5}\ln x\right) + c_2 \cos\left(\dfrac{1}{5}\ln x\right)\right]$

13. $y = c_1 x^{-5} + c_2 x^{-1} + c_3 x^4,$

   $y = -\dfrac{1}{4}x^{-5} + \dfrac{41}{20}x^{-1} + \dfrac{1}{5}x^4$

15. $y_p(x) = x,\ y = c_1 x^3 + c_2 x^5 + x$

17. $y = c_1 + c_2 x + (-2c_1 + 2c_2)x^2 + \left(-\dfrac{8}{3}c_1 + 2c_2\right)x^3 +$

   $\left(-2c_1 + \dfrac{4}{3}c_2\right)x^4 + \left(-\dfrac{16}{15}c_1 + \dfrac{2}{3}c_2\right)x^5 + \cdots$

19. $y = c_1 + c_2 x + \left(\dfrac{3}{2}c_1 - c_2\right)x^2 + \left(\dfrac{7}{6}c_2 - c_1 + \dfrac{1}{6}\right)x^3 +$

   $\left(\dfrac{7}{8}c_1 - \dfrac{5}{6}c_2\right)x^4 + \left(\dfrac{1}{20} - \dfrac{1}{2}c_1 + \dfrac{61}{120}c_2\right)x^5 + \cdots$

21. $y = c_1 + c_2 x - \dfrac{3}{2}c_1 x^2 - \dfrac{1}{6}c_2 x^3 - \dfrac{5}{8}c_1 x^4 - \dfrac{1}{8}c_2 x^5 + \cdots$

23. $y = c_1 x^{-8/3}\left(1 - \dfrac{1}{5}x + \dfrac{1}{20}x^2 + \dfrac{1}{60}x^3 + \dfrac{1}{960}x^4 + \dfrac{1}{33600}x^5 + \cdots\right) +$

   $c_2\left(1 + \dfrac{1}{11}x + \dfrac{1}{308}x^2 + \dfrac{1}{15708}x^3 + \dfrac{1}{1256640}x^4 + \dfrac{1}{144513600}x^5 + \cdots\right)$

25. $y = c_1 x^{-2} + c_2 x^{1/2}$

27. $y = c_1 x^7\left(1 + \dfrac{1}{8}x^2 + \dfrac{1}{160}x^4 + \dfrac{1}{5760}x^6 + \dfrac{1}{322560}x^8 + \cdots\right) +$

   $c_2 x \ln x\left(-1800x^6 - 225x^8 - \dfrac{45}{4}x^{10} - \dfrac{5}{16}x^{12} - \dfrac{5}{896}x^{14} + \cdots\right) +$

   $c_2 x\left(-86400 + 21600x^2 - 5400x^4 + \dfrac{1125}{8}x^8 + \dfrac{351}{32}x^{10} + \cdots\right)$

29. $y = F(2, -5, 1; x)$

C H A P T E R   7

Exercises 7.1

1. $\dfrac{21}{s^2}$

3. $\dfrac{2}{s-1}$

5. $\dfrac{4}{s^2+4}$

7. $\dfrac{e^{-s}}{s}$

9. $\dfrac{1}{s^2+1}(e^{-s\pi/2}+s)$

11. $\dfrac{1}{s^2}(e^{-3s}(2s+1)+s-1)$

13. $-\dfrac{2e^{-10s}}{s}+\dfrac{1}{s}$

15. $\dfrac{k}{s^2+k^2}$

17. $\dfrac{28}{s-1}$

19. $\dfrac{-18}{s-3}$

21. $\dfrac{1}{s^2(s^2+4)}=\dfrac{1}{4s^2}-\dfrac{1}{4(s^2+4)}$

23. $\dfrac{s}{s^2+25}$

25. $\dfrac{2(s^2+2)}{s(s^2+4)}$

27. $\dfrac{s^2+5s+25}{s(s^2+25)}$

29. (a) $\dfrac{\sqrt{2}}{2}\left(\dfrac{s+1}{s^2+1}\right)$; (b) $\dfrac{\sqrt{2}}{2}\left(\dfrac{s-1}{s^2+1}\right)$;

(c) $\dfrac{1}{2}\left(\dfrac{\sqrt{3}s-1}{s^2+1}\right)$; (d) $\dfrac{1}{2}\left(\dfrac{s+\sqrt{3}}{s^2+1}\right)$

31. (a) $\mathscr{L}\{t^{-1/2}\}=\sqrt{\pi/s}$; (b) $\mathscr{L}\{t^{1/2}\}=\dfrac{\sqrt{\pi}}{2s^{3/2}}$

32. (b) $f(t)=e^{bt}\sin t,\ b>0$, is of exponential order b but $\lim\limits_{t\to\infty}(e^{bt}\sin t)(e^{-bt})=\lim\limits_{t\to\infty}\sin t$ does not exist.

Exercises 7.2

1. $\dfrac{-16}{s-2}$

3. $\dfrac{5040}{s^8}$

5. $\dfrac{2}{(s+3)^3}$

7. $\dfrac{120}{(s+4)^6}$

9. $\dfrac{s^2-9}{(s^2+9)^2}$

11. $\dfrac{6s}{(s^2+1)^2}$

13. $\dfrac{14s}{(s^2-49)^2}$

15. $\dfrac{1}{2}\left[\dfrac{1}{s-6}+\dfrac{1}{s-8}\right]=\dfrac{s-7}{(s-7)^2-1}$

17. $\dfrac{s+2}{(s+2)^2+16}$

19. $\dfrac{7}{(s+2)^2+49}$

21. $\dfrac{s-5}{(s-5)^2+49}$

23. $\dfrac{1}{s+1}-\dfrac{1}{s+3}=\dfrac{2}{(s+2)^2-1}$

25. $\dfrac{-128}{(s^2+16)^2}+\dfrac{8}{s^2+16}=\dfrac{8s^2}{(s^2+16)^2}$

27. $\dfrac{48s(s^2-4)}{(s^2+4)^4}$

29. $\dfrac{s}{(s^2+9)^2}$

35. $\dfrac{1}{2}\left[\dfrac{s}{s^2+4k^2}+\dfrac{1}{s}\right]=\dfrac{s^2+2k^2}{s(s^2+4k^2)}$

Exercises 7.3

1. t

3. te^{2t}

5. $\dfrac{1}{2}t^2e^{-6t}$

7. $4te^{3t}$

9. $\cos 4t$

11. $\dfrac{2}{9}e^{-2t} + \dfrac{7}{9}e^{7t}$

13. $\dfrac{1}{2} + \dfrac{1}{2}e^{4t}$

15. $\dfrac{1}{2}e^{6t} + \dfrac{1}{2}e^{8t}$

17. $-\dfrac{1}{8}e^{-2t} + \dfrac{1}{8}e^{6t}$

19. $e^t \sin t$

21. $e^t \cos 7t$

23. $e^{-2t} \cos 2t$

25. $e^{7t} \cos t$

27. $\dfrac{1}{4}t - \dfrac{1}{8}\sin 2t$

29. $-t \cos 2t$

31. $\dfrac{1}{2}t^2 \cos 5t$

33. $\dfrac{1}{4}t^2(e^t + e^{-t})$

35. $\dfrac{1}{4}t^2(e^{7t} + e^{-7t})$

37. $-2\cos(\sqrt{6}t) + \dfrac{1}{2}\sin 2t$

39. $e^t \cos 3t + \dfrac{\sqrt{2}}{2}e^{6t}\sin(\sqrt{2}t)$

41. **(a)** $-\dfrac{ab(b^2 - a^2)}{(s^2 + a^2)(s^2 + b^2)}, \ -\dfrac{s(b^2 - a^2)}{(s^2 + a^2)(s^2 + b^2)};$

(b) $\dfrac{1}{b^2 - a^2}\left[\dfrac{1}{a}\sin at - \dfrac{1}{b}\sin bt\right],$

$\dfrac{1}{b^2 - a^2}[\cos at - \cos bt]$

43. **(a)** $\displaystyle\lim_{s\to\infty} \dfrac{s}{4 - s} = -1,$ **(b)** $\displaystyle\lim_{s\to\infty} \dfrac{3s}{s + 1} = 3,$

(c) $\displaystyle\lim_{s\to\infty} \dfrac{s^2}{4s + 10} = +\infty,$ **(d)** $\displaystyle\lim_{s\to\infty} \dfrac{5s^3}{s^2 + 1} = +\infty.$ The
inverse Laplace transform does not exist for any of
these functions.

Exercises 7.4

1. $y = \dfrac{3}{5}e^{-8t} - \dfrac{8}{5}e^{-3t}$

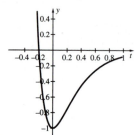

3. $y = -\dfrac{3}{7}e^{-5t} - \dfrac{4}{7}e^{2t}$

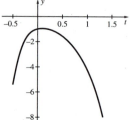

5. $y = 2 \cos 2t$

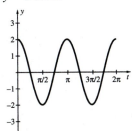

7. $y = \dfrac{11}{42}e^{-3t} + \dfrac{13}{6}e^{3t} - \dfrac{10}{7}e^{4t}$

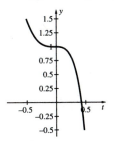

9. $y = -4e^{-t} - \cos 2t - 2 \sin 2t$

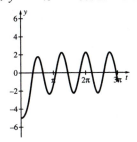

11. $y = \dfrac{1}{156}(2 \cos 2t + 154e^{6t} \cos 2t + 3 \sin 2t - 465e^{6t} \sin 2t)$

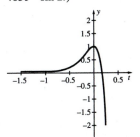

13. $y = -\dfrac{47}{70}e^{-3t} - \dfrac{1}{10}e^{2t} - \dfrac{53}{238}e^{4t} - \dfrac{1}{170}\cos t + \dfrac{13}{170}\sin t$

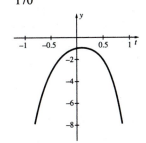

15. $y = \dfrac{3}{2}t^2$

17. $y = t^2$

19. (a) $Y'(s) + s^2 Y(s) = s$
    (b) $s^2 Y''(s) + 2sY'(s) - (s^2 + n(n+1))Y(s) - 1 = 0$

Exercises 7.5

1. $\dfrac{-28}{se^{3s}}$

3. $\dfrac{3 - e^{4s}}{se^{8s}}$

5. $\dfrac{-7 - 7e^s + e^{5s}}{se^{7s}}$

7. $\dfrac{-42e^{-4s}}{s - 1}$

9. $\dfrac{-12}{e^{2s}(s^2 - 1)}$

11. $\dfrac{-14}{e^{2\pi s/3}(s^2 + 1)}$

13. $\dfrac{e^s - s - 1}{s^2 e^s (1 - e^{-2s})}$

15. $\dfrac{e^s + 2}{s(e^{2s} + e^s + 1)}$

17. $e^{-\pi s}$

19. $e^{-s} + e^{-2s}$

21. $\dfrac{e^{-2\pi s}}{s^2 + 1} + e^{-\pi s/2}$

23. $-3\mathcal{U}(t - \pi)$

25. $2\mathcal{U}(t - 1) - 3\mathcal{U}(t - 4)$

27. $-3\mathcal{U}(t - 6) + 3\mathcal{U}(t - 5) - 4\mathcal{U}(t - 3)$

29. $e^{t-4}\mathcal{U}(t - 4)$

31. $\cos(2t - 6)\mathcal{U}(t - 3)$

33. $\dfrac{1}{3}(e^{5t-25} - e^{2t-10})\mathcal{U}(t - 5)$

35. $2[-t + (4 - 2t)\mathcal{U}(t - 2) + (12 - 2t)\mathcal{U}(t - 6) +$
$(20 - 2t)\mathcal{U}(t - 10) + (28 - 2t)\mathcal{U}(t - 14) + \cdots]$

37. $\cosh(t - 4)\mathcal{U}(t - 4)$

39. $t - (t - 4)\mathcal{U}(t - 4) + (t - 8)\mathcal{U}(t - 8) -$
$(t - 12)\mathcal{U}(t - 12) + (t - 16)\mathcal{U}(t - 16) + \cdots$

41. $\dfrac{1}{4}[\sin 4t - \sin(4t - 12)\mathcal{U}(t - 3) + \sin(4t - 24)\mathcal{U}(t -$
$6) - \sin(4t - 36)\mathcal{U}(t - 9) + \cdots]$

43. $\left(\dfrac{1}{8}\sin 2t - \dfrac{1}{4}t\right) +$

$3\left(\dfrac{1}{2} - \dfrac{1}{4}t + \dfrac{1}{8}\sin(2t - 4)\right)\mathcal{U}(t - 2) +$

$9\left(1 - \dfrac{1}{4}t + \dfrac{1}{8}\sin(2t - 8)\right)\mathcal{U}(t - 4) +$

$27\left(\dfrac{3}{2} - \dfrac{1}{4}t + \dfrac{1}{8}\sin(2t - 12)\right)\mathcal{U}(t - 6) + \cdots$

45. $y(t) = \dfrac{1}{3}e^{-3t}(-1 + e^{3t} - e^{6}\mathcal{U}(2 - t) + e^{3t}\mathcal{U}(2 - t))$

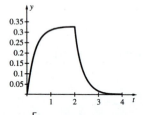

47. $y = \left[-\dfrac{\pi}{4 + \pi^2}e^{t-1} + \dfrac{\pi}{36 + \pi^2}e^{3t-3}\right.$

$+ \dfrac{32\pi}{(4 + \pi^2)(36 + \pi^2)}\cos\left(\dfrac{\pi}{2}(t - 1)\right)$

$+ \left.\dfrac{48 - 4\pi^2}{(4 + \pi^2)(36 + \pi^2)}\sin\left(\dfrac{\pi}{2}(t - 1)\right)\right]\mathcal{U}(t - 1)$

$- \dfrac{8 + \pi^2}{2(4 + \pi^2)}e^t + \dfrac{48 + \pi^2}{2(36 + \pi^2)}e^{3t}$

$+ \dfrac{48 - 4\pi^2}{(4 + \pi^2)(36 + \pi^2)}\cos\left(\dfrac{\pi}{2}t\right)$

$- \dfrac{32\pi}{(4 + \pi^2)(36 + \pi^2)}\sin\left(\dfrac{\pi}{2}t\right)$

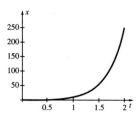

49. $x(t) = \sin(t - \pi)\mathcal{U}(t - \pi) + 1 - \cos t$

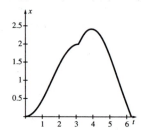

51. $x(t) = \mathcal{U}(t - \pi)(e^{\pi-t} - e^{2\pi-2t})$

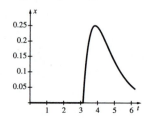

53. $x(t) = -\dfrac{1}{3}e^{2\pi-2t}\mathcal{U}(t - \pi)\sin 3t + \dfrac{3}{40}\sin t -$
$\dfrac{1}{40}\cos t - \dfrac{1}{120}e^{-2t}\sin 3t + \dfrac{1}{40}e^{-2t}\cos 3t$

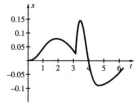

55. $x(t) = -\dfrac{1}{3}\mathcal{U}(t - \pi)\sin 3t + \dfrac{1}{8}\cos t - \dfrac{1}{8}\cos 3t$

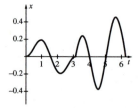

57. $x(t) = -\mathcal{U}(t-1)e^{2-2t} + \mathcal{U}(t-1)e^{1-t} + e^{-2t} + te^{-t} - e^{-t}$

59. $x = 2 + \dfrac{1}{9}(1 - e^{9\pi - 9t}) \cdot \mathcal{U}(t - \pi)$

61. $x(t) = 3\mathcal{U}(t - \pi)e^{3\pi - 3t}\sin(t - \pi)$

63. $x(t) = 2e^{-3t}\sin t + 3\mathcal{U}(t - \pi)e^{3\pi - 3t}\sin(t - \pi)$

Exercises 7.6

1. $\dfrac{1}{3}t^3$

3. $t - 1 + e^{-t}$

5. $-\dfrac{3}{32}t + \dfrac{1}{4}t^3 + \dfrac{3}{128}\sin 4t$

7. $\dfrac{1}{s(s+1)}$

9. $\dfrac{1}{s^2(s^2+1)}$

11. $\dfrac{1}{s^2(s-1)}$

13. $t - 1 + e^{-t}$

15. $\dfrac{1}{2}t\sin t$

17. $\dfrac{1}{96}t^3 - \dfrac{1}{256}t + \dfrac{1}{1024}\sin 4t$

19. $\dfrac{1}{116}e^{10t} - \dfrac{1}{116}\cos 4t - \dfrac{5}{232}\sin 4t$

21. $-\dfrac{1}{200}t\cos 10t + \dfrac{1}{2000}\sin 10t$

23. $g(t) = \sin t$

25. $h(t) = \dfrac{5}{3}\cos(\sqrt{2}t) - \dfrac{2\sqrt{2}}{3}\sin(\sqrt{2}t) + \dfrac{10}{3}e^{-2t}$

27. $y(t) = \left(\dfrac{1}{10}t^5 + \dfrac{1}{4}t^4\right)e^{2t}$

35. $\dfrac{d^2 y}{dt^2} - 4\dfrac{dy}{dt} + 4y = 2t^3 e^{2t} + 3t^2 e^{2t}$, $y(0) = 0$, $y'(0) = 0$

Exercises 7.7

1. $x(t) = \cos 2t$

3. $x(t) = -2e^{-3t} + 3e^{-2t}$

5. $x(t) = e^{-2t}\left(\cos 3t + \dfrac{2}{3}\sin 3t\right)$

7. $x(t) = -\dfrac{1}{40}e^{-3t} + \dfrac{1}{20}e^{-2t} - \dfrac{1}{40}\cos t + \dfrac{1}{40}\sin t$

9. $x(t) = \dfrac{2}{9}e^{-2t}\sin^2\left(\dfrac{3}{2}t\right)$

11. $x(t) = \dfrac{1 + e^{\pi}}{4e^{2t}} - \dfrac{2(1 + e^{3\pi/2})}{13e^{3t}} +$

$\left(\dfrac{2}{13}e^{3\pi/2 - 3t} - \dfrac{1}{4}e^{\pi - 2t} - \dfrac{5}{52}\cos 2t + \dfrac{1}{52}\sin 2t\right)\mathcal{U}(\pi/2 - t)$

13. $x(t) = -e^{-2t} + 2te^{-2t} + (t-1)^2 e^{-2t}\mathcal{U}(1 - t)$

15. $x(t) = \dfrac{1}{6}e^{-2t}(\pi - \pi\mathcal{U}(\pi - t) + t\mathcal{U}(\pi - t))\sin 3t$

17. $x(t) = \dfrac{1}{3}e^{-t} + \dfrac{1}{9}e^{-4t}(1 - e^3) +$

$\dfrac{1}{9}e^{-4t}(e^3 - 4e^{3t} + 3te^{3t})\mathcal{U}(1 - t)$

21. $x(t) = -\dfrac{1}{3}e^{4-2t}\sin(6 - 3t)\mathcal{U}(t - 2)$

23. $Q(t) = 2 - 2e^{-t} - 2(1 - e^{1-t})\mathcal{U}(t - 1)$,
$I(t) = 2e^{-t} - 2(1 - e^{1-t})\delta(t - 1) - 2e^{1-t}\mathcal{U}(t - 1)$

25. $I(t) = \dfrac{1}{2}(1 - e^{-t}) - \dfrac{1}{2}(1 - e^{-(t-1)})\mathcal{U}(t - 1) +$

$\dfrac{1}{2}(1 - e^{-(t-2)})\mathcal{U}(t - 2) - \dfrac{1}{2}(1 - e^{-(t-3)})\mathcal{U}(t - 3) + \cdots$

27. $I(t) = (t - 1 + e^{-t}) - (t - 1)\mathcal{U}(t - 1) +$
$(t - 3 + e^{-(t-2)})\mathcal{U}(t - 2) - (t - 3)\mathcal{U}(t - 3) +$
$(t - 5 + e^{-(t-4)})\mathcal{U}(t - 4) - (t - 5)\mathcal{U}(t - 5) + \cdots$

29. $I(t) = 100te^{-3t}$

31. $I(t) = \dfrac{2}{3}\sqrt{3}e^{-t/2}\sin\left(\dfrac{\sqrt{3}}{2}t\right) -$

$\dfrac{2}{3}\sqrt{3}e^{-t/2+\pi/2}\sin\left(\dfrac{\sqrt{3}}{2}(t-\pi)\right)\mathcal{U}(t-\pi)$

33. (a) $I(t) = e^{-4t+4}\mathcal{U}(t-1)$; **(b)** $I(t) = e^{-4t} + e^{-4t+4}\mathcal{U}(t-1)$

35. $I(t) = t - 1 + e^{-t} - (t - 2 + e^{-t+1})\mathcal{U}(t-1)$

37. $I(t) = e^{2-t}\mathcal{U}(t-2) + e^{6-t}\mathcal{U}(t-6)$

39. $x(t) = 200 + \dfrac{127150}{13}e^{-5t} + \dfrac{250}{13}\cos t - \dfrac{1250}{13}\sin t$,

bounded

41. $x(t) = 200 + \dfrac{63650}{13}e^{-5t} - \dfrac{1250}{13}\cos t - \dfrac{250}{13}\sin t$;

bounded

43. $x(t) = -250 + 5350e^{2t} - 200\sin t - 100\cos t$,
unbounded

Chapter 7 Review Exercises

1. $\dfrac{1}{s} - \dfrac{1}{s^2}$

3. $\dfrac{e^{5s} - 1}{se^{5s}}$

5. $\dfrac{5(s^5 + 24)}{s^6}$

7. $(s - 2)^{-2}$

9. $6(s - 1)^{-4}$

11. $\dfrac{s^2 - 9}{(s^2 + 9)^2}$

13. $\dfrac{s + 5}{s^2 + 10s + 34}$

15. $e^{-3\pi s/2}$

17. $\dfrac{2(3 - 2e^{3s})}{se^{7s}}$

19. $\dfrac{-42e^{5-s}}{s - 5}$

21. $\dfrac{2}{s^3 e^{2s}} + \dfrac{4}{s^2 e^{2s}} + \dfrac{4}{se^{2s}}$

23. $\dfrac{e^{2s} - 2e^s + 1}{s^2 e^{2s}(1 - e^{-2s})} = \dfrac{e^s - 1}{s^2(e^s + 1)}$

25. $\dfrac{\pi e^s + \pi}{e^s(s^2 + \pi^2)(1 - e^{-2s})} = \dfrac{\pi e^s}{(e^s - 1)(s^2 + \pi^2)}$

45. $x(t) = -10000 + 17500e^t + 2500\cos t + 2500\sin t + (10000 - 10000e^{t-5} + 2500e^{t-5}\cos 5 - 2500\cos 5\cos(t - 5) + 2500e^{t-5}\sin 5 - 2500\sin 5\cos(t - 5) - 2500\cos 5\sin(t - 5) + 2500\sin 5\sin(t - 5))\mathcal{U}(t - 5)$

47. $x(t) = -5000\Big\{1 - \dfrac{5}{2}e^t + \dfrac{1}{2}(\cos t - \sin t) + \dfrac{1}{2}(-2 + 2e^{t-1} + e^{t-1}\cos 1 - \cos t - e^{t-1}\sin 1 + \sin t)\mathcal{U}(t - 1) + \left(1 - \dfrac{3}{2}e^{t-2} + \dfrac{1}{2}(\cos(t - 2) - \sin(t - 2))\right)\mathcal{U}(t - 2) + \cdots\Big\}$

49. $x(t) = \dfrac{c}{k} + e^{-kt}\left(x_0 - \dfrac{c}{k}\right)$; $\displaystyle\lim_{t\to\infty} x(t) = \dfrac{c}{k}$

51. $x(t) = x_0 e^{-kt} + \dfrac{c}{k}(1 - e^{-kt}) - \dfrac{c}{k}(1 - e^{-k(t-1)})\mathcal{U}(t - 1) + \dfrac{c}{k}(1 - e^{-k(t-2)})\mathcal{U}(t - 2) + \cdots$

27. $-2\sinh 5t$

29. $t^5 - 3$

31. $e^{-5t}\cos t - 2$

33. $7\mathcal{U}(t - 7) + 6\mathcal{U}(t - 3)$

35. $8\cos(t - 3)\mathcal{U}(t - 3)$

37. $\dfrac{1}{27}e^{-3t}(9t^2 e^{3t} - 6te^{3t} + 2e^{3t} - 2)$

39. $y(t) = e^{-3t}\sin t$

41. $y(t) = 6t^2 + 1$

43. $y(t) = -e^{-4t} + 2e^{-2t} + \left(\dfrac{1}{8} + \dfrac{1}{8}e^{4-4t} - \dfrac{1}{4}e^{2-2t}\right)\mathcal{U}(t - 1)$

45. $x(t) = \dfrac{1}{8}\cos t - \dfrac{1}{8}\cos 3t - \dfrac{1}{3}\sin(3t)\mathcal{U}(t - \pi)$

47. $g(t) = 1 + \sin t - \cos t$

51. (a) $\dfrac{s}{s^2 - 1}$ **(b)** $\dfrac{1}{s^2 - 1}$

(c) $\dfrac{s^3}{s^4 + 4}$ **(d)** $\dfrac{s^2 + 2}{s^4 + 4}$

(e) $\dfrac{s^2 - 2}{s^4 + 4}$ **(f)** $\dfrac{2s}{s^4 + 4}$

53. $x(t) = -2\sin\dfrac{t}{2} + \dfrac{1}{15}\cos(2t)\mathcal{U}(t - \pi) - \dfrac{1}{15}\sin\left(\dfrac{t}{2}\right)\mathcal{U}(t - \pi) - \dfrac{1}{15}\cos(2t) + \dfrac{1}{15}\cos\left(\dfrac{t}{2}\right)$

55. $Q(t) = -220\left(\dfrac{1}{10000} - \dfrac{1}{10000}\cos(100t - 200)\right)\mathcal{U}(t - 2) + \dfrac{11}{500} - \dfrac{11}{500}\cos(100t)$

57. $I(t) = 100\sqrt{2}e^{-t}\sin(2\sqrt{2}t)$

59. $x(t) = 10000e^{2t} + 100e^{2t-2}\mathcal{U}(t - 1)$

61. $y(x) = \dfrac{w}{24}(x^4 - 2x^3 + x^2)$

C H A P T E R   8

Exercises 8.1

1.
$$\begin{cases} x(t) = \dfrac{1}{3} + c_1e^{3t} + c_2e^{-4t} \\[2mm] y(t) = \dfrac{5}{3} - \dfrac{5}{2}c_1e^{3t} + c_2e^{-4t} \end{cases}$$

3.
$$\begin{cases} x(t) = \dfrac{7}{34}\cos t + \dfrac{11}{34}\sin t + c_1e^{4t} - c_2e^{-t} \\[2mm] y(t) = -\dfrac{15}{34}\cos t - \dfrac{9}{34}\sin t + \dfrac{3}{2}c_1e^{4t} + c_2e^{-t} \end{cases}$$

5.
$$\begin{cases} x(t) = \dfrac{2}{9} - \dfrac{1}{4}e^t + c_1e^{-3t} + 2tc_2e^{-3t} \\[2mm] y(t) = \dfrac{1}{3} + c_2e^{-3t} \end{cases}$$

7.
$$\begin{cases} x(t) = -2c_3\sin\sqrt{6}t - 2c_4\cos\sqrt{6}t + c_1\sin t + c_2\cos t \\ y(t) = c_3\sin\sqrt{6}t + c_4\cos\sqrt{6}t + 2c_1\sin t + 2c_2\cos t \end{cases}$$

9.
$$\begin{cases} x(t) = (c_2 + c_1)\sin t + c_1\cos t - c_3e^t - c_4te^t \\ y(t) = c_2\cos t + (-c_2 - 2c_1)\sin t + c_3e^t + c_4te^t \end{cases}$$

11.
$$\begin{cases} x(t) = -t - 1 + c_1e^t - c_4te^t + c_2e^{-t} + c_3te^{-t} \\ y(t) = -t + 1 + c_4e^t + (3c_3 + 2c_2)e^{-t} + 2c_3te^{-t} \end{cases}$$

13.
$$\begin{cases} x(t) = -c_1 + c_2e^{3t} - 2c_3e^{-3t} \\ y(t) = c_1 + c_2e^{3t} + c_3e^{-3t} \\ z(t) = -\dfrac{5}{2}c_1 + c_2e^{3t} + c_3e^{-3t} \end{cases}$$

15.
$$\begin{cases} x(t) = -\dfrac{3}{10}\cos t - \dfrac{1}{10}\sin t - 2c_1e^{2t} + \\ \qquad\qquad (3c_3 - 2c_2)e^{-t} - 2c_3te^{-t} \\[2mm] y(t) = \dfrac{1}{2}\sin t + (3c_2 - 4c_3)e^{-t} + 3c_3te^{-t} \\[2mm] z(t) = \dfrac{2}{5}\cos t + \dfrac{3}{10}\sin t + c_1e^{2t} + c_2e^{-t} + c_3te^{-t} \end{cases}$$

17.
$$\begin{cases} x(t) = 2te^t - \dfrac{10}{3} + 3e^t + c_1e^t - 2c_2e^{2t} - c_3e^{3t} \\[2mm] y(t) = -2te^t + \dfrac{5}{6} - 3e^t + (1 - c_1)e^t + c_2e^{2t} + c_3e^{3t} \\[2mm] z(t) = -4te^t + \dfrac{10}{3} - 6e^t - 2c_1e^t + 4c_2e^{2t} + 4c_3e^{3t} \end{cases}$$

19.
$$\begin{cases} x(t) = -\dfrac{1}{2}e^{3t} + \dfrac{1}{2}e^{-t} \\[2mm] y(t) = -\dfrac{1}{2}e^{3t} - \dfrac{1}{2}e^{-t} \end{cases}$$

21.
$$\begin{cases} x(t) = -20 + 20e^t - 20te^t \\ y(t) = -30 + 30e^t - 20te^t \end{cases}$$

23.
$$\begin{cases} x(t) = -t + \dfrac{1}{6}t^3 + \sin t \\[2mm] y(t) = \dfrac{1}{6}t^3 \end{cases}$$

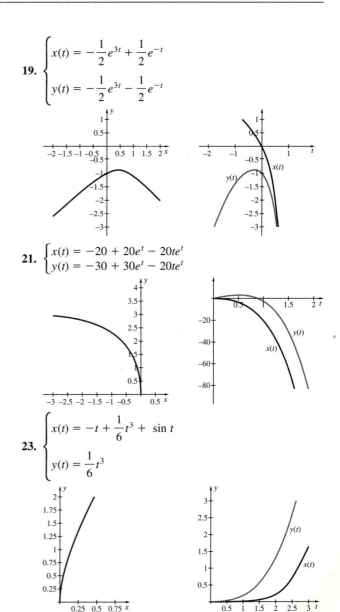

29. $k = -4$

31. $c \neq 4$

Exercises 8.2

1. $\begin{pmatrix} -3 & 5 \\ 3 & 2 \end{pmatrix}$

3. $\begin{pmatrix} -5 & 21 \\ 1 & -3 \end{pmatrix}$

5. $\begin{pmatrix} 12 & 15 & 5 \\ 50 & 8 & 31 \\ 26 & -22 & -15 \end{pmatrix}$

7. $\begin{pmatrix} -8 & 30 & 16 \\ -30 & -18 & 0 \\ -8 & 30 & -2 \end{pmatrix}$

9. $\mathbf{AB} = \begin{pmatrix} 22 & -5 & -13 \\ 22 & -24 & -17 \\ -31 & -4 & 37 \end{pmatrix}$ and

$\mathbf{BA} = \begin{pmatrix} 1 & 9 & 16 & -35 \\ -25 & 20 & 20 & 5 \\ 0 & -9 & -8 & -28 \\ -27 & -11 & -22 & 22 \end{pmatrix}$

11. $\mathbf{AB} = \begin{pmatrix} -1 & -11 \\ -2 & -28 \end{pmatrix}$ and $\mathbf{BA} = \begin{pmatrix} -2 & -3 \\ -16 & -27 \end{pmatrix}$

13. $\mathbf{AB} = \begin{pmatrix} -1 & -12 & -4 \\ 9 & 22 & -3 \\ -10 & -36 & -9 \end{pmatrix}$ and

$\mathbf{BA} = \begin{pmatrix} -1 & -4 & 11 \\ -22 & 25 & -6 \\ -6 & 14 & -12 \end{pmatrix}$

15. $|\mathbf{A}| = 17$

17. $|\mathbf{A}| = 0$

19. $|\mathbf{A}| = 10$

21. $\begin{pmatrix} 1 & \frac{1}{2} \\ -1 & 0 \end{pmatrix}$

23. $\begin{pmatrix} 2 & 1 & -1 \\ -3 & -1 & 2 \\ 0 & 0 & -\frac{1}{2} \end{pmatrix}$

24. $\lambda_1 = -5,\ \mathbf{v}_1 = \begin{pmatrix} -1 \\ 1 \end{pmatrix};\ \lambda_2 = -3,\ \mathbf{v}_2 = \begin{pmatrix} -1 \\ 2 \end{pmatrix}$

25. $\lambda_1 = -5,\ \mathbf{v}_1 = \begin{pmatrix} 1 \\ 1 \end{pmatrix};\ \lambda_2 = -4,\ \mathbf{v}_2 = \begin{pmatrix} 1 \\ 2 \end{pmatrix}$

27. $\lambda_1 = \lambda_2 = -3;\ \mathbf{v}_1 = \begin{pmatrix} 1 \\ 0 \end{pmatrix};\ \mathbf{v}_2 = \begin{pmatrix} 0 \\ 1 \end{pmatrix}$

29. $\lambda_1 = \lambda_2 = -5;\ \mathbf{v} = \begin{pmatrix} 2 \\ 1 \end{pmatrix}$

31. $\lambda_1 = -2 + 2i,\ \mathbf{v}_1 = \begin{pmatrix} 1 \\ 1 - 2i \end{pmatrix};\ \lambda_2 = -2 - 2i,$

$\mathbf{v}_2 = \begin{pmatrix} 1 \\ 1 + 2i \end{pmatrix}$

33. $\lambda_1 = -1,\ \mathbf{v}_1 = \begin{pmatrix} 1 \\ -2 \\ -2 \end{pmatrix};\ \lambda_2 = -2,\ \mathbf{v}_2 = \begin{pmatrix} 1 \\ -\frac{3}{2} \\ -1 \end{pmatrix};$

$\lambda_3 = 3,\ \mathbf{v}_3 = \begin{pmatrix} 3 \\ -3 \\ 1 \end{pmatrix}$

35. $\begin{pmatrix} 2e^{2t} \\ -5e^{-5t} \end{pmatrix};\ \begin{pmatrix} \dfrac{e^{2t}}{2} + c_1 \\ \dfrac{e^{-5t}}{-5} + c_2 \end{pmatrix}$

37. $\begin{pmatrix} -\sin t & \cos t - t \sin t \\ \cos t & \sin t + t \cos t \end{pmatrix};\ \begin{pmatrix} \sin t & \cos t + t \sin t \\ -\cos t & \sin t - t \cos t \end{pmatrix}$

39. $\begin{pmatrix} 4e^{4t} \\ -3 \sin 3t \\ 3 \cos 3t \end{pmatrix};\ \begin{pmatrix} \frac{1}{4}e^{4t} + c_1 \\ \frac{1}{3} \sin 3t + c_2 \\ -\frac{1}{3} \cos 3t + c_3 \end{pmatrix}$

Exercises 8.3

1. $\begin{pmatrix} x' \\ y' \end{pmatrix} = \begin{pmatrix} 1 & 4 \\ 2 & -1 \end{pmatrix} \begin{pmatrix} x \\ y \end{pmatrix}$

3. $\begin{pmatrix} x' \\ y' \end{pmatrix} = \begin{pmatrix} 2 & 1 \\ 1 & -5 \end{pmatrix} \begin{pmatrix} x \\ y \end{pmatrix}$

5. $\begin{pmatrix} x' \\ y' \end{pmatrix} = \begin{pmatrix} 1 & 0 \\ 1 & 1 \end{pmatrix} \begin{pmatrix} x \\ y \end{pmatrix} + \begin{pmatrix} e^t \\ 0 \end{pmatrix}$

7. $\begin{pmatrix} x' \\ y' \end{pmatrix} = \begin{pmatrix} 0 & 1 \\ -1 & 0 \end{pmatrix} \begin{pmatrix} x \\ y \end{pmatrix} + \begin{pmatrix} 0 \\ 2 \sin t \end{pmatrix}$

9. Lin. indep.

11. Lin. indep.

13. Lin. indep.

15. Yes

17. Yes

19. No

21. $\begin{cases} x' = y \\ y' = 3y - 4x \end{cases}$

23. $\begin{cases} x' = y \\ y' = -16x + t \sin t \end{cases}$

25. $\begin{cases} y' = z \\ z' = w \\ w' = x - 3y - 6z - 3w \end{cases}$

27. $\begin{cases} x' = y \\ y' = -x - \mu[\frac{1}{3}y^2 - 1]y \end{cases}$

29. $\begin{cases} x(t) = 30e^{4t} - 24e^{3t} \\ y(t) = 20e^{4t} - 24e^{3t} \end{cases}$

31. $\begin{cases} X' = Z \\ Z' = \dfrac{1}{m}(C_M\omega^a W(Z^2 + W^2)^{(b-1)/2} - \frac{1}{2}C_D\rho AX'\sqrt{Z^2 + W^2}) \\ Y' = W \\ W' = -g - \dfrac{1}{m}(C_M\omega^a Z(Z^2 + W^2)^{(b-1)/2} + \\ \qquad\qquad \frac{1}{2}C_D\rho AW\sqrt{Z^2 + W^2}) \end{cases}$

Exercises 8.4

1. $\begin{cases} x(t) = c_1e^{15t} + 2c_2e^{-4t} \\ y(t) = -\dfrac{7}{5}c_1e^{15t} + c_2e^{-4t} \end{cases}$

3. $\begin{cases} x(t) = c_1e^t + c_2e^{5t} \\ y(t) = 5c_1e^t + c_2e^{5t} \end{cases}$

5. $\begin{cases} x(t) = 3c_1e^{7t} \\ y(t) = c_1e^{7t} + c_2e^{-8t} \end{cases}$

7. $\begin{cases} x(t) = -c_1e^{-11t} + 3c_2e^{-4t} \\ y(t) = 2c_1e^{-11t} + c_2e^{-4t} \end{cases}$

9. $\begin{cases} x(t) = c_1e^{-12t} - 2c_2e^{-4t} \\ y(t) = \dfrac{3}{2}c_1e^{-12t} + c_2e^{-4t} \end{cases}$

11. $\begin{cases} x(t) = c_1e^{-8t} + (2c_2 + 2c_1)te^{-8t} \\ y(t) = c_2e^{-8t} + (-2c_2 - 2c_1)te^{-8t} \end{cases}$

13. $\begin{cases} x(t) = 2c_1\cos 4t + 2c_2\sin 4t \\ y(t) = -c_1\sin 4t + c_2\cos 4t \end{cases}$

15. $\begin{cases} x(t) = c_1e^{-2t}\sin 3t + c_2e^{-2t}\cos 3t \\ y(t) = (-3c_2 - 2c_1)e^{-2t}\sin 3t + (-2c_2 + 3c_1)e^{-2t}\cos 3t \end{cases}$

17. $\begin{cases} x(t) = c_1e^{4t} + c_2e^{-t} \\ y(t) = c_3e^{-2t} \\ z(t) = -5c_2e^{-t} \end{cases}$

19. $\begin{cases} x(t) = 4c_1e^{-4t} + c_2e^{-3t} + 23c_3te^{-3t} \\ y(t) = c_3e^{-3t} \\ z(t) = c_1e^{-4t} - 5c_3e^{-3t} \end{cases}$

21. $\begin{cases} x(t) = -\dfrac{7}{4}c_2e^{3t} + c_1e^{-t} - c_3te^{-t} \\ y(t) = c_2e^{3t} + c_3e^{-t} \\ z(t) = \dfrac{3}{2}c_2e^{3t} - 2c_1e^{-t} + 2c_3te^{-t} \end{cases}$

23. $\begin{cases} x(t) = -13c_1e^{-t} + e^{-2t}\left[\left(-\dfrac{3}{5}c_2 + \dfrac{4}{5}c_3\right)\cos 4t \right. \\ \qquad\qquad \left. + \left(-\dfrac{4}{5}c_2 - \dfrac{3}{5}c_3\right)\sin 4t\right] \\ y(t) = 17c_1e^{-t} \\ z(t) = 24c_1e^{-t} + e^{-2t}(c_2\cos 4t + c_3\sin 4t) \end{cases}$

25. $\begin{cases} x(t) = c_1e^t + e^t\left[\left(-c_2 - \dfrac{3}{2}c_3\right)\cos 2t + \left(\dfrac{3}{2}c_2 - c_3\right)\sin 2t\right] \\ y(t) = e^t(-c_3\cos 2t + c_2\sin 2t) \\ z(t) = e^t(c_2\cos 2t + c_3\sin 2t) \end{cases}$

27. $\begin{cases} x(t) = 4e^{2t} - 4e^t \\ y(t) = 4e^{2t} \end{cases}$

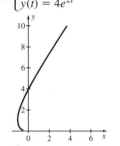
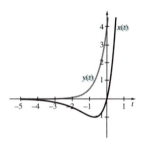

29. $\begin{cases} x(t) = 8e^{4t} \\ y(t) = 16te^{4t} \end{cases}$

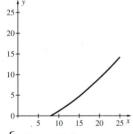
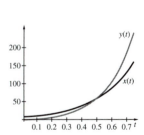

31. $\begin{cases} x(t) = -8\sin 4t \\ y(t) = 8\cos 4t \end{cases}$

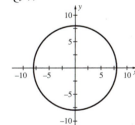
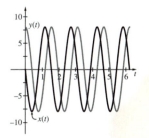
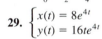
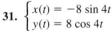

33. $\begin{cases} x(t) = e^{2t} - 2e^{3t} \\ y(t) = 2e^{t} - 2e^{2t} + 2e^{3t} \\ z(t) = -2e^{2t} + 2e^{3t} \end{cases}$

35. a and **b** are linearly independent so $\begin{vmatrix} a_1 & b_1 \\ a_2 & b_2 \end{vmatrix} =$

$a_1 b_2 - a_2 b_1 \neq 0$. Thus,

$W(e^{\alpha t}(\mathbf{a} \cos \beta t - \mathbf{b} \sin \beta t), \ e^{\alpha t}(\mathbf{a} \cos \beta t + \mathbf{b} \sin \beta t))$

$= \begin{vmatrix} e^{\alpha t}(a_1 \cos \beta t - b_1 \sin \beta t) & e^{\alpha t}(a_1 \cos \beta t + b_1 \sin \beta t) \\ e^{\alpha t}(a_2 \cos \beta t - b_2 \sin \beta t) & e^{\alpha t}(a_2 \cos \beta t + b_2 \sin \beta t) \end{vmatrix}$

$= (a_1 b_2 - a_2 b_1)e^{2\alpha t} \sin(2\beta t) \neq 0$

37. (e)

39. (a)

41. (d)

Exercises 8.5

1. $\mathbf{X}(t) = c_1 \begin{pmatrix} -3 \\ 1 \end{pmatrix} e^{-7t} + c_2 \begin{pmatrix} 1 \\ 0 \end{pmatrix} e^{-5t} + \begin{pmatrix} \frac{6}{35} \\ \frac{1}{7} \end{pmatrix} t + \begin{pmatrix} \frac{173}{1225} \\ -\frac{1}{49} \end{pmatrix}$

or $\begin{cases} x(t) = -3c_1 e^{-7t} + c_2 e^{-5t} + \dfrac{6}{35}t + \dfrac{173}{1225} \\ y(t) = c_1 e^{-7t} + \dfrac{1}{7}t - \dfrac{1}{49} \end{cases}$

3. $\begin{cases} x(t) = \dfrac{1}{6}t - \dfrac{1}{36} - c_2 e^{-6t} \\ y(t) = \dfrac{1}{6}t^2 - \dfrac{1}{12}t + \dfrac{1}{54} + c_1 e^{-6t} + c_2 t e^{-6t} \end{cases}$

5. $\begin{cases} x(t) = -\dfrac{2}{143}e^{11t} + \dfrac{3}{13}e^{2t} - 4c_2 e^{-2t} + c_1 e^{-11t} \\ y(t) = -\dfrac{1}{26}e^{2t} + \dfrac{7}{143}e^{11t} + c_2 e^{-2t} + 2c_1 e^{-11t} \end{cases}$

7. $\begin{cases} x(t) = \dfrac{7}{16}e^{-4t} + \dfrac{43}{85}\cos t + \dfrac{19}{85}\sin t + c_1 e^{4t} + 7c_2 e^{-2t} \\ y(t) = -\dfrac{1}{16}e^{-4t} + \dfrac{9}{85}\cos t + \dfrac{2}{85}\sin t + c_1 e^{4t} + c_2 e^{-2t} \end{cases}$

9. $\begin{cases} x(t) = e^{-6t}\left(c_2 \cos 4t - c_1 \sin 4t - \dfrac{1}{4}\right) \\ y(t) = e^{-6t}\left(c_1 \cos 4t + c_2 \sin 4t + \dfrac{1}{4}\right) \end{cases}$

11. $\begin{cases} x(t) = e^{3t}\left[c_1\left(-\dfrac{1}{4}\cos 2t - \dfrac{3}{4}\sin 2t\right) + c_2\left(-\dfrac{3}{4}\cos 2t + \dfrac{1}{4}\sin 2t\right)\right] \\ \quad + \dfrac{1}{8}e^{3t}(10 - \cos 2t - 12t\cos 2t + 3\sin 2t + \\ \qquad\qquad\qquad\qquad\qquad\qquad\qquad 4t\sin 2t) \\ y(t) = e^{3t}[c_1 \cos 2t + c_2 \sin 2t] + \\ \quad \dfrac{1}{2}e^{3t}(-3 + 4t\cos 2t - \sin 2t) \end{cases}$

13. $\mathbf{X}(t) = \begin{pmatrix} \sin t & \cos t \\ -\cos t & \sin t \end{pmatrix}\begin{pmatrix} c_1 \\ c_2 \end{pmatrix} + \begin{pmatrix} t\cos t - \ln|\cos t|\sin t \\ t\sin t + \ln|\cos t|\cos t \end{pmatrix}$

15. $\begin{cases} x(t) = c_1 \sin t - c_2 \cos t - 2\sin t \ln|\sin t| + \\ \qquad\qquad\qquad\qquad 2\sin t \ln|\cos t + 1| \\ y(t) = c_1 \cos t + c_2 \sin t - 2 + 2\cos t \ln|\cos t + 1| - \\ \qquad\qquad\qquad\qquad 2\cos t \ln|\sin t| \end{cases}$

17. $\mathbf{X}(t) = \begin{pmatrix} -2e^{-4t} & -4e^{-t} & 8e^{2t} \\ -3e^{-4t} & -3e^{-t} & 9e^{2t} \\ 2e^{-4t} & e^{-t} & 4e^{2t} \end{pmatrix}\begin{pmatrix} c_1 \\ c_2 \\ c_3 \end{pmatrix} +$

$\begin{pmatrix} -\dfrac{83}{16} + \dfrac{11}{4}t - t^2 \\ -\dfrac{125}{32} + \dfrac{15}{8}t - t^2 \\ \dfrac{11}{16} - \dfrac{3}{4}t \end{pmatrix}$

19.
$$
\begin{cases}
x(t) = -\dfrac{1}{36}te^{-2t} + \dfrac{1}{8}t - \dfrac{1}{32} - \dfrac{1}{12}t^2 e^{-2t} - \dfrac{1}{216}e^{-2t} - \\
\qquad \dfrac{7}{16}e^{-4t} + 2c_2 e^{-2t} - \dfrac{8}{11}c_3 e^{4t} \\[4pt]
y(t) = \dfrac{1}{18}te^{-2t} - \dfrac{5}{432}e^{-2t} + \dfrac{9}{32}t + \dfrac{1}{64} - \dfrac{3}{16}te^{-4t} + \\
\qquad \dfrac{1}{24}t^2 e^{-2t} - \dfrac{15}{128}e^{-4t} + \left(c_1 + \dfrac{5}{16}\right)e^{-4t} \\
\qquad + \left(-\dfrac{1}{12} - c_2\right)e^{-2t} - \dfrac{29}{11}c_3 e^{4t} \\[4pt]
z(t) = -\dfrac{3}{16}te^{-4t} + \dfrac{5}{32}t - \dfrac{5}{64} - \dfrac{31}{432}e^{-2t} + \dfrac{7}{36}te^{-2t} - \\
\qquad \dfrac{15}{128}e^{-4t} - \dfrac{1}{24}t^2 e^{-2t} + c_1 e^{-4t} + c_2 e^{-2t} + c_3 e^{4t}
\end{cases}
$$

21.
$$
\begin{cases}
x(t) = -t + \dfrac{8}{5} - \dfrac{1}{4}e^t + c_1 e^{5t} + c_2 e^{-t} + c_3 te^{-t} \\[4pt]
y(t) = \dfrac{3}{5} - \dfrac{1}{4}e^t + c_1 e^{5t} + \left(c_2 - \dfrac{3}{2}c_3\right)e^{-t} + c_3 te^{-t} \\[4pt]
z(t) = \dfrac{1}{4}e^t - \dfrac{12}{5} + t + c_1 e^{5t} + \left(-2c_2 + \dfrac{5}{4}c_3\right)e^{-t} - \\
\qquad 2c_3 te^{-t}
\end{cases}
$$

23. $\mathbf{X}(t) = $
$$
\begin{pmatrix}
-7e^{4t} & e^{-2t}\left(\dfrac{3}{4}\cos t - \dfrac{1}{4}\sin t\right) \\[8pt]
-2e^{4t} & e^{-2t}\left(-\dfrac{1}{2}\cos t + \dfrac{1}{4}\sin t\right) \\[8pt]
e^{4t} & e^{-2t}\cos t
\end{pmatrix}
$$
$$
\begin{pmatrix}
e^{-2t}\left(\dfrac{1}{4}\cos t + \dfrac{3}{4}\sin t\right) \\[8pt]
e^{-2t}\left(-\dfrac{1}{4}\cos t - \dfrac{1}{2}\sin t\right) \\[8pt]
e^{-2t}\sin t
\end{pmatrix}
\begin{pmatrix} c_1 \\ c_2 \\ c_3 \end{pmatrix} +
$$
$$
\begin{pmatrix}
\dfrac{23}{80} + \dfrac{71}{1369}e^{4t} - \dfrac{1}{4}t + \dfrac{28}{37}te^{4t} \\[8pt]
-\dfrac{43}{200} - \dfrac{59}{1369}e^{4t} + \dfrac{3}{10}t + \dfrac{8}{37}te^{4t} \\[8pt]
\dfrac{91}{400} + \dfrac{48}{1369}e^{4t} - \dfrac{1}{20}t - \dfrac{4}{37}te^{4t}
\end{pmatrix}
$$

25.
$$
\begin{cases}
x(t) = t - 1 + e^{-t} \\
y(t) = 0
\end{cases}
$$

27.
$$
\begin{cases}
x(t) = -\dfrac{2}{3}e^{-t} - \dfrac{1}{2}e^t + \dfrac{1}{6}e^{5t} \\[4pt]
y(t) = \dfrac{2}{3}e^{-t} + \dfrac{1}{4}e^t + \dfrac{1}{12}e^{5t}
\end{cases}
$$

29.
$$
\begin{cases}
x(t) = \dfrac{20}{13} + e^{2t}\left(-\dfrac{7}{13}\cos 3t + \dfrac{1}{39}\sin 3t\right) \\[4pt]
y(t) = -\dfrac{30}{13} + e^{2t}\left(\dfrac{4}{13}\cos 3t + \dfrac{31}{39}\sin 3t\right)
\end{cases}
$$

31. (b)

33. (a)

35. (c)

Exercises 8.6

1. $\begin{cases} x(t) = e^{-7t} - e^{5t} \\ y(t) = 3e^{-7t} + e^{5t} \end{cases}$

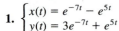

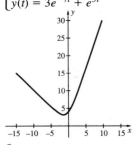

3. $\begin{cases} x(t) = 4e^t \sin 2t + 2e^t \cos 2t \\ y(t) = -4e^t \sin 2t \end{cases}$

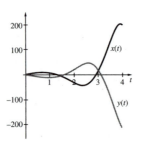

5. $\begin{cases} x(t) = -\dfrac{21}{2}e^t + 8e^{6t} + \dfrac{5}{2}e^{-3t} \\ y(t) = -3e^t - 2e^{6t} + 5e^{-3t} \\ z(t) = 15e^t \end{cases}$

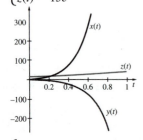

7. $\begin{cases} x(t) = \dfrac{5}{8}te^{4t} + \dfrac{3}{64}e^{4t} - \dfrac{3}{64}e^{-4t} \\ y(t) = \dfrac{5}{8}te^{4t} - \dfrac{5}{64}e^{4t} + \dfrac{5}{64}e^{-4t} \end{cases}$

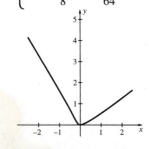

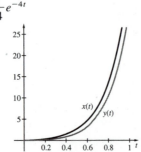

9. $\begin{cases} x(t) = \dfrac{2}{5}e^t + \dfrac{3}{5}e^{-4t} \\ y(t) = \dfrac{2}{5}e^t - \dfrac{2}{5}e^{-4t} \end{cases}$

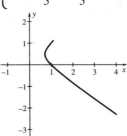

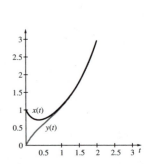

11. $\begin{cases} x(t) = -\dfrac{3}{10}e^{-t} - \dfrac{2}{5}e^{2t} - \dfrac{2}{35}e^{4t} - \dfrac{17}{70}e^{-3t} \\ y(t) = \dfrac{3}{10}e^{2t} + \dfrac{1}{5}e^{-t} + \dfrac{1}{70}e^{4t} + \dfrac{17}{35}e^{-3t} \end{cases}$

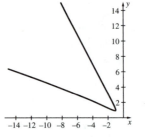

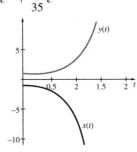

13. $\begin{cases} x(t) = -\dfrac{1}{6}\cos 2t + \dfrac{1}{6}\cos 4t + \dfrac{1}{6}\sin 2t - \dfrac{5}{6}\sin 4t \\ y(t) = -\dfrac{1}{4}\cos 2t - \dfrac{3}{4}\cos 4t + \dfrac{1}{6}\sin 2t - \dfrac{7}{12}\sin 4t \end{cases}$

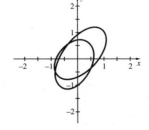

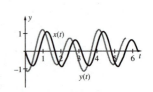

15.
$$\begin{cases} x(t) = \dfrac{1}{2}(-\pi \cos t + \pi \cos t\, \mathcal{U}(\pi - t) - \\ \qquad t \cos t\, \mathcal{U}(\pi - t) + \sin t\, \mathcal{U}(\pi - t)) \\ y(t) = \dfrac{1}{2} \sin t(\pi - \pi \mathcal{U}(\pi - t) + t\, \mathcal{U}(\pi - t)) \end{cases}$$

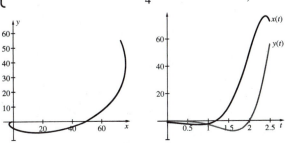

17.
$$\begin{cases} x(t) = -e^{2t} \cos 2t + \left(-\dfrac{1}{2} - \dfrac{1}{2}e^{2t} \sin 2t + \dfrac{1}{2}e^{2t} \cos 2t\right) \\ \qquad + (1 + e^{2(t-1)} \sin 2(t-1) - \\ \qquad\qquad e^{2(t-1)} \cos 2(t-1))\mathcal{U}(t-1) \\ \qquad + \left(-\dfrac{1}{2} - \dfrac{1}{2}e^{2(t-2)} \sin 2(t-2) + \\ \qquad\qquad \dfrac{1}{2}e^{2(t-2)} \cos 2(t-2)\right)\mathcal{U}(t-2) \\ y(t) = -\dfrac{1}{2}e^{2t} \sin 2t + \left(-\dfrac{1}{4} + \dfrac{1}{4}e^{2t} \sin 2t + \dfrac{1}{4}e^{2t} \cos 2t\right) \\ \qquad + \left(\dfrac{1}{2} - \dfrac{1}{2}e^{2(t-1)} \sin 2(t-1) + \\ \qquad\qquad \dfrac{1}{2}e^{2(t-1)} \cos 2(t-1)\right)\mathcal{U}(t-1) \\ \qquad + \left(-\dfrac{1}{4} + \dfrac{1}{4}e^{2(t-2)} \sin 2(t-2) + \\ \qquad\qquad \dfrac{1}{4}e^{2(t-2)} \cos 2(t-2)\right)\mathcal{U}(t-2) \end{cases}$$

19.
$$\begin{cases} x(t) = \dfrac{-1}{2}e^{-5t} + \dfrac{3}{2}e^{-3t} - \\ \qquad \left[-\dfrac{2}{5} - \dfrac{3}{5}e^{-5(t-3)} + e^{-3(t-3)}\right]\mathcal{U}(t-3) - \\ \qquad \left[\dfrac{1}{5} + \dfrac{3}{10}e^{-5(t-2)} - \dfrac{1}{2}e^{-3(t-2)}\right]\mathcal{U}(t-2) - \\ \qquad \left[\dfrac{-1}{5} + \dfrac{3}{10}e^{-5(t-1)} - \dfrac{1}{2}e^{-3(t-1)}\right]\mathcal{U}(t-1) + \cdots \\ y(t) = \dfrac{-1}{2}e^{-5t} + \dfrac{1}{2}e^{-3t} - \\ \qquad \left[\dfrac{4}{15} - \dfrac{3}{5}e^{-5(t-3)} + \dfrac{1}{3}e^{-3(t-3)}\right]\mathcal{U}(t-3) \\ \qquad + \left[\dfrac{2}{15} - \dfrac{3}{10}e^{-5(t-2)} + \dfrac{1}{6}e^{-3(t-2)}\right]\mathcal{U}(t-2) \\ \qquad + \left[\dfrac{2}{15} - \dfrac{3}{10}e^{-5(t-1)} + \dfrac{1}{6}e^{-3(t-1)}\right]\mathcal{U}(t-1) + \cdots \end{cases}$$

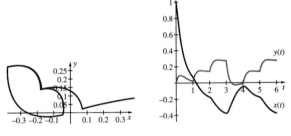

21.
$$\begin{cases} x(t) = t + 1 + \dfrac{\sqrt{2}}{2}\sin\sqrt{2}t + \cos\sqrt{2}t - \sin t \\ y(t) = 1 + \sqrt{2}\sin\sqrt{2}t - \cos\sqrt{2}t + \cos t \end{cases}$$

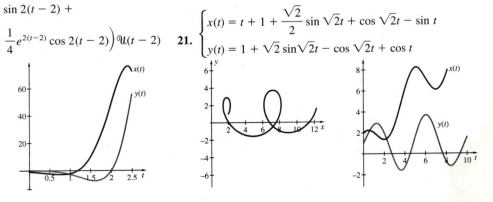

23.
$$\begin{cases} x(t) = \dfrac{1}{10}\cos t + \dfrac{15}{8} - \dfrac{1}{4}t + \dfrac{13}{40}e^{2t} - \dfrac{3}{10}e^{-2t} \\ y(t) = \dfrac{1}{10}\sin t - \dfrac{3}{4} - \dfrac{13}{20}e^{2t} - \dfrac{3}{5}e^{-2t} \end{cases}$$

25.
$$\begin{cases} x(t) = -\dfrac{1}{2} - \dfrac{1}{3}e^{t} + 2\sin t - 3\cos t - \dfrac{4\sqrt{2}}{3}\sin\sqrt{2}t + \dfrac{29}{6}\cos\sqrt{2}t \\ y(t) = -\dfrac{1}{2}t - \dfrac{3}{2}\sin t - \cos t \end{cases}$$

Exercises 8.7

1. $\lambda_1 = -11$, $\lambda_2 = 1$; saddle

3. $\lambda_1 = \lambda_2 = 2$; degenerate unstable node

5. $\lambda_1 = -12$, $\lambda_2 = -8$; stable node

7. $\lambda_1 = 3 + \sqrt{46}$, $\lambda_2 = 3 - \sqrt{46}$; saddle

9. $\lambda_1 = \dfrac{-5 + i\sqrt{191}}{2}$, $\lambda_2 = \dfrac{-5 - i\sqrt{191}}{2}$; stable spiral

11. $\lambda_1 = i\sqrt{41}$, $\lambda_2 = -i\sqrt{41}$; center

13. $(2, -1)$: $\lambda_1 = -6$, $\lambda_2 = 1$, saddle; $(8, -1)$: $\lambda_1 = 6$, $\lambda_2 = 1$, unstable node

15. $(0, 0)$: $\lambda_1 = -\sqrt{2}$, $\lambda_2 = \sqrt{2}$, saddle; $(1, 1)$: $\lambda_1 = -1 + i$, $\lambda_2 = -1 - i$, stable spiral

17. $(0, 2)$: $\lambda_1 = 2$, $\lambda_2 = 4$, unstable node; $(0, -2)$: $\lambda_1 = -2$, $\lambda_2 = -4$, stable node $(2, 0)$: $\lambda_1 = -2\sqrt{2}$, $\lambda_2 = 2\sqrt{2}$, saddle; $(-2, 0)$: $\lambda_1 = -2\sqrt{2}$, $\lambda_2 = 2\sqrt{2}$, saddle

19. $(1, 1)$: $\lambda_1 = 2$, $\lambda_2 = -2$, saddle; $(-1, -1)$: $\lambda_1 = 1 + i\sqrt{3}$, $\lambda_2 = 1 - i\sqrt{3}$, unstable spiral

21. $(0, 0)$: $\lambda_1 = a_1$, $\lambda_2 = -a_2$, saddle; $\left(\dfrac{a_1}{b_1}, 0\right)$: $\lambda_1 = -a_1$, $\lambda_2 = -\left(a_2 - \dfrac{c_2 a_1}{b_1}\right)$, stable node $\left(-\dfrac{a_2}{c_2}, \dfrac{a_1 c_2 + a_2 b_1}{c_1 c_2}\right)$: $\lambda = \dfrac{a_2 b_1}{c_2} \pm \sqrt{\left(\dfrac{a_2 b_1}{c_2}\right)^2 - \dfrac{4a_2(a_1 c_2 + a_2 b_1)}{b_1}}$,

unstable node if $\left(\dfrac{a_2 b_1}{c_2}\right)^2 - \dfrac{4a_2(a_1 c_2 + a_2 b_1)}{b_1} \geq 0$; unstable spiral if $\left(\dfrac{a_2 b_1}{c_2}\right)^2 - \dfrac{4a_2(a_1 c_2 + a_2 b_1)}{b_1} < 0$

23. (e)
25. (d)
27. (c)
31. $y = 0$, stable
33. $y = 0$, unstable; and $y = 2$, stable

35. $y = 1$, stable; and $y = 4$, unstable

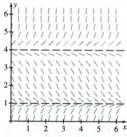

Exercises 8.8

1. $x(1) \approx 8.64479$, $y(1) \approx 8.29263$

3. $x(1) \approx -7.2362$, $y(1) \approx -1.5998$

5. $x(1) \approx -0.113115$, $y(1) \approx -1.06576$

7. $x(1) \approx -326.204$, $y(1) \approx 608$

9. $x(1) \approx 0.164251$, $y(1) \approx -0.504587$

11. $x(1) \approx 1.52319$, $y(1) \approx 0.419975$

13. $\begin{cases} x' = y \\ y' = -2x - 3y; \text{ Euler's method yields} \\ x(0) = 0, y(0) = -3 \end{cases}$
$x(1) \approx -0.723913$, $y(1) \approx 0.40179$; exact solution is
$x(t) = 3e^{-2t} - 3e^{-t}$ so $x(1) = 3e^{-2}(1 - e) \approx$
-0.697632.

15. $\begin{cases} x' = y \\ y' = -9x; \text{ Euler's method yields} \\ x(0) = 0, y(0) = 3 \end{cases}$
$x(1) \approx 0.346313$, $y(1) \approx -4.49743$; exact solution is
$x(t) = \sin 3t$ so $x(1) = \sin 3 \approx 0.14112$.

17. $\begin{cases} x' = y \\ y' = -\dfrac{1}{t^2}(16x + ty); \text{ Euler's method yields} \\ x(1) = 0, y(1) = 4 \end{cases}$
$x(2) \approx 0.354942$, $y(2) \approx -2.90834$; exact solution is
$x(t) = \sin(4 \ln t)$ so $x(2) = \sin(4 \ln 2) \approx 0.360687$.

19. (See 13) Runge-Kutta yields $x(1) \approx -0.697621$,
$y(1) \approx 0.291602$

21. (See 15) Runge-Kutta yields $x(1) \approx 0.141307$,
$y(1) \approx -2.96975$

23. (See 17) Runge-Kutta yields $x(1) \approx 0.360845$,
$y(1) \approx -1.86541$

Chapter 8 Review Exercises

1. $\{-4, 11\}$, $\left\{ \begin{pmatrix} -2 \\ 1 \end{pmatrix}, \begin{pmatrix} 1 \\ 2 \end{pmatrix} \right\}$

3. $\{-3, -2\}$, $\left\{ \begin{pmatrix} -1 \\ 1 \end{pmatrix}, \begin{pmatrix} -2 \\ 1 \end{pmatrix} \right\}$

5. $\{1, 2\}$, $\left\{ \begin{pmatrix} -2 \\ 1 \\ 0 \end{pmatrix}, \begin{pmatrix} 1 \\ 0 \\ 1 \end{pmatrix} \right\}$

7. $\{2, 3, 4\}$, $\left\{ \begin{pmatrix} -1 \\ -1 \\ 2 \end{pmatrix}, \begin{pmatrix} 0 \\ -1 \\ 1 \end{pmatrix}, \begin{pmatrix} 1 \\ -1 \\ 2 \end{pmatrix} \right\}$

9. $\{-4, -2 \pm 3i\}$, $\left\{ \begin{pmatrix} 1 \\ 0 \\ 0 \end{pmatrix}, \begin{pmatrix} -\frac{9}{13} + \frac{6}{13}i \\ i \\ 1 \end{pmatrix}, \begin{pmatrix} -\frac{9}{13} - \frac{6}{13}i \\ -i \\ 1 \end{pmatrix} \right\}$

11. $\mathbf{X} = \begin{pmatrix} -e^{6t} & 4e^t \\ e^{6t} & e^t \end{pmatrix} \begin{pmatrix} c_1 \\ c_2 \end{pmatrix}$

13. $\begin{cases} x = 3e^{2t} \\ y = 2e^{-t} + e^{2t} \end{cases}$

15. $\mathbf{X} = \begin{pmatrix} 3 \sin \frac{t}{2} - \cos \frac{t}{2} & \sin \frac{t}{2} + 3 \cos \frac{t}{2} \\ \sin \frac{t}{2} & \cos \frac{t}{2} \end{pmatrix} \begin{pmatrix} c_1 \\ c_2 \end{pmatrix}$

17. $\mathbf{X}' = \begin{pmatrix} -4 & -1 \\ 5 & -2 \end{pmatrix} \mathbf{X}$,
$\mathbf{X} = e^{-3t} \begin{pmatrix} \sin 2t - 2 \cos 2t & -\cos 2t \\ -5 \sin 2t & \cos 2t - 2 \sin 2t \end{pmatrix} \begin{pmatrix} c_1 \\ c_2 \end{pmatrix}$

19. $\mathbf{X} = \begin{pmatrix} -2 \cos 4t & 2 \sin 4t \\ \sin 4t & \cos 4t \end{pmatrix} \begin{pmatrix} c_1 \\ c_2 \end{pmatrix}$

21. $\begin{cases} x(t) = e^{-t}(2 \cos 3t + \sin 3t) \\ y(t) = 3e^{-t}(\cos 3t + 3 \sin 3t) \end{cases}$

23. $\mathbf{X} = e^{7t} \begin{pmatrix} 1 & 1 + t \\ 1 & t \end{pmatrix} \begin{pmatrix} c_1 \\ c_2 \end{pmatrix}$

25. $\begin{cases} x(t) = c_1 e^{-t} + 9c_2 e^{7t} + \dfrac{5}{4} e^{7t} - 18te^{7t} \\ y(t) = c_1 e^{-t} + c_2 e^{7t} + \dfrac{5}{4} e^{7t} - 2te^{7t} \end{cases}$

27. $\begin{cases} x = 1 - c_1 e^{-t} + 2c_2 e^{5t} \\ y = -1 + c_1 e^{-t} + c_2 e^{5t} \end{cases}$

29. $\begin{cases} x = \cos t - t \sin t \\ y = \sin t + t \cos t \end{cases}$

31. $\begin{cases} x = e^t - 1 \\ y = -\dfrac{1}{2} + 2e^t - \dfrac{3}{2} e^{2t} \end{cases}$

33. $\begin{cases} x = -\dfrac{2}{3} + e^t - \dfrac{1}{3} e^{3t} + \\ \qquad 2 \left(\dfrac{1}{3} + \dfrac{1}{6} e^{3t-6} - \dfrac{1}{2} e^{t-2} \right) \mathcal{U}(t - 2) \\ y = -\dfrac{1}{3} + \dfrac{1}{3} e^{3t} - \left(-\dfrac{1}{3} + \dfrac{1}{3} e^{3t-6} \right) \mathcal{U}(t - 2) \end{cases}$

C H A P T E R   9

Exercises 9.1

1. (a)
$$\begin{cases} Q(t) = \dfrac{3\sqrt{5}}{5}e^{-5t/3}\sin\left(\dfrac{\sqrt{5}}{3}t\right) \\[2mm] I(t) = -\sqrt{5}e^{-5t/3}\sin\left(\dfrac{\sqrt{5}}{3}t\right) + e^{-5t/3}\cos\left(\dfrac{\sqrt{5}}{3}t\right) \end{cases}$$

(b)
$$\begin{cases} Q(t) = \dfrac{1}{3}e^{-t} + \dfrac{7\sqrt{5}}{15}e^{-5t/3}\sin\left(\dfrac{\sqrt{5}}{3}t\right) - \\[2mm] \qquad\qquad \dfrac{1}{3}e^{-5t/3}\cos\left(\dfrac{\sqrt{5}}{3}t\right) \\[2mm] I(t) = -\dfrac{1}{3}e^{-t} - \dfrac{2\sqrt{5}}{3}e^{-5t/3}\sin\left(\dfrac{\sqrt{5}}{3}t\right) + \\[2mm] \qquad\qquad \dfrac{4}{3}e^{-5t/3}\cos\left(\dfrac{\sqrt{5}}{3}t\right) \end{cases}$$

3. (a)
$$\begin{cases} Q(t) = 10^{-6}e^{-t}\cos(\sqrt{2}t) \\ I_2(t) = 10^{-6}e^{-t}\sin(\sqrt{2}t) \end{cases}$$

(b)
$$\begin{cases} Q(t) = \dfrac{2}{9}e^{-t/2} + \dfrac{4\sqrt{2}}{9}e^{-t}\sin(\sqrt{2}t) - \\[2mm] \qquad\qquad \dfrac{1999991}{9000000}e^{-t}\cos(\sqrt{2}t) \\[2mm] I_2(t) = \dfrac{8}{9}e^{-t/2} - \dfrac{1999991}{4500000\sqrt{2}}e^{-t}\sin(\sqrt{2}t) - \\[2mm] \qquad\qquad \dfrac{8}{9}e^{-t}\cos(\sqrt{2}t) \end{cases}$$

5. (a)
$$\begin{cases} Q(t) = 10^{-6}e^{-2t}(1+t) \\ I_2(t) = 10^{-6}te^{-2t} \Longrightarrow \\ I(t) = -10^{-6}e^{-2t}(1+2t) \Longrightarrow I_1(t) = -10^{-6}e^{-2t}(1+t) \end{cases}$$

(b)
$$\begin{cases} Q(t) = \dfrac{135}{2} - \dfrac{135}{2}e^{-2t} - 45te^{-2t} \\[2mm] I_2(t) = \dfrac{45}{2} - \dfrac{45}{2}e^{-2t} - 45te^{-2t} \end{cases} \Longrightarrow$$

$$I(t) = 90e^{-2t}(1+t) \Longrightarrow I_1(t) = \dfrac{45}{2}e^{-2t}(3 + e^{2t} + 2t)$$

7. (a)
$$\begin{cases} Q(t) = 10^{-6}e^{-t} - 2\cdot 10^{-6}t^2e^{-t} \\ I_2(t) = 10^{-6}te^{-t} \\ I_3(t) = 2\cdot 10^{-6}t^2e^{-t} \end{cases}$$

(b)
$$\begin{cases} Q(t) = 10^{-6}e^{-t} + te^{-t} - 2\cdot 10^{-6}t^2e^{-t} \\ I_2(t) = 10^{-6}te^{-t} \\ I_3(t) = -te^{-t} + 2\cdot 10^{-6}t^2e^{-t} \end{cases}$$

9. (a)
$$\begin{cases} Q(t) = 90 - te^{-t} - 90e^{-t} \\ I_2(t) = e^{-t} \\ I_3(t) = -90 + te^{-t} + 90e^{-t} \end{cases}$$

(b)
$$\begin{cases} Q(t) = -te^{-t} + 45e^{-t} + 45\sin t - 45\cos t \\ I_2(t) = e^{-t} \\ I_3(t) = te^{-t} - 45e^{-t} - 45\sin t + 45\cos t \end{cases}$$

11. (a)
$$\begin{cases} Q(t) = 90 - \dfrac{4}{3}e^{-t/4} - \dfrac{266}{3}e^{-t} \\[2mm] I_2(t) = e^{-t/4} \\[2mm] I_3(t) = -90 + \dfrac{4}{3}e^{-t/4} + \dfrac{266}{3}e^{-t} \end{cases} \quad ;$$

(b)
$$\begin{cases} Q(t) = -45\cos t + 45\sin t - \dfrac{4}{3}e^{-t/4} + \dfrac{139}{3}e^{-t} \\[2mm] I_2(t) = e^{-t/4} \\[2mm] I_3(t) = 45\cos t - 45\sin t + \dfrac{4}{3}e^{-t/4} - \dfrac{139}{3}e^{-t} \end{cases}$$

13.
$$\begin{cases} \dfrac{dx}{dt} = y \\[2mm] \dfrac{dy}{dt} = -9x \end{cases}, \quad \lambda = \pm 3i; \text{ undamped}$$

15.
$$\begin{cases} \dfrac{dx}{dt} = y \\[2mm] \dfrac{dy}{dt} = -9x - 10y \end{cases}, \quad \lambda_1 = -1,\ \lambda_2 = -9; \text{ overdamped}$$

17.
$$\begin{cases} \dfrac{dx}{dt} = y \\[2mm] \dfrac{dy}{dt} = -50x - 10y \end{cases},$$
$$\lambda_1 = -5 + 5i,\ \lambda_2 = -5 - 5i; \text{ underdamped}$$

19.
$$\begin{cases} \dfrac{dx}{dt} = y \\[2mm] \dfrac{dy}{dt} = -25x - 10y \end{cases}, \quad \lambda_1 = -5,\ \lambda_2 = -5;$$
critically damped

21. (13) $\begin{cases} x(t) = \cos 3t \\ y(t) = -3 \sin 3t \end{cases}$;

(15) $\begin{cases} x(t) = \dfrac{9}{8}e^{-t} - \dfrac{1}{8}e^{-9t} \\ y(t) = -\dfrac{9}{8}e^{-t} + \dfrac{9}{8}e^{-9t} \end{cases}$

23. $\lambda = -\dfrac{c}{2m} \pm \dfrac{1}{2m}\sqrt{c^2 - 4km}$, overdamped if $c^2 > 4km$; critically damped if $c^2 = 4km$, underdamped if $c^2 < 4km$.

Exercises 9.2

1. $\begin{cases} x(t) = \dfrac{3}{2} - \dfrac{1}{2}e^{-t} \\ y(t) = \dfrac{3}{2} + \dfrac{1}{2}e^{-t} \end{cases}$, $\lim_{t\to\infty}(x(t), y(t)) = \left(\dfrac{3}{2}, \dfrac{3}{2}\right)$

3. $\begin{cases} x(t) = \dfrac{8}{3} - \dfrac{8}{3}e^{-3t/2} \\ y(t) = \dfrac{4}{3} + \dfrac{8}{3}e^{-3t/2} \end{cases}$, $\lim_{t\to\infty}(x(t), y(t)) = \left(\dfrac{8}{3}, \dfrac{4}{3}\right)$

5. $\begin{cases} x(t) = 1 + 3e^{-15t/4} \\ y(t) = 4 - 3e^{-15t/4} \end{cases}$, $\lim_{t\to\infty}(x(t), y(t)) = (1, 4)$

7. $\{x(t) = 10 - 10e^{-t/5}, y(t) = 10 - 10e^{-t/5} - 2te^{-t/5}\}$, $\lim_{t\to\infty} x(t) = \lim_{t\to\infty} y(t) = 10$, $x(t)$

9. $\left\{x(t) = 4e^{-t/5}, y(t) = 4e^{-t/5} + \dfrac{4}{5}te^{-t/5}\right\}$,

$\lim_{t\to\infty} x(t) = \lim_{t\to\infty} y(t) = 0$, $x(t)$

11. $\{x(t) = 300 - 300e^{-t/20}, y(t) = 150 + 150e^{-t/10} - 300e^{-t/20}\}$, (b) $\lim_{t\to\infty} x(t) = 300$, $\lim_{t\to\infty} y(t) = 150$

13. $\left\{x(t) = \dfrac{325}{6} - \dfrac{125}{12}e^{-3t/25} - \dfrac{175}{4}e^{-t/25}, y(t) = \dfrac{200}{3} + \dfrac{125}{6}e^{-3t/25} - \dfrac{175}{2}e^{-t/25}\right\}$; $\lim_{t\to\infty} x(t) = \dfrac{325}{6}$,

$\lim_{t\to\infty} y(t) = \dfrac{200}{3}$; $x(t) = y(t)$ at approx. $t = 29.1$ min.

15. System: $\begin{cases} \dfrac{dx}{dt} = 2y - 2x + 1 \\ \dfrac{dy}{dt} = -3y + x + 1 \end{cases}$

$\left\{x(t) = \dfrac{5}{4} - \dfrac{23}{12}e^{-4t} + \dfrac{8}{3}e^{-t}, y(t) = \dfrac{3}{4} + \dfrac{23}{12}e^{-4t} + \dfrac{4}{3}e^{-t}\right\}$, $\lim_{t\to\infty} x(t) = \dfrac{5}{4}$, $\lim_{t\to\infty} y(t) = \dfrac{3}{4}$; Max. $x = 4$, $y \approx 1$

17. System: $\begin{cases} \dfrac{dx}{dt} = -\dfrac{x}{10} + 10 \\ \dfrac{dy}{dt} = \dfrac{x}{10} - \dfrac{y}{10} \\ \dfrac{dz}{dt} = \dfrac{y}{10} - \dfrac{z}{10} \end{cases}$

$\begin{cases} x(t) = 100 - 100e^{-t/10} \\ y(t) = 100 - 100e^{-t/10} - 10te^{-t/10} \\ z(t) = 100 - 100e^{-t/10} - 10te^{-t/10} - \dfrac{1}{2}t^2 e^{-t/10} \end{cases}$

$\lim_{t\to\infty} x(t) = \lim_{t\to\infty} y(t) = \lim_{t\to\infty} z(t) = 100$

19. $\mathbf{X}(t) = \begin{pmatrix} x(t) \\ y(t) \end{pmatrix} = \begin{pmatrix} \dfrac{25}{3}e^{4t} + \dfrac{5}{3}e^{-2t} \\ 25e^{4t} - 5e^{-2t} \end{pmatrix}$

21. $\mathbf{X}(t) = \begin{pmatrix} x(t) \\ y(t) \end{pmatrix} = \begin{pmatrix} 15e^{2t} - 10e^{-t} \\ 10e^{-t} \end{pmatrix}$

23. $\mathbf{X}(t) = \begin{pmatrix} x(t) \\ y(t) \\ z(t) \end{pmatrix} = \begin{pmatrix} -\dfrac{67}{7}e^{-10t} + \dfrac{171}{10}e^{7t} + \dfrac{663}{70}e^{-3t} \\ 5e^{-10t} + \dfrac{171}{10}e^{7t} - \dfrac{221}{10}e^{-3t} \\ \dfrac{52}{7}e^{-10t} + \dfrac{171}{10}e^{7t} + \dfrac{663}{70}e^{-3t} \end{pmatrix}$;

$x(1) \approx z(1) \approx 18752.89$

25. $\mathbf{X}(t) = \begin{pmatrix} x(t) \\ y(t) \\ z(t) \end{pmatrix} = \begin{pmatrix} \dfrac{43}{7} + \dfrac{46}{21}e^{7t} - \dfrac{1}{3}e^{-8t} \\ \dfrac{4}{3}e^{7t} + \dfrac{2}{3}e^{-8t} \\ -\dfrac{43}{7} + \dfrac{136}{21}e^{7t} - \dfrac{1}{3}e^{-8t} \end{pmatrix}$;

$z(t) \approx 7095.86$

29. $\begin{cases} x(t) = x_0 e^{-at} \\ y(t) = y_0 e^{-at} + ax_0 te^{-at} \\ z(t) = x_0 + y_0 + z_0 - (x_0 + y_0)e^{-at} - ax_0 te^{-at} \end{cases}$;

$\lim_{t\to\infty} x(t) = 0$, $\lim_{t\to\infty} y(t) = 0$, $\lim_{t\to\infty} z(t) = x_0 + y_0 + z_0$

33. $\begin{cases} x(t) = 7e^{-6t} \\ y(t) = \dfrac{47}{5}e^{-t} - \dfrac{42}{5}e^{-6t} \quad ; \ \lim\limits_{t\to\infty} z(t) = 16 \\ z(t) = 16 - \dfrac{47}{5}e^{-t} + \dfrac{7}{5}e^{-6t} \end{cases}$

35. $\begin{cases} x(t) = e^{-4t} \\ y(t) = 3e^{-2t} - 2e^{-4t} \quad ; \ \lim\limits_{t\to\infty} z(t) = 3 \\ z(t) = 3 - 3e^{-2t} + e^{-4t} \end{cases}$

37. $\begin{cases} x(t) = 2 \\ y(t) = 2 - 2e^{-t} \quad ; \\ z(t) = t + 1 + e^{-t} \end{cases}$
$\lim\limits_{t\to\infty} x(t) = 2, \ \lim\limits_{t\to\infty} y(t) = 2, \ \lim\limits_{t\to\infty} z(t) = \infty$

39. $\begin{cases} x(t) = 10 - 2e^{-t} \\ y(t) = 10 - 8e^{-t} - 2te^{-t} \quad ; \\ z(t) = 10t - 8 + 10e^{-t} + 2te^{-t} \end{cases}$
$\lim\limits_{t\to\infty} x(t) = 10, \ \lim\limits_{t\to\infty} y(t) = 10, \ \lim\limits_{t\to\infty} z(t) = \infty$

Exercises 9.3

1. $\begin{cases} x(t) = \dfrac{2}{5}\cos t - \dfrac{2}{5}\cos\sqrt{6}t \\ y(t) = \dfrac{4}{5}\cos t + \dfrac{1}{5}\cos\sqrt{6}t \end{cases}$

3. $\begin{cases} x(t) = \dfrac{1}{3}\cos t + \dfrac{2}{3}\cos 2t \\ y(t) = \dfrac{2}{3}\cos t - \dfrac{2}{3}\cos 2t \end{cases}$

5. $\begin{cases} x(t) = \dfrac{4}{5}\cos\sqrt{3}t + \dfrac{1}{5}\cos\dfrac{t}{\sqrt{2}} \\ y(t) = -\dfrac{2}{5}\cos\sqrt{3}t + \dfrac{2}{5}\cos\dfrac{t}{\sqrt{2}} \end{cases}$

7. $\begin{cases} x(t) = \dfrac{1}{4} + \dfrac{1}{18}\sin 2t + \dfrac{7}{12}\cos 2t + \dfrac{1}{18}\sin t + \\ \qquad \dfrac{1}{6}\cos t - \dfrac{1}{6}t\cos t \\ y(t) = \dfrac{1}{4} - \dfrac{1}{18}\sin 2t - \dfrac{7}{12}\cos 2t + \dfrac{4}{9}\sin t + \\ \qquad \dfrac{1}{3}\cos t - \dfrac{1}{3}t\cos t \end{cases}$

9. $\begin{cases} x(t) = \dfrac{7}{18}\cos t + \dfrac{11}{18}\cos 2t + \dfrac{1}{12}t\sin 2t \\ y(t) = \dfrac{7}{9}\cos t - \dfrac{7}{9}\cos 2t - \dfrac{1}{12}t\sin 2t \end{cases}$

11. $\begin{cases} \theta_1(t) = -\dfrac{1}{8}\sqrt{3}\sin\dfrac{2t}{\sqrt{3}} + \dfrac{1}{8}\sin 2t \\ \theta_2(t) = -\dfrac{1}{4}\sqrt{3}\sin\dfrac{2t}{\sqrt{3}} - \dfrac{1}{4}\sin 2t \end{cases}$

13. $\begin{cases} \theta_1(t) = \dfrac{1}{2}\cos\dfrac{2t}{\sqrt{3}} + \dfrac{1}{2}\cos 2t \\ \theta_2(t) = \cos\dfrac{2t}{\sqrt{3}} - \cos 2t \end{cases}$

15. $\begin{cases} \theta_1(t) = \dfrac{\sqrt{3}}{4}\sin\dfrac{2t}{\sqrt{3}} + \dfrac{1}{4}\sin 2t \\ \theta_2(t) = \dfrac{\sqrt{3}}{2}\sin\dfrac{2t}{\sqrt{3}} - \dfrac{1}{2}\sin 2t \end{cases}$

17. $\begin{cases} \theta_1(t) = -\dfrac{1}{4}\cos\dfrac{2t}{\sqrt{3}} + \dfrac{1}{4}\cos 2t \\ \theta_2(t) = -\dfrac{1}{2}\cos\dfrac{2t}{\sqrt{3}} - \dfrac{1}{2}\cos 2t \end{cases}$

19. $\begin{cases} x(t) = \dfrac{b+d}{2\omega_0}\sin\omega_0 t + \dfrac{a+c}{2}\cos\omega_0 t + \\ \qquad \dfrac{(b-d)\sqrt{m}}{2\sqrt{2k+m\omega_0^2}}\sin T + \dfrac{a-c}{2}\cos T \\ y(t) = \dfrac{b+d}{2\omega_0}\sin\omega_0 t + \dfrac{a+c}{2}\cos\omega_0 t + \\ \qquad \dfrac{(d-b)\sqrt{m}}{2\sqrt{2k+m\omega_0^2}}\sin T + \dfrac{c-a}{2}\cos T \end{cases}$,
$T = \dfrac{t\sqrt{2k+k\,\omega_0^2}}{\sqrt{m}}$

21. $\begin{cases} x(t) = \dfrac{1}{2\omega_0}\sin\omega_0 t + \dfrac{\sqrt{m}}{2\sqrt{2k+m\omega_0^2}}\sin\left(\dfrac{t\sqrt{2k+k\omega_0^2}}{\sqrt{m}}\right) \\ y(t) = \dfrac{1}{2\omega_0}\sin\omega_0 t - \dfrac{\sqrt{m}}{2\sqrt{2k+m\omega_0^2}}\sin\left(\dfrac{t\sqrt{2k+k\omega_0^2}}{\sqrt{m}}\right) \end{cases}$

23. $\mathbf{X}'(t) = \begin{pmatrix} 0 & 1 & 0 & 0 \\ -(k_1 + k_2)/m_1 & 0 & k_2/m_2 & 0 \\ 0 & 0 & 0 & 1 \\ k_2/m_2 & 0 & -(k_2 + k_3)/m_2 & 0 \end{pmatrix}\mathbf{X}(t)$

Exercises 9.4

1. $-a \ln|y| + by - c \ln|x| + dx = C$

3. (a) $(2k, k)$ stable spiral; **(b)** $(2k, k)$ center

5. $\begin{cases} \dfrac{dx}{dt} = y \\ \dfrac{dy}{dt} = -\sin x \end{cases}$, $(k\pi, 0)$, $k = 0, \pm 1, \pm 2, \ldots,$

(vertical position of pendulum)

7. $\begin{cases} \dfrac{dx}{dt} = y \\ \dfrac{dy}{dt} = -k^2 x \end{cases}$; $\dfrac{dy}{dx} = \dfrac{dy/dt}{dx/dt} = \dfrac{-k^2 x}{y}$; $\dfrac{1}{2}y^2 + \dfrac{k^2}{2}x^2 = C$;

$(0, 0)$ center

9. $\begin{cases} \dfrac{dx}{dt} = y \\ \dfrac{dy}{dt} = -\dfrac{gM_1}{x^2} + \dfrac{gM_2}{(R-x)^2} \end{cases}$; $(x_0, y_0) =$

$\left(\dfrac{R(M_1 + \sqrt{M_1 M_2})}{M_1 - M_2}, 0 \right) = \left(\dfrac{R\sqrt{M_1}}{\sqrt{M_1} - \sqrt{M_2}}, 0 \right)$; saddle

11. If $\dfrac{dx}{dt} = y - F(x) = y - \displaystyle\int_0^x f(u)\,du$ and $\dfrac{dy}{dt} = -g(x)$,

then $\dfrac{d^2x}{dt^2} = \dfrac{dy}{dt} - f(x)\dfrac{dx}{dt} = -g(x) - f(x)\dfrac{dx}{dt}$ which is

equivalent to $\dfrac{d^2x}{dt^2} + f(x)\dfrac{dx}{dt} + g(x) = 0$.

13. In each case, no limit cycle in the xy-plane:
(a) $f_x(x, y) + g_y(x, y) = 3x^2 + 1 + x^2 + 1 = 4x^2 + 1 > 0$;
(b) $f_x(x, y) + g_y(x, y) = -1 - 1 = -2 < 0$;
(c) $f_x(x, y) + g_y(x, y) = y^2 + 8 > 0$.

15. $\begin{cases} \dfrac{du}{d\theta} = y \\ \dfrac{dy}{d\theta} = \alpha - u - ku^2 \end{cases}$; $y = 0$

and $\alpha - u - ku^2 = 0 \Rightarrow u = \dfrac{1 \pm \sqrt{1 - 4k\alpha}}{2k}$. We

choose $u = \dfrac{1 - \sqrt{1 - 4k\alpha}}{2k}$ because it is closer to $u = 0$

than $u = \dfrac{1 + \sqrt{1 - 4k\alpha}}{2k}$, and we are considering a

small change in the orbit. $J\left(\dfrac{1 + \sqrt{1 - 4k\alpha}}{2k}, 0 \right) =$

$\begin{pmatrix} 0 & 1 \\ -\sqrt{1 - 4k\alpha} & 0 \end{pmatrix} \Rightarrow \lambda^2 + \sqrt{1 - 4k\alpha} = 0$; center.

17. Differentiating $\dfrac{1}{2}m(x)\left(\dfrac{dx}{dt}\right)^2 + V(x) = E$ with respect to

t yields $\dfrac{1}{2}m'(x)\left(\dfrac{dx}{dt}\right)^3 + m(x)\dfrac{dx}{dt}\dfrac{d^2x}{dt^2} + V'(x)\dfrac{dx}{dt} =$

$\dfrac{dx}{dt}\left[\dfrac{1}{2}m'(x)\left(\dfrac{dx}{dt}\right)^2 + m(x)\dfrac{d^2x}{dt^2} + V'(x)\right] = 0$, so

$$\dfrac{1}{2}m'(x)\left(\dfrac{dx}{dt}\right)^2 + m(x)\dfrac{d^2x}{dt^2} + V'(x) = 0$$

$$\dfrac{1}{\sqrt{m(x)}}\left[\dfrac{1}{2}m'(x)\left(\dfrac{dx}{dt}\right)^2 + m(x)\dfrac{d^2x}{dt^2} + V'(x)\right] = 0$$

$$\dfrac{1}{2}\dfrac{m'(x)}{\sqrt{m(x)}}\left(\dfrac{dx}{dt}\right)^2 + \sqrt{m(x)}\dfrac{d^2x}{dt^2} + \dfrac{V'(x)}{\sqrt{m(x)}} = 0$$

$$\dfrac{d}{dt}\left[\sqrt{m(x)}\dfrac{dx}{dt}\right] + \dfrac{V'(x)}{\sqrt{m(x)}} = 0$$

Notice that $\dfrac{du}{dx} = \sqrt{m(x)}$, so the equation is

$$\dfrac{d}{dt}\left[\dfrac{du}{dx}\dfrac{dx}{dt}\right] + \dfrac{V'(x)}{\sqrt{m(x)}} = 0.$$

Notice also that $\dfrac{du}{dt} = \dfrac{du}{dx}\dfrac{dx}{dt}$, so we have

$$\dfrac{d}{dt}\left[\dfrac{du}{dt}\right] + \dfrac{V'(x)}{\sqrt{m(x)}} = 0 \text{ or } \dfrac{d^2u}{dt^2} + \dfrac{V'(x)}{\sqrt{m(x)}} = 0.$$

19. Paths: circles; equil. pt.: center; agree

21. (a) center if $x = 1$; saddle if $x = -1$; **(b)** saddle if $x = 0$; center if $x = 1$, $x = -1$.

23. $\begin{cases} \dfrac{d\theta}{dt} = y \\ \dfrac{dy}{dt} = -\dfrac{FR}{I}\text{sgn}(y) \end{cases}$; equilibrium points: $(\theta, 0)$;

if $\dfrac{d\theta}{dt} > 0$, then we integrate $I\dfrac{d\theta}{dt}\dfrac{d}{d\theta}\left(\dfrac{d\theta}{dt}\right) = -FR$

with respect θ to obtain the parabolas $\dfrac{1}{2}I\left(\dfrac{d\theta}{dt}\right)^2 =$

$-FR\theta + C$, $\dfrac{d\theta}{dt} > 0$. Similar calculations follow

for $\dfrac{d\theta}{dt} < 0$.

Chapter 9 Review Exercises

1. (a)
$$\begin{cases} Q(t) = \dfrac{1}{500000}\left(\dfrac{3}{2}e^{-2t/3} - e^{-t}\right) \\[4mm] I(t) = \dfrac{1}{500000}(-e^{-2t/3} + e^{-t}) \end{cases}$$

(b)
$$\begin{cases} Q(t) = -\dfrac{4500001}{500000}e^{-t} - 3te^{-t} + \dfrac{9000003}{1000000}e^{-2t/3} \\[4mm] I(t) = \dfrac{3000001}{500000}e^{-t} + 3te^{-t} + \dfrac{3000001}{1000000}e^{-2t/3} \end{cases}$$

3. (a)
$$\begin{cases} Q(t) = \dfrac{1}{1000000}e^{-t}\cos t \\[4mm] I_2(t) = \dfrac{1}{1000000}e^{-t}\sin t \end{cases};$$

(b)
$$\begin{cases} Q(t) = 60 + 60e^{-t}\sin t - \dfrac{59999999}{1000000}e^{-t}\cos t \\[4mm] I_2(t) = 60 - \dfrac{59999999}{1000000}e^{-t}\sin t - 60e^{-t}\cos t \end{cases}$$

5. (a) $\{x(t) = e^{-t} + te^{-t},\ y(t) = -te^{-t}\}$ (critically damped);
(b) $\{x(t) = \cos 2t,\ y(t) = -2\sin 2t\}$ (undamped)

7. $\left\{ x(t) = \dfrac{15}{2} - \dfrac{5}{2}e^{-t},\ y(t) = \dfrac{15}{2} + \dfrac{5}{2}e^{-t} \right\}$

9. $\{x(t) = 15e^{3t} + 5e^{t},\ y(t) = 15e^{3t} - 5e^{t}\}$

11.
$$\begin{cases} x(t) = -\dfrac{5}{3}e^{-5t} - \dfrac{128}{51}e^{-14t} + \dfrac{139}{17}e^{3t} \\[4mm] y(t) = \dfrac{7}{6}e^{-5t} - \dfrac{64}{51}e^{-14t} + \dfrac{139}{34}e^{3t} \\[4mm] z(t) = -\dfrac{1}{6}e^{-5t} + \dfrac{208}{51}e^{-14t} + \dfrac{139}{34}e^{3t} \end{cases}$$

13.
$$\begin{cases} x(t) = -\dfrac{8}{25}\cos\sqrt{6}t + \dfrac{8}{25}\cos t + \dfrac{t}{5}\sin t \\[4mm] y(t) = \dfrac{4}{25}\cos\sqrt{6}t - \dfrac{4}{25}\cos t + \dfrac{2t}{5}\sin t \end{cases};\ \text{Resonance}$$

15.
$$\begin{cases} \theta_1(t) = \dfrac{1}{8}\sin 2t - \dfrac{1}{4}\cos 2t - \dfrac{\sqrt{3}}{8}\sin\dfrac{2t}{\sqrt{3}} + \\[4mm] \qquad\qquad \dfrac{1}{4}\cos\dfrac{2t}{\sqrt{3}} \\[4mm] \theta_2(t) = -\dfrac{1}{4}\sin 2t + \dfrac{1}{2}\cos 2t - \dfrac{\sqrt{3}}{4}\sin\dfrac{2t}{\sqrt{3}} + \\[4mm] \qquad\qquad \dfrac{1}{2}\cos\dfrac{2t}{\sqrt{3}} \end{cases}$$

17. $J(x, y) = \begin{pmatrix} 0 & 1 \\ -\dfrac{g}{L}\cos x & -b \end{pmatrix};$

$J(n\pi, 0) = \begin{pmatrix} 0 & 1 \\ -\dfrac{g}{L}\cos n\pi & -b \end{pmatrix};$

n is odd: $\lambda = \dfrac{-b \pm \sqrt{b^2 + 4\,g/L}}{2}$ $(\lambda_2 < 0 < \lambda_1)$ real;

n is even: $\lambda = \dfrac{-b \pm \sqrt{b^2 - 4\,g/L}}{2}$ (complex

conjugate pair or two negative real.)

Index